EXACTLY SOLVABLE MODELS OF STRONGLY CORRELATED ELECTRONS

ADVANCED SERIES IN MATHEMATICAL PHYSICS

Published*

Vol. 29: *Bombay Lectures on Highest Weight Representations of Infinite Dimensional Lie Algebras (Second Edition)*
by Victor G Kac, Ashok K Raina and Natasha Rozhkovskaya

Vol. 28: *An Introduction to the Mathematical Structure of Quantum Mechanics: A Short Course for Mathematicians (Second Edition)*
by F. Strocchi

Vol. 27: *An Introduction to the Mathematical Structure of Quantum Mechanics: A Short Course for Mathematicians*
by F Strocchi

Vol. 26: *Soliton Equations and Hamiltonian Systems (Second Edition)*
by L A Dickey

Vol. 25: *Basic Methods in Soliton Theory*
by I Cherednik

Vol. 24: *The Mathematical Beauty of Physics*
edited by J M Drouffe and J B Zuber

Vol. 23: *Quantum Theory and Global Anomalies*
by R A Baadhio

Vol. 22: *W-Symmetry*
by P Bouwknegt and K Schoutens

Vol. 21: *Multidimensional Hypergeometric Functions and Representation Theory of Lie Algebras and Quantum Groups*
by A Varchenko

Vol. 20: *The State of Matter*
edited by M Aizenman and H Araki

Vol. 19: *Under the Spell of the Gauge Principle*
by G 't Hooft

Vol. 18: *Exactly Solvable Models and Strongly Correlated Electrons*
by V Korepin and F H L Ebler

*For the complete list of volumes in this series, please visit
www.worldscientific.com/series/asmp

Advanced Series in Mathematical Physics
Vol. 18

EXACTLY SOLVABLE MODELS OF STRONGLY CORRELATED ELECTRONS

Editors
Vladimir E. Korepin
Fabian H. L. Eβler
Institute for Theoretical Physics
State University of New York at Stony Brook

World Scientific
Singapore • New Jersey • London • Hong Kong

Published by

World Scientific Publishing Co. Pte. Ltd.
5 Toh Tuck Link, Singapore 596224
USA office: 27 Warren Street, Suite 401-402, Hackensack, NJ 07601
UK office: 57 Shelton Street, Covent Garden, London WC2H 9HE

British Library Cataloguing-in-Publication Data
A catalogue record for this book is available from the British Library.

The editors and publisher are grateful to the authors concerned and to publishers of the following journals for their permission to reproduce the articles found in this volume:

American Institute of Physics (*J. Maths. Phys. and Sov. Phys. JETP*);
American Physical Society (*Phys. Rev. B and Phys. Rev. Letts.*);
Elsevier Science Publishers (*Nucl. Phys. B and Phys. Lett.*);
Institute of Physics Publishing (*J. Phys. A, J. Phys. C and J. Phys: Cond. Mat.*);
Physical Society of Japan (*J. Phys. Soc. Jpn.*);
Plenum Publishing Corp. (*J. Stat. Phys.*);
Progress of Theoretical Physics (*Prog. Theor. Phys.*) and
Springer-Verlag (*Z. Phys. B*)

Advanced Series in Mathematical Physics — Vol. 18
EXACTLY SOLVABLE MODELS OF STRONGLY CORRELATED ELECTRONS

ISBN-13 978-981-02-1534-7
ISBN-10 981-02-1534-7

PREFACE

In recent years there has been a strong increase of interest in strongly correlated electronic systems in the Condensed Matter community. The reason for this development is that certain phenomena like high-T_c superconductivity cannot seem to be explained in the framework of weak interactions, that is mean-field theory and perturbation around it. There are mainly two different approaches in dealing with strong correlations: numerical studies and exact analytical solutions. Analytical solutions provide a complete and unambiguous picture of the dynamics of the models under consideration. The analytical approach is mainly limited to one spatial dimension, but there are indications that two-dimensional systems may share certain features with their one-dimensional analogs (see e.g. P. W. Anderson in Ref. 1; for a modified Hubbard model it was recently shown that the ground state structure is identical in one and two dimensions[2]). This volume is devoted to exact solutions of models of strongly correlated electrons in one spatial dimension by means of the Bethe *Ansatz*.

The first such exactly solved model is the non-relativistic continuum model of electrons with local interaction, solved by C. N. Yang in 1967[3] (see also Ref. 4). In the solution of the model the nested Bethe *Ansatz*, which is the basis for all exact solutions of electronic models in one dimension, was discovered.

The most important model of strongly correlated electrons, and the central topic of this volume, is the Hubbard model. The first chapter is devoted to its exact solution in one space and one time dimension, which started with E. H. Lieb's and F. Y. Wu's work of 1968. The chapter starts with reprints dealing with the construction of eigenstates of the Hubbard hamiltonian and the determination of the ground state and excitation spectrum at zero temperature. The next few topics are the study of magnetic properties in an external magnetic field, the thermodynamics of the model, and the asymptotic behaviour of correlation functions. The chapter closes with reprints on transport properties of the model.

The second chapter is devoted to the t-J model at the supersymmetric point $J = \pm 2t$. The t-J model first appeared as an effective model describing the strong coupling limit of the Hubbard model for $H \ll t$). Later the model was reinvented to describe copper-oxide planes in high-T_c superconductors. Nowadays the t-J model is the most thoroughly studied model of strongly correlated electrons after the Hubbard model. The organization of Chapter II is similar to that of Chapter I.

The third chapter is about other solvable models of strongly correlated electrons. The first half of the chapter is devoted to various electronic models with local interaction — the nonrelativistic continuum model of electrons with local interaction, the Luttinger model, and models which were constructed recently in relation to high-T_c superconductivity. The second part of the chapter deals with models with long-range interactions, which have recently attracted much attention in relation to fractional statistics.

The reprinted papers are complemented by a list of some 230 references at the end of the volume. In this list reprinted papers are marked by a bullet (•). When referring to reprinted papers in the comments we first give the chapter number and then the number of the reprint within the chapter, e.g. [repr.III.A.2] would refer to the paper by M. Takahashi reprinted in part A of chapter III under the number 2.

The central topic of this volume is lattice models of strongly correlated electrons. We did not include reprints on models describing interactions of electrons with impurities (see the excellent reviews in Refs. 5, 6) or other excitations in this volume. We also did not include papers dealing with purely mathematical aspects of integrable models. There are many excellent books dealing with these issues[7–16] and we refer the interested reader to them. We also did not include any reviews on the Hubbard model.

Due to republication fees charged by certain journals we were not able to include a significant number of papers which we had originally planned to get reprinted. We are very sorry about this unfortunate development but are confident that the volume in its current form will still be helpful for researchers and students.

While working on this volume we have benefitted greatly from discussions with L. D. Faddeev, F. D. M. Haldane, A. I. Larkin, E. H. Lieb, B. S. Shastry, E. K. Sklyanin, B. Sutherland, F. Woynarovich, and C. N. Yang. We would like to take this opportunity to sincerely thank them for their help and support.

Vladimir E. Korepin　　Fabian H. L. Eßler

Stony Brook, July 1993

CONTENTS

Preface v

I. The One-Dimensional Hubbard Model 1

[1] E. H. Lieb and F. Y. Wu: *Absence of Mott transition in an exact solution of the short-range one-band model in one dimension*
Phys. Rev. Lett. **20** (1968) 1445. 9

[2] A. A. Ovchinnikov: *Excitation spectrum of the one-dimensional Hubbard model*
Sov. Phys. JETP **30** (1970) 1160. 13

[3] C. F. Coll: *Excitation spectrum in the one-dimensional Hubbard model*
Phys. Rev. B **9** (1974) 2150. 17

[4] F. Woynarovich: *Excitations with complex wavenumbers in a Hubbard chain: I. States with one pair of complex wavenumbers*
J. Phys. C **15** (1982) 85. 26

[5] F. Woynarovich: *Excitations with complex wavenumbers in a Hubbard chain: II. States with several pairs of complex wavenumbers*
J. Phys. C **15** (1982) 97. 38

[6] F. Woynarovich: *Spin excitations in a Hubbard chain*
J. Phys. C **16** (1983) 5293. 51

[7] F. Woynarovich: *Low-energy excited states in a Hubbard chain with on-site attraction*
J. Phys. C **16** (1983) 6593. 63

[8] A. Klümper, A. Schadschneider, J. Zittartz: *A new method for the excitations of the one-dimensional Hubbard model*
Z. Phys. B **78** (1990) 99. 75

[9] M. Takahashi: *Magnetization curve for the half-filled Hubbard model*
Prog. Theor. Phys. **42** (1969) 1098. 86

[10] K. Penc and F. Woynarovich: *Novel magnetic properties of the Hubbard chain with an attractive interaction*
Z. Phys. B **85** (1991) 269. 95

[11] M. Takahashi: *One-Dimensional Hubbard model at finite temperature*
Prog. Theor. Phys. **47** (1972) 69. 107

[12] M. Takahashi: *Low temperature specific heat of one-dimensional Hubbard model*
Prog. Theor. Phys. **52** (1974) 103. 121

[13] F. H. L. Eßler, V. E. Korepin and K. Schoutens: *Completeness of the SO(4) extended Bethe* ansatz *for the one-dimensional Hubbard model*
Nucl. Phys. B **384** (1992) 431. 133

[14] B. S. Shastry: *Decorated star-triangle relations and exact integrability of the one-dimensional Hubbard model*
J. Stat. Phys. **50** (1988) 57. 161

[15] T. Koma: *An extension of the thermal Bethe* ansatz
Prog. Theor. Phys. **83** (1990) 655. 184

[16] H. Tsunetsugu: *Temperature dependence of spin correlation length of half-filled one-dimensional Hubbard model*
J. Phys. Soc. Japan **60** (1991) 1460. 190

[17] A. M. Finkel'shteĭn: *Correlation functions in the one-dimensional Hubbard model*
Sov. Phys. JETP Lett. **25** (1977) 73. 194

[18] H. J. Schulz: *Correlated fermions in one dimension*
Int. J. Mod. Phys. B **5** (1991) 57. 198

[19] J. M. P. Carmelo and A. A. Ovchinnikov: *Generalization of the Landau liquid concept: example of the Luttinger liquid*
J. Phys. Cond. Mat. **3** (1991) 757. 216

[20] F. Woynarovich and H. P. Eckle: *Finite-size corrections for the low-lying states of a half-filled Hubbard chain* J. Phys. A **20** (1987) L443. 225

[21] N. M. Bogoliubov and V. E. Korepin: *The role of quasi-one-dimensional structures in high-T_c superconductivity* Int. J. Mod. Phys. B **3** (1989) 427. 232

[22] F. Woynarovich: *Finite-size effects in a non-half-filled Hubbard chain* J. Phys. A **22** (1989) 4243. 245

[23] H. Frahm and V. E. Korepin: *Correlation functions of the one-dimensional Hubbard model in a magnetic field* Phys. Rev. B **43** (1991) 5653. 259

[24] B. S. Shastry and B. Sutherland: *Twisted boundary conditions and effective mass in Heisenberg-Ising and Hubbard rings* Phys. Rev. Lett. **65** (1990) 243. 269

[25] M. Fowler and N. Yu: *Persistent current of a Hubbard ring threaded with a magnetic flux* Phys. Rev. B **45** (1992) 11795. 273

II. The Supersymmetric t-J Model 283

[1] B. Sutherland: *Model for a multicomponent quantum system* Phys. Rev. B **12** (1975) 3795. 287

[2] P. Schlottmann: *Integrable narrow-band model with possible relevance to heavy-fermion systems* Phys. Rev. B **36** (1987) 5177. 298

[3] P. A. Bares and G. Blatter: *Supersymmetric t-J model in one dimension: separation of spin and charge* Phys. Rev. Lett. **64** (1990) 2567. 307

[4] S. Sarkar: *The supersymmetric t-J model in one dimension* J. Phys. A **24** (1991) 1137. 311

[5] N. Kawakami and S.-K. Yang: *Luttinger liquid properties of highly-correlated electron systems in one dimension* J. Phys. C **3** (1991) 5983. 326

[6] A. Förster and M. Karowski: *Algebraic properties of the Bethe* ansatz *for* an *spl(2,1)-supersymmetric t-J model* Nucl. Phys. B **396** (1993) 611. 352

III. Other Models of Strongly Correlated Electrons 381

A. Models with Local Interaction 382

[1] C. N. Yang: *Some exact results for the many-body problem in one dimension with repulsive delta-function interaction* Phys. Rev. Lett. **19** (1967) 1312. 384

[2] M. Takahashi: *One-dimensional electron gas with delta-function interaction at finite temperature* Prog. Theor. Phys. **46** (1971) 1388. 388

[3] D. C. Mattis, E. H. Lieb: *Exact solution of a many-fermion system and its associated boson field* J. Math. Phys. **6** (1965) 304. 407

[4] F. D. M. Haldane: *'Luttinger liquid theory' of one-dimensional quantum fluids: I. Properties of the Luttinger model and their extension to the general 1D interacting spinless Fermi gas* J. Phys. C **14** (1981) 2585. 416

[5] F. D. M. Haldane: *Demonstration of the "Luttinger Liquid" character of Bethe-ansatz-soluble mdoels of 1-D quantum fluids* Phys. Lett. **81** A (1981) 153. 441

[6] R. Z. Bariev: *Integrable spin chain with two- and three-particle interactions* J. Phys. A **24** (1991) L549. 444

[7] R. Z. Bariev, A. Klümper, A. Schadschneider and J. Zittartz: *Exact solution of a one-dimensional model of hole superconductivity* *J. Phys.* A **26** (1993) 1249. 449

[8] F. H. L. Eßler, V. E. Korepin and K. Schoutens: *New exactly solvable model of strongly correlated electrons motivated by high T_c superconductivity* *Phys. Rev. Lett.* **68** (1992) 2960. 458

B. Models with long-range interaction **462**

[1] B. Sutherland: *Quantum many-body problem in one dimension: ground state* *J. Math. Phys.* **12** (1971) 246. 464

[2] B. Sutherland: *Quantum many-body problem in one dimension: thermodynamics* *J. Math. Phys.* **12** (1971) 251. 469

[3] Y. Kuramoto, H. Yokoyama: *Exactly soluble supersymmetric t-J model with long-range exchange and transfer* *Phys. Rev. Lett.* **67** (1991) 1338. 475

[4] N. Kawakami: *Asymptotic Bethe ansatz: Application to the one-dimensional t-J model with long-range exchange and transfer* *Phys. Rev.* B **45** (1992) 7525. 479

References **483**

I

The one-dimensional Hubbard model

I. THE ONE-DIMENSIONAL HUBBARD MODEL

The Hubbard model is the basis of the modern theory of strongly correlated electrons. It was invented in 1963 as a short-range, one-band model for the correlation effects of electrons in a partially filled energy band[17–20]. The hamiltonian describes electrons that hop between the Wannier states of neighbouring lattice sites, and which interact if they occupy the same site (in which case they must have opposite spins due to the Pauli principle).

Of particular interest in the initial study of the model were the investigation of a Mott (conductor-insulator) transition at finite values of the coupling U, and the determination of the magnetic properties in an external magnetic field. More recently the Hubbard model has drawn a great deal of attention in relation to high-T_c superconductivity[1].

This history of the exact solution of the one-dimensional Hubbard model starts with E. H. Lieb's and F. Y. Wu's seminal paper of 1968[21] [repr.I.1]. Using the formulation of the nested Bethe *Ansatz* invented in [repr.III.A.1] they reduced the problem of diagonalizing the hamiltonian to solving a set of coupled nonlinear equations, known as the Bethe *Ansatz* or Lieb-Wu equations. These equations are the basis for most subsequent exact studies of the model. Lieb and Wu also determined the ground state of the system (and its energy) and gave a discussion of the nature of the excitation spectrum for the half-filled band. They showed that the Hubbard model at half-filling is an insulator for all positive values of the coupling U. Lieb and Wu explicitly considered the positive U (repulsive) Hubbard model, in which the ground state is described by means of a set of two coupled integral equations.

The excitation spectrum was studied by A. A. Ovchinnikov[22], G. V. Uimin and S. V. Fomichev[23], C. F. Coll[24], F. Woynarovich[25–28], T. C. Choy and W. Young[29], A. Klümper, A. Schadschneider and J. Zittartz[30,31], and N. Kawakami and A. Okiji[32].

In [repr.I.2.] A. A. Ovchinnikov obtained the correct dispersion for the spin-triplet excitations over the half-filled ground state. He also studied some spin-singlet excitations. G. V. Uimin and S. V. Fomichev extended Ovchinnikov's results to less than half-filling. They also studied the magnetic susceptibility at less than half-filling. C. F. Coll determined the one-particle type and spin-wave type excitations (which only involve real spectral parameters) for strictly less than half-filling [repr.I.3] (the excitation spectrum at half-filling is qualitatively different). He found a linear dispersion relation of the spin-wave excitation for small momenta and evaluated the corresponding spin-wave velocity. T. C. Choy and W. Young re-examined the triplet spin-wave excitations. F. Woynarovich gave a very detailed analysis of the charge excitations (which involve complex spectral parameters) at half-filling [repr.I.4, I.5], and of the spin excitations [repr.I.6]. In [repr.I.4] he studied a so-called Λ'-string† spin-singlet excitation over the ground state, found that this type of excitation has a gap, and determined the dispersion relation. This paper also contains a very clear exposition of the Bethe *Ansatz* wave functions. In [repr.I.5] and [repr.I.6] he extended his results to the case of spin-singlet (charge) excitations with many Λ' strings and spin-triplet and higher spin-multiplet excitations respectively. He also studied the ground state and low-lying excitations in the attractive half-filled case[28] [repr.I.7] (we find his discussion of the quantum numbers of the excitations in the attractive case somewhat confusing‡). A. Klümper, A. Schadschneider and J. Zittartz developed a different method to solve the nested Bethe *Ansatz* equations at half-filling, rederiving all known results [repr.I.8]. The last two authors extended this method to small and almost half-filling. N. Kawakami and A. Okiji investigated a charged hole excitation in the half-filled band.

The magnetic properties at zero temperature were studied by M. Takahashi[34–36], H. Shiba[37], G. V. Uimin and S. V. Fomichev (see above), T. B. Bahder, K. Penc and F. Woynarovich[38–40], and J. M. P. Carmelo, P. Horsch, P. A. Bares and A. A. Ovchinnikov[41]. M. Takahashi found the magnetization curves and the magnetic susceptibility at zero temperature for the case of the half-filled band for both positive and negative values of the coupling constant U [repr.I.9]. H. Shiba extended Takahashi's results to arbitrary band fillings for the repulsive case (positive U). T. B. Bahder and F. Woynarovich studied the magnetization curves for arbitrary filling and negative U. In [repr.I.10] F. Woynarovich and K. Penc derived exact analytic expressions for magnetization and magnetic susceptibility at arbitrary filling (and negative U). J. M. P. Carmelo, P. Horsch, P. A. Bares and A. A. Ovchinnikov studied magnetic properties of the Hubbard model by means of the Landau-Luttinger approach (see [repr.I.19]).

The thermodynamics of the model were investigated by M. Takahashi[42,43], T. Koma[44–46], by N. Kawakami, T. Usuki and A. Okiji[47], and by J. M. P. Carmelo, P. Horsch, P. A. Bares and A. A. Ovchinnikov[48]. In [repr.I.11] M. Takahashi derived an infinite set of coupled integral equations that determine energy, entropy, density and magnetization for fixed temperature, chemical potential and magnetic field. He solved this infinite set of integral equations in the limits of infinite and zero temperature and coupling U. In the first step of his derivation he classified all solutions of the Lieb-Wu equations by deriving the adequate form of the string

†In the terminology of [repr.I.11], where this type of solution to the Lieb-Wu equations was first discussed.

‡A *complete* set of excitations is given by the scattering states of a gapless charge doublet (without spin) and a spin-doublet with a gap (and no charge)[33].

hypothesis for the Hubbard model. In his treatment of the thermodynamics (see also Ref. 49) the use of the string hypothesis[8,49–52] is necessary. T. Koma initiated a different approach to determine thermodynamic properties, which does not make use of the string hypothesis and which is discussed below. In [repr.I.12] M. Takahashi used his results obtained in [repr.I.11] to compute the low-temperature specific heat. N. Kawakami, T. Usuki and A. Okiji performed numerical studies of the magnetic susceptibility, compressibility, and specific heat. J. M. P. Carmelo, P. Horsch, P. A. Bares and A. A. Ovchinnikov studied the low-temperature thermodynamics of a Hubbard chain in the presence of a magnetic field by means of the Landau-Luttinger approach[53,54]. They derived explicit expressions for the specific heat for a nearly half-filled band and for values of the magnetic field close to the ferromagnetic saturated state. They also evaluated the "renormalized charge and hole masses" which determine the exponential decay of correlation functions related to the development of gaps in the spectrum of the spin-triplet excitations for large magnetic fields, and of the charge excitations at half-filling. The compressibility at zero temperature was calculated by T. Usuki, N. Kawakami and A. Okiji[55].

The issue of whether the eigenstates of the hamiltonian found by Lieb and Wu actually form a complete set of states was resolved by F. H. L. Eßler, V. E. Korepin and K. Schoutens[56–58] [repr.I.13]. Based on the representation of the $SO(4)$ symmetry of the Hubbard model[59,60] in terms of spin $SU(2)$ and η-pairing $SU(2)$[61–63] they proved that the Bethe *Ansatz* does not provide a complete set of states. They proved a lowest weight theorem with respect to the $SO(4)$-symmetry for the Bethe *Ansatz* states, and constructed a complete set of states by acting with the $SO(4)$ raising operators on the Bethe *Ansatz* states.

J. M. P. Carmelo and D. Baeriswyl studied the ground state properties of the Hubbard model by means of a perturbative expansion in powers of $1/U$ for large U for arbitrary band filling[64]. A comparison of the energy gap for the half-filled band with the gap-equation of BCS theory was performed by A. L. Kholodenko and A. L. Beyerlein[65].

It is quite a general feature of Bethe *Ansatz* solvable one-dimensional *quantum* models that their respective hamiltonians can be embedded in commuting families of transfer matrices describing two-dimensional *classical* integrable models. The commuting transfer matrices depend on a spectral parameter, and upon a formal power series expansion in this parameter. The commutativity property of the transfer matrices translates into the existence of an infinite number of conservation laws, one of which is by construction the hamiltonian of the one-dimensional quantum model.

The Hubbard model appeared to be an exception to this rule for many years until B. S. Shastry succeeded in constructing a covering classical model and its commuting family of transfer matrices[66–68]. In [repr.I.14] he constructed the R-matrix and showed it to obey a Yang-Baxter equation. This R-matrix has the unusual feature that it depends on the difference *and* the sum of the spectral variables in the Yang-Baxter equation. [repr.I.14] also contains explicit expressions for some higher conservation laws of the Hubbard model and a conjecture for the eigenvalues of the transfer matrix. Subsequently R. Z. Bariev reported a Bethe *Ansatz* solution for the diagonal-to-diagonal transfer matrix problem of the classical model and used this to compute the free energy and finite size corrections[69,70].

Preceeding Shastry's work were other attempts to construct a covering two-dimensional classical vertex model starting from the Hubbard hamiltonian. M. Barma and B. S. Shastry[71] used the Trotter product formula to transform the partition function of the Hubbard model into the partition function of a two-dimensional classical spin model, for which the Hubbard hamiltonian is the logarithmic derivative of the transfer matrix (at a special value of the spectral parameter). However, their transfer matrices do *not* commute for generic values of the spectral parameters and it is not known whether there exists a commuting family describing the Barma-Shastry model. This of course raises the question of the integrability of this model, which to the best of our knowledge has not been completely clarified[72]. However, R. Z. Bariev[73] reported a diagonalization of the diagonal-to-diagonal transfer matrix of the Barma-Shastry model by means of a Bethe *Ansatz*. This point mystifies us to some extent because to the best of our knowledge the existence of a commuting family of transfer matrices is essential for the applicability of the Bethe *Ansatz*. Further studies were performed by T. T. Truong, K. D. Schotte and S. Iwabuchi[74,75].

In [repr.I.15] T. Koma used Bariev's results to construct the so-called thermal Bethe *Ansatz*[44,45,76–81] for the Hubbard model[46]. Using the equivalence of the partition functions of the Hubbard model and the classical two-dimensional vertex model of Barma and Shastry, Koma managed to describe the thermodynamics of the Hubbard model by a system of coupled algebraic equations, without making use of the string hypothesis needed in Takahashi's approach to the thermodynamics. Combining Koma's results with an extension of the thermal Bethe *Ansatz* method developed by M. Takahashi[82,83], H. Tsunetsugu evaluated the temperature dependence of the spin correlation length for the half-filled band[84] [repr.I.16]. In this method the spin correlation length is determined by a set of coupled algebraic equations, which have to be solved numerically.

At zero temperature, correlation functions decay as a power of the distance (in the gapless situation). The corresponding powers are called critical exponents or conformal dimensions. They are the main objects of interest for the large-distance asymptotics of correlation functions at zero temperature. They were studied by employing a variety of different methods.

In 1977 A. M. Finkel'shteĭn derived the critical exponents for several correlators for small coupling U and a half-filled band[85] [repr.I.17]. He used a bosonization technique to derive an effective hamiltonian describing low-lying excitations close to the Fermi surface. His results turn out to be exact for all values of U.

In the limit $U \to \infty$ the Bethe *Ansatz* wave functions take an especially simple form[86] (see also [repr.I.5]). This fact allows for direct studies of asymptotic properties of correlation functions. M. Ogata, T. Sugiyama and H. Shiba[86–88] performed numerical studies of momentum distribution and spin-correlation functions in the limit $U \to \infty$. An analytic derivation of the asymptotics for spin-spin correlators in the limit $U \to \infty$ at zero magnetic field and quarter filling was obtained by A. Parola and S. Sorella[89].

Another method for the calculation of correlators at zero temperature is the so-called *Luttinger-liquid* approach invented by F. D. M. Haldane[90–94], which is essentially based on the fact that the Hubbard model, as well as many other critical one-dimensional models, are certain realizations of the Gaussian model[95]. By generalizing Haldane's method H. J. Schulz[96,97] [repr.I.18] and independently Y. Ren and P. W. Anderson[98] obtained the expressions for the critical exponents of the charge and spin correlators in zero magnetic field. The singularities of the electron momentum distribution function (which are connected to the long-distance asymptotics of correlation functions by Fourier transform) were studied by M. Brecht, J. Voigt and H. Büttner[99] in the framework of Luttinger liquid theory.

An extension of F. D. M. Haldane's description to the case of finite magnetic fields is provided by the Landau-Luttinger-liquid approach of J. M. P. Carmelo, P. Horsch and A. A. Ovchinnikov[53,100]. In Ref. 100 the critical exponents were obtained in the framework of a generalized Landau-liquid theory, confirming the results of H. Frahm and V.E. Korepin[101,102] (see below). The concept of a Landau-Luttinger liquid was introduced by J. M. P. Carmelo and A. A. Ovchinnikov in [repr.I.19][53], where the spectral properties and asymptotic behaviour of correlation functions of the one-dimensional Hubbard model were reformulated in terms of interacting charge and spin pseudoparticles.

The modern version of the Luttinger-liquid approach is related to conformal quantum field theory[103]. It was shown in Refs. 104–106 that the critical exponents can be extracted from the finite size corrections to the spectrum of the hamiltonian. In Bethe *Ansatz* solvable models it is possible to evaluate the finite size corrections exactly[107–110]. The corrections for the low-lying states in the repulsive half-filled Hubbard model were obtained by F. Woynarovich and H. P. Eckle[111] [repr.I.20]. The finite size corrections and critical exponents in the attractive Hubbard model were determined by N. M. Bogoliubov and V. E. Korepin[112,113] [repr.I.21]. Unlike in the repulsive case there is only one branch of gapless excitations in the attractive case (note that the expression for the correlation length ζ [describing the exponential decay of the correlator of Fermi fields] given in the paragraph between formulas (7) and (8) is incorrect).

Expressions for the finite size corrections in a general solvable model were given A. Izergin, V. E. Korepin and N. Yu. Reshetikhin in Ref. 114. Using these results F. Woynarovich treated the case of the non-half-filled Hubbard model[115] [repr.I.22]. The derivation of the critical exponents from the finite-size corrections in the repulsive case for arbitrary values of the external magnetic field, the density, and the coupling U was performed by H. Frahm and V. E. Korepin[101,102] [repr.I.23]. They were the first to evaluate the critical exponents for finite magnetic fields. They found that the conformal dimensions depend strongly on the external magnetic field and the other system parameters like density and coupling constant. In Ref. 102 an expression that determines the conformal dimensions for any (gapless) model solvable by a nested Bethe Ansatz was obtained[†]. They also investigated the singularities of the electron momentum distribution function. The case of zero magnetic field was also studied by N. Kawakami and S. K. Yang[117]. Logarithmic amplitude corrections to the asymptotical behaviour of the correlators were calculated by T. Giamarchi and H. J. Schulz[118]. Other methods for determining the critical exponents, based on perturbation theory and a renormalization group analysis, can be found in Refs. 119–121.

The transport properties of the one-dimensional Hubbard model were studied by B. Sutherland and B. S. Shastry[122], M. Fowler and N. Yu[123], J. M. P. Carmelo, P. Horsch and D. Campbell[124], and by M. J. Martins and R. M. Fye[125]. In [repr.I.24] B. S. Shastry and B. Sutherland showed that the boundary energy of a fermionic lattice system with twisted boundary conditions can be identified with the inverse effective charge- and spin-carrying masses (or stiffnesses). They evaluated the charge- and spin-stiffnesses for the Hubbard model and showed that these quantities can be used as a probe for a Mott-Hubbard transition. The twisting of boundary conditions can be viewed as the effect of a transverse ("two-dimensional") magnetic field.

[†]An application of this formula to other models can be found in [repr.II.5] and Ref. 116

N. Yu and M. Fowler studied the persistent current in Hubbard rings as a function of the flux through the ring and the coupling U [repr.I.25]. They found that the persistent current is a periodic function of the flux and showed that the ground state is always a spin-singlet. J. M. P. Carmelo, P. Horsch and D. K. Campbell investigated the transport properties of the Hubbard model in the framework of the Landau-Luttinger-liquid approach. They evaluated the charge and spin dynamical form factors for arbitrary momentum in a magnetic field[124]. M. J. Martins and R. M. Fye evaluated the charge and spin stiffnesses for the repulsive and attractive Hubbard models with toroidal boundary conditions.

VOLUME 20, NUMBER 25 PHYSICAL REVIEW LETTERS 17 JUNE 1968

ABSENCE OF MOTT TRANSITION IN AN EXACT SOLUTION OF THE SHORT-RANGE, ONE-BAND MODEL IN ONE DIMENSION

Elliott H. Lieb* and F. Y. Wu
Department of Physics, Northeastern University, Boston, Massachusetts
(Received 22 April 1968)

The short-range, one-band model for electron correlations in a narrow energy band is solved exactly in the one-dimensional case. The ground-state energy, wave function, and the chemical potentials are obtained, and it is found that the ground state exhibits no conductor-insulator transition as the correlation strength is increased.

The correlation effect of electrons in a partially filled energy band has been a subject of interest for many years.[1-4] A realistic model which takes this correlation into consideration, and which is hopefully amenable to mathematical treatment, is the short-range, one-band model considered by a number of authors.[2-5] In this model, one pictures the electrons in a narrow energy band hopping between the Wannier states of neighboring lattice sites, with a repulsive interaction energy between two electrons of opposite spins occupying the same lattice site. The central problems of interest have been (a) the possible existence of a "Mott transition" between conducting and insulating states as the strength of the interaction is increased, and (b) the magnetic nature (ferromagnetic or antiferromagnetic) of the ground state. While previous treatments of this model have always been approximate, we have succeeded in solving the model exactly in the one-dimensional case. Our exact result shows that the Mott transition does occur in the ground state of the one-dimensional model. Furthermore, a general theorem of Lieb and Mattis[6] on one-dimensional systems tells us that the ground state is necessarily antiferromagnetic.

It may be argued that the absence of a Mott transition in one dimension is irrelevant for the study of real three-dimensional systems because of the folkloristic dictum that there are never any phase transitions in one dimension with short-range interactions. In actual fact, the dictum is only true for nonzero temperature; the ground state is another matter. Generally speaking, when a Hamiltonian is considered to be a function of some parameter, U (which in our case is the electron-electron repulsion), singularities with respect to U usually do appear in the ground-state wave function, energy, polarizability, etc., even in one dimension. A good example of this is the one-dimensional Heisenberg chain (to which the present model is very close) which, when considered as a function of the anisotropy parameter, does have two singularities in the ground state and, presumably, no singularities for nonzero temperatures.[7,8]

Consider a crystal (one-, two-, or three-dimensional) of N_a lattice sites with a total of $N \leq 2N_a$ electrons. We suppose that the electrons can hop between the Wannier states of neighboring lattice sites, and that each site is capable of accommodating two electrons of opposite spins, with an interaction energy $U>0$. The Hamiltonian to consider is then[2-5]

$$H = T\sum_{\langle ij\rangle}\sum_{\sigma} c_{i\sigma}^{\dagger} c_{j\sigma} + U\sum_{i} c_{i\uparrow}^{\dagger} c_{i\uparrow} c_{i\downarrow}^{\dagger} c_{i\downarrow}, \quad (1)$$

where $c_{i\sigma}^{\dagger}, c_{i\sigma}$ are, respectively, the creation and annihilation operators for an electron of spin σ in the Wannier state at the ith lattice site, and the sum

$$\sum_{\langle ij\rangle}$$

is restricted to nearest-neighbor sites.

First of all, it can be shown that the energy spectrum of H is invariant under the replacement of T by $-T$.[9] Therefore, for simplicity we shall take, in appropriate units, $T=-1$. Since the numbers M of down-spin electrons and M' of up-spin electrons are good quantum numbers $(M+M'=N)$, we may designate the ground-state energy of H by $E(M,M';U)$. It is then easy to derive the following relations [by considering holes instead of particles in (1)]:

$$\begin{aligned} E(M,M';U) &= -(N_a-M-M')U + E(N_a-M, N_a-M';U) \\ &= MU + E(M, N_a-M'; -U) \\ &= M'U + E(N_a-M, M'; -U). \end{aligned} \quad (2)$$

Without loss of generality, therefore, we may take

$$S_z \equiv \tfrac{1}{2}(N-2M) \geq 0 \text{ and } N \leq N_a$$

(less than half-filled band).

VOLUME 20, NUMBER 25 PHYSICAL REVIEW LETTERS 17 JUNE 1968

It can similarly be shown that the maximum energy $G(M,M';U)$ is related to the ground-state energy by

$$G(M,M';U)=M'U-E(N_a-M,M';U).$$

Therefore, a knowledge of the ground-state energies also tells us about the maximum energies.

For a one-dimensional system, the lattice sites can be numbered consecutively from 1 to N_a. Let $f(x_1,x_2,\cdots,x_M,x_{M+1},\cdots,x_N)$ represent the amplitude in ψ for which the down spins are at the sites $x_1,\cdots,x_M$, and the up spins at $x_{M+1},\cdots,x_N$. Then the eigenvalue equation $H\psi=E\psi$ leads to

$$-\sum_{i=1}^{N}\sum_{s=\pm 1} f(x_1,\cdots,x_i+s,\cdots,x_N) + U\sum_{i<j}\delta(x_i-x_j)f(x_1\cdots x_N) = Ef(x_1\cdots x_N), \tag{3}$$

where it is understood that we require a solution of the form

$$f(x_1,x_2,\cdots,x_M|x_{M+1},x_{M+2},\cdots,x_N) \tag{4}$$

which is antisymmetric in the first M and the last $N-M$ variables.

In each region defined by $1\leqslant x_{Q1}\leqslant x_{Q2}\leqslant\cdots x_{QN}\leqslant N$, we make the following Ansatz for f:

$$f(x_1,\cdots,x_M|x_{M+1},\cdots,x_N) = \sum_P [Q,P]\exp(i\sum_{j=1}^{N} k_{Pj}x_{Qj}), \tag{5}$$

where $P=(P1,P2,\cdots,PN)$ and $Q=(Q1,Q2,\cdots,QN)$ are two permutations of the numbers $(1,2,\cdots,N)$, $\{k_1,k_2,\cdots k_N\}$ is a set of N unequal real numbers, and $[Q,P]$ is a set of $N!\times N!$ coefficients to be determined.

The coefficients $[Q,P]$ are not all independent. The condition of single valuedness (or continuity) of f and the requirement that (5) be a solution of (3) lead to the following:

$$E=-2\sum_{j=1}^{N}\cos k_j \tag{6}$$

and, for all Q and P, the coefficients $[Q,P]$ must be chosen to satisfy the relations

$$[Q,P]=Y_{nm}^{ab}[Q,P']. \tag{7}$$

In (7), Y_{nm}^{ab} is an operator defined by

$$Y_{nm}^{ab}=\frac{-\frac{1}{2}iU}{\sin k_n-\sin k_m+\frac{1}{2}iU}+\frac{\sin k_n-\sin k_m}{\sin k_n-\sin k_m+\frac{1}{2}iU}P^{ab}, \tag{8}$$

where, for $j=i+1$,

$Qi=a=Q'j,\quad Qj=b=Q'i,$

$Qk=Q'k$ for all $k\neq i,j$;

$Pi=m=P'j,\quad Pj=n=P'i,$

$Pk=P'k$ for all $k\neq i,j$;

and P^{ab} is an operator which exchanges $Qi=a$ and $Qj=b$.

It is fortunate that the Ansatz (5) and the algebraic consistency conditions (7) and (8) have, in essence, appeared before in the study of the one-dimensional delta-function gas for particles in a continuum. The first solution of that problem was for bosons (symmetric f) by Lieb and Liniger[10] but this case is not relevant here, besides which the consistency conditions there are trivial to solve. The two-component fermion case was solved by McGuire[11] for $M=1$, but again (7) is trivial because of translational invariance. The next development was the solution of the case $M=2$ by Flicker and Lieb[12] by an inelegant algebraic method which could not be easily generalized. However, the case $M=2$ is the simplest one which displays the full difficulty of the problem. Shortly thereafter, Gaudin[13] published the solution of the general-M problem. The method of his brilliant solution did not appear for some time and is now available as his thesis.[14] In the meantime, Yang[15] also discovered the method of solution (essentially the same as Gaudin's) and published it with considerable detail. Here, we have followed Yang's notation with slight modification.

The important point is that our Eqs. (7) and (8) are the same as for the continuum gas except for the replacement of k by $\sin k$ in the latter. This has no effect on the beautiful algebraic analysis which finally leads to the following condi-

VOLUME 20, NUMBER 25 PHYSICAL REVIEW LETTERS 17 JUNE 1968

tions which determine the set $\{k_{1,2},\cdots,k_N\}$:

$$N_a k_j = 2\pi I_j + \sum_{\beta=1}^{M} \theta(2\sin k_j - 2\Lambda_\beta),\quad j=1,2,\cdots,N, \tag{9}$$

where the Λ's are a set of real numbers related to the k's through

$$-\sum_{j=1}^{N} \theta(2\Lambda_\alpha - 2\sin k_j) = 2\pi J_\alpha - \sum_{\beta=1}^{M} \theta(\Lambda_\alpha - \Lambda_\beta),\quad \alpha=1,2,\cdots,M, \tag{10}$$

$$\theta(p) \equiv -2\tan^{-1}(2p/U),\quad -\pi \leq \theta < \pi, \tag{11}$$

and I_j = integers (or half-odd integers) for M = even (or odd), J_α = integers (or half-odd integers) for M' = odd (or even). An immediate consequence is

$$\sum_{j=1}^{N} k_j = \frac{1}{N_a}\Big(\sum_j I_j + \sum_\alpha J_\alpha\Big). \tag{12}$$

For the ground state, J_α and I_j are consecutive integers (or half-odd integers) centered around the origin and satisfying $\sum_j k_j = 0$.

In the limit of $N\to\infty$, $N_a\to\infty$, $M\to\infty$ with the ratios N/N_a, M/N_a kept finite, the real numbers k and Λ are distributed continuously between $-Q$ and $Q \leq \pi$ and $-B$ and $B \leq \infty$, with density functions $\rho(k)$ and $\sigma(\Lambda)$, respectively. Equations (9) and (10) then lead to the coupled integral equations for the distribution function $\rho(k)$ and $\sigma(\Lambda)$:

$$2\pi\rho(k) = 1 + \cos k \int_{-B}^{B} \frac{8U\sigma(\Lambda)d\Lambda}{U^2 + 16(\sin k - \Lambda)^2}, \tag{13}$$

$$\int_{-Q}^{Q} \frac{8U\rho(k)dk}{U^2 + 16(\Lambda - \sin k)^2} = 2\pi\sigma(\Lambda) + \int_{-B}^{B} \frac{4U\sigma(\Lambda')d\Lambda'}{U^2 + 4(\Lambda - \Lambda')^2}, \tag{14}$$

where Q and B are determined by the conditions

$$\int_{-Q}^{Q}\rho(k)dk = N/N_a, \tag{15}$$

$$\int_{-B}^{B}\sigma(\Lambda)d\Lambda = M/N_a. \tag{16}$$

The ground-state energy (6) now becomes

$$E = -2N_a\int_{-Q}^{Q}\rho(k)\cos k\,dk. \tag{17}$$

We have established the following:

(a) Equations (13)-(16) have a unique solution which is positive for all allowed B and Q.

(b) M/N is a monotonically increasing function of B reaching a maximum of $\frac{1}{2}$ at $B=\infty$. This is the antiferromagnetic case, $S_z = 0$, and corresponds to the absolute ground state.

(c) N/N_a is a monotonically increasing function of Q, reaching a maximum of 1 (half-filled band) at $Q=\pi$.

For $B=\infty$ and $Q=\pi$, (13)-(16) can be solved in closed form by Fourier transforms with the result

$$\sigma(\Lambda) = (2\pi)^{-1}\int_0^\infty \operatorname{sech}(\tfrac{1}{4}\omega U) \times \cos(\omega\Lambda)J_0(\omega)d\omega, \tag{18}$$

$$\rho(k) = (2\pi)^{-1} + \pi^{-1}\cos k \times \int_0^\infty \frac{\cos(\omega\sin k)J_0(\omega)d\omega}{1+\exp(\frac{1}{2}\omega U)}, \tag{19}$$

$$E \equiv E(\tfrac{1}{2}N_a, \tfrac{1}{2}N_a; U) = -4N_a\int_0^\infty \frac{J_0(\omega)J_1(\omega)d\omega}{\omega[1+\exp(\frac{1}{2}\omega U)]}, \tag{20}$$

where J_0 and J_1 are Bessel functions.

To investigate whether the ground state is conducting or insulating, we compute the chemical

VOLUME 20, NUMBER 25 PHYSICAL REVIEW LETTERS 17 JUNE 1968

potentials μ_+ and μ_- as defined in a forthcoming paper by Mattis[16]:

$$\mu_+ \equiv E(M+1,M;U)-E(M,M;U),$$
$$\mu_- \equiv E(M,M;U)-E(M-1,M;U). \qquad (21)$$

If μ_+ and μ_- are equal, the system has the property of a conductor. If, on the other hand, we find $\mu_+ > \mu_-$, then the system shares the property of an insulator. We can compute μ_- directly from (9) and (10) by replacing $M \to M-1$ and $N \to N-1$, while letting all the k's, Λ's, and their distribution functions change slightly. The procedure is quite similar to the calculation of the excitation spectrum of the continuum gas.[10] If $N<\frac{1}{2}N_a$, we can compute μ_+ in the same way and thereby find that $\mu_+=\mu_-$ for all U. If, however, N is exactly $\frac{1}{2}N_a$, then we must compute μ_+ by using the first line of (2) which tells us that

$$\mu_+ = U-\mu_- \text{ (half-filled band).} \qquad (22)$$

The calculation of μ_- can be done in closed form for a half-filled band with the result

$$\mu_- - 2 = -4\int_0^\infty \frac{J_1(\omega)d\omega}{\omega[1+\exp(\frac{1}{2}\omega U)]}$$
$$= -4\sum_{n=1}^\infty (-1)^n[(1+\tfrac{1}{4}n^2U^2)^{\frac{1}{2}}-\tfrac{1}{2}nU]. \qquad (23)$$

It can be established from (22) and (23) that, indeed, $\mu_+>\mu_-$ for $U>0$, and

$$\lim_{U\to 0} \mu_\pm = 0.$$

Therefore, we conclude that the ground state for a half-filled band is insulating for any nonzero U, and conducting for $U=0$. That is, there is no Mott transition for nonzero U. This absence of a Mott transition is also reflected by the fact that the ground-state energy and the ground-state wave function are analytic in U on the real axis (except at the origin).

We have also investigated the excitation spectrum $E(p)$ for a given total momentum $\sum_j k_j = p$ and a given value of S_z. Just as in the case of a continuum gas for which the spectrum can be regarded as consisting of several elementary excitations,[10,15] we find three types of excitations: (I) a "hole" state in the Λ distribution, (II) a "hole" state in the k distribution, and (III) a "particle" state in the k distribution. While the $S_z=0$ spin-wave state may have any of these three types of spectra, the $S_z=1$ spin-wave state is always associated with the type-I spectrum. The type-I excitation has the lowest energy and is characterized by a double periodicity similar to that of an antiferromagnetic chain.[7] In the limit $U\to 0$, it goes over to $E(p)=|\sin p|$, while the type-II and -III spectra have the identical limiting form $E(p)=|2\sin(\frac{1}{2}p)|$. Detailed discussions of these matters will be given elsewhere.

We are grateful to Dr. D. C. Mattis for helpful advice and suggestions and E. L. would like to thank Dr. J. Zittartz for interesting him in the problem.

---

*Work partially supported by National Science Foundation Grant No. GP-6851.

[1]J. H. Van Vleck, Rev. Mod. Phys. 25, 220 (1953).

[2]M. C. Gutzwiller, Phys. Rev. Letters 10, 159 (1963), and Phys. Rev. 134, A923 (1964), and 137, A1726 (1965).

[3]J. Hubbard, Proc. Roy. Soc. (London), Ser. A 276, 238 (1963), and 277, 237 (1964).

[4]G. Kemeny, Ann. Phys. (N.Y.) 32, 69, 404 (1965).

[5]See also the review article by C. Herring, in Magnetism, edited by G. T. Rado and H. Suhl (Academic Press, Inc., New York, 1966), Vol. IV, Chap. 10.

[6]E. Lieb and D. C. Mattis, Phys. Rev. 125, 164 (1962). The proof of the theorem given in this reference can be adopted to our Hamiltonian (1).

[7]J. des Cloizeaux and M. Gaudin, J. Math. Phys. 7, 1384 (1966).

[8]C. N. Yang and C. P. Yang, Phys. Rev. 150, 321, 327 (1966).

[9]The proof assumes A and B sublattices and uses the unitary transformation

$$\exp[i\pi\sum_\sigma \sum_{i\in A} c_{i\sigma}^\dagger c_{i\sigma}]$$

which changes $c_{i\sigma}$ to $-c_{i\sigma}$ and $c_{i\sigma}^\dagger$ to $-c_{i\sigma}^\dagger$ for $i\in A$. This transformation does not change the number operator.

[10]E. Lieb and W. Liniger, Phys. Rev. 130, 1605 (1963); E. Lieb, Phys. Rev. 130, 1616 (1963).

[11]J. B. McGuire, J. Math. Phys. 6, 432 (1965).

[12]M. Flicker, thesis, Yeshiva University, 1966 (unpublished); M. Flicker and E. Lieb, Phys. Rev. 161, 179 (1967).

[13]M. Gaudin, Phys. Letters 24A, 55 (1967).

[14]M. Gaudin, thesis, University of Paris, 1967 (unpublished).

[15]C. N. Yang, Phys. Rev. Letters 19, 1312 (1967).

[16]D. Mattis, to be published.

SOVIET PHYSICS JETP VOLUME 30, NUMBER 6 JUNE, 1970

EXCITATION SPECTRUM IN THE ONE-DIMENSIONAL HUBBARD MODEL

A. A. OVCHINNIKOV

L. Ya. Karpov Physical-Chemical Institute

Submitted June 12, 1969

Zh. Eksp. Teor. Fiz. 57, 2137–2143 (December, 1969)

The spectrum of the lowest (quasi-homopolar) excitations in the one-dimensional Hubbard model are investigated within the framework of the exact method developed in articles[8-10]. The excitations are classified according to spin and momentum. The singlet states are states of the bound type. It is shown that both singlet and triplet excitations start from zero, i.e., they do not have a gap. The magnitude of the gap is determined for the spectrum of quasi-ionic states to which an optical transition is possible. Its dependence on the parameter characterizing the electron interaction is investigated.

1. INTRODUCTION

IN order to describe the metal-dielectric transition associated with an increase of the repulsion between electrons, Hubbard[1] proposed a model of a Fermi lattice gas having an interaction of the electrons only at one center. In the case of a one-dimensional cyclic chain, the corresponding Hamiltonian has the following form:

$$H = \sum_{m,n}^{N} \sum_{\sigma} T_{m,n} a_{m\sigma}^{+} a_{n\sigma} + \frac{\gamma}{2} \sum_{n}^{N} \sum_{\sigma} a_{n-\sigma}^{+} a_{n-\sigma} a_{n\sigma}^{+} a_{n\sigma}, \qquad (1)$$

where $a_{n\sigma}^{+}$ and $a_{n\sigma}$ denote the creation and annihilation operators for an electron with spin σ in atom n; all $T_{m,n} = 0$ except $T_{n\pm1,n} = -\beta$ $(\beta > 0)$.

The Hamiltonian (1) was used in article[2] in order to explain the appearance of a gap in the optical spectrum of long polymers with conjugated bonds. In this connection it was shown, within the framework of the generalized Hartree-Fock method, that an excited state to which an optical transition is possible is separated from the ground state by a gap for arbitrary values of the parameter γ. For a suitable choice of γ it was possible to obtain agreement with the experimentally observed dependence of the magnitude of the first transition on the length of the chain. In addition to the excitations of the indicated type, the Hamiltonian (1) has below a gap a set of singlet and triplet quasi-homopolar excitations.[3,4] Here, as shown in the work by Kohn[5] and Bulaevskiĭ[3], an optical transition to these states is forbidden or very weak. Meanwhile these states play a major role in the determination of the physical and chemical properties of long systems with conjugated bonds. For example, the fact that the spectrum of the triplet excitations starts from zero leads, for infinitely long chains, to an appreciable paramagnetism of these molecules.[6]

The goal of the present article is a determination of the spectrum of the lowest quasi-homopolar excitations of the Hamiltonian (1) and their classification. We shall use the exact expression for the wave function of the Hamiltonian which was obtained in articles[7-9], where Bethe's idea[10] was extended.

Let us consider an eigenfunction of the Hamiltonian (1) with the number of electrons equal to the number of sites, i.e., N, and with the z-component of the total spin equal to zero (we shall assume N to be even). We shall seek it in the form

$$\Psi_Q(n_1, n_2, \ldots, n_N) = \sum_P [Q, P] \exp\Big\{ i \sum_{j=1}^{N} k_{P_j} n_{Q_j} \Big\},$$
$$1 \leqslant n_{Q_1} \leqslant n_{Q_2} \leqslant \ldots \leqslant n_{Q_N} \leqslant N. \qquad (2)$$

Here $k_1, k_2, \ldots, k_N$ denotes the set of quasimomenta for which the equation will be written down; $(Q_1, Q_2, \ldots, Q_N)$ and $(P_1, P_2, \ldots, P_N)$ denote permutations among the coordinates and momenta respectively. The summation in (2) is carried out over all permutations of the momenta k_i; the $[Q, P]$ are coefficients which simultaneously depend on Q and P and which are represented by a square matrix of order $N! \times N!$, which must be determined. The Schrödinger equation gives the following relation between these coefficients:

$$[Q, P] = Y_{nm}^{ab} [Q, P'], \qquad (3)$$

where the operator Y_{nm}^{ab} has the form[10]

$$Y_{nm}^{ab} = -\frac{i\gamma/2 + (\sin k_n - \sin k_m) P^{ab}}{\sin k_n - \sin k_m + i\gamma/2};$$
$$Q_i = a = Q_j', \quad Q_j = b = Q_i',$$
$$P_i = m = P_j', \quad P_j = n = P_i', \qquad (4)$$

$Q_k = Q_k'$, $P_k = P_k'$ for $k \neq i, j$ and the operator P^{ab} interchanges the sites Q_i and Q_j. In this connection the characteristic energy of the system is expressed in terms of the quasimomenta k_i in the following way:

$$E = -2\beta \sum_{j=1}^{N} \cos k_j. \qquad (5)$$

By successively applying the operator Y_{mn}^{ab}, one can express any arbitrary coefficient $[Q, P]$ in terms of (a vector of dimension $N!$) the coefficient $[Q, I]$, where I denotes the identity permutation among the momenta $k_1, k_2, \ldots, k_N$.

Utilization of the conditions for the cyclic nature and symmetry of the wave function leads to a system of equations for the coefficients $[Q, I]$. Omitting the subsequent calculations which are rather completely given in the article by Yang,[9] let us write down the transcendental equations for the quasimomenta k_i arising upon the solution of this system

$$Nk_j = 2\pi I_j + \sum_{\beta=1}^{N/2} \varphi(j\beta), \tag{6a}$$

$$\sum_{j=1}^{N} \varphi(j\alpha) = 2\pi J_\alpha + \sum_{\beta=1}^{N/2} \psi(\beta\alpha) + \pi, \tag{6b}$$

$$e^{i\varphi(j\beta)} = \frac{\sin k_j - \Lambda_\beta + ic/2}{\sin k_j - \Lambda_\beta - ic/2}, \tag{7a}$$

$$e^{i\psi(\beta\alpha)} = \frac{\Lambda_\beta - \Lambda_\alpha + ic}{\Lambda_\beta - \Lambda_\alpha - ic}, \quad c = \frac{\gamma}{2\beta}. \tag{7b}$$

Here Λ_α ($\alpha = 1, 2, \ldots, N/2$) denotes a set of numbers, all of which are different, and which in general may be complex. The phases $\psi(\alpha\beta)$ and $\varphi(j\beta)$ are determined so that

$$-\pi < \operatorname{Re}\psi(\alpha\beta),\ \operatorname{Re}\varphi(j\beta) < \pi,$$

I_j ($j = 1, 2, \ldots, N$) and J_α ($\alpha = 1, 2, \ldots, N/2$) are integers; they label the eigenstates of the system. For example, the total momentum Q of the system is expressed in terms of them in the following manner:

$$Q = \sum_{j=1}^{N} k_j = \frac{2\pi}{N}\Big(\sum_{j=1}^{N} I_j + \sum_{\alpha=1}^{N/2} J_\alpha\Big). \tag{8}$$

2. SPECTRUM OF THE TRIPLET EXCITATIONS

Let us consider the solution of the system of Eqs. (6) and (7) in the limit $\gamma \to \infty$. As is well-known, in this limit all eigenstates of the Hamiltonian (1) are divided into groups of almost degenerate states: homopolar, ionic, doubly ionic, etc. The first group consists of 2^N states with almost zero energy. The splitting of the energy levels among this group is described by the Heisenberg spin Hamiltonian. The second group consists of $2^N N$ states with energy $\sim\gamma$. A lowest excited state, to which an optical transition is possible, is found among this group. The third group contains $N(N-1)2^{N-1}$ states with energy $\sim 2\gamma$ and so forth. We will primarily be interested in the first group of states. Since the excited states of the spin Hamiltonian are well-known, then this makes it possible to classify the quasihomopolar states of the Hamiltonian (1) according to spin and momentum.

As $\gamma \to \infty$, Eqs. (6a), (6b) and (7a), (7b) go over into the following system of equations:

$$Nk_j = 2\pi I_j + \sum_{\beta=1}^{N/2} p_\beta, \quad \xi_\alpha = \operatorname{ctg}\frac{p_\alpha}{2} = -\frac{2\Lambda_\alpha}{c},$$

$$Np_\beta = 2\pi J_\beta + \sum_{\alpha(\neq\beta)} \psi(\alpha\beta), \quad 0 < p_\beta < 2\pi,$$

$$\operatorname{ctg}\frac{\psi(\beta\alpha)}{2} = \frac{1}{2}(\xi_\alpha - \xi_\beta). \tag{9}$$

This system agrees with the system of equations for the case of the spin Hamiltonian.[4] For the ground state of the system it is necessary to choose J_α and I_j in the following way:

$$J_\alpha = 1, 3, 5, \ldots, N-1, \tag{10}$$

$$I_j = -N/2, -N/2+1, \ldots, N/2-1. \tag{11}$$

For the quasi-homopolar levels all k_j are real, and for convenience one can reduce them to the interval $(-\pi, \pi)$.

In order to determine the excited triplet states, following[11] let us choose J_α in the form

$$J_\alpha = 0, 2, 4, \ldots, 2n-2, 2n+1, \ldots, N-1, \tag{12}$$

where n is a certain number which determines the total quasimomentum of the system. The solution of Eqs. (6) and (7) is obtained by changing to a continuous distribution of the numbers k_j and Λ_α. In this connection one can use the formal equation $\rho(k) = dj/dk_j$ for the density of the numbers k_j in the interval $(-\pi, \pi)$ and $\sigma(\Lambda) = d\alpha/d\Lambda_\alpha$ for the density of the numbers Λ_α over the entire axis $(-\infty, \infty)$. Carrying out the required differentiation in Eqs. (6) and (7) under the conditions (11) and (12), we obtain the following system of equations for the triplet states:

$$\rho(k) = \frac{1}{2\pi} + \frac{\cos k}{2\pi}\int_{-\infty}^{\infty} \frac{4c\sigma(\Lambda)\,d\Lambda}{c^2 + 4(\Lambda - \sin k)^2}, \tag{13}$$

$$\int_{-\pi}^{\pi} \frac{4c\rho(k)\,dk}{c^2 + 4(\Lambda - \sin k)^2} = 2\pi\sigma(\Lambda) + \int_{-\infty}^{\infty} \frac{2c\sigma(\Lambda')\,d\Lambda'}{c^2 + (\Lambda - \Lambda')^2} + \frac{2\pi}{N}\delta(\Lambda - \Lambda_n),$$

$$E = -2N\beta\int_{-\pi}^{\pi} \rho(k)\cos k\,dk. \tag{14}$$

Here Λ_n is equal to its own unperturbed value, i.e., it is obtained from the solution of Eqs. (6) and (7) by utilization of the numbers J_α and I_j, just as for the ground state (10) and (11). Taking the Fourier transform of the function $\sigma(\Lambda)$, one can easily obtain an expression for $\rho(k)$ and $\sigma(\Lambda)$. Omitting this calculation, we cite the answer for the energy of the triplet states

$$E_t(q) = E_0 + 2\beta\int_0^\infty \frac{J_1(\omega)\cos\omega\Lambda_n\,d\omega}{\omega\,\operatorname{ch}(\omega c/2)}. \tag{15}$$

Here E_0, the energy of the ground state which was first determined by Lieb and Wu,[10] is given by

$$E_0 = -4N\beta\int_0^\infty \frac{J_1(\omega)J_0(\omega)\,d\omega}{\omega(1 + e^{\omega c})},$$

$J_0(\omega)$ and $J_1(\omega)$ are Bessel functions. The quantity Λ_n is expressed in terms of the quasimomentum of the system $q = 2\pi n/N$ in the following way:

$$q = \frac{\pi}{2} + \int_0 \frac{J_0(\omega)}{\omega}\frac{\sin\omega\Lambda_n}{\operatorname{ch}(\omega c/2)}\,d\omega. \tag{16}$$

The system (15) and (16) parametrically determines the $E_t(q)$ dependence. The function $E_t(q)$ possesses a double periodicity and reaches a maximum at $q = \pi/2$. If $\gamma \to \infty$

$$\varepsilon_t(q) = E_t(q) - E_0 \cong (4\pi\beta^2/2\gamma)\,|\sin q|,$$

which agrees with the expression for the triplet excitations[11] in the Heisenberg model with an exchange integral equal to $4\beta^2/\gamma$.

3. SPECTRUM OF THE SINGLET EXCITATIONS

As was shown in[4] the lowest singlet states of an antiferromagnetic Heisenberg chain necessarily belong to the bound state type, i.e., they correspond to complex momenta in the spin system. Our calculation of the spectrum of the singlet quasi-homopolar excitations of the Hamiltonian (1) will be entirely based on an analogy with a similar calculation for the spin Hamiltonian.

Let us choose sets of numbers I_j and J_α in the

following way. Let us leave the set I_j unchanged, as given by Eq. (11), but

$$J_\alpha = 1, 3, \ldots, 2\beta_1 - 1, 2\beta_1 - 1, \ldots, 2\beta_2 - 3, 2\beta_2 + 1, \ldots, N - 1. \quad (17)$$

According to[4] two complex-conjugate numbers $\Lambda_a = \lambda + i\kappa$ and $\Lambda_b = \lambda - i\kappa$ will correspond to two identical numbers J_{β_1}. We note that the total quasimomentum of such a system will be determined in the following way:

$$q = 2\pi (J_{\beta_1} - J_{\beta_2}) / N. \quad (18)$$

One can choose all remaining Λ_α to be real. From the imaginary part of Eq. (6c) for $\alpha = a$ we have $\kappa = c/2$. Changing to a continuous distribution of the numbers Λ_α and k_j and introducing the corresponding densities according to the formulas of the preceding Section, we obtain the following system of equations:

$$\rho(k) = \frac{1}{2\pi} + \frac{\cos k}{2\pi} \int_{-\infty}^{\infty} \frac{4c\sigma(\Lambda)\, d\Lambda}{c^2 + 4(\Lambda - \sin k)^2} - \frac{1}{2N\pi} T(k),$$

$$T(k) = \left[2\pi\delta(\sin k - \Lambda) - \frac{2c}{c^2 + (\sin k - \lambda)^2}\right] \cos k$$

$$+ \cos k \sum_{m=1,2} \left[\frac{4c}{c^2 + 4(\bar{\Lambda}_{\beta_m} - \sin k)^2} - 2\pi\delta(\sin k - \bar{\Lambda}_{\beta_m})\right]; \quad (19)$$

$$\int_{-\pi}^{\pi} \frac{4c\rho(k)\, dk}{c^2 + 4(\sin k - \Lambda)^2} = 2\pi\sigma(\Lambda) + \int_{-\infty}^{\infty} \frac{2c\sigma(\Lambda')\, d\Lambda'}{c^2 + (\Lambda - \Lambda')^2} + \frac{D(\Lambda)}{N},$$

$$D(\Lambda) = -4\pi\delta(\Lambda - \lambda) + \frac{4c}{4(\Lambda - \lambda)^2 + c^2} + \frac{12c}{4(\lambda - \Lambda)^2 + 9c^2}$$

$$+ \sum_{m=1,2} \left[2\pi\delta(\Lambda - \bar{\Lambda}_{\beta_m}) - \frac{2c}{c^2 + (\Lambda - \bar{\Lambda}_{\beta_m})^2}\right]. \quad (20)$$

In connection with the derivation of these equations we added to the system of real numbers Λ_α two additional numbers $\bar{\Lambda}_{\beta_1}$ and $\bar{\Lambda}_{\beta_2}$ which satisfy the same equations as the number Λ_α for $J_{\beta_1} = 2\beta_1 - 1$, $J_{\beta_2} = 2\beta_2 - 1$. The function $\sigma(\Lambda)$ is represented out of the density of real numbers Λ_α together with the two additional numbers $\bar{\Lambda}_{\beta_1}$ and $\bar{\Lambda}_{\beta_2}$.

The solution of the system of equations is obtained by transition to the Fourier transform for the function $\sigma(\Lambda)$. Omitting the calculations, we write down an expression for the energy of the singlet quasi-homopolar excitations

$$E_s(q) = E_0 + 8\beta \int_0^{\infty} \frac{d\omega J_1(\omega)}{\omega\, \mathrm{ch}(\omega c/2)} (\cos \omega\bar{\Lambda}_{\beta_1} + \cos \omega\bar{\Lambda}_{\beta_2} - \cos \omega\lambda). \quad (21)$$

In this connection, just as in[4], the following restriction is imposed on $\bar{\Lambda}_{\beta_1}$, $\bar{\Lambda}_{\beta_2}$, and λ:

$$|\bar{\Lambda}_{\beta_1}| > 1, \quad |\bar{\Lambda}_{\beta_2}| > 1, \quad |\lambda| > 1. \quad (22)$$

The condition for solvability of the system of equations for the number Λ_α (here it is required that $\Lambda_\alpha \neq \Lambda_\beta$ for $\alpha \neq \beta$) at once gives the equation

$$\lambda = \bar{\Lambda}_{\beta_2}. \quad (23)$$

The real part of Eq. (6b) for $\alpha = a$ together with Eq. (23) leads to the relation

$$2\beta_1 = \beta_2. \quad (24)$$

Finally Eqs. (18) and (24) make it possible to relate the total momentum q of the system to Λ_{β_1}:

$$|q| = \pi - 2 \int_0^{\infty} \frac{d\omega J_0(\omega) \sin \omega\bar{\Lambda}_{\beta_1}}{\omega\, \mathrm{ch}(\omega c/2)}. \quad (25)$$

Equation (25) together with the equation which follows from (21) and (23),

$$\varepsilon_s(q) = 8\beta \int_0 \frac{d\omega J_1(\omega) \cos \omega\bar{\Lambda}_{\beta_1}}{\omega\, \mathrm{ch}(\omega c/2)} \quad (26)$$

give the parametric dependence of the energy of the singlet excitations on the quasimomenta. Here one should keep in mind the limiting condition $|\bar{\Lambda}_{\beta_1}| > 1$. It leads to the result that the singlet excitation spectrum has a termination point at

$$q_0 = \pi - 2 \int_0^{\infty} d\omega \frac{J_0(\omega) \sin \omega}{\omega\, \mathrm{ch}(\omega c/2)}. \quad (27)$$

For $\gamma \to \infty$ the value $|q_0| = \pi/2$. If $\gamma = 0$ then $q_0 = 0$, which indicates the absence of bound states in this limit. For small q the spectra of singlet and triplet excitations have identical slopes:

$$\varepsilon_{t,s}(q) = |q| \frac{2\beta I_1(\pi/c)}{I_0(\pi/c)}, \quad (28)$$

where I_1 and I_0 are Bessel functions of imaginary argument. For large values of q the singlet levels always lie above the triplet levels. For sufficiently large but not infinite values of N, the energy of the first triplet level tends to zero in the following way:

$$\varepsilon_t(N) = \frac{4\pi\beta}{N} \frac{I_1(2\pi\beta/\gamma)}{I_0(2\pi\beta/\gamma)} \quad (29)$$

Let us make several remarks about the energy of the singlet quasi-ionic states. A strong optical transition takes place precisely to these states. The quasi-ionic states possess a nonvanishing current. The energy of the lowest current state and, consequently, the gap in the optical spectrum in the one-dimensional Hubbard model were calculated in the article by Lieb and Wu.[10] For its determination they obtained an energy $E_+ = E_0 + \mu_+$ for the ground state of the system containing N + 1 electrons and an energy $E_- = E_0 + \mu_-$ for the ground state of the system containing N − 1 electrons. The gap in the spectrum of the quasi-ionic states is then determined in the following way:

$$\Delta E = E_+ - E_- = \mu_+ - \mu_-. \quad (30)$$

In order to determine the spectrum of the quasi-ionic states it is necessary to determine the energy of a system containing N + 1 or N − 1 electrons and having a total momentum q. This computation is quite

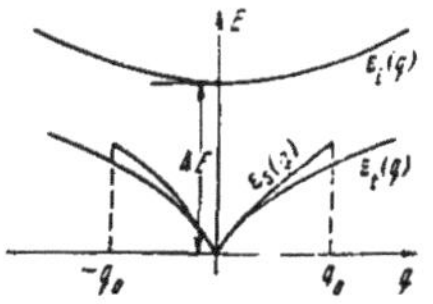

Different types of excitations of the system. $\epsilon_S(q)$ is the spectrum of singlet homopolar excitations for small q, as given by Eqs. (25) and (26); q_0 given by Eq. (27) is the point of termination of the spectrum; $\epsilon_t(q)$ is the spectrum for the homopolar triplet excitations which are described by Eqs. (15) and (16); $\epsilon_i(q)$ is the spectrum for the ionic excitations, and ΔE given by Eq. (33) is the gap in the spectrum of the ionic states.

analogous to the one given in the text. Without giving it in detail, in the figure we show the general form of the spectra for the lowest excited states. Different types of excitations of the system. $\epsilon_s(q)$ is the spectrum of singlet homopolar excitations for small q, as given by Eqs. (25) and (26); q_0 given by Eq. (27) is the point of termination of the spectrum; $\epsilon_t(q)$ is the spectrum for the homopolar triplet excitations which are described by Eqs. (15) and (16); $\epsilon_i(q)$ is the spectrum for the ionic excitations, and ΔE given by Eq. (33) is the gap in the spectrum of the ionic states.

Lieb and Wu[10] arrived at the following expression for the gap ΔE:

$$\Delta E = \gamma - 4\beta + 8\beta \sum_{n=1}^{\infty} (-1)^n [(1 + c^2 n^2)^{1/2} - cn]. \tag{31}$$

It is possible to give a more convenient expression for ΔE. For this purpose let us represent the series in (31) in terms of an integral along a contour C_0 which encompasses the real axis from c to ∞:

$$\sum_{n=1}^{\infty} (-1)^n [(1 + c^2 n^2)^{1/2} - nc] = \frac{1}{2ic} \int_{C_0} \frac{dz}{\operatorname{sh}(\pi z/c)} (\sqrt{z^2+1} - z). \tag{32}$$

Deforming the contour C_0 until it coincides with the imaginary axis, we can represent ΔE in the form

$$\Delta E = \frac{16\beta^2}{\gamma} \int_1^{\infty} \frac{\sqrt{y^2-1}\, dy}{\operatorname{sh}(\pi y/c)}. \tag{33}$$

For $\gamma \to \infty$ the gap is given by $\Delta E \approx \gamma - 4\beta + (8\beta^2/\gamma) \ln 2 + \ldots$. If the strength of the electron interaction is decreased, i.e., if $\gamma \to 0$, then

$$\Delta E \approx 8\pi^{-1} \sqrt{\gamma\beta}\, e^{-2\pi\beta/\gamma}. \tag{34}$$

We note that to within the pre-exponential factor this expression agrees with the expression given in article[2] for the gap as $\gamma \to 0$.

In conclusion the author thanks Ya. B. Zel'dovich, I. M. Khalatnikov, I. M. Lifshitz, and E. Lieb (USA) for interesting discussions of this work.

[1] J. Hubbard, Proc. Roy. Soc. (London), Ser. A276, 238 (1963), and 277, 237 (1964).
[2] I. A. Misurkin and A. A. Ovchinnikov, ZhETF Pis. Red. 4, 248 (1966) [JETP Lett. 4, 167 (1966)].
[3] L. N. Bulaevskiĭ, Zh. Eksp. Teor. Fiz. 51, 230 (1966) Sov. Phys.-JETP 24, 154 (1967).
[4] A. A. Ovchinnikov, Zh. Eksp. Teor. Fiz. 56, 1354 (1969) [Sov. Phys.-JETP 29, 727 (1969)].
[5] W. Kohn, Phys. Rev. 133, A171 (1964).
[6] L. A. Blyumenfel'd, A. A. Berlin, A. A. Slinkin, and A. É. Kalmanson, Zhurn. strukt. khim. 1, 1031 (1960).
[7] M. Gaudin, Phys. Letters 24A, 55 (1967).
[8] C. N. Yang, Phys. Rev. Letters 19, 1312 (1967).
[9] E. H. Lieb and F. Y. Wu, Phys. Rev. Letters 20, 1445 (1968).
[10] H. Bethe, Z. Physik 71, 205 (1931).
[11] J. des Cloizeaux and J. J. Pearson, Phys. Rev. 128, 2131 (1962).

Translated by H. H. Nickle
246

PHYSICAL REVIEW B VOLUME 9, NUMBER 5 1 MARCH 1974

Excitation spectrum of the one-dimensional Hubbard model*

Cornelius F. Coll, III
Department of Physics, University of California, Los Angeles, California 90024
(Received 13 September 1973)

We have extended calculations of the zero-temperature excitation spectrum of the one-dimensional Hubbard model to the case where the number of electrons is less than the number of sites in the chain. The results are computed as a function of the ratio U/t, where U represents the on-site Coulomb repulsion and t is the transfer integral, assumed to be nonzero only for nearest neighbors. Exact calculations are made for the energy and momentum of excitations having single-particle character. Unlike the situation for the half-filled band, we find no gap in the excitation spectrum. We have also considered excitations of the spin-wave type. These are shown to vary linearly with momentum for small momentum. The group velocity for small momentum is found to be inversely proportional to the magnetic susceptibility.

I. INTRODUCTION

There has been much interest recently in systems which for some purposes may be considered one dimensional. For example, the system $Cu(NH_3)_4SO_4 \cdot H_2O$ exhibits magnetic properties which are reasonably described by the linear-chain Heisenberg Hamiltonian.[1] Our concern in this work is with the Hubbard model for a linear chain, i.e., a model for interacting itinerant electrons. This model has been used to analyze results of studies of the salt N-methylphenazinium-tetracyanoquinodimethan (NMP-TCNQ).[2] Here the half-filled band is the appropriate model. Other TCNQ salts may be described as more or less than half-filled bands; e.g., in quinolinium-$TCNQ_2$ there presumably exists one electron per two TCNQ molecules and thus a quarter-filled band. It is to these types of materials that we hope the results of this work will prove applicable.

The Hubbard Hamiltonian can be written as[3]

$$H = -\sum_{\substack{i,j\\ \sigma}} t_{ij} C^\dagger_{i\sigma} C_{j\sigma} + U \sum_i n_{i\uparrow} n_{i\downarrow} , \qquad (1.1)$$

where t_{ij} is the hopping integral, assumed to be nonzero only for (i,j) nearest neighbors. We consider a one-dimensional crystal of N_a lattice sites with a total of $N \leqslant 2N_a$ electrons. Since the numbers M of spin-down electrons and M' of spin-up electrons are good quantum numbers, we can classify states of the system by, say, the numbers N and M. At zero temperature the model Hamiltonian is characterized by the parameters $u = U/t$, the ratio of the Coulomb interaction energy and the nearest-neighbor hopping integral, and the electron density N/N_a.

Lieb and Wu gave[4] an exact solution for the lowest energy state of the Hubbard model for fixed M/N_a. For the half-filled-band case $[(N/N_a)=1]$ they derived an analytic expression for the ground-state energy as a function of u. Shiba considered the ground-state energy for arbitrary electron density and gave numerical results for various values of u.[5] In addition, he calculated the lowest energy as a function of magnetization and was thereby able to obtain numerical results for the magnetic susceptibility for arbitrary electron density.

For the half-filled band the spectrum of the lowest excitations was considered by Ovchinnikov.[6] He found $S=1$ excitations of spin-wave character having a double periodicity similar to that of the antiferromagnetic chain.[7] He also investigated the spectrum of "quasi-ionic" states, i.e., states of $(N \pm 1)$ electrons with total momentum q. For the case of the half-filled band there is a gap in the spectrum of the quasi-ionic states.

In this work we investigate some of the low-lying excited states of the system for arbitrary electron density. We find excitations of spin-wave character whose group velocity at long wavelengths is inversely proportional to the magnetic susceptibility. For the spectrum of quasi-ionic states we find no gap in the spectrum for $N/N_a < 1$. The results are derived for arbitrary $N/N_a < 1$ and u. We give numerical results for the quarter-filled band:

$$N/N_a = \tfrac{1}{2} .$$

II. EQUATIONS DETERMINING DISTRIBUTION FUNCTIONS $\rho(k)$ AND $\sigma(\Lambda)$

For the Hamiltonian of Eq. (1.1) it was shown by Lieb and Wu[4] that the energy and momentum of a system of N electrons, M of which have down spin, is given by

$$E = -2t \sum_{j=1}^{N} \cos k_j , \qquad (2.1a)$$

$$p = \sum_{j=1}^{N} k_j . \qquad (2.1b)$$

The "momenta" k_j are determined by the equation

$$N_a k_j = 2\pi I_j + \sum_{\beta=1}^{M} \theta(2\sin k_j - 2\Lambda_\beta), \quad j = 1, \ldots, N \tag{2.2}$$

where the Λ's are a set of numbers related to the k's by

$$-\sum_{j=1}^{N} \theta(2\Lambda_\alpha - 2\sin k_j) = 2\pi J_\alpha - \sum_{\beta=1}^{M} \theta(\Lambda_\alpha - \Lambda_\beta), \quad \alpha = 1, \ldots, M \tag{2.3}$$

and

$$\theta(x) = -2\tan^{-1}(2x/u) . \tag{2.4}$$

In these equations I_j and J_α are integers (or half-odd integers) which we consider as the quantum numbers describing the state of the system.

From Eqs. (2.2)–(2.4) we see that the momentum p can be conveniently written as

$$p = \frac{2\pi}{N_a}\left(\sum_{j=1}^{N} I_j + \sum_{\alpha=1}^{M} J_\alpha\right) . \tag{2.5}$$

We are interested in solutions to Eqs. (2.2) and (2.3) for real k's and Λ's. It was shown by Ovchinnikov[6] for the half-filled band that there exist singlet ($S = 0$) excitations of the system for which some of the k's and Λ's are complex. We hope to return to a study of this case in a future work.

We begin by writing the equations

$$N_a(k_{j+1} - k_j)\left(1 + \cos k_j \frac{1}{N_a}\sum_{\beta=1}^{M} \frac{8u}{u^2 + 16(\sin k_j - \Lambda_\beta)^2}\right) = 2\pi(I_{j+1} - I_j) , \tag{2.6a}$$

$$N_a(\Lambda_{\alpha+1} - \Lambda_\alpha)\left(\frac{1}{N_a}\sum_{j=1}^{N}\frac{8u}{u^2 + 16(\sin k_j - \Lambda_\alpha)^2} - \frac{1}{N_a}\sum_{\beta=1}^{M}\frac{4u}{u^2 + 4(\Lambda_\alpha - \Lambda_\beta)^2}\right) = 2\pi(J_{\alpha+1} - J_\alpha) , \tag{2.6b}$$

where we have used a Taylor expansion for the function $\theta(x)$, since $(k_{j+1} - k_j) \sim O(1/N_a)$, $(\Lambda_{\alpha+1} - \Lambda_\alpha) \sim O(1/N_a)$, and we are interested eventually in the large-N_a limit.

We introduce two functions $\rho(k)$ and $\sigma(\Lambda)$ defined at the points k_j and Λ_α, respectively, by

$$\frac{1}{N_a \rho(k_j)} = k_{j+1} - k_j , \tag{2.7a}$$

$$\frac{1}{N_a \sigma(\Lambda_\alpha)} = \Lambda_{\alpha+1} - \Lambda_\alpha . \tag{2.7b}$$

By means of these functions the sums in Eq. (6) may be approximated in the limit of a large system by an integral,

$$\frac{1}{N_a}\sum_{j=1}^{N} f(k_j) \to \int_{-Q}^{Q} \rho(k) f(k)\, dk , \tag{2.8a}$$

$$\frac{1}{N_a}\sum_{\beta=1}^{M} g(\Lambda_\beta) \to \int_{-B}^{B} \sigma(\Lambda) g(\Lambda)\, d\Lambda , \tag{2.8b}$$

where $\rho(k)$ and $\sigma(\Lambda)$ obey the normalization conditions (in the limit N, M, $N_a \to \infty$, N/N_a, M/N fixed)

$$\int_{-Q}^{Q} \rho(k)\, dk = N/N_a , \tag{2.9a}$$

$$\int_{-B}^{B} \sigma(\Lambda)\, d\Lambda = M/N_a . \tag{2.9b}$$

For a large system the meaning of $\rho(k)$ and $\sigma(\Lambda)$ is that $N_a\rho(k)dk$ is the number of k's in $(k, k+dk)$; $N_a\sigma(\Lambda)d\Lambda$ is the number of Λ's in $(\Lambda, \Lambda + d\Lambda)$. One immediate result of Eqs. (2.8) for a large system is, from Eq. (2.1),

$$E/N_a = -2t \int_{-Q}^{Q} dk \cos k\, \rho(k) .$$

In the following sections we proceed to determine the equations satisfied by $\rho(k)$ and $\sigma(\Lambda)$ for particular choices of I_j, J_α. We repeat results found previously[5] for the ground state, since we will need some of the results for the later investigations. Generally speaking, we will find that the distribution functions determining the energy of various excited states can be written in the form

$$\rho(k) = \rho_0(k) + (1/N_a)\rho_1(k) ,$$

$$\sigma(\Lambda) = \sigma_0(\Lambda) + (1/N_a)\sigma_1(\Lambda) ,$$

where $\rho_0(k)$ and $\sigma_0(\Lambda)$ are the *ground*-state distribution functions. Since we are interested in the excitation energy, i.e., the difference in energies of the excited state and the ground state, we see that the excitation energy is determined by $\rho_1(k)$ and $\sigma_1(\Lambda)$.

III. GROUND STATE

As shown by Lieb and Wu,[4] we take for the ground state

$$I_{j+1} - I_j = 1 , \tag{3.1a}$$

$$J_{\alpha+1} - J_\alpha = 1 . \tag{3.1b}$$

Substituting these expressions into Eqs. (2.6) and changing to a continuous distribution of the numbers k_j and Λ_α, we find the following equations for the distribution functions $\rho(k)$ and $\sigma(\Lambda)$:

$$2\pi\rho_0(k) = 1 + \cos k \int_{-\infty}^{\infty} d\Lambda\, \sigma_0(\Lambda) \frac{8u}{u^2 + 16(\sin k - \Lambda)^2} , \tag{3.2a}$$

$$\int_{-Q_0}^{Q_0} dk\, \rho_0(k) \frac{8u}{u^2 + 16(\sin k - \Lambda)^2} = 2\pi\sigma_0(\Lambda) + \int_{-\infty}^{\infty} d\Lambda'\, \sigma_0(\Lambda') \frac{4u}{u^2 + 4(\Lambda - \Lambda')^2} . \tag{3.2b}$$

We denote the solutions of these equations with a subscript 0. The distribution functions satisfy the subsidiary conditions

$$\int_{-Q_0}^{Q_0} \rho_0(k)\, dk = N/N_a , \tag{3.3a}$$

$$\int_{-\infty}^{\infty} \sigma_0(\Lambda)\, d\Lambda = \tfrac{1}{2}(N/N_a)\ , \qquad (3.3b)$$

i.e., $M/N = \frac{1}{2}$. These equations have been solved previously for the half-filled band by Lieb and Wu[4] and for arbitrary N/N_a by Shiba.[5] We write the solutions here since we will make reference to them later.

By introducing the Fourier transform $\sigma_0(\omega) = \int_{-\infty}^{\infty} \sigma_0(\Lambda) e^{-i\omega\Lambda}\, d\Lambda$ one can show that $\rho_0(k)$ satisfies the following integral equation:

$$\rho_0(k) = \frac{1}{2\pi} + \cos k \int_{-Q_0}^{Q_0} dk'\, R\left(\frac{4}{u}(\sin k - \sin k')\right) \times \rho_0(k')\ , \qquad (3.4)$$

where $R(x)$ is defined as

$$R(x) = \frac{1}{4\pi}\int_{-\infty}^{\infty} dy\, \frac{e^{ixy/2}}{1 + e^{|y|}}\ . \qquad (3.5)$$

The solution for $\sigma_0(\Lambda)$ can then be written in terms of $\rho_0(k)$ as

$$\sigma_0(\Lambda) = \frac{1}{u}\int_{-Q_0}^{Q_0} dk\, \rho_0(k)\, \mathrm{sech}\left(\frac{2\pi}{u}(\Lambda - \sin k)\right)\ . \qquad (3.6)$$

For the half-filled band ($N/N_a = 1$; $Q = \pi$) Lieb and Wu[4] found an analytic expression for $\rho_0(k)$. For $N/N_a \neq 1$ one can solve for $\rho_0(k)$ numerically and then use this result in Eq. (3.5) to find $\sigma_0(\Lambda)$.

From Eqs. (3.4)–(3.6) we can gain some insight into the nature of the solutions for $\rho_0(k)$ and $\sigma_0(\Lambda)$. From Eq. (3.4) we see that $\rho_0(k)$ is an even function of k; in addition, using the fact that R is an even function of its argument we see that $\rho_0(k)$ has a maximum at $k = 0$. Physically this is what we expect by looking at the expression for the ground-state energy:

$$E_0 = -2tN_a \int_{-Q_0}^{Q} \rho_0(k) \cos k\, dk\ . \qquad (3.7)$$

From this we see that the energy is minimized if $\rho_0(k)$ is largest for small k.

In Fig. 1 we show numerical results for $\rho_0(k)$ for various values of $u = U/t$ for the quarter-filled band. From this figure we see the effect of increasing u on the ground-state distribution. The resulting effect on the ground-state energy can be seen in the work of Shiba.[5]

From Eq. (3.6) we see that $\sigma_0(\Lambda)$ is an even function of Λ, it has a maximum at $\Lambda = 0$ and decreases exponentially for large Λ.

IV. SPIN-WAVE STATE

This is the state which Lieb and Wu[4] classify as the "hole in the Λ distribution" state. We choose the integers I_j as in the ground state and take

$$J_{\alpha+1} - J_\alpha = 1 + \delta_{\alpha,\alpha_0}\ . \qquad (4.1)$$

Substituting these expressions into Eqs. (2.6) and changing to a continuous distribution for the numbers k_j and Λ_α, we find for the distribution function $\rho(k)$ and $\sigma(\Lambda)$ the equations

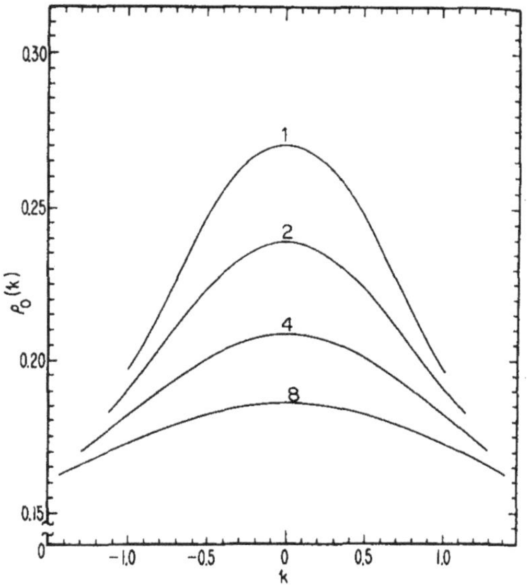

FIG. 1. Ground-state distribution function $\rho_0(k)$ for the quarter-filled band. Individual curves are labeled with the values of $u = U/t$. The cutoff momentum is determined by the normalization condition, $\int_{-Q_0}^{Q_0} \rho_0(k)dk = N/N_a$

$$2\pi\rho(k) = 1 + \cos k \int_{-\infty}^{\infty} d\Lambda\, \sigma(\Lambda) \frac{8u}{u^2 + 16(\sin k - \Lambda)^2}\ , \qquad (4.2a)$$

$$\int_{-Q}^{Q} dk\, \rho(k) \frac{8u}{u^2 + 16(\sin k - \Lambda)^2} = 2\pi\sigma(\Lambda) + \int_{-\infty}^{\infty} d\Lambda'\, \sigma(\Lambda') \frac{4u}{u^2 + 4(\Lambda - \Lambda')^2} + \frac{2\pi}{N_a}\delta(\Lambda - \Lambda_0)\ , \qquad (4.2b)$$

where Λ_0 is the value of Λ for which $\alpha = \alpha_0$. The limit of the k integration, Q, is in general different from that in the ground state for a given N/N_a. If we denote the corresponding limiting momentum in the ground state by Q_0 then Q will be related to Q_0 by the condition that the ratio N/N_a is fixed. Also note that in Eqs. (4.2) the distribution functions depend explicitly on Λ_0. In general, we would expect the limits of the Λ integration in Eqs. (4.2) to be different from those in the ground state. One can show, using manipulations of the form employed by Shiba,[5] that for purposes of calculating the excitation energy this limit may be taken to be as in the ground state.

Introduce the distribution functions $\rho_1(k)$ and $\sigma_1(\Lambda)$ by

$$\rho(k) = \rho_0(k) + (1/N_a)\rho_1(k)\ , \qquad (4.3a)$$

$$\sigma(\Lambda) = \sigma_0(\Lambda) + (1/N_a)\sigma_1(\Lambda)\ , \qquad (4.3b)$$

where $\rho_0(k)$ and $\sigma_0(\Lambda)$ are the ground-state distribution functions for *fixed* Q. The distribution functions $\rho_1(k)$ and $\sigma_1(\Lambda)$ then satisfy the equations

$$2\pi\rho_1(k)=\cos k\int_{-\infty}^{\infty}d\Lambda\,\sigma_1(\Lambda)\frac{8u}{u^2+16(\sin k-\Lambda)^2}\,, \tag{4.4a}$$

$$\int_{-Q}^{Q}dk\,\rho_1(k)\frac{8u}{u^2+16(\sin k-\Lambda)^2}=2\pi\delta(\Lambda-\Lambda_0)+2\pi\sigma_1(\Lambda)+\int_{-\infty}^{\infty}d\Lambda'\,\sigma_1(\Lambda')\frac{4u}{u^2+4(\Lambda-\Lambda')^2}\,. \tag{4.4b}$$

An integral equation for $\rho_1(k)$ can be obtained by introducing the Fourier transform $\sigma_i(\omega)=\int_{-\infty}^{\infty}d\Lambda\times\sigma_1(\Lambda)e^{-i\omega\Lambda}$. This leads to the equation

$$\rho_1(k)=\frac{-1}{u}\cos k\,\mathrm{sech}\left(\frac{2\pi}{u}(\sin k-\Lambda_0)\right)+\frac{4}{u}\cos k\times\int_{-Q}^{Q}dk'\,R\left(\frac{4}{u}(\sin k-\sin k')\right)\rho_1(k')\,. \tag{4.5}$$

It is useful to note that the inhomogeneous term in the integral equation is the solution for $\rho_1(k)$ for the half-filled band.

An integral equation for $\sigma_1(\Lambda)$ can be found by substituting Eq. (4.4a) into Eq. (4.4b):

$$2\pi\sigma_1(\Lambda)=-2\pi\delta(\Lambda-\Lambda_0)-\int_{-\infty}^{\infty}d\Lambda'\,S_Q(\Lambda,\Lambda')\sigma_1(\Lambda')\,, \tag{4.6}$$

where the kernel $S_Q(\Lambda,\Lambda')$ is given by[5]

$$S_Q(\Lambda,\Lambda')=\frac{4u}{u^2+4(\Lambda-\Lambda')^2}-\int_{-Q}^{Q}\frac{dk}{2\pi}\cos k\times\frac{8u}{u^2+16(\sin k-\Lambda)^2}\times\frac{8u}{u^2+16(\sin k-\Lambda')^2}\,. \tag{4.7}$$

We can write the formal solution for $\sigma_1(\Lambda)$ in terms of the resolvent kernel $\mathcal{S}_Q(\Lambda,\Lambda')$ as

$$\sigma_1(\Lambda)=-\delta(\Lambda-\Lambda_0)+\mathcal{S}_Q(\Lambda,\Lambda_0)\,, \tag{4.8}$$

where the resolvent kernel is defined by the equations

$$2\pi\mathcal{S}_Q(\Lambda,\Lambda')=S_Q(\Lambda,\Lambda')-\int_{-\infty}^{\infty}d\Lambda''\,\mathcal{S}_Q(\Lambda,\Lambda'')S_Q(\Lambda'',\Lambda')=S_Q(\Lambda,\Lambda')-\int_{-\infty}^{\infty}d\Lambda''\,S_Q(\Lambda,\Lambda'')\mathcal{S}_Q(\Lambda'',\Lambda')\,. \tag{4.9}$$

The energy of the spin-wave state is given by

$$E=-2tN_a\int_{-Q}^{Q}\rho(k)\cos k\,dk\,, \tag{4.10}$$

or, in terms of $\rho_1(k)$, by

$$E=E_0(Q)-2t\int_{-Q}^{Q}\rho_1(k)\cos k\,dk\,, \tag{4.11}$$

where $E_0(Q)$ is the ground-state energy for *fixed* Q. The excitation energy ϵ is then the difference between this energy and the ground-state energy for a fixed density N/N_a:

$$\epsilon(\Lambda_0)=-2t\int_{-Q}^{Q}\rho_1(k,\Lambda_0)\cos k\,dk+E_0(Q)-E_0(Q_0)\,. \tag{4.12}$$

We have explicitly indicated the dependence on Λ_0 in Eq. (4.12).

To relate Q to Q_0 we use the condition that the density of electrons is fixed; i.e., the distribution function $\rho(k)$ must satisfy the normalization condition

$$\int_{-Q}^{Q}\rho(k)\,dk=\frac{N}{N_a}=\frac{N_0(Q_0)}{N_a} \tag{4.13}$$

or

$$\frac{N_0(Q)}{N_a}+\frac{1}{N_a}\int_{-Q}^{Q}\rho_1(k)\,dk=\frac{N_0(Q_0)}{N_a}\,. \tag{4.14}$$

Equation (4.14) gives us the relation between Q and Q_0. To $O(1/N_a)$ we find

$$Q-Q_0=\left(\frac{1}{N_a}\frac{\partial N_0(Q_0)}{\partial Q_0}\right)^{-1}\left(\frac{-1}{N_a}\int_{-Q_0}^{Q_0}\rho_1(k)\,dk\right)\,. \tag{4.15}$$

In order to calculate the excitation energy correctly we must include all terms of $O(1)$. We can neglect higher-order terms for a large system. Thus for a very large system we find for the excitation energy

$$\epsilon(\Lambda_0)=-2t\int_{-Q_0}^{Q_0}dk\,\rho_1(k,\Lambda_0)\cos k-\mu\int_{-Q_0}^{Q_0}dk\,\rho_1(k,\Lambda_0)\,, \tag{4.16}$$

where we have defined μ as

$$\mu=\left(\frac{1}{N_a}\frac{\partial E_0(Q_0)}{\partial Q_0}\right)\left(\frac{1}{N_a}\frac{\partial N_0(Q_0)}{\partial Q_0}\right)^{-1}\,. \tag{4.17}$$

It is understood that in Eq. (4.16) $\rho_1(k)$ is now to be found as a solution to Eq. (4.5) with $Q=Q_0$.

To complete the calculation of the dispersion relation for the spin-wave state we must find how the momentum is related to the parameter Λ_0. From Eqs. (2.5) and (4.1) it follows that the momentum p is given by

$$p/2\pi=\int_{\Lambda_0}^{\infty}\sigma(\Lambda)\,d\Lambda\,. \tag{4.18}$$

Following the analogous treatment by des Cloizeaux and Pearson,[7] we simplify this equation by replacing $\sigma(\Lambda)$ by $\sigma_0(\Lambda)$. The omitted terms are of order $1/N_a$ and can be neglected. So, combining Eqs. (4.18) and (3.6), we find

$$\frac{p}{2\pi}=\frac{1}{\pi}\int_{-Q_0}^{Q_0}dk\,\rho_0(k)\tan^{-1}\left[\exp\left(-\frac{2\pi}{u}(\Lambda_0-\sin k)\right)\right]\,. \tag{4.19}$$

Equations (4.16) and (4.19) determine the parametric dependence of ϵ on the momentum p.

We can obtain some general properties of the spin-wave dispersion relations by examining the behavior of $\epsilon(\Lambda_0)$ and $p(\Lambda_0)$ as a function of Λ_0. From Eq. (4.5) we see that $\rho_1(-k,-\Lambda_0)=\rho_1(k,\Lambda_0)$.

This implies through Eq. (4.16) that ϵ is an even function of Λ_0. One can also show from Eq. (19), using the fact that $\rho_0(k)=\rho_0(-k)$, that $[p-\frac{1}{2}\pi(N/N_a)]$ is an odd function of Λ_0. This implies that as a function of momentum ϵ is symmetric about $p=\frac{1}{2}\pi\times(N/N_a)$. Further, one can show that as $\Lambda_0\rightarrow\infty$ both p and ϵ approach zero. We will show further in Sec. IV A that in the region of small momentum the excitation energy ϵ varies linearly with momentum.

The procedure followed to calculate the dispersion relation is to solve Eq. (4.5) for $\rho_1(k)$ numerically. The value of μ is also determined numerically in the manner demonstrated by Shiba.[5] Then the excitation energy and the momentum [using numerical results for $\rho_0(k)$] are calculated as a function of Λ_0. These results for various values of U/t are shown in Fig. 2 for $N/N_a=\frac{1}{2}$.

A. Spin-wave velocity

We would like to examine the dispersion relation for small values of the momentum. From Eq. (4.19) we see that small momentum corresponds to large values of the parameter Λ_0. Thus for large Λ_0

$$p/2\pi\rightarrow(1/\pi)e^{(-2\pi/u)\Lambda_0}\int_{-Q_0}^{Q_0}dk\,\rho_0(k)e^{(2\pi/u)\sin k}\,. \tag{4.20}$$

One can show that $1/2\pi\int_{-Q_0}^{Q_0}\rho_0(k)e^{(2\pi/u)\sin k}=I_{Q_0}^{(0)}(u)$, where the function $I_{Q_0}^{(n)}(u)$ were introduced by Shiba in the calculation of the magnetic susceptibility.[5] Thus for small momentum

$$p/2\pi=2e^{(-2\pi/u)\Lambda_0}I_{Q_0}^{(0)}(u)\,. \tag{4.21}$$

From Eq. (4.5) we see that for large Λ_0, $\rho_1(k)$ has the asymptotic form

$$\rho_1(k)=(-2/u)\cos k\,e^{(-2\pi/u)\Lambda_0}\psi(k)\,, \tag{4.22}$$

where $\psi(k)$ has been introduced by Shiba[5] and, as shown there, satisfies the equation

$$\psi(k)=e^{(2\pi/u)\sin k}+\int_{-Q_0}^{Q_0}dk'\cos k'\times\left[\frac{4}{u}R\left(\frac{4}{u}(\sin k-\sin k')\right)\right]\psi(k')\,. \tag{4.23}$$

The functions $I_{Q_0}^{(n)}(u)$ are written in terms of $\psi(k)$ as

$$I_{Q_0}^{(n)}(u)=\int_{-Q_0}^{Q_0}\frac{dk}{2\pi}\cos^n k\,\psi(k)\,. \tag{4.24}$$

Therefore for large Λ_0, $\epsilon(\Lambda_0)$ has the form

$$\epsilon(\Lambda_0)\rightarrow\frac{4t}{u}2\pi\left(I_{Q_0}^{(2)}(u)+\frac{u}{2t}I_{Q_0}^{(1)}(u)\right)e^{(-2\pi/u)\Lambda_0}\,. \tag{4.25}$$

Thus, if we define the velocity v_s as $\lim_{p\rightarrow 0}[\epsilon(p)/p)]$, we have

$$v_s=\pi\frac{4t^2}{u}\left(\frac{I_Q^{(2)}(u)}{I_Q^{(0)}(u)}+\frac{\mu}{2t}\frac{I_Q^{(1)}(u)}{I_Q^{(0)}(u)}\right)\,. \tag{4.26}$$

Comparing this result with the work of Shiba[5] we see that v_s is inversely proportional to the magnetic suceptibility:

$$v_s=\frac{2t}{\pi}\left(\frac{\chi}{N_a\mu_B^2}\right)^{-1}\,. \tag{4.27}$$

This relationship was pointed out by Takahashi[8] for the case of the half-filled band. We see that the relationship is valid for arbitrary electron density.

B. Atomic limit of spin-wave frequency

As $U/t\rightarrow\infty$ we may approximate the distribution function $\rho_1(k)$ by

$$\rho_1(k)\underset{u\rightarrow\infty}{\longrightarrow}-\frac{1}{u}\cos k\,\mathrm{sech}\left(\frac{2\pi}{u}\Lambda_0\right)\,. \tag{4.28}$$

Then for large u

$$\epsilon(\Lambda_0)\underset{u\rightarrow\infty}{\longrightarrow}\frac{2t}{u}\mathrm{sech}\left(\frac{2\pi}{u}\Lambda_0\right)\left(\int_{-Q_0}^{Q_0}\cos^2 k\,dk+\frac{\mu}{2t}\int_{-Q_0}^{Q_0}dk\cos k\right)\,. \tag{4.29}$$

For large u the momentum p assumes the simple form

$$p(\Lambda_0)=\frac{N}{N_a}\sin^{-1}\left(\mathrm{sech}\,\frac{2\pi}{u}\Lambda_0\right)\,. \tag{4.30}$$

Therefore in the large-u limit

$$\epsilon(p)=\pi\frac{4t^2}{U}\left(\int_{-Q}^{Q}\frac{dk}{2\pi}\cos^2 k+\frac{\mu}{2t}\int_{-Q}^{Q}\frac{dk}{2\pi}\cos k\right)\times\sin\frac{p}{\rho}\,, \tag{4.31}$$

where $\rho=N/N_a$ is the electron density. For $u\rightarrow\infty$ we have $Q_0\rightarrow\pi(N/N_a)$ and $\mu/2t\rightarrow-\cos\pi(N/N_a)$. So to first order in t/U we have for the dispersion relation

$$\epsilon(p)=\frac{\pi}{2}\left(\frac{4t^2}{U}\right)\rho\left(1-\frac{\sin 2\pi\rho}{2\pi\rho}\right)\sin\left(\frac{p}{\rho}\right)\,. \tag{4.32}$$

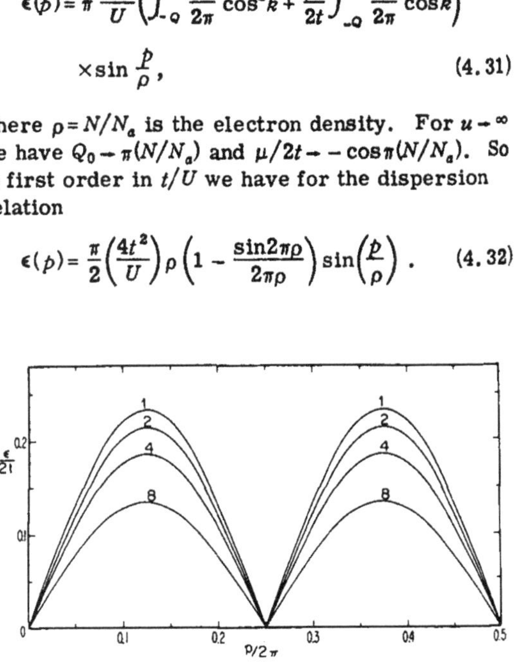

FIG. 2. Spin-wave energy for the quarter-filled band. Individual curves are labeled with the value of $u=U/t$.

Note that for $\rho = 1$ this agrees with result of des Cloizeaux and Pearson[7] for the Heisenberg antiferromagnetic chain if we identify $J = 4t^2/U$.

V. HOLE STATE

For this state, which is classified by Lieb and Wu[4] as the "hole in the k distribution" state we choose the integers J_α as in the ground state and take

$$I_{j+1} - I_j = 1 + \delta_{j,n} \ . \tag{5.1}$$

With these expressions for the integers J_α and I_j, and proceeding to a continuous distribution for the numbers k_j and Λ_α, we find the equations

$$2\pi\rho(k) = \frac{-2\pi}{N_a}\,\delta(k-k_0) + 1 + \cos k \int_{-\infty}^{\infty} d\Lambda\, \sigma(\Lambda) \times \frac{8u}{u^2 + 16(\sin k - \Lambda)^2}\,, \tag{5.2a}$$

$$\int_{-Q}^{Q} dk\, \rho(k) \frac{8u}{u^2 + 16(\sin k - \Lambda)^2} = 2\pi\sigma(\Lambda) + \int_{-\infty}^{\infty} d\Lambda'\, \sigma(\Lambda') \frac{4u}{u^2 + 4(\Lambda - \Lambda')^2}\,, \tag{5.2b}$$

where k_0 is the value of k for which $j = n$ and $|k_0| \leq Q_0$ will be determined from the ground-state distribution.

We again introduce distribution functions $\rho_1(k)$ and $\sigma_1(\Lambda)$ defined by Eqs. (4.3) and find that these distribution functions satisfy the equations

$$2\pi\rho_1(k) = -2\pi\delta(k-k_0) + \cos k \int_{-\infty}^{\infty} d\Lambda\, \sigma_1(\Lambda) \times \frac{8u}{u^2 + 16(\sin k - \Lambda)^2}\,, \tag{5.3a}$$

$$\int_{-Q}^{Q} dk\, \rho_1(k) \frac{8u}{u^2 + 16(\sin k - \Lambda)^2} = 2\pi\sigma_1(\Lambda) + \int_{-\infty}^{\infty} d\Lambda'\, \sigma_1(\Lambda') \frac{4u}{u^2 + 4(\Lambda - \Lambda')^2}\,. \tag{5.3b}$$

We isolate the δ-function term in the equation for the distribution function by writing

$$\rho_1(k) = -\delta(k-k_0) + \rho_1'(k)\ . \tag{5.4}$$

Then the equations determining the distribution functions are given by

$$2\pi\rho_1'(k) = \cos k \int_{-\infty}^{\infty} d\Lambda\, \sigma_1(\Lambda) \frac{8u}{u^2 + 16(\sin k - \Lambda)^2}\,, \tag{5.5a}$$

$$\int_{-Q}^{Q} dk\, \rho_1'(k) \frac{8u}{u^2 + 16(\sin k - \Lambda)^2} = 2\pi\sigma_1(\Lambda) + \int_{-\infty}^{\infty} d\Lambda'\, \sigma_1(\Lambda') \frac{4u}{u^2 + 4(\Lambda - \Lambda')^2} + \frac{8u}{u^2 + 16(\sin k_0 - \Lambda)^2}\,. \tag{5.5b}$$

We construct an integral equation for $\rho_1'(k)$ by introducing the Fourier transform of $\sigma_1(\Lambda)$, as we did in Sec. IV. We find

$$\rho_1'(k) = -\frac{4}{u}\cos k\, R\left(\frac{4}{u}(\sin k - \sin k_0)\right) + \frac{4}{u}\cos k \times \int_{-Q}^{Q} dk'\, R\left(\frac{4}{u}(\sin k - \sin k')\right) \rho_1'(k')\ . \tag{5.6}$$

The energy of the hole state relative to the ground state is found to be

$$\epsilon = 2t\cos k_0 - 2t\int_{-Q}^{Q} \rho_1'(k)\cos k\, dk + E_0(Q) - E_0(Q_0)\,, \tag{5.7}$$

where again Q_0 is the limiting momentum for the ground-state distribution for a fixed electron density N/N_a and $E_0(Q)$ is the ground-state energy for fixed Q.

Q is determined in terms of Q_0 by the requirement that the distribution functions $\rho(k)$ describe the same electron density as the ground-state distribution function. Using this requirement and keeping all terms in ϵ of first order we find for the excitation energy the result

$$\epsilon(k_0) = 2t\cos k_0 - 2t\int_{-Q_0}^{Q_0} \rho_1'(k)\cos k\, dk + \mu\left(1 - \int_{-Q_0}^{Q_0} dk\, \rho_1'(k)\right)\,, \tag{5.8}$$

where μ is as defined in Eq. (4.17). Again it is understood that the distribution function $\rho_1'(k)$ in Eq. (5.8) is a solution to Eq. (5.6) with $Q = Q_0$ there.

The momentum is related to the parameter k_0 by the equation

$$\frac{p}{2\pi} = \int_{k_0}^{Q_0} \rho(k)\, dk \simeq \int_{k_0}^{Q_0} \rho_0(k)\, dk\,, \tag{5.9}$$

where we may, with sufficient accuracy in the limit of a large system, treat the approximation in Eq. (5.9) as an equality.

After some manipulation we can write an expression for p in terms of $\rho_0(k)$ as

$$\frac{p}{2\pi} = \frac{1}{2}\frac{N}{N_a} - \frac{k_0}{2\pi} - \int_{-Q_0}^{Q_0} dk\, \rho_0(k) \times F\left(\frac{4}{u}(\sin k_0 - \sin k)\right)\,. \tag{5.10}$$

where $F(x)$ is defined in terms of $R(x)$ as

$$F(x) = \int^{x} dx'\, R(x')\ . \tag{5.11}$$

From Eqs. (5.8) and (5.10) we can examine some general features of the dispersion relation. From Eq. (5.6) we see that $\rho_1'(-k, -k_0) = \rho_1'(k, k_0)$. This implies through Eq. (5.8) that ϵ is an even function of k_0. From Eq. (5.10), using the fact that $F(x)$ is an odd function of its argument, we find that $p - \pi(N/N_a)$ is an odd function of k_0. This implies that as a function of momentum ϵ is symmetric about $p=\pi N/N_a$.

One can also show that $-[\rho_1'(k,Q_0)+\rho_1'(-k,Q_0)]$ satisfies the same equation as $[1/\rho_0(Q_0)]\partial\rho_0(k)/\partial Q_0$. This implies that we can write

$$\frac{\mu}{2t}=-\left(\frac{\cos Q_0-\int_{-Q_0}^{Q_0}dk\cos k\rho_1'(k,Q_0)}{1-\int_{-Q_0}^{Q_0}dk\,\rho_1'(k,Q_0)}\right). \quad (5.12)$$

From this relation and Eq. (5.8) it is evident that

$$\epsilon(k_0)\xrightarrow[k_0\to\pm Q_0]{}0 .$$

One can also show by integrating Eq. (3.2a) that p vanishes in the same limit.

The dispersion curves are found as a function of U/t by solving Eq. (6) for $\rho_1'(k)$ numerically and using numerical results for $\rho_0(k)$.

VI. PARTICLE STATE

Here we have in mind removing an electron from the highest occupied momentum in the ground state Q_0 and placing it in a momentum state $k_0>Q_0$. We choose the integers J_α as in the ground state. The numbers I_j are chosen as follows:

$$I_{j+1}-I_j=1,\quad j=1,\ldots,N-2 \quad (6.1a)$$

$$I_N-I_{N-1}\gg 1 . \quad (6.1b)$$

So we find from Eqs. (2.6)

$$2\pi\rho(k_j)=1+\cos k_j\sum_{\beta=1}^{m}\frac{8u}{u^2+16(\sin k_j-\Lambda_\beta)^2},\quad j=1,\ldots,N-2 \quad (6.2)$$

$$\frac{1}{N_a}\sum_{j=1}^{N}\frac{8u}{u^2+16(\sin k_j-\Lambda_\beta)^2}=2\pi\sigma(\Lambda_\alpha)+\frac{1}{N_a}\sum_{\beta=1}^{m}\frac{4u}{u^2+4(\Lambda_\alpha-\Lambda_\alpha)^2} . \quad (6.3)$$

Separating from Eq. (6.3) the term with $k_N=k_0$ and proceeding to the limit of a large system, we find for the distribution function $\rho(k)$ and $\sigma(\Lambda)$ the equations

$$2\pi\rho(k)=1+\cos k\int_{-\infty}^{\infty}d\Lambda\,\sigma(\Lambda)\frac{8u}{u^2+16(\sin k-\Lambda)^2}, \quad (6.4a)$$

$$\int_{-Q}^{Q}dk\,\rho(k)\frac{8u}{u^2+16(\sin k-\Lambda)^2}=2\pi\sigma(\Lambda)+\int_{-\infty}^{\infty}d\Lambda'\,\sigma(\Lambda')\frac{4u}{u^2+4(\Lambda-\Lambda')^2}-\frac{1}{N_a}\times\frac{8u}{u^2+16(\sin k_0-\Lambda)^2} . \quad (6.4b)$$

It must be recognized that in Eqs. (6.4) $\rho(k)$ is the distribution function for $N-1$ electrons; i.e., it satisfies the normalization condition

$$\int_{-Q}^{Q}dk\,\rho(k)=\frac{N-1}{N_a} . \quad (6.5)$$

We introduce the distribution functions $\rho_1(k)$ and $\sigma_1(\Lambda)$ defined by Eq. (4.3). These distribution functions satisfy the equations

$$2\pi\rho_1(k)=\cos k\int_{-\infty}^{\infty}d\Lambda\,\sigma_1(\Lambda)\frac{8u}{u^2+16(\sin k-\Lambda)^2}, \quad (6.6a)$$

$$\int_{-Q}^{Q}dk\,\rho_1(k)\frac{8u}{u^2+16(\sin k-\Lambda)^2}=2\pi\sigma_1(\Lambda)+\int_{-\infty}^{\infty}d\Lambda'\frac{\sigma_1(\Lambda')4u}{u^2+4(\Lambda-\Lambda')^2}-\frac{8u}{u^2+16(\sin k_0-\Lambda)^2} . \quad (6.6b)$$

We find an integral equation for $\rho_1(k)$ alone by again introducing the Fourier transform of $\sigma_1(\Lambda)$:

$$\rho_1(k)=\frac{4}{u}\cos k\,R\left(\frac{4}{u}(\sin k-\sin k_0)\right)+\frac{4}{u}\cos k\times\int_{-Q}^{Q}dk'R\left(\frac{4}{u}(\sin k-\sin k')\right)\rho_1(k') . \quad (6.7)$$

Notice the similarity between the equation and Eq. (5.6). The differences are that here $|k_0|\geq Q_0$ and the inhomogeneous term has the opposite sign.

The energy of the particle state relative to that of the ground state is given by

$$\epsilon=-2t\cos k_0-2t\int_{-Q}^{Q}\cos k\rho_1(k)\,dk+E_0(Q)-E_0(Q_0) . \quad (6.8)$$

If we relate Q to Q_0 by the condition that the density of electrons is fixed, we find

$$\epsilon(k_0)=-2t\cos k_0-2t\int_{-Q_0}^{Q_0}dk\cos k\rho_1(k,k_0)-\mu\left(1+\int_{-Q_0}^{Q_0}dk\rho_1(k,k_0)\right) . \quad (6.9)$$

By manipulations similar to those mentioned in Sec. V we can show that

$$\mu=-2t\left(\frac{\cos Q_0+\int_{-Q_0}^{Q_0}dk\cos k\rho_1(k,Q_0)}{1+\int_{-Q_0}^{Q_0}dk\,\rho_1(k,Q_0)}\right). \quad (6.10)$$

From this relation it is evident that $\epsilon(k_0)\to 0$ as $k_0\to Q_0$.

The momentum of the particle state is related to the parameter k_0 by the equation

$$\frac{p}{2\pi}=\int_{Q_0}^{k_0}\rho_0(k)\,dk,\quad k_0\geq Q_0 . \quad (6.11)$$

We can rewrite this in terms of the function F [Eq. (5.11)] as

$$\frac{p}{2\pi}=\frac{-1}{2}\frac{N}{N_a}+\frac{k_0}{2\pi}+\int_{-Q_0}^{Q_0}dk\,\rho_0(k)F\left(\frac{4}{u}(\sin k_0-\sin k)\right) . \quad (6.12)$$

From Eqs. (6.9) and (6.12) we can examine the general properties of ϵ as a function of momentum. Since $\rho_1(k,k_0)=\rho_1(-k,-k_0)$ we see that $\epsilon(k_0)$ is an even function of k_0. Likewise, we see from (6.12) that $p+\pi(N/N_a)$ is an odd function of k_0. This implies that as a function of momentum $\epsilon(p)$ is symmetric about $\pm\pi N/N_a$.

The dispersion curves are found as a function of U/t by solving numerically Eq. (6.7) for $\rho_1(k)$ and using numerical results for $\rho_0(k)$.

VII. PARTICLE-HOLE EXCITATIONS

We can combine the results of Secs. V and VI to calculate the energy and momentum of those states which arise by removing one electron from a momentum level k_0 occupied in the ground state and placing it in a momentum level p_0 not occupied in the ground state. The energy and momentum of this state, denoted by $\epsilon(k_0,p_0)$ and $p(k_0,p_0)$, respectively, will depend parametrically on the quantities k_0 and p_0, the momenta of the "hole" and the "electron," respectively. The energy $\epsilon(k_0,p_0)$ is defined as

$$\epsilon(k_0,p_0)=[E(N+1,p_0)-E_0(N)]-[E_0(N)-E(N-1,k_0)]\ . \tag{7.1}$$

In Eq. (7.1) $E_0(N)$ is the ground-state energy for N particles, $E(N+1,p_0)$ is the energy of that state which arises by adding one electron with momentum p_0 to the ground state of N electrons, and $E(N-1,k_0)$ is the energy of that state which arises by removing an electron with momentum k_0 from the ground state of N electrons. As k_0 (or p_0) $\rightarrow Q_0$, $[E_0(N)-E(N-1,k_0)]$ $\{[E(N+1,p_0)-E_0(N)]\}$ $\rightarrow \mu_-$ (μ_+). The parameters μ_+ and μ_- were introduced by Lieb and Wu.[4] The explicit expression for $\epsilon(k_0,p_0)$ is

$$\epsilon(k_0,p_0)=2t\cos k_0-2t\cos p_0-\mu\int_{-Q_0}^{Q_0}[\rho_h(k,k_0)+\rho_p(k,p_0)]\,dk-2t\int_{-Q_0}^{Q_0}[\rho_h(k,k_0)+\rho_p(k,p_0)]\cos k\,dk\ , \tag{7.2}$$

where $\rho_h(k,k_0)$ and $\rho_p(k,p_0)$ satisfy Eqs. (5.6) and (6.7), respectively. The quantity μ is defined in Eq. (4.17).

We have calculated the energy $\epsilon(k_0,p_0)$ and momentum $p(k_0,p_0)$ by using numerical results for the distribution functions ρ_h and ρ_p. We then find a band of states for the "particle-hole" excitations. We show in Fig. 3 results for the quarter-filled band for various values of U/t.

VIII. SUMMARY AND CONCLUSIONS

Using the formalism of Lieb and Wu[4] we have been able to write down the integral equations determining the dispersion relation for excitations having either single-particle or spin-wave character. We have restricted attention in this investigation to a class of states in which the parameters k_j and Λ_α of Eqs. (2.2) and (2.3) are real. This is an extension of the work of Ovchinnikov[6] to the situation in which the number of electrons is less than the number of sites in the chain. The results are found as numerical solutions to a set of coupled integral equations. Analytic results are given for some limiting cases.

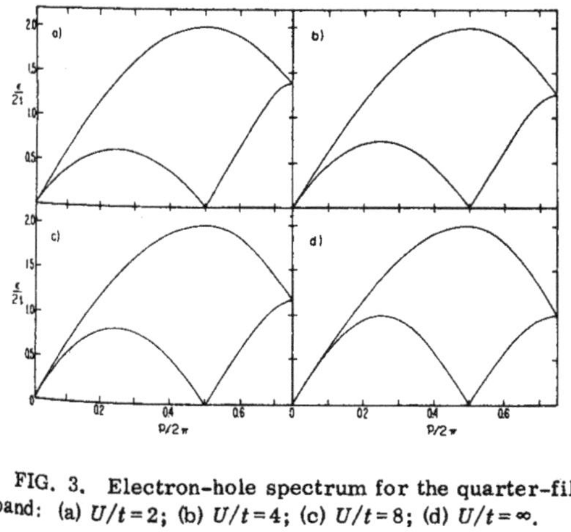

FIG. 3. Electron-hole spectrum for the quarter-filled band: (a) $U/t=2$; (b) $U/t=4$; (c) $U/t=8$; (d) $U/t=\infty$.

For the case of the single-particle excitations we have been able to demonstrate that there is no gap in the spectrum for $N/N_a<1$, unlike the result for the half-filled band ($N/N_a=1$). Thus, according to the criterion of Lieb and Wu,[4] the system has the properties of a conductor regardless of the magnitude of U/t. As a consequence one would expect a linear term ($\sim\gamma T$) in the specific heat from thermal excitation of these modes. We have graphically displayed the shape of the single-particle band for $N/N_a<1$, and the numerical results indicate that the shape is relatively insensitive to the magnitude of U/t, at least if $U/t\geq 1$.

Several interesting features emerge from the investigation of the excitations of the spin-wave type. We find that the period of the spin-wave excitation energy is incommensurate with the periodicity of the lattice unless N_a/N equals an integer. For small momentum the energy varies linearly with momentum. This is the type of behavior one associates with antiferromagnetic systems. Therefore one expects[2] a linear contribution ($\sim\alpha_M T$) to the low-temperature specific heat from thermal excitation of these modes. This contribution may be difficult to isolate experimentally from the contribution expected from the single-particle excitations. It has also been shown that the slope of the spin-wave dispersion curve for small momentum is a measure of the inverse of the static magnetic susceptibility. This relationship, first noted for the half-filled band by Takahashi,[8] is seen to be valid for arbitrary electron density.

In attempting to apply the results of these calculations to the interpretation of the experimental results one is beset by at least two difficulties. First, one has only partial knowledge of the spectrum of low-lying states. Assuredly there are other modes, not enumerated, which need to be considered. In fact, Ovchinnikov[6] showed that for the half-filled band there exist spin-wave bound states which certainly contribute to the low-temperature thermodynamic properties of the system. Second, as emphasized in Ref. 2, there is the uncertainty of the contribution of each mode to the thermodynamic properties; i.e., we don't know the spectral weight function.

ACKNOWLEDGMENTS

The author would like to thank Professor P. Pincus for many useful discussions throughout the course of this work. In addition, the author is grateful to R. Luskin for advice concerning the numerical calculations.

*Work supported in part by the National Science Foundation under Contract No. GH35689X.

[1]R. B. Griffiths, Phys. Rev. 135, A659 (1964).

[2]A. J. Epstein, S. Etemad, A. F. Garito, and A. J. Heeger, Phys. Rev. B 5, 952 (1972).

[3]J. Hubbard, Proc. R. Soc. A 276, 238 (1963).

[4]E. H. Lieb and F. Y. Wu, Phys. Rev. Lett. 20, 1445 (1968).

[5]H. Shiba, Phys. Rev. B 6, 930 (1972).

[6]A. A. Ovchinnikov, Zh. Eksp. Teor. Fiz. 57, 2137 (1969) [Sov. Phys.-JETP 30, 1160 (1970)].

[7]J. des Cloizeaux and J. J. Pearson, Phys. Rev. 128, 2131 (1962).

[8]M. Takahashi, Prog. Theor. Phys. 43, 1619 (1970).

J. Phys. C: Solid State Phys., 15 (1982) 85–96. Printed in Great Britain

Excitations with complex wavenumbers in a Hubbard chain: I. States with one pair of complex wavenumbers

F Woynarovich
Central Research Institute for Physics, H-1525 Budapest, POB 49, Hungary

Received 26 August 1980, in final form 23 June 1981

Abstract. Those excited states of a half-filled 1D Hubbard chain that are connected with electron pairs occupying the same sites are studied. It is argued that these states should be described by solutions of the Lieb–Wu equations in which some of the wavenumbers are complex. Solutions of this type have been found that correspond to states in which the spin part is not excited. The energy–momentum dispersion is also calculated. The gap in the spectrum of the singlet excitations equals the discontinuity of the chemical potential calculated by Lieb and Wu.

1. Introduction

The one-dimensional Hubbard model, being a non-trivial but exactly treatable model for interacting spin $\frac{1}{2}$ fermions, is of great theroretical interest. It describes electrons that can hop between the Wannier states of neighbouring sites in a chain and exhibit repulsion if two of them (with opposite spins) occupy the same site. The Hamiltonian of the model is

$$\hat{H} = t\sum_{i=1}^{N}\sum_{\sigma}(c_{i\sigma}^{+}c_{i+1\sigma} + c_{i+1\sigma}^{+}c_{i\sigma}) + U\sum_{i=1}^{N} n_{i\uparrow}n_{i\downarrow}. \tag{1.1}$$

Here N is the number of sites on the chain; $c_{i\sigma}^{+}$, $c_{i\sigma}$ and $n_{i\sigma}$ are the creation, annihilation and number operators, respectively, for an electron with spin σ in the Wannier state centred around the site i. The problem is uniquely defined by imposing periodic boundary conditions on the system.

In the exact solution of the model the first step, providing a base for all further work, was made by Lieb and Wu (1968). These authors, starting from Yang's solution for the continuum version of the model (Yang 1967), showed that the diagonalisation of (1.1) is equivalent to solving a set of coupled non-linear equations. They calculated the ground-state energy of the system for a half-filled band, and the gap in the spectrum of the one-particle type excitations at this band filling (half-filled band ± one particle). The equations set up by Lieb and Wu provided the basis for the Ovchinnikov (1970) calculation of the lower edge of the continuum of the triplet excitations of a half filled band. Calculations for the singlet excitations were also given in the same paper. Coll (1974) determined the spin-wave-type and one-particle-type excitations for general band filling. The $T = 0$ magnetic properties of the model were worked out by Takahashi (1969) and by Shiba (1972); in particular Takahashi found the magnetisation curve for the half-filled band and, extending this work, Shiba gave the magnetic susceptibility for an arbitrary concentration of electrons.

The aim of the present work is to study those excitations of a Hubbard chain that are connected with charge rearrangement. In a non-half-filled band two kinds of such excitations exist. The first kind, which have been described by Coll (1974), differ from the ground state only in the momentum distribution of the electrons. The number of excitations of this type disappears as the band filling approaches $\frac{1}{2}$; a fact that suggests that in all these states the electrons occupy different lattice sites (at least in the large-U limit). The other type of 'charge excitations' is connected with electron pairs occupying the same lattice sites. Our aim is to find a way to describe such excitations. For the sake of simplicity, the half-filled band is studied first, since in this case charge excitations of the first kind do not exist.

The paper is organised as follows. In § 2, after introducing the general formalism an analysis of the wavefunction leads us to argue that complex wavenumbers have to be used to describe the states in question. In § 3 we give the solutions of the Lieb–Wu equations, in which the k-set has two complex ks. We restrict our study to states whose spin degrees of freedom are not excited.

2. The Lieb–Wu equations; some properties of the eigenstates

2.1. The Lieb–Wu equations and the eigenfunctions

Using Yang's method, Lieb and Wu showed that the finding of an eigenstate of the Hamiltonian (1.1) (with $t = -1$) is equivalent to solving the system of equations

$$Nk_j = 2\pi\mathcal{J}_j - \sum_{\beta=1}^{M} 2\tan^{-1}\frac{4}{U}(\sin k_j - \lambda_\beta) \qquad (j = 1, 2, \ldots N_e) \tag{2.1}$$

$$\sum_{j=1}^{N_e} 2\tan^{-1}\frac{4}{U}(\lambda_\alpha - \sin k_j) = 2\pi\mathcal{F}_\alpha + \sum_{\beta=1}^{M} 2\tan^{-1}\frac{2}{U}(\lambda_\alpha - \lambda_\beta) \qquad (\alpha = 1, 2, \ldots M). \tag{2.2}$$

Here N_e and M are the number of electrons and the number of down spins, respectively. It is supposed that $N_e \leqslant N$ and $M \leqslant N_e - M$. (Any state for which these conditions do not hold can be transformed to the required form by introducing holes instead of the electrons and/or changing the spin directions.) The parameters $\mathcal{J}$ are integers (or half odd-integers) if M is even (or odd) and the parameters $\mathcal{F}$ are integers (or half odd-integers) if $N_e - M$ is odd (or even). In this system of equations the ks and λs are the unknowns and the parameters $\mathcal{J}$ and $\mathcal{F}$ are the actual quantum numbers specifying the state. Only those solutions for which all the ks and λs are different are meaningful. A special difficulty arises because it is not clear for which sets of $\mathcal{J}$ and $\mathcal{F}$ can a meaningful solution of equations (2.1) and (2.2) be found. Thus the solving of (2.1) and (2.2) represents a two-fold task: the need to find the appropriate $\mathcal{J}$ and $\mathcal{F}$ sets and the need to find the corresponding ks and λs.

The energy belonging to a state corresponding to a solution of equations (2.1)–(2.2) is

$$E = -\sum_{j=1}^{N_e} 2\cos k_j. \tag{2.3}$$

The momentum is

$$p = \sum_{j=1}^{N_e} k_j \tag{2.4a}$$

which, by summing (2.1) and (2.2), yields

$$p = \frac{2\pi}{N}\left(\sum_{j=1}^{N_e} \mathcal{I}_j + \sum_{\alpha=1}^{M} \mathcal{F}_\alpha\right). \tag{2.4b}$$

If the reasoning that led to (2.1)–(2.2) is followed the wavefunction can be constructed. One finds that the amplitude of finding the electrons at positions $n_1, n_2, n_3, \ldots, n_{N_e}$ with spins $\sigma_1, \sigma_2, \sigma_3, \ldots, \sigma_{N_e}$ is (up to a normalisation factor)

$$f(n_1\sigma_1, n_2\sigma_2, \ldots n_{N_e}\sigma_{N_e})$$

$$= \sum_P (-1)^Q(-1)^P \exp\left(i\sum_{j=1}^{N_e} k_{P_j} n_{Q_j}\right) \varphi_P(\sigma_{Q1}, \sigma_{Q2}, \ldots \sigma_{QN_e}) \tag{2.5}$$

where the permutation Q is defined by the condition

$$1 \leq n_{Q1} \leq n_{Q2} \leq \ldots \leq n_{QN_e} \leq N \tag{2.6}$$

and the summation is extended over all permutations P of the ks. The function φ_P is given in the form

$$\varphi_P(\sigma_{Q1}, \sigma_{Q2}, \ldots \sigma_{QN_e}) = \sum_\pi A(\lambda_{\pi 1}, \lambda_{\pi 2}, \ldots \lambda_{\pi M})\left(\prod_{l=1}^{M} F_P(\lambda_{\pi l}; y_l)\right) \tag{2.7}$$

with

$$F_P(\lambda; y) = \left(\prod_{j=1}^{y-1} \frac{i(\sin k_{Pj} - \lambda) - U/4}{i(\sin k_{Pj} - \lambda) + U/4}\right) \frac{1}{i(\sin k_{P_y} - \lambda) + U/4} \tag{2.8}$$

and

$$\frac{A(\ldots \lambda_{\pi i}, \lambda_{\pi i+1} \ldots)}{A(\ldots \lambda_{\pi i+1}, \lambda_{\pi i} \ldots)} = \frac{i(\lambda_{\pi i+1} - \lambda_{\pi i}) - U/2}{i(\lambda_{\pi i+1} - \lambda_{\pi i}) + U/2} \tag{2.9}$$

where the ys are the positions of the down spins in the series $\sigma_{Q1}, \sigma_{Q2}, \ldots, \sigma_{QN_e}$ in increasing order

$$1 \leq y_1 < y_2 < \ldots < y_M \leq N_e. \tag{2.10}$$

In connection with the wavefunction (2.5)–(2.10) it should be noted that it is uniquely defined even if Q is not: if for example n_i and n_j are equal, there are two permutations, Q and $Q' = QP_{ij}$ which arrange the spatial coordinates into non-decreasing order but the value of f does not depend on the choice of Q or Q' in (2.5).

2.2. The ground state and simple excitations

According to Lieb and Wu the ground state is characterised by the parameter set in which both the $\mathcal{I}_j$s and $\mathcal{J}_\alpha$s are consecutive integers (or half odd-integers) centred around the origin. The ground state is a singlet (if N_e is even) with $M = N_e/2$ or a doublet, $S^z = \pm\frac{1}{2}$, $S^2 = \frac{3}{4}$ (if N_e is odd).

The simplest excited states can be obtained by making small changes in the $\mathcal{I}$ set, the $\mathcal{J}$ set or in both. The simplest excitations with one spin turned over are triplet spin waves and are described by $\mathcal{J}$ sets in which one of the $\mathcal{J}_\alpha$s of the ground state set is missing ('hole in the λ distribution' in the Lieb and Wu classification). The simplest excitations connected with the $\mathcal{J}$ distribution may be 'hole-', 'particle-' or 'particle–hole'-type ones; they are all described by certain well defined modifications of the ground state $\mathcal{I}$ set. The 'hole'-type excitations, in accordance with Coll's description, are defined by $\mathcal{I}$ sets in which one $\mathcal{I}_j$ is removed from the bulk of the set and another one is added to the set at one of its ends. In the case of 'particle'-type excitations the change in the $\mathcal{I}$ set is performed by removing one $\mathcal{I}_j$ from one of the ends of the set and adding to it another $\mathcal{I}_{j'}$ that is outside the region covered by the ground state set. The 'particle–hole' type of excitation is, as the name suggests, a combination of the other two types.

All the above described excitations are connected with real sets of ks and λs, and are discussed in detail in Coll's paper. Here, we would just point out that for a half-filled band ($N_e = N$) neither of the excitations connected with the k distribution and discussed above can exist. This can be seen as follows. All $\mathcal{I}_j$ can be taken as satisfying $-N/2 \leqslant \mathcal{I}_j \leqslant N/2$, because changing k_j by 2π does not affect anything. In the above region there are N different integers or half odd-integers, thus the ground state $\mathcal{I}$ set for N electrons just covers this region: one can displace no $\mathcal{I}$ outside the ground state $\mathcal{I}$ set. As a consequence, the excited states of the system of $N_e = N$ electrons connected with the k distribution cannot be described in the framework given by Coll.

2.3. Eigenstates with real ks in the large-U limit

Looking at (2.2) one sees that in the large-U limit all $\lambda_\alpha - \sin k_j$ must be of the order of U. As for real k_js, $|\sin k_j| \leqslant 1$, the λ_αs must be proportional to U. Thus the $\sin k_j$ terms can be neglected in relation to the λ_αs and the limiting values of the λ_αs must satisfy the equations

$$N_e 2\tan^{-1}\frac{4\lambda_\alpha}{U} = 2\pi\mathcal{J}_\alpha + \sum_{\beta=1}^{M} 2\tan^{-1}\frac{1}{2}\left(\frac{4\lambda_\alpha}{U} - \frac{4\lambda_\beta}{U}\right). \tag{2.11}$$

These are essentially the secular equations of an isotropic Heisenberg chain with N_e sites (see e.g. Griffiths 1964) independently of the actual values of the k_js. (In fact, the substitution $2\tan^{-1} 4\lambda_\alpha/U = \pi - k_\alpha$ leads to the form of equations used in the literature of the Heisenberg chain.) The limiting values of the wavenumbers in the same approximation are (see (2.1))

$$k_j = \frac{2\pi}{N}\mathcal{I}_j + \frac{1}{N}\sum_{\beta=1}^{M} 2\tan^{-1}\frac{4\lambda_\beta}{U}. \tag{2.12}$$

They are essentially the ks of a non-interacting spinless Fermi system with the modification that all ks are displaced by $1/N$ times the total momentum of the given state of the Heisenberg chain.

It is not hard to see that also the energy and the wavefunction separate into two more or less independent parts. The energy becomes the sum of the kinetic energy of the spinless Fermi system and the energy of the Heisenberg chain; for the latter the effective coupling is proportional to $1/U$; the wavefunction factorises into the product of the wavefunctions of the Fermi system and the Heisenberg chain. In this limiting process the ground state goes over to the ground state of both the Fermi system and the Heisenberg chain. Excitations connected with the $\mathcal{F}$ set correspond to the excitations of the Heisenberg chain while the excitations generated by the $\mathcal{I}$ set and described by real ks will be the excitations of the Fermi system. This is in good agreement with the fact that the number of excited states connected with the $\mathcal{I}$ set and described by real ks disappears as the number of electrons approaches the number of sites.

Because in a spinless Fermi system all particles occupy different sites, in all states described by real ks the amplitude of finding electron pairs occupying the same sites must disappear as $U \to \infty$. All the states in which this amplitude does not vanish must be described by k-sets containing complex wavenumbers too. The energy of these states is expected to have a term proportional to U; that is these states are important if U is of the order of unity, but they are also important if the band filling is near to $\frac{1}{2}$ being the only excitations connected with the charge distribution. The goal of the present study is to find the solutions of (2.1)–(2.2) corresponding to these states.

3. Singlet states with one pair of complex wavenumbers

3.1. The Lieb–Wu equations for singlet states with one pair of complex wavenumbers

To describe $S^z = 0$ states in a half-filled band we need to introduce $N/2$ λs. For this, it is supposed that the number of sites and thus the number of electrons is even. This restriction makes no difference with regard to the nature of the excitations since if N were odd, both the ground state and the excited states analogous to those discussed here would belong to $S^z = \pm\frac{1}{2}$, in which case both for even and odd N the excitations themselves can be regarded as singlet ones.

In possession of two complex ks it is clear that each must be the complex conjugate of the other, otherwise the momentum would not be real. We will denote them as $\kappa \pm \mathrm{i}\chi$. The equations for them can be written in the form

$$N(\kappa \pm \mathrm{i}\chi) = 2\pi\mathcal{I} - \sum_{\beta=1}^{N/2-1} 2\tan^{-1}\frac{4}{U}[\sin(\kappa \pm \mathrm{i}\chi) - \lambda_\beta]$$
$$- 2\tan^{-1}\frac{4}{U}[\sin(\kappa \pm \mathrm{i}\chi) - \Lambda]. \tag{3.1}$$

Here we have separated one λ (the one denoted by Λ) from the others. We have to do this for the following reason. We may suppose that the presence of an excitation will modify the λ distribution by terms of the order of $1/N$ only. If so, the RHS of (3.1) can be estimated by means of the ground-state distribution of the λs. It turns out that at any value of the κ and χ the difference between the two sides of (3.1) is of the order of N; that is the equation cannot be satisfied unless there is one term on the RHS that is in itself of the order of N. This term is separated. To satisfy the imaginary part of (3.1), the connection between κ, χ and Λ must be

$$\sin(\kappa \pm \mathrm{i}\chi) = \Lambda \mp \mathrm{i}U/4 + \mathrm{O}(\mathrm{e}^{-\eta N}) \tag{3.2}$$

where

$$\eta = \chi + \mathrm{Im}\left(\frac{1}{N}\sum_{\beta=1}^{N/2-1} 2\tan^{-1}\frac{4}{U}[\sin(\kappa + i\chi) - \lambda_\beta]\right) \tag{3.3}$$

and $O(e^{-\eta N})$ denotes the terms of the order of $e^{-\eta N}$. Since η is a positive number of the order of unity, in the large-N limit the $O(e^{-\eta N})$ terms can be neglected in (3.2), and the remaining equation yields

$$\kappa = \sin^{-1}[\tfrac{1}{2}\{[(U/4)^2 + (\Lambda + 1)^2]^{1/2} - [(U/4)^2 + (\Lambda - 1)^2]^{1/2}\}] \qquad (\cos\kappa < 0) \tag{3.4a}$$

$$\chi = \cosh^{-1}[\tfrac{1}{2}\{[(U/4)^2 + (\Lambda + 1)^2]^{1/2} + [(U/4)^2 + (\Lambda - 1)^2]^{1/2}\}] \qquad (\chi > 0). \tag{3.4b}$$

By means of (3.2) the real part of (3.1) can be written in a simpler form. It is not hard to check that the identity

$$\mathrm{Re}\, 2\tan^{-1}(4/U)\,[\sin(\kappa + i\chi) - \lambda] = (\pi/2)\,\mathrm{sign}(\Lambda - \lambda) + \tan^{-1}[(2/U)(\Lambda - \lambda)] \tag{3.5}$$

holds up to terms exponentially small in N. With (3.5) the real part of (3.1) takes the form

$$N\kappa = 2\pi\mathcal{I} - \frac{\pi}{2}\sum_{\beta=1}^{N/2-1}\mathrm{sign}(\Lambda - \lambda_\beta) - \sum_{\beta=1}^{N/2-1}\tan^{-1}\frac{2}{U}(\Lambda - \lambda_\beta)$$
$$- \mathrm{Re}\, 2\tan^{-1}\frac{4}{U}[\sin(\kappa + i\chi) - \Lambda]. \tag{3.6}$$

In the following this equation together with (3.5) will be used.

The real ks are defined by the equations

$$Nk_j = 2\pi\mathcal{I}_j - \sum_{\beta=1}^{N/2-1} 2\tan^{-1}\frac{4}{U}(\sin k_j - \lambda_\beta) - 2\tan^{-1}\frac{4}{U}(\sin k_j - \Lambda). \tag{3.7}$$

For the $\mathcal{I}_j$ set, depending on the parity of $N/2$, one has to choose one of the sets

$$-\tfrac{1}{2}(N-2), -\tfrac{1}{2}(N-4), \ldots, \tfrac{1}{2}(N-2), \tfrac{1}{2}N \qquad (N/2 \text{ even}) \tag{3.8a}$$

$$-\tfrac{1}{2}(N-1), -\tfrac{1}{2}(N-3), \ldots, \tfrac{1}{2}(N-3), \tfrac{1}{2}(N-1) \qquad (N/2 \text{ odd}) \tag{3.8b}$$

with two holes left in it. Equation (3.7) additionally defines ks to the $\mathcal{I}$ parameters left out of the $\mathcal{I}_j$ set, we will denote these ks as k_l and k_m.

The equations for the λs are

$$\sum_{j\neq l,m} 2\tan^{-1}\frac{4}{U}(\Lambda - \sin k_j) + 4\,\mathrm{Re}\tan^{-1}\frac{4}{U}[(\Lambda - \sin(\kappa + i\chi)]$$
$$= 2\pi\mathcal{F} + \sum_{\beta=1}^{N/2-1} 2\tan^{-1}\frac{2}{U}(\Lambda - \lambda_\beta) \tag{3.9}$$

and

$$\sum_{j\neq l,m} 2\tan^{-1}\frac{4}{U}(\lambda_\alpha - \sin k_j) + 4\,\mathrm{Re}\tan^{-1}\frac{4}{U}[\lambda_\alpha - \sin(\kappa + i\chi)]$$
$$= 2\pi\mathcal{F}_\alpha + \sum_{\beta=1}^{N/2-1} 2\tan^{-1}\frac{2}{U}(\lambda_\alpha - \lambda_\beta) + 2\tan^{-1}\frac{2}{U}(\lambda_\alpha - \Lambda). \tag{3.10}$$

This second equation (3.10) combined with (3.5) is equivalent to

$$\sum_{j\neq l,m} 2\tan^{-1}\frac{4}{U}(\lambda_\alpha - \sin k_j) = 2\pi\mathcal{F}'_\alpha + \sum_{\beta=1}^{N/2-1} 2\tan^{-1}\frac{2}{U}(\lambda_\alpha - \lambda_\beta)$$

$$\mathcal{F}'_\alpha = \mathcal{F}_\alpha - \tfrac{1}{2}\,\mathrm{sign}(\lambda_\alpha - \Lambda). \tag{3.11}$$

It is interesting to notice that from (3.10) both the Λ and the $\kappa + i\chi$ drop out and the remaining equations (3.11) are formally the same as the corresponding equations for a system of $N-2$ electrons. Based on this analogy one may suppose that this is the equation that defines the spin state of the $N-2$ electrons described by the real ks. Thus to avoid having the spin degrees of freedom excited, we choose, for the $\mathcal{F}'$ set, the same set that would correspond to the ground state of a system of $N-2$ electrons, i.e. the set

$$-\tfrac{1}{2}(N/2-2),\ -\tfrac{1}{2}(N/2-4), \ldots, \tfrac{1}{2}(N/2-2). \tag{3.12}$$

3.2. Solution of the system (3.6), (3.7), (3.9) and (3.11)

In the following we suppose that in the large-N limit, as in the case of the ground state, both the real k set and the λ set can be described by their density functions†. If we suppose that the number of k_js (λ_αs) in the interval $(k; k+dk)$ $((\lambda; \lambda + d\lambda))$ is $N\rho(k)dk$ $(N\sigma(\lambda)d\lambda)$ equations (3.7) and (3.13) lead to the equations

$$\rho(k) = \frac{1}{2\pi} + \frac{2\cos k}{2\pi}\int_{-\infty}^{\infty}\frac{U/4}{(U/4)^2 + (\sin k - \lambda)^2}\sigma(\lambda)\,d\lambda + \frac{2\cos k}{2\pi N}\frac{U/4}{(U/4)^2 + (\sin k - \Lambda)^2} \tag{3.13}$$

$$2\int_{-\pi}^{\pi}\frac{U/4}{(U/4)^2 + (\lambda - \sin k)^2}\rho^*(k) = 2\pi\sigma(\lambda) + 2\int_{-\infty}^{\infty}\frac{U/2}{(U/2)^2 + (\lambda - \lambda')^2}\sigma(\lambda')\,d\lambda' \tag{3.14}$$

with

$$\rho^*(k) = \rho(k) - (1/N)\,\delta(k - k_l) - (1/N)\,\delta(k - k_m) \tag{3.15}$$

($\rho(k)$ being the density of ks defined by the whole (3.8) set contains additionally k_l and k_m). This system can be solved by Fourier transformation giving

$$\sigma(\lambda) = \sigma_0(\lambda) - \frac{1}{NU}\left(\frac{1}{\cosh[(\lambda - \sin k_l)\,2\pi/U]} + \frac{1}{\cosh[(\lambda - \sin k_m)\,2\pi/U]}\right) \tag{3.16}$$

$$\rho(k) = \rho_0(k) + \frac{1}{2\pi N}\cos k\frac{U/4}{(U/4)^2 + (\sin k - \Lambda)^2} - \frac{\cos k}{2\pi N}\int_0^\infty \frac{e^{-\omega U/4}}{\cosh \omega U/4} \times \{\cos[\omega(\sin k - \sin k_l)] + \cos[\omega(\sin k - \sin k_m)]\}\,d\omega \tag{3.17}$$

† A description of the k and λ sets by their density functions is very plausible, but it is not established in a strict mathematical sense. The problem is that one must be certain that the error introduced by turning the sums into integrals is much less than the $1/N$ terms that are present due to the excitations. Replacing $(1/N)\,\Sigma_k$ by $\int dk\rho(k)$ introduces an error of the order of $1/N^2$ but the replacement of $(1/N)\,\Sigma_\lambda$ by $\int d\lambda\sigma(\lambda)$ may introduce a larger error because the integration interval is infinite.

where $\sigma_0(\lambda)$ and $\rho_0(k)$ are the ground-state distributions

$$\sigma_0(\lambda) = \frac{1}{2\pi}\int_0^\infty \frac{J_0(\omega)\cos\omega\lambda}{\cosh\omega U/4}\,\mathrm{d}\omega \tag{3.18}$$

$$\rho_0(k) = \frac{1}{2\pi} + \frac{1}{2\pi}\cos k \int_0^\infty \frac{\mathrm{e}^{-\omega U/4}}{\cosh\omega U/4} J_0(\omega)\cos(\omega\sin k_0)\,\mathrm{d}\omega \tag{3.19}$$

with $J_0(\omega)$ being the zeroth-order Bessel function.

Having $\rho(k)$ at hand, (3.6) and (3.9) can be reduced: on the LHS of (3.9) the sum over the k_js can be replaced by an integral over the k variable with density $\rho^*(k)$, and the integral can be evaluated giving

$$\begin{aligned}2Nk(\Lambda) &- 2\tan^{-1}\frac{4}{U}(\Lambda - \sin k_l) - 2\tan^{-1}\frac{4}{U}(\Lambda - \sin k_m)\\ &+ 4\,\mathrm{Re}\tan^{-1}(\Lambda - \sin(\kappa + \mathrm{i}\chi))\\ &= 2\pi\mathcal{F} + \sum_{\beta=1}^{N/2-1} 2\tan^{-1}\frac{2}{U}(\Lambda - \lambda_\beta)\end{aligned} \tag{3.20}$$

with

$$\begin{aligned}k(\Lambda) = \sin^{-1}&[\![\tfrac{1}{2}\{[(U/4)^2 + (\Lambda+1)^2]^{1/2} - [(U/4)^2 + (\Lambda-1)^2]^{1/2}\}]\!]\\ &(\cos k(\Lambda) > 0).\end{aligned} \tag{3.21}$$

Comparing (3.6) and (3.20) one obtains that

$$\begin{aligned}2\pi\Big[N\,\mathrm{sign}\,\Lambda &- \Big(\mathcal{F} + 2\mathcal{J} - \tfrac{1}{2}\sum_{\beta=1}^{N/2-1}\mathrm{sign}(\Lambda - \lambda_\beta)\Big)\Big]\\ &= 2\tan^{-1}\frac{4}{U}(\Lambda - \sin k_l) + 2\tan^{-1}\frac{4}{U}(\Lambda - \sin k_m).\end{aligned} \tag{3.22}$$

This equation may have two non-equivalent solutions

$$\Lambda = \tfrac{1}{2}(\sin k_l + \sin k_m) \qquad \text{if the LHS is 0} \tag{3.23a}$$

$$\Lambda \to \infty \qquad \text{if the LHS is } 2\pi. \tag{3.23b}$$

It can be seen that k_l and k_m enable all the other unknowns to be expressed: (3.23a) or (3.23b) gives the connection between Λ, k_l and k_m, and through these three variables all κ, χ, $\sigma(\lambda)$ and $\rho(k)$ can be given by (3.4a), (3.4b), (3.16) and (3.17), respectively. The equations defining k_l and k_m are obtained from (3.7) through eliminating the λ_αs by means of $\sigma(\lambda)$ and by eliminating Λ using either (3.23a) or (3.23b):

$$\begin{aligned}k_{l(m)} &+ \int_0^\infty \frac{\mathrm{e}^{-\omega U/4}}{\omega\cosh\omega U/4}\mathcal{J}_0(\omega)\sin(\omega\sin k_{l(m)})\,\mathrm{d}\omega = \frac{2\pi}{N}\mathcal{J}_{l(m)}\\ &+ \frac{1}{N}\int_0^\infty \frac{\mathrm{e}^{-\omega 3U/4}}{\omega\cosh\omega U/4}\sin[\omega(\sin k_{l(m)} - \sin k_{m(l)})]\,\mathrm{d}\omega\end{aligned} \tag{3.24a}$$

$$\begin{aligned}k_{l(m)} &+ \int_0^\infty \frac{\mathrm{e}^{-\omega U/4}}{\omega\cosh\omega U/4}\mathcal{J}_0(\omega)\sin(\omega\sin k_{l(m)}) = \frac{2\pi}{N}(\mathcal{J}_{l(m)} + \tfrac{1}{2})\\ &+ \frac{1}{N}\int_0^\infty \frac{\mathrm{e}^{-\omega U/4}}{\omega\cosh\omega U/4}\sin[\omega(\sin k_{l(m)} - \sin k_{m(l)})]\,\mathrm{d}\omega.\end{aligned} \tag{3.24b}$$

To characterise the above described excited states we have to fix three prameters, $\mathcal{J}_l$, $\mathcal{J}_m$, and the LHS of (3.22), that is $\mathcal{F} + 2\mathcal{J}$. It seems that we have one parameter i.e. $\mathcal{J}$, which can take any value without affecting the state. This is not so. It can be checked that if we calculate the exponentially small correction to κ and χ this parameter will also be fixed: for the phase of the terms denoted by $O(e^{-\eta N})$ in (3.2), (3.1) gives a solution only at a given value of $\mathcal{J}$.

In connection with the states corresponding to the solution given by (3.23b) and (3.24b) we note the following: in these states both the real k and normal λ sets are exactly the same as those for states of $N-2$ electrons in which the spin part is in its ground state. If we substitute these ks and λs together with a complex k pair satisfying (3.4a) and (3.4b) into the wavefunction, and take it to the $\Lambda \to \infty$ limit, we obtain a wavefunction that is the same as that corresponding to the state obtained from the $N-2$ electron state through acting on it by the operator $\Sigma_{n=1}^{N}[(-1)^n c_{n\uparrow}^+ c_{n\downarrow}^+]$. As the commutator of this operator with $\hat{H}$ of (1.1) is a number (U), solution (3.23b), (3.24b) indeed corresponds to an eigenstate of the system.

3.3. The energy and momentum

The energy of the states is easily calculated by means of $\rho^*(k)$:

$$E = -N\int_{-\pi}^{\pi} 2\cos k\,\rho(k)\,dk - 4\cos\kappa\cosh\chi$$

$$= E_0 + \varepsilon(k_l) + \varepsilon(k_m) + U \tag{3.25}$$

with E_0 and $\varepsilon(k)$ given by

$$E_0 = -2N\int_0^{\infty} \frac{e^{-\omega U/4} J_0(\omega)\,J_1(\omega)}{\omega\cosh\omega U/4}\,d\omega \tag{3.26}$$

$$\varepsilon(k) = 2\cos k + 2\int_0^{\infty} \frac{e^{-\omega U/4}}{\omega\cosh\omega U/4} J_1(\omega)\cos(\omega\sin k)\,d\omega. \tag{3.27}$$

The momentum, according to (2.2b), (3.8), (3.11) and (3.12) is

$$p = \frac{2\pi}{N}\left(-I_l - I_m + \sum_{\beta=1}^{N/2-1} \tfrac{1}{2}\,\mathrm{sign}(\lambda_\beta - \Lambda) + \mathcal{F} + 2\mathcal{J}\right) + p_0. \tag{3.28}$$

Here p_0, the momentum of the ground state, is π if $N/2$ is even and 0 if $N/2$ is odd. This, combined with (3.23) and (3.24), yields

$$p - p_0 = -p(k_l) - p(k_m) \tag{3.29}$$

where we understand

$$p(k) = k + \int_0^{\infty} \frac{e^{-\omega U/4}}{\omega\cosh\omega U/4} J_0(\omega)\sin(\omega\sin k)\,d\omega. \tag{3.30}$$

The energy–momentum curves defined by (3.27) and (3.30) for different values of U are shown in figure 1. Figure 2 displays the continuum of the excitations discussed in this section. For the sake of comparison in the same figure we have also plotted the energy of a spin wave. It can be seen that at the given value of U ($U = 2$) the energy band of the spin waves and the band of the excitations with one complex k pair overlap, i.e.

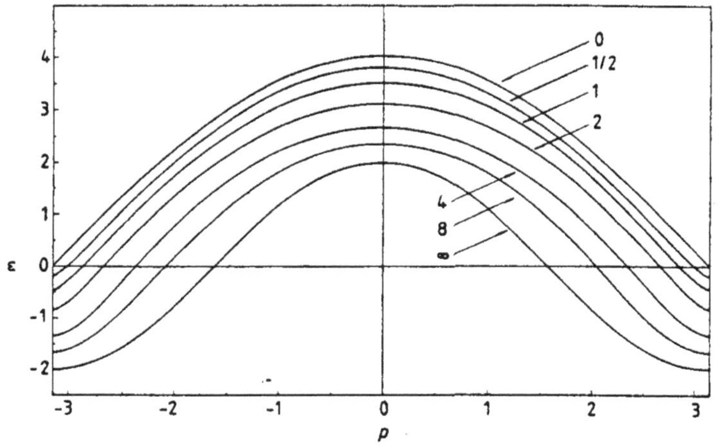

Figure 1. $\varepsilon(p)$ dispersions for different values of U. The individual curves are labelled with the values of U.

both kinds of excitations have the same order of magnitude energy. For increasing values of U this overlap gradually disappears, since the lower edge of the band of the excitations with complex wavenumbers increases roughly like $U - 4$, while the maximum energy of a spin wave decreases like $1/U$.

3.4. Comments

The number of states described above is $N(N-1)$ since there are $N(N-1)/2$ choices of the $\mathcal{J}_l$ and $\mathcal{J}_m$ parameters and for each choice we have two solutions (3.23a) and (3.23b).

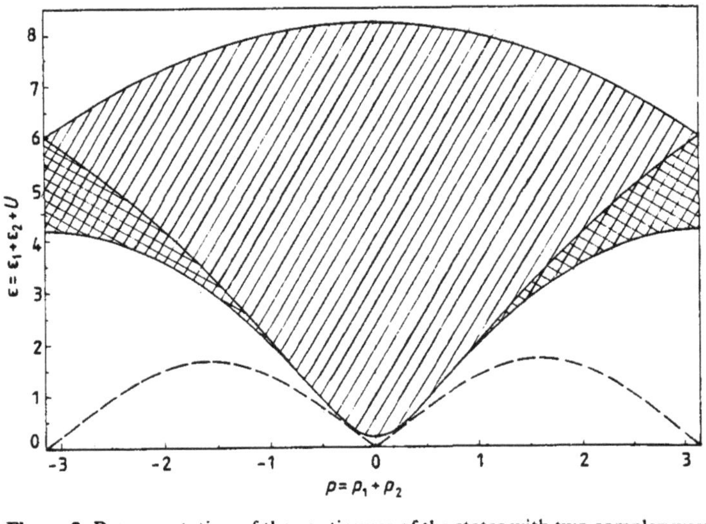

Figure 2. Representation of the continuum of the states with two complex wavenumbers in the energy–momentum plane for $U = 2$. Cross-hatched areas represent degenerate states: to one $(\varepsilon; p)$ point there are two non-equivalent p_1, p_2 pairs for which $p_1 + p_2 = p$ and $\varepsilon(p_1) + \varepsilon(p_2) + U = \varepsilon$. The broken curve shows the lower edge of the continuum of the triplet spin waves.

On the other hand, it can be verified that these states in the large U limit correspond to states in which one site is doubly occupied and one site is empty, and the Heisenberg chain of the spins belonging to the singly occupied sites is in its ground state. As the number of these states is also $N(N-1)$ we have found all of them.

It is interesting to note that the state with lowest energy is the one in which $k_l \simeq k_m \simeq \pi$. Here the energy is

$$(E - E_0)_{\min} = U - 4 + 4 \int_0^\infty \frac{e^{-\omega U/4}}{\omega \cosh \omega U/4} J_1(\omega)\, d\omega \tag{3.31}$$

which is exactly the same as

$$\mu_+ - \mu_- = (E_0(N+1) - E_0(N)) - (E_0(N) - E_0(N-1)) \tag{3.32}$$

($E_0(N \pm 1)$ being the ground state energy of a system with $N \pm 1$ electrons) calculated by Lieb and Wu. In other words the gap calculated through the one particle excitations coincides with the gap in the spectrum of particle number conserving charge excitations. A detailed discussion of the results of this chapter is given in the second part of the present work (Woynarovich 1981), where solutions corresponding to an arbitrary number of complex pairs are found for the case of an arbitrary number of electrons.

4. Summary

The main points of the present study can be summarised as follows.

(i) Based on the analysis of the $U \to \infty$ limiting form of the eigenfunctions of the 1D Hubbard chain it is shown that all those states in which the amplitude of finding electron pairs occupying the same sites is finite (even if U is large) can be described by solutions of the Lieb–Wu equations in which some of the wavenumbers are complex.

(ii) Solutions with one pair of complex wavenumbers, corresponding to states in which the spin part is in its ground state (singlet) are found. These states are characterised by three parameters (k_l, k_m and Λ) and these parameters are coupled to each other by (3.23) and (3.24). In the energy and momentum only the parameters k_l and k_m appear explicitly ((3.25), (3.27), (3.29) and (3.30)). In the spectrum a gap is found which is of the same magnitude as that calculated by Lieb and Wu for the one-particle-type excitations.

Acknowledgments

It is a pleasure to acknowledge many helpful and illuminating discussions with Drs A Sütö and P Fazekas.

References

Coll C F 1974 *Phys. Rev.* B **9** 2150
Griffiths R B 1964 *Phys. Rev.* **133** A768
Lieb E M and Wu F Y 1968 *Phys. Rev. Lett.* **20** 1445

Ovchinnikov A A 1970 *Sov. Phys.–JETP* **30** 1160
Shiba H 1972 *Phys. Rev.* B **6** 930
Takahashi M 1969 *Prog. Theor. Phys.* **42** 1098
Woynarovich F 1982 *J. Phys. C: Solid State Phys.* **15** 97–109
Yang C N 1967 *Phys. Rev. Lett.* **19** 1312

J. Phys. C: Solid State Phys., 15 (1982) 97–109. Printed in Great Britain

Excitations with complex wavenumbers in a Hubbard chain: II. States with several pairs of complex wavenumbers

F Woynarovich
Central Research Institute for Physics, H-1525 Budapest, POB 49, Hungary

Received 26 August 1980, in final form 23 June 1981

Abstract. States of the 1D Hubbard chain with several pairs of complex wavenumbers are studied. The original set of Lieb–Wu equations is replaced by an equivalent set in which only real wavenumbers appear. These equations are reduced for the states in which the spin degrees of freedom are not excited. The energy–momentum dispersion for these states is also found.

1. Introduction

In the first part of the present work (Woynarovich 1982, hereafter referred to as I) those eigenstates of the 1D Hubbard Hamiltonian

$$\hat{H} = -\sum_{i=1}^{N}\sum_{\sigma}(c_{i+1\sigma}^{+}c_{i\sigma} + c_{i\sigma}^{+}c_{i+1\sigma}) + U\sum_{i=1}^{N} n_{i\uparrow}n_{i\downarrow} \tag{1.1}$$

have been studied that correspond to states in which the amplitude of finding electron pairs occupying the same site does not vanish even if U is large. It has been established that these states are to be described by such solutions of the Lieb and Wu (1968) equations

$$Nk_j = 2\pi\mathcal{I}_j - \sum_{\beta=1}^{N_e/2-S^z} 2\tan^{-1}\frac{4}{U}(\sin k_j - \lambda_\beta) \tag{1.2}$$

$$\sum_{j=1}^{N_e} 2\tan^{-1}\frac{4}{U}(\lambda_\alpha - \sin k_j) = 2\pi\mathcal{F}_\alpha + \sum_{\beta=1}^{N_e/2-S^z} 2\tan^{-1}\frac{2}{U}(\lambda_\alpha - \lambda_\beta) \tag{1.3}$$

in which some of the wavenumbers k are complex. Solutions with one pair of complex wavenumbers corresponding to states of the half-filled band ($N_e = N$) in which the spin part is not excited were discussed in I.

In this work we intend to generalise our results in two directions: we look for states with several pairs of complex wavenumbers, and at the same time we do not fix the band filling; the number of electrons can be less than the number of sites. We wish to derive a system of equations in which, instead of the parameters of all electrons, parameters referring only to the excitations appear. As a first step (§ 2) we deduce from the original

0022-3719/82/010097 + 13 \$01.50

set of Lieb–Wu equations an equivalent set that contains real wavenumbers only. In § 3 the equations will be further reduced to describe states in which the spin part is not excited. The resulting equations will provide a connection between parameters that can be attributed to the excitations only. The energy and momentum of these states will be found too. It will be seen that both the energy and the momentum of the excited states in question can be expressed as the sums of energies and momenta of a set of quasiparticles.

Throughout the present paper the formulae of I will be prefixed with a roman I.

2. Equations for states with several complex wavenumbers

2.1. Elimination of the complex wavenumbers from the Lieb–Wu equations

In this section we deal with states described by a k set containing several pairs of complex wavenumbers. We do not fix the band filling. It can be less than half, that is $N_e \leqslant N$. The number of electrons needed to make the band half-filled is denoted by H: $H = N - N_e$; the number of complex k pairs is L and the number of down spins is M ($M \geqslant L$). The spin state will not be fixed.

It is supposed that, similarly to the case discussed in I, for each complex k pair there is one λ among the M λs, for which

$$\sin(\kappa_n^{(\pm)} \pm i\chi_n^{(\pm)}) = \Lambda_n \mp iU/4 + O(e^{-\eta_n^{(\pm)}N}) \tag{2.1}$$

(λs coupled to complex k pairs by (2.1) will be denoted by Λ to distinguish them from 'normal' λs). These equations are the generalisations of (I. 3.2) allowing for the possibility of Λ being complex (in a complex k pair the $\kappa_n^{(+)} + i\chi_n^{(+)}$ and the $\kappa_n^{(-)} - i\chi_n^{(-)}$ are the complex conjugates of each other only if Λ_n is real; if $\Lambda_n = \Lambda_m^*$, then $\kappa_n^{(+)} + i\chi_n^{(+)} = (\kappa_m^{(-)} - i\chi_m^{(-)})^*$). Substituting (2.1) into (1.2):

$$N(\kappa_n^{(\pm)} \pm i\chi_n^{(\pm)}) = 2\pi\mathcal{I}_n^{(\pm)} - \sum_{\beta=1}^{M-L} 2\tan^{-1}\frac{4}{U}[\sin(\kappa_n^{(\pm)} \pm i\chi_n^{(\pm)}) - \lambda_\beta]$$
$$- \sum_{m=1}^{L} 2\tan^{-1}\frac{4}{U}[\sin(\kappa_n^{(\pm)} \pm i\chi_n^{(\pm)}) - \Lambda_m]. \tag{2.2}$$

One finds that the terms denoted by $O(e^{-\eta_n^{(\pm)}N})$ are indeed exponentially small in N, if the numbers defined as

$$\eta_n^{(\pm)} = \chi_n^{(\pm)} \pm \frac{1}{N}\mathrm{Im}\Big\{\sum_{\beta} 2\tan^{-1}\frac{4}{U}[\sin(\kappa_n^{(\pm)} \pm i\chi_n^{(\pm)}) - \lambda_\beta]$$
$$+ \sum_{m\neq n} 2\tan^{-1}\frac{4}{U}[\sin(\kappa_n^{(\pm)} \pm i\chi_n^{(\pm)}) - \Lambda_m]\Big\} \tag{2.3}$$

are positive and of the order of unity. Whereas in the case discussed in I (one pair of complex ks, spin part not excited) η could be estimated without solving the system completely, now, as we are dealing with a more general case, we cannot tell anything about the $\eta_n^{(\pm)}$s. We suppose, however, that all the $\eta_n^{(\pm)}$s are positive and of the order of unity. Since the following treatment will be consistent only if this supposition is correct, after solving the equations one has to check the validity of this assumption.

If the terms $O(e^{-\eta_n^{(\pm)}N})$ are neglected, the solutions of (2.1) are given in the form

$$\begin{aligned}
\kappa_n^{(\pm)} &= \sin^{-1}[\![\tfrac{1}{2}\{[(U/4 \mp \operatorname{Im}\Lambda_n)^2 + (\operatorname{Re}\Lambda_n + 1)^2]^{1/2} \\
&\qquad - [(U/4 \mp \operatorname{Im}\Lambda_n)^2 + (\operatorname{Re}\Lambda_n - 1)^2]^{1/2}\}]\!] \\
\chi_n^{(\pm)} &= \cosh^{-1}[\![\tfrac{1}{2}\{[(U/4 \mp \operatorname{Im}\Lambda_n)^2 + (\operatorname{Re}\Lambda_n + 1)^2]^{1/2} \\
&\qquad + [(U/4 \mp \operatorname{Im}\Lambda_n)^2 + (\operatorname{Re}\Lambda_n - 1)^2]^{1/2}\}]\!] \\
\chi_n^{(\pm)} &> 0 \qquad \operatorname{sign}(\cos\kappa_n^{(\pm)}) = -\operatorname{sign}(U/4 \mp \operatorname{Im}\Lambda_n).
\end{aligned} \tag{2.4}$$

The equations for the real ks are

$$\begin{aligned}
Nk_j = 2\pi\mathcal{I}_j &- \left(\sum_{\beta=1}^{M-L} 2\tan^{-1}\frac{4}{U}(\sin k_j - \lambda_\beta) + \sum_{m=1}^{L} 2\tan^{-1}\frac{4}{U}(\sin k_j - \Lambda_m)\right)_{\text{discont.}} \\
&- \left(\sum_{\beta=1}^{M-L} 2\tan^{-1}\frac{4}{U}(\sin k_j - \lambda_\beta) + \sum_{m=1}^{L} 2\tan^{-1}\frac{4}{U}(\sin k_j - \Lambda_m)\right)_{\text{cont.}}
\end{aligned} \tag{2.5}$$

where we have split $2\tan^{-1}(4/U)(\sin k - \lambda)$ into two parts:

$$\begin{aligned}
\left(2\tan^{-1}\frac{4}{U}(\sin k - \lambda)\right)_{\text{cont.}} &= \tan^{-1}\frac{\sin k - \operatorname{Re}\lambda}{U/4 + \operatorname{Im}\lambda} + \tan^{-1}\frac{\sin k - \operatorname{Re}\lambda}{U/4 - \operatorname{Im}\lambda} \\
&\quad + \frac{1}{2i}\ln\frac{(U/4 + \operatorname{Im}\lambda)^2 + (\sin k - \operatorname{Re}\lambda)^2}{(U/4 - \operatorname{Im}\lambda)^2 + (\sin k - \operatorname{Re}\lambda)^2}
\end{aligned} \tag{2.6a}$$

$$\left(2\tan^{-1}\frac{4}{U}(\sin k - \lambda)\right)_{\text{discont.}} = \begin{cases} \pi\operatorname{sign}(\sin k - \operatorname{Re}\lambda) & \text{if } |\operatorname{Im}\lambda| > U/4 \\ 0 & \text{if } |\operatorname{Im}\lambda| < U/4. \end{cases} \tag{2.6b}$$

In (2.5) it is more convenient to introduce an $\mathcal{I}'$ set instead of the $\mathcal{I}$ set with the definition

$$\mathcal{I}_j' = \mathcal{I}_j - \frac{1}{2\pi}\left(\sum_{\beta=1}^{M-L} 2\tan^{-1}\frac{4}{U}(\sin k_j - \lambda_\beta) + \sum_{m=1}^{L} 2\tan^{-1}\frac{4}{U}(\sin k_j - \Lambda_m)\right)_{\text{discont.}} \tag{2.7}$$

Because both the λ and Λ sets are assumed to consist of real numbers and complex conjugate pairs, then $\mathcal{I}_j'$ is an integer if $\mathcal{I}_j$ is an integer (that is if M is even) and $\mathcal{I}_j'$ is half an odd-integer otherwise. For the $\mathcal{I}'$ set we choose $N - 2L - H$ different numbers from one of the sets

$$-\tfrac{1}{2}(N-1), -\tfrac{1}{2}(N-3), \ldots, \tfrac{1}{2}(N-3), \tfrac{1}{2}(N-1) \tag{2.8a}$$

$$-\tfrac{1}{2}(N-2), -\tfrac{1}{2}(N-4), \ldots, \tfrac{1}{2}(N-2), \tfrac{1}{2}N. \tag{2.8b}$$

(The particular set to be used is decided by the parities of the numbers N and M.) Equation (2.5) additionally defines ks to the $2L + H$ $\mathcal{I}$'s left out from (2.8a) or (2.8b). We will denote these ks by the index h (for 'hole'). The density of the ks satisfying (2.5) is

$$\begin{aligned}
\rho(k) = \frac{1}{2\pi} + \frac{2\cos k}{2\pi N}&\left(\sum_{\beta=1}^{M-L}\frac{U/4}{(U/4)^2 + (\sin k - \lambda_\beta)^2}\right. \\
&\left. + \sum_{m=1}^{L}\frac{U/4}{(U/4)^2 + (\sin k - \Lambda_m)^2}\right).
\end{aligned} \tag{2.9}$$

As this $\rho(k)$ also contains the k_h variables, the density of the k_js is

$$\rho^*(k) = \rho(k) - \frac{1}{N}\sum_{h=1}^{H+2L} \delta(k - k_h). \tag{2.10}$$

It should be noted that the introduction of $\rho(k)$ in the form (2.9) makes sense only if it is positive, but whether it is positive for all k or not depends on the distribution of the λs and Λs. As we do not know these distributions, let us suppose for the time being that $\rho(k)$ of (2.9) is positive; of course this must be checked for any given solution.

Although the equations for the normal λs and Λs have a common form in the original system (1.3), in our case it is worthwhile to treat them differently. Substituting (2.1) into the equations for the normal λs:

$$\begin{aligned}
&\sum_{j\neq h} 2\tan^{-1}\frac{4}{U}(\lambda_\alpha - \sin k_j) + \sum_{m=1}^{L} 2\tan^{-1}\frac{4}{U}[\lambda_\alpha - \sin(\kappa_m^{(+)} + \mathrm{i}\chi_m^{(+)})]\\
&\qquad + \sum_{m=1}^{L} 2\tan^{-1}\frac{4}{U}[\lambda_\alpha - \sin(\kappa_m^{(-)} - \mathrm{i}\chi_m^{(-)})]\\
&\qquad = 2\pi\mathcal{F}_\alpha + \sum_{\beta=1}^{M-L} 2\tan^{-1}\frac{2}{U}(\lambda_\alpha - \lambda_\beta)\\
&\qquad + \sum_{m=1}^{L} 2\tan^{-1}\frac{2}{U}(\lambda_\alpha - \Lambda_m)
\end{aligned} \tag{2.11}$$

and using the identity

$$\begin{aligned}
&2\tan^{-1}(4/U)(\lambda_\alpha - \Lambda_m - \mathrm{i}\,U/4) + 2\tan^{-1}(4/U)(\lambda_\alpha - \Lambda_m + \mathrm{i}\,U/4)\\
&\qquad = 2\tan^{-1}(2/U)(\lambda_\alpha - \Lambda_m) + \pi\,\mathrm{sign}[\mathrm{Re}(\lambda_\alpha - \Lambda_m)]
\end{aligned} \tag{2.12}$$

one has (up to exponentially small terms)

$$\sum_{j\neq h} 2\tan^{-1}\frac{4}{U}(\lambda_\alpha - \sin k_j) = 2\pi\mathcal{F}'_\alpha + \sum_{\beta=1}^{M-L} 2\tan^{-1}\frac{2}{U}(\lambda_\alpha - \lambda_\beta) \tag{2.13}$$

with

$$\mathcal{F}'_\alpha = \mathcal{F}_\alpha - \sum_{m=1}^{L} \tfrac{1}{2}\mathrm{sign}[\mathrm{Re}(\lambda_\alpha - \Lambda_m)]. \tag{2.14}$$

The $\mathcal{F}'_\alpha$ parameters are integers if $N_e - (M - L)$ is odd, otherwise they are half odd-integers.

The equations for the Λ_ns coupled to the complex wavenumbers

$$\begin{aligned}
&\sum_{j\neq h} 2\tan^{-1}\frac{4}{U}(\Lambda_n - \sin k_j) + \sum_{m=1}^{L} 2\tan^{-1}\frac{4}{U}[\Lambda_n - \sin(\kappa_m^{(+)} + \mathrm{i}\chi_m^{(+)})]\\
&\qquad + \sum_{m=1}^{L} 2\tan^{-1}\frac{4}{U}[\Lambda_n - \sin(\kappa_m^{(-)} - \mathrm{i}\chi_m^{(-)})]
\end{aligned}$$

$$= 2\pi\mathcal{F}_n + \sum_{\beta=1}^{M-L} 2\tan^{-1}\frac{2}{U}(\Lambda_n - \lambda_\beta)$$

$$+ \sum_{m=1}^{L} 2\tan^{-1}\frac{2}{U}(\Lambda_n - \Lambda_m) \tag{2.15}$$

need a different treatment. In the first term of the LHS we replace the continuous part of the sum over the real ks by an integral over the ks with density $\rho^*(k)$. In this integral only the constant and the δ-function terms of $\rho^*(k)$ contribute, and it can be evaluated in closed form giving

$$\frac{1}{N}\sum_{j\neq h}\left(2\tan^{-1}\frac{4}{U}(\Lambda_n - \sin k_j)\right)_{\text{cont.}} \simeq -\frac{1}{N}\sum_{h=1}^{H+2L}\left(2\tan^{-1}\frac{4}{U}(\Lambda_n - \sin k_h)\right)_{\text{cont.}}$$
$$+ \operatorname{sign}(U/4 - \operatorname{Im}\Lambda_n)\sin^{-1}[\![\tfrac{1}{2}\{[(U/4 - \operatorname{Im}\Lambda_n)^2 + (\operatorname{Re}\Lambda_n + 1)^2]^{1/2}$$
$$- [(U/4 - \operatorname{Im}\Lambda_n)^2 + (\operatorname{Re}\Lambda_n - 1)^2]^{1/2}\}]\!]$$
$$+ \operatorname{sign}(U/4 + \operatorname{Im}\Lambda_n)\sin^{-1}[\![\tfrac{1}{2}\{[(U/4 + \operatorname{Im}\Lambda_n)^2 + (\operatorname{Re}\lambda_n + 1)^2]^{1/2}$$
$$- [(U/4 + \operatorname{Im}\Lambda_n)^2 + (\operatorname{Re}\Lambda_n - 1)^2]^{1/2}\}]\!] - \mathrm{i}(\chi_n^{(+)} - \chi_n^{(-)}). \tag{2.16}$$

This identity holds up to terms of the order of $1/N^2$ (also calculating the discontinuous part by means of $\rho^*(k)$ would introduce an error of the order of $1/N$). Substituting (2.16) into (2.15), and summing (2.15) and (2.2) (this latter twice: once with the upper and once with the lower signs) one obtains

$$\sum_{h=1}^{H+2L} 2\tan^{-1}\frac{4}{U}(\Lambda_n - \sin k_h) = 2\pi\mathcal{F}'_n + \sum_{m=1}^{L} 2\tan^{-1}\frac{2}{U}(\Lambda_n - \Lambda_m) \tag{2.17}$$

with

$$\mathcal{F}'_n = N\operatorname{sign}(\operatorname{Re}\Lambda_n) - \mathcal{F}_n - \mathcal{J}_n^{(+)} - \mathcal{J}_n^{(-)} + \sum_{m\neq n}\operatorname{sign}[\operatorname{Re}(\Lambda_n - \Lambda_m)]$$
$$+ \tfrac{1}{2}\sum_{\beta}\operatorname{sign}[\operatorname{Re}(\Lambda_n - \lambda_\beta)] + \sum_{\text{all }k}\tfrac{1}{4}\left[\operatorname{sign}\left(\frac{U}{4} - |\operatorname{Im}\Lambda_n|\right) - 1\right]$$
$$\times [\operatorname{sign}(\operatorname{Re}\Lambda_n) - \operatorname{sign}(\operatorname{Re}\Lambda_n - \sin k)]$$
$$+ \frac{1}{\pi}\operatorname{Re}\tan^{-1}\frac{4}{U}(\mathrm{i}\operatorname{Im}\Lambda_n). \tag{2.18}$$

The parameters $\mathcal{F}'_n$ are integers if $N_e - L$ is odd; otherwise they are half odd-integers.

To summarise the result of this subsection, we conclude that the actual system to be solved is the system of (2.5), (2.13) and (2.17). The solution of these equations determines ((through (2.4)) a solution of the original Lieb–Wu equations with an error of the order of $1/N^2$ if the $\eta_n^{(\pm)}$s defined by (2.3) are positive and of the order of unity, and the $\rho(k)$ function given by (2.9) is positive.

2.2. *Symmetry of the system (2.5), (2.13) and (2.17)*

The system (2.5), (2.13) and (2.17) is entirely symmetrical in the variables k_j, λ_α and k_h, Λ_n. Changing the role of these variables one obtains another solution of the system. In this subsection we wish to show that this symmetry is present in some form in the momentum and energy as well.

Let us define the complex wavenumbers $\kappa_\alpha^{(+)} + \mathrm{i}\chi_\alpha^{(+)}$ and $\kappa_\alpha^{(-)} - \mathrm{i}\chi_\alpha^{(-)}$ for the variables λ_α, analogously to (2.1), and calculate

$$\sum_{\mathrm{all}\,k} k + \sum_{\alpha=1}^{M-L} [(\kappa_\alpha^{(+)} + i\chi_\alpha^{(+)}) + (\kappa_\alpha^{(-)} - i\chi_\alpha^{(-)})] + \sum_{n=1}^{L} [(\kappa_n^{(+)} + i\chi_n^{(+)}) + (\kappa_n^{(-)} - i\chi_n^{(-)})]. \tag{2.19}$$

Summing (2.5) for all indices j and h, and evaluating the expression obtained in this way by means of (2.16) (with the difference that now, as the sum is over all ks, the first term on the RHS of (2.16) is omitted) one obtains that the value of (2.1) is (up to $n2\pi$)

$$\frac{2\pi}{N}\sum_{\mathrm{all}\,\mathcal{J}'} \mathcal{J}' = \begin{cases} 0 & \text{if } N-M \text{ is odd} \\ \pi & \text{if } N-M \text{ is even.} \end{cases} \tag{2.20}$$

If we calculate the sum

$$\sum_{\mathrm{all}\,k} (-2\cos k) - \sum_{\alpha=1}^{M-L} 2[\cos(\kappa_\alpha^{(+)} + \mathrm{i}\chi_\alpha^{(+)}) + \cos(\kappa_\alpha^{(-)} - \mathrm{i}\chi_\alpha^{(-)})] - \sum_{n=1}^{L} 2[\cos(\kappa_n^{(+)} + \mathrm{i}\chi_n^{(+)}) + \cos(\kappa_n^{(-)} - \mathrm{i}\chi_n^{(-)})] \tag{2.21}$$

we find that this is equal to MU. Thus, in some sense the states in which the roles of the variables k_j, λ_α and k_h, Λ_n are interchanged are 'complementers': both the sum of their momenta and the sum of their energies are constants that depend only on the number of the down spins, but not on the structure of the states.

The above-established complementarity can be used to calculate highly excited states of the Hubbard chain or to calculate the low-energy states of a chain with negative U. From this, for example, one knows that the highest-energy state of N electrons is the one in which all wavenumbers are complex and the distribution of the Λ_ns is the same as the distribution of the λ_αs in the ground state. Because the Hubbard Hamiltonian with negative U is essentially the same as a Hubbard Hamiltonian with positive U but with an overall minus sign, the above state is also the ground state of a Hubbard chain with negative U.

This complementarity is probably connected with the property of the Hubbard Hamiltonian that if we introduce holes instead of the up spin electrons, then

$$\hat{H} \to U\sum_{i=1}^{N} n_{i\downarrow} - \hat{H}' \tag{2.22}$$

where $\hat{H}'$ has the same structure as $\hat{H}$. Taking into account the parallelism that in the complementary states the parameters λ_α describing the spin part change roles with the parameters Λ_n connected with the charge distribution, and that the transformation which connects $\hat{H}$ and $\hat{H}'$ introduces doubly occupied or empty sites instead of the singly occupied ones (uncompensated spins) and *vice versa*, the above-suspected connection seems very probable.

2.3. The $U \to \infty$ limit of the wavefunction

Taking the $U \to \infty$ limit in the wavefunction (see expressions (I. 2.5)–(I. 2.10)) one finds that some of the amplitudes diverge. As in the normalised wavefunction only the terms with the strongest divergence will give finite contributions, picking out the most divergent terms we can separate those configurations that can be realised even if U is very large. In this way we get the result that for large U only those configurations remain in which the number of doubly occupied sites is equal to the number of complex k pairs. In the amplitude of these configurations only those permutations P and π in which the $\kappa_n^{(+)} + i\chi_n^{(+)}$ and $\kappa_n^{(-)} - i\chi_n^{(-)}$ wavenumber pair belongs to one doubly occupied site, and the Λ_n belongs to the down spin at this site give contributions. Using the fact that for large U all the sin k_js can be neglected compared with the λ_αs and Λ_ns, which are of the order of U, and also using (2.1) the amplitude of the configurations in question can be given as

$$(-1)^Q\left[\exp\left(i\sum_{l=1}^{L} n_l^{\mathrm{d}}\right)\varphi_1(y_1^{\mathrm{d}}, y_2^{\mathrm{d}}, \ldots, y_L^{\mathrm{d}})\right]$$
$$\times\left[\left(\sum_P(-1)^P\exp\left(i\sum_{j=1}^{N-H-2L}k_{Pj}n_{Qj}^{\mathrm{s}}\right)\right)\varphi_2(y_1^{\mathrm{s}}, y_2^{\mathrm{s}}, \ldots, y_{M-L}^{\mathrm{s}})\right]. \qquad (2.23)$$

Here the permutation Q arranges the coordinates $n_1, n_2, \ldots, n_{N-H}$ into non-decreasing order with the restriction that from two equal coordinates that of the electron with down spin must come first. The n_{Qj}^{s}s refer to singly occupied sites, the n_l^{d}s to doubly occupied ones and P goes over all permutations of the real wavenumbers. The functions φ_1 and φ_2 are essentially the Heisenberg eigenfunctions:

$$\varphi_2 = \sum_\pi A(\lambda_{\pi 1}, \lambda_{\pi 2}, \ldots, \lambda_{\pi M-L})\left(\frac{(i\lambda_{\pi 1} + U/4)}{(i\lambda_{\pi 1} - U/4)}\right)^{y_1^{\mathrm{s}}}\cdots\left(\frac{(i\lambda_{\pi M-L} + U/4}{(i\lambda_{\pi M-L} - U/4)}\right)^{y_{M-L}^{\mathrm{s}}}$$

$$A(\ldots, \lambda_{\pi i}, \lambda_{\pi i+1}, \ldots)/A(\ldots, \lambda_{\pi i+1}, \lambda_{\pi i}, \ldots)$$
$$= [i(\lambda_{\pi i+1} - \lambda_{\pi i}) - U/2]/[i(\lambda_{\pi i+1} - \lambda_{\pi i}) + U/2]. \qquad (2.24)$$

The numbers y^{s} are the coordinates of the down spins in the chain of singly occupied sites in increasing order. φ_1 is formally the same as φ_2 with the difference that U must be replaced by $-U$, the λ_αs by the Λ_ns, and the numbers y^{d} are the coordinates of the doubly occupied sites in the chain containing only the doubly and unoccupied sites. The amplitude of the configurations in which the number of doubly occupied sites is more or less than L vanishes at least like $1/U$ as $U \to \infty$.

To understand (2.23) let us consider a configuration in which the first $N - H - 2L$ sites in the chain are singly occupied, and the remaining $H + 2L$ sites are the empty or doubly occupied ones. In this configuration the electrons cannot move (except for the last one) as either the Pauli principle or the large onsite repulsion prevents it. Although in this configuration there is no direct interaction between the electrons, through an intermediate state with energy U neighbouring electrons can see each other's spins, and electrons with different spins can change position, i.e. the spins can move in the same way as do those in a Heisenberg chain; the distribution of the spins will correspond to the eigenstates of the Heisenberg Hamiltonian. The situation with the empty and doubly occupied sites is similar: neighbouring sites can observe each others occupancy through an intermediate state of relative energy $-U$; moreover the same intermediate state makes it possible for an empty and doubly occupied site to change position. Thus the distribution of the empty and doubly occupied sites will be the same as the distribution of up and down spins in a Heisenberg chain. It is clear that neither the spin distribution nor the relative distribution of the empty and doubly occupied sites does change if the chain of singly occupied sites is 'diluted' by empty and doubly occupied sites making possible also direct propagation for the electrons.

In connection with the $U \to \infty$ limiting form of the wavefunction we have to note that in general we have no reason for regarding the states with complex wavenumbers as some sort of bound states. This notion is right only if the chain consists mainly of empty and doubly occupied sites; then the electrons occupying the same site can be regarded as bound ones. Otherwise the neighbours of a doubly occupied site are most probably singly occupied ones, one of the electrons of the two can move, and the situation is rather similar to having one free electron hopping from a site to a site already singly occupied.

3. Singlet states with only charge excitations present

3.1. Solution for the distribution of the normal λs

The system of equations reduced from the original Lieb–Wu equations may have many solutions depending on the choice of the parameters $\mathcal{F}'_\alpha$. In this section we want to deal with those solutions in which the spin degrees of freedom are not excited. For the sake of simplicity we suppose that these states are singlet, i.e. the number of uncompensated spins, $N - H - 2L$, is even. It is not hard to see ((2.23), (2.24) and the $U \to \infty$ form of (2.13)) that to have the states in question, we have to choose the $\mathcal{F}'_\alpha$ set in (2.13) that is characteristic of the ground state of $N - H - 2L$ electrons, that is, the set

$$-\tfrac{1}{2}[(N-H-2L)/2-1],\ -\tfrac{1}{2}[(N-H-2L)/2-3],\ \ldots,\ \tfrac{1}{2}[(N-H-2L)/2-1]. \tag{3.1}$$

With this choice of $\mathcal{F}'_\alpha$s, the density of the λs must satisfy the equation

$$2\int_{-\pi}^{\pi} \frac{U/4}{(U/4)^2 + (\lambda - \sin k)^2}\,\rho^*(k)\,\mathrm{d}k = 2\pi\,\sigma(\lambda) + 2\int_{-\infty}^{\infty} \frac{U/2}{(U/2)^2 + (\lambda-\lambda')^2}\,\sigma(\lambda')\,\mathrm{d}\lambda'. \tag{3.2}$$

The solution of (3.2) is easily obtained by Fourier transformation:

$$\sigma(\lambda) = \frac{1}{2\pi}\int_0^{\infty} \frac{F_0(\omega)\cos(\omega\lambda)}{\cosh(\omega U/4)}\,\mathrm{d}\omega - \frac{1}{NU}\sum_{h=1}^{H+2L} \frac{1}{\cosh[(\lambda - \sin k_h)\,2\pi/U]}. \tag{3.3}$$

This $\sigma(\lambda)$ allows us to eliminate the λ_βs from $\rho(k)$ of (2.9), with the result

$$\rho(k) = \frac{1}{2\pi}\left(1 + \cos k \int_0^\infty \frac{e^{-\omega U/4}}{\cosh(\omega U/4)} F_0(\omega)\cos(\omega \sin k)\,d\omega\right)$$
$$-\frac{\cos k}{2\pi N}\int_0^\infty \frac{e^{-\omega U/4}}{\cosh(\omega U/4)} \sum_{h=1}^{H+2L} \cos(\omega(\sin k - \sin k_h))\,d\omega$$
$$+\frac{2\cos k}{2\pi N}\sum_{m=1}^{L} \frac{U/4}{(U/4)^2 + (\sin k - \Lambda_m)^2}. \tag{3.4}$$

By means of $\sigma(\lambda)$ from (2.5), equations for the variables k_h can be obtained too:

$$N\left(k_h + \int_0^\infty \frac{e^{-\omega U/4}}{\cosh(\omega U/4)} \frac{F_0(\omega)\sin(\omega \sin k_h)}{\omega}\,d\omega\right)$$
$$= 2\pi \mathcal{J}_h + \sum_{h'=1}^{H+2L} \int_0^\infty \frac{e^{-\omega U/4}}{\cosh(\omega U/4)} \frac{\sin[\omega(\sin k_h - \sin k_{h'})]}{\omega}\,d\omega$$
$$-\sum_{m=1}^{L} 2\tan^{-1}\frac{4}{U}(\sin k_h - \Lambda_m). \tag{3.5}$$

As by the k_h and Λ_m variables all the other unknowns are determined, the problem is reduced to the solving of (2.17) and (3.5). This system is very similar in structure to the original Lieb–Wu equations ((2.17) is formally the same as (1.3) while (3.5) is the analogue of (1.2)). The main differences are that it contains only the parameters of the excitations (instead of the parameters of all electrons) and that the variables k_h, being the positions of the holes in the distribution of the real ks, are always real.

3.2. *Energy and momentum*

The energy is calculated by the formula

$$E = -N\int_{-\pi}^{\pi} 2\cos k\rho''(k)\,dk - \sum_{m=1}^{L} 2[\cos(\kappa_m^{(+)} + i\chi_m^{(+)}) + \cos(\kappa_m^{(-)} - i\chi_m^{(-)})] \tag{3.6}$$

which yields

$$E = E_0 + \sum_{h=1}^{H+2L} \varepsilon(k_h) + LU \tag{3.7}$$

with E_0 being the ground-state energy of $N(!)$ electrons

$$E_0 = -2N\int_0^\infty \frac{e^{-\omega U/4} F_0(\omega)F_1(\omega)}{\cosh(\omega U/4)}\frac{d\omega}{\omega} \tag{3.8}$$

and

$$\varepsilon(k) = 2\cos k + 2\int_0^\infty \frac{e^{-\omega U/4} F_1(\omega)}{\cosh(\omega U/4)}\cos(\omega \sin k)\frac{d\omega}{\omega}. \tag{3.9}$$

The momentum evaluated by means of the formulae (I. 2.4), (2.7), (2.8), (2.14), (2.18), (3.1) and (3.5), is (up to an integer multiply of 2π)

$$p = \sum_{h=1}^{H+2L} -p(k_h) + \psi \tag{3.10}$$

with

$$p(k) = k + \int_0^\infty \frac{e^{-\omega U/4} F_0(\omega) \sin(\omega \sin k)}{\cosh(\omega U/4)} \frac{d\omega}{\omega} \tag{3.11}$$

and

$$\psi = \pi[1 + N + (N - H)/2] \tag{3.12}$$

(we wish to mention that the appearance of ψ is not connected with the presence of the complex wavenumbers. It is connected with the fact that even the ground state momentum of a Heisenberg chain can be 0 or π, depending on the parity of the number of sites, and that even the ground state momentum of a half-filled Hubbard chain can be 0 to π depending on the parity of $N/2$).

It is worthwhile to compare the (3.7), (9.10) energy–momentum dispersion with that for states with more than N electrons. A state with $N + H'$ electrons can be obtained by acting on a state with $N - H'$ electrons by the operator†

$$\exp\left(-i \sum_{n=1}^{N} \pi(n + \tfrac{1}{2})(c^+_{n\downarrow} c^+_{n\uparrow} + c_{n\uparrow} c_{n\downarrow})\right).$$

This operation introduces holes instead of the particles, and changes the energy by $H'U$ and the momentum by $\pi H'$. According to this, the energy and the momentum of a state with H' extra electrons and with L pairs of complex wavenumbers is

$$\begin{aligned} E &= E_0 + \sum_{h=1}^{H'+2L} \varepsilon(k_h) + (H' + L)U \\ p &= \sum_{h=1}^{H'+2L} -p(k_h) + \pi[1 + N + (N + H')/2]. \end{aligned} \tag{3.13}$$

Comparing (3.7), (3.10) with (3.13) we see that the energy momentum dispersion is very much like the energy of a state with $L + H$ (L) holes in a subband with dispersion $-\varepsilon(k(p))$ and L ($L + H'$) particles in another band with dispersion $\varepsilon(k(p)) + U$. Thus, the form of the energy momentum dispersion suggests that if pairs of complex wavenumbers are introduced instead of real ks it acts like exciting a number of carriers from one band to the other. This picture, however, reflects only the apparent additivity of the energy and the momentum, and gives the right coefficient of U. The excitations, if we treat them as quasiparticles, should be regarded as interacting ones. This is reflected in the fact that the momenta of the quasiparticles are not free parameters; they are connected with the actual quantum numbers through the system of equations (2.17) and (3.5).

In figure 1 we have represented the energy bands as functions of U for a half-filled band ($H = 0$) for one, two and three pairs of complex wavenumbers. The lower and

† The state with H' extra electrons can be constructed in this way only if N is even. If N is odd, the transformation between particles and holes changes the periodic boundary condition into antiperiodic or changes the sign of the kinetic energy.

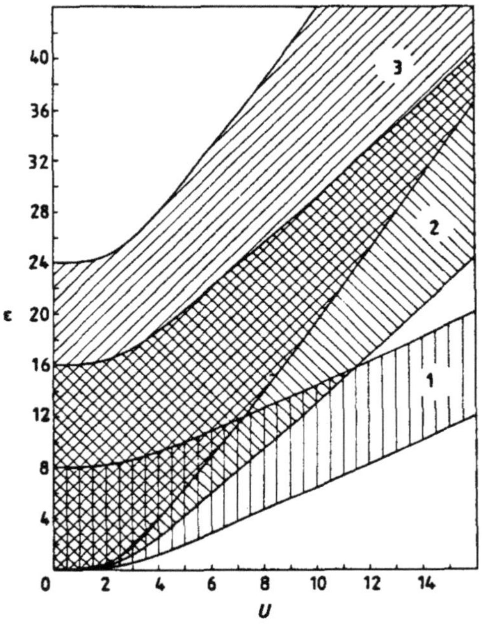

Figure 1. Energy bands for excitations with one, two, and three pairs of complex wavenumbers as functions of U. The different bands are differently shaded; the numbers in them indicate the numbers of complex wavenumber pairs.

upper edges of the bands can be obtained through (3.7) and (3.9) taking all the k_hs as π and 0 respectively.

3.3. *A special solution for the* Λ_n*s*

The equations (2.17) and (3.5) are highly non-linear but there is one case when they can be replaced by a linear integral equation. This is the case when $H = 0$, L is macroscopic (comparable to N) and we choose for the $\mathcal{F}'_n$ set the numbers

$$-\tfrac{1}{2}(L-1),\ -\tfrac{1}{2}(L-3),\ \ldots,\ \tfrac{1}{2}(L-1). \tag{3.14}$$

With this choice of $\mathcal{F}'_n$s all Λ_ns will be real, and the number of Λ_ns between Λ and $\Lambda + \mathrm{d}\Lambda$ can be given as $(2L)\xi(\Lambda)\mathrm{d}\Lambda$, where

$$\xi(\Lambda) = \frac{1}{2L}\sum_{h=1}^{2L}\frac{1}{U}\frac{1}{\cosh[(\Lambda - \sin k_n)2\pi/U]}. \tag{3.15}$$

Combining this with (3.3) one finds that the density of all λs and Λs is the same as the density of λs in the ground state:

$$\sigma(\lambda) + \frac{2L}{N}\xi(\lambda) = \frac{1}{2\pi}\int_0^\infty \frac{F_0(\omega)}{\cosh(\omega U/4)}\cos(\omega\lambda)\,\mathrm{d}\omega = \sigma_0(\lambda). \tag{3.16}$$

Using the $\xi(\Lambda)$ to evaluate the sum over the Λ_ms in (3.5), we find that

$$p(k_h) = (2\pi/N)\,\mathcal{I}_h \tag{3.17}$$

i.e. in this special case the quasiparticles are not interacting. We should emphasise that this holds only if L is large, i.e. the error introduced by using $\xi(\Lambda)$ in the summations (for which $1/L$ is an upper limit) is sufficiently small and if we choose (3.14) to characterise the system.

3.4. On the conditions

Any solution of the system (2.5), (2.13) and (2.17) is meaningful only if, on substituting the λs and Λs into (2.3), all of the $\eta_n^{(\pm)}$s are positive and of the order of unity; and $\rho(k)$ of (2.9) is positive. We are not able to show in general that these conditions are satisfied for all solutions, but in two cases: one in which the number of excitations is small compared with N (i.e. the number of Λs is not macroscopic, and at the same time the spin part is near to its ground state). The other case is when although the number of Λs is macroscopic, the system is near to the state described in § 3.3. In both cases the sums in (2.3) can be estimated by integrating over the λs (all λs and Λs) using the ground-state density. (The error of this estimation is of the order of $(\ln N)/N$). This estimation shows that both $\eta_n^{(+)}$ and $\eta_n^{(-)}$ are definitely positive and do not vanish as $N \rightarrow \infty$. $\rho(k)$ can be estimated in the same way with an error of the order of $1/N$; it is also positive for all k. It is also true that for a small number of complex k pairs for each pair there must exist a Λ satisfying (2.1). (If a complex k pair without Λ existed, then for that pair $\eta^{(+)}$ and $\eta^{(-)}$ should be zero; this is however, according to the above-described estimation, impossible.)

4. Summary

In the present work we have investigated those eigenstates of the 1D Hubbard model for which in the wavenumber set there are several pairs of complex wavenumbers. Our results are the following.

(i) It has been shown that for states with several pairs of complex wavenumbers the solving of the Lieb–Wu equations is equivalent to solving (2.5), (2.13) and (2.17). This system is very similar in structure to the original Lieb–Wu equations, with the difference that there are no complex wavenumbers. Auxiliary variables appearing instead of the complex ks are the positions of the holes in the distribution of the real wavenumbers.

(ii) A complementarity between different solutions of the system (2.5), (2.13) and (2.17) has been established. By one solution of the system two eigenstates can be described. In the complementary states the parameters connected with the charge and spin degrees of freedom change roles. This complementarity can be used to find the eigenvalues and eigenstates of the Hubbard Hamiltonian with negative U too.

(iii) It has been shown that for large U, in a state with L complex k pairs, there are L doubly occupied sites. In this limit the distribution of the empty and doubly occupied sites is the same as the distribution of the up and down spins in a spin $\frac{1}{2}$ Heisenberg chain. The distribution of the spins belonging to the singly occupied sites is also described by an eigenfunction of the Heisenberg Hamiltonian.

(iv) To describe states in which the spin degrees of freedom are not excited, the (2.5), (2.13) and (2.17) system has been reduced to a simpler one ((2.17) and (3.5)). This system is the analogue of the original Lieb–Wu equations with the difference that it contains the parameters of the excitations only. For these states the energy and the momentum can be given as the sum of the energies and the momenta of quasiparticles.

These quasiparticles are, however, interacting ones. The energy bands belonging to different numbers of complex k pairs are given.

We have concentrated on the 'charge excitations'. To isolate charge rearrangement effects we examined in detail those states in which the spin part was in its ground state. Similar calculations can also be performed for those states in which the spin part is excited too. Preliminary results show that, as expected, the presence of spin excitations does not effect drastically the results connecting the states studied so far, just in addition a new type of 'elementary excitations' must be introduced.

Acknowledgments

I am very grateful to Drs A Sütõ and P̂ Fazekas for the many valuable discussions.

Note added in proof. The derivation of the system (2.5), (2.13) and (2.17) presented in § 2.1 is correct only if there are no λ_α and Λ_n with $|\text{Re}\ \lambda(\Lambda)| \leqslant 1$ and $|U/4 - |\text{Im}\ \lambda(\Lambda)|| \ll 1/N$. It can be checked that a solution of the system (2.5), (2.13) and (2.17) generates a solution of the original Lieb–Wu equations even if some of the λ and Λ have imaginary parts very near to $U/4$. In this case:

(*a*) For the N different I' equation (2.5) has more than N different solutions. Among these solutions to each λ and Λ pair satisfying the above inequalities there is a pair of real k with $\sin k^{(1,2)} = \text{Re}\ \lambda(\Lambda)$. This k should be ignored, and the k_j and k_h are given by the remaining N solutions.

(*b*) Two of the four k coupled to a complex Λ pair satisfying the above inequalities and defined by (2.4) are real and equal. Instead of these k the two different solutions of $\sin k^{(1,2)} = \text{Re}\ \Lambda$ are to be used.

Since in the $\rho(k)$ of (2.9) the real k-pairs connected with the λ and Λ in question appear as δ-like peaks with opposite signs, $\rho(k)$ automatically takes care of the alterations connected with the Λ. The only point where care must be taken is the calculation of the spin excitations, thus no results and conclusions of the other sections are affected.

References

Lieb E M and Wu F Y 1968 *Phys. Rev. Lett.* **20** 1445
Woynarovich F 1982 *J. Phys. C: Solid State Phys.* **15** 85–96

J. Phys. C: Solid State Phys., **16** (1983) 5293–5304. Printed in Great Britain

Spin excitations in a Hubbard chain

F Woynarovich
Central Research Institute for Physics H-1525 Budapest, POB 49, Hungary

Received 15 February 1983

Abstract. The spin excitations of a Hubbard chain are calculated on the basis of the Lieb–Wu equations. A system of equations, which contains the parameters of the excitation only, is derived. It is found that both the energy and the momentum of the spin excitations depend only on the positions of the holes in the real λ distribution. The energy–momentum dispersion for a half-filled band is given. The structure and degeneracy of the lowest energy excitations are also discussed.

1. Introduction

As is well known, the one-dimensional Hubbard model can be treated exactly by the Bethe *ansatz*. Lieb and Wu (1968) showed that the diagonalisation of the Hamiltonian can be reduced to solving a set of coupled non-linear equations and they also found the solution corresponding to the ground state of this system. Following this work, Ovchinnikov (1970) calculated the triplet spin waves for a half-filled band while Coll (1974) calculated the spin-wave-type and one-particle-type excitations for general band filling. Recently Choy and Young (1982) re-examined the triplet spin wave excitations and the present author analysed the charge excitations for a half-filled chain (Woynarovich 1982a, b). In the latter paper (Woynarovich 1982b) the original Lieb–Wu equations were replaced by an equivalent system in which the parameters corresponding to the charge and spin degrees of freedom were separated. This system has also been solved for the case when the spin part was not excited. In the present work we intend to complete these studies by giving those solutions of our equations which correspond to states which also include spin excitations (singlet, triplet and higher multiplet excitations).

This paper is divided into two parts. In § 2, after reviewing the formalism, we reduce our system to another one establishing connections among the parameters of the excitations only. In § 3, we give an account of the low-energy spin excitations of a half-filled chain and also describe the structure of the spin excitations for a band which is less than half-filled.

Throughout this study we will use the results of our previous paper extensively (Woynarovich 1982b, hereafter referred to as II).

2. Equations for the excitations

2.1. Basic formalism

It is well known (Lieb and Wu 1968), that the diagonalisation of the 1D Hubbard Hamiltonian with periodic boundary conditions

$$\hat{H} = -\sum_{i=1}^{N}\sum_{\sigma}(c_{i+1\sigma}^{+}c_{i\sigma} + c_{i\sigma}^{+}c_{i+1\sigma}) + U\sum_{i=1}^{N}\hat{n}_{i\uparrow}\hat{n}_{i\downarrow} \qquad c_{N+1\sigma} = c_{1\sigma} \tag{2.1}$$

($c_{i\sigma}^{+}(c_{i\sigma})$ creates (destroys) an electron with spin σ at the site i; $\hat{n}_{i\sigma}$ is the number operator of the electrons with spin σ at the site i) is equivalent to solving the system of equations

$$Nk_l = 2\pi I_l - \sum_{\alpha=1}^{M} 2\tan^{-1}\frac{4}{U}(\sin k_l - \lambda_\alpha) \tag{2.2a}$$

$$\sum_{l=1}^{N_e} 2\tan^{-1}\frac{4}{U}(\lambda_\alpha - \sin k_l) = 2\pi J_\alpha + \sum_{\alpha'=1}^{M} 2\tan^{-1}\frac{2}{U}(\lambda_\alpha - \lambda_{\alpha'}). \tag{2.2b}$$

Here the parameter sets I_l and J_α consist of integer or half odd-integer numbers, the parities of the numbers $2I_l$ and $2J_\alpha$ being the same as those of the numbers M and $(N_e - M - 1)$ respectively. A solution of equation (2.2a) and (2.2b) corresponds to an $S^z = \frac{1}{2}N_e - M$ $(N \geqslant N_e \geqslant 2M)$ eigenstate of N_e electrons with an energy and a momentum

$$E = \sum_{l=1}^{N_e}(-2\cos k_l) \tag{2.3a}$$

$$p = \sum_{l=1}^{N_e} k_l = \frac{2\pi}{N}\left(\sum_{l=1}^{N_e} I_l + \sum_{\alpha=1}^{M} J_\alpha\right) \tag{2.3b}$$

respectively.

In II we were searching for such solutions of equations (2.2a) and (2.2b) in which the wavenumber set k_l also contains complex wavenumbers, especially pairs of wavenumbers of the form

$$k_n^{\mp} = \sin^{-1}(\Lambda_n \mp \mathrm{i}U/4) + \mathrm{O}(\exp(-\eta_n^{\mp}N)) \tag{2.4}$$

(with the set of Λ_n being a subset of the λ_α of equation (2.2b)). We found that the $k_n^{\mp}$ pairs of (2.4) drop out of equations (2.2a) and (2.2b), but instead of these the positions of the holes (k_h) left in the real $k(k_j)$ distribution appear and one has to solve the system

$$Nk_{j(h)} = 2\pi I_{j(h)} - \sum_{\beta=1}^{M-L} 2\tan^{-1}\frac{4}{U}(\sin k_{j(h)} - \lambda_\beta) - \sum_{n=1}^{L} 2\tan^{-1}\frac{4}{U}(\sin k_{j(h)} - \Lambda_n) \tag{2.5a}$$

$$\sum_{j=1}^{N_e-2L} 2\tan^{-1}\frac{4}{U}(\lambda_\beta - \sin k_j) = 2\pi J'_\beta + \sum_{\beta'=1}^{M-L} 2\tan^{-1}\frac{2}{U}(\lambda_\beta - \lambda_{\beta'}) \tag{2.5b}$$

$$\sum_{h=1}^{H+2L} 2\tan^{-1}\frac{4}{U}(\Lambda_n - \sin k_h) = 2\pi J'_n + \sum_{n'=1}^{L} 2\tan^{-1}\frac{2}{U}(\Lambda_n - \Lambda_{n'}). \tag{2.5c}$$

Here H is the number of electrons needed to make the band half-filled ($H = N - N_e$), L is the number of $k^{\pm}$ pairs, and the parities of the numbers $2J'_\beta$ and $2J'_n$ are the same as those of the numbers $N_e - (M - L) - 1$ and $N_e - L - 1$ respectively. Equation (2.5*a*) (at fixed λ_β and Λ_n sets) is an equation with one unknown and has (for appropriately chosen *I*s) *N* non-equivalent real solutions, provided that for no λ or Λ

$$|\mathrm{Re}\,\lambda(\Lambda)| < 1 \qquad \text{and}\ ||\mathrm{Im}\,\lambda(\Lambda)| - U/4| \lessdot 1/N.$$

If for a λ or Λ

$$|\mathrm{Re}\,\lambda(\Lambda)| < 1 \qquad \text{and}\ ||\mathrm{Im}\,\lambda(\Lambda)| - U/4| \lessdot 1/N,$$

then in addition to the above real solutions (2.4*a*) also has real solutions of the form

$$k = \sin^{-1}(\mathrm{Re}\,\lambda(\Lambda)) + \mathrm{O}(\mathrm{Im}\,\lambda(\Lambda) - U/4). \tag{2.6}$$

(These *k*s are the special cases of (2.4)). The elements of the sets k_j and k_h are given by the *N* non-equivalent real roots of (2.5*a*), which are not of the form (2.6). Then the set λ_α of (2.2*a*) and (2.2*b*) is the union of the sets λ_β of (2.5*b*) and Λ_n of (2.5*c*), while the set k_l of (2.2*a*) and (2.2*b*) is given by the union of the set k_j of (2.5*a*) and the set of *k*s defined as

$$k_n^\pm = \sin^{-1}(\Lambda_n \mp \mathrm{i}U/4); \qquad \pm\,\mathrm{Im}\,k_n^\pm > 0$$
$$\text{if}\ ||\mathrm{Im}\,\Lambda_n| - U/4| > 1/N \ \text{or}\ |\mathrm{Re}\,\Lambda_n| > 1 \tag{2.7a}$$

$$k_n^+ = \sin^{-1}(\Lambda_n - \mathrm{i}U/4); \qquad k_n^- = \sin^{-1}(\mathrm{Re}\,\Lambda_n); \quad \mathrm{Im}\,k_n^+ > 0$$
$$k_{n'}^+ = \pi - k_n^-; \quad k_{n'}^- = (k_n^+)^* \tag{2.7b}$$
$$\text{if}\ |\mathrm{Im}\,\Lambda_n + U/4| < 1/N,\ |\mathrm{Re}\,\Lambda_n| < 1 \ \text{and}\ \Lambda_n = \Lambda_{n'}^*.$$

It is also possible to give those I_l and J_α parameters which define the above solution for (2.2*a*) and (2.2*b*). Since, however, these parameters may appear explicitly only in the momentum, it is more useful to determine the momentum in terms of the new parameters

$$p = \frac{2\pi}{N}\Bigg[\sum_j I_j + \sum_\beta J'_\beta - \sum_n J'_n$$
$$+ \frac{1}{2}\sum_{|\mathrm{Im}\,\Lambda_n| > U/4}\Bigg(\sum_j \mathrm{sgn}(\mathrm{Re}\,\Lambda_h - \sin k_j) + \sum_h \mathrm{sgn}(\mathrm{Re}\,\Lambda_n - \sin k_h)\Bigg)\Bigg]. \tag{2.8}$$

Equations (2.5*a*) to (2.5*c*) make it possible to treat the charge and spin degrees of freedom separately: in II it was shown (using an argument based on the $U \to \infty$ limit of both the equations and the wavefunction) that the set Λ_n is characteristic of the charge distribution (the relative distribution of doubly occupied and empty sites), while the distribution of the uncompensated spins is described by the set λ_β only. In II the λ_β distribution, characteristic for those states in which the spin degrees of freedom are not excited, was also determined. In this paper we intend to describe those solutions which correspond to states where spin excitations are also present. In the first step we will take the set k_j as a given one and we will reduce equation (2.5*b*) to a simpler one which contains the parameters of the spin excitations only. In this procedure we will determine the distribution of the real λ_βs as well. In the second step we will eliminate both the real λ_βs from equation (2.5*a*) and the k_js from the distribution of the real λ_βs. In this way, we will obtain the set of equations determining the parameters of the charge and spin excitations and we will also find the distributions of the k_js and the real λ_βs.

2.2. Reduction of equation (2.5b)

In a state with no spin excitations present, all the λ_βs are real and their number is $N_e/2 - L$ (for N_e even). To describe states with excited spins, we have to suppose that the number of λ_βs is less than $N_e/2 - L$ and/or some of the λ_βs are complex. To find such solutions for equation (2.5*b*), the same strategy can be used that was applied to reduce the secular equations of a Heisenberg chain (Woynarovich 1982c, Babelon *et al* 1982). The essence of this method is that the distribution of the real λ_βs can be calculated supposing that the complex λ_βs are given. By means of this distribution the real λ_βs can be eliminated from the equations of the complex λ_βs. In this way one obtains a set of equations which contains the complex λ_βs and the holes in the real λ_β distribution only. Since this method is described in sufficient detail in the papers referred to above, we give here the main steps only.

(i) Suppose that the number of complex λs (later on, the complex λs will be denoted by λ_μ) with $0 < |\mathrm{Im}\,\lambda_\mu| < U/2$ and $U/2 < |\mathrm{Im}\,\lambda_\mu|$ is $2m_1$ and $2m_2$ respectively. Then we have to introduce $N_e - 2M + 2m_1 + 4m_2$ holes in the real λ_β distribution (later on, the positions of the holes will be denoted by λ_η). Solving the equations of the real λ_βs (equation (2.5*b*) with a real λ_β on the LHS) for the density of these variables yields

$$\sigma(\lambda) = \frac{1}{2\pi N}\int_{-\infty}^{\infty} \frac{\exp(\mathrm{i}\omega\lambda)}{2\cosh(\omega U/4)}\left(\sum_j \exp(-\mathrm{i}\omega\sin k_j) - \sum_\eta \exp(|\omega|U/4 - \mathrm{i}\omega\lambda_\eta)\right)\mathrm{d}\omega$$
$$- \frac{1}{2\pi N}\int_{-\infty}^{\infty} \frac{\exp(\mathrm{i}\omega\lambda + |\omega|U/4)}{2\cosh(\omega U/4)}\sum_\mu f(\omega, \lambda_\mu, U/2)\,\mathrm{d}\omega. \tag{2.9}$$

Here we have used the notation

$$f(\omega, \lambda_\mu, U/2) = \frac{1}{2\pi}\int_{-\infty}^{\infty} \exp(-\mathrm{i}\omega\lambda)2\frac{U/2}{(U/2)^2 + (\lambda - \lambda_\mu)^2}\,\mathrm{d}\omega. \tag{2.10}$$

(ii) From the equations for the complex λs (equation (2.5*b*) with a λ_μ in the LHS) the real λs can be eliminated by means of $\sigma(\lambda)$. The equations obtained in this way can be satisfied (at least if the number of excitations i.e. the number of λ_ηs is small compared with $N_e - 2L$) only if to each λ_μ with $|\mathrm{Im}\,\lambda_\mu| < U/2$ there exists another $\lambda_{\mu'}$ for which

$$|\mathrm{Im}\,\lambda_{\mu'}| < U/2 \qquad \text{and} \qquad \lambda_\mu - \lambda_{\mu'} = \pm\mathrm{i}U/2 + \mathrm{O}(\exp(-N_e + 2L)).$$

This makes it convenient to represent the set λ_μ by a set of auxiliary variables λ'_μ in the following way:

(α) one λ'_μ with $|\mathrm{Im}\,\lambda'_\mu| < U/4$ represents a pair of λ_μs with $|\mathrm{Im}\,\lambda^\pm_\mu| < U/2$:

$$\lambda^\pm_\mu = \lambda'_\mu \pm \mathrm{i}U/4 \tag{2.11a}$$

(β) a λ'_μ with $|\mathrm{Im}\,\lambda'_\mu| > U/4$ represents a single λ_μ with $|\mathrm{Im}\,\lambda_\mu| > U/2$

$$\lambda_\mu = \lambda'_\mu + \mathrm{i}U/4\,\mathrm{sgn}(\mathrm{Im}\,\lambda'_\mu). \tag{2.11b}$$

The number of the variables λ'_μ is $m_1 + 2m_2$ and substituting them into the equations of the λ_μs it turns out that they have to satisfy the equations

$$\sum_\eta 2\tan^{-1}\frac{4}{U}(\lambda'_\mu - \lambda_\eta) = 2\pi J''_\mu + \sum_{\mu'} 2\tan^{-1}\frac{2}{U}(\lambda'_\mu - \lambda'_{\mu'}). \tag{2.12}$$

Here the J''_μs are integers if $N_e - 2M - (m_1 + 2m_2)$ is odd and they are half odd integers otherwise. Their connections with the parameters J'_μ are

$$J''_\mu = J'_\mu + \frac{1}{2}\sum_\beta \operatorname{sgn}(\operatorname{Re}\lambda'_\mu - \lambda_\beta) + \frac{1}{2}\sum_\eta \operatorname{sgn}(\operatorname{Re}\lambda'_\mu - \lambda_\eta)$$

$$-\frac{1}{2}\sum_j \operatorname{sgn}(\operatorname{Re}\lambda'_\mu - \sin k_j) + \frac{1}{2}\sum_{|\operatorname{Im}\lambda_{\mu'}|<U/4} \operatorname{sgn}(\operatorname{Re}(\lambda'_\mu - \lambda'_{\mu'})) \tag{2.13a}$$

and

$$J''_\mu = J^+_\mu + J^-_\mu - \frac{1}{2}\sum_j \operatorname{sgn}(\operatorname{Re}\lambda'_\mu - \sin k_j) + \frac{1}{2}\sum_{\mu'\neq\mu} \operatorname{sgn}(\operatorname{Re}(\lambda'_\mu - \lambda'_{\mu'})). \tag{2.13b}$$

Here, (2.13*a*) is for $|\operatorname{Im}\lambda'_\mu| > U/4$, while (2.13*b*) is for λ'_μs with $|\operatorname{Im}\lambda'_\mu| < U/4$.

(iii) With the introduction of the λ'_μs the last term in $\sigma(\lambda)$ of (2.9) can be simplified:

$$-\frac{1}{2\pi N}\int_{-\infty}^{\infty} \frac{\exp(i\omega\lambda + |\omega|U/4)}{2\cosh(\omega U/4)} \sum_{\lambda_\mu} f(\omega, \lambda_\mu, U/2)$$

$$= -\frac{1}{2\pi N}\sum_{\lambda'_\mu} 2\frac{(U/4)}{(U/4)^2 + (\lambda - \lambda'_\mu)^2}. \tag{2.14}$$

The equations for the λ_ηs (equation (2.5*b*) with a λ_η substituted instead of λ_β) can be given (after eliminating the λ_βs by means of $\sigma(\lambda)$) in the form

$$\sum_j \int_{-\infty}^{\infty} \frac{\exp[i\omega(\lambda_\eta - \sin k_j)]}{2\cosh(\omega U/4)}\frac{d\omega}{i\omega} = 2\pi J'_\eta - \sum_{\eta'}\int_{-\infty}^{\infty} \frac{\exp[i\omega(\lambda_\eta - \lambda_{\eta'}) - |\omega|U/4]}{2\cosh(\omega U/4)}\frac{d\omega}{i\omega}$$

$$+ \sum_\mu 2\tan^{-1}\frac{4}{U}(\lambda_\eta - \lambda'_\mu). \tag{2.15}$$

2.3. *Equations for the parameters of the excitations*

As a next step we calculate the density of the variables k_j. Since some of the λ'_μ can be real, meaning that the corresponding $\lambda^\pm_\mu$ have imaginary parts very near to $+U/4$ or $-U/4$, we have to be careful to filter out the additional real solutions of equation (2.5*a*) (those of the form (2.6)). This can be done by splitting the $\tan^{-1}$ functions into continuous and discontinuous parts in the following manner:

$$\left(2\tan^{-1}\frac{4}{U}(\sin k - \lambda_\mu)\right)_{\text{cont.}} = \tan^{-1}\frac{\sin k - \operatorname{Re}\lambda_\mu}{U/4 + \operatorname{Im}\lambda_\mu} + \tan^{-1}\frac{\sin k - \operatorname{Re}\lambda_\mu}{U/4 - \operatorname{Im}\lambda_\mu}$$

$$+\frac{1}{2i}\ln\frac{(U/4 + \operatorname{Im}\lambda_\mu)^2 + (\sin k - \operatorname{Re}\lambda_\mu)^2}{(U/4 - \operatorname{Im}\lambda_\mu)^2 + (\sin k - \operatorname{Re}\lambda_\mu)^2} \tag{2.16a}$$

$$(2\tan^{-1}\frac{4}{U}(\sin k - \lambda_\mu))_{\text{discont.}} = \begin{cases}\pi\operatorname{sgn}(\sin k - \operatorname{Re}\lambda_\mu) & \text{if } |\operatorname{Im}\lambda_\mu| > U/4 \\ 0 & \text{if } |\operatorname{Im}\lambda_\mu| < U/4.\end{cases} \tag{2.16b}$$

This splitting is to be used for λ_μs with $||\operatorname{Im}\lambda_\mu| - U/4| > 1/N$. If $||\operatorname{Im}\lambda_\mu| - U/4| \ll 1/N$, then for a $\sin k_j$ or a $\sin k_h$ one term in (2.16*a*) is $+\pi/2$ or $-\pi/2$ with an accuracy of $O(|\operatorname{Im}\lambda_\mu| - U/4)$, but this is not true for the corresponding k of (2.6). Thus for a λ_μ with $|\operatorname{Im}\lambda_\mu| \sim U/4$ we take

$$\operatorname{Re}(2\tan^{-1}4U^{-1}(\sin k - \lambda_\mu))_{\text{cont.}} = \tan^{-1}2U^{-1}(\sin k - \operatorname{Re}\lambda_\mu) \tag{2.16c}$$

$$(2\tan^{-1}4U^{-1}(\sin k - \lambda_\mu))_{\text{discont.}} = \tfrac{1}{2}\pi\operatorname{sgn}(\sin k - \operatorname{Re}\lambda_\mu). \tag{2.16d}$$

With these definitions equation (2.5*a*) can be written in the form

$$Nk_{j(h)} = 2\pi I'_{j(h)} - \sum_{\text{all }\lambda}\left(2\tan^{-1}\frac{4}{U}(\sin k_{j(h)} - \lambda)\right)_{\text{cont.}} - \sum_{n}\left(2\tan^{-1}\frac{4}{U}(\sin k_{j(h)} - \Lambda_n)\right)_{\text{cont.}} \tag{2.17}$$

where

$$I'_{j(h)} = I_{j(h)} - \frac{1}{2}\sum_{\mu}\operatorname{sgn}(\sin k_{j(h)} - \operatorname{Re}\lambda'_\mu) - \frac{1}{2}\sum_{|\operatorname{Im}\Lambda_n|>U/4}\operatorname{sgn}(\sin k_{j(h)} - \operatorname{Re}\Lambda_n). \tag{2.18}$$

The set of possible non-equivalent real solutions of equation (2.17) coincide (with an exponential accuracy) with those real solutions of (2.5*a*) which are not of the form (2.6) i.e. with the union of the k_j and k_h sets. Taking $N_e - 2L$ different integers (or half odd integers) between $+N/2$ and $-N/2$ we find that the density of the k_js is

$$\rho(k) = \frac{1}{2\pi} - \frac{1}{2\pi N}\sum_h \delta(k - k_h) + \frac{\cos k}{2\pi N}\left(\sum_{\text{all }\lambda} g(\sin k - \lambda) + \sum_{\text{all }\Lambda} g(\sin k - \Lambda)\right) \tag{2.19}$$

where

$$g(x) = \frac{\mathrm{d}}{\mathrm{d}x}\left(2\tan^{-1}\frac{4}{U}x\right)_{\text{cont.}}. \tag{2.20}$$

Using $\rho(k)$ to evaluate the sum over k_j in (2.9) we have

$$\sigma(\lambda) = \frac{1}{2\pi}\int_0^\infty \frac{J_0(\omega)\cos\omega\lambda}{\cosh(\omega U/4)}\,\mathrm{d}\omega - \frac{1}{N}\sum_\eta(\delta(\lambda - \lambda_\eta) - f_1(\lambda - \lambda_\eta)) - \frac{1}{NU}\sum_h \frac{1}{\cosh[(\lambda - \sin k_h)2\pi/U]} - \frac{1}{2\pi N}\sum_\mu 2\frac{U/4}{(U/4)^2 + (\lambda - \lambda'_\mu)^2} \tag{2.21}$$

where $J_0(\omega)$ is the zeroth-order Bessel function and we have used the notation

$$f_1(x) = \frac{1}{2\pi}\int_0^\infty \frac{\exp(-\omega U/4)\cos\omega x}{\cosh(\omega U/4)}\,\mathrm{d}\omega. \tag{2.22}$$

In turn, by means of $\sigma(\lambda)$ the coefficient of the $\cos k$ term in $\rho(k)$ of (2.19) can be evaluated yielding

$$\rho(k) = \frac{1}{2\pi}\left(1 + \cos k\int_0^\infty \frac{\exp(-\omega U/4)J_0(\omega)\cos(\omega\sin k)}{\cosh(\omega U/4)}\,\mathrm{d}\omega\right) - \frac{1}{N}\sum_h\{\delta(k - k_h) + \cos k f_1(\sin k - \sin k_h)\} - \frac{\cos k}{NU}\sum_\eta \frac{1}{\cosh[(\sin k - \lambda_\eta)2\pi/U]} + \frac{\cos k}{2\pi N}\sum_n 2\frac{(U/4)^2}{(U/4)^2 + (\sin k - \Lambda_n)^2}. \tag{2.23}$$

In a similar way the k_js and the real λ_βs can be eliminated from (2.15) and (2.5*a*) respectively with the results

$$N\int_0^\infty \frac{J_0(\omega)\sin(\omega\lambda_\eta)}{\cosh(\omega U/4)}\frac{d\omega}{\omega} = 2\pi J'_\eta - \sum_{\eta'} F_1(\lambda_\eta - \lambda_{\eta'})$$

$$+ \sum_h 2\tan^{-1}\{\tanh[(\lambda_\eta - \sin k_h)\pi/U]\} + \sum_\mu 2\tan^{-1}\frac{4}{U}(\lambda_\eta - \lambda'_\mu) \qquad (2.24)$$

$$N\left(k_h + \int_0^\infty \frac{J_0(\omega)\exp(-\omega U/4)\sin(\omega\sin k_h)}{\cosh(\omega U/4)}\frac{d\omega}{\omega}\right) = 2\pi I''_h + \sum_{h'} F_1(\sin k_h - \sin k_{h'})$$

$$+ \sum_\eta 2\tan^{-1}\{\tanh[(\sin k_h - \lambda_\eta)\pi/U]\} - \sum_n 2\tan^{-1}\frac{4}{U}(\sin k_h - \Lambda_n). \qquad (2.25)$$

Here

$$F_1(x) = \int_0^\infty \frac{\exp(-\omega U/4)\sin(\omega x)}{\cosh(\omega U/4)}\frac{d\omega}{\omega} \qquad (2.26)$$

and

$$I''_h = I_h - \frac{1}{2}\sum_\mu \mathrm{sgn}(\sin k_h - \mathrm{Re}\,\lambda'_\mu). \qquad (2.27)$$

As a result, for the complete description of an eigenstate, the system of equations (2.24)–(2.25) together with the equations (2.5*c*) and (2.12) should be solved. This system may seem more complicated in structure than the original system of Lieb–Wu equations, but it contains the parameters of the excitations only.

2.4. *Energy and momentum*

To calculate the energy, according to (2.3*a*) one has to sum up the contributions of the complex *k*s (which are given by (2.7*a*)–(2.7*b*)) and those of the k_js. These latter can be done using the density (2.23). As a result one obtains

$$E = E_0 + \sum_h \varepsilon_c(k_h) + LU + \sum_\eta \varepsilon_s(\lambda_\eta). \qquad (2.28)$$

Here E_0 is the ground-state energy of *N* electrons

$$E_0 = -2N\int_0^\infty \frac{\exp(-\omega U/4)J_0(\omega)J_1(\omega)}{\omega\cosh(\omega U/4)}\,d\omega \qquad (2.29)$$

($J_1(\omega)$ being the first-order Bessel function) and $\varepsilon_c(k_h)$ and $\varepsilon_s(\lambda_\eta)$ are given by

$$\varepsilon_c(k) = 2\cos k + 2\int_0^\infty \frac{\exp(-\omega U/4)J_1(\omega)\cos(\omega\sin k)}{\omega\cosh(\omega U/4)} \qquad (2.30)$$

$$\varepsilon_s(\lambda) = 2\int_0^\infty \frac{J_1(\omega)\cos(\omega\lambda)}{\omega\cosh(\omega U/4)}\,d\omega. \qquad (2.31)$$

The momentum should be calculated according to (2.8). Following carefully the redefinitions of the *I* and $\mathcal{J}$ parameters ((2.13 (2.18) (2.27)) one obtains that

$$p = \frac{2\pi}{N}\left(\sum_j I'_j + \sum_h I'_h\right) - \frac{2\pi}{N}\left(\sum_h I''_h + \sum_\eta J'_\eta + \sum_\mu J''_\mu + \sum_n J'_n\right) \qquad (2.32)$$

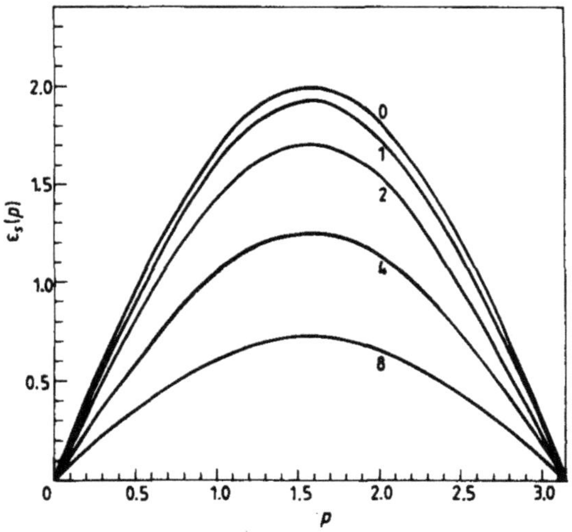

Figure 1. The energy–momentum dispersion for the spin excitations of a half-filled chain. The individual curves are labelled by the value of U.

which, summing up equations (2.5*c*), (2.12), (2.24) and (2.25), yields

$$p = \pi(N - N_e/2 + 1) - \sum_h p_c(k_h) + \sum_\eta p_s(\lambda_\eta) \bmod 2\pi \qquad (2.33)$$

where

$$p_c(k) = k + \int_0^\infty \frac{\exp(-\omega U/4) J_0(\omega) \sin(\omega \sin k)}{\cosh(\omega U/4)} \frac{d\omega}{\omega} \qquad (2.34)$$

and

$$0 < p_s(\lambda) = \pi/2 - \int_0^\infty \frac{J_0(\omega) \sin(\omega\lambda)}{\cosh(\omega U/4)} \frac{d\omega}{\omega} < \pi \qquad (2.35)$$

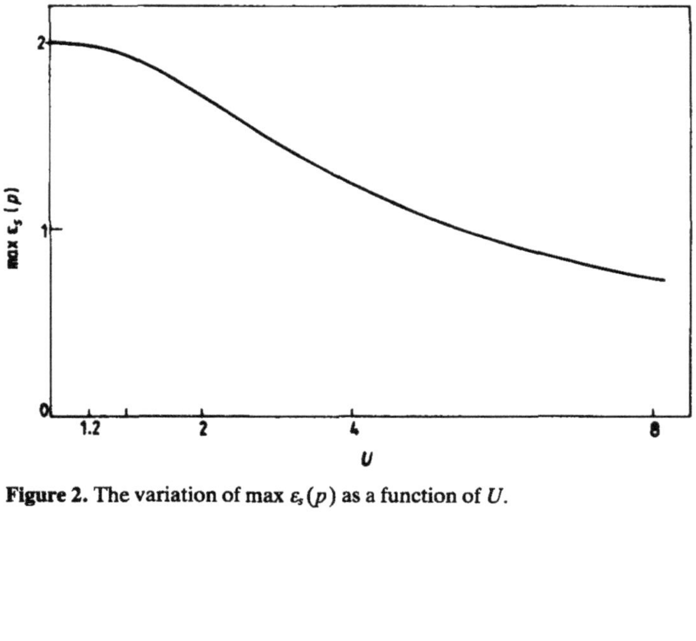

Figure 2. The variation of max $\varepsilon_s(p)$ as a function of U.

The $\varepsilon_c(p_c)$ dispersion is given in II. The $\varepsilon_s(p_s)$ dispersion for different values of U is given in figure 1. The variation of max $\varepsilon_s(p)$ as a function of U is displayed in figure 2.

3. Spin excitations

3.1. Spin excitations of a half-filled chain

For a complete description of an eigenstate the system (2.5*c*), (2.12), (2.24) and (2.25) should be solved. If the band is macroscopically half-filled ($H \ll N$), and both the charge and spin parts are near to their ground states ($N - N_e + 2L \ll N$ and $N_e - 2M + 2(m_1 + 2m_2) \ll N$) this sytem becomes simpler: since the only terms on the RHS of equations (2.24) and (2.25) comparable with N are the $2\pi J'_\eta$ and $2\pi I''_h$ ones, the λ_η and k_h variables can be treated as free parameters (up to an error of the order of $1/N$) and only equations (2.5*c*) and (2.12) should be solved for the Λ_ns and λ'_μs. These variables, however, do not appear explicitly in the energy–momentum dispersion, thus to give an account for the excitations it is sufficient to establish what kind of solutions exist for equations (2.5*c*) and (2.12) for a given k_h and λ_η set. The charge excitations (those connected with the k_hs and Λ_ns) were described in II, here we will discuss the spin excitations in more detail.

It is convenient to classify the excited states according to the number of λ_ηs (in analogy with a Heisenberg chain). The number of λ_ηs is always even if N_e is even and it is always odd if N_e is odd. For the sake of simplicity, we will suppose that N_e is even.

The simplest excitations are those with two λ_ηs. In one class of them there is no λ'_μ, these are the $S^z = 1$ triplet excitations analysed by Choy and Young (1982). In the other class there is one real λ'_μ (fixed by the values of the λ_ηs: $\lambda'_\mu = (\lambda_{\eta_1} + \lambda_{\eta_2})/2$) corresponding to a two-string. These states are singlets. The energy–momentum dispersion for the two classes are the same:

$$E_s^{(2)} = \varepsilon_s(p_{\eta_1}) + \varepsilon_s(p_{\eta_2}). \tag{3.1}$$

The excitations with four λ_ηs can be grouped into four classes. In the first there is no λ'_μ, the states of this class are of $S^z = 2$, $S^2 = 6$. In the second class there is one real λ'_μ which (at fixed λ_ηs), according to the three different choices for J''_μ, assume three different values. These states are triplets with one two-string. In the third class there are two real λ'_μs which are fixed by the λ_ηs. In the fourth class, depending on the actual values of the λ_ηs, there are either two real λ'_μs or there is a complex conjugate pair of λ'_μs. The states of the third class are singlets with two two-strings, while the states of the fourth class are singlets either with two two-strings (real λ'_μs) or with one quartet (if $|\mathrm{Im}\ \lambda'_\mu| < U/4$; $\lambda^{\pm}_{\mu_1} = \lambda'_{\mu_1} \pm \mathrm{i}U/4$, $\lambda^{\pm}_{\mu_2} = \lambda'_{\mu_2} \pm \mathrm{i}U/4 = \lambda'^{*}_{\mu_1} \pm \mathrm{i}U/4$: these complex λ configurations can be regarded as two two-strings with complex conjugate centres) or singlets with one complex λ_μ pair for which $|\mathrm{Im}\ \lambda_\mu| > U/2$ (if $|\mathrm{Im}\ \lambda'_\mu| > U/4$). The energy–momentum dispersion for all the four classes are

$$E_s^{(4)} = \sum_{\eta=1}^{4} \varepsilon_s(p_\eta). \tag{3.2}$$

Taking into account that the degeneracy of $S^2 = 6$ states is fivefold while the $S^2 = 2$ states are threefold degenerate, the (3.2) dispersion is sixteenfold degenerate.

Introducing more and more λ_ηs, the situation becomes less and less transparent. In any case, however, for $\mathcal{H}(\mathcal{H} \ll N_e)$ λ_ηs we expect different solutions for equation (2.12) in which the number of λ'_μs is m as much as

$$\binom{\mathcal{H}}{m} - \binom{\mathcal{H}}{m-1}.$$

These solutions should correspond to $S^z = \mathcal{H}/2 - m$ and $S^2 = (\mathcal{H}/2 - m)(\mathcal{H}/2 - m + 1)$ states and the degeneracy of the

$$E_s^{(\mathcal{H})} = \sum_{\eta=1}^{\mathcal{H}} \varepsilon_s(p_\eta) \tag{3.3}$$

dispersion should be $2^{\mathcal{H}}$.

3.2. On the spin excitations of a less-than-half-filled band

The fact that the energy (2.31) and the momentum (2.35) depend on the λ_ηs, i.e. on the parameters of the spin excitations only, would suggest at first sight that the spin excitations of a half-filled and a non-half-filled band are the same. This is, however, not true: although the structure of the spin excitations is the same, the energy and the momentum carried by the λ_ηs are different from those given by (2.31) and (2.35).

The ground state of $N_e(N - N_e \sim N_e)$ electrons is characterised by a certain k_h set. Introducing a small number of spin excitations, although the connection between the λ_ηs and J_ηs, due to the k_hs, will be different from that in the half-filled case, it remains true that the λ_ηs are free parameters (up to an error of the order of $1/N$, equation (2.24)). Thus the possible λ_η and λ'_μ sets characteristic for the spin excitations are the same for both the half-filled and the non-half-filled band.

The form of the energy–momentum dispersion implicitly depends on the band filling: the introduction of a small number of λ_ηs changes each element of the set of k_hs in the ground state by a term which is linear in $2\tan^{-1}\{\tanh[(\sin k_h - \lambda_\eta)\pi/U]\}$ and is of the order of $1/N$ (equation (2.25)). If the number of k_hs is of the order of N these modifications sum up to a finite value both in the energy and in the momentum. This is most apparently seen in the momentum: in the ground state according to equation (2.25):

$$-\sum_h p_c(k_h) = -\frac{2\pi}{N}\sum_h I'_h \tag{3.4a}$$

while in a state with spin excitations also present

$$-\sum_h p_c(k_h) = -\frac{2\pi}{N}\sum_h I'_h - \frac{1}{N}\sum_{h,\eta} 2\tan^{-1}\{\tanh[(\sin k_h - \lambda_\eta)\pi/U]\}. \tag{3.4b}$$

Here in the second term the ground-state values can be taken for the k_hs and this term can be joined to the contribution of the λ_ηs. Thus the momentum to be attributed to a single spin excitation is

$$p_s(\lambda_\eta) - \frac{1}{N}\sum_h \tan^{-1}\{\tanh[(\sin k_h - \lambda_\eta)\pi/U]\}. \tag{3.5}$$

This momentum changes by a value of $(N_e/N)\pi$ while λ_η runs from $+\infty$ to $-\infty$.

The change in the energy can not be expressed explicitly but it is not hard to see that it has the form:

$$\delta E = \sum_{h,\eta} \frac{1}{N} \delta\varepsilon_h \tan^{-1}\{\tanh[\sin k_h - \lambda_\eta)\pi/U]\}. \tag{3.6}$$

This implies that the energy of the spin part is of the form

$$E_s = \sum_{\eta} \varepsilon_s'(\lambda_\eta) \tag{3.7}$$

also for a general band filling but ε_s' is different from (2.31).

4. Summary and comments

In the present work we have studied the spin excitations of a Hubbard chain. The basis of our study has been the system of equations (2.5*a*)–(2.5*c*) which was derived from the original Lieb–Wu equations in an earlier work and in which the charge and spin degrees of freedom were separated.

We have reduced our basic equations by eliminating the larger part of the parameter set to obtain a system (equations (2.5*c*), (2.12), (2.24), (2.25)) which couples the parameters corresponding to the excitations only. In this way we have established that the spin excitations are characterised by the set of holes in the real λ distribution and by a set of auxiliary variables, this latter representing the set of complex λs. We have also found that both the energy and the momentum of the excited states depend explicitly on the positions of the holes in the real λ distribution only, but they do not depend on the actual values of the complex λs.

The lowest energy spin excitations have been described in more detail: the energy–momentum dispersion, the number of possible states and the structures of the parameters describing these states have been given for the cases when the number of holes is two and four. We have also established that the structures of the spin excitations of a half-filled and a less-than-half-filled band are the same, the only difference between the two cases is in the functional form of the energy–momentum dispersion.

Finally we note that similarly to the Heisenberg model, in the low-energy excitations in the λ set the maximal length of strings is two. Instead of the longer strings, quartets (two two-string with complex conjugate centres for which the modulus of the imaginary parts is not larger than $U/4$) and wide pairs (complex conjugate λ pairs with $|\mathrm{Im}\,\lambda| > U/2$) appear. All these have been expected on the basis of the analogy between the spin part of the Hubbard model and the isotropic Heisenberg chain. As far as the complex Λs are concerned, in the low-energy excited states of a macroscopically half-filled band they have no special configurations in the complex Λ plane. This is due to the fact that in these states the number of k_hs is small and equation (2.5*c*) determines the Λ set directly (but not a set of auxiliary variables of the type (2.11*a*)–(2.11*b*) like equation (2.12)). In a macroscopically less-than-half-filled band ($H \sim N$) the complex-Λ configurations are of strictly string form.

Acknowledgments

I am very grateful to Drs J Sólyom and A Sütö for a critical reading of the manuscript and for their valuable comments.

References

Babelon O, de Vega H J and Viallet C M 1983 *Nucl. Phys.* B **220** 13–34
Choy T C and Young W 1982 *J. Phys. C: Solid State Phys.* **15** 521–8
Coll C F 1974 *Phys. Rev.* B **9** 2150
Lieb E M and Wu F Y 1968 *Phys. Rev. Lett.* **20** 1445
Ovchinnikov A A 1970 *Sov. Phys.–JETP* **30** 1160
Woynarovich F 1982a *J. Phys. C: Solid State Phys.* **15** 85–96
—— 1982b *J. Phys. C: Solid State Phys.* **15** 97–107
—— 1983c *J. Phys. A: Math. Gen.* **15** 2985–96

J. Phys. C: Solid State Phys., **16** (1983) 6593–6604. Printed in Great Britain

Low-energy excited states in a Hubbard chain with on-site attraction

F Woynarovich
Central Research Institute for Physics, H-1525 Budapest, POB 49, Hungary

Received 27 May 1983

Abstract. The structure and energy of the eigenstates of a 1D Hubbard Hamiltonian with negative U is studied. Solutions for the Lieb–Wu equations corresponding to $U < 0$ are constructed from those for $U > 0$. It is found that the ground state is continuous, while the ground-state energy, although continuous, is not analytic at the $U = 0$ point. In the excited states, in contrast to the $U > 0$ case, both kinds of excitation are connected with the charge distribution and the state of the spins does not affect the energy explicitly. It is also found that the distribution of parameters can change discontinuously in the excited states as U crosses zero.

1. Introduction

The aim of the present work is to study the ground state and the low-energy excited states of a Hubbard chain with on-site attraction, i.e. we intend to study the eigenvalues and eigenstates of the Hamiltonian

$$\hat{H}(U) = -\sum_{i=1}^{N}\sum_{\sigma}(c_{i\sigma}^{+}c_{i+1\sigma} + c_{i+1\sigma}^{+}c_{i\sigma}) + U\sum_{i=1}^{N}\hat{n}_{i\uparrow}\hat{n}_{i\downarrow} \qquad (N+1\equiv 1) \tag{1.1}$$

for the case when $U < 0$. In (1.1) $c_{i\sigma}^{+}(c_{i\sigma})$ creates (destroys) an electron with spin σ at the site i and $\hat{n}_{i\sigma} = c_{i\sigma}^{+}c_{i\sigma}$.

The problem for $U > 0$ has been already worked out to a large extent. Lieb and Wu (1968) reduced the diagonalisation of (1.1) to the solution of the coupled system of equations

$$Nk_l = 2\pi I_l - \sum_{\alpha=1}^{M} 2\tan^{-1}\frac{4}{U}(\sin k_l - \lambda_\alpha) \qquad l = 1, \dots, N_e \tag{1.2a}$$

$$\sum_{l=1}^{N_e} 2\tan^{-1}\frac{4}{U}(\lambda_\alpha - \sin k_l) = 2\pi\mathcal{J}_\alpha + \sum_{\alpha'=1}^{M} 2\tan^{-1}\frac{2}{U}(\lambda_\alpha - \lambda_{\alpha'})$$

$$\alpha = 1, \dots, M \tag{1.2b}$$

where N_e and M are the numbers of electrons and the number of down-spin electrons respectively and the quantum numbers I_l and $\mathcal{J}_\alpha$ are appropriately chosen integers or half odd-integers ($2I_l = M (\mathrm{mod}\ 2)$, $2\mathcal{J}_\alpha = N_e - M + 1\ (\mathrm{mod}\ 2)$). They also calculated the solution of (1.2a)–(1.2b) corresponding to the ground state of a half-filled chain ($N_e = N$). Based on this work, Coll (1974) calculated the spin-wave-type and one-particle-type excitations for a general band filling. Recently Choy and Young (1982) re-examined the triplet spin wave excitations and the present author studied the excitations connected with the charge (Woynarovich 1982a, b) and spin (Woynarovich 1983) degrees of freedom of a macroscopically half-filled band. The $T = 0$ magnetic properties have been worked out by Takahashi (1969) and Shiba (1972).

Studying the $U < 0$ case does not, in principle, require a complete new calculation since there are many possibilities of reducing the problem with $U < 0$ to the one with $U > 0$:

(i) As is well known, by introducing holes instead of the up-spin electrons $\hat{H}(-U)$ can be transformed into $\hat{H}(U)$ (the form of this transformation will be used later), and thus the spectrum of $\hat{H}(U < 0)$ can be given without performing new calculations. The problem of this method is, however, that in this way one does not have the wavefunction in an explicit form and it is hard to find the meaning of the parameters appearing in the spectrum.

(ii) Another possibility, which also allows one to construct the wavefunction, is provided by the facts that equations (1.2a)–(1.2b) are also the secular equations for $U < 0$ and if the sets $\{k_l\}$ and $\{\lambda_\alpha\}$ satisfy equations (1.2a)–(1.2b) for U then the sets $\{k_l + \pi\}$ and $\{-\lambda\}$ give a solution for $-U$. To take advantage of these facts, however, one should know the solutions of (1.2a)–(1.2b) corresponding to high-energy states: owing to the shifting of all momenta by π, the energy, which is

$$E = -\sum 2 \cos k \tag{1.3}$$

changes its sign, and the low-energy solutions for $U > 0$ generate high-energy solutions for $U < 0$ and vice versa. A tool for overcoming these difficulties is provided by the complementary properties of the solutions for equations (1.2a)–(1.2b) (Woynarovich 1982b, referred to as II hereafter). From a solution of equations (1.2a)–(1.2b) another solution can be constructed trivially, and if one of these solutions corresponds to a low-energy state, then the other describes a high-energy one. This way it is possible to connect, through the complementary solutions and the property of the Lieb–Wu equations mentioned in point (ii), the low-energy solutions for $U > 0$ with those for $U < 0$. Actually, it will turn out that in this way from a solution for $U < 0$ one obtains the same solution for $U < 0$ as one would obtain applying the transformation mentioned in point (i).

In the present paper we would like to describe the ground state and excitations for $U < 0$. In the first part (§ 2) we rederive the properties of the complementary solutions for the Lieb–Wu equations in an exact way and give the connection between the low-energy states for $U > 0$ and $U < 0$. The second part of the paper (§ 3) is devoted to describing the ground state and excited states, and analysing their continuity at $U = 0$.

2. Relation between the eigenstates in the repulsive and attractive cases

2.1. *Complementary solutions for equations (1.2a)–(1.2b)*

The complementary properties of the solutions for the Lieb–Wu equations has been derived earlier (in II) by means of techniques permittable only in the $N \to \infty$ limit. Now we will show that the complementary solutions exist for any finite N.

Let us first consider equation (1.2*a*). At a fixed λ_α set this is an equation with a single unknown. For the different choices of I it has $N + 2M$ non-equivalent solutions. This is easily seen, if we write equation (1.2*a*) in a polynomial form:

$$P(x) = 0 \tag{2.1a}$$

where $P(x)$ is a polynomial of the order of $N + 2M$

$$P(x) = x^N \prod_{\alpha=1}^{M} [x^2 - 2\mathrm{i}(\lambda_\alpha + \mathrm{i}U/4)x - 1] - \prod_{\alpha=1}^{M} [x^2 - 2\mathrm{i}(\lambda_\alpha - \mathrm{i}U/4)x - 1] \tag{2.1b}$$

of the variable

$$x = \exp(\mathrm{i}k). \tag{2.1c}$$

N_e of the $N + 2M$ roots for equation (2.1*a*) constitutes the set of x_ls (k_ls). Let us denote the rest of the roots by $x_g(k_g)$. Now we show that if $\{k_l\}$ and $\{\lambda_\alpha\}$ satisfy equations (1.2*a*)–(1.2*b*), then $\{k_g\}$ and $\{\lambda_\alpha\}$ is also a solution (for which the $2\mathcal{F}_\alpha$s are also equal to $N_e - M + 1 \pmod 2$). Since the k_gs satisfy equation (2.1*a*), and thus equation (1.2*a*), we have to show that they also satisfy equation (1.2*b*). To do this we write equation (1.2*b*) in the form

$$\sum_l \frac{1}{\mathrm{i}} \ln \frac{x_l^2 - 2\mathrm{i}(\lambda_\alpha - \mathrm{i}U/4)x_l - 1}{x_l^2 - 2\mathrm{i}(\lambda_\alpha + \mathrm{i}U/4)x_l - 1} = \pi + \sum_{\alpha'} \frac{1}{\mathrm{i}} \ln \frac{\lambda_\alpha - \lambda_{\alpha'} - \mathrm{i}U/2}{\lambda_\alpha - \lambda_{\alpha'} + \mathrm{i}U/2} \pmod{2\pi} \tag{2.2}$$

and evaluate the sum on the LHS by means of residues:

$$\sum_l \frac{1}{\mathrm{i}} \ln \frac{x_l^2 - 2\mathrm{i}(\lambda_\alpha - \mathrm{i}U/4)x_l - 1}{x_l^2 - 2\mathrm{i}(\lambda_\alpha + \mathrm{i}U/4)x_l - 1} = \sum_l \frac{1}{2\pi\mathrm{i}} \oint_{C_l} \frac{1}{\mathrm{i}} \ln \frac{x^2 - 2\mathrm{i}(\lambda_\alpha - \mathrm{i}U/4)x - 1}{x^2 - 2\mathrm{i}(\lambda_\alpha + \mathrm{i}U/4)x - 1} \left(\frac{\mathrm{d}}{\mathrm{d}x} \ln P(x)\right) \mathrm{d}x \tag{2.3a}$$

where the contour C_l encircles the root x_l. Deforming the contour in (2.3*a*) to encircle the roots x_g and the cuts of the integrand we have

$$\sum_l \frac{1}{\mathrm{i}} \ln \frac{x_l^2 - 2\mathrm{i}(\lambda_\alpha - \mathrm{i}U/4)x_l - 1}{x_l^2 - 2\mathrm{i}(\lambda_\alpha + \mathrm{i}U/4)x_l - 1} = - \sum_g \frac{1}{\mathrm{i}} \ln \frac{x_g^2 - 2\mathrm{i}(\lambda_\alpha - \mathrm{i}U/4)x_g - 1}{x_g^2 - 2\mathrm{i}(\lambda_\alpha + \mathrm{i}U/4)x_g - 1} + \frac{2}{\mathrm{i}} \sum_{\alpha'} \ln \frac{\lambda_\alpha - \lambda_{\alpha'} - \mathrm{i}U/4}{\lambda_\alpha - \lambda_{\alpha'} + \mathrm{i}U/4} + \pi N \pmod{2\pi}. \tag{2.3b}$$

Comparing (2.2) and (2.3*b*) and rewriting the logarithmic terms in $\tan^{-1}$ form we have

$$\sum_g 2\tan^{-1}\frac{4}{U}(\lambda_\alpha - \sin k_g) = 2\pi\mathcal{F}_\alpha + \sum_{\alpha'} 2\tan^{-1}\frac{2}{U}(\lambda_\alpha - \lambda_{\alpha'}) \tag{2.4}$$

where the $2\mathcal{F}_\alpha N_e - M + 1 \pmod 2$.

Let us denote the eigenstate of $\hat{H}(U)$ corresponding to $\{k_l\}$ and $\{\lambda_\alpha\}$ by $|k_l, \lambda_\alpha, U\rangle$ and the state built up according to the same prescription but using $\{k_g\}$ and $\{\lambda_\alpha\}$ by $|k_g, \lambda_\alpha, U\rangle$. The $|k_g, \lambda_\alpha, U\rangle$ describes a state of $\tilde{N}_e = N + 2M - N_e$ electrons, and this state is an eigenstate of $\hat{H}(U)$ if the $\{k_g\}$, $\{\lambda_\alpha\}$ satisfy a system analogous to (1.2*a*)–(1.2*b*). According to the former paragraphs, this is true if $2\mathcal{F}_\alpha = \tilde{N}_e - M + 1 \pmod 2$ i.e. if N is even (otherwise $|k_g, \lambda_\alpha, U\rangle$ fails to satisfy periodic boundary conditions). In the following we will take N to be even, thus both $|k_l, \lambda_\alpha, U\rangle$ and $|k_g, \lambda_\alpha, U\rangle$ are eigenstates of the chain, and these are the states which we call complementary ones.

Using the properties of the $P(x)$ polynomial it is easy to see that the momenta of the complementary states complete each other to $\pi(N+M+1)(\mathrm{mod}\ 2\pi)$ while the energies add up to MU. For the sum of the momenta

$$p+\bar{p}=\sum_l k_l+\sum_g k_g=\frac{1}{\mathrm{i}}\ln\left(\prod_{\text{all roots}} x\right). \tag{2.5a}$$

The product of all the roots of $P(x)$ is $(-1)^{2M+N}$ times the zeroth-order term in $P(x)$ i.e.

$$p+\bar{p}=\frac{1}{\mathrm{i}}\ln\left((-1)^{2M+N}(-1)^{M+1}\right)=\pi(N+M+1)\ (\mathrm{mod}\ 2\pi). \tag{2.5b}$$

The sum of the energies is

$$E+\bar{E}=-\sum_l 2\cos k_l-\sum_g 2\cos k_g=-\sum_{\text{all roots}} x-\sum_{\text{all roots}}\frac{1}{x}. \tag{2.6a}$$

These terms can be expressed by the coefficients of the $(N+2M-1)$th-order and first-order terms in $P(x)$ yielding

$$E+\bar{E}=MU. \tag{2.6b}$$

2.2. *Construction of complementary solutions*

If we want to describe a state $|k_l, \lambda_\alpha, U\rangle$ which is near to the ground state of a half-filled band, we may suppose that the λ distribution deviates from that of the ground state only by small terms. If this is so, as can be checked, $2M$ of the possible $N+2M$ k-values satisfying equation (1.2a) are of the form

$$k_\alpha^\pm \simeq \sin^{-1}(\lambda_\alpha \mp \mathrm{i}U/4) \pm \mathrm{Im}\, k_\alpha^\pm > 0,$$

$$\mathrm{sign}(\cos(\mathrm{Re}\, k_\alpha^\pm)) = -\mathrm{sign}(U/4 \mp \mathrm{Im}\,\lambda_\alpha) \tag{2.7a}$$

(if $||\mathrm{Im}\,\lambda_\alpha|-U/4| \gg 1/N$ or $|\mathrm{Re}\,\lambda_\alpha|>1$), or

$$k_\alpha^+ \simeq \sin^{-1}(\mathrm{Re}\,\lambda_\alpha) \qquad k_\alpha^- \simeq \sin^{-1}(\mathrm{Re}\,\lambda_\alpha + \mathrm{i}U/2)$$

$$-\mathrm{Im}\, k_\alpha^- > 0 \qquad \cos(\mathrm{Re}\, k_\alpha^-) < 0 \qquad k_{\alpha'}^+ \simeq k_\alpha^{-*} \qquad k_{\alpha'}^- = \pi - k_\alpha^+ \tag{2.7b}$$

(if $|\mathrm{Im}\,\lambda_\alpha - U/4| \leqslant 1/N$, $|\mathrm{Re}\,\lambda_\alpha|<1$ and $\lambda_{\alpha'}=\lambda_\alpha^*$).

These equalities are correct up to terms exponentially small in N. (Note that (2.7b) is the refinement of (2.7a) for the case when $|\mathrm{Im}\,\lambda| \simeq U/4$ and (2.7a) would yield two equal real ks for λ and λ^*.) These $2M$ ks, although in special cases some of them can turn out to be real (see (2.7b)), we will call complex ones. The rest of the ks are always real, and in the large-N limit, they are disturbed continuously in the region $(-\pi, \pi)$. These ks will be referred to as real ones.

In II it has been shown that if $\{k_l\}$ is such that if a k_α^+ (k_α^-) is an element of it then k_α^- (k_α^+) is also in $\{k_l\}$, then equations (1.2b) can be replaced by the equations

$$\sum_j 2\tan^{-1}\frac{4}{U}(\lambda_\beta-\sin k_j)=2\pi\mathcal{F}_\beta'+\sum_{\beta'}2\tan^{-1}\frac{2}{U}(\lambda_\beta-\lambda_{\beta'}) \tag{2.8}$$

and

$$\sum_h 2\tan^{-1}\frac{4}{U}(\Lambda_m-\sin k_h)=2\pi\mathcal{F}_m'+\sum_{m'}2\tan^{-1}\frac{2}{U}(\lambda_m-\lambda_{m'}). \tag{2.9}$$

Here k_j denote the real elements of $\{k_l\}$ and the λ_βs are those elements in $\{\lambda_\alpha\}$ which are not associated with complex k_ls. The k_h refer to the real elements of $\{k_g\}$ and the Λ_ms are those λ_αs which are connected with the complex k_ls (are not connected with complex k_gs). $2\mathcal{F}'_\beta = N_e - (M - L) - 1 \pmod 2$ and $2\mathcal{F}'_m = N_e - L - 1 \pmod 2$, L being the number of complex k pairs in $\{k_l\}$. Now this equation can be obtained easily from equations (1.2*b*) and (2.4) respectively, by using the equality

$$2\tan^{-1}\frac{4}{U}(\lambda - \sin k_\alpha^+) + 2\tan^{-1}\frac{4}{U}(\lambda - \sin k_\alpha^+)$$
$$= 2\tan^{-1}\frac{2}{U}(\lambda - \lambda_\alpha) + \pi\,\mathrm{sign}(\mathrm{Re}(\lambda - \lambda_\alpha)) \qquad \lambda \neq \lambda_\alpha \tag{2.10}$$

which is valid up to an exponentially small error.

Thus to find a solution for equations (1.2*a*)–(1.2*b*) one has to solve the system (2.8) and (2.9) simultaneously with the equation

$$Nk_{j(h)} = 2\pi I_{j(h)} - \sum_\beta 2\tan^{-1}\frac{4}{U}(\sin k_{j(h)} - \lambda_\beta) - \sum_m 2\tan^{-1}\frac{4}{U}(\sin k_{j(h)} - \Lambda_m). \tag{2.11}$$

This system is simpler than equations (1.2*a*)–(1.2*b*), since here the k_js and k_hs are always real (being the N solutions which are not of the form (2.7*a*)–(2.7*b*)), and once it is solved, it provides an easy way to construct the complementary solutions: for one

$$\{k_l\} = \{k_j; k_m^\pm\} \qquad \{\lambda_\alpha\} = \{\lambda_\beta; \Lambda_m\} \tag{2.12a}$$

and for the other

$$\{k_g\} = \{k_h; k_\beta^\pm\} \qquad \{\lambda_\alpha\} = \{\lambda_\beta; \Lambda_m\}. \tag{2.12b}$$

2.3. Relation between the structures of the complementary states

It is apparent that the variable sets $\{k_j, \lambda_\beta\}$ and $\{k_h, \Lambda_m\}$ change role in the complementary states. As can be seen in the $U \to \infty$ limiting form of the wavefunction (see II) this means that in the complementary states the charge and spin degrees of freedom are interchanged. To be more quantative let us consider the unitary transformation

$$\hat{U}_1 = \exp(\mathrm{i}\hat{S}_1)$$
$$\hat{S}_1 = (\pi/2)\sum_{n=1}^{N/2}[(c_{n\uparrow} - c^+_{N-n+1\uparrow})(c^+_{n\uparrow} - c_{N-n+1\uparrow})$$
$$+ (c_{n\downarrow} + \mathrm{i}c_{N-n+1\downarrow})(c^+_{n\downarrow} - \mathrm{i}c^+_{N-n+1\downarrow})$$
$$+ (2n-1)(c_{n\downarrow}c^+_{n\downarrow} + c_{N-n+1\downarrow}c^+_{N-n+1\downarrow})]. \tag{2.13}$$

With this transformation

$$\hat{U}_1 c_{n\uparrow}\hat{U}_1^{-1} = c^+_{N-n+1\uparrow} \qquad \hat{U}_1 c_{n\downarrow}\hat{U}_1^{-1} = (-1)^{N+n+1}c_{N-n+1\downarrow} \tag{2.14}$$

i.e. this transformation reflects the chain through its centre, and changes the up-spin electrons to holes. Applying this transformation to $\hat{H}(U)$ and $\hat{p}$ we find that

$$\hat{U}_1\hat{H}(U)\hat{U}_1^{-1} = U\sum_{i=1}^N \hat{n}_{i\downarrow} - \hat{H}(U) \qquad \hat{U}_1\hat{p}\hat{U}_1^{-1} = \pi(N+1) + \pi\sum_{i=1}^N \hat{n}_{i\downarrow} - \hat{p} \tag{2.15}$$

thus

$$\hat{H}(U)\hat{U}_1^{-1}|k_l, \lambda_\alpha, U\rangle = \bar{E}\hat{U}_1^{-1}|k_l, \lambda_\alpha, U\rangle \tag{2.16a}$$

and

$$\hat{p}\hat{U}_1^{-1}|k_l, \lambda_\alpha; U\rangle = \bar{p}\hat{U}_1^{-1}|k_l, \lambda_\alpha, U\rangle. \tag{2.16b}$$

This suggests that $\hat{U}_1^{-1}|k_l, \lambda_\alpha, U\rangle$ and $|k_g, \lambda_\alpha, U\rangle$ are identical or degenerate. As can be checked $\hat{U}_1^{-1}|k_l, \lambda_\alpha, U\rangle$ and $|k_g, \lambda_\alpha, U\rangle$ coincide in the $U \to \infty$ limit; thus, supposing continuity in U, we may conclude that

$$|k_g, \lambda_\alpha, U\rangle = \hat{U}_1^{-1}|k_l, \lambda_\alpha, U\rangle. \tag{2.17}$$

Now the charge–spin symmetry between the complementary states can be understood on the basis of (2.17): if at any site there is an up- or down-spin electron, $\hat{U}_1$ changes it to an empty or doubly occupied site respectively, while if a site is empty or doubly occupied, $\hat{U}_1$ puts to this site an up-spin electron or taking off the up-spin electron creates an uncompensated down-spin electron. Thus $\hat{U}_1$ is a transformation between the charge and spin degrees of freedom.

In connection with (2.17) we would like to note the following also: the fact that the RHS is an eigenstate of $\hat{H}(U)$, is a symmetry of $\hat{H}(U)$, and, if the solutions of the Lieb–Wu equations yield a complete set, $\hat{U}_1^{-1}|k_l, \lambda_\alpha, U\rangle$ should also be of the Bethe *ansatz* form. What is interesting is that $|k_l, \lambda_\alpha, U\rangle$ and $\hat{U}_1^{-1}|k_l, \lambda_\alpha, U\rangle$ have the same λ set and for the momenta they share the roots of the same equation.

2.4. The low-energy states of (1.1) with negative U

According to (ii) of the Introduction, solutions for equations (1.2*a*)–(1.2*b*) with $-U$, corresponding to low-energy states, can be obtained from the high-energy solutions for these equations with U, simply by the transformation $k \to k + \pi$; $\lambda \to -\lambda$. Thus, if $\{k_l\}$ and $\{\lambda_\alpha\}$ is a low-energy solution for U, its complementary solution $\{k_g\}$ and $\{\lambda_\alpha\}$ is a high-energy one, and $\{k_g + \pi\}$, $\{-\lambda_\alpha\}$ is a low-energy solution for equations (1.2*a*)–(1.2*b*) with $-U$. The energy and momentum belonging to this state (according to (2.5*b*) and (2.6*b*)) are

$$\hat{H}(-U)|k_g + \pi; -\lambda_\alpha; -U\rangle = \Big(-\sum_g 2\cos(k_g + \pi)\Big)|k_g + \pi; -\lambda_\alpha; -U\rangle$$

$$= (-MU + E)|k_g + \pi; -\lambda_\alpha; -U\rangle \tag{2.18a}$$

$$\hat{p}|k_g + \pi; -\lambda_\alpha; -U\rangle = \Big(\sum_g (k_g + \pi)\Big)|k_g + \pi; -\lambda_\alpha; -U\rangle$$

$$= ((M + 1 - N_e)\pi - p)|k_g + \pi; -\lambda_\alpha; -U\rangle \tag{2.18b}$$

where E and p are the energy and momentum defined by

$$\hat{H}(U)|k_l; \lambda_\alpha; U\rangle = E|k_l; \lambda_\alpha; U\rangle \tag{2.19a}$$

$$\hat{p}|k_l; \lambda_\alpha; U\rangle = p|k_l; \lambda_\alpha; U\rangle. \tag{2.19b}$$

It is not hard to see that applying the operator which transforms $\hat{H}(-U)$ into $\hat{H}(U)$ leads to the same result. If $\hat{U}_2$ is the trasnformation which shifts all momenta by π,

$$\hat{U}_2 = \exp(i\hat{S}_2) \qquad \hat{S}_2 = \pi\sum_{m=1}^{N}\sum_\sigma m\hat{n}_{m\sigma} \tag{2.20}$$

then

$$\hat{U}_2\hat{H}(-U)\hat{U}_2^{-1} = -\hat{H}(U) \qquad \hat{U}_2\hat{p}\hat{U}_2^{-1} = \pi\sum_{i=1}^{N}\sum_{\sigma}\hat{n}_{i\sigma} + \hat{p} \tag{2.21}$$

which combined with (2.15) yields

$$(\hat{U}_1\hat{U}_2)\hat{H}(-U)(\hat{U}_1\hat{U}_2)^{-1} = H(U) - U\sum_{i=1}^{N}\hat{n}_{i\downarrow}$$

$$(\hat{U}_1\hat{U}_2)\hat{p}(\hat{U}_1\hat{U}_2)^{-1} = \pi - \pi\sum_{i=1}^{N}\hat{n}_{i\uparrow} - \hat{p}. \tag{2.22}$$

On the other hand, applying $\hat{U}_2^{-1}$ on both sides of (2.17) and using the explicit form of the wavefunction (see equations (2.5–10) of Woynarovich 1982a) one has

$$(\hat{U}_1\hat{U}_2)^{-1}|k_l;\lambda_\alpha;U\rangle = |k_g + \pi; -\lambda_\alpha; -U\rangle. \tag{2.23}$$

Equations (2.22) and (2.23) also yield equations (2.18*a*)–(2.18*b*). We note, however, that equation (2.22) in itself would only give the spectrum, and (2.23) gives the explicit form of the corresponding eigenstates.

3. The low-energy states of a macroscopically half-filled band as functions of U

3.1. *The ground state*

The ground state of a half-filled band ($N_e = N$) with positive U has already been described by Lieb and Wu (1968). In this state the number of down spins is $M = N/2$, and all the k_ls and λ_αs are real and their distributions are described by the densities

$$\rho_0(k) = \frac{1}{2\pi}\left(1 + \cos k\int_0^\infty \frac{\exp(-\omega U/4)}{\cosh(\omega U/4)} J_0(\omega)\cos(\omega\sin k)\,\mathrm{d}\omega\right) \tag{3.1}$$

$$\sigma_0(\lambda) = \frac{1}{2\pi}\int_0^\infty \frac{J_0(\omega)\cos(\omega\lambda)}{\cosh(\omega U/4)}\,\mathrm{d}\omega. \tag{3.2}$$

The ground-state energy per site is

$$E_0(U>0) = -2\int_0^\infty \frac{\exp(-\omega U/4)}{\cosh(\omega U/4)} J_0(\omega)J_1(\omega)\,\mathrm{d}\omega \tag{3.3}$$

with $J_0(\omega)$ and $J_1(\omega)$ being the zeroth-and first-order Bessel functions.

It is clear that the ground state of a half-filled band with negative U is to be constructed (by the transformation $(\hat{U}_1\hat{U}_2)^{-1}$) from the ground state for $U' = |U|$ (this is not true if the band is not half filled). Thus, the λ set again consists of $M = N/2$ real λs distributed according to (3.2) ($\sigma_0(\omega)$ is an even function), but since all of the N real roots of equation (1.2*a*) belong to the ground state with U', the k set of its complementary solution consists of the $2M = N$ complex roots

$$k_\alpha^\pm = \sin^{-1}(\lambda_\alpha \mp \mathrm{i}U/4) \qquad \pm\mathrm{Im}\,k_\alpha^\pm > 0. \tag{3.4}$$

Thus the k set of the ground state for $U<0$ is $\{k_\alpha^\pm + \pi\}$.

As far as the structure of the ground state is concerned, we know that for $U \to +\infty$ all sites are singly occupied, and the spins are distributed in the same way as they are distributed in the ground state of a Heisenberg chain. In the opposite limit ($U \to -\infty$), according to the spin–charge symmetry discussed in § 2.3, half of the sites are empty and the other half are doubly occupied, and the distribution of empty and doubly occupied sites is the same as the distribution of up and down spins in the ground state of a Heisenberg chain. The transition between these two opposite limits is continuous: to see this we have to note that for $U \to +0$ all $k_\alpha^\pm$ become real, and since

$$\lim_{U\to+0} \sigma_0(\lambda) = \begin{cases} \dfrac{1}{2\pi}\dfrac{1}{(1-\lambda^2)^{1/2}} & |\lambda|<1 \\ 0 & |\lambda|>1 \end{cases} \tag{3.5}$$

the density of the limiting values of the $k_\alpha^\pm$ s is

$$\rho\left(\lim_{U\to+0} k^\pm(\lambda)\right) = \pi^{-1}\,St(|k|-\pi/2). \tag{3.6}$$

Hence, the distribution of the momenta in the $U \to -0$ limit is

$$\pi^{-1}\,St(\pi/2-|k|) \tag{3.7}$$

that is, exactly the same as the $U \to +0$ limit of (3.1).

The ground-state energy for $U<0$, according to (2.19*b*), is

$$E_0(U<0) = E_0(|U|) - |U|/2. \tag{3.8}$$

It is not hard to establish that both E_0 and $\partial E_0/\partial U$ are continuous functions of U even at $U=0$, but the calculation of the higher derivatives of E_0 at $U=0$ encounters difficulties (after the second differentiation with respect to U in E_0 of (3.3), the order of integration and the $U \to 0$ limit can not be changed). What we were able to establish is

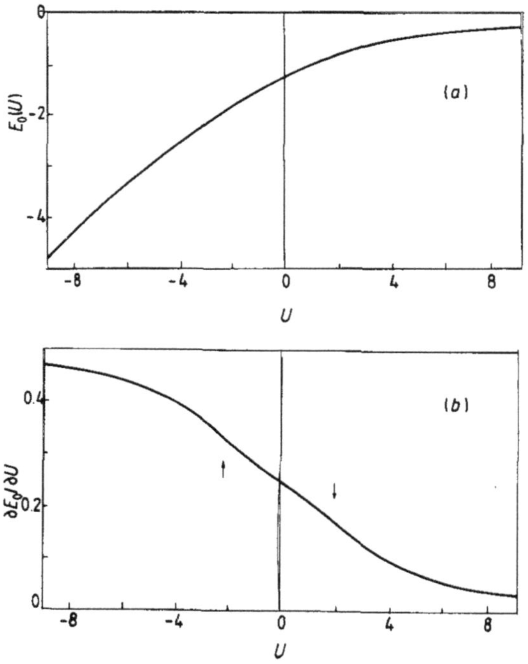

Figure 1. (*a*) The ground-state energy E_0, and (*b*) its derivative with respect to U as a function of U.

that E_0 as a function of U must have a singularity at $U = 0$ since it can not be expanded into a convergent power series around this point. The form we have found (valid for both positive and negative U) is

$$E_0(U) = -(4/\pi) + (U/4) + \sum_{n=1}^{M} a_n(U/4)^{2n} + \mathcal{O}(U^{2M+1})$$

$$a_n = 8\frac{2^{2n+1}-1}{(2\pi)^{2n+1}}\zeta(2n+1)\sum_{k=0}^{2n-1}\frac{(-1)^k}{k!(2n-1-k)!}\frac{\Gamma(\frac{1}{2}+k)\Gamma(\frac{3}{2}+(2n-1-k))}{\Gamma(\frac{1}{2}-k)\Gamma(\frac{3}{2}-(2n-1-k))}. \tag{3.9}$$

(Here ζ is Riemann's ζ-function and $\mathcal{O}(U^{2m+1})$ denotes terms for which $\mathcal{O}(U^{2M+1})$ $U^{-(2M+1)}$ is bounded in a finite interval containing $U = 0$). Since the coefficients a_n generate a divergent series for all finite values of U, the expansion (3.9) is an asymptotic one. E_0 is plotted as a function of U in figure 1(a).

It is also worth examining the first derivative of E_0 with respect to U (figure 1(b)). This quantity gives the expectation value of $\hat{n}_{i\uparrow}\hat{n}_{1\downarrow}$, i.e. the number of doubly occupied sites divided by the chain length in the ground state. For $U \to \pm\infty$, $\partial E_0/\partial U$ approaches its limiting values (0 and $\frac{1}{2}$ respectively) as $1/U^2$ and for small values of U it is linear. The transition between the different regions takes place through inflections situated at about $|U| = 2,1$. What is interesting is that this is about the same as the value where the gap starts to grow drastically with U.

3.2. The excited states

Excitations connected with the charge degrees of freedom can be obtained by: (i) allowing for N_e to be smaller than N, $N_e = N - H$ (introducing 'holes'); and (ii) introducing complex wavenumbers instead of a number of real ones. Spin excitations can be introduced by: (iii) taking $M < N_e/2$ (this leads to a total magnetisation $S^z = N_e/2 - M > 0$); and (iv) allowing for some of the λs being complex. As has been shown (II and Woynarovich 1983) to find solutions of equations (1.2a)–(1.2b) corresponding to such excited states the equation

$$\sum_{\eta}^{\mathcal{H}+2\mathcal{L}} 2\tan^{-1}\frac{4}{U}(\lambda'_\mu - \lambda_\eta) = 2\pi\mathcal{F}_\mu + \sum_{\mu'} 2\tan^{-1}\frac{2}{U}(\lambda'_\mu - \lambda'_{\mu'}) \tag{3.10}$$

together with equation (2.9) should be solved. Here the λ_ηs are the holes in the distribution of real λ_βs, their number is $\mathcal{H} + 2\mathcal{L}$, where $\mathcal{H} = N_e - 2M = 2S^z$, and $\mathcal{L}$ is the number of auxiliary variables λ'_μ which generate the complex elements of $\{\lambda_\beta\}$ (denoted by λ_μ):

$$\begin{aligned} \lambda_\mu^\pm &= \lambda'_\mu \pm \mathrm{i}U/4 && \text{if } |\mathrm{Im}\,\lambda'_\mu| < U/4 \\ \lambda_\mu &= \lambda'_\mu + (iU/4)\mathrm{sign}(\mathrm{Im}\,\lambda'_\mu) && \text{if } |\mathrm{Im}\,\lambda'_\mu| > U/4. \end{aligned} \tag{3.11}$$

In equation (2.9) for the number of k_hs $H + 2L$ should be taken, where L is the number of complex k pairs (the number of Λ_ms). In as much as both $H + 2L$ and $\mathcal{H} + 2\mathcal{L}$ are non-macroscopic (much less than N) the positions of the holes in the distribution of real λ_βs(λ_η) and the holes in the k_j distribution (k_h) are free parameters. Once equations (2.9) and (3.10) are solved, the distribution of the k_js and real λ_βs can be given by

$$\rho(k) = \rho_0(k) - (1/N)\sum_{h}^{H+2L}(\delta(k-k_h) + \cos k f(\sin k - \sin k_h))$$

$$-\frac{\cos k}{NU}\sum_{\eta}^{\mathcal{H}+2\mathcal{L}}\frac{1}{\cosh((\sin k-\lambda_\eta)2\pi/U)}$$

$$+\frac{\cos k}{2\pi N}\sum_{m}^{L}2\frac{(U/4)^2}{(U/4)^2+(\sin k-\Lambda_m)^2} \tag{3.12}$$

and

$$\sigma(\lambda)=\sigma_0(\lambda)-(1/N)\sum_{\eta}^{\mathcal{H}+2\mathcal{L}}(\delta(\lambda-\lambda_\eta)-f(\lambda-\lambda_\eta))$$

$$-\frac{1}{NU}\sum_{h}^{H+2L}\frac{1}{\cosh((\lambda-\sin k_h)2\pi/U)}$$

$$-\frac{1}{2\pi N}\sum_{\mu}^{\mathcal{L}}2\frac{(U/4)}{(U/4)^2+(\lambda-\lambda'_\mu)^2} \tag{3.13}$$

with

$$f(x)=\frac{1}{2\pi}\int_0^\infty\frac{\exp(-\omega U/4)\cos(\omega x)}{\cosh(\omega U/4)}\,\mathrm{d}\omega. \tag{3.14}$$

The energy and momentum of the corresponding state is

$$E=NE_0(U)+\sum_{h}^{H+2L}\varepsilon_c(k_h)+LU+\sum_{\eta}^{\mathcal{H}+2\mathcal{L}}\varepsilon_s(\lambda_\eta) \tag{3.15}$$

$$p=\pi(N/2+H/2+1)-\sum_{h}^{H+2L}p_c(k_h)+\sum_{\eta}^{\mathcal{H}+2\mathcal{L}}p_s(\lambda_\eta) \tag{3.16}$$

with

$$\varepsilon_c=2\cos k+2\int_0^\infty\frac{\exp(-\omega U/4)J_1(\omega)\cos(\omega\sin k)}{\omega\cosh(\omega U/4)}\,\mathrm{d}\omega$$

$$p_c=k+\int_0^\infty\frac{\exp(-\omega U/4)J_0(\omega)\sin(\omega\sin k)}{\omega\cosh(\omega U/4)}\,\mathrm{d}\omega \tag{3.17}$$

$$\varepsilon_s=2\int_0^\infty\frac{J_1(\omega)\cos(\omega\lambda)}{\omega\cosh(\omega U/4)}\,\mathrm{d}\omega$$

$$p_s=\pi/2-\int_0^\infty\frac{J_0(\omega)\sin(\omega\lambda)}{\omega\cosh(\omega U/4)}\,\mathrm{d}\omega. \tag{3.18}$$

In the $U<0$ eigenstates, generated by the above $U>0$ excited states, the $\{k_h+\pi\}$ is the set of real k and the set of complex k is given by $\{k_\beta^\mp+\pi\}$. The $-\Lambda_m$s and $-\lambda_\beta$s describe (in the $U\to-\infty$ limit) the distribution of the uncompensated spins (the spins of electrons with real wavenumbers) and the distribution of empty and doubly occupied sites respectively. The total number of electrons is $N-\mathcal{H}$ while the magnetisation is $\frac{1}{2}H$. The energy and momentum of such a state are easily obtained from (2.18*a*)–(2.18*b*) and (3.15)–(3.16):

$$E=NE_0(U<0)+|U|(\mathcal{H}+H+2L)/2+\sum_{h}^{H+2L}\varepsilon_c(k_h)+\sum_{\eta}^{\mathcal{H}+2\mathcal{L}}\varepsilon_s(\lambda_\eta) \tag{3.19}$$

$$p = \pi(N - \mathcal{H}/2) + \sum_{h}^{H+2L} p_c(k_h) - \sum_{\eta}^{\mathcal{H}+2\mathcal{L}} p_s(\lambda_\eta). \tag{3.20}$$

Formally this dispersion is (apart from the signs of the momenta) the same as that for $U > 0$ (3.15)–(3.16), but in principle it has a very different feature: while for $U > 0$ ε_c and ε_s could be attributed to the excitations in the charge and spin degrees of freedom, respectively, for $U < 0$ both ε_c and ε_s are connected with the charge distribution. ε_s is due to the deviation of the distribution of empty and doubly occupied sites (i.e. the bound pairs) from that of the ground state, while ε_c can be regarded as the kinetic energy of the non-bounded electrons (those which occupy their sites alone). The details of the state of uncompensated spins do not appear explicitly in (3.19): since in the ground state all electrons are bound, i.e. all spins are compensated, to excite the spin degrees of freedom, some of the bound pairs must be broken up, and (3.18) contains only the energy needed for the pair-breaking and the 'kinetic' energy of the resulting 'free' particles.

3.3. On the continuity of the excited states at $U = 0$

The $U > 0$ and $U < 0$ excited states described by the same sets of λ_η and k_h in general are not the continuation of each other through $U = 0$ (for $N_e \neq N/2 + M$ even the numbers of particles are not the same in the two states). Finding the continuation of one state from $U > 0$ to $U < 0$ encounters difficulties, and the possibility that the parameter sets (k, λ distribution) is not continuous at $U = 0$ can not be ruled out.

Let us consider the most simple triplet states of $N/2$ electrons ($H = 0$, $\mathcal{H} = 2$, $L = \mathcal{L} = 0$). The $U \to +0$ limit of the λ distribution is

$$\lim_{U \to +0} \sigma_0 - (\tfrac{1}{2}N)(\delta(\lambda - \lambda_{\eta_1}) + \delta(\lambda - \lambda_{\eta_2})) \tag{3.21}$$

and the same limit of the excitation energy is

$$2\sin(p(\lambda_{\eta_1})) + 2\sin(p(\lambda_{\eta_2})). \tag{3.22}$$

The $U < 0$ continuation of such a state should be sought among those states which are obtained through (2.23) from the $U > 0$ states with $H' = 2$; $\mathcal{H}' = 0$. If in these states L' and $\mathcal{L}'$ are 0, the limit of the energy is

$$4\cos(p(k_{h_1})/2) + 4\cos(p(k_h)/2). \tag{3.23}$$

Since (3.32) and (3.33) do not overlap, the energy can be continuous through $U = 0$ only if $\mathcal{L}'$ is at least one for $U < 0$. If so, for $U < 0$, some of the λ_βs are complex. Even in the simplest possible case (i.e. if we suppose that $\mathcal{L}' = 1$, then $\lambda'_\mu = (\lambda_{\eta_1} + \lambda_{\eta_2})/2$ and the corresponding $\lambda^{\pm}_\mu$ became real as $U \to 0$) the limit of the λ distribution is

$$\begin{aligned}\lim \sigma_0 &- (1/2N)(\delta(\lambda - \lambda_{\eta_1}) + \delta(\lambda - \lambda_{\eta_2})) + (1/N)\delta(\lambda - \lambda'_\mu) \\ &- (1/2N)(\delta(\lambda - \sin k_{\eta_1}) + \delta(\lambda - \sin k_{\eta_2}))\end{aligned} \tag{3.24}$$

which is different from (3.21) and it is easy to see that looking for the $U < 0$ continuation, for a state with $H = 0$, $\mathcal{H} = 2$, $U > 0$ in more complicated form ($L' > 0$, $\mathcal{L}' > 1$) leads to larger differences in the limiting λ distributions. Thus we conclude, that even if we suppose that the energy of a given state is continuous as U crosses zero, the parameter sets describing the state are not necessarily (in the present example are not) continuous.

We would like to note that the above conclusion concerns only the behaviour of the parameters in the $N \to \infty$ limit: for any finite N all parameters are expected to be continuous (equations (1.2*a*)–(1.2*b*)) in the form (2.1) and (2.2) are continuous in U) but in the regime $|U| < 1/N$ the structure of the parameter sets can change significantly (a relatively small change, shifting all λ in different directions by $1/N$, may result in the discontinuities demonstrated above).

4. Summary

In the present work we have studied the ground state and the low-energy excited states of a Hubbard chain with an on-site attraction. The main tool of this study has been the complementary property of the solutions for the Lieb–Wu equations.

In the first part we rederived the properties of the complementary states in an exact way, and we have found that the existence of the complementary solutions of the Lieb–Wu equations is a reflection of the particle–hole symmetry in the Hubbard model. Since, to obtain the complementary states, only the up-spin particles should be replaced by holes, in the complementary states the spin distribution and charge distribution change role. Using the complementary solutions we have described how to find the low-energy solutions of the Lieb–Wu equations if $U < 0$.

We have given the ground state and ground-state energy for $U < 0$. We have found that the parameters describing the ground state are continuous, and the ground-state energy is continuous, although not an analytic function of U at $U = 0$.

The structure and excitation energy of the low-lying states is also given. We have found that although the energy–momentum dispersion of the excitations is the same as that for $U > 0$, it has a different meaning. In contrast to the $U > 0$ case, where the two kinds of excitations could be connected with the charge and spin degrees of freedom, for $U < 0$ both kinds of excitations are connected with the charge distribution: one with the distribution of the bound pairs, the other with the motion of the 'free' electrons. The state of the uncompensated spins does not affect the energy explicitly.

Finally, we have demonstrated in a simple example that by taking the $N \to \infty$ limit first and then the $U \to 0$ limit, the parameter sets describing the excited states are not continuous for all states in $U = 0$.

Acknowledgment

I am very grateful for Drs J Sólyom and A Sütö for their continuous interest and critical reading of the manuscript.

References

Choy T C and Young W 1982 *J. Phys. C: Solid State Phys.* **15** 521–8
Coll C F 1974 *Phys. Rev.* B **9** 2150
Lieb E M and Wu F Y 1968 *Phys. Rev. Lett.* **20** 1445
Shiba H 1972 *Phys. Rev.* B **6** 950
Takahashi M 1969 *Prog. Theor. Phys.* **42** 1098
Woynarovich F 1982a *J. Phys. C: Solid State Phys.* **15** 85–96
—— 1982b *J. Phys. C: Solid State Phys.* **15** 87–109
—— 1983 *J. Phys. C: Solid State Phys.* **16**

Z. Phys. B – Condensed Matter 78, 99–109 (1990)

Condensed Zeitschrift für Physik B Matter

A new method for the excitations of the one-dimensional Hubbard model*,**

A. Klümper, A. Schadschneider, and J. Zittartz
Institut für Theoretische Physik, Universität zu Köln, Federal Republic of Germany

Received July 14, 1989

We present a new method to solve the nested Bethe Ansatz equations for the one-dimensional Hubbard model. First the "spin problem" is solved completely by using inversion/functional relations and an equivalence to the eight-vertex model. Secondly we solve the "electronic problem", i.e. we determine the ground state of the half-filled band and all finite energy excitations relative to this case. Apart from the calculation of the energy bands we also derive basic equations for the characteristic parameters which determine the excited states.

I. Introduction

The one-dimensional Hubbard model is one of the few exactly solvable models for spin-$\frac{1}{2}$ fermions. Lieb and Wu [1] showed that the Hamiltonian can be diagonalized by a Bethe Ansatz. They derived a nested set of nonlinear equations, also called nested Bethe Ansatz equations, which determine the eigenstates and the corresponding energies and momenta. From these equations they calculated the ground state of the system for the half-filled band case. Subsequently various types of excitations relative to this ground state have been investigated. Spin excitations were considered in [2–4], while charge excitations which involve complex wave numbers have been studied in some detail by Woynarovich [5, 6].

Common to all these investigations is that the two nested sets of Bethe Ansatz equations are solved simultaneously by setting up integral equations for density distributions of the Bethe Ansatz variables. This procedure is very cumbersome, as it also requires some preknowledge of the location of the variables, and makes the classification of the excited states a difficult problem.

In this paper we present a new method for the solution of the Bethe Ansatz equations by splitting them up into two problems which can be treated separately. In Sect. II the basic equations for the two problems which we call "spin problem" and "electronic problem", respectively, are derived. The "spin problem" is then completely solved in Sect. III for all energy eigenstates in the thermodynamic limit. Here we make essentially use of inversion/functional relations for characteristic excitation functions. Similar inversion relations have been successfully used recently in solving corresponding Bethe Ansatz equations for the eight-vertex model [7, 8]. In fact we shall show very briefly in Sect. III and Appendix A that the present problem is a special case of the eight-vertex model problem.

After having solved the "spin problem" it turns out that the remaining "electronic problem" can be treated easily and systematically for the ground state of the half-filled band and for all finite energy excitations relative to this case. In Sect. IV we treat subsequently first spin excitations characterized by gapless spin-wave dispersions, secondly so-called "hole excitations", and thirdly "charge excitations". Apart from recovering all results on excitations known so far we obtain a complete classification of the eigenstates. The states are characterized by sets of parameters v_α and λ_α which satisfy basic equations derived in Appendices A and B. Section V contains a summary.

We believe that our new method to solve Bethe Ansatz equations will also prove useful for other models which can be solved by the Bethe Ansatz.

---

* Dedicated to Professor W. Brenig on the occasion of his 60th birthday

** Work performed within the research program of the Sonderforschungsbereich 341, Köln-Aachen-Jülich

II. Model and Bethe Ansatz equations

The simplest model describing hopping of electrons on a one-dimensional chain plus repulsion of two electrons on the same site due to the Coulomb interaction is the Hubbard model with Hamiltonian

$$H=-t\sum_{j=1}^{L}\sum_{\sigma=\pm 1}(c_{j+1\sigma}^{+}c_{j\sigma}+c_{j\sigma}^{+}c_{j+1\sigma})+U\sum_{j=1}^{L}n_{j\uparrow}n_{j\downarrow}. \quad (1)$$

$c_{j\sigma}^{+}$ ($c_{j\sigma}$) creates (annihilates) an electron with spin σ on site j and $n_{j\sigma}=c_{j\sigma}^{+}c_{j\sigma}$ is the corresponding number operator. As usual periodic boundary conditions are imposed, $c_{L+1\sigma}=c_{1\sigma}$, where L is the number of sites. t is the hopping matrix element and U is the interaction energy of two electrons occupying the same site. Because of various symmetries [1] of the Hamiltonian (1) it is sufficient to consider the model for $t=+1$, $U\geqq 0$, and to restrict the number of electrons N and the number of electrons with down spin N_-, which are both conserved, to

$$N\leqq L, \quad N_-\leqq\frac{N}{2}. \quad (2)$$

For convenience we assume N to be even in the following.

In 1968 Lieb and Wu [1] could explicitly construct the eigenstates of (1) under the restriction (2) via a double Bethe Ansatz (BA). Their method traces back to [9] and is described in detail in [10, 11]. The integrability of a many particle system like (1) requires that two particles interact only on the same site and that their scattering matrix satisfies the factorization equation (or Yang-Baxter-, Star-Triangle-Equation).

In each fundamental region characterized by a fixed sequence of the particle coordinates, $1\leqq x_{Q_1}\leqq\ldots\leqq x_{Q_N}\leqq L$ where Q is a permutation of $\{1,\ldots,N\}$, a superposition of plane waves with N "wave numbers" k_j is a suitable ansatz for the wave function

$$\psi(x_1,\sigma_1;\ldots;x_N,\sigma_N)=\sum_P A_{P,Q}(\sigma_1,\ldots,\sigma_N)\exp\left(i\sum_{j=1}^{N}k_{P_j}x_{Q_j}\right). \quad (3)$$

The sum is over all permutations P of $\{1,\ldots,N\}$ and $A_{P,Q}$ are functions of the spin variables. If no two particle coordinates are equal the ansatz (3) satisfies $H|\psi\rangle=E|\psi\rangle$ with energy and momentum

$$E=-2\sum_{j=1}^{N}\cos k_j, \qquad P=\sum_{j=1}^{N}k_j. \quad (4)$$

The requirement that $H|\Psi\rangle=E|\psi\rangle$ is also satisfied on the boundaries of each fundamental region (where at least two particles scatter) imposes a set of linear equations for the coefficients $A_{P,Q}$. Other equations arise from the symmetry property of fermions as well as from the periodic boundary conditions. This requires the diagonalization of an appropriate operator T acting on the functions $A_{I,I}$:

$$T\cdot A_{I,I}=e^{iLk_j}\cdot A_{I,I}. \quad (5)$$

Actually T is related to the transfer matrix of an inhomogeneous six-vertex model. This problem again can be solved by a suitable Bethe Ansatz with N_- "moments" v_α which leads to the equations*

$$\prod_{j=1}^{N}\frac{v_\alpha-i\sin k_j-U/4}{v_\alpha-i\sin k_j+U/4}=-\prod_{\beta=1}^{N_-}\frac{v_\alpha-v_\beta-U/2}{v_\alpha-v_\beta+U/2}, \quad \alpha=1,\ldots,N_-. \quad (6a)$$

Equating the corresponding eigenvalue of T with e^{iLk_j} (see (5)) yields

$$e^{iLk_j}=\prod_{\alpha=1}^{N_-}\frac{i\sin k_j-v_\alpha-U/4}{i\sin k_j-v_\alpha+U/4}, \quad j=1,\ldots,N. \quad (6b)$$

The two sets of Eqs. (6) constitute the nested Bethe Ansatz equations for the one-dimensional Hubbard model. It is generally assumed that all possible solutions of these equations provide the complete solution of the energy eigenvalue problem.

In the following we cite (6a) as the 'spin problem' because it arises from the diagonalization of the operator T acting on the space of all spins. We regard (6a) as a set of equations determining the variables v_α as functions of the "wave functions" k_j. Equations (6b) will be called the 'electronic problem' because it will be used to determine the "wave numbers" k_j once the "spin variables" v_α are known (as functions of the k_j).

Usually (see [1]) the logarithmic form of (6) is used to derive a set of coupled integral equations which determines the densities $\rho(k)$, $\sigma(v)$ of the wave numbers k_j, v_α, respectively, in the thermodynamic

* Our parameters v_α are related to the parameters Λ_α in [1] by $v_\alpha=i\Lambda_\alpha$

limit. We shall not follow this procedure here. Instead we consider an alternative version of Eqs. (6) which will turn out to be more convenient. First we introduce the functions

$$\Phi(v)=\prod_{j=1}^{N}(v-\mathrm{i}\sin k_j), \tag{7a}$$

$$q(v)=\prod_{\beta=1}^{N_-}(v-v_\beta), \tag{7b}$$

depending on the complex variable v. The BA equations then read

$$\mathrm{e}^{\mathrm{i}Lk_j}=\frac{q\left(\mathrm{i}\sin k_j-\frac{U}{4}\right)}{q\left(\mathrm{i}\sin k_j+\frac{U}{4}\right)}, \tag{8a}$$

$$\frac{\Phi\left(v_\alpha+\frac{U}{4}\right)}{\Phi\left(v_\alpha-\frac{U}{4}\right)}=-\frac{q\left(v_\alpha+\frac{U}{2}\right)}{q\left(v_\alpha-\frac{U}{2}\right)}. \tag{8b}$$

The key step now is to observe that (8 b) guarantees the analyticity of a function $\Lambda(v)$ which is defined – ad hoc – by*

$$\Lambda(v)\,q(v)=\Phi\left(v-\frac{U}{4}\right)q\left(v+\frac{U}{2}\right)+\Phi\left(v+\frac{U}{4}\right)q\left(v-\frac{U}{2}\right). \tag{9}$$

This is clear because $\Lambda(v)$ is the ratio of two analytic functions where the zeros v_β of the denominator are cancelled by zeros of the numerator (see (8 b)).

Inserting $v=\mathrm{i}\sin k_j+\frac{U}{4}$ into (9) and observing that $\Phi(\mathrm{i}\sin k_j)=0$, (8 a) can be written as

$$\mathrm{e}^{\mathrm{i}Lk_j}=\frac{\Lambda\left(\mathrm{i}\sin k_j+\frac{U}{4}\right)}{\Phi\left(\mathrm{i}\sin k_j+\frac{U}{2}\right)}. \tag{10}$$

The problem can now be stated in principal as to find all analytic functions $\Lambda(v)$ which satisfy (9) ("spin problem") and then to solve (10) for all variables k_j ("electronic problem"). Energy and momentum finally can be calculated from (4). We remark that (7), (9) are the eigenvalue equations for an inhomogeneous six-vertex model which can be treated very much in the same manner as the homogeneous six-vertex model (or the eight-vertex model [7, 8, 12]). The program is carried out in the following sections by using the method of inversion/functional relations [7, 8].

* Here $\Lambda(v)$ seems only to be an auxiliary function. But indeed it has the meaning of an eigenvalue of the vertex model transfer matrix T mentioned before

III. Solution of the spin problem

In this section we solve the "spin problem", i.e. we determine the functions $\Lambda(v)$ in terms of a given distribution of wave numbers k_j. There are two possibilities to do this. The first possibility is to solve the above equations directly which we shall do in the following. The second possibility is to view the present problem as a particular scaling limit of the corresponding problem for the transfer matrix eigenvalues of the eight-vertex model. This second method relies heavily on results obtained in [7, 8] and is briefly discussed in Appendix A.

We shall solve the "spin problem" under the condition that the number of electrons N is a macroscopic parameter, i.e N/L tends to a non-zero value in the thermodynamic limit, and that the magnetization $M=N/2-N_-$ remains finite in this limit. For the distribution of the wave numbers k_j we require that most of them should be real while a finite number of them may be complex. As we shall see this includes the energy ground state for each N as well as all states of finite energy excitations.

With a view to (7 a) it is easily seen that $\Phi\left(v+\frac{U}{4}\right)=O(\mathrm{e}^{-N})$ for v out of some neighbourhood of $v=-\frac{U}{4}$ where $O(\mathrm{e}^{-N})$ means a quantity which decreases exponentially with $N\to\infty$. Then it follows from (9) that we have

$$\begin{aligned}\Lambda(v)&=\Phi\left(v-\frac{U}{4}\right)\frac{q(v+U/2)}{q(v)}\cdot[1+O(\mathrm{e}^{-N})],\\ \Lambda\left(v+\frac{U}{2}\right)&=\Phi\left(v+\frac{3U}{4}\right)\frac{q(v)}{q(v+U/2)}\cdot[1+O(\mathrm{e}^{-N})].\end{aligned} \tag{11}$$

Obviously (11) gives rise to the following fundamental inversion relation for $\Lambda(v)$:

$$\Lambda(v)\,\Lambda\left(v+\frac{U}{2}\right)=\Phi\left(v-\frac{U}{4}\right)\Phi\left(v+\frac{3U}{4}\right)\cdot(1+O(\mathrm{e}^{-N})). \tag{12}$$

This functional equation for all possible $\Lambda(v)$-functions has been derived for some neighbourhood of

$v=-\frac{U}{4}$. We remark, but we omit the detailed proof, that the validity can be extended to the region $-\frac{U}{2}<\mathrm{Re}(v)<0$.

Now we are going to solve (12) for all possible $\Lambda(v)$-functions. This has the advantage that we need not determine the distribution of the variables v_α which characterize the "spin states" $q(v)$ (7b) beforehand as the right hand side of (12) obviously depends only on the universal function $\Phi(v)$, but is independent of $q(v)$. The investigation of the "spin states" $q(v)$ can thus be postponed, this problem will be considered in Appendix A.

There are various solutions of (12) corresponding to the ground state with magnetization $M=0$ and to the excited states with arbitrary M. We first look for the largest function $\Lambda_0(v)$, which will turn out later on to correspond to the ground state and thus has $M=0$, and then for the next-largest functions. From the investigation of the corresponding problem for the eight-vertex model [7, 8] it is reasonable to assume that the function $\Lambda_0(v)$ is analytic *and* non-zero in the "physical regime" $-\frac{U}{4}<\mathrm{Re}(v)<\frac{U}{4}$. The fundamental relation (12) together with this property and the asymptotic behaviour obtained from (9), $\Lambda(v)\simeq v^N$ for $v\to\pm i\infty$, uniquely determines $\Lambda_0(v)$ up to a sign. The final result is

$$\Lambda_0(v)=(-1)^{N/2}\prod_{l=1}^{N} z(v-\mathrm{i}\sin k_l), \tag{13}$$

where

$$z(v):=\left(v+\frac{U}{4}\right)\phi\left(v-\frac{U}{4}\right)$$

$$\phi(v):=\prod_{n=1}^{\infty}\frac{4n-2-4v/U}{4n-2+4v/U}\cdot\frac{4n+4v/U}{4n-4v/U} =\frac{\Gamma\left(\frac{1}{2}+\frac{v}{U}\right)\Gamma\left(1-\frac{v}{U}\right)}{\Gamma\left(\frac{1}{2}-\frac{v}{U}\right)\Gamma\left(1+\frac{v}{U}\right)}. \tag{14}$$

We omit the detailed derivation, but mention that (12) can be checked easily and that the ultimate sign $(-1)^{N/2}$ has been determined from the limit $U\to\infty$. $\left(\text{Remark that } N_-=\frac{N}{2} \text{ for } M=0.\right)$

The next-largest functions $\Lambda(v)$ are characterized by the excitation function $l(v)$ in the thermodynamic limit

$$l(v):=\lim_{L\to\infty}\frac{\Lambda(v)}{\Lambda_0(v)} \tag{15}$$

which satisfies the functional equation (see (12))

$$l(v)\,l\left(v+\frac{U}{2}\right)=1. \tag{16}$$

In contrast to $\Lambda_0(v)$ the excitation function exhibits zeros in the physical regime $-\frac{U}{4}<\mathrm{Re}(v)<\frac{U}{4}$. Relation (16) together with the asymptotic behaviour $\lim_{v\to\pm i\infty} l(v)=1$ now implies

$$l(v)=\pm\prod_{\alpha=1}^{\nu}\tan\left(\frac{\pi}{U}(v-\Theta_\alpha)\right) \tag{17}$$

where the Θ_α are the zeros of $l(v)$. A detailed derivation of the result (17) has been given in [7] and in particular in Sect. (IV.3) of [13]. As in [13] ν turns out to be an even number and each Θ_α is a parameter which can take arbitrary imaginary values. From $\lim_{v\to\pm i\infty} l(v)=1$ we determine the sign in (17). This yields the final result

$$l(v)=(-1)^{\nu/2}\prod_{\alpha=1}^{\nu}\tan\left(\frac{\pi}{U}(v-i\theta_\alpha)\right), \tag{18}$$

where now all $\theta_\alpha=\frac{1}{i}\Theta_\alpha$ are real parameters.

Equations (13) and (18) constitute the solution of the "spin problem". One should observe that $\Lambda_0(v)$ explicitly depends on the wave numbers k_j whereas $l(v)$ is completely independent of them. We have already remarked that the magnetization of the state corresponding to $\Lambda_0(v)$ is $M=0$. In Appendix A we investigate the degeneracies of all functions $l(v)$ (18) and classify the excitations. It will turn out that M is not determined uniquely but can take the values $M=0, 1, \ldots, \nu/2$. The case $\nu=2$ for instance comprises one singlet and three triplet bands (which are degenerate). As a further result a classification scheme for higher ν is presented.

IV. The electronic problem

In the last section we have determined the functions $\Lambda(v)$ in dependence of the wave numbers k_j. These functions enter (10) from which finally all wave numbers have to be calculated. First we insert Λ_0 from (13) into (10) and we get the set of equations

$$e^{iLk_j}=(-1)^{N/2}\cdot\prod_{l=1}^{N}\phi(\mathrm{i}\sin k_j-\mathrm{i}\sin k_l). \tag{19}$$

Secondly we use (15) in order to take into account also spin excitations. In this case we obtain the set of equations

$$e^{iLk_j}=(-1)^{N/2}\cdot\left[\prod_{l=1}^{N}\phi(\mathrm{i}\sin k_j-\mathrm{i}\sin k_l)\right]\cdot l\left(\mathrm{i}\sin k_j+\frac{U}{4}\right). \tag{20}$$

In the following we first derive an integral equation for the density of wave numbers k_j in the ground state of the Hubbard chain with N electrons. In case of the "half filled" band, i.e. $N=L$, the integral equation can be solved trivially. Then we treat various excitations above the ground state: spin excitations, hole excitations where $N<L$, and double occupations.

a. The ground state

The ground state is characterized by N real wave numbers k_j centered around the origin [1]. Their actual positions are determined by (19). We therefore introduce the density $\rho(k)$ of wave numbers in the interval $[-K, K]$ which is filled densely in the thermodynamic limit

$$\rho(k_j)=\frac{1}{L(k_{j+1}-k_j)}. \tag{21}$$

K $(-K)$ is the maximum (minimum) value and the k_j are assumed to increase with increasing index. Taking the logarithm of (19) we obtain

$$\mathrm{i}Lk_j=2\pi\mathrm{i}\cdot I_j+\sum_{l=1}^{N}\ln\phi(\mathrm{i}\sin k_j-\mathrm{i}\sin k_l) \tag{22}$$

where the I_j are consecutive integers (half-odd integers) for $\frac{N}{2}$ even (odd). By taking the difference for consecutive k_{j+1} and k_j, this leads to an integral equation for $\rho(k)$ in the thermodynamic limit $L\to\infty$

$$2\pi\rho(k)+\cos k\int_{-K}^{K}(\ln\phi)'(\mathrm{i}\sin k-\mathrm{i}\sin\tilde{k})\,\rho(\tilde{k})\,\mathrm{d}\tilde{k}=1. \tag{23}$$

This equation is easily solved in the case $K=\pi$ where we denote the density by $\rho_0(k)$. First the substitution $k\to\pi-k$ shows that

$$\rho_0(k)+\rho_0(\pi-k)=\frac{1}{\pi}. \tag{24}$$

A first consequence of this equation is $\int_{-\pi}^{\pi}\rho_0(k)\,\mathrm{d}k=1$ (ρ is 2π-periodic!). According to the definition (21) this implies that $K=\pi$ corresponds to $N=L$, i.e. the case of the half-filled band. Splitting the integration interval into two parts $\left[-\frac{\pi}{2},\frac{\pi}{2}\right]$ and $\left[\frac{\pi}{2},\frac{3\pi}{2}\right]$ and substituting in the second one $\tilde{k}\to\pi-\tilde{k}$ we immediately find

$$\rho_0(k)=\frac{1}{2\pi}-\frac{\cos k}{2\pi^2}\int_{-\pi/2}^{\pi/2}(\ln\phi)'(\mathrm{i}\sin k-\mathrm{i}\sin\tilde{k})\,\mathrm{d}\tilde{k}. \tag{25}$$

Using (14) explicitly this can be rewritten as [14]

$$\rho_0(k)=\frac{1}{2\pi}+\frac{\cos k}{\pi}\int_0^{\infty}\frac{J_0(\omega)\cos(\omega\sin k)}{1+\exp\left(\frac{\omega U}{2}\right)}\,\mathrm{d}\omega \tag{26}$$

where J_0 denotes the Bessel function. Equation (26) is the well-known result of [1].

The corresponding ground state energy follows from (4) and is given by

$$\begin{aligned}E_0&=-2L\int_{-\pi}^{\pi}\rho_0(k)\cos k\,\mathrm{d}k\\&=\frac{LU}{2\pi^2}\sum_{n=1}^{\infty}(-1)^n\int_{-\pi/2}^{\pi/2}\int_{-\pi/2}^{\pi/2}\frac{4n\cos^2 k}{\left(\frac{nU}{2}\right)^2+(\sin k-\sin\tilde{k})^2}\,\mathrm{d}\tilde{k}\,\mathrm{d}k\\&=-4L\int_0^{\infty}\frac{J_0(\omega)\,J_1(\omega)}{\omega\left(1+\exp\left(\frac{\omega U}{2}\right)\right)}\,\mathrm{d}\omega\end{aligned} \tag{27}$$

in terms of Bessel functions. For illustration E_0 is plotted in Fig. 1 as a function of the interaction ener-

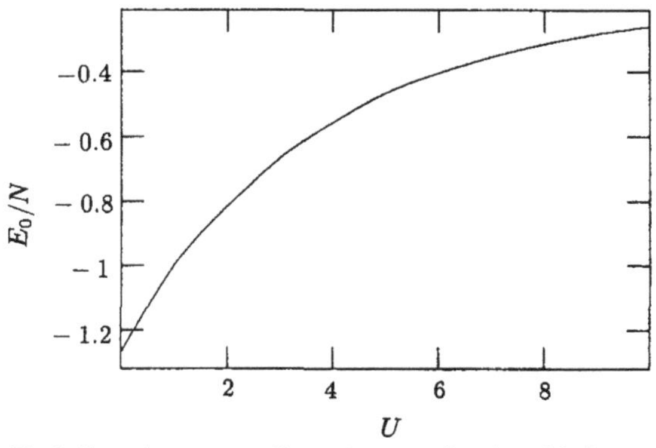

Fig. 1. Groundstate energy E_0 per electron as function of the interaction U according to (27)

gy U. The corresponding momentum P_0 (4) of the ground state is 0 or π depending on whether $N/2$ is odd or even.

b. Spin excitations

Now we determine the "spin excitations" above the ground state of the half filled band, $N=L$. Instead of (19) we have to consider (20). As for the ground state we look for purely real wave numbers, now denoted by $\hat{k}_j$. Proceeding as before we take the logarithm of (20) and obtain

$$iL\hat{k}_j = 2\pi i \cdot \hat{I}_j + \sum_{l=1}^{L} \ln \phi(i \sin \hat{k}_j - i \sin \hat{k}_l) + \ln l\left(i \sin \hat{k}_j + \frac{U}{4}\right). \tag{28}$$

We assume that the difference between the wave numbers $\hat{k}_j$ and k_j of the (low-lying) excited states and the ground state, respectively, is $O\left(\frac{1}{L}\right)$. Introducing the quantities

$$\begin{aligned} \chi(k_j) &= L(\hat{k}_j - k_j), \\ \sigma(k) &= \chi(k) \cdot \rho_0(k), \end{aligned} \tag{29}$$

choosing $\hat{I}_j = I_j$, and subtracting (22) from (28) we obtain the following integral equation for $\sigma(k)$

$$2\pi\sigma(k) + \int_{-\pi}^{\pi} (\ln \phi)'(i \sin k - i \sin \hat{k}) \cos \hat{k}\, \sigma(\hat{k})\, d\hat{k} = -i \ln l\left(i \sin k + \frac{U}{4}\right). \tag{30}$$

The substitution $k \to \pi - k$ shows that

$$\sigma(k) = \sigma(\pi - k). \tag{31}$$

The integral in (30) is again reduced to the interval $-\frac{\pi}{2} \leqq \hat{k} \leqq \frac{\pi}{2}$ and is easily seen to vanish by using (31). Thus the solution of (30) is simply given by

$$\sigma(k) = \frac{1}{2\pi i} \ln l\left(i \sin k + \frac{U}{4}\right). \tag{32}$$

The excitation energy is from (4)

$$\begin{aligned} E - E_0 &= -2\sum_{j=1}^{L} (\cos \hat{k}_j - \cos k_j) = 2\int_{-\pi}^{\pi} \sigma(k) \sin k\, dk \\ &= \frac{2}{\pi} \int_{-\pi/2}^{\pi/2} (\ln l)'\left(i \sin k + \frac{U}{4}\right) \cos^2 k\, dk. \end{aligned}$$

The latter result follows by partial integration. Using (18) explicitly we can write

$$E - E_0 = \sum_{\alpha=1}^{\nu} \varepsilon_S(\theta_\alpha) \tag{33a}$$

with

$$\varepsilon_S(\theta) := \frac{4}{U} \int_{-\pi/2}^{\pi/2} \frac{\cos^2 k}{\cosh\left(\frac{2\pi}{U}(\sin k - \theta)\right)}\, dk. \tag{33b}$$

The corresponding momentum (4) is

$$\begin{aligned} P - P_0 &= \sum_{j=1}^{L} (\hat{k}_j - k_j) = \int_{-\pi}^{\pi} \sigma(k)\, dk \\ &= \frac{1}{\pi i} \int_{-\pi/2}^{\pi/2} \ln l\left(i \sin k + \frac{U}{4}\right) dk. \end{aligned}$$

Again we can write

$$P - P_0 = \sum_{\alpha=1}^{\nu} p_S(\theta_\alpha) \tag{34a}$$

with

$$p_S(\theta) := \frac{\pi}{2} + \frac{2}{\pi} \int_{-\pi/2}^{\pi/2} \arctan\left[\tanh\left(\frac{\pi}{U}(\sin k - \theta)\right)\right] dk. \tag{34b}$$

These are the results found in [2, 3] after lengthy calculations. In Fig. 2 the dispersion (33a), (34a) is

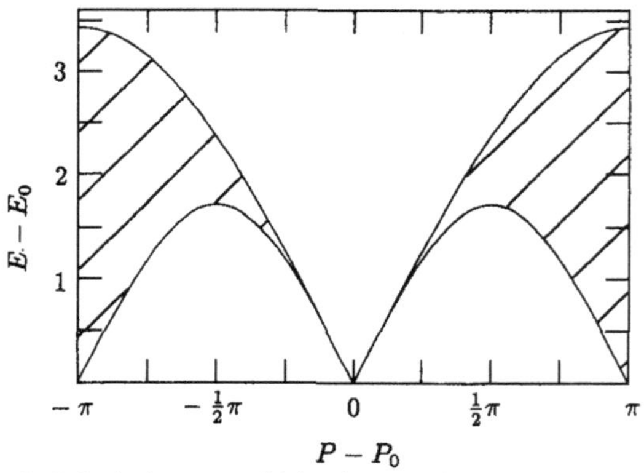

Fig. 2. Excitation energy (33a) as function of total momentum (34a) for the $\nu=2$ spin-wave continuum with interaction $U=2$

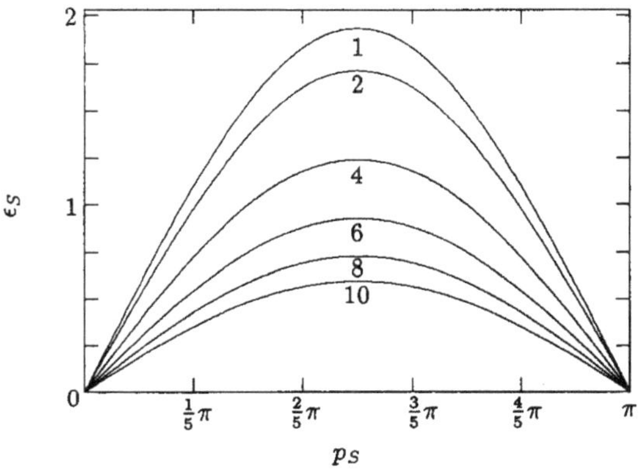

Fig. 3. The spin-wave dispersion ε_S (33b) as function of momentum p_S (34b) for different values of U. This corresponds to the lower curve in Fig. 2

shown for $U=2$ and $v=2$. In Fig. 3 the dispersion (33b), (34b) which corresponds to one half of the lower curve in Fig. 2 is plotted for different values of U. The strictly positive values of ε_S rigorously prove that the ground state is characterized by $\Lambda_0(v)$ and that the excited states are determined from the next-largest functions $\Lambda(v)$.

The spin excitations (33), (34) may be viewed as spin waves above the singlet ground state. Their number is always even as $v=2,4,\ldots$ and the corresponding spin wave bands are 2^v-fold degenerate as we argue in Appendix A. For each v they comprise various multiplets of the magnetization M.

c. *Hole excitations*

Here we calculate energy and momentum of another type of excitations from the ground state of the half filled band ($N=L$). We allow for a slightly different number of particles, i.e. $H:=L-N$ may be positive, but finite in the thermodynamic limit. To treat these hole excitations (apart from spin excitations) we must solve (19) for N real wave numbers, now denoted by $\hat{k}_j$,

$$e^{iL\hat{k}_j}=(-1)^{\frac{L-H}{2}}\cdot\prod_{l=1}^{L-H}\phi(i\sin\hat{k}_j-i\sin\hat{k}_l). \tag{35}$$

This equation has L solutions for the $N=L-H$ variables $\hat{k}_j$. Denoting also the "empty" hole positions $k_1^h,\ldots,k_H^h$ by $\hat{k}$ (thus we have a set of L wave numbers $\hat{k}_j$) (35) can be rewritten as

$$e^{iL\hat{k}_j}=(-1)^{L/2}\cdot\left[\prod_{l=1}^{L}\phi(i\sin\hat{k}_j-i\sin\hat{k}_l)\right]l_h(i\sin\hat{k}_j), \tag{36}$$

where

$$l_h(v):=\prod_{\alpha=1}^{H}[i\phi(v-i\sin k_\alpha^h)]^{-1}. \tag{37}$$

Comparing with (20) we see that the present problem is almost the same as the problem of spin excitations treated before, the difference being only that the spin excitation function l in (20) is replaced by the hole excitation function l_h (37). Thus we can proceed as in b). The distribution $\sigma(k)$ (29), (32) is now given by

$$\sigma^h(k)=\frac{1}{2\pi i}\ln l_h(i\sin k). \tag{38}$$

Energy and momentum are again derived from (4):

$$\begin{aligned}E-E_0&=-2\sum_{j=1}^{L-H}\cos\hat{k}_j+2\sum_{j=1}^{L}\cos k_j\\&=-2\sum_{j=1}^{L}(\cos\hat{k}_j-\cos k_j)+2\sum_{\alpha=1}^{H}\cos k_\alpha^h\\&=\frac{2}{\pi}\int_{-\pi/2}^{\pi/2}(\ln l_h)'(i\sin k)\cos^2 k\,dk+2\sum_{\alpha=1}^{H}\cos k_\alpha^h\\&=\sum_{\alpha=1}^{H}\left[\varepsilon_h(k_\alpha^h)-\frac{U}{2}\right].\end{aligned} \tag{39a}$$

The hole dispersion ε_h $\left(\text{including a shift } \frac{U}{2} \text{ to render } \varepsilon_h \text{ positive}\right)$ is given by

$$\begin{aligned}\varepsilon_h(k^h):=&\frac{32}{\pi U}\int_{-\pi/2}^{\pi/2}\sum_{n=1}^{\infty}(-1)^{n+1}\\&\cdot\frac{n\cos^2 k}{(2n)^2+\left(\frac{4}{U}(\sin k-\sin k^h)\right)^2}dk\\&+2\cos k^h+\frac{U}{2}.\end{aligned} \tag{39b}$$

The corresponding total momentum is

$$\begin{aligned}P-P_0&=\sum_{j=1}^{L-H}\hat{k}_j-\sum_{j=1}^{L}k_j=\sum_{j=1}^{L}(\hat{k}_j-k_j)-\sum_{\alpha=1}^{H}k_\alpha^h\\&=\frac{1}{\pi i}\int_{-\pi/2}^{\pi/2}\ln l_h(i\sin k)\,dk-\sum_{\alpha=1}^{H}k_\alpha^h\\&=\sum_{\alpha=1}^{H}p_h(k_\alpha^h)-H\cdot\frac{\pi}{2},\end{aligned} \tag{40a}$$

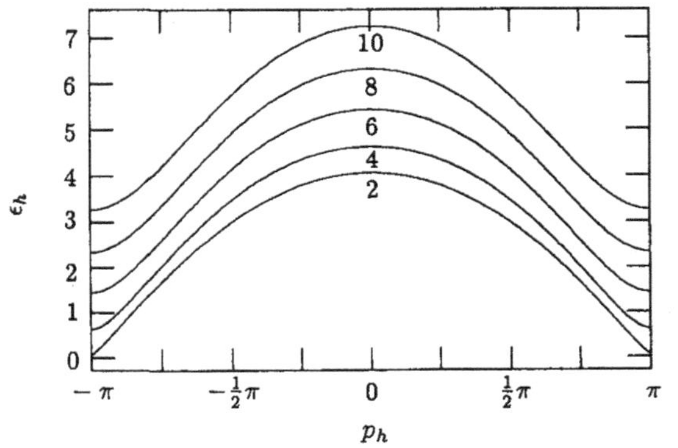

Fig. 4. The hole dispersion ε_h (39b) as function of momentum p_h (40b) for different values of U. It exhibits a gap at $p_h = \pi$

such that

$$p_h(k^h) := -k^h - \frac{2}{\pi} \int_{-\pi/2}^{\pi/2} \sum_{n=1}^{\infty} (-1)^n \cdot \arctan\left(\frac{2}{nU}(\sin k - \sin k^h)\right) dk. \tag{40b}$$

Inverting $p_h(k^h)$ and inserting $k^h(p)$ into $\varepsilon_h(k^h)$ yields the dispersion $\varepsilon_h(p)$. This can only be done numerically, the result for different values of U is represented in Fig. 4. In contrast to the spin wave dispersion the hole dispersion (39b) exhibits a gap $\varepsilon_h(\pi) > 0$ at $k^h = \pi$.

In [1] Lieb and Wu introduced the two chemical potentials $\mu_\pm$

$$\begin{aligned} \mu_+ &:= E_0(L+1) - E_0(L), \\ \mu_- &:= E_0(L) - E_0(L-1), \end{aligned} \tag{41}$$

where $E_0(N)$ is the ground state energy for N particles. They concluded from the nonvanishing difference, $\mu_+ - \mu_- > 0$, that the ground state of the half-filled band, $N = L$, is insulating for all $U > 0$ (with a Mott transition at $U = 0$). This is easily rederived from the above results. First it follows from the symmetries of the Hamiltonian (1) mentioned before that $\mu_+ = U - \mu_-$ [1]. μ_- is the energy gain by creating one hole, i.e. $\mu_- = \frac{U}{2} - \min \varepsilon_h(k^h) = \frac{U}{2} - \varepsilon_h(\pi)$. Thus we obtain $\mu_+ - \mu_- = 2\varepsilon_h(\pi)$ which is strictly positive for $U > 0$.

By calculating the hole excitations (39a) we have obtained the energy eigenvalues of those states which differ from the half-filled band case by adding a finite number H of holes to the system. To derive the full spectrum of energies which differ from the ground state energy of the half-filled band by a finite amount in the thermodynamic limit we only need to consider one more class of excitations, namely solutions of the basic equations (19), (20) where some wave numbers k_j take on complex values. This problem is considered in the following.

d. Excitations with double occupation

States with complex wave numbers k_j have been investigated in [5, 6]. Their energies scale with $E - E_0 \simeq U$, $U \to \infty$, relative to the ground state energy of the half-filled band which means that we have real (rather than virtual) double occupations of individual lattice sites in these states. This explains the terminology.

In Appendix B we derive the conditions under which complex wave numbers are allowed in (19), (20). We consider μ pairs $k_1^\pm, \ldots, k_\mu^\pm$ which satisfy (B.1), i.e.

$$\mathrm{i} \sin k_\alpha^\pm = \lambda_\alpha \pm \frac{U}{4}. \tag{42}$$

To treat the Eqs. (19) for the remaining $L - 2\mu$ real wave numbers $\hat{k}_j$ we then proceed as follows. Starting from the half-filled band case, $N = L$, we first create 2μ holes $k_1^h, \ldots, k_{2\mu}^h$ and then we add the complex pairs (thus $N = L$ is unchanged). As in the treatment of hole excitations before we count the "empty" hole positions twice, once among the set of real $\hat{k}_j$ and then separately by "subtraction". Equation (19) can then be written as

$$\mathrm{e}^{\mathrm{i}L\hat{k}_j} = (-1)^{L/2} \cdot \left[\prod_{l=1}^{L} \phi(\mathrm{i} \sin \hat{k}_j - \mathrm{i} \sin \hat{k}_l)\right] \cdot \bar{T}(\mathrm{i} \sin \hat{k}_j), \tag{43}$$

where the "excitation" function $\bar{T}$ is now

$$\bar{T}(v) := \left[\prod_{\alpha=1}^{2\mu} \mathrm{i}\phi(v - \mathrm{i} \sin k_\alpha^h)\right]^{-1} \cdot \prod_{\alpha=1}^{\mu} \frac{v - \mathrm{i} \sin k_\alpha^+}{v - \mathrm{i} \sin k_\alpha^-}. \tag{44}$$

The first product describes the holes (compare with (37)), the second product describes the complex pairs and we have used (B.3) to simplify the expression.

Comparing with (20) or (36) we see that again we can proceed as in b) to calculate energy and momentum. With a view to (39a) we obtain

$$E - E_0 = \frac{2}{\pi} \int_{-\pi/2}^{\pi/2} (\ln \bar{T})'(\mathrm{i} \sin k) \cos^2 k \, dk + 2 \sum_{\alpha=1}^{2\mu} \cos k_\alpha^h - 2 \sum_{\alpha=1}^{\mu} (\cos k_\alpha^+ + \cos k_\alpha^-). \tag{45}$$

Then using the "hole results" (39a) and (39b) to treat the first part of (44) in (45) and using (B.2) to treat the second product of (44) in (45) we easily derive the final result

$$E-E_0=\sum_{\alpha=1}^{2\mu}\varepsilon_h(k_\alpha^h). \tag{46}$$

Similarly the momentum is

$$P-P_0=\sum_{\alpha=1}^{2\mu}p_h(k_\alpha^h). \tag{47}$$

The quantities ε_h and p_h have been defined in (39b), (40b). As obviously $\varepsilon_h\simeq\frac{U}{2}$ for $U\to\infty$, the energy difference $E-E_0\simeq\mu U$ $(\mu=1,2,\dots)$ proves that real double occupations of lattice sites characterize the eigenstates which was mentioned before. In contrast to the spin excitations of IV.b) which are gapless, the present excitation bands are separated from the ground state by the gaps $2\mu\varepsilon_h(\pi)$ $(\mu=1,2,\dots)$. They may also be viewed as charge excitations because the states may carry a charge current. Finally we remark that the excitations always involve an even number of "ε-particles" $(2\mu=2,4,\dots)$.

V. Summary

In the foregoing we have presented a new method to solve the one-dimensional Hubbard model. Starting point is the set of Bethe Ansatz equations (6) consisting of the "spin problem" (6a) and the "electronic problem" (6b). By introducing the characteristic Λ-functions (9) and investigating their analytic properties we could determine all possible Λ-functions in the thermodynamic limit by using the fundamental inversion relations (12), (16). The complete solution of the "spin problem" in Sect. III is contained in Eqs. (13) and (18). The remaining "electronic problem" (Sect. IV) is described by (19), (20) for the ground state and excitations.

The method of solution then proceeds by first determining the ground state of the half filled band, which serves as reference state. We remark that the calculation of the ground state distribution $\rho_0(k)$ (25) is trivial in our method and energy (27) and momentum are obtained easily. However, the main advantage of our new method is then shown in the treatment of all possible excitations of finite energy in the thermodynamic limit. There are three types of excitations (and this exhausts all possibilities), namely spin excitations (IV.b)), hole excitations (IV.c)), and excitations with double occupation (IV.d)) which are characterized by complex wave numbers.

The spin excitations are spin-wave "particles" which occur in even numbers ν, the corresponding bands are 2^ν-fold degenerate (Appendix A) and the spin wave dispersion (33b) is gapless.

The hole excitations treated in IV.c) refer to states where the particle numbers N differ from the half-filled band case, $N=L$, by an arbitrary number H of holes which, however, must remain finite for $L\to\infty$. The characteristic energies and momenta of these excitations are given by (39) and (40).

The excitations with double occupation are treated in IV.d) and are characterized by complex wave numbers $k^\pm$ which must satisfy the conditions (42) (also Appendix B). The corresponding (charged) "particles" have the dispersion (39b), (40b) and contribute to energy and momentum with expressions (46), (47). The number of ε_h-particles again is even in all excitations, but their energy-bands are separated from the ground state by a finite gap – in contrast to the spin waves.

We have seen by the way of solution in Sect. IV that the different excitations contribute independently to the total energy and total momentum. This means now that the most general excitation is a simple superposition of independent contributions of the form

$$E-E_0=\sum_{\alpha=1}^{\nu}\varepsilon_S(\theta_\alpha)+\sum_{\alpha=1}^{2\mu}\varepsilon_h(k_\alpha^h)+\sum_{\alpha=2\mu+1}^{H-2\mu}\left[\varepsilon_h(k_\alpha^h)-\frac{U}{2}\right], \tag{48}$$

$$P-P_0=\sum_{\alpha=1}^{\nu}p_S(\theta_\alpha)+\sum_{\alpha=1}^{2\mu}p_h(k_\alpha^h)+\sum_{\alpha=2\mu+1}^{H-2\mu}\left[p_h(k_\alpha^h)-\frac{\pi}{2}\right], \tag{49}$$

with ν spin waves (from (33), (34)), 2μ "charged particles" (from (46), (47)) which correspond to μ pairs of complex $k^\pm$, and additional $H-2\mu$ holes (from (39), (40)). The ν parameters θ_α and the $H(\geqq 2\mu)$ parameters k_α^h are free continuous variables in the thermodynamic limit, they give rise to the energy-bands.

Although the excitations contribute to (48), (49) by a simple superposition, this does not mean that the corresponding energy eigenstate is likewise a simple product state. The eigenstates depend and are classified by the moments v_α (occurring in (7b)) and the λ_α parameters (42) which characterize the complex pairs $k^\pm$ (B.1). The v_α parameters (or their associated string parameters χ_α) are all correlated to one another and satisfy the set of Eqs. (A.2). The λ_α parameters also are all correlated to one another and satisfy the set of Eqs. (B.5). This shows that the one-dimensional Hubbard model describes a strongly correlated Fermi system.

Appendix A

We give a brief survey over the distributions of v_α-parameters corresponding to the functions $\Lambda_0(v)$ (13) and $l(v)$ (18). $\Lambda_0(v)$ is characterized by $N/2$ purely imaginary numbers v_α. For classifying the spin excitations $l(v)$ there are two possibilities. First one could try to determine the v_α-distributions, or the "spin states" $q(v)$, directly from the basic equations (7) and (9). Secondly one can use the intimate relationship to an (inhomogeneous) eight-vertex model. We choose the latter way since we have studied the eight-vertex model in some detail in [7, 8] and several results can be taken over immediately.

To set up the relationship one has to perform a certain scaling limit $\eta \to 0$ of some relations used in [8]. Substituting $v \to \eta v$, $v_\alpha \to \eta v_\alpha$, and $\lambda \to \eta \frac{U}{4}$ in the Eqs. (1), (2), (4) of [8] and letting $\eta \to 0$ one recovers the Eqs. (7) and (9) of the present paper. [In addition the function $\Phi(v)$ should be defined in (2) of [8] as a product of $h(\ldots)$ functions (Eq. (3) of [8]) rather than a simple power. This slight alteration, however, does not change the analysis and results of [8].] The scaling procedure actually recovers (7), (9) with $N_- = N/2$, the general case $N_- \leqq N/2$ will be treated below.

In order to calculate excited states we looked in [7, 8] for v_α-distributions similar to the state corresponding to $\Lambda_0(v)$. While most v_α remain on the imaginary axis, a finite number of them is allowed to take on "complex" values. In the strip $-2\lambda \leqq \mathrm{Re}(v) \leqq 2\lambda$ (here $-U/2 \leqq \mathrm{Re}(v) \leqq U/2$) all complex v_α must occur pairwise as so-called 2-strings with a fixed distance 2λ (here $U/2$). Outside this strip there is no such requirement and complex v_α occur as singles or 1-strings.

As in [8] it is convenient to introduce string variables χ_α by defining the complex v_α's of a 2-string or a 1-string, respectively, as

$$\begin{aligned} &\chi \pm \frac{U}{4} \quad \left(-\frac{U}{4} \leqq \mathrm{Re}(\chi) \leqq \frac{U}{4}, 2-\text{string}\right),\\ &\chi + \frac{U}{4} \quad \left(\frac{U}{4} \leqq \mathrm{Re}(\chi), 1-\text{string}\right), \qquad\qquad (A.1)\\ &\chi - \frac{U}{4} \quad \left(\mathrm{Re}(\chi) \leqq -\frac{U}{4}, 1-\text{string}\right), \end{aligned}$$

which one obtains from the appropriate definitions in [8] by scaling $\chi \to \eta \chi$. The string variables are determined by the set of Eqs. (55) of [8]. Performing the scaling limit (also setting $\Theta \to \eta \Theta$ in (55) of [8]) then yields the following set of equations

$$\prod_{j=1}^{\nu} \frac{\chi_\alpha - \Theta_j + U/4}{\chi_\alpha - \Theta_j - U/4} = -\prod_{\beta=1}^{\nu_1} \frac{\chi_\alpha - \chi_\beta + U/2}{\chi_\alpha - \chi_\beta - U/2}, \qquad (A.2)$$

where $\nu_1 = \frac{\nu}{2}$. From this set of ν_1 equations, one for each string variable χ_α, one can determine the positions χ_α in terms of given excitation parameters $\Theta = i\theta$ occurring in (17), (18).

This is our final result as long as $N_- = N/2$. The number of solutions of (A.2) for fixed $\nu (= 2, 4, \ldots)$ and given Θ_j determines the degeneracy of the excitations. However, in this way we only find the singlet excitations, as $M = 0$ for $N_- = \frac{N}{2}$.

To treat triplets, quintets etc. one must allow for $N_- < N/2$. In this case the appropriate relationship to an eight-vertex model is set up by scaling only N_- variables v_α in (1), (2), (4) of [8] with η whereas $\nu_2 := N/2 - N_-$ complex v_α's are kept finite. Of course, also the corresponding ν_2 string variables χ remain finite in the scaling limit and it is easily seen from (55) of [8] that they do not contribute to the right hand side of (A.2) after scaling. Thus (A.2) is also valid for the case $\nu_1 = \frac{\nu}{2} - \nu_2$ $(\nu_2 > 0)$.

The magnetization of the state corresponding to a particular distribution of v_α is just $M = \nu_2$. This state has maximum weight of the associated multiplet [15] which therefore has dimension $2\nu_2 + 1$. This together with (A.2) enables one to classify the excitations. For instance $\nu = 2$ allows (ν_1, ν_2) to be (0, 1) or (1, 0). In the first case (A.2) admits one solution $\chi = (\Theta_1 + \Theta_2)/2$ whereas in the second case there are no equations to be satisfied. Hence there is 1 singlet and 1 triplet for $\nu = 2$. If $\nu = 4$ three possible choices for (ν_1, ν_2) exist: (2, 0), (1, 1), (0, 2). In the first case (A.2) admits 2 solutions, in the second case 3 solutions, whereas in the last case there are no equations. Therefore one has 2 singlets, 3 triplets, and 1 quintet for $\nu = 4$. In both cases the degeneracy of the excitation function $l(v)$ is 2^ν, 4 for $\nu = 2$ and 16 for $\nu = 4$. As in [8] we conjecture that the result is true for general ν.

Appendix B

We investigate the conditions under which complex wave numbers k_j are allowed in the basic equations (19), (20). We consider one particular k^+ with positive imaginary part and the corresponding equation (19),

$e^{iLk^+} = \dots$. Since the left hand side vanishes exponentially for $L \to \infty$ like $O(e^{-L})$, so must the right hand side. As $\phi\left(\frac{U}{2}\right) = 0$ (according to (14)), we conclude that there should be another wave number k^- such that $i(\sin k^+ - \sin k^-) = \frac{U}{2} + O(e^{-L})$. Thus we may write

$$i \sin k^\pm = \lambda \pm \frac{U}{4} + O(e^{-L}). \tag{B.1}$$

Each pair $k^\pm$ is therefore characterized by one parameter λ. (Vice versa the same reasoning can be applied to k^- and its Eq. (19), $e^{iLk^-} = \dots$.)

In Sect. IV.d two integrals involving the quantities (B.1) as parameters are needed. These are

$$-\frac{U}{\pi} \int_{-\pi/2}^{\pi/2} \frac{\cos^2 k}{(\sin k - \sin k^+)(\sin k - \sin k^-)} dk = 2(\cos k^+ + \cos k^-) + U$$

$$\frac{i}{\pi} \int_{-\pi/2}^{\pi/2} \ln\left(\frac{\sin k - \sin k^+}{\sin k^- - \sin k}\right) dk = (k^+ + k^-) \bmod 2\pi. \tag{B.2}$$

Furthermore we note the relation

$$\phi(v - i \sin k^+) \cdot \phi(v - i \sin k^-) = -\frac{v - i \sin k^+}{v - i \sin k^-} \tag{B.3}$$

which follows from (B.1) and (14) explicitly.

If we consider several complex pairs $k_1^\pm, \dots, k_\mu^\pm$ with parameters $\lambda_1, \dots, \lambda_\mu$ we need equations to determine these parameters. We shall briefly indicate how these equations are obtained. We consider the most general excitation of the half-filled band, $N = L$. First we have spin excitations (discussed in IV.b)) characterized by the l-functions (18), secondly we insert H holes with "empty" positions $k_1^h, \dots, k_H^h$, thirdly we consider μ complex pairs $k^\pm$. For each pair we have to create two holes beforehand as a complex k replaces a real k_j. Thus we have $H \geqq 2\mu$ to satisfy our general condition $N \leqq L$ for the excitation case. The Eqs. (20) for a particular pair $k^\pm$ (with parameter λ) are then written as

$$e^{iLk^\pm} = \prod_{j=1}^{L} [i\phi(i \sin k^\pm - i \sin k_j)] \cdot \prod_{\alpha=1}^{\mu} i\phi(i \sin k^\pm - i \sin k_\alpha^+)\, i\phi(i \sin k^\pm - i \sin k_\alpha^-) \cdot \left[\prod_{\beta=1}^{H} i\phi(i \sin k^\pm - i \sin k_\beta^h)\right]^{-1} \cdot l\left(i \sin k^\pm + \frac{U}{4}\right) \tag{B.4}$$

where the L real wave numbers k_j again include also the H hole positions k_β^h. We multiply both Eqs. (B.4) and evaluate the products on the right hand side by using formulas of the foregoing sections (for instance $l^+ \cdot l^- = 1$ because of (16)). Omitting details (which can be found in [14]) we arrive at

$$\prod_{\beta=1}^{H} \frac{i \sin k_\beta^h - \lambda - U/4}{i \sin k_\beta^h - \lambda + U/4} = -\prod_{\alpha=1}^{\mu} \frac{\lambda_\alpha - \lambda - U/2}{\lambda_\alpha - \lambda + U/2}. \tag{B.5}$$

This set of equations, one for each λ, determines the λ_α parameters in terms of the hole positions k_β^h which therefore may be taken as free parameters.

References

1. Lieb, E.H., Wu, F.Y.: Phys. Rev. Lett. **20**, 1445 (1968)
2. Ovchinnikov, A.A.: Sov. Phys. JETP **30**, 1160 (1970)
3. Choy, T.C., Young, W.: J. Phys. C**15**, 521 (1982)
4. Coll, C.F.: Phys. Rev. B**9**, 2150 (1974)
5. Woynarovich, F.: J. Phys. C**15**, 85, 97 (1982)
6. Woynarovich, F.: J. Phys. C**16**, 5293 (1983)
7. Klümper, A., Zittartz, J.: Z. Phys. B – Condensed Matter **71**, 495 (1988)
8. Klümper, A., Zittartz, J.: Z. Phys. B – Condensed Matter **75**, 371 (1989)
9. Yang, C.N.: Phys. Rev. Lett. **19**, 1312 (1967)
10. Sutherland, B.: In: Lecture Notes in Physics. Vol. 242: Exactly solvable problems in condensed matter and relativistic field theory. Shastry, B.S., Jha, S.S., Singh, V. (eds.). Berlin, Heidelberg, New York, Tokyo: Springer 1985
11. Tsvelick, A.M., Wiegmann, P.B.: Adv. Phys. **32**, 435 (1983)
12. Baxter, R.J.: Exactly solved models in statistical mechanics. London: Academic Press 1982
13. Klümper, A., Schadschneider, A., Zittartz, J.: Z. Phys. B – Condensed Matter **76**, 247 (1989)
14. Schadschneider, A.: Diplom-Arbeit, Universität zu Köln 1988 (unpublished)
15. Takhtajan, L.A.: In: Lecture Notes in Physics. Vol. 242: Exactly solvable problems in condensed matter and relativistic field theory. Shastry, B.S., Jha, S.S., Singh, V. (eds.). Berlin, Heidelberg, New York, Tokyo: Springer 1985

A. Klümper, A. Schadschneider, J. Zittartz
Institut für Theoretische Physik
Universität zu Köln
Zülpicher Strasse 77
D-5000 Köln 41
Federal Republic of Germany

1098

Progress of Theoretical Physics, Vol. 42, No. 5, November 1969

Magnetization Curve for the Half-Filled Hubbard Model

Minoru TAKAHASHI

Department of Physics, University of Tokyo, Tokyo

(Received June 3, 1969)

The lowest energies of the one-dimensional half-filled Hubbard model as a function of the total spin are computed from the exact solution of one-dimensional interacting fermions. The lowest energies are used to calculate the magnetization curve at zero temperature. In the case of repulsive interaction our work is an extension of Griffiths' theory for the Heisenberg model to the itinerant electron model of antiferromagnet. In the case of attractive interaction it is shown that the magnetic susceptibility in zero field vanishes.

§ 1. Introduction

The half-filled Hubbard model has properties similar to the antiferromagnetic Heisenberg model. Griffiths[1] calculated the magnetization curve for the Heisenberg linear chain at zero temperature. He used the lowest energies computed from Hulthen's integral equation as a function of the total spin. In this paper we apply his method to the one-dimensional Hubbard model.[2] We assume that the electrons can hop between the Wannier states of neighbouring lattice sites. The Hamiltonian is

$$\mathcal{H} = -T\sum_{\langle ij\rangle}\sum_{\sigma} c_{i\sigma}^{\dagger}c_{j\sigma} + I\sum_{i=1}^{N_a} c_{i\uparrow}^{\dagger}c_{i\uparrow}c_{i\downarrow}^{\dagger}c_{i\downarrow}\,; \quad T>0\,, \tag{1}$$

where N_a is the number of the lattice sites, $c_{i\sigma}^{\dagger}$ and $c_{i\sigma}$ are, respectively, the creation and annihilation operators of an electron of spin σ in the Wannier states at the i-th lattice site. In the case of finite N_a the eigenvalue problem of this Hamiltonian can be reduced to a set of coupled algebraic equations. From these equations Lieb and Wu[3] derived coupled integral equations which determine the exact lowest energy of the infinite system in the case $I>0$.

In § 2 we compute the lowest energies of the half-filled system with positive I using their integral equations. In § 3 we derive coupled integral equations and compute the lowest energies in the case $I<0$. In § 4 magnetization curve and magnetic susceptibility at zero temperature are calculated from the results obtained in §§ 2 and 3. We show that the magnetic susceptibility is given by $(g^2\mu^2/\pi^2)I/4T^2$ for $I>0$ and zero for $I<0$ where μ is the Bohr magneton and g the electron g factor.

§ 2. The Lieb-Wu integral equations

Let m and m' be the densities of up-spin electron and down-spin electron, respectively:

$$m=\sum_{i=1}^{N_a} c_{i\uparrow}^{\dagger}c_{i\uparrow}/N_a\,, \quad m'=\sum_{i=1}^{N_a} c_{i\downarrow}^{\dagger}c_{i\downarrow}/N_a\,. \tag{2}$$

For the sake of simplicity we define the normalized lowest energy as a function of m, m' and $U=I/T$ as

$$f(m, m'; U)=\lim_{N_a\to\infty}\frac{1}{N_a T}(\text{the lowest eigenvalue of } \mathcal{H} \text{ for fixed } m \text{ and } m'). \tag{3}$$

Lieb and Wu[3] derived integral equations which determine the lowest energies of the Hamiltonian (1) for $I>0$. Their equations contain two functions $\rho(k)$ and $\sigma(\Lambda)$:

$$2\pi\rho(k)=1+\cos k\int_{-B}^{B}\frac{8U\sigma(\Lambda)d\Lambda}{U^2+16(\Lambda-\sin k)^2}\,,$$

$$2\pi\sigma(\Lambda)+\int_{-B}^{B}\frac{4U\sigma(\Lambda')d\Lambda'}{U^2+4(\Lambda-\Lambda')^2}=\int_{-Q}^{Q}\frac{8U\rho(k)dk}{U^2+16(\Lambda-\sin k)^2}\,. \tag{4}$$

Here parameters B and Q are determined from the conditions

$$n=\int_{-Q}^{Q}\rho(k)dk \quad \text{and} \quad m=\int_{-B}^{B}\sigma(\Lambda)d\Lambda\,, \tag{5}$$

where n is the electron density. The normalized lowest energy is given by

$$f(m, n-m; U)=-2\int_{-Q}^{Q}\cos k\rho(k)dk\,. \tag{6}$$

In the half-filled case (i.e. $n=1$), we find that $Q=\pi$. Then we can eliminate $\rho(k)$ in Eqs. (4) and (6) and obtain the integral equation for the half-filled case:

$$2\pi\sigma(\Lambda)+\int_{-B}^{B}\frac{4U\sigma(\Lambda')d\Lambda'}{U^2+4(\Lambda-\Lambda')^2}=g_0(\Lambda)\,, \tag{7}$$

$$f(\tfrac{1}{2}-s, \tfrac{1}{2}+s; U)=-\int_{-B}^{B}g_1(\Lambda)\sigma(\Lambda)d\Lambda\,, \tag{8}$$

where

$$g_0(\Lambda)=\frac{1}{2\pi}\int_{-\pi}^{\pi}\frac{8Udk}{U^2+16(\Lambda-\sin k)^2}\,, \quad g_1(\Lambda)=\int_{-\pi}^{\pi}\frac{dk}{\pi}\frac{8U\cos^2 k}{U^2+16(\Lambda-\sin k)^2}\,.$$

It is fortunate that the equation of this type has appeared in the study of an antiferromagnetic linear chain. Here we use Griffiths' method with slight modification.

By Fourier transform of (7) one obtains

$$\sigma(\Lambda)=\sigma_0(\Lambda)+\int_{|\Lambda'|>B}\frac{4}{U}R\Big(\frac{4}{U}(\Lambda-\Lambda')\Big)\sigma(\Lambda')d\Lambda'\,, \tag{9}$$

where

$$R(x)=\frac{1}{2\pi}\int_{-\infty}^{\infty}\frac{e^{i\omega x}d\omega}{e^{2|\omega|}+1}=\frac{1}{4\pi}\int_{-\infty}^{\infty}\frac{\operatorname{sech}(\pi y/2)\,dy}{1+(x+y)^2},$$

$$\sigma_0(\Lambda)=\frac{1}{U}\int_{-\pi}^{\pi}\frac{dk}{2\pi}\operatorname{sech}\frac{2\pi(\Lambda-\sin k)}{U}.$$

It is clear that $\sigma_0(\Lambda)$ is the solution for $B=\infty$. Magnetization and energy are given by

$$s=-s_z=\frac{1}{2}-\int_{-B}^{B}\sigma(\Lambda)\,d\Lambda=\int_{B}^{\infty}\sigma(\Lambda)\,d\Lambda \tag{10a}$$

and

$$f(\tfrac{1}{2}-s,\tfrac{1}{2}+s;U)-f(\tfrac{1}{2},\tfrac{1}{2};U)=-\int_{-B}^{B}\sigma(\Lambda)g_1(\Lambda)\,d\Lambda+\int_{-\infty}^{\infty}\sigma_0(\Lambda)g_1(\Lambda)\,d\Lambda$$

$$=4\pi\int_{B}^{\infty}\sigma_1(\Lambda)\sigma(\Lambda)\,d\Lambda, \tag{10b}$$

respectively, where

$$\sigma_1(\Lambda)=\frac{2}{U}\int_{-\pi}^{\pi}\frac{dk}{2\pi}\cos^2 k\operatorname{sech}\frac{2\pi(\Lambda-\sin k)}{U}.$$

Putting $x=4\Lambda/U$, $\alpha=4B/U$ and $p(x)=(U/2)\,e^{2\pi B/U}\sigma(4(x+\alpha)/U)$, we obtain for $B\gg\mathrm{Max}(U,1)$

$$p(x)-\int_{0}^{\infty}\{R(x-y)+R(x+y+2\alpha)\}p(y)\,dy=e^{-\pi x/2}+O(e^{-3\pi x/2-\pi\alpha}), \tag{11}$$

$$s=\frac{1}{2}e^{-\pi\alpha/2}\int_{0}^{\infty}p(x)\,dx=e^{-\pi\alpha/2}\Big[a_0+O\Big(\frac{1}{\alpha}\Big)\Big],$$

and

$$\delta f=\frac{4\pi}{U}e^{-\pi\alpha}\int_{0}^{\infty}e^{-\pi x/2}p(x)\,dx=\frac{4}{U}e^{-\pi\alpha}\Big[b_0+O\Big(\frac{1}{\alpha^2}\Big)\Big].$$

C. N. Yang and C. P. Yang[4] proved the relation $b_0/a_0^2=\pi^2/2$. Then we obtain

$$\delta f=\frac{2\pi^2}{U}s^2\Big[1+O\Big(\frac{1}{\ln|s|}\Big)\Big], \tag{12}$$

for $|s|\ll\mathrm{Min}(e^{-2\pi/U},e^{-2\pi})$.

When $B\ll\mathrm{Min}(1,U)$, we can expand f and m as series in powers of B by the use of Eq. (7). Eliminating B in the equations obtained we find a power series expansion of function f with respect to the variable m for $0\leqq m\ll\mathrm{Min}(U,1)$:

$$f(m,1-m;U)=-4\Big(\sqrt{1+\Big(\frac{U}{4}\Big)^2}-\frac{U}{4}\Big)m+\frac{\pi^2}{6}\frac{1}{\sqrt{1+(U/4)^2}}m^3+O(m^4). \tag{13}$$

At $U=0+$ we can solve (7) and (8) analytically. The result is

$$f(\tfrac{1}{2}-s,\ \tfrac{1}{2}+s;0+) = -\frac{4}{\pi}\cos \pi s\,, \tag{14}$$

which coincides with the result obtained from the band theory of electrons.

§ 3. Case of attractive interaction

Coupled integral equations (4), (5) and (6) are not applicable to the case $I<0$. But if one exchanges particles and holes in down spin-band, one obtains

$$f(x, y; U) = f(x, 1-y; -U) + xU\,. \tag{15}$$

From this relation we have derived the integral equations which determine the lowest energy in the case $I<0$. For the sake of clarity we replace $\rho(k)$ by $\rho(\pi-k)$. Our integral equations are

$$2\pi\rho(k) = 1-\cos k \int_{-B}^{B} \frac{8|U|\sigma(\Lambda)d\Lambda}{U^2+16(\Lambda-\sin k)^2} \tag{16a}$$

and

$$2\pi\sigma(\Lambda) + \int_{-B}^{B} \frac{4|U|\sigma(\Lambda')d\Lambda'}{U^2+4(\Lambda-\Lambda')^2} = \mathrm{Re}\frac{2}{\sqrt{1-(\Lambda-Ui/4)^2}} - \int_{-Q}^{Q} \frac{8|U|\rho(k)dk}{U^2+16(\Lambda-\sin k)^2}\,. \tag{16b}$$

The parameters B and Q are determined by the conditions

$$n = 2\int_{-B}^{B}\sigma(\Lambda)d\Lambda + \int_{-Q}^{Q}\rho(k)dk \quad \text{and} \quad m = \int_{-B}^{B}\sigma(\Lambda)d\Lambda\,. \tag{17}$$

Thus the lowest energy is given by

$$f = -4\int_{-B}^{B}\mathrm{Re}\sqrt{1-\left(\Lambda-\frac{Ui}{4}\right)^2}\,\sigma(\Lambda)d\Lambda - 2\int_{-Q}^{Q}\cos k\rho(k)dk\,. \tag{18}$$

These equations correspond to Gaudin's integral equations[5] for electron gas with attractive interaction of a delta function type in the continuum. One can also derive these equations from the theory of the finite system directly. But we shall not give the details here. In Eqs. (16), (17) and (18), we can regard $\rho(k)$ and $\sigma(\Lambda)$ as the distribution functions of the unpaired electrons and electron pairs, respectively. One can easily prove the following properties.

i) $m=0,\ n=1$ at $Q=\pi$.

ii) $n=1$ at $B=\infty$.

iii) $m/n=1/2$ at $Q=0$.

For $B=\infty$ and $Q=0$, i.e. the ground state of the half-filled case, we can solve these equations analytically with the results

$$\sigma(\Lambda) = \frac{1}{|U|}\int_{-\pi}^{\pi}\frac{dk}{2\pi}\,\mathrm{sech}\,\frac{2\pi(\Lambda-\sin k)}{U},$$

$$\rho(k) = \rho_0(k) \equiv \frac{1}{2\pi} - \frac{\cos k}{2\pi}\int_{-\pi}^{\pi} dk\,\frac{4}{|U|}R\left(\frac{4(\sin k - \sin k')}{U}\right)$$

and

$$f(\tfrac{1}{2}, \tfrac{1}{2};\, U) = -\frac{|U|}{2} - 4\int_0^{\infty}\frac{J_0(\omega)J_1(\omega)\,d\omega}{\omega(1+\exp|U\omega/2|)}, \tag{19}$$

where J_0 and J_1 are Bessel functions. For $B=\infty$, $\sigma(\Lambda)$ can be eliminated in the above equations and consequently one obtains

$$\rho(k) = \rho_0(k) + \cos k\int_{-Q}^{Q}\frac{4}{|U|}R\left(\frac{4(\sin k - \sin k')}{U}\right)\rho(k')\,dk' \tag{20a}$$

or

$$\rho(k) = \frac{1}{2\pi} - \cos k\int_{\pi>|k'|>Q}\frac{4}{|U|}R\left(\frac{4(\sin k-\sin k')}{U}\right)\rho(k')\,dk', \tag{20b}$$

$$s = \frac{1}{2}\int_{-Q}^{Q}\rho(k)\,dk, \tag{21}$$

and

$$f(\tfrac{1}{2}-s, \tfrac{1}{2}+s;\, U) = f(\tfrac{1}{2}, \tfrac{1}{2};\, U) + \int_{-Q}^{Q}\rho_1(k)\rho(k)\,dk, \tag{22}$$

where

$$\rho_1(k) = -2\cos k + \frac{|U|}{2} + 2\int_{-\pi}^{\pi}\frac{4}{|U|}R\left(\frac{4(\sin k-\sin k')}{U}\right)\cos^2 k'\,dk'.$$

From this we obtain

$$f(\tfrac{1}{2}-s, \tfrac{1}{2}+s;\, U) = f(\tfrac{1}{2}, \tfrac{1}{2};\, U + 2\rho_1(0)s + \frac{1}{3}\frac{\rho_1''(0)}{\{\rho_0(0)\}^3}s^3 + O(s^4) \tag{23}$$

at $0 \leqq s \ll \mathrm{Min}(U^2, \frac{1}{2})$. We can easily show that $\rho_1(0)>0$. This means that the lowest energy curve has a cusp at $s_z=0$.

From Eq. (20b), we obtain a power series expansion of f at $0 \leqq m \ll \mathrm{Min}(\frac{1}{2}, |U|)$ as follows:

$$f(m, 1-m;\, U) = -(4+|U|)m + \frac{8\pi^2}{3}m^3 + O(m^4). \tag{24}$$

§ 4. Magnetization curve and magnetic susceptibility

We discuss the case where a uniform magnetic field H is applied in the direction of the spin quantization axis. The Zeeman energy

$$\mathcal{H}_z=\tfrac{1}{2}g\mu H\sum_{i=1}^{N_a}(c_{i\uparrow}^{\dagger}c_{i\uparrow}-c_{i\downarrow}^{\dagger}c_{i\downarrow})$$

commutes with the Hamiltonian (1). Under the condition that the magnetization per site is $s=-s_z$, the lowest energy of the system is given by

$$E_{\text{total}}=N_a[Tf(\tfrac{1}{2}-s,\tfrac{1}{2}+s;U)-g\mu sH]. \tag{25}$$

This is minimized by the s-value which satisfies

$$g\mu H=T\frac{d}{ds}f(\tfrac{1}{2}-s,\tfrac{1}{2}+s;U). \tag{26}$$

Therefore magnetic susceptibility in zero field becomes

$$\chi=g\mu\frac{ds}{dH}\bigg|_{H=0}=g^2\mu^2\times\left(T\frac{d^2}{ds^2}f(\tfrac{1}{2}-s,\tfrac{1}{2}+s;U)\bigg|_{s=0}\right)^{-1}. \tag{27}$$

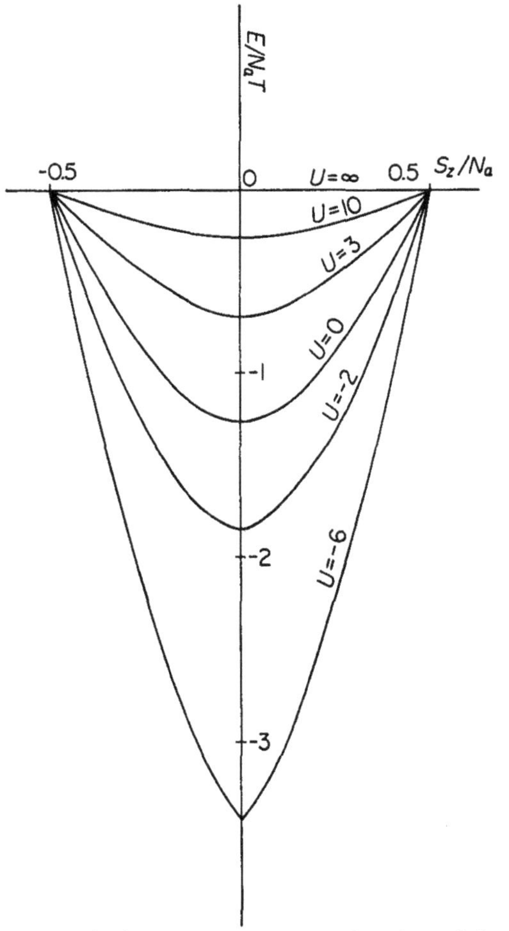

Fig. 1. The lowest energies as a function of the magnetization.

The integral equations (7) and (8) are approximated by a set of 40 coupled linear algebraic equations and Eqs. (20) ~ (22) are approximated by a set of 20 coupled linear algebraic equations. They were solved with the use of HITAC 5020 at the computer center, University of Tokyo. Figure 1 shows the value of f calculated as a function of s_z. The derivative df/ds yields the magnetization curve.

The magnetic susceptibility at zero temperature and in zero field can be calculated from Eqs. (12), (14) and (23), in terms of the relation (27). For $I>0$ we obtain

$$\chi=\frac{g^2\mu^2}{\pi^2}\cdot\frac{I}{4T^2}. \tag{28}$$

This is consistent with Griffiths' result[1] for the antiferromagnetic linear chain of spin $\frac{1}{2}$ described by the Hamiltonian $\mathcal{H}=J\sum S_i S_{i+1}$.

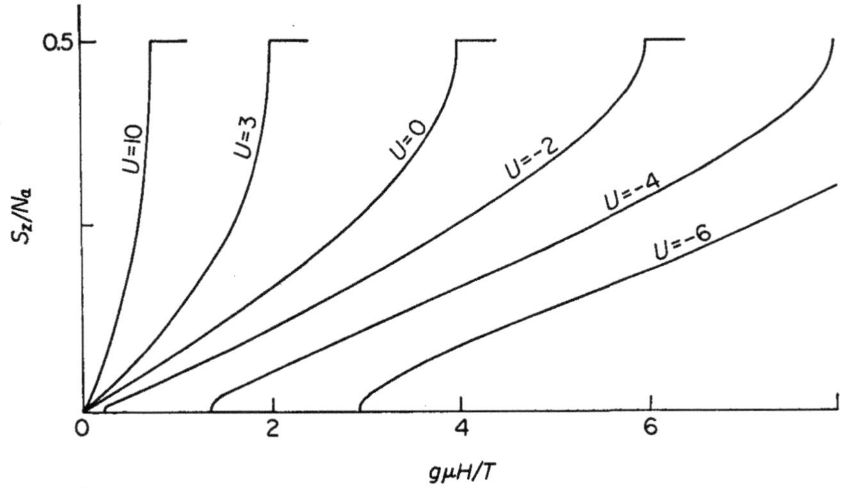

Fig. 2. The magnetization as a function of magnetic field for the one-dimensional half-filled Hubbard model at zero temperature.

Taking the canonical transformation*) of (1) in the half-filled case we obtain an effective Hamiltonian

$$\mathcal{H}_{\text{eff}}=4T^2/I\sum\left(s_i s_{i+1}-\frac{1}{4}\right)+O\left(\frac{T^3}{I^2}\right).$$

Then we see that the Heisenberg model is equivalent to the half-filled Hubbard model in the limit $U=\infty$. The difference is that the result (28) holds even at $U=$ finite.

For $I=0$ we obtain easily from (14)

$$\chi=\frac{g^2\mu^2}{4\pi T}\,. \tag{29}$$

For $I<0$ no magnetization appears under the magnetic field in the range

$$H<H_c\equiv 2T\rho_1(0)/g\mu\,. \tag{30}$$

Then the magnetic susceptibility in zero field vanishes. A peculier feature of the magnetization curve is that it has an infinite slope at $H=H_c+0$.

§ 5. Discussion

The discontinuity of the magnetic susceptibility at $I=0$ suggests that the perturbational treatment of the interaction term $I\sum c_{i\uparrow}^\dagger c_{i\uparrow}c_{i\downarrow}^\dagger c_{i\downarrow}$ is not a good approximation at least in the half-filled case. This is also justified by the fact that the ground state energy is not analytic at $I=0$. The analytic properties of the ground state energy as a function of U has been investigated in reference 6).

*) See for example, W. Kohn, Phys. Rev. **133** (1964), A 171, §3.

Acknowledgement

The auther would like to express his sincere thanks to Professor M. Suzuki for valuable discussions at the final stage of the present work.

References

1) R. B. Griffiths, Phys. Rev. **133** (1964), A 768.
2) J. Hubbard, Proc. Roy. Soc. **A276** (1963), 238; **277** (1964), 237.
3) E. H. Lieb and F. Y. Wu, Phys. Rev. Letters **20** (1968), 1445.
4) C. N. Yang and C. P. Yang, Phys. Rev. **150** (1966), 327.
5) M. Gaudin, Phys. Letters **24A** (1967), 55; thesis, University of Paris, 1967 (unpublished).
6) M. Takahashi, thesis, University of Tokyo, 1969 (unpublished).

Erratum

Magnetization Curve for the Half-Filled Hubbard Model

Minoru TAKAHASHI

Prog. Theor. Phys. 42 (1969), 1098

In the calculations from Eq. (11) to Eq. (12) the factors of modified Bessel functions are missed. Then Eq. (12) should read

$$\delta f = 2\pi s^2\left[1+O\left(\frac{1}{\ln|s|}\right)\right] I_1\left(\frac{2\pi}{U}\right) \Big/ I_0\left(\frac{2\pi}{U}\right).$$

Therefore expression of magnetic susceptibility for $I>0$ in §1 and in Eq. (28) should be replaced by $g^2\mu^2 I_0(2\pi/U)/4\pi T I_1(2\pi/U)$. This continues to Eq. (29) at the limit $U\to+0$.

Z. Phys. B – Condensed Matter 85, 269–280 (1991)

Condensed Zeitschrift für Physik B Matter

Novel magnetic properties of the Hubbard chain with an attractive interaction

F. Woynarovich and K. Penc

Central Research Institute for Physics, P.O.B. 49, H-1525 Budapest 114, Hungary

Received June 17, 1991

The magnetic properties of an attractive Hubbard chain are considered. Based on the Bethe Ansatz equations of the problem, exact analytic expressions are derived for the magnetization and susceptibility. These formulae can be evaluated after solving certain "derivatives" of the Bethe Ansatz equations. These derivative equations are also given. We give the magnetization and susceptibility curves for several values of the interaction-strength and bandfilling. We find that the susceptibility at the onset of magnetization (at the critical field) is *finite* for all bandfillings, except for the cases of half filled and empty bands, and in the limit of vanishing interaction. We argue that the finiteness of the initial susceptibility is due to the fermion-like behavior of the bound pairs. We also give the gap (what is equal to the critical field) and the initial susceptibility as functions of the interaction-strength and bandfilling for the cases of nearly half filled and almost empty bands as a function of the interaction, and in the weak coupling limit as a function of the bandfilling. To our knowledge, this is the first Bethe Ansatz calculation for the gap in this latter limit.

I. Introduction

The one-dimensional (1-d) Hubbard model is one of the simplest model of an interacting electron gas on a lattice. It has attracted a great deal of interest, since – in spite of its simplicity – it is able to give a correct account of the characteristic features of a strongly correlated electron system. Among the related models it is special and more intensively studied, since it can be treated exactly [1] in terms of the Bethe Ansatz. Recently a new impetus has been given to the study of the Hubbard chain [2] by the hope, that it may have relevance in the understanding of the high T_c superconductivity.

The Hamiltonian of a Hubbard chain of length N in a magnetic field h is given by

$$\hat{H} = -\sum_{i=1}^{N} \sum_{\sigma} (c_{i,\sigma}^{+} c_{i+1,\sigma} + \text{h.c.}) + U \sum_{i=1}^{N} n_{i,\uparrow} n_{i,\downarrow} - \frac{h}{2} \sum_{i=1}^{N} (n_{i,\uparrow} - n_{i,\downarrow}). \tag{1.1}$$

Here the hopping integral (the coefficient of the first term) is taken to be unity, and $c_{i,\sigma}^{+}$, $c_{i,\sigma}$ and $n_{i,\sigma}$ are the creation, destruction and number operators of an electron with spin σ at the site i. Lieb and Wu have shown that the diagonalization of this Hamiltonian is equivalent to solving a set of nonlinear equations [1], the so called Lieb-Wu equations, which are the Bethe Ansatz equations (BAE) of the problem. Most of the theoretical work concerning the zero temperature properties of the model has been based on the different solutions of these equations. The spectrum of the excitations has been investigated by Ovchinnikov [3], Coll [4], Choy and Young [5], one of the present authors [6–8], and by Schadschneider, Klümper, and Zittartz [9, 10]. Also the zero-temperature magnetic properties have been studied by several authors. Takahashi [11] gave the magnetization curve for the half-filled band for both positive and negtive U. Extending this work, Shiba [12] calculated the zero-field magnetic susceptibility for an arbitrary concentration of electrons, but for repulsive (positive) U only. The magnetization curves for different electron densities at attractive (negative) U have been first studied by Bahder and Woynarovich [13]. The present study is an extension of this latter work, and it also completes a recent comment [14] on a work of Lee and Schlottmann [15].

As it is well known, in the ground state of the attractive Hubbard chain the electrons form singlet bound pairs [8], so the system has no magnetic response below a critical field. Above the critical field bound pairs are broken up and "free" (interacting but not bound) electrons with uncompensated spins are created. With increasing magnetic field the number of pairs is decreased while the number of uncompensated spins is increased. At a second critical field, which we call saturation-field, also the last pairs are broken up, and all the spins are aligned in one direction. A special character is given to this behavior by the fact, that in this system not only the free electrons, but also the bound pairs follow Fermi-like distribution (although they are bosons).

To describe the above behavior quantitatively, one has to solve the BAE, calculate the energy and minimize it with respect to the magnetization. All the earlier studies [11–13, 15] performed this program *numerically*. In the present work we derive *exact analytic* expressions for the magnetization and the susceptibility: we express these quantities by the different densities, dressed energies and dressed charges taken at the Fermi-points. The quantities entering into the magnetization and susceptibility expressions can be given after solving a set of equations derived from the BAE. This way one can discuss the magnetization and the susceptibility analytically, and in the most interesting limits (high and low density of electrons; or vanishing interaction) the critical fields as well as the susceptibility at the critical fields can be given even in closed form.

The major result of this analysis is that the susceptibility at the onset of magnetization – contrary to the naïve expectations – is always *finite* except in the above mentioned limits. This novel behavior – as we argue – is a consequence of the Fermi like distribution of the bound pairs. In a recent work based on numerical study Lee and Schlottmann [15] claimed that the initial susceptibility is infinite for all bandfillings. By giving the explicit susceptibility curves we demonstrate, that as the critical field is approached from above, the susceptibility shows a several order of magnitude enhancement which in a numerical analysis can be easily mistaken for a divergence.

The paper is organized as follows. In Chapt. II we summarize the Lieb-Wu equations on which we base our analysis, and explain the formalism used in the paper. Chapter III is devoted to the derivation of the magnetization, and the susceptibility. The behavior at the onset of magnetization is discussed in detail in Chapt. IV while the saturation field and susceptibility are given in Chapt. V.

II. Lieb-Wu equations and formalism

The Lieb-Wu equations

In the Bethe Ansatz solution of the Hubbard model the electrons are characterized by a set of wavenumbers k_j and in addition to this the down spins have parameters λ_α called rapidities. These quantities are connected by a set of nonlinear equations derived by Lieb and Wu [1]

$$N k_j = 2\pi I_j - \sum_{\alpha=1}^{M} 2\tan^{-1}\frac{\sin k_j - \lambda_\alpha}{U/4}, \tag{2.1a}$$

$$\sum_{j=1}^{N_e} 2\tan^{-1}\frac{\lambda_\alpha - \sin k_j}{U/4} = 2\pi J_\alpha + \sum_{\beta=1}^{M} 2\tan^{-1}\frac{\lambda_\alpha - \lambda_\beta}{U/2}. \tag{2.1b}$$

Here N_e is the number of electrons, M is the number of down spins, i.e. the magnetization $S=(N_e/2-M)$, and the quantum numbers I_j and J_α are integers or half-odd-integers depending on the parities of N_e and M. The energy and the momentum of the corresponding state is given by

$$E = -\sum_{j=1}^{N_e} 2\cos k_j - h(N_e/2 - M), \tag{2.2}$$

$$P = \sum_{j=1}^{N_e} k_j. \tag{2.3}$$

Equations (2.1), (2.2) and (2.3) hold regardless of the sign of U, nevertheless the structure of the solutions is very different for $U>0$ and $U<0$. Here we treat the simplest solutions with nonzero magnetization for negative U. In the corresponding states there are a number of singlet bound pairs and a number of electrons with uncompensated up-spins. The bound pairs are characterized [8] by a pair of complex wavenumbers $k_\alpha^\pm$ and a rapidity λ_α. These three quantities are connected through the relation

$$\sin k_\alpha^\pm = \lambda_\alpha \pm i u \tag{2.4}$$

which is accurate up to a correction exponentially small in N, and where

$$u = |U|/4. \tag{2.5}$$

The electrons with uncompensated spins are described by real k_j satisfying (2.1.a) which we write in the form

$$2\pi I_j' = N k_j - \sum_{\alpha=1}^{M} 2\tan^{-1}\frac{\sin k_j - \lambda_\alpha}{u}. \tag{2.6a}$$

Using the relation (2.4) the $k_\alpha^\pm$ can be eliminated from the system (2.1 a, b) and we have

$$2\pi J_\alpha' = 2N\,\mathrm{Re}(\sin^{-1}(\lambda_\alpha - iu)) - \sum_{j=1}^{N_e'} 2\tan^{-1}\frac{\lambda_\alpha - \sin k_j}{u} - \sum_{\beta=1}^{M} 2\tan^{-1}\frac{\lambda_\alpha - \lambda_\beta}{2u} \tag{2.6b}$$

with N_e' being the the number of electrons with uncompensated up spins, i.e. the number of real k_j: $N_e' = N_e - 2M = 2S$. The energy of the state given by (2.6a, b) is found by substituting (2.4) into (2.2):

$$E = -\sum_j \left(2\cos k_j + \frac{h}{2}\right) - \sum_\alpha 4\,\mathrm{Re}\sqrt{1-(\lambda_\alpha - iu)^2}. \tag{2.7}$$

At a given S for the lowest energy state I_j' and J_α' are consecutive integers or half-odd-integers centered around the origin (the fact that all J_α' should be different [1] gives a fermionic character to the bound pairs). In the thermodynamic limit $N_e/N \to n$, $S/N \to s$ the k and λ will be distributed in the intervals $(Q; -Q)$ and $(B; -B)$ with densities $\varrho_1(k)$ and $\varrho_2(\lambda)$, respectively (ϱ_1 is defined so that $N\varrho_1(k)\mathrm{d}k$ gives the number of k_j in the interval

$(k; k+\mathrm{d}k)$, and a similar definition holds for ϱ_2). The equations for these densities can be derived by a standard procedure from (2.6a, b):

$$\rho_1(k)=\frac{1}{2\pi}-\frac{1}{2\pi}\cos k \int_{-B}^{B} K_1(\sin k-\lambda')\rho_2(\lambda')\mathrm{d}\lambda' \quad (2.8\mathrm{a})$$

and

$$\rho_2(\lambda)=\frac{1}{2\pi}\,\mathrm{Re}\frac{2}{\sqrt{1-(\lambda-\mathrm{i}u)^2}} -\frac{1}{2\pi}\int_{Q}^{Q} K_1(\lambda-\sin k')\rho_1(k')\mathrm{d}k' -\frac{1}{2\pi}\int_{-B}^{B} K_2(\lambda-\lambda')\rho_2(\lambda')\mathrm{d}\lambda' \quad (2.8\mathrm{b})$$

with

$$K_\nu(x)=2\frac{\nu u}{(\nu u)^2+x^2}. \quad (2.9)$$

The integration limits (cut-offs) are determined by the conditions

$$\int_{-Q}^{Q} \rho_1(k)=2s, \qquad \int_{-B}^{B} \rho_2(\lambda)=\frac{n}{2}-s. \quad (2.10)$$

The energy density in this thermodynamic limit is

$$\frac{E}{N}=-\int_{-Q}^{Q}\left(2\cos k+\frac{h}{2}\right)\rho_1(k) -\int_{-B}^{B} 4\,\mathrm{Re}\sqrt{1-(\lambda-\mathrm{i}u)^2}\,\rho_2(\lambda). \quad (2.11)$$

Equations (2.8)–(2.11) give the starting point of our analysis. In connection with them we note two things. First: they can be derived also from the $U>0$ equations using the particle-hole symmetry of the system [8, 13]. Second: a different but equivalent system of equations can be obtained by taking the $T\to 0$ limit of the finite temperature Bethe Ansatz equations [15].

Formalism

Since equations of the type (2.8a, b) with different inhomogeneous parts play a central role in the following analysis, we introduce a short-hand vector-matrix notation for them:

$$\mathbf{x}(k,\lambda)=\mathbf{x}_0(k,\lambda)+\mathbf{K}(k,\lambda|k',\lambda')\otimes\mathbf{x}(k',\lambda') \quad (2.12)$$

or just $\mathbf{x}=\mathbf{x}_0+\mathbf{K}\otimes\mathbf{x}$. Here $\mathbf{x}$ and $\mathbf{x}_0$ are column-vectors

$$\mathbf{x}(k,\lambda)=\begin{pmatrix}x_1(k)\\ x_2(\lambda)\end{pmatrix}, \qquad \mathbf{x}_0(k,\lambda)=\begin{pmatrix}x_{1,0}(k)\\ x_{2,0}(\lambda)\end{pmatrix}, \quad (2.13)$$

$\mathbf{K}$ is a matrix with integral operator elements

$$\mathbf{K}(k,\lambda|k',\lambda')=\frac{1}{2\pi}\begin{pmatrix}0 & -\cos k\,K_1(\sin k-\lambda')\\ -K_1(\lambda-\sin k') & -K_2(\lambda-\lambda')\end{pmatrix}, \quad (2.14)$$

and the product $\otimes$ means the usual matrix product *and* integration over the common variables, from $-Q$ to Q over the k type and from $-B$ to B over the λ type ones, respectively. An equation of the type (2.12) is solved by the vector

$$\mathbf{x}(k,\lambda)=\mathbf{x}_0(k,\lambda)+\mathbf{R}(k,\lambda|k',\lambda')\otimes\mathbf{x}_0(k',\lambda') \quad (2.15)$$

if the resolvent operator [16] $\mathbf{R}$ satisfies the relation

$$\mathbf{K}(k,\lambda|k',\lambda')+\mathbf{K}(k,\lambda|k'',\lambda'')\otimes\mathbf{R}(k'',\lambda''|k',\lambda') =\mathbf{R}(k,\lambda|k',\lambda'). \quad (2.16)$$

In the analysis also the following type of equations is of great importance:

$$\mathbf{y}(k,\lambda)=\mathbf{y}_0(k,\lambda)+\mathbf{K}^{\mathrm{T}}(k,\lambda|k',\lambda')\otimes\mathbf{y}(k',\lambda') \quad (2.17)$$

where $\mathbf{K}^{\mathrm{T}}$ is the transposed of $\mathbf{K}$, i.e. $\mathbf{K}^{\mathrm{T}}_{ij}(k,\lambda|k',\lambda')=\mathbf{K}_{ji}(k',\lambda'|k,\lambda)$. Using the symmetries of $\mathbf{K}$ it is not hard to convince ourselves, that the resolvent of (2.17) is the transposed of $\mathbf{R}$, i.e. if (2.16) is satisfied, then also

$$\mathbf{K}^{\mathrm{T}}(k,\lambda|l',\lambda')+\mathbf{K}^{\mathrm{T}}(k,\lambda|k'',\lambda'')\otimes\mathbf{R}^{\mathrm{T}}(k'',\lambda''|k',\lambda') =\mathbf{R}^{\mathrm{T}}(k,\lambda|k',\lambda') \quad (2.18)$$

holds, and

$$\mathbf{y}(k,\lambda)=\mathbf{y}_0(k,\lambda)+\mathbf{R}^{\mathrm{T}}(k,\lambda|k',\lambda')\otimes\mathbf{y}_0(k',\lambda'). \quad (2.19)$$

An important consequence of (2.15)–(2.19) is that the "scalar-product" of the vectors $\mathbf{y}_0$ and $\mathbf{x}$ has the property

$$(\mathbf{y}_0^{\mathrm{T}}(k,\lambda)\otimes\mathbf{x}(k,\lambda))=(\mathbf{x}_0^{\mathrm{T}}(k,\lambda)\otimes\mathbf{y}(k,\lambda)). \quad (2.20)$$

(Also here the upper index "T" means transposition, i.e. it makes a row-vector out of a column-vector and vice versa.)

Dressed energy and dressed charge

In our notation (2.8a, b) read

$$\rho=\rho_0+\mathbf{K}\otimes\rho \quad (2.21)$$

with

$$\rho(k,\lambda)=\begin{pmatrix}\rho_1(k)\\ \rho_2(\lambda)\end{pmatrix}, \qquad \rho_0(k,\lambda)=\frac{1}{2\pi}\begin{pmatrix}1\\ \mathrm{Re}\left(2/\sqrt{1-(\lambda-\mathrm{i}u)^2}\right)\end{pmatrix}. \quad (2.22)$$

If we introduce the vector

$$\epsilon_0(k,\lambda)=\begin{pmatrix}-(2\cos k+h/2)\\ -4\,\mathrm{Re}\sqrt{1-(\lambda-\mathrm{i}u)^2}\end{pmatrix}, \quad (2.23)$$

(2.11) can be written in a product form

$$E/N=(\epsilon_0^{\mathrm{T}}\otimes\rho), \quad (2.24)$$

which by (2.20) is equivalent to

$$E/N=(\rho_0^{\mathrm{T}}\otimes\epsilon), \quad (2.25)$$

where the vector ϵ is called the dressed energy, and it satisfies the equation

$$\epsilon=\epsilon_0+\mathbf{K}^{\mathrm{T}}\otimes\epsilon. \tag{2.26}$$

Defining the vectors $\xi^{(1)}$ and $\xi^{(2)}$ by

$$\xi^{(1,2)}=\xi_0^{(1,2)}+\mathbf{K}^{\mathrm{T}}\otimes\xi^{(1,2)} \tag{2.27}$$

with

$$\xi_0^{(1)}=\begin{pmatrix}1\\0\end{pmatrix} \quad\text{and}\quad \xi_0^{(2)}=\begin{pmatrix}0\\1\end{pmatrix}, \tag{2.28}$$

(2.10) read

$$2s=(\xi_0^{(1)\mathrm{T}}\otimes\rho)=(\rho_0^{\mathrm{T}}\otimes\xi^{(1)}),$$
$$\frac{n}{2}-s=(\xi_0^{(2)\mathrm{T}}\otimes\rho)=(\rho_0^{\mathrm{T}}\otimes\xi^{(2)}). \tag{2.29}$$

We call the 2×2 matrix ξ for which the first column is $\xi^{(1)}$ and the second one is $\xi^{(2)}$, i.e. which has the elements $\xi_{ij}=\xi_i^{(j)}$ the dressed charge matrix [17]. (In other works [18, 19] its transposed is defined as the dressed charge.)

III. Magnetization and susceptibility

Calculation of the magnetization

The magnetization is a constant of motion for the Hamiltonian (1.1), so one has to calculate the energy at a fixed magnetization s, and than minimize it with respect to s at a fixed particle density n. Since the energy depends directly on the cut-offs Q and B only, to minimize it one has to take the derivatives of E/N given by (2.25) with respect to Q and B, and use (2.29) to find out the derivatives of Q and B with respect to s at fixed n.

The energy density (2.25) depends on Q and B directly through the integration limits in the $\otimes$ product, and also implicitly (i.e. through the Q and B depence of the dressed energy ϵ). Thus:

$$\frac{\partial E/N}{\partial Q}=2\rho_{1,0}(Q)\epsilon_1(Q)+\left(\rho_0^{\mathrm{T}}\otimes\frac{\partial\epsilon}{\partial Q}\right),$$
$$\frac{\partial E/N}{\partial B}=2\rho_{2,0}(B)\epsilon_2(B)+\left(\rho_0^{\mathrm{T}}\otimes\frac{\partial\epsilon}{\partial \mathrm{B}}\right) \tag{3.1}$$

where it has been already used that ρ and ϵ are even functions. By taking the derivative (2.26) with respect to Q and B we find that

$$\frac{\partial\epsilon}{\partial Q}=\epsilon_1(Q)\mathbf{r}_0^{(1)}+\mathbf{K}^{\mathrm{T}}\otimes\frac{\partial\epsilon}{\partial Q},$$
$$\frac{\partial\epsilon}{\partial B}=\epsilon_2(B)\mathbf{r}_0^{(2)}+\mathbf{K}^{\mathrm{T}}\otimes\frac{\partial\epsilon}{\partial \mathrm{B}} \tag{3.2}$$

with

$$\mathbf{r}_0^{(1)}=-\frac{1}{2\pi}\begin{pmatrix}0\\K_1(\lambda-\sin Q)\cos Q+K_1(\lambda+\sin Q)\cos Q\end{pmatrix},$$
$$\mathbf{r}_0^{(2)}=-\frac{1}{2\pi}\begin{pmatrix}K_1(\sin k-B)+K_1(\sin k+B)\\K_2(\lambda-B)+K_2(\lambda+B)\end{pmatrix}. \tag{3.3}$$

Since these equations are linear, and $\mathbf{r}_0^{(1,2)}(k,\lambda)$ are closely related to the columns of $\mathbf{K}^{\mathrm{T}}(k,\lambda|\pm Q,\pm B)$, using (2.18) the solution of (3.2) can be expressed through the columns of $\mathbf{R}^{\mathrm{T}}(k,\lambda|\pm Q,\pm B)$, and after a straight-forward algebraic manipulation one arrives at

$$\frac{\partial E/N}{\partial Q}=2\rho_1(Q)\epsilon_1(Q),\quad \frac{\partial E/N}{\partial B}=2\rho_2(B)\epsilon_2(B). \tag{3.4}$$

The same way one can calculate the derivatives of n and s with respect to Q and B using (2.29), and one finds

$$2\frac{\partial s}{\partial Q}=2\rho_1(Q)\xi_{11}(Q),\quad \frac{1}{2}\frac{\partial n}{\partial Q}-\frac{\partial s}{\partial Q}=2\rho_1(Q)\xi_{12}(Q),$$
$$2\frac{\partial s}{\partial B}2\rho_2(B)\xi_{21}(B),\quad \frac{1}{2}\frac{\partial n}{\partial B}-\frac{\partial s}{\partial B}=2\rho_2(B)\xi_{22}(B). \tag{3.5}$$

These equations yield

$$\left.\frac{\partial Q}{\partial s}\right|_{n=\mathrm{const}}=\frac{\zeta_2(B)}{2\rho_1(Q)\det\xi(Q,B)},$$
$$\left.\frac{\partial B}{\partial s}\right|_{n=\mathrm{const}}=\frac{-\zeta_1(Q)}{2\rho_2(B)\det\xi(Q,B)} \tag{3.6}$$

where the vector ζ is a linear combination of $\xi^{(1)}$ and $\xi^{(2)}$

$$\zeta(k,\lambda)=\xi^{(1)}(k,\lambda)+2\xi^{(2)}(k,\lambda). \tag{3.7}$$

With (3.4) and (3.6) one has

$$\left.\frac{\partial E/N}{\partial s}\right|_{n=\mathrm{const}}=\frac{\epsilon_1(Q)\zeta_2(B)-\epsilon_2(B)\zeta_1(Q)}{\det\xi(Q,B)}, \tag{3.8}$$

i.e. at the optimal value of magnetization

$$\epsilon_1(Q)\zeta_2(B)-\epsilon_2(B)\zeta_1(Q)=0. \tag{3.9}$$

Since the vector ϵ contains h linearly, from (3.9) one can express h explicitly. Defining the vector ε by the equation

$$\varepsilon=\varepsilon_0+\mathbf{K}^{\mathrm{T}}\otimes\varepsilon \tag{3.10}$$

with

$$\varepsilon_0(k,\lambda)=\begin{pmatrix}-2\cos k\\-4\,Re\sqrt{1\text{-}(\lambda-iu)^2}\end{pmatrix}, \tag{3.11}$$

it is clear through (2.23) and (2.28), that

$$\epsilon=\varepsilon-\frac{h}{2}\xi^{(1)}. \tag{3.12}$$

Substituting this into (3.9) yields

$$h=\frac{\varepsilon_1(Q)\zeta_2(B)-\varepsilon_2(B)\zeta_1(Q)}{\det\xi(Q,B)}. \tag{3.13}$$

Equation (3.13) together with (2.29) are a parametric representation of the magnetization s as a function of h at fixed n: Equation (2.29) can be used to calculate

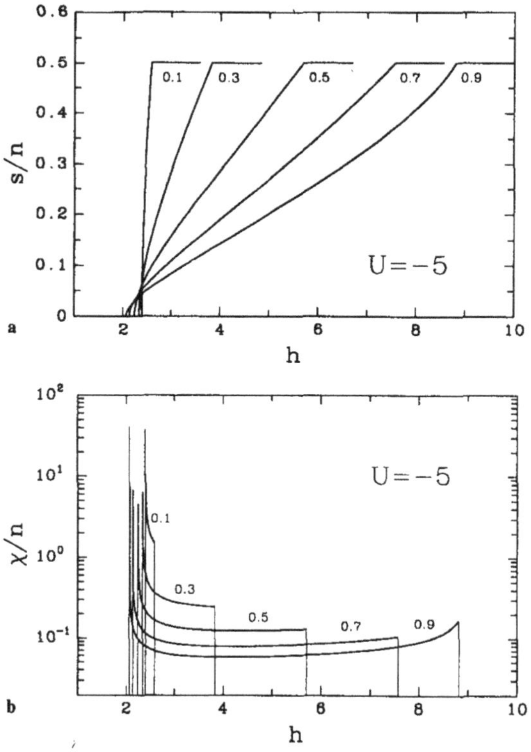

Fig. 1. Magnetization **a** and susceptibility **b** versus magnetic field for $U=-5$, at different bandfillings. The individual curves are labelled by the value of the bandfilling

Q and B as functions of s at fixed n, and then (3.13) gives the magnetic field h for which s is the optimal magnetization. This is the way as Figs. 1a and 2a which give the magnetization versus magnetic field at different bandfillings for $U=-5$ and $U=-1$ were obtained. All curves start at some critical value of the magnetic field h_c and reach $s/n=1/2$ at some saturation field h_s. Since at h_c there are no free particles yet, $s=0$ and according to (2.10) ((2.29)) h_c corresponds to $Q=0$. At h_s all the electrons are free, i.e. $n=2s$, thus for h_s $B=0$.

Derivation of the susceptibility

To calculate the susceptibility, we take the derivative of (3.13) with respect to Q and B, and then using (3.6) we obtain $\partial s/\partial h$. Taking the derivatives of ε with respect to the implicite Q and B dependence is done the same way as it has been done with ϵ (see (3.2)–(3.3)), leading to

$$\frac{\partial \varepsilon(k,\lambda)}{\partial Q}=\varepsilon_1(Q)\mathbf{r}^{(1)}(k,\lambda), \qquad \frac{\partial \varepsilon(k,\lambda)}{\partial B}=\varepsilon_2(B)\mathbf{r}^{(2)}(k,\lambda). \tag{3.14}$$

To calculate the derivative of ε with respect to its arguments we take the derivative of the first row of (3.10)

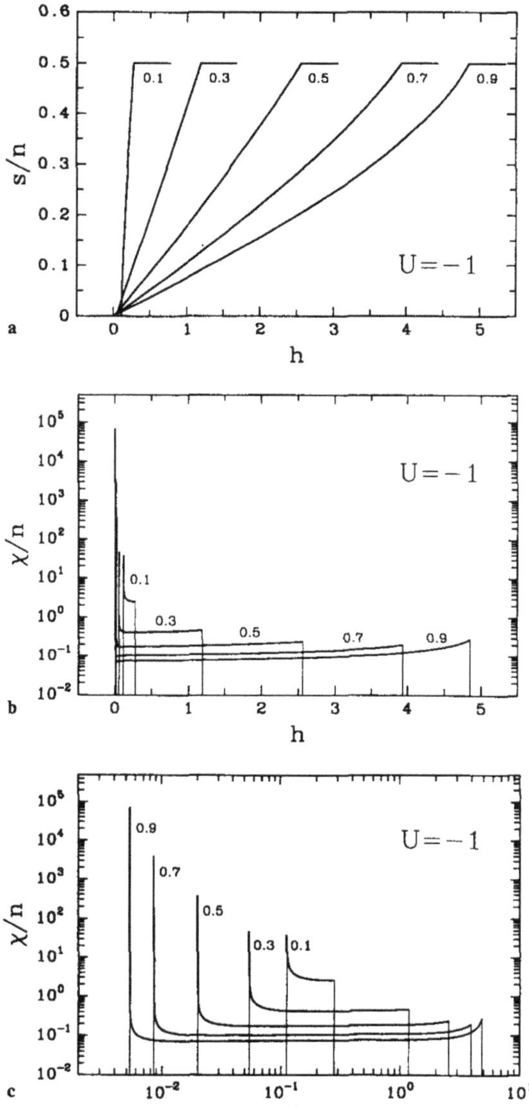

Fig. 2a–c. The same as Fig. 1 but for $U=-1$. **c** repeats **b** with a logarithmic scale for h

with respect to k, and that of the second row with respect to λ. After integration by parts we get

$$\varepsilon'=\mathbf{v}+\varepsilon_1(Q)\mathbf{u}+\varepsilon_2(B)\mathbf{w} \tag{3.15}$$

where

$$\varepsilon'=\begin{pmatrix}\partial\varepsilon_1(k)/\partial k\\ \partial\varepsilon_2(\lambda)/\partial\lambda\end{pmatrix} \tag{3.16}$$

and the vectors $\mathbf{v}$, $\mathbf{u}$ and $\mathbf{w}$ are defined by equations analogues to (2.12) with $\mathbf{x}_0$ replaced by

$$\mathbf{v}_0=\begin{pmatrix} 2\sin k \\ \mathrm{Re}\left(4(\lambda-iu)/\sqrt{1-(\lambda-iu)^2}\right)\end{pmatrix}, \tag{3.17}$$

$$\mathbf{u}_0=\frac{1}{2\pi}\begin{pmatrix} 0 \\ K_1(\lambda-\sin Q)-K_1(\lambda+\sin Q)\end{pmatrix} \tag{3.18}$$

and

$$\mathbf{w}_0=\frac{1}{2\pi}\begin{pmatrix} \cos k K_1(\sin k-B)-\cos k K_1(\sin k+B) \\ K_2(\lambda-B)-K_2(\lambda+B)\end{pmatrix} \tag{3.19}$$

respectively. This way the derivatives of the components of ε can be given:

$$\begin{aligned}
\frac{\partial \varepsilon_1(Q)}{\partial Q}&=v_1(Q)+\varepsilon_1(Q)u_1(Q)+\varepsilon_2(B)w_1(Q)+\varepsilon_1(Q)r_1^{(1)}(Q),\\
\frac{\partial \varepsilon_2(B)}{\partial Q}&=\varepsilon_1(Q)r_2^{(1)}(B),\\
\frac{\partial \varepsilon_1(Q)}{\partial B}&=\varepsilon_2(B)r_1^{(2)}(Q),\\
\frac{\partial \varepsilon_2(B)}{\partial B}&=v_2(B)+\varepsilon_1(Q)u_2(B)+\varepsilon_2(B)w_2(B)+\varepsilon_2(B)r_2^{(2)}(B).
\end{aligned} \tag{3.20}$$

In a similar way we can calculate the derivatives of $\xi^{(1,2)}$, and after a straight-forward calculation we obtain

$$\frac{\partial h}{\partial Q}=\frac{v_1(Q)\zeta_2(B)}{\det \xi(Q,B)},\quad \frac{\partial h}{\partial B}=-\frac{v_2(B)\zeta_1(Q)}{\det \xi(Q,B)} \tag{3.21}$$

which by (3.6) yields

$$\left.\frac{ds}{dh}\right|_{n=\mathrm{const}}=\chi$$

$$\chi=\left(\frac{v_1(Q)(\zeta_2(B))^2}{2\rho_1(Q)(\det\xi(Q,B))^2}+\frac{v_2(B)(\zeta_1(Q))^2}{2\rho_2(B)(\det\xi(Q,B))^2}\right)^{-1}. \tag{3.22}$$

This formula can be somewhat simplified using the Fermi-velocities defined through the low energy excitations [17]

$$v_{F1}=\frac{1}{2\pi}\frac{v_1(Q)}{\rho_1(Q)}\quad\text{and}\quad v_{F2}=\frac{1}{2\pi}\frac{v_2(B)}{\rho_2(B)}: \tag{3.23}$$

$$\chi=\frac{1}{2\pi}\left(\frac{v_{F1}(\zeta_2(B))^2}{2(\det\xi(Q,B))^2}+\frac{v_{F2}(\zeta_1(Q))^2}{2(\det\xi(Q,B))^2}\right)^{-1}. \tag{3.24}$$

This expression can be obtained also by using the low energy part of the spectrum [17].

The susceptibility curves of Figs. 1b and 2b were obtained through the formula (3.22).

IV. Behavior at h_c

The critical field

As it has been already mentioned, at the critical field there are no unpaired electrons yet, thus $Q=0$, i.e.:

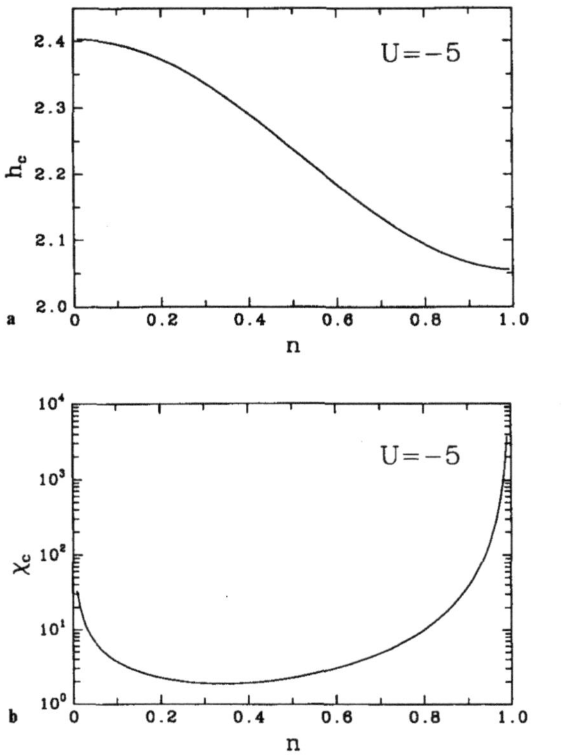

Fig. 3. Critical field **a** and susceptibility **b** versus bandfilling for $U=-5$

$$h_c=\left.\frac{\varepsilon_1(Q)\zeta_2(B)-\varepsilon_2(B)\zeta_1(Q)}{\det\xi(Q,B)}\right|_{Q=0} \tag{4.1}$$

(note that $Q=0$ not only in the arguments of the functions, but also in the equations defining the different functions). The critical field is practically the energy needed to break up the first pair (up to a constant factor), so it is the gap in the spin excitations. It is plotted as a function of the bandfilling for different values of U on Figs. 3a, 4a, 6a and 7a. Its behavior has been discussed in more detail in [13].

Finiteness of the initial susceptibility

The critical value of the susceptibility (χ at h_c) can be directly calculated through (3.22) by setting $Q=0$ (just as for h_c). Since $v_1(Q=0)=0$ (both v are odd functions for any Q and B)

$$\chi_c=\left.\frac{2\rho_2(B)(\det\xi(Q,B))^2}{v_2(B)(\zeta_1(Q))^2}\right|_{Q=0} \tag{4.2}$$

This expression exactly coincides with the one given in [14]. It is a very important observation that this quantity

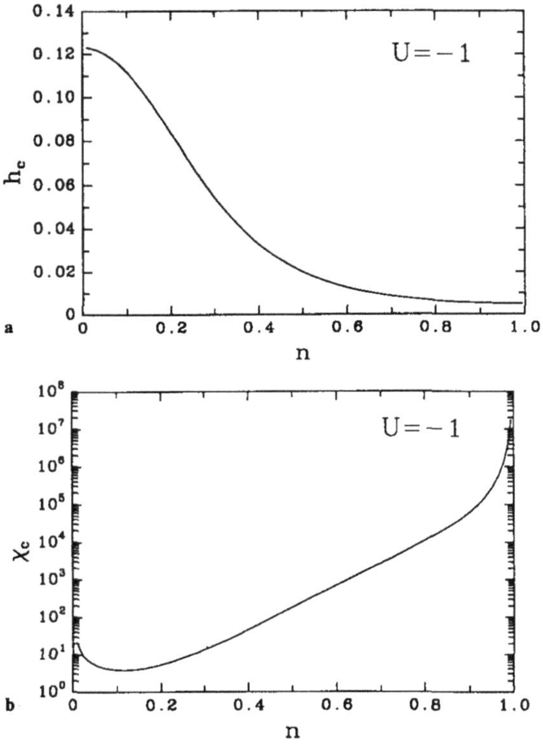

Fig. 4a, b. The same as Fig. 3 but for $U=-1$

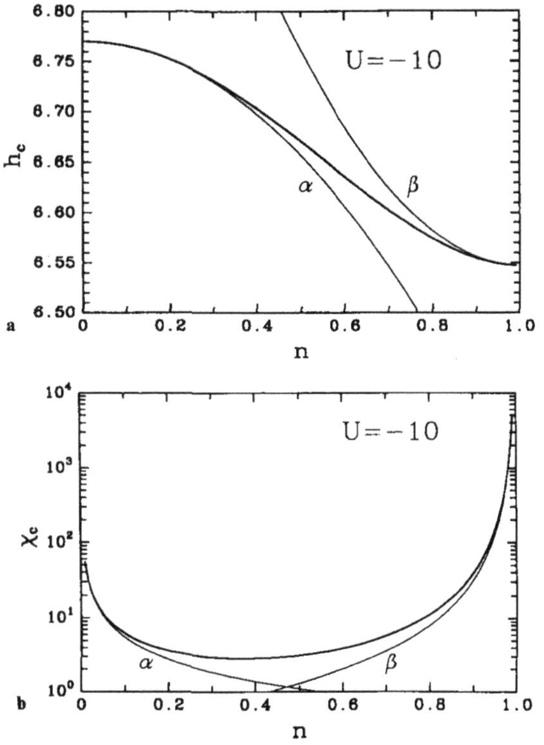

Fig. 6. Critical field a and susceptibility b versus bandfilling for a large value of the interaction: $U=-10$. The unlabelled (thick) curves give the exact values, the curves labelled by "α" correspond to the approximative functions (4.7) and (4.8), and the curves labelled by "β" represent (4.9) and (4.10)

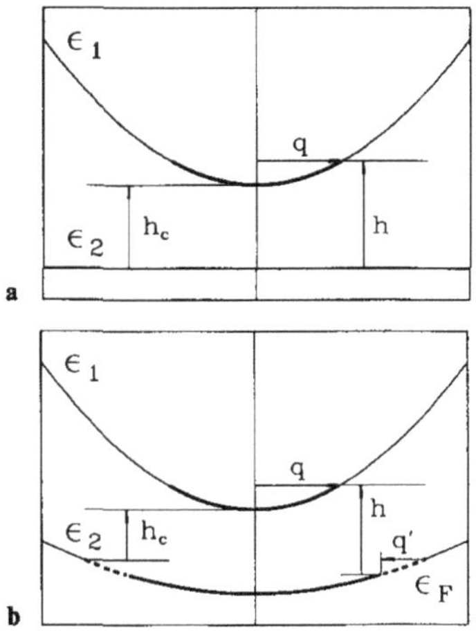

Fig. 5a, b. Schematic representations of the relations between the different energy bands and magnetic field to explain the initial susceptibility. On both parts ϵ_1 corresponds to the band of electrons with uncompenseated spins, and ϵ_2 represents the the energy of the bound pairs. a Corresponds to the case when the bound pairs condensate into the same level (i.e. ϵ_2 is the top or the bottom of a band, or it is the energy of a Bose-Einstein condensate). b Corresponds to the case when the bound pairs behave like fermions and fill in a band with a dispersion. q is the Fermi-momentum of the free electrons at h and q' is the change in the "Fermi-momentum" of the pairs as the magnetic field increases from h_c to h

is always *finite* except for the cases of half-filled and empty bands, and for the $U \nearrow 0$ limit. This has been noticed in [13] and it is proved in a strict sense in [14]. Here we give the plots of χ_c versus the bandfilling at different values of U (Figs. 3b, 4b, 6b and 7b).

The finiteness of χ_c – what is a most unusual property since at h_c a square-root like singularity might have been expected – can be understood on the basis of the fermion-like behavior of the bound pairs. These bound pairs, being formed of two fermions, should be bosons. They, however, do not form a Bose-Einstein condensate, but rather follow a Fermi-like distribution, and this makes the magnetization curves to start linearly. This is explained by means of Fig. 5a, b. (In this explanation we concentrate on the effect of the dispersion and distribution of the bound pairs and neglect all other effect which do not change the picture.) If the pairs would condensate into one state, or would fill in states of the same energy (like the top or the bottom of a band), the situation would be like on Fig. 5a, where the band ϵ_1 is for the unpaired electrons, and the level ϵ_2 is for the pairs. From this figure we can see that

$$h-h_c \propto q^2, \qquad (4.3)$$

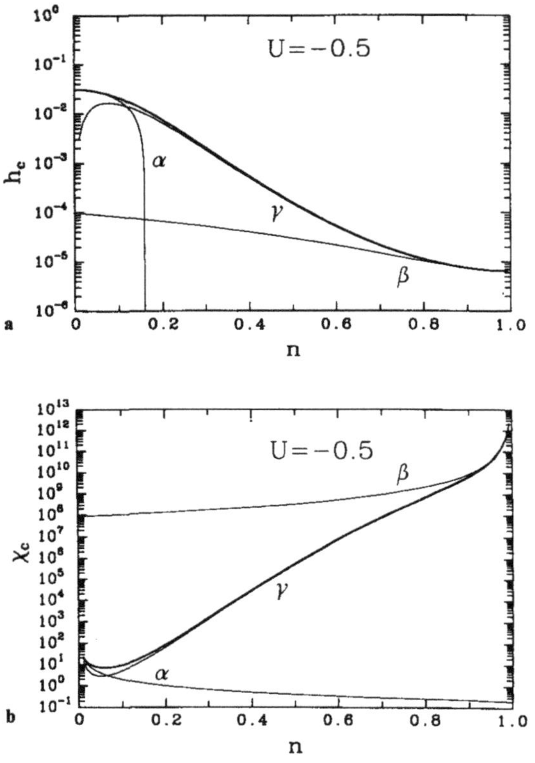

Fig. 7a, b. The same as Fig. 6 but for a small interaction: $U=-0.5$. The curves "γ" represent (4.11) and (4.12)

where q is the "Fermi-momentum" of the free electrons at h, and since $s \propto q$

$$s \propto \sqrt{h-h_c}. \tag{4.4}$$

The situation in the attractive Hubbard model is, however, different. The band ϵ_2 has a dispersion and the pairs fill it in up to a certain "Fermi-level" as depicted on Fig. 5b. This figure shows that

$$h-h_c \propto a q^2 + b q', \tag{4.5}$$

where q is the same as on the "a" part of the figure and q' represents the change in the "Fermi-momentum" of the pairs while the magnetic field increases from h_c to h. Since $q \propto q' \propto s$, for $h \searrow h_c$

$$s \propto h-h_c. \tag{4.6}$$

Recently Lee and Schlottmann [15] calculated the magnetization curves. Based on their *numerical* findings they claimed, that all the magnetization curves start with infinite slopes. Our susceptibility curves (Figs. 1b, 2b–c, 8b) show very strong, several order of magnitude increases as $h \searrow h_c$. In a merely numerical analysis it is very easy to misinterpret this enhancement as a singularity, and we think that happened in [15].

h_c and χ_c in the limiting cases

As it can be seen on Figs. 2b, 3b, 5b, 6b and 8b, and as it is also pointed out in [14], χ_c diverges only in the limiting cases $n \to 1$ and $n \to 0$ (for all (negative) values of U), and if $U \to 0$ (for all values of n). Actually these are also the cases when h_c and χ_c can be evaluated in closed form. The cases $n \to 0$ and $n \to 1$ have already been treated by different methods in [13]. Evaluating the formulae (4.1) and (4.2) reproduce the results obtained there. The results concerning the $U=0$ limit are new.

The low density limit: $n \to 0$. In this case, as it is easy to see from (2.10), $B \to 0$ and one can expand all quantities – by expanding the equations – in terms of B. The resulting equations can be solved, and yield

$$h_c(n \to 0) \cong (\sqrt{U^2+16}-4) - \frac{(\pi n)^2}{2\sqrt{U^2+16}} \tag{4.7}$$

and

$$\chi_c(n \to 0) \cong \frac{2}{(2\pi)^2}\sqrt{U^2+16}\,\frac{1}{n}. \tag{4.8}$$

These approximations are valid for all U. (4.7) and (4.8) give the curves labelled by "α" on the Figs. 6 and 7.

The nearly half-filled-band case: $n \to 1$. The integration of (2.8b) from $-\infty$ to ∞ with B set to ∞ shows that for $n=1$ $B=\infty$, i.e., the limit $n \to 1$ corresponds to $B \to \infty$ for any finite U. If so, to solve the BAE the Wiener-Hopf technique can be applied leading to

$$h_c(n \to 1) \cong h_c(n=1) + (1-n)^2 \pi \frac{I_1(2\pi/|U|)}{(I_0(2\pi/|U|))^2} \tag{4.9a}$$

with

$$h_c(n=1) = |U| - 4 + 4\int_0^\infty \frac{e^{-|\omega||U|/4} J_1(\omega)}{\omega \cosh \omega|U|/4}\, d\omega \tag{4.9b}$$

and

$$\chi_c(n \to 1) \cong \frac{1}{4\pi(1-n)^2}\,\frac{(I_0(2\pi/|U|))^3}{I_1(2\pi/|U|)}. \tag{4.10}$$

Here J_1 is a Bessel function, I_0 and I_1 are Bessel functions of imaginary arguments, and according to [13] the approximations hold as long as $(1-n) \ll 4 I_0(2\pi/|U|)$ $\exp(-2\pi/|U|)/\pi e$. (4.9) and (4.10) give the curves labelled by "β" in Figs. 6 and 7.

The small coupling limit: $U \to 0$. It is known that for $U \to 0$, $B \to \sin n\pi/2$. (This can bee seen by taking the $U \to 0$ limit of (2.8b).) For $n>0$, B is finite, but the width of the kernel (what is actually u) tends to zero, and for this the Wiener-Hopf technique can be applied also in this case. The technical details are given in the Appendix, here we give the results only:

$$h_c(|U| \ll 1) \cong \frac{8}{\pi}\sqrt{|U| \sin^3 \frac{\pi n}{1}}\, e^{-\frac{2\pi \sin \pi n/2}{|U|}} \tag{4.11}$$

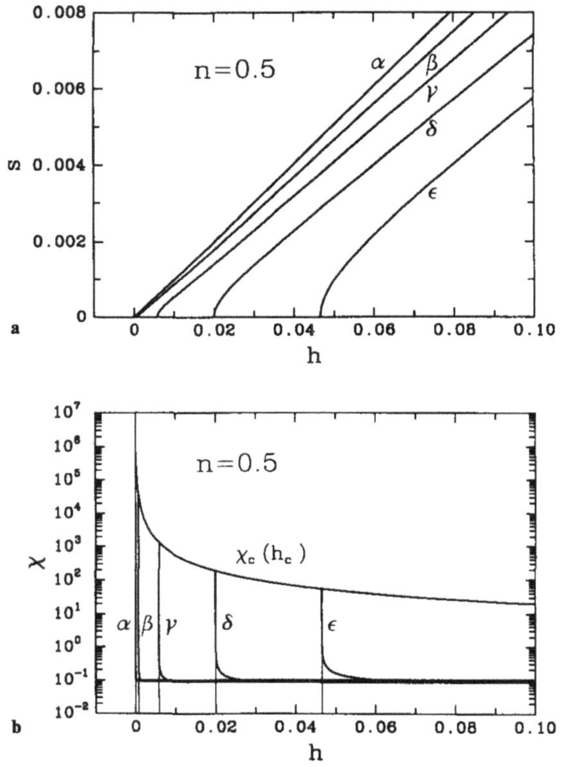

Fig. 8. Magnetization **a** and susceptibility **b** for small values of h for a quarter-filled band. The curves α, β, γ, δ and ϵ correspond to $U=-0.4$, -0.6, -0.8, -1.0 and -1.2, respectively. The sharp increase in χ as $h \searrow h_c$ is apparent

and

$$\chi_c(|U| \ll 1) \cong \frac{|U|}{16\pi \sin^2 \pi n} \, e^{\frac{4\pi \sin \pi n/2}{|U|}}. \tag{4.12}$$

(4.11) and (4.12) are valid for any n provided $|U|$ is sufficiently small, i.e. if $\sin(\pi n/2)/|U| \gg 1$. The curves labelled by "γ" on Fig. 7a, b represent (4.11) and (4.12). The fit to the unlabelled exact curves is remarkably good in a large range of n. We note that expanding (4.11) and (4.12) around $n=1$, and taking the $|U| \ll 1$ limit of (4.9) and (4.10) give the same result.

To our knowledge this work is the first Bethe Ansatz calculation for the gap in the magnetic excitations in the non-half-filled band case. Earlier Larkin and Sak calculated this gap by solving the renormalization group equations and using the exact gap of the half-filled band case as a "boundary condition". Equation (4.11) exactly confirms their result [20].

In connection with the $U \to 0$ limit we want to draw attention to a strange singular behavior. When $U=0$ the susceptibility is finite, when $U<0$ χ_c is again finite, but

$$\lim_{U \to -0} \chi_c(U) = \infty, \qquad 0<n<1 \tag{4.13}$$

i.e.

$$\lim_{U \to -0} \chi_c(U) \neq \chi_c(U=0). \tag{4.14}$$

This strange behaviour of χ_c and $\chi(h)$ near h_c in the $u \to 0$ limit (what is a manifestation of the essential singularity of the model at $U=0$ reported in several works) is shown on Fig. 8b.

V. Behavior at saturation

In all the cases the magnetisation saturates at $s=n/2$ at a finite field h_s. Since at this field also the last pair is broken up, $B=0$, and the problem becomes equivalent to a free fermion problem. Accordingly, all the quantities can be calculated in a closed form. As a result on gets (like in [13])

$$h_s = -\frac{2}{\pi} \cos \pi n \left[2\pi - 2 \tan^{-1}\left(\frac{\sin \pi n}{|U|/4} \right) \right] + \frac{\sqrt{U^2+16}}{2} + \frac{|U|}{\pi} \left[\frac{\sqrt{U^2+16}}{|U|} \tan^{-1}\left(\frac{|U|}{\sqrt{U^2+16}} \cot \pi n \right) + \pi n \right] \tag{5.1}$$

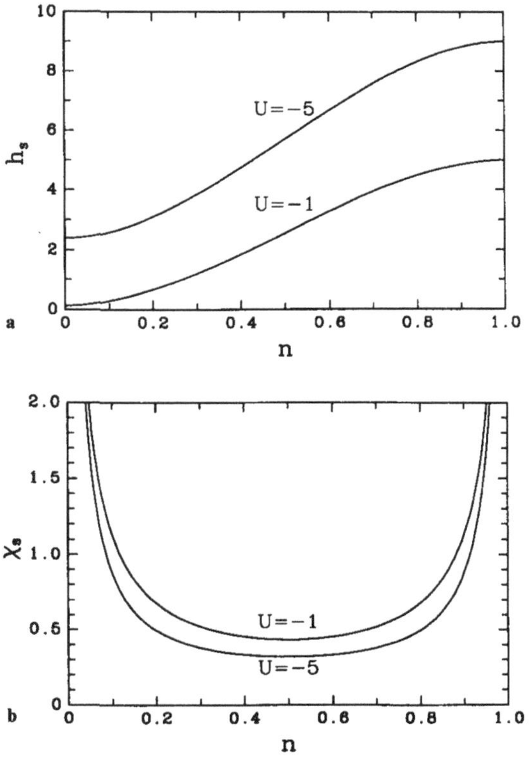

Fig. 9. The saturation field **a** and the susceptibility **b** versus bandfilling at different values of U

and

$$\chi_s=\left[\frac{2}{\pi}\sin\pi n\left(2\pi-2\tan^{-1}\left(\frac{\sin\pi n}{|U|/4}\right)\right)\right]^{-1} \tag{5.2}$$

These quantities are represented on Fig. 9a, b.

We are very grateful to Prof. J. Zittartz and the Institut für Theoretische Physik, Universität zu Köln – where a significant part of this work has been completed – for the valuable discussions and kind hospitality. Also the financial support from SFB341 is gratefully acknowledged.

Appendix

In this Appendix we give some details of the calculation of h_c and χ_c in the $U\to 0$ limit. First we show, how to reduce the problem to the solutions of Wiener-Hopf equations, then we give the B as a function of the bandfilling, finally we give the h_c and χ_c.

Reduction to Wiener-Hopf equations

As it has been seen, all the important distribution functions are determined by equations of the type (2.12) or (2.17). For $Q=0$ both (2.12) and (2.17) reduce to the same one

$$x_2(\lambda)=x_{2,0}(\lambda)-\frac{1}{2\pi}\int_{-B}^{B}K_2(\lambda-\lambda')x_2(\lambda')\,d\lambda', \tag{A.1}$$

and the calculation of quantities of the type $x_1(Q=0)$ reduce to the evaluation of expressions

$$x_1(0)=x_{1,0}(0)-\frac{1}{2\pi}\int_{-B}^{B}K_1(\lambda')x_2(\lambda')\,d\lambda'. \tag{A.2}$$

In treating (A.1) first we suppose that the Fourier-transform of $x_{2,0}$ exists, and it is finite [21]. Then also the Fourier-transform of x_2 exists and also that is finite. If so, (A.1) can be written in the form

$$x_2(\lambda)=x^*_{2,0}(\lambda)+\frac{1}{2\pi}\int_{B}^{\infty}K^*_2(\lambda-\lambda')x_2(\lambda')\,d\lambda' + \frac{1}{2\pi}\int_{-\infty}^{-B}K^*_2(\lambda-\lambda')x_2(\lambda')\,d\lambda' \tag{A.3}$$

with

$$x^*_{2,0}(\lambda)=\int_{-\infty}^{\infty}\frac{\tilde{x}_{2,0}(\omega)e^{i\omega\lambda}}{1+\tilde{K}_2(\omega)}\,d\omega,$$

$$K^*_2(\lambda)=\int_{-\infty}^{\infty}\frac{\tilde{K}_2(\omega)e^{i\omega\lambda}}{1+\tilde{K}_2(\omega)}\,d\omega \tag{A.4}$$

the tilde denoting fourier-transform. After introducing

$$x(\eta)=x_2(\eta+B) \quad\text{and}\quad I_x(\eta)=x^*_{2,0}(\eta+B) \tag{A.5}$$

we have

$$x(\eta)=I_x(\eta)+\frac{1}{2\pi}\int_0^{\infty}K^*_2(\eta-\eta')x(\eta')\,d\eta' \pm\frac{1}{2\pi}\int_0^{\infty}K^*_2(\eta+\eta'+2B)x(\eta')\,d\eta' \tag{A.6}$$

provided $x_2(\lambda)$ is an even/odd function. Following Yang and Yang [22] we look for the solution in form of a series

$$x(\eta)=\sum_{l=1}^{\infty}x^{(l)}(\eta). \tag{A.7}$$

This solves (A.6) if

$$x^{(l)}(\eta)=I_x^{(l)}(\eta)+\frac{1}{2\pi}\int_0^{\infty}K^*_2(\eta-\eta')x^{(l)}(\eta')\,d\eta' \tag{A.8}$$

with

$$I_x^{(l)}(\eta)=\begin{cases} I_x(\eta), & \text{if } l=1\\ \pm\frac{1}{2\pi}\int_0^{\infty}K^*_2(\eta+\eta'+2B)x^{(l-1)}(\eta'), & \text{if } l>1. \end{cases} \tag{A.9}$$

We have to note here, that this series expansion is useful if each term is much smaller than the previous one. This is verified, if $B\gg 1$ ($n\to 1$ case) or $u\ll 1$ (the present case): $K^*_2(2B)$ can be estimated to be of the order of (u/B), i.e. $I_x^{(l)}$ (and so $x^{(l)}$) are approximately (u/B)-times smaller than $x^{(l-1)}$. (The essential difference between the $u\ll 1$ and $B\gg 1$ cases, which leads to the different results is in the analytic properties of the inhomogeneous parts $I_x(\eta)$.) Equations (A.8) are of the Wiener-Hopf type and can be solved by standard methods (for more details see for eg. [22]).

Quantities of the type $x_1(0)$ in principle can be calculated by (A.2), if $x_2(\lambda)$ is given. One has to keep in mind, however, that the $x_2(\lambda)$ obtained the way described above, due to the truncation of the series (A.7) after $l=1$ or 2, is more accurate around B and for $\lambda>B$, and it is less accurate for $\lambda\ll B$. So one has to convert the integral over the region $(-B;B)$ in (A.2) into an integral over $B<\lambda<\infty$. This we do in the following way. Multiplying (A.1) by $K_{2\nu-1}$ (for K see (2.9)) and integrating over λ we obtain

$$\int_{-B}^{B}K_{2\nu-1}(\lambda)x_2(\lambda) = -2\int_{B}^{\infty}K_{2\nu-1}(\lambda)x_2(\lambda) + \int_{-\infty}^{\infty}K_{2\nu-1}(\lambda)x_{2,0}(\lambda) - \int_{-B}^{B}K_{2\nu+1}(\lambda)x_2(\lambda) \tag{A.10}$$

provided x_2 is an even function. Applying this relation to (A.2) successively, we arrive at

$$x_1(0)=\left[x_{1,0}(0)-\frac{1}{2\pi}\sum_{\nu=0}^{\infty}(-1)^\nu\int_{-\infty}^{\infty}K_{2\nu+1}(\lambda)x_{2,0}(\lambda)\right]+\frac{1}{\pi}\sum_{\nu=0}^{\infty}(-1)^\nu\int_{B}^{\infty}K_{2\nu+1}(\lambda)x_2(\lambda),\tag{A.11}$$

what, using the relation

$$\sum_{\nu=0}^{\infty}(-1)^\nu K_{2\nu+1}(\lambda)=\frac{\pi}{2u}\frac{1}{\cosh\frac{\lambda\pi}{2u}}\tag{A.12}$$

leads to

$$x_1(0)=\left[x_{1,0}(0)-\frac{1}{4u}\int_{-\infty}^{\infty}\frac{x_{2,0}(\lambda)}{\cosh\frac{\lambda\pi}{2u}}\right]+\frac{1}{2u}\int_{B}^{\infty}\frac{x_2(\lambda)}{\cosh\frac{\lambda\pi}{2u}}.\tag{A.13}$$

Using the explicit form of the solution of (A.8) a very important relation can be derived from (A.13):

$$x_1(0)=\frac{2}{\pi}\sqrt{\frac{\pi}{e}}\,e^{-\frac{B\pi}{2u}}x_2(B)+\frac{4u}{\pi^4}\sqrt{\frac{\pi}{e}}\,e^{-\frac{B\pi}{2u}}\frac{x'_{2,0}(B)}{\sqrt{2}}+\left[x_{1,0}(0)-\frac{1}{4u}\int_{-\infty}^{\infty}\frac{x_{2,0}(\lambda)}{\cosh\frac{\lambda\pi}{2u}}\right]+o\left(ue^{-\frac{B\pi}{2u}}\right)\tag{A.14}$$

where $x'_{2,0}$ is the derivative of $x_{2,0}$ and $o(u\exp(-\frac{B\pi}{2u}))$ means terms decaying faster than $u\exp(-\frac{B\pi}{2u})$ as $u\to 0$. Another important relation what we shall use and can be obtained from the explicit form of the solution for (A.8) is

$$\lim_{u\to 0}x_2(B)=\frac{1}{\sqrt{2}}\lim_{u\to 0}x_{2,0}(B).\tag{A.15}$$

The particle number

According to (2.10) at $Q=0$ (i.e. for $s=0$) the density is given by

$$\int_{-B}^{B}\rho_2(\lambda)=\frac{n}{2}\tag{A.16}$$

with ρ_2 determined (see (2.8.b)) through

$$\rho_2(\lambda)=\frac{1}{2\pi}\mathrm{Re}\frac{2}{\sqrt{1-(\lambda-iu)^2}}-\frac{1}{2\pi}\int_{-B}^{B}K_2(\lambda-\lambda')\rho_2(\lambda')d\lambda'.\tag{A.17}$$

Integrating (A.17) from $-\infty$ to ∞ (A.16) can be written in the form

$$\int_{B}^{\infty}\rho_2(\lambda)=\tfrac{1}{2}(1-n).\tag{A.18}$$

Solving (A.17) by the method outlined above, and calculating (A.18) up to the required accuracy (one has to solve (A.8) for $l=1$ and 2) one has

$$\frac{1}{2}(1-n)=\left(\frac{1}{2}-\frac{k_0}{\pi}\right)-\frac{u}{\pi^2\cos k_0}\left(1+\ln\frac{\pi\cos^2k_0\sin k_0}{2u}\right),\tag{A.19}$$

where k_0 is defined through $B=\sin k_o$, i.e.:

$$B=\sin\frac{\pi n}{2}-\frac{u}{\pi}\left(1+\ln\frac{\pi\cos^2\frac{\pi n}{2}\sin\frac{\pi n}{2}}{2u}\right).\tag{A.20}$$

The critical field h_c

According to (4.1)

$$h_c=\left.\frac{\varepsilon_1(Q)\zeta_2(B)-\varepsilon_2(B)\zeta_1(Q)}{\det\xi(Q,B)}\right|_{Q=0}.\tag{A.21}$$

First of all, it is not hard to see from (2.28) and (2.29), that if $Q=0$, than $\xi_{21}=\xi_2^{(1)}=0$ and $\xi_{11}=\xi_1^{(1)}=1$, and so, due to (3.7) and the definitions of the matrix ξ and the vector ζ

$$\det\xi(0,B)=\xi_{22}(B)=\tfrac{1}{2}\zeta_2(B).\tag{A.22}$$

The $\zeta_2(\lambda)$ and $\zeta_1(0)$ are determined by equations of the type (A.1) and (A.2) with $x_{1,0}(0)$ and $x_{2,0}(\lambda)$ replaced by

$$\zeta_{1,0}(0)=1\quad\text{and}\quad\zeta_{2,0}(\lambda)=2,\tag{A.23}$$

respectively. Applying (A.14) to calculate $\zeta_1(0)$ we find that the quantity in the square brackets is zero, and since $\zeta'_{2,0}=0$ we have

$$\zeta_1(0)=\frac{2}{\pi}\sqrt{\frac{\pi}{e}}\,e^{-\frac{B\pi}{2u}}\zeta_2(B).\tag{A.24}$$

Also the quantities $\varepsilon_2(\lambda)$ and $\varepsilon_1(0)$ are determined by (A.1) and (A.2), but now $x_{1,0}(0)$ and $x_{2,0}(\lambda)$ are to be replaced by

$$\varepsilon_{1,0}(0)=-2\quad\text{and}\quad\varepsilon_{2,0}(\lambda)=-4\,\mathrm{Re}\sqrt{1-(\lambda-iu)^2},\tag{A.25}$$

respectively. Substituting these into (A.14) the expression in the squarebrackets gives

$$\left[\varepsilon_{1,0}(0)-\frac{1}{4u}\int_{-\infty}^{\infty}\frac{\varepsilon_{2,0}(\lambda)}{\cosh\frac{\lambda\pi}{2u}}\right]\cong\frac{8\sqrt{u}}{\pi}e^{-\frac{\pi}{2u}}.\tag{A.26}$$

Since this decays exponentially faster then the other terms in (A.14) (this is due to $B<1$), we can neglect it and we have

$$\varepsilon_1(0)=\frac{2}{\pi}\sqrt{\frac{\pi}{e}}\,e^{-\frac{B\pi}{2u}}\varepsilon_2(B)$$

$$+\frac{4u}{\pi^2}\sqrt{\frac{\pi}{e}}\,e^{-\frac{B\pi}{2u}}\frac{4}{\sqrt{2}}\frac{B}{\sqrt{1-B^2}}. \tag{A.27}$$

Substituting (A.22), (A.24) and (A.27) into (A.21) we obtain

$$h_c=\frac{32u}{\pi^2\sqrt{2}}\sqrt{\frac{\pi}{e}}\,e^{-\frac{B\pi}{2u}}\frac{B}{\sqrt{1-B^2}}. \tag{A.28}$$

This, after substituting (A.20) for B yields

$$h_c=\frac{8}{\pi}\sqrt{|U|\sin^3\frac{\pi n}{2}}\,e^{-\frac{2\pi\sin\pi n/2}{|U|}} \tag{A.29}$$

The critical susceptibility χ_c

To evaluate the expression

$$\chi_c=\left.\frac{2\rho_2(B)(\det\xi(Q,B))^2}{v_2(B)(\zeta_1(Q))^2}\right|_{Q=0} \tag{A.30}$$

in addition to the quantities we have already, we need to calculate $\rho_2(B)$ and $v_2(B)$. For both quantities we may use their $u\to 0$ limiting values. According to (A.15)

$$\rho_2(B)\cong\frac{1}{\sqrt{2}}\rho_{2,0}(B)=\frac{1}{2\pi\sqrt{2}}\frac{1}{\sqrt{1-B^2}} \tag{A.31}$$

and (see also the $u\to 0$ limit of (3.17))

$$v_2(B)\cong\frac{1}{\sqrt{2}}v_{2,0}(B)=2\sqrt{2}\frac{B}{\sqrt{1-B^2}}. \tag{A.32}$$

Combining (A.22), (A.24), (A.31) and (A.32) yields

$$\chi_c=\frac{e}{64B}\,e^{\frac{B\pi}{u}}, \tag{A.33}$$

i.e.

$$\chi_c=\frac{|U|}{16\pi\sin^2\pi n}\,e^{\frac{4\pi\sin\pi n/2}{|U|}}. \tag{A.34}$$

References

1. Lieb, E.H., Wu, F.Y.: Phys. Rev. Lett. **20**, 1445 (1968)
2. Toward the theoretical understanding of the high T_c superconductivity, workshop, Trieste 1988. Int. J. Modern Phys. B**1**, Nr. 5 (1988)
3. Ovchinnikov, A.A.: Zh. Eksp. Teor. Fiz. **57**, 2137 (1969) [Sov. Phys.-JETP **30**, 116] (1970)
4. Coll, C.F.: Phys. Rev. B**9**, 2150 (1974)
5. Choy, T.C., Young, W.: J. Phys. C**15**, 521 (1982)
6. Woynarovich, F.: J. Phys. C**15**, 85, 97 (1982)
7. Woynarovich, F.: J. Phys. C**16**, 5293 (1983)
8. Woynarovich, F.: J. Phys. C**16**, 6593 (1983)
9. Klümper, A., Schadschneider, A., Zittartz, J.: Z. Phys. B – Condensed Matter **78**, 99 (1990)
10. Schadschneider, A., Zittartz, J.: Z. Phys. B – Condensed Matter **82**, 387 (1991)
11. Takahashi, M.: Prog. Theor. Phys. **42**, 1098 (1969)
12. Shiba, H.: Phys. Rev. B**6**, 930 (1972)
13. Bahder, T.B., Woynarovich, F.: Phys. Rev. B**33**, 2114 (1986)
14. Woynarovich, F.: Phys. Rev. B**43**, 11448 (1991)
15. Lee, K-J-B., Schlottmann, P.: Phys. Rev. B**40**, 9104 (1989)
16. See e.g. Bronstein, I.N., Semendyayev, K.A.: Handbook of Mathematics, New York: Van Nostrand Reinhold Company, Thun Frankfurt/M: Deutsch
17. Woynarovich, F.: J. Phys. A**22**, 4243 (1989)
18. Izergin, A.G., Korepin, V.E., Reshetikhin, N.Yu.: J. Phys. A**22**, 2615 (1989)
19. Frahm, H., Korepin, V.E.: Phys. Rev. B**42**, 10553 (1990)
20. Larkin, A.I., Sak, J.: Phys. Rev. Lett. **39**, 1025 (1977)
21. This is a convenient but not necessarry condition, and all the important relations like (A.14) and (A.15) are valid also when $x_{2,0}$ tends to a finite constant if $|\lambda|\to\infty$: one may introduce a cut-off at some very large Λ for $x_{2,0}$, and at the end the cut-off can be taken to infinity
22. Yang, C.N., Yang, C.P.: Phys. Rev. **150**, 327 (1966)

Progress of Theoretical Physics, Vol. 47, No. 1, January 1972

One-Dimensional Hubbard Model at Finite Temperature

Minoru TAKAHASHI

Department of Physics, University of Tokyo, Tokyo

(Received June 5, 1971)

The method of integral equations with infinitely many unknowns which were used for the one-dimensional Heisenberg model and electron gas is applied to the one-dimensional Hubbard model. From these equations one can calculate energy, entropy, density and magnetization at given temperature, chemical potential and magnetic field.

§ 1. Introduction

Lieb and Wu[1] solved exactly the ground state energy of the one-dimensional Hubbard model. Using their formulation, the author calculated the magnetization curve and magnetic susceptibility at zero temperature in the half-filled case.[2] In this paper we derive the integral equations for the one-dimensional Hubbard model at finite temperature. In the problem of an electron gas[3] we considered the distribution functions of magnon bound states, unpaired electrons and electron pairs. Besides these functions we must treat the distribution functions of bound states of pairs in the problem of the one-dimensional Hubbard model.

In a recent paper Sokoloff[4] gave the exact solution for the case of an infinite repulsive interaction. It will be shown that our equations give the same results as those of Sokoloff in the limit of infinite repulsion.

§ 2. Theory of finite system

We consider the Hamiltonian

$$\mathcal{H} = -\sum_{\langle ij\rangle}\sum_{\sigma} c^{\dagger}_{i\sigma}c_{j\sigma} + 4U\sum_{i=1}^{N} n_{i\uparrow}n_{i\downarrow} - \sum_{i=1}^{N}\mu_0 H(n_{i\uparrow}-n_{i\downarrow}). \tag{2·1}$$

The wave function is

$${}_M\Phi_1(x_1, x_2, \cdots, x_{N-M}|x_{N-M+1}, \cdots, x_N) = \sum_P [Q, P]\exp(i\sum_{j=1}^{N} k_{Pj}x_{Qj}), \tag{2·2}$$

at $x_{Q1}\leqq x_{Q2}\leqq x_{Q3}\cdots\leqq x_{QN}$. The energy eigenvalue is

$$E = -2\sum_{j=1}^{N}\cos k_j\,. \tag{2·3}$$

One can easily show that the wave function (2·2) is an eigenstate of (2·1) if we put

$$[Q, P] = \varepsilon(Q_1)\varepsilon(Q_2)\sum_R A_R \prod_{j=1}^{M} F_P(\Lambda_{Rj}, y_j), \tag{2·4}$$

with

$$F_P(\Lambda, y) = \prod_{j=1}^{y-1} \frac{\sin k_{Pj} - \Lambda + iU}{\sin k_{P(j+1)} - \Lambda - iU}, \tag{2·5}$$

and

$$A_R = \prod_{i<j} \left(\frac{\Lambda_{Ri} - \Lambda_{Rj} - 2iU}{\Lambda_{Ri} - \Lambda_{Rj} + 2iU} \right). \tag{2·6}$$

Here we have introduced parameters $\Lambda_1, \Lambda_2, \cdots, \Lambda_M$, $y_1\ y_2, \cdots, y_M$ are coordinates of $x_{N-M+1}, \cdots, x_N$ along the chain, Q_1 and Q_2 signify the orders of $1, 2, \cdots, N-M$ and $N-M+1, \cdots, N$ in the permutation Q, $\varepsilon(Q_i)$ is the parity of the permutation Q_i. The periodic boundary condition

$${}_M\Phi_1(x_1, x_2, \cdots, x_i, \cdots, x_N) = {}_M\Phi_1(x_1, \cdots, x_i + N_a, \cdots, x_N), \quad i = 1, 2, \cdots, N, \tag{2·7}$$

is satisfied if we put

$$\exp\,(ik_j N_a) = \prod_{\alpha=1}^{M} e\left(\frac{\sin k_j - \Lambda_\alpha}{U}\right), \tag{2·8a}$$

$$\prod_{j=1}^{N} e\left(\frac{\Lambda_\alpha - \sin k_j}{U}\right) = -\prod_{\beta=1}^{M} e\left(\frac{\Lambda_\alpha - \Lambda_\beta}{2U}\right), \tag{2·8b}$$

where we put $e(x) \equiv (x+i)/(x-i)$. For excited states k's and Λ's are not necessarily real numbers. To specify the all eigenstates of Hamiltonian (2·1) we use the following two conjectures.

Conjecture 1. A complex k_j belongs to a pair of electrons or a bound state of pairs. For a bound state of n-pairs there are $n-\Lambda$'s which have the same real part and the imaginary parts as $(n-1)iU$, $(n-3)iU, \cdots -(n-1)iU$ within the accuracy of $O(\exp -\delta N_a)$, where δ is some positive number. And $2n-k$'s belong to this bound state and take the values

$$\begin{aligned}
k_\alpha^1 &= \pi - \sin^{-1}(\Lambda_\alpha'^n + niU), \\
k_\alpha^2 &= \sin^{-1}(\Lambda_\alpha'^n + (n-2)iU), \\
k_\alpha^3 &= \pi - k_\alpha^2, \\
k_\alpha^4 &= \sin^{-1}(\Lambda_\alpha'^n + (n-4)iU), \\
k_\alpha^5 &= \pi - k_\alpha^4, \\
&\vdots \\
k_\alpha^{2n-2} &= \sin^{-1}(\Lambda_\alpha'^n - (n-2)iU), \\
k_\alpha^{2n-1} &= \pi - k_\alpha^{2n-2}, \\
k_\alpha^{2n} &= \pi - \sin^{-1}(\Lambda_\alpha'^n - niU),
\end{aligned} \tag{2·9a}$$

within the accuracy of $O(\exp(-\delta N_a))$. Here we put $-\pi/2 < \mathrm{Re}\,\sin^{-1}(x) \leq \pi/2$.

We denote the k's and Λ's which belong to the bound state of n-pairs as

$$k_\alpha^{n,l}, \qquad l=1,2,\cdots,2n,$$

and

$$\Lambda_\alpha'^{nj}=\Lambda_\alpha'^{n}+(n+1-2j)Ui, \qquad j=1,2,\cdots,n. \tag{2·9b}$$

And we define $\Lambda_\alpha'^{n}$ as

$$\Lambda_\alpha'^{n}\equiv \mathrm{Re}\sin^{-1}k_\alpha^{n,l}\equiv \mathrm{Re}\,\Lambda_\alpha^{n,l}.$$

Conjecture 2. Complex Λ belongs to a bound state of pairs or bound state of Λ's. In the latter case the real parts of Λ's are the same and the imaginary parts are $(n-1)iU$, $(n-3)iU$, $\cdots$, $-(n-1)iU$ within the accuracy of $O(\exp -\delta N)$.

We denote these as $\Lambda_\alpha^{n,j}$ and real part of these as Λ_α^{n};

$$\Lambda_\alpha^{n,j}=\Lambda_\alpha^{n}+(n+1-2j)iU. \tag{2·10}$$

We write unpaired k's as k_j. One can specify the eigenstates of Hamiltonian by k_j's, Λ_α^{n}'s and $\Lambda_\alpha'^{n}$'s. The equations for these are derived from Eqs. (2·8a) and (2·9b).

$$e^{ik_jN_a}=\prod_{(n,\alpha)}e\left(\frac{\sin k_j-\Lambda_\alpha^{n}}{nU}\right)\prod_{(n,\alpha)}e\left(\frac{\sin k_j-\Lambda_\alpha'^{n}}{nU}\right), \tag{2·11a}$$

$$\exp(iN_a\sum_{l=1}^{2n}k_\alpha^{n,l})=\exp\{-N_a(\sin^{-1}(\Lambda_\alpha'^{n}+inU)+\sin^{-1}(\Lambda_\alpha'^{n}-inU))\}$$

$$=-\prod_{j=1}^{N-2M'}e\left(\frac{\Lambda_\alpha'^{n}-\sin k_j}{nU}\right)\prod_{(m,\beta)}E_{nm}\left(\frac{\Lambda_\alpha'^{n}-\Lambda_\beta'^{m}}{U}\right), \tag{2·11b}$$

$$\prod_{j=1}^{N-2M'}e\left(\frac{\Lambda_\alpha^{n}-\sin k_j}{nU}\right)=-\prod_{(m,\beta)}E_{nm}\left(\frac{\Lambda_\alpha^{n}-\Lambda_\beta^{m}}{U}\right), \tag{2·11c}$$

where

$$E_{nm}(x)\equiv\begin{cases} e\left(\frac{x}{|n-m|}\right)e^2\left(\frac{x}{|n-m|+2}\right)e^2\left(\frac{x}{|n-m|+4}\right)\cdots e^2\left(\frac{x}{n+m-2}\right)e\left(\frac{x}{n+m}\right) & \text{for } n\neq m,\\ e^2\left(\frac{x}{2}\right)e^2\left(\frac{x}{4}\right)\cdots e^2\left(\frac{x}{2n-2}\right)e\left(\frac{x}{2n}\right) & \text{for } n=m. \end{cases}$$

Taking the logarithm of these equations we have the following equations:

$$k_jN_a=2\pi I_j-\sum_{n=1}^{\infty}\sum_{\alpha=1}^{M_n}\theta\left(\frac{\sin k_j-\Lambda_\alpha^{n}}{nU}\right)-\sum_{n=1}^{\infty}\sum_{\alpha=1}^{M_n'}\theta\left(\frac{\sin k_j-\Lambda_\alpha'^{n}}{nU}\right), \tag{2·12a}$$

$$N_a(\sin^{-1}(\Lambda_\alpha'^{n}+inU)+\sin^{-1}(\Lambda_\alpha'^{n}-inU))$$

$$=2\pi J_\alpha'^{n}+\sum_{j=1}^{N-2M'}\theta\left(\frac{\Lambda_\alpha'^{n}-\sin k_j}{nU}\right)+\sum_{(m,\beta)}\Theta_{nm}\left(\frac{\Lambda_\alpha'^{n}-\Lambda_\beta'^{m}}{U}\right), \tag{2·12b}$$

$$\sum_{j=1}^{N-2M'} \theta\left(\frac{\Lambda_\alpha^n - \sin k_j}{nU}\right) = 2\pi J_\alpha^n + \sum_{(m,\beta)} \Theta_{nm}\left(\frac{\Lambda_\alpha^n - \Lambda_\beta^m}{U}\right), \tag{2·12c}$$

where

$$\theta(x) \equiv 2\tan^{-1} x,$$

$$\Theta_{nm}(x) \equiv \begin{cases} \theta\left(\frac{x}{|n-m|}\right) + 2\theta\left(\frac{x}{|n-m|+2}\right) + \cdots + 2\theta\left(\frac{x}{n+m-2}\right) + \theta\left(\frac{x}{n+m}\right) & \text{for } n \neq m, \\ 2\theta\left(\frac{x}{2}\right) + \cdots + 2\theta\left(\frac{x}{2n-2}\right) + \theta\left(\frac{x}{2n}\right) & \text{for } n = m. \end{cases}$$

Here I_j, J_α^n, $J_\alpha'^n$ should satisfy the following conditions

$$I_j = \begin{cases} \text{integer}, & (\sum_{i=1}^{\infty}(M_i + M_i') \text{ even}) \\ \text{half odd integer}, & (\sum_{i=1}^{\infty}(M_i + M_i') \text{ odd}) \end{cases}$$

$$J'^n = \begin{cases} \text{integer}, & (N_a - (N - M_n') \text{ odd}) \\ \text{half odd integer}, & (N_a - (N - M_n') \text{ even}) \end{cases}$$

$$J^n = \begin{cases} \text{integer}, & (N - M_n \text{ odd}) \\ \text{half odd integer}, & (N - M_n \text{ even}) \end{cases}$$

$$|J_\alpha'^n| < \tfrac{1}{2}(N_a - N + 2M' - \sum_{m=1}^{\infty} t_{nm} M_m'),$$

$$|J_\alpha^n| < \tfrac{1}{2}(N - 2M' - \sum_{m=1}^{\infty} t_{nm} M_m), \qquad t_{nm} = 2\,\mathrm{Min}(n, m) - \delta_{nm}.$$

Total momentum and total energy are given by

$$K = (2\pi \sum_{j=1}^{N-2M'} I_j - 2\pi \sum_{(n,\alpha)} J_\alpha'^n)/N_a + \pi \prod_{n=1}^{\infty} (-1)^{M_{2n}'}, \tag{2·13a}$$

$$E = \sum_{j=1}^{N-2M'} (-2\cos k_j - \mu_0 H) + \sum_{(n,\alpha)} 4\,\mathrm{Re}\sqrt{1 - (\Lambda_\alpha'^n - niU)^2} + 2\mu_0 H \sum_{n=1}^{\infty} n M_n, \tag{2·13b}$$

where M' is defined by

$$M' \equiv \sum_{n=1}^{\infty} n M_n'. \tag{2·13c}$$

The number of up-spin and down-spin electrons are given by

$$N_\downarrow = \sum_{n=1}^{\infty} n M_n + M', \tag{2·13d}$$

$$N_\uparrow = N - N_\downarrow. \tag{2·13e}$$

§ 3. The integral equations

We put the distribution functions of k, Λ^n and Λ'^n as $\rho(k)$, $\sigma_n(\Lambda)$ and $\sigma_n'(\Lambda)$, respectively, and corresponding distribution functions of holes as $\rho^h(k)$, $\sigma_n^h(\Lambda)$ and $\sigma_n'^h(\Lambda)$. The integral equations for these functions are derived from (2·12):

$$\frac{1}{2\pi}=\rho(k)+\rho^h(k)-\cos k\left(\sum_{n=1}^{\infty}\int_{-\infty}^{\infty}\frac{nU}{\pi}\cdot\frac{(\sigma_n'(\Lambda)+\sigma_n(\Lambda))d\Lambda}{n^2U^2+(\sin k-\Lambda)^2}\right), \tag{3·1a}$$

$$\int_{-\pi}^{\pi}\frac{nU}{\pi}\frac{\rho(k)dk}{(nU)^2+(\sin k-\Lambda)^2}=\sigma_n^h(\Lambda)+\sum_{m=1}^{\infty}A_{nm}\sigma_m(\Lambda), \tag{3·1b}$$

$$\frac{1}{\pi}\operatorname{Re}\frac{1}{\sqrt{1-(\Lambda-Ui)^2}}-\int_{-\pi}^{\pi}\frac{nU}{\pi}\frac{\rho(k)dk}{(nU)^2+(\sin k-\Lambda)^2}=\sigma_n'^h(\Lambda)+\sum_{m=1}^{\infty}A_{nm}\sigma_m'(\Lambda), \tag{3·1c}$$

where A_{nm} is an operator defined by

$$A_{nm}f(x)=\delta_{nm}f(x)+\frac{d}{dx}\int_{-\infty}^{\infty}\Theta_{nm}\left(\frac{x-x'}{U}\right)f(x')\frac{dx'}{2\pi}.$$

The thermodynamic potential $\Omega\equiv E-AN-TS$ is given by

$$\begin{aligned}\Omega/N_a=&\int_{-\pi}^{\pi}[(-2\cos k-A-\mu_0H)\rho(k)-T\{(\rho+\rho^h)\ln(\rho+\rho^h)-\rho\ln\rho-\rho^h\ln\rho^h\}]dk\\&+\sum_{n=1}^{\infty}\int_{-\infty}^{\infty}[2n\mu_0H\sigma_n(\Lambda)-T\{(\sigma_n+\sigma_n^h)\ln(\sigma_n+\sigma_n^h)-\sigma_n\ln\sigma_n-\sigma_n^h\ln\sigma_n^h\}\\&+(4\operatorname{Re}\sqrt{1-(\Lambda-nUi)^2}-2nA)\sigma_n'(\Lambda)\\&-T\{(\sigma_n'+\sigma_n'^h)\ln(\sigma_n'+\sigma_n'^h)-\sigma_n'\ln\sigma_n'-\sigma_n'^h\ln\sigma_n'^h\}]d\Lambda\,.\end{aligned} \tag{3·2}$$

The condition $\delta\Omega=0$ gives the equation for $\zeta\equiv\rho^h/\rho$, $\eta_n\equiv\sigma_n^h/\sigma_n$ and $\eta_n'\equiv\sigma_n'^h/\sigma_n'$ as follows:

$$\begin{aligned}\ln\zeta(k)=&\frac{-2\cos k-\mu_0H-A}{T}\\&+\sum_{n=1}^{\infty}\int_{-\infty}^{\infty}\frac{nU}{\pi}\cdot\frac{1}{(nU)^2+(\sin k-\Lambda)^2}\{\ln(1+\eta_n'^{-1}(\Lambda))-\ln(1+\eta_n^{-1}(\Lambda))\}d\Lambda\,,\end{aligned} \tag{3·3a}$$

$$\begin{aligned}\ln(1+\eta_n(\Lambda))+&\int_{-\pi}^{\pi}\frac{dk\cos k}{\pi}\frac{nU}{(nU)^2+(\sin k-\Lambda)^2}\ln(1+\zeta^{-1}(k))\\&=\frac{2n\mu_0H}{T}+\sum_{m=1}^{\infty}A_{nm}\ln(1+\eta_m^{-1}(\Lambda)),\end{aligned} \tag{3·3b}$$

$$\begin{aligned}\ln(1+\eta_n'(\Lambda))+&\int_{-\pi}^{\pi}\frac{dk\cos k}{\pi}\frac{nU}{(nU)^2+(\sin k-\Lambda)^2}\ln(1+\zeta^{-1}(k))\\&=\frac{4\operatorname{Re}\sqrt{1-(\Lambda-nUi)^2}-2nA}{T}+\sum_{m=1}^{\infty}A_{nm}\ln(1+\eta_m'^{-1}(\Lambda)).\end{aligned} \tag{3·3c}$$

Equations (3·1) are rewritten as

$$\frac{1}{2\pi}=(1+\zeta(k))\rho(k)-\cos k\left\{\sum_{n=1}^{\infty}\int_{-\infty}^{\infty}\frac{nU}{\pi}\cdot\frac{(\sigma_n(\Lambda)+\sigma_n'(\Lambda))d\Lambda}{(nU)^2+(\sin k-\Lambda)^2}\right\}, \tag{3·4a}$$

$$\int_{-\pi}^{\pi}\frac{nU}{\pi}\frac{\rho(k)dk}{(nU)^2+(\sin k-\Lambda)^2}=\eta_n(\Lambda)\sigma_n(\Lambda)+\sum_{m=1}^{\infty}A_{nm}\sigma_m(\Lambda), \tag{3·4b}$$

$$\frac{1}{\pi}\operatorname{Re}\frac{1}{\sqrt{1-(\Lambda-nUi)^2}}-\int_{-\pi}^{\pi}\frac{nU}{\pi}\cdot\frac{\rho(k)dk}{(nU)^2+(\sin k-\Lambda)^2}$$
$$=\eta_n'(\Lambda)\sigma_n'(\Lambda)+\sum_{m=1}^{\infty}A_{nm}\sigma_m(\Lambda). \tag{3·4c}$$

Substituting Eqs. (3·3) and (3·4) into (3·2) we have an expression for thermodynamic potential:

$$\Omega/N_a=-T\int_{-\pi}^{\pi}\ln(1+\zeta^{-1}(k))\frac{dk}{2\pi}-T\sum_{n=1}^{\infty}\int_{-\infty}^{\infty}\ln(1+\eta_n'^{-1}(\Lambda))\operatorname{Re}\frac{1}{\sqrt{1-(\Lambda-nUi)^2}}\frac{d\Lambda}{\pi}, \tag{3·5a}$$

and this can be transformed as

$$=E_0-A-T\left\{\int_{-\pi}^{\pi}\rho_0(k)\ln(1+\zeta(k))dk+\int_{-\infty}^{\infty}\sigma_0(\Lambda)\ln(1+\eta_1(\Lambda))d\Lambda\right\}, \tag{3·5b}$$

where E_0, $\rho_0(k)$ and $\sigma_0(\Lambda)$ are energy, $\rho(k)$ and $\sigma_1(\Lambda)$ at $T=0$, $H=0$ and $A=2U$. According to Lieb and Wu,

$$E_0=-4\int_0^{\infty}\frac{J_0(\omega)J_1(\omega)d\omega}{1+\exp(2U\omega)},$$

$$\sigma_0(\Lambda)=\int_{-\pi}^{\pi}\frac{1}{4U}\operatorname{sech}\frac{\pi(\Lambda-\sin k)}{2U}\frac{dk}{2\pi},$$

$$\rho_0(k)=\frac{1}{2\pi}+\cos k\int_{-\infty}^{\infty}\frac{d\Lambda}{\pi}\frac{U}{U^2+(\Lambda-\sin k)^2}\sigma_0(\Lambda).$$

Equations (3·3) can be replaced by the following integral equations:

$$\ln\zeta=-\frac{2\cos k}{T}+\int_{-\infty}^{\infty}\frac{d\Lambda}{4U}\operatorname{sech}\frac{\pi(\Lambda-\sin k)}{2U}\left[\frac{4}{T}-\operatorname{Re}\sqrt{1-(\Lambda-Ui)^2}+\ln\left(\frac{1+\eta_1'}{1+\eta_1}\right)\right], \tag{3·6a}$$

$$\ln\eta_1=s*\left\{\ln(1+\eta_2)-\int_{-\pi}^{\pi}\delta(\Lambda-\sin k)\ln(1+\zeta^{-1})\cos k dk\right\}, \tag{3·6b}$$

$$\ln\eta_1'=s*\left\{\ln(1+\eta_2')-\int_{-\pi}^{\pi}\delta(\Lambda-\sin k)\ln(1+\zeta)\cos k dk\right\}, \tag{3·6c}$$

$$\ln\eta_n=s*\ln\{(1+\eta_{n-1})(1+\eta_{n+1})\},\qquad n=2,3,\cdots, \tag{3·6d}$$

$$\ln\eta_n'=s*\ln\{(1+\eta_{n-1}')(1+\eta_{n+1}')\},\qquad n=2,3,\cdots, \tag{3·6e}$$

with asymptotic conditions

$$\lim_{n\to\infty}\frac{1}{n}\ln\eta_n=\frac{2\mu_0 H}{T}, \tag{3·6f}$$

$$\lim_{n\to\infty}\frac{1}{n}\ln\eta_n'=\frac{4U-2A}{T}. \tag{3·6g}$$

Here s^* is an operator defined by

$$s^*f(\Lambda)=\int_{-\infty}^{\infty}d\Lambda'\frac{1}{4U}\operatorname{sech}\frac{\pi(\Lambda-\Lambda')}{2U}f(\Lambda').$$

§ 4. Special cases

1) *Limit* $U\to\infty$

In this case the equations become

$$\eta_n'=\infty,$$

$$\ln\zeta=\frac{-2\cos k-\mu_0 H-A}{T},$$

$$\ln(1+\eta_n(\Lambda))=\frac{2n\mu_0 H}{T}+\sum_{m=1}^{\infty}A_{nm}\ln(1+\eta_m^{-1}(\Lambda)),$$

and

$$\sigma_n'(\Lambda)=0,$$

$$\rho(k)=\left(2\pi\right)^{-1}\left\{\exp\left(-\frac{2\cos k+\mu_0 H+A}{T}\right)+1\right\}^{-1},$$

$$\left(\frac{N}{N_a}\right)\frac{1}{\pi}\cdot\frac{nU}{(nU)^2+\Lambda^2}=\eta_n(\Lambda)\sigma_n(\Lambda)+\sum_{m=1}^{\infty}A_{nm}\sigma_m(\Lambda).$$

The solution of these equations are

$\eta_n=f^2(n)-1, \qquad f(n)=(z^{n+1}-z^{-n-1})/(z-z^{-1}),$

$z=\exp(-\mu_0 H/T),$

$$\sigma_n(\Lambda)=\frac{1}{\pi f(1)}\left(\frac{N}{N_a}\right)\left\{\frac{1}{f(n-1)f(n)}\cdot\frac{nU}{(nU)^2+\Lambda^2}-\frac{1}{f(n)f(n+1)}\cdot\frac{(n+2)U}{(n+2)^2U^2+\Lambda^2}\right\}.$$

The magnetization is

$$S_z/N_a=\frac{1}{2}\frac{N}{N_a}-\sum_{n=1}^{\infty}n\int_{-\infty}^{\infty}\sigma_n(\Lambda)d\Lambda=\frac{1}{2}\frac{N}{N_a}\tanh\left(\frac{\mu_0 H}{T}\right).$$

Therefore the magnetization of the one-dimensional Hubbard model behaves as that of $\frac{1}{2}$-spins which are free each other when U is infinity. This fact was already noted by Sokoloff.[4]

2) *Limit* $U\to 0$

We define $\rho_-(\Lambda)$, $\rho_+(\Lambda)$, $\zeta_-(\Lambda)$ and $\zeta_+(\Lambda)$ as

$$\rho_-(\Lambda)\equiv\frac{1}{\sqrt{1-\Lambda^2}}\rho(\sin^{-1}\Lambda),\qquad \zeta_-(\Lambda)\equiv\zeta(\sin^{-1}\Lambda),$$

$$\rho_+(\Lambda)\equiv\frac{1}{\sqrt{1-\Lambda^2}}\rho(\pi-\sin^{-1}\Lambda),\qquad \zeta_+(\Lambda)\equiv\zeta(\pi-\sin^{-1}\Lambda),$$

where $|\sin^{-1}\Lambda|\leqq\pi/2$, $|\Lambda|\leqq 1$. Then integral equations (3·3) are reduced to a set of equations

$$\ln\zeta_\pm=\frac{\pm2\sqrt{1-\Lambda^2}-\mu_0H-A}{T}+\sum_{n=1}^{\infty}\{\ln(1+\eta_n'^{-1})-\ln(1+\eta_n^{-1})\},\tag{4·1a}$$

$$\ln(1+\eta_n)+\ln(1+\zeta_-^{-1})-\ln(1+\zeta_+^{-1})=\frac{2n\mu_0H}{T}+\sum_{m=1}^{\infty}a_{nm}\ln(1+\eta_m^{-1}),\tag{4·1b}$$

$$\ln(1+\eta_n')+\ln(1+\zeta_-^{-1})-\ln(1+\zeta_+^{-1})=\frac{4\sqrt{1-\Lambda^2}-2nA}{T}+\sum_{m=1}^{\infty}a_{nm}\ln(1+\eta_m'^{-1})\tag{4·1c}$$

where $a_{nm}\equiv2\,\mathrm{Min}(n,m)$. It is clear that functions $\rho_\pm$, $\zeta_\pm$, η_n, η_n', σ_n and σ_n' have physical meaning only at $|\Lambda|\leqq1$. The solutions (4·1b) and (4·1c) are

$$\eta_n=f^2(n)-1,\qquad f(n)=\frac{az^n-a^{-1}z^{-n}}{z-z^{-1}},\qquad z=\exp\left(-\frac{\mu_0H}{T}\right),\tag{4·2a}$$

$$\eta_n'=f'^2(n)-1,\qquad f'(n)=\frac{bw^n-b^{-1}w^{-n}}{w-w^{-1}},\qquad w=\exp\left(\frac{A}{T}\right),\tag{4·2b}$$

with the conditions

$$f^2(0)=\frac{1+\zeta_+^{-1}}{1+\zeta_-^{-1}},\tag{4·3a}$$

$$f'^2(0)=\exp\left(\frac{4\sqrt{1-\Lambda^2}}{T}\right)\frac{1+\zeta_+^{-1}}{1+\zeta_-^{-1}}.\tag{4·3b}$$

Equations (4·1a) are rewritten as

$$\zeta_+=f'(1)/f(1),\qquad \zeta_-=\exp\left(-\frac{4\sqrt{1-\Lambda^2}}{T}\right)f'(1)/f(1).$$

Substituting these into (4·3a) and (4·3b) we have

$$f(-1)+f'(-1)=0\qquad\text{and}\qquad xf(0)=f'(0),\tag{4·4}$$

where

$$x=\exp\left(\frac{2\sqrt{1-\Lambda^2}}{T}\right).$$

From these algebraic equations we have

$$a=\sqrt{\frac{x^2+xz(w+w^{-1})+z^2}{x^2+xz^{-1}(w+w^{-1})+z^{-2}}},\qquad b=\sqrt{\frac{x^{-2}+x^{-1}w(z+z^{-1})+w^{-2}}{x^{-2}+x^{-1}w^{-1}(z+z^{-1})+w^{-2}}}.\tag{4·5}$$

Equations (3·4a), (3·4b) and (3·4c) become

$$\frac{1}{2\pi}\frac{1}{\sqrt{1-\Lambda^2}}=(1+\zeta_\pm)\rho_\pm\pm\sum_{n=1}^{\infty}(\sigma_n+\sigma_n'),\tag{4·6a}$$

$$\rho_++\rho_-=\eta_n\sigma_n\pm\sum_{m=1}^{\infty}a_{nm}\sigma_m,\tag{4·6b}$$

$$\frac{1}{\pi}\frac{1}{\sqrt{1-\Lambda^2}}-\rho_+-\rho_-=\eta_n'\sigma_n'+\sum_{m=1}^{\infty}a_{nm}\sigma_m'.\tag{4·6c}$$

Equations (4·6b) and (4·6c) are easily solved and the results are

$$\sigma_n=\frac{f(0)(\rho_++\rho_-)}{a+a^{-1}}\left\{\frac{1}{f(n-1)f(n)}-\frac{1}{f(n)f(n+1)}\right\},\tag{4·7a}$$

$$\sigma_n'=\frac{f'(0)(\pi^{-1}(1-\Lambda^2)^{-1/2}-\rho_+-\rho_-)}{b+b^{-1}}\left\{\frac{1}{f'(n-1)f'(n)}-\frac{1}{f'(n)f'(n+1)}\right\}.\tag{4·7b}$$

Substituting these into Eqs. (4·6b), we have a set of coupled equations for ρ_+ and ρ_-. The momentum distribution of electrons are given by

$$\rho_++\sum_{n=1}^{\infty}(n+1)\sigma_n'=\rho_++\left(\frac{1}{\pi}\frac{1}{\sqrt{1-\Lambda^2}}-\rho_+-\rho_-\right)\frac{1}{b+b^{-1}}\left(b+\frac{1}{f'(1)}\right)\tag{4·8a}$$

at $\pi\geqq|k|\geqq\pi/2$ and

$$\rho_-+\sum_{n=1}^{\infty}(n-1)\sigma_n'=\rho_-+\left(\frac{1}{\pi}\frac{1}{\sqrt{1-\Lambda^2}}-\rho_+-\rho_-\right)\frac{1}{b+b^{-1}}\left(b-\frac{1}{f'(1)}\right)\tag{4·8b}$$

at $|k|\leqq\pi/2$. A simple algebraic calculation gives

$$(4·8a)=\frac{1}{2\pi\sqrt{1-\Lambda^2}}\left\{\left(1+\exp\frac{2\sqrt{1-\Lambda^2}+\mu_0H-A}{T}\right)^{-1}+\left(1+\exp\frac{2\sqrt{1-\Lambda^2}-\mu_0H-A}{T}\right)^{-1}\right\},$$

$$(4·8b)=\frac{1}{2\pi\sqrt{1-\Lambda^2}}\left\{\left(1+\exp\frac{-2\sqrt{1-\Lambda^2}+\mu_0H-A}{T}\right)^{-1}+\left(1+\exp\frac{-2\sqrt{1-\Lambda^2}-\mu_0H-A}{T}\right)^{-1}\right\}.$$

Thus we deduce that the number of quasi-momenta between k and $k+dk$ is

$$\frac{1}{2\pi}\left(\left(1+\exp\frac{-2\cos k-\mu_0H-A}{T}\right)^{-1}+\left(1+\exp\frac{-2\cos k+\mu_0H-A}{T}\right)^{-1}\right)dk.\tag{4·9}$$

This result is the well-known momentum distribution of fermions at finite temperature.

3) *Limit* $T\to 0$

We put $\varepsilon_n(\Lambda)\equiv T\ln\eta_n(\Lambda)$, $\varepsilon_n'(\Lambda)\equiv T\ln\eta_n'(\Lambda)$ and $\kappa(k)\equiv T\ln\zeta(k)$. Substituting (3·3a), Eq. (3·3c) is transformed as

$$\ln\eta_n'(\Lambda)=\frac{n(4U-2\Lambda)}{T}-\int_{-\pi}^{\pi}\frac{dk\cos k}{\pi}\frac{nU}{n^2U^2+(\Lambda-\sin k)^2}\ln(1+\zeta(k))$$
$$+\sum_m T_{nm}\ln(1+\eta_m'^{-1}(\Lambda)),$$

where

$$T_{nm}\equiv A_{nm}-\delta_{nm}[0].$$

As $\Lambda\leqq 2U$ we have

$$\varepsilon_n'(\Lambda)\geqq 0\,,\qquad n=1,2\cdots.$$

And from (3·3b) we have

$$\varepsilon_n(\Lambda)=2(n-1)\mu_0H+[n-1]T\ln\left(1+\exp\frac{\varepsilon_1(\Lambda)}{T}\right)$$
$$+T\sum_{m=2}^{\infty}T_{n-1,m-1}\ln\left(1+\exp-\frac{\varepsilon_m(\Lambda)}{T}\right),\qquad n=2,3,\cdots.$$

Therefore $\varepsilon_2(\Lambda)$, $\varepsilon_3(\Lambda)\cdots$ are always positive. So in the limit $T\to 0$ we have a set of equations in spite of (3·3):

$$\kappa(k)=-2\cos k-\mu_0H-A+\int_{-\infty}^{\infty}\frac{U}{\pi}\frac{1}{U^2+(\sin k-\Lambda)^2}\varepsilon_1^-(\Lambda)d\Lambda\,,\tag{4·10a}$$

$$\varepsilon_1(\Lambda)=\int_{-\pi}^{\pi}\frac{dk\cos k}{\pi}\cdot\frac{U}{U^2+(\sin k-\Lambda)^2}\kappa^-(k)+2\mu_0H-[2]\varepsilon_1^-(\Lambda),\tag{4·10b}$$

where the suffices '+' and '−' mean

$$f^+(k)\equiv\begin{cases}f(k) & \text{at } f(k)>0\,,\\ 0 & \text{at } f(k)\leqq 0\,,\end{cases}\qquad f^-(k)\equiv\begin{cases}0 & \text{at } f(k)\geqq 0\,,\\ f(k) & \text{at } f(k)<0\,.\end{cases}$$

In the Appendix we shall prove that (4·10a) and (4·10b) have a set of solutions $\varepsilon_1(\Lambda)$ and $\kappa(k)$ which are monotonically increasing functions of Λ^2 and k^2. So ε_1 and κ are negative in the regions $[B,-B]$ and $[Q,-Q]$. Equations (3·4) can be written as

$$\rho(k)=\frac{1}{2\pi}+\cos k\int_{-B}^{B}\frac{U}{\pi}\cdot\frac{\sigma_1(\Lambda)d\Lambda}{U^2+(\Lambda-\sin k)^2}\quad\text{at } |k|\leqq Q\,,$$
$$\rho(k)=0\quad\text{at } \pi\geqq|k|>Q\,,\tag{4·11a}$$

$$\int_{-Q}^{Q}\frac{U}{\pi}\cdot\frac{\rho(k)dk}{U^2+(\sin k-\Lambda)^2}=\sigma_1(\Lambda)+\frac{2U}{\pi}\int_{-B}^{B}\frac{\sigma_1(\Lambda')d\Lambda'}{4U^2+(\Lambda-\Lambda')^2}\quad\text{at } |\Lambda|\leqq B\,,$$
$$\sigma(\Lambda)=0\quad\text{at } |\Lambda|\geqq B\,.\tag{4·11b}$$

These equations coincide with those of Lieb and Wu.[1)]

4) *Limit* $T \to \infty$

We go to the limit $T \to \infty$ with $\mu_0 H/T$ and A/T kept finite. In this limit the term $-2\cos k/T$ and $4\,\mathrm{Re}\sqrt{1-(\Lambda-Ui)^2}/T$ in (3·3a) and (3·3c) are neglected. So we see that ζ, η_n and η_n' are all constants. The solutions of (3·3) are

$$\zeta = \frac{w+w^{-1}}{z+z^{-1}}, \qquad \eta_n = g^2(n) - 1, \qquad \eta_n' = g'^2(n) - 1,$$

where

$$g(n) \equiv \frac{z^{n+1} - z^{-n-1}}{z - z^{-1}}, \qquad g'(n) \equiv \frac{w^{n+1} - w^{-n-1}}{w - w^{-1}},$$

$$z \equiv \exp\left(-\frac{\mu_0 H}{T}\right), \qquad w \equiv \exp\left(\frac{A-2U}{T}\right).$$

And the solutions of (3·4) are

$$\sigma_n = \frac{1}{\pi}\left(\frac{1}{z+z^{-1}+w+w^{-1}}\right)\left\{\frac{1}{g(n-1)\,g(n)}\,\mathrm{Re}\,\frac{1}{\sqrt{1-(\Lambda-nUi)^2}} - \frac{1}{g(n)\,g(n+1)}\,\mathrm{Re}\,\frac{1}{\sqrt{1-(\Lambda-(n+2)\,Ui)^2}}\right\}, \tag{4·12a}$$

$$\sigma_n' = \frac{1}{\pi}\left(\frac{1}{z+z^{-1}+w+w^{-1}}\right)\left\{\frac{1}{g'(n-1)\,g'(n)}\,\mathrm{Re}\,\frac{1}{\sqrt{1-(\Lambda-nUi)^2}} - \frac{1}{g'(n)\,g'(n+1)}\,\mathrm{Re}\,\frac{1}{\sqrt{1-(\Lambda-(n+2)\,Ui)^2}}\right\}, \tag{4·12b}$$

$$\rho = \frac{1}{1+\zeta}\left[\frac{1}{2\pi} + \frac{1}{\pi}\cdot\frac{\cos k}{(z+z^{-1})(w+w^{-1})}\,\mathrm{Re}\frac{1}{\sqrt{1-(\sin k - 2Ui)^2}}\right]. \tag{4·12c}$$

So one obtains

$$n_\downarrow = \sum_{n=1}^{\infty} n \int_{-\infty}^{\infty} \{\sigma_n(\Lambda) + \sigma_n'(\Lambda)\}\,d\Lambda = \frac{1}{1+z^{-1}w^{-1}}, \tag{4·12d}$$

$$n_\uparrow = \int_{-\pi}^{\pi} \rho(k)\,dk + \sum_{n=1}^{\infty} 2n \int_{-\infty}^{\infty} \sigma_n'(\Lambda)\,d\Lambda - n_\downarrow = \frac{1}{1+zw^{-1}}. \tag{4·12e}$$

This result coincides with the first term of high-temperature expansion of thermodynamic potential by usual method.

§ 5. Discussion

In these series of papers[3),5)] we derived non-linear integal equations for thermodynamics of the Heisenberg model, electron gas and Hubbard model in one dimension. Our integral equations do not contradict known exact results of these models. A weak point of our theory is that we used some conjectures to derive these equations. A more rigorous derivation is desired.

Appendix

Let us consider the series of functions $\{\kappa^{(n)}(k)\}$, $\{\varepsilon_1^{(n,m)}(\Lambda)\}$ and $\{\varepsilon_1^{(n)}(\Lambda)\}$ which are defined by

$$\kappa^{(1)} = -2\cos k - \mu_0 H - A\,, \tag{A1}$$

$$\varepsilon_1^{(n,1)} = 2\mu_0 H\,, \tag{A2}$$

$$\varepsilon_1^{(n,m+1)}(\Lambda) = \mu_0 H + \int_{-\pi}^{\pi} \frac{dk\cos k}{4U}\,\mathrm{sech}\frac{\pi(\Lambda - \sin k)}{2U}\,\kappa^{(n)-}(k) + \int_{-\infty}^{\infty}\frac{d\Lambda'}{U}\,R\Big(\frac{\Lambda-\Lambda'}{U}\Big)\varepsilon^{(n,m)+}(\Lambda'), \tag{A3}$$

$$\varepsilon_1^{(n)} = \lim_{m\to\infty} \varepsilon^{(n,m)}, \tag{A4}$$

$$\kappa^{(n+1)}(k) = -2\cos k - \mu_0 H - A + \int_{-\infty}^{\infty}\frac{U}{\pi}\,\frac{\varepsilon_1^{(n)-}(\Lambda)\,d\Lambda}{U^2 + (\sin k - \Lambda)^2}\,. \tag{A5}$$

By the method of mathematical induction we easily prove the following lemmas:

lemma 1

a) $\varepsilon_1^{(n,m)} \geqq \varepsilon_1^{(n,m+1)}$,

b) $\varepsilon_1^{(n,m)} \geqq -2\int \frac{dk\cos^2 k}{4U}\,\mathrm{sech}\frac{\pi(\Lambda-\sin k)}{2U}$.

From this lemma we see that the limit (A4) exists.

lemma 2

a) $\varepsilon_1^{(n,m)} \geqq \varepsilon_1^{(n+1,m)}$,

b) $\kappa^{(n)} \geqq \kappa^{(n+1)}$,

c) $\varepsilon_1^{(n)} \geqq \varepsilon_1^{(n+1)}$.

[Proof] It is clear from (A1) and (A5) that $\kappa^{(1)} \geqq \kappa^{(2)}$. Then from (A2) and (A3) we have $\varepsilon_1^{(1,m)} \geqq \varepsilon_1^{(2,m)}$. So we have $\varepsilon_1^{(1)} > \varepsilon_1^{(2)}$.

If $\varepsilon_1^{(n-1)} \geqq \varepsilon_1^{(n)}$ we have $\kappa^{(n)} \geqq \kappa^{(n+1)}$, $\varepsilon_1^{(n,m)} \geqq \varepsilon_1^{(n+1,m)}$ and $\varepsilon_1^{(n)} \geqq \varepsilon_1^{(n+1)}$.

[Q.E.D.]

lemma 3

a) $\varepsilon_1^{(n,m)}(\Lambda)$ is an MIF of Λ^2,

b) $\kappa^{(n)}(k)$ is an MIF of k^2 at $0 \leqq k^2 \leqq (\pi/2)^2$.

[Proof] We rewrite the second term of right-hand side of (A3) as

$$\int_{-\infty}^{\infty} d\Lambda' \frac{1}{4U} \operatorname{sech} \frac{\pi(\Lambda-\Lambda')}{2U} K^{(n)}(\Lambda'),$$

where

$$K^{(n)}(\Lambda') \equiv \int_{-\pi}^{\pi} dk \cos k \delta(\Lambda' - \sin k) \kappa^{(n)-}(k).$$

We see that $K^{(n)}(\Lambda')$ is a continuous function and takes the values zero, $-4\sqrt{1-\Lambda^2}$ or $\kappa^{(n)}(\sin^{-1}\Lambda)$. So if $\kappa^{(n)}(k)$ is an MIF of k^2 at $0 \leqq k^2 \leqq (\pi/2)^2$, $K^{(n)}(\Lambda')$ is an MIF of Λ'^2. $\kappa^{(1)}(k)$ is an MIF of k^2 at $0 \leqq k^2 \leqq (\pi/2)^2$. Then $\varepsilon_1^{(1,m+1)}$ is an MIF of Λ^2 if $\varepsilon^{(1,m)}$ is an MIF of Λ^2. $\varepsilon_1^{(1,1)}$ is an MIF of Λ^2. Using (A5) we have $\kappa^{(2)}$ is an MIF of k^2 at $0 \leqq k^2 \leqq (\pi/2)^2$.

If $\kappa^{(n)}$ is an MIF of k^2 at $0 \leqq k^2 \leqq (\pi/2)^2$, $\varepsilon^{(n,m)}$ and $\varepsilon^{(n)}$ are MIF of Λ^2. So we have $\kappa^{(n+1)}$ is an MIF of k^2 at $0 \leqq k^2 \leqq (\pi/2)^2$. [Q.E.D.]

lemma 4

$\kappa^{(n+1)}(k)$ is an MIF of k^2 at $(\pi/2)^2 \leqq k^2 \leqq \pi^2$.

[Proof] $\varepsilon_1^{(n)}$ satisfies the equation

$$\varepsilon_1^{(n)} = \mu_0 H + \int_{-\pi}^{\pi} \frac{dk \cos k}{4U} \operatorname{sech} \frac{\pi(\Lambda - \sin k)}{2U} (-2\cos k)$$

$$- \int_{-\pi}^{\pi} \frac{dk \cos k}{4U} \operatorname{sech} \frac{\pi(\Lambda - \sin k)}{2U} \kappa^{(n)+} + \int \frac{d\Lambda'}{U} R\left(\frac{\Lambda - \Lambda'}{U}\right) \varepsilon_1^{(n)+}(\Lambda').$$

Substituting this into (A5) we have

$$\kappa^{(n+1)}(k) = \left\{-2\cos k + 2\int_{-\pi}^{\pi} \frac{dk' \cos^2 k'}{U} R\left(\frac{\sin k - \sin k'}{U}\right)\right\}$$

$$- \int_{-\pi}^{\pi} \frac{dk \cos k'}{U} R\left(\frac{\sin k - \sin k'}{U}\right) \kappa^{(n)+}(k)$$

$$- \int_{-\infty}^{\infty} \frac{d\Lambda}{4U} \operatorname{sech} \frac{\pi(\Lambda - \sin k)}{2U} \varepsilon_1^{(n)+}(\Lambda) - A.$$

We see easily that the all terms of r.h.s. are MIF's of k^2 at $(\pi/2)^2 \leqq k^2 \leqq \pi^2$ if $\kappa^{(n)+}$ is an MIF of k^2 at $(\pi/2)^2 \leqq k^2 \leqq \pi^2$. From (A1) we see that $\kappa^{(1)}$ satisfies this condition. So by the mathematical induction we have lemma 4. [Q.E.D.]

If we define $\varepsilon_1(\Lambda)$ and $\kappa(k)$ by

$$\varepsilon_1(\Lambda) \equiv \lim_{n\to\infty} \varepsilon_1^{(n)}(\Lambda), \qquad \kappa(k) \equiv \lim_{n\to\infty} \kappa^{(n)}(k),$$

these functions satisfy the equations

$$\varepsilon_1(\Lambda) = \mu_0 H + \int_{-\pi}^{\pi} \frac{dk \cos k}{4U} \operatorname{sech} \frac{\pi(\Lambda - \sin k)}{2U} \kappa^-(k) + \int_{-\infty}^{\infty} \frac{d\Lambda'}{U} R\left(\frac{\Lambda - \Lambda'}{U}\right) \varepsilon_1^+(\Lambda'),$$

$$\kappa(k) = -2\cos k - \mu_0 H - A + \int_{-\infty}^{\infty} \frac{U}{\pi} \cdot \frac{\varepsilon_1^-(\Lambda) d\Lambda}{U^2 + (\Lambda - \sin k)^2}.$$

These equations are equivalent to (4·10a) and (4·10b). So we see that equations (4·10a) and (4·10b) have solutions $\varepsilon_1(\Lambda)$ and $\kappa(k)$ which are monotonically increasing functions of Λ^2 and k^2, respectively.

References

1) E. H. Lieb and F. Y. Wu, Phys. Rev. Letters **20** (1968), 1445.
2) M. Takahashi, Prog. Theor. Phys. **42** (1969), 1098; **43** (1970), 860, 1619.
3) M. Takahashi, Prog. Theor. Phys. **46** (1971), 1388.
4) J. B. Sokoloff, Phys. Rev. **B2** (1970), 779.
5) M. Takahashi, Prog. Theor. Phys. **46** (1971), 401.

Progress of Theoretical Physics, Vol. 52, No. 1, July 1974

Low-Temperature Specific-Heat of One-Dimensional Hubbard Model

Minoru TAKAHASHI

Department of Physics, College of General Education Osaka University, Osaka

(Received February 18, 1974)

Low-temperature specific heat per site (C) of one-dimensional Hubbard model is investigated by the method of non-linear integral equations. For the half-filled case we show $\lim_{H\to 0}\lim_{T\to 0} C/T=\pi I_0(\pi/2U)/(6I_1(\pi/2U))$, where T is temperature, H is magnetic field, U is the coupling constant, and I_0 and I_1 are modified Bessel functions. Although this equation yeilds $\lim_{T,H\to 0} C/T=\pi/6$ in the limit $U\to 0+$, the true value of $\lim_{T,H\to 0} C/T$ at $U=0$ is $\pi/3$. This means that $\lim_{T,H\to 0} C/T$ is a discontinuous function of U at $U=0$. This discontinuity disappears when the band is not half filled.

§ 1. Introduction

Low-temperature behavior of Hubbard model is interesting physically, and difficult to treat rigorously. The one-dimensional case of this model has been investigated by many physicists. Its thermodynamic potential density is difined by

$$\omega(U,T,A,H) = -T\lim_{N_a\to\infty}\{\ln(\mathrm{Tr}\exp(-T^{-1}(\mathcal{H}-A\sum_{i=1}^{N_a}(n_{i\uparrow}+n_{i\downarrow})))/N_a\}, \tag{1·1a}$$

where $\mathcal{H}$ is the Hamiltonian:

$$\mathcal{H}=-\sum_{i=1}^{N_a}\sum_{\sigma}(c_{i\sigma}^\dagger c_{i+1\sigma}+c_{i+1\sigma}^\dagger c_{i\sigma})+4U\sum_{i=1}^{N_a}n_{i\uparrow}n_{i\downarrow}-\mu_0 H\sum_{i=1}^{N_a}(n_{i\uparrow}-n_{i\downarrow}),$$

$$c_{N_a+1\sigma}\equiv c_{1\sigma}, \qquad n_{i\sigma}\equiv c_{i\sigma}^\dagger c_{i\sigma}. \tag{1·1b}$$

Here we have following symmetry relations through appropriate unitary transformations:

$$\begin{aligned}\omega(U,T,A,H)&=\omega(U,T,A,-H)=4U-2A+\omega(U,T,4U-A,H)\\&=\mu_0H-A+\omega(-U,T,\mu_0H-2U,\mu_0^{-1}(A-2U)).\end{aligned} \tag{1·2}$$

The first identity is obtained by changing up-spin and down-spin, the second by changing the creation and annihilation operators and the third by changing the creation and annihilation operators in the up-spin band. If we know the value of ω in the region $U\geq 0$, $H\geq 0$ and $A\leq 2U$, we easily obtain the value of ω outside of this region through the relations (1·2). Then we restrict ourselves to calculate ω in this region. Other thermodynamic quantities such as energy and entropy per site (e, s), specific heat per site $(C_{H,A})$ and densities of up-spin and down-spin electrons $(n_\uparrow, n_\downarrow)$ are obtained by the differentiations of ω:

$$n_\uparrow + n_\downarrow = \frac{\partial\omega}{\partial A}, \qquad n_\uparrow - n_\downarrow = \frac{1}{\mu_0}\frac{\partial\omega}{\partial H}, \qquad e = -T^2\frac{\partial}{\partial T}\left(\frac{\omega}{T}\right) + \frac{\partial\omega}{\partial A}A,$$

$$S = -\frac{\partial\omega}{\partial T}, \qquad C_{H,A} = -T\frac{\partial^2\omega}{\partial T^2}, \qquad \chi = -\frac{\partial^2\omega}{\partial H^2}. \tag{1·3}$$

In a previous paper[1] the author derived a set of non-linear integral equations for the calculation of thermodynamic potential density ω. We used Bethe ansatz, which was first applied to this model by Lieb and Wu,[2] and some assumptions on the distributions of quasi-momenta k and parameters Λ on the complex plane. Recently Shiba and Pincus[3] calculated the energy levels of this model in the case of finite atomic numbers (such as six or five) and thermodynamic quantities. Their method is not useful to investigate the low-temperature properties of the model in the thermodynamic limit. For example, magnetic susceptibility of the finite system becomes zero or infinity in the limit of zero temperature. But this is not valid in the thermodynamic limit because magnetic susceptibility has finite values at $T=0$ in the half-filled state.[4),5)]

In the following sections we investigate the low-temperature behavior of this system, using the set of integral equations given in Ref. 1), and come to the conclusion that in the half-filled case low-temperature specific heat is proportional to temperature and coefficient is given analytically:

$$\lim_{H\to 0}\lim_{T\to 0} C/T = \pi I_0(\pi/2U)/(6I_1(\pi/2U)).$$

It should be noted that this is inversely proportional to the magnon velocity[6] at $T=0$:

$$v = 2I_1(\pi/2U)/I_0(\pi/2U),$$

and proportional to the magnetic susceptibility[5] at $T=0$:

$$\chi = \mu_0^2 I_0(\pi/2U)/(\pi I_1(\pi/2U)).$$

§ 2. Integral equations

The eigenvalue problem of one-dimensional Hubbard model described by the Hamiltonian (1·1b) can be treated by the method of Bethe's hypothesis. According to Lieb and Wu, we must solve a set of equations for N quasi-momenta k and M parameters Λ where N is the number of fermions and M is the number of down-spin fermions,

$$e^{ik_jN_a} = -\prod_{\alpha=1}^{M}\left(\frac{k_j - \Lambda_\alpha - 2iU}{k_j - \Lambda_\alpha + 2iU}\right), \qquad j = 1, 2, \cdots, N,$$

$$\prod_{j=1}^{N}\left(\frac{\Lambda_\alpha - k_j + iU}{\Lambda_\alpha - k_j - iU}\right) = -\prod_{\beta=1}^{M}\left(\frac{\Lambda_\alpha - \Lambda_\beta + 2iU}{\Lambda_\alpha - \Lambda_\beta - 2iU}\right), \qquad \alpha = 1, 2, \cdots, M.$$

In the previous paper[1] the author assumed that the k's and Λ's form bound states on the complex plane, and derived a set of non-linear integral equations for the distribution of the bound states at given temperature T, magnetic field H and chemical potential A:

$$\ln\zeta(k) = \kappa_0(k)/T + \int_{-\infty}^{\infty} s(\Lambda - \sin k)\ln((1+\eta_1'(\Lambda))/(1+\eta_1(\Lambda)))d\Lambda\,, \quad (2\cdot1\text{a})$$

$$\ln\eta_1(\Lambda) = s^*\ln(1+\eta_2(\Lambda)) - \int_{-\pi}^{\pi} dk\cos k\, s(\Lambda-\sin k)\ln(1+\zeta^{-1}(k)), \quad (2\cdot1\text{b})$$

$$\ln\eta_1'(\Lambda) = s^*\ln(1+\eta_2'(\Lambda)) - \int_{-\pi}^{\pi} dk\cos k\, s(\Lambda-\sin k)\ln(1+\zeta(k)), \quad (2\cdot1\text{c})$$

$$\ln\eta_n(\Lambda) = s^*\ln(1+\eta_{n-1}(\Lambda))(1+\eta_{n+1}(\Lambda)), \qquad n=2,3,\cdots, \quad (2\cdot1\text{d})$$

$$\ln\eta_n'(\Lambda) = s^*\ln(1+\eta_{n-1}'(\Lambda))(1+\eta_{n+1}'(\Lambda)), \qquad n=2,3,\cdots, \quad (2\cdot1\text{e})$$

$$\lim_{n\to\infty}\frac{\ln\eta_n}{n} = \frac{2\mu_0 H}{T}\,, \quad (2\cdot1\text{f})$$

$$\lim_{n\to\infty}\frac{\ln\eta_n'}{n} = \frac{4U-2A}{T}\,, \quad (2\cdot1\text{g})$$

where $s(\Lambda) \equiv \text{sech}(\pi x/2U)/4U$, $f^*g \equiv \int_{-\infty}^{\infty} f(\Lambda-\Lambda')g(\Lambda')d\Lambda'$,

$$\kappa_0(k) = -2\cos k - 4\int_{-\infty}^{\infty} s(\Lambda-\sin k)(\text{Re}\sqrt{1-(\Lambda-Ui)^2})d\Lambda\,. \quad (2\cdot1\text{h})$$

Function $\zeta(k)$ is the ratio of hole density and particle density of unbound quasi-momenta. Function $\eta_n(\Lambda)$ is that of n-th order bound state of Λ. Function $\eta_n'(\Lambda)$ is that of bound state of the $n\Lambda$'s and $2nk$'s. Thermodynamic potential per site is given by

$$\omega(T,A,H) = -T\int_{-\pi}^{\pi}\ln(1+\zeta^{-1}(k))\frac{dk}{2\pi} - T\sum_{n=1}^{\infty}\int_{-\infty}^{\infty}\ln(1+\eta_n^{-1}(\Lambda))$$
$$\times\text{Re}\frac{1}{\sqrt{1-(\Lambda-nUi)^2}}\cdot\frac{d\Lambda}{\pi} \quad (2\cdot2\text{a})$$
$$= E_0 - A - T\left\{\int_{-\pi}^{\pi}\rho_0(k)\ln(1+\zeta(k))dk + \int_{-\infty}^{\infty}\sigma_0(\Lambda)\ln(1+\eta_1(\Lambda))d\Lambda\right\}, \quad (2\cdot2\text{b})$$

where E_0, $\rho_0(k)$, $\sigma_0(\Lambda)$ are the ground state energy per site, distribution function of the k's and that of the Λ's at $T=0$, $A=2U$, $\mu_0 H=0$, respectively:

$$\sigma_0(\Lambda) \equiv \int_{-\pi}^{\pi} s(\Lambda-\sin k)\frac{dk}{2\pi}\,, \quad (2\cdot2\text{c})$$

$$\rho_0(k) \equiv \frac{1}{2\pi} + \cos k\int_{-\infty}^{\infty} d\Lambda a_1(\Lambda-\sin k)\sigma_0(\Lambda), \quad (2\cdot2\text{d})$$

$$a_n(\Lambda) \equiv \frac{nU}{\pi(\Lambda^2+(nU)^2)}\,, \quad (2\cdot2\text{e})$$

$$E_0 \equiv -2\int_{-\pi}^{\pi} \cos k\, \rho_0(k)dk\,. \tag{2·2f}$$

One should note that Eqs. (2·1) and (2·2) are valid only at $U \geq 0$, $A \leq 2U$ and $\mu_0 H \geq 0$. The other cases can be treated through Eqs. (1·2).

From Eqs. (2·1c), (2·1e) and (2·1g) we have

$$\ln(1+\eta_n') \geq 2n(2U-A)/T\,, \qquad n=1,2,3,\cdots.$$

At $2U-A \gg T$, we can replace $\ln \eta_n'$ by $\ln(1+\eta_n')$ in Eqs. (2·1c), (2·1e) and (2·1g) and obtain

$$\ln(1+\eta_n') = 2n(2U-A)/T + \int_{-\pi}^{\pi} a_n(\Lambda - \sin k)\ln(1+\zeta(k))\cos k\, dk$$
$$+O(\exp(-(2U-A)/T)), \qquad n=1,2,\cdots. \tag{2·3a}$$

Substituting case $n=1$ of this equation into (2·1a), we have

$$\kappa(k) = \kappa_0(k) + 2U - A + T\int_{-\pi}^{\pi} R(\sin k - \sin k')\ln(1+\exp(\kappa(k')/T))\cos k' dk'$$
$$-T\int_{-\infty}^{\infty} s(\Lambda - \sin k)\ln(1+\exp(\varepsilon_1(\Lambda)/T))d\Lambda + O(T\exp(-(2U-A)/T)), \tag{2·4a}$$

where $R \equiv s * a_1$, $\kappa = T\ln\zeta$, $\varepsilon_1 = T\ln\eta_1$.

At $2U-A = O(T)$, function $\kappa(k)$ is always negative. Then the last term of (2·1c) is of the order of $T^{1/2}\exp(\kappa^{(0)}(\pi)/T)$ at low temperatures. Then we have

$$1+\eta_j' = (\mathrm{sh}\{(j+1)(2U-A)/T\}/\mathrm{sh}\{(2U-A)/T\})^2 + O(T^{1/2}\exp(\kappa^{(0)}(\pi)/T)), \tag{2·3b}$$

where $\kappa^{(0)}$ is κ at zero temperature (hereafter we put (0) for the functions at zero temperature). Substituting this into (2·1a), we have

$$\kappa(k) = \kappa_0(k) + T\ln(2\,\mathrm{ch}\{(2U-A)/T\}) - T\int_{-\infty}^{\infty} s(\Lambda - \sin k)$$
$$\times \ln\left(1+\exp\frac{\varepsilon_1(\Lambda)}{T}\right)d\Lambda + O(T^{3/2}\exp(\kappa^{(0)}(\pi)/T)). \tag{2·4b}$$

At $\mu_0 H \gg T$, we have

$$\ln(1+\eta_n) = a_{n-1}^* \ln(1+\eta_1) + 2(n-1)\mu_0 H/T\,, \qquad n=2,3,\cdots. \tag{2·3c}$$

Substituting this into (2·1b), we have

$$\varepsilon_1(\Lambda) = TR^*\ln(1+\exp(\varepsilon_1(\Lambda)/T)) + \mu_0 H - T\int_{-\pi}^{\pi} dk\cos k\, s(\Lambda - \sin k)$$
$$\times \ln(1+\exp(-\kappa(k)/T)) + O(T\exp(2\mu_0 H/T)). \tag{2·4c}$$

Equations (2·4a) and (2·4c) are transformed as follows:

$$\kappa(k) = -2\cos k - A - \mu_0 H - T\int_{-\infty}^{\infty} a_1(\sin k - \Lambda)\ln(1+\exp(-\varepsilon_1(\Lambda)/T))d\Lambda\,, \tag{2·5a}$$

$$\varepsilon_1(\Lambda) = T\int_{-\infty}^{\infty} a_2(\Lambda-\Lambda')\ln(1+\exp(-\varepsilon_1(\Lambda')/T))d\Lambda'$$
$$-T\int_{-\pi}^{\pi} a_1(\Lambda - \sin k)\ln(1+\exp(-\kappa(k)/T))\cos k\, dk + 2\mu_0 H\,. \tag{2·5b}$$

From Eq. (2·2a) we have

$$\omega(T, A, H) = -T\int_{-\pi}^{\pi} \ln(1+\exp(-\kappa(k)/T))\frac{dk}{2\pi}\,. \tag{2·5c}$$

Here we have neglected the terms which are of the order of $e^{-2\mu_0 H/T}$ or $e^{-(4U-2A)/T}$. Equations (2·4) or (2·5) are useful to obtain thermodynamic potential at $2U-A \gg T$ and $2\mu_0 H \gg T$.

As shown in Fig. 1, (A, H) plane is devided into several regions by the low-temperature properties. The number of fermions per site n has the following properties at zero temperature:

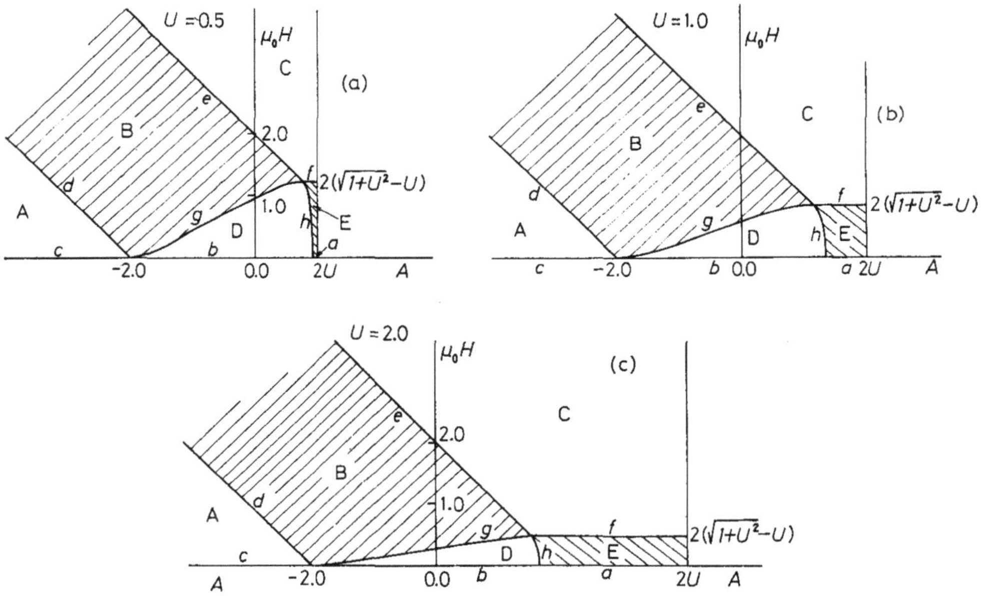

Fig. 1. Characteristic regions of low-temperature specific heat for various values of U. On lines d, e, f, g and h, low-temperature specific heat is proportional to $T^{1/2}$. In regions B, D and E, it is proportional to T. In regions A and C, it is proportional to $T^{-3/2}\exp(-\alpha/T)$.
a) $U=0.5$
b) $U=1.0$
c) $U=2.0$

$n=1$ at regions C and E,

$0<n<1$ at regions B and D,

$n=0$ at region A.

On lines a, b and c, magnetization is zero.

§ 3. Case $\mu_0 H \gg T$

a) $A \leq -2-\mu_0 H$ (*Region A*)

In this region, density of fermions is zero at zero temperature. From Eq. (2·5a), we have

$$\kappa(k) = -2\cos k - A - \mu_0 H - T\exp(-2\mu_0 H/T).$$

Substituting this into (2·5c), we obtain

$$\omega(T,A,H) = -\pi^{-1}T^{3/2}\int_0^\infty \ln\left(1+\exp\left(\frac{2+A+\mu_0 H}{T}\right)e^{-x^2}\right)dx.$$

b) $\varepsilon_1^{(0)}(0) \geq 0$, $A > -2-\mu_0 H$ (*Region B*)

Here the number of fermions per site, n, satisfies $1>n>0$. At zero temperature all fermions have up-spin. From Eq. (2·5c) we have

$$\omega(T,A,H)-\omega(0,A,H) = -T\int_{-\pi}^{\pi}\ln(1+\exp(-|\kappa(k)|/T))\frac{dk}{2\pi} - \int_{-Q}^{Q}\delta\kappa(k)\frac{dk}{2\pi},$$

where Q and $-Q(Q>0)$ are zeroes of $\kappa^{(0)}(k)$. From Eqs. (2·5a) and (2·5b) we obtain

$$\delta\kappa = -T\int_{-\infty}^{\infty} a_1(\sin k - \Lambda)\ln(1+\exp(-\varepsilon_1(\Lambda)/T))d\Lambda,$$

$$\varepsilon_1(\Lambda) = -2\int_{-Q}^{Q} a_1(\Lambda - \sin k)\cos^2 k\, dk + 2\mu_0 H + O(T^2) + O(T^{3/2}\exp(-\varepsilon_1^{(0)}/T)).$$

Then we have

$$\omega(T,A,H) = \omega(0,A,H) - \frac{T^2}{2\pi}\frac{1}{2\sin Q}\cdot\frac{\pi^2}{3} - T^{3/2}2g(0)\sqrt{\frac{2}{\varepsilon_1^{(0)\prime\prime}(0)}}\int_0^\infty \ln(1+\exp(-\varepsilon_1^{(0)}(0)/T)\cdot e^{-x^2})dx,$$

where

$$g(\Lambda) = \int_{-Q}^{Q} a_1(\Lambda - \sin k)\frac{dk}{2\pi}.$$

c) $\mu_0 H \geq 2(\sqrt{1+U^2}-U)$, $A \geq 2-\mu_0 H$ (*Region C*)

At zero temperature, density of fermion is one, and all fermions have up-spin. Substituting Eq. (2·5a) into Eq. (2·5c), we have

$$\omega(T,A,H) = -A-\mu_0 H - T\int_{-\pi}^{\pi} \ln(1+\exp(\kappa(k)/T))\frac{dk}{2\pi}$$
$$-T\int_{-\infty}^{\infty} 2\left(\mathrm{Re}\frac{1}{\sqrt{1-(\Lambda-Ui)^2}}\right)\ln(1+\exp(-\varepsilon_1(\Lambda)/T))d\Lambda\,. \qquad (3\cdot1)$$

From Eqs. (2·5a) and (2·5b), we obtain

$$\kappa(k) = -2\cos k - A - \mu_0 H - O(T^{3/2}\lambda),$$
$$\varepsilon_1(\Lambda) = -4\,\mathrm{Re}(\sqrt{1-(\Lambda-Ui)^2}-U) + 2\mu_0 H + O(T^{3/2}\lambda) + O(T^{3/2}\mu),$$
$$\lambda \equiv \exp(-(4(\sqrt{1+U^2}-U)-2\mu_0 H)/T), \qquad \mu = \exp((2-A-\mu_0 H)/T).$$

Substituting these into Eq. (3·1), we have

$$\omega(T,A,H) = -A-\mu_0 H - \pi^{-1}T^{3/2}\int_0^{\infty}\ln(1+\mu e^{-x^2})dx$$
$$-4T^{3/2}(\sqrt{1+U^2}-U)(1+U^2)^{-1/4}\int_0^{\infty}\ln(1+\lambda e^{-x^2})dx\,. \qquad (3\cdot2)$$

On the boundary of this region we have

$$\omega = \begin{cases} -A-\mu_0 H - \pi^{-1}T^{3/2}\zeta\left(\frac{3}{2}\right)\left(1-\frac{1}{\sqrt{2}}\right)\frac{\sqrt{\pi}}{2} & \text{at } \mu_0 H = 2(\sqrt{1+U^2}-U), \\ -A-\mu_0 H - 4T^{3/2}(\sqrt{1+U^2}-U)(1+U^2)^{-1/4}\zeta\left(\frac{3}{2}\right)\left(1-\frac{1}{\sqrt{2}}\right)\frac{\sqrt{\pi}}{2} & \\ & \text{at } A = 2-\mu_0 H. \end{cases} \qquad (3\cdot3)$$

d) $\varepsilon_1^{(0)}(0)<0$, $\kappa^{(0)}(\pi)>0$ (*Region D*)

From Eq. (2·5c), we have

$$\omega(T,A,H) - \omega(0,A,H) = -\frac{\pi T^2}{6\kappa'(Q)} + \int_{-Q}^{Q}\frac{dk}{2\pi}\delta\kappa(k)\,. \qquad (3\cdot4)$$

Function $\delta\kappa(k)$ is determined by

$$\delta\kappa(k) = \int_{-B}^{B} a_1(\sin k-\Lambda)\delta\varepsilon_1(\Lambda)d\Lambda - \frac{\pi^2 T^2}{6\varepsilon_1'(B)}\{a_1(\sin k - B) + a_1(\sin k + B)\}, \qquad (3\cdot5\mathrm{a})$$

$$\delta\varepsilon(\Lambda) + \int_{-B}^{B} a_2(\Lambda-\Lambda')\delta\varepsilon(\Lambda')d\Lambda' = \int_{-Q}^{Q} dk\cos k\, a_1(\sin k-\Lambda)\delta\kappa(k)$$
$$-\frac{\pi^2 T^2\cos Q}{6\kappa'(Q)}\{a_1(\sin Q-\Lambda) + a_1(\sin Q+\Lambda)\}$$

$$+\frac{\pi^2T^2}{6\varepsilon_1'(B)}\{a_2(B-\Lambda)+a_2(B+\Lambda)\}, \tag{3·5b}$$

where Q and B are zeroes of $\kappa^{(0)}(k)$ and $\varepsilon_1^{(0)}(\Lambda)$, respectively. From these equations we obtain

$$\omega(T,A,H)-\omega(0,A,H)=-\frac{\pi^2T^2}{3}\left[\frac{\sigma_1^{(0)}(B)}{\varepsilon_1^{(0)\prime}(B)}+\frac{\rho^{(0)}(Q)}{\kappa^{(0)\prime}(Q)}\right]+O(T^3), \tag{3·6a}$$

where $\rho^{(0)}(k)$ and $\sigma_1^{(0)}(\Lambda)$ are the distribution functions of k and Λ at zero temperature and determined by

$$\rho_1^{(0)}(k)=\frac{1}{2\pi}+\cos k\int_{-B}^{B}a_1(\Lambda-\sin k)\sigma_1^{(0)}(\Lambda)d\Lambda, \tag{3·6b}$$

$$\sigma_1^{(0)}(\Lambda)+\int_{-B}^{B}a_2(\Lambda-\Lambda')\sigma_1^{(0)}(\Lambda')d\Lambda'=\int_{-Q}^{Q}a_1(\Lambda-\sin k)\rho^{(0)}(k)dk. \tag{3·6c}$$

The equations for $\sigma_1^{(0)}$ and $\varepsilon_1^{(0)\prime}$ are written as

$$\sigma_1^{(0)}(\Lambda)-\int_{|\Lambda|>B}R(\Lambda-\Lambda')\sigma_1^{(0)}(\Lambda')d\Lambda'=\int_{-Q}^{Q}s(\Lambda-\sin k)\rho^{(0)}(k)dk, \tag{3·6d}$$

$$\varepsilon_1^{(0)\prime}(\Lambda)-\int_{|\Lambda|>B}R(\Lambda-\Lambda')\varepsilon_1^{(0)\prime}(\Lambda')d\Lambda'=\int_{-Q}^{Q}s(\Lambda-\sin k)\kappa^{(0)\prime}(k)dk. \tag{3·6e}$$

The right-hand sides of these equations are

$$\exp\left(-\frac{\pi|\Lambda|}{2U}\right)(2U)^{-1}\int_{-Q}^{Q}dk\exp\left(-\frac{\pi\sin k}{2U}\right)\rho^{(0)}(k)$$

and

$$\text{sign}(\Lambda)\exp\left(-\frac{\pi|\Lambda|}{2U}\right)(2U)^{-1}\int_{-Q}^{Q}dk\exp\left(-\frac{\pi\sin k}{2U}\right)\kappa^{(0)\prime}(k)$$

at $|\Lambda|\gg 1$, U. Then we have

$$\frac{\sigma_1^{(0)}(B)}{\varepsilon_1^{(0)\prime}(B)}=\frac{\int_{-Q}^{Q}dk\exp(\pi\sin k/2U)\rho^{(0)}(k)}{\int_{-Q}^{Q}dk\exp(\pi\sin k/2U)\kappa^{(0)\prime}(k)}+O(B^{-2})$$

and

$$\omega(T,A,H)-\omega(0,A,H)=-\frac{\pi^2T^2}{3}\left\{\frac{\rho^{(0)}(Q)}{\kappa^{(0)\prime}(Q)}+\frac{\int_{-Q}^{Q}dk\exp(\pi\sin k/2U)\rho^{(0)}(k)}{\int_{-Q}^{Q}dk\exp(\pi\sin k/2U)\kappa^{(0)\prime}(k)}+O(\{\ln(\mu_0H)\}^{-2})\right\}, \tag{3·7}$$

when μ_0H is very small. From this equation we obtain

$$\lim_{H\to0}\lim_{T\to0}C_{A,H}/T=\frac{2\pi^2}{3}\left\{\frac{\rho^{(0)}(Q)}{\kappa^{(0)\prime}(Q)}+\frac{\int_{-Q}^{Q}dk\exp(\pi\sin k/2U)\rho^{(0)}(k)}{\int_{-Q}^{Q}dk\exp(\pi\sin k/2U)\kappa^{(0)\prime}(k)}\right\}. \tag{3·8}$$

e) $2(\sqrt{1+U^2}-U)>\mu_0 H\gg T$, $\kappa^{(0)}(\pi)\leq 0$ *(Region E)*

From Eq. (2·2b) we have

$$\omega(T,A,H)-\omega(0,A,H)=-2T^{3/2}\rho_0(\pi)\sqrt{\frac{2}{-\kappa(\pi)}}\int_0^\infty \ln\left(1+\exp\left(\frac{\kappa(\pi)}{T}\right)\cdot e^{-x^2}\right)dx$$
$$-T\int_{-\infty}^{\infty}\sigma_0(\Lambda)\ln\left(1+\exp\left(-\frac{|\varepsilon_1(\Lambda)|}{T}\right)\right)d\Lambda-T\int_{|\Lambda|>B}\sigma_0(\Lambda)\delta\varepsilon_1(\Lambda)d\Lambda+O(T^4), \tag{3·9}$$

where $\delta\varepsilon\equiv\varepsilon-\varepsilon^{(0)}$. From Eq. (2·4b) we have

$$\delta\varepsilon(\Lambda)-\int_{|\Lambda'|>B}R(\Lambda-\Lambda')\delta\varepsilon_1(\Lambda')d\Lambda'$$
$$=-\pi^2T^2(R(\Lambda-B)$$
$$+R(\Lambda+B))/(6\varepsilon_1'(B))$$
$$+O((B-B')^2),$$

where B and B' are zeroes of ε_1 and $\varepsilon_1^{(0)}$, respectively. Summing the second and the third terms of r.h.s. of (3·9), we have

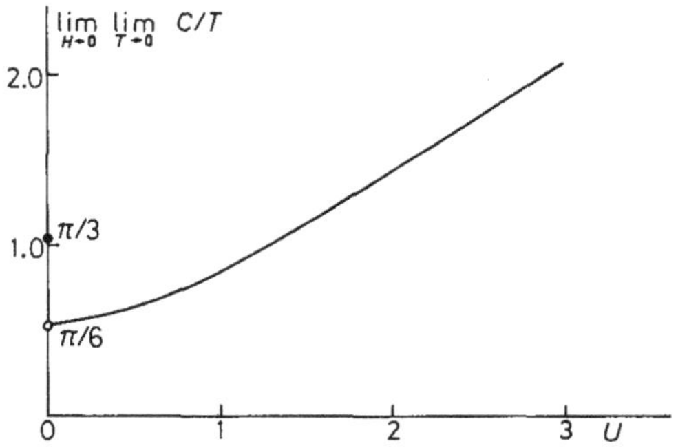

Fig. 2. Coefficient of T-linear low-temperature specific heat in the half-filled case ($A=2U$), and $\mu_0H=0$.

$$-\frac{\pi^2T^2}{3}\frac{\sigma_1^{(0)}(B)}{\varepsilon^{(0)\prime}(B)}+O(T^3), \tag{3·10}$$

where $\sigma_1^{(0)}$ and $\varepsilon_1^{(0)\prime}$ are determined by

$$\sigma_1^{(0)}(\Lambda)-\int_{|\Lambda'|>B}R(\Lambda-\Lambda')\sigma_1^{(0)}(\Lambda')d\Lambda'=\sigma_0(\Lambda), \tag{3·11a}$$

$$\varepsilon_1^{(0)\prime}(\Lambda)-\int_{|\Lambda'|>B}R(\Lambda-\Lambda')\varepsilon_1^{(0)\prime}(\Lambda')d\Lambda'=2\int_{-\pi}^{\pi}s(\Lambda-\sin k)\sin k\,dk\,. \tag{3·11b}$$

At $|\Lambda|\gg\max.(1,1/U)$, r.h.s. of (3·11a) and (3·11b) are $(2U)^{-1}I_0(\pi/2U)e^{-\pi|\Lambda|/2U}$ and $\mathrm{sign}(\Lambda)\cdot 2\pi U^{-1}I_1(\pi/2U)e^{-\pi|\Lambda|/2U}$, respectively. Then we have

$$\frac{\sigma_1^{(0)}(B)}{\varepsilon_1^{(0)\prime}(B)}=\frac{I_0(\pi/2U)}{4\pi I_1(\pi/2U)}+O(B^2)$$

and

$$\omega(T,A,H)-\omega(O,A,H)=-2T^{3/2}\rho_0(\pi)\sqrt{\frac{2}{-\kappa^{(0)\prime}(\pi)}}$$
$$\times\int_0^\infty\ln(1+\exp(\kappa^{(0)}(\pi)/T)e^{-x^2})-\frac{\pi T^2}{12}\cdot\frac{I_0(\pi/2U)}{I_1(\pi/2U)}$$
$$+O((\ln\mu_0H)^{-2})+O(T^3). \tag{3·12}$$

The coefficient of T-linear specific heat at $A=2U$ is

$$\lim_{H\to 0}\lim_{T\to 0} C_{A,H}/T=\frac{\pi}{6}\cdot\frac{I_0(\pi/2U)}{I_1(\pi/2U)}.$$

This value is shown in Fig. 2 as a function of U.

§ 4. Case $\mu_0 H=O(T)$

a) $\kappa^{(0)}(\pi)\leq 0$ *(near line a)*

From Eqs. (2·1) we have

$$\ln\eta_1(\Lambda)=s*\ln(1+\eta_2(\Lambda))-\frac{2}{T}\int_{-\pi}^{\pi}dk\cos^2 k\, s(\Lambda-\sin k)+O\left(T^{1/2}\exp\frac{\kappa^{(0)}(\pi)}{T}\right),$$

$$\ln\eta_n(\Lambda)=s*\ln(1+\eta_{n-1}(\Lambda))(1+\eta_{n+1}(\Lambda)),\qquad n=2,3,\cdots,$$

$$\lim_{n\to\infty}\frac{\ln\eta_n}{n}=\frac{2\mu_0 H}{T}.$$

Thus Eq. (2·2b) can be written as follows:

$$\omega(T,A,H)=E_0-A-2T^{3/2}\rho_0(\pi)\sqrt{\frac{2}{-\kappa^{(0)''}(\pi)}}\int_0^\infty\ln\left(1+\exp\frac{\kappa^{(0)}(\pi)}{T}\cdot e^{-x^2}\right)dx$$
$$-\frac{T^2}{2}\cdot\frac{I_0(\pi/2U)}{I_1(\pi/2U)}C\left(\frac{2\mu_0 H}{T}\right)+O(T^4),\tag{4·1}$$

where $C(y)$ is determined by

$$C(y)=\int_{-\infty}^{\infty}e^{-\pi x/2}\ln(1+\eta_1(x))dx,$$

$$\ln\eta_1(x)=-e^{-\pi x/2}+\int_{-\infty}^{\infty}\frac{1}{4}\operatorname{sech}\frac{\pi(x-x')}{2}\ln(1+\eta_2(x'))dx',$$

$$\ln\eta_n(x)=\int_{-\infty}^{\infty}\frac{1}{4}\operatorname{sech}\frac{\pi(x-x')}{2}\ln(1+\eta_{n-1}(x'))(1+\eta_{n+1}(x'))dx',$$

$$n=2,3,\cdots,$$

$$\lim_{n\to\infty}\frac{\ln\eta_n}{n}=y.\tag{4·2}$$

b) $\kappa^{(0)}(\pi)>0$ *(near line b)*

From Eq. (2·2b) we have

$$\omega(T,A,H)-\omega(0,A,H)=\int_{-Q}^{Q}\frac{dk}{2\pi}\delta\kappa(k)-\frac{\pi}{12}\cdot\frac{T^2}{\kappa^{(0)'}(Q)}+O(T^3).\tag{4·3}$$

The equation for $\delta\kappa\equiv\kappa-\kappa^{(0)}$ is

$$\delta\kappa(k)-\int_{-Q}^{Q}R(\sin k-\sin k')\delta\kappa(k')\cos k'dk'$$

$$= -T\int_{-\infty}^{\infty} s(\Lambda-\sin k)\ln(1+\eta_1(\Lambda))d\Lambda$$

$$+\frac{\pi^2T^2}{6\kappa^{(0)}(Q)}(R(\sin k-\sin Q)+R(\sin k+\sin Q))+O(T^4). \qquad (4\cdot4)$$

After some calculations we obtain

$$\omega(T,A,H)-\omega(0,A,0)=-\frac{\pi^2T^2\rho^{(0)}(Q)}{3\kappa^{(0)\prime}(Q)}-2\pi T^2C\left(\frac{2\mu_0H}{T}\right)$$

$$\times\left(\int_{-Q}^{Q}\exp\left(\frac{\sin k}{U}\right)\rho^{(0)}(k)dk\Big/\int_{-Q}^{Q}\exp\left(\frac{\sin k}{U}\right)\kappa^{(0)\prime}(k)dk\right)+O(T^3), \quad (4\cdot5)$$

where $\rho^{(0)}$, $\kappa^{(0)}$ and $C(y)$ are defined in Eqs. (3·6) and (4·2). Functions similar to $C(y)$ defined in (4·2) appeared in the investigation of the low-temperature specific heat of Heisenberg-Ising ring at $|\Delta|\leq 1$.[7] From the result of numerical calculation in Ref. 7) we conjecture

$$C(0)=\pi/6 \quad\text{and}\quad C''(0)=1/2\pi\,. \qquad (4\cdot6)$$

If these equations are true, we obtain

$$\lim_{T\to0}\lim_{H\to0}C/T=\lim_{H\to0}\lim_{T\to0}C/T\,,$$

$$\lim_{T\to0}\lim_{H\to0}\chi=\lim_{H\to0}\lim_{T\to0}\chi\,.$$

§ 5. Discussions and summary

From the theory of non-interacting fermions, thermodynamic potential per site at $U=0$ is

$$\omega(T,A,H)=\frac{1}{2\pi}\left\{\int_{-\pi}^{\pi}\ln(1+\exp(-2\cos k-\mu_0H-A/T))dk\right.$$

$$\left.+\int_{-\pi}^{\pi}\ln(1+\exp(-2\cos k+\mu_0H-A/T))dk\right\}. \qquad (5\cdot1)$$

From this equation we obtain

$$\lim_{H\to0}\lim_{T\to0}C/T=\pi/3$$

at $A=2U=0$ and $\mu_0H=0$. This value differs from $\lim_{U\to0}\lim_{H\to0}\lim_{T\to0}C/T=\pi/6$. One can interpret this discontinuity of the coefficient of T-linear specific heat at $U=0$ as follows. In the half-filled case at $U>0$ one-particle excitation spectrum has a energy gap $-\kappa^{(0)}(\pi)=2U-2+4\int_0^\infty d\omega\, J_1(\omega)/\omega(1+e^{2U\omega})$. Then this excitation does not contribute to the coefficient of T-linear specific heat. But at $U=0$, gap is zero and this excitation does contribute to the coefficient. In the case $n<1$ one finds no such discontinuity, because both magnon excitation and one-particle excitation contribute to the coefficient of T-linear specific heat.

References

1) M. Takahashi, Prog. Theor. Phys. **47** (1972), 69.
2) E. Lieb and F. Y. Wu, Phys. Rev. Letters **20** (1968), 1445.
3) H. Shiba and P. A. Pincus, Phys. Rev. **B5** (1972), 1966.
4) M. Takahashi, Prog. Theor. Phys. **42** (1969), 1098; **43** (1970), 860, 1619.
5) H. Shiba, Phys. Rev. **B6** (1972), 930.
6) A. A. Ovchinikov, Zhur. Eksp. i Theoret. Fiz. **57** (1969), 2137.
7) M. Takahashi, Prog. Theor. Phys. **50** (1973), 1519, and to be published.

Nuclear Physics B 384 (1992) 431–458
North-Holland

NUCLEAR
PHYSICS B

Completeness of the SO(4) extended Bethe ansatz for the one-dimensional Hubbard model

Fabian H.L. Eßler, Vladimir E. Korepin and Kareljan Schoutens

Institute for Theoretical Physics, State University of New York at Stony Brook, Stony Brook, NY 11794-3840, USA

Received 16 December 1991
Accepted for publication 30 June 1992

We show how to construct a complete set of eigenstates of the hamiltonian of the one-dimensional Hubbard model on a lattice of even length L. This is done by using the nested Bethe ansatz *and* the SO(4) symmetry of the model. We discuss in detail how the counting of independent eigenstates is carried out.

1. Introduction

An important ingredient in the search for a theory of high-T_c superconductors is the analysis of strongly correlated electron systems (see for example ref. [1]). A prototype model for these is the Hubbard model. Especially interesting is the η-pairing mechanism proposed by Yang in refs. [2–4].

The one-dimensional Hubbard model has been known to be exactly solvable since the work of Lieb and Wu of 1968 [5]. In their paper, a large set of eigenfunctions of the hamiltonian were found by using the nested Bethe ansatz [6]. However, the issue of whether this set of eigenfunctions is actually *complete* has not been considered until recently. In a recent paper [7],we used the SO(4) symmetry of the Hubbard model (which has been explored in refs. [2–4,8,9]) to show that the Bethe ansatz is not complete (see below for some comments on this result).

In this paper we will show that the SO(4) structure can be used to extend the set of Bethe-ansatz eigenstates to a complete set of eigenstates of the one-dimensional Hubbard model. This result was first announced in ref. [10]. Here we provide a detailed account of the derivations involved.

The Hubbard model describes electrons, which can hop along a one-dimensional lattice and which interact with coupling constant U if two of them occupy the same site. The hamiltonian of the Hubbard model on a periodic one-dimensional lattice of even, finite length L is given by (the small modifications in the potential term as compared to ref. [5] are such that the SO(4) symmetry becomes explicit, see refs. [8,2,9,3])

$$H = -\sum_{i=1}^{L} \sum_{\sigma=1,-1} \left(c_{i,\sigma}^{\dagger} c_{i+1,\sigma} + c_{i+1,\sigma}^{\dagger} c_{i,\sigma} \right)$$

$$+ U \sum_{i=1}^{L} \left(n_{i,1} - \tfrac{1}{2} \right) \left(n_{i,-1} - \tfrac{1}{2} \right). \tag{1.1}$$

Here the $c_{i,\sigma}$ (σ takes the values ± 1) are canonical Fermi operators on the lattice ($i = 1, \ldots, L$ labels the lattice sites), with anti-commutation relations given by $\{c_{i,\sigma}^{\dagger}, c_{j,\tau}\} = \delta_{i,j}\delta_{\sigma,\tau}$. They act in a Fock space with the pseudo vacuum $|0\rangle$ defined by $c_{i,\sigma}\ |0\rangle = 0$. The operator $n_{i,\sigma} = c_{i,\sigma}^{\dagger} c_{i,\sigma}$ is the number operator for electrons with spin σ on site i. U is the coupling constant and can be either positive (repulsive case) or negative (attractive case). For later convenience we define

$$u = \frac{U}{2i}. \tag{1.2}$$

The analysis by Lieb and Wu in ref. [5] resulted in a large number of eigenstates of the hamiltonian, which are characterized by momenta k_i and rapidities Λ_α, where $i = 1, 2, \ldots, N_e$ and $\alpha = 1, 2, \ldots, M$ for an eigenstate with a total number of N spin-up and M spin-down electrons. Our convention throughout the paper will be: N = number of spin-up electrons; M = number of spin-down electrons; $N_e = N + M$ = total number of electrons.

We will now discuss the Bethe wave functions in the form as given by Woynarovich [11], which is equivalent to the form found by Lieb and Wu. The nested Bethe ansatz provides us with the following set of eigenstates with M spins down and N spins up,

$$|\Psi_{M,N}\rangle = \sum_{1 \leqslant x_k \leqslant L} \psi_{-1,\ldots,-1,1,\ldots,1}(x_1, \ldots, x_{N_e}) \prod_{j=1}^{M} c_{x_j,-1}^{\dagger} \prod_{i=M+1}^{N_e} c_{x_i,1}^{\dagger}\ |0\rangle$$

$$= \sum_{1 \leqslant x_k \leqslant L} \psi_{\sigma_1,\sigma_2,\ldots,\sigma_{N_e}}(x_1, \ldots, x_{N_e}) \prod_{j=1}^{N_e} c_{x_j,\sigma_j}^{\dagger}\ |0\rangle, \tag{1.3}$$

where we have put $\sigma_1 = \ldots = \sigma_M = -1$, $\sigma_{M+1} = \ldots = \sigma_{N_e} = 1$.

The Bethe-ansatz wave functions explicitly depend on the relative ordering of the x_i. We represent this dependence by a permutation Q of N_e elements, which is such that $1 \leqslant x_{Q_1} \leqslant x_{Q_2} \leqslant \ldots \leqslant x_{Q_{N_e}} \leqslant L$. In the sector Q the general Bethe wave function for M spins down and N spins up reads

$$\psi_{\sigma_1,\sigma_2,\ldots,\sigma_{N_e}}(x_1,\ \ldots,\ x_{N_e})$$

$$= \sum_{P\in S_{N_e}} \operatorname{sgn}(Q)\ \operatorname{sgn}(P)\ \exp\left(i\sum_{j=1}^{N_e} k_{P_j}x_{Q_j}\right)\phi(y_1,\ \ldots,\ y_M\,|\,P). \tag{1.4}$$

The P-summation extends over all permutations of N_e elements and sgn(π) is the sign of the permutation π (π = Q, P). The amplitudes $\phi(y_1,\ \ldots,\ y_M\,|\,P)$ are f the form

$$\phi(y_1,\ \ldots,\ y_M\,|\,P) = \sum_{\pi\in S_M} A_\pi \prod_{l=1}^{M} F_P(\Lambda_{\pi_l}, y_l), \tag{1.5}$$

with

$$F_P(\Lambda_j,\ y) = \left(\prod_{i=1}^{y-1}\frac{e_-^{(j)}(P_i)}{e_+^{(j)}(P_i)}\right)\frac{1}{e_+^{(j)}(P_y)}$$

$$= \left(\prod_{i=1}^{y-1}\frac{\sin(k_{P_i})-\Lambda_j-U/4i}{\sin(k_{P_i})-\Lambda_j+U/4i}\right)\left(\frac{1}{\sin(k_{P_y})-\Lambda_j+U/4i}\right), \tag{1.6}$$

where we defined

$$e_\pm^{(j)}(i) = \sin(k_i) - \Lambda_j \pm \tfrac{1}{2}u \tag{1.7}$$

and

$$\frac{A_\pi}{A_{(t,t+1)\pi}} = \frac{\Lambda_{\pi_{t+1}} - \Lambda_{\pi_t} - u}{\Lambda_{\pi_{t+1}} - \Lambda_{\pi_t} + u}. \tag{1.8}$$

By $\pi = (\pi_1,\ \pi_2,\ \ldots,\ \pi_t,\ \pi_{t+1},\ \ldots,\ \pi_M)$ we denote a permutation of M elements (spin-down electrons) and $(t,\ t+1)\pi = (\pi_1,\ \pi_2,\ \ldots,\ \pi_{t+1},\ \pi_t,\ \ldots,\ \tau_M)$. A solution of eq. (1.8) is given by

$$A_\pi = \prod_{1\leqslant l<k\leqslant M}\left(\frac{\Lambda_{\pi_l} - \Lambda_{\pi_k} + u}{\Lambda_{\pi_l} - \Lambda_{\pi_k}}\right). \tag{1.9}$$

The amplitudes $\phi(y_1,\ \ldots,\ y_M\,|\,P)$ depend on $\sigma_1,\ \ldots,\ \sigma_{N_e}$ and on Q through the numbers $y_1,\ \ldots,\ y_M$, which are defined to be the positions of the down spins among the spins in the series $\sigma_{Q_1},\ \sigma_{Q_2},\ \ldots,\ \sigma_{Q_{N_e}}$ in increasing order, i.e.,

$$1 \leqslant y_1 < y_2 < y_3 < \ldots < y_M \leqslant N_e. \tag{1.10}$$

For example, for one spin down and one spin up (and $\sigma_1 = -1$, $\sigma_2 = 1$) we have the two cases $y = 1$ (if the spin down is to the left, which holds in the Q = (id) sector) and $y = 2$ (if the spin down is to the right, which holds for Q = (21)).

As we already indicated in eq. (1.3), we will choose the notation such that the M down spins are at the positions $x_1, \dots, x_M$, i.e., $\sigma_1 = \dots = \sigma_M = -1$ and $\sigma_{M+1} = \dots = \sigma_{N_e} = 1$.

We see that all solutions are characterised by N_e momenta $\{k_j \mid j = 1, \dots, N_e\}$ of charged spinless excitations (holons), and M rapidities $\{\Lambda_k \mid k = 1, ,\dots, M\}$ of spin waves (spinons).

Imposing periodic boundary conditions on the Bethe-ansatz wave functions leads to the following equations for the parameters k_i and Λ_α,

$$\exp(ik_jL) = \prod_{\alpha=1}^{M} \frac{\sin(k_j) - \Lambda_\alpha - U/4i}{\sin(k_j) - \Lambda_\alpha + U/4i} \qquad (j = 1, 2, \dots, N_e),$$

$$\prod_{j=1}^{N_e} \frac{\sin(k_j) - \Lambda_\alpha - U/4i}{\sin(k_j) - \Lambda_\alpha + U/4i} = -\prod_{\beta=1}^{M} \frac{\Lambda_\beta - \Lambda_\alpha - U/2i}{\Lambda_\beta - \Lambda_\alpha + U/2i} \qquad (\beta = 1, 2, \dots, M). \tag{1.11}$$

Energy and momentum, i.e., the eigenvalues of the hamiltonian (1.1) and the logarithm of the translation operator, of the system in a state corresponding to a solution of (1.11) are

$$E_{N_e} = -2\sum_{i=1}^{N_e} \cos(k_i) + \tfrac{1}{2}U\left[\tfrac{1}{2}L - N_e\right],$$

$$P = \sum_{i=1}^{N_e} k_i. \tag{1.12}$$

The second term in the expression for the energy is due to the shift of $n_{j,\sigma}$ by $\frac{1}{2}$ in eq. (1.1).

Because of the antisymmetry of the product over $c^\dagger$'s under interchange of any two of them, the wave functions $\psi_{\sigma_1,\dots,\sigma_{N_e}}(x_1, \dots, x_{N_e})$ can be (and have been) chosen to be completely antisymmetric under the simultaneous exchange $x_k \leftrightarrow x_j$ and $\sigma_k \leftrightarrow \sigma_j$, i.e.,

$$\psi_{\sigma_1,\dots,\sigma_j,\dots,\sigma_k,\dots,\sigma_{N_e}}(x_1, \dots, x_j, \dots, x_k, \dots, x_{N_e})$$
$$= -\psi_{\sigma_1,\dots,\sigma_k,\dots,\sigma_j,\dots,\sigma_{N_e}}(x_1, \dots, x_k, \dots, x_j, \dots, x_{N_e}). \tag{1.13}$$

We now define "regular" Bethe-ansatz states (for finite L), to be denoted by $|\psi_{M,N}\rangle$, by the properties that $N-M\geqslant 0$ (non-negative third component of the spin), $N_e \leqslant L$ (less than or equal to half filling), and that all Λ_α and all k_j are finite. Bethe-ansatz states with $N-M<0$ and/or $N_e>L$ can be obtained from the regular Bethe-ansatz states by using simple symmetry operations, i.e., reflection of the third component of the spin and particle/hole correspondence, which commute with the hamiltonian [12]. The model is invariant under spin rotations, with the corresponding SU(2) generators given by

$$\zeta=\sum_{i=1}^{L} c_{i,1}^{\dagger}c_{i,-1},\qquad \zeta^{\dagger}=(\zeta)^{\dagger},\qquad \zeta_z=\tfrac{1}{2}\sum_{i=1}^{L}(n_{i,-1}-n_{i,1}). \tag{1.14}$$

(Note that ζ_z equals *minus* the third component of the total spin.) For even L the model has a second SU(2) invariance, generated by [3]

$$\eta=\sum_{i=1}^{L}(-1)^{i}c_{i,1}c_{i,-1},\quad \eta^{\dagger}=\sum_{i=1}^{L}(-1)^{i}c_{i,-1}^{\dagger}c_{i,1}^{\dagger},\quad \eta_z=\tfrac{1}{2}\sum_{i=1}^{L}(n_{i,-1}+n_{i,1})-\tfrac{1}{2}L. \tag{1.15}$$

The raising operator $\eta^\dagger$ of this second SU(2) creates a pair of two opposite-spin electrons on the same site, with momentum π. Combining the two SU(2)'s, which commute with the hamiltonian and with one another, leads to an SO(4) invariance of the one-dimensional Hubbard model for even lattice lengths [9,3]. For a discussion of the theoretical and possible experimental consequences of the existence of this symmetry, which also exists in the Hubbard model in two or three dimensions, we refer to the papers [3,4,9,13].

In a previous paper [7], we established the following remarkable property of the regular Bethe-ansatz eigenstates of the hamiltonian: they are all *lowest-weight states* of the SO(4) algebra (1.14), (1.15), i.e.,

$$\eta\,|\psi_{M,N}\rangle=0,\qquad \zeta\,|\psi_{M,N}\rangle=0. \tag{1.16}$$

This shows that acting with the raising operators $\eta^\dagger$ and $\zeta^\dagger$ on $|\psi_{M,N}\rangle$ leads to new eigenstates of the hamiltonian, which are not in the regular Bethe ansatz. In this way, every regular Bethe-ansatz state $|\psi_{M,N}\rangle$ is the lowest-weight state in a multiplet of states, which form a representation of SO(4). Since

$$\eta_z\,|\psi_{M,N}\rangle=\tfrac{1}{2}(N_e-L)\,|\psi_{M,N}\rangle,\qquad \zeta_z\,|\psi_{M,N}\rangle=\tfrac{1}{2}(M-N)\,|\psi_{M,N}\rangle, \tag{1.17}$$

a state $|\psi_{M,N}\rangle$ has spin $\eta=\frac{1}{2}(L-N_e)$ with respect to the η-pairing SU(2) algebra and spin $\zeta=\frac{1}{2}(N-M)$ with respect to the ζ SU(2) algebra. The dimension of the corresponding SO(4) multiplet is therefore given by

$$\dim_{M,N}=(2\eta+1)(2\zeta+1)=(L-N_e+1)(N-M+1). \tag{1.18}$$

The states in this multiplet are of the form

$$|\psi_{M,N}^{\alpha,\beta}\rangle = (\eta^\dagger)^\alpha (\zeta^\dagger)^\beta |\psi_{M,N}\rangle. \tag{1.19}$$

By symmetry, all the states that are highest- or lowest-weight states with respect to one of the SU(2) algebras are again given by the Bethe ansatz (although in general they are outside the *regular* Bethe ansatz). All other states are *not* given by the Bethe ansatz, which shows that for this model the Bethe ansatz is not complete. The simplest example of a state that is outside the Bethe ansatz is $\eta^\dagger|0\rangle$, which describes a single η-pair of momentum π. The fact that this state is outside the Bethe ansatz was explicitly confirmed in ref. [7].

It is the main purpose of this paper to show that, if one counts the number of eigenstates that are related to the regular Bethe-ansatz states by the SO(4) symmetry, one finds 4^L, which is precisely the correct dimension of the Hilbert space of the model *. Thus we will conclude that the Bethe ansatz together with the SO(4) structure leads to a complete set of eigenstates of the one-dimensional Hubbard model.

The paper is organized as follows. In sect. 2 we discuss in some detail the nature of the solutions of the Bethe equations for the Hubbard model. In appendix A we show that the so-called Λ and $k-\Lambda$ strings give wave functions that describe bound states. In sect. 3 we will then count eigenstates and prove completeness. An explicit construction of the $\frac{1}{2}(L+2)(L-1)$ solutions of the Bethe equations in the sector with one spin-up and one spin-down electron (as opposed to the somewhat indirect construction used in the general proof) is presented in appendix B.

2. Solutions of the Bethe equations for the Hubbard model

Let us focus on the Bethe equations (1.11), which express the fact that the Bethe-ansatz wave functions (1.4) satisfy periodic boundary conditions.

Counting regular Bethe-ansatz states means counting inequivalent solutions of eqs. (1.11) while taking into account the "regularity conditions" $N-M \geqslant 0$ and $N_e \leqslant L$. Following Takahashi [14], we will first distinguish different types of solutions $\{k_i, \Lambda_\alpha\}$ of eq. (1.11). The idea is that for a solution $\{k_i, \Lambda_\alpha\}$, the set of all the k_i's and Λ_α's can be split into (three) different kinds of subsets (strings), which are:

(1) a single real momentum k_i;

(2) m Λ_α's combine into a string-type configuration (Λ-strings); this includes the case $m=1$, which is just a single real Λ_α **;

* There are 4 possible electron configurations per lattice site (spin up, spin down, both spin up and spin down, and empty site), thus the corresponding direct product Hilbert space is 4^L-dimensional.

** These correspond to bound states of spin waves (magnons).

(3) $2m$ k_i's and m Λ_α's combine into a different string-type configuration ($k-\Lambda$-strings) *.

For large lattices ($1 \ll L$), almost all the string configurations are close to "idealized" string solutions where both the k's and the Λ's are assigned imaginary parts according to a "equal-spacing" prescription [14]. For a Λ-string of length m the rapidities involved are

$$\Lambda_\alpha^{m,j} = \Lambda_\alpha^m - \tfrac{1}{2}(m+1-2j)u \qquad (\Lambda_\alpha^m \text{ real}, \qquad j=1,2,\ldots,m). \tag{2.1}$$

The k's and Λ's involved in a $k-\Lambda$-string are

$$\begin{aligned}
k_\alpha^1 &= \pi - \sin^{-1}(\Lambda_\alpha'^m - \tfrac{1}{2}mu),\\
k_\alpha^2 &= \sin^{-1}[\Lambda_\alpha'^m - \tfrac{1}{2}(m-2)u],\\
k_\alpha^3 &= \pi - k_\alpha^2,\\
k_\alpha^4 &= \sin^{-1}[\Lambda_\alpha'^m - \tfrac{1}{2}(m-4)u],\\
k_\alpha^5 &= \pi - k_\alpha^4, \ldots,\\
k_\alpha^{2m-2} &= \sin^{-1}[\Lambda_\alpha'^m + \tfrac{1}{2}(m-2)u],\\
k_\alpha^{2m-1} &= \pi - k_\alpha^{2m-2},\\
k_\alpha^{2m} &= \pi - \sin^{-1}(\Lambda_\alpha'^m + \tfrac{1}{2}mu).
\end{aligned} \tag{2.2}$$

and

$$\Lambda_\alpha'^{m,j} = \Lambda_\alpha'^m - \tfrac{1}{2}(m+1-2j)u \qquad (\Lambda_\alpha'^m \text{ real}, \qquad j=1,2,\ldots,m). \tag{2.3}$$

Eqs. (2.1)–(2.3) are valid up to exponential corrections of order $O(\exp(-\delta L))$, where δ is real and positive (and depends on the specific string under consideration). In appendix A we discuss the wave functions corresponding to some of the the string configurations (2) and (3) and show that they correspond to bound states.

Let us now consider a solution that splits into M_m Λ-strings of lengths m, M_n' $k-\Lambda$-strings of length n (containing $2n$ k_i's and n Δ_α's) and M_e additional single

* The case $m=1$ describes a "Cooper pair" of electrons.

k_i's. Clearly, we have

$$N_e = M_e + 2\sum_{m=1}^{\infty} mM'_m, \qquad M = \sum_{m=1}^{\infty} m(M_m + M'_m). \tag{2.4}$$

How many solutions of this type exist?

The idea is that each of the strings in a solution can be characterised by the position of its center (a real number), which we denote as in eq. (2.1) by Λ^m_α, $\alpha = 1, 2, \dots, M_m$, for the length-$m$ Λ-strings, by Λ'^m_α, $\alpha = 1, 2, \dots, M'_m$ for the length-m $k-\Lambda$-strings [as in eq. (2.3)] and which is simply equal to k_j for the unpaired momenta k_j, $j = 1, 2, \dots, M_e$. Because of the periodic boundary conditions, these parameters will have to be chosen from a discrete set.

Following ref. [14], we now write the following equations for the parameters k_j, Λ^m_α and Λ'^m_α. They follow from eq. (1.11) and the form of the "idealized" string solutions which we discussed above (we write $M' = \sum_{m=1}^{\infty} mM'_m$),

$$k_j L = 2\pi I_j - \sum_{n=1}^{\infty}\sum_{\alpha=1}^{M_n} \theta\left(\frac{\sin k_j - \Lambda^n_\alpha}{nU}\right) - \sum_{n=1}^{\infty}\sum_{\alpha=1}^{M'_n} \theta\left(\frac{\sin k_j - \Lambda'^n_\alpha}{nU}\right),$$

$$\sum_{j=1}^{N_e-2M'} \theta\left(\frac{\Lambda^n_\alpha - \sin k_j}{nU}\right) = 2\pi J^n_\alpha + \sum_{m=1}^{\infty}\sum_{\beta=1}^{M_m} \Theta_{nm}\left(\frac{\Lambda^n_\alpha - \Lambda^m_\beta}{U}\right),$$

$$L\left[\sin^{-1}(\Lambda'^n_\alpha + in\tfrac{1}{4}U) + \sin^{-1}(\Lambda'^n_\alpha - in\tfrac{1}{4}U)\right]$$

$$= 2\pi J'^n_\alpha + \sum_{j=1}^{N_e-2M'} \theta\left(\frac{\Lambda'^n_\alpha - \sin k_j}{nU}\right) + \sum_{m=1}^{\infty}\sum_{\beta=1}^{M'_m} \Theta_{nm}\left(\frac{\Lambda'^n_\alpha - \Lambda'^m_\beta}{U}\right), \tag{2.5}$$

where

$$\theta(x) = 2\tan^{-1}(4x),$$

$$\Theta_{nm}(x) = \begin{cases} \theta\left(\frac{4x}{|n-m|}\right) + 2\theta\left(\frac{4x}{|n-m|+2}\right) + \dots + 2\theta\left(\frac{4x}{n+m-2}\right) \\ \quad + \theta\left(\frac{4x}{n+m}\right) & (\text{for } n \neq m) \\ 2\theta\left(\frac{2x}{1}\right) + \dots + 2\theta\left(\frac{2x}{n-1}\right) + \theta\left(\frac{2x}{n}\right) & (\text{for } n = m). \end{cases} \tag{2.6}$$

The I_j, J^n_α and J'^n_α are integer or half odd integer according to the following prescriptions: I_j is integer (half odd integer) if $\sum_m (M_m + M'_m)$ is even (odd); the J^n_α are integer (half odd integer) if $N_e - M_n$ is odd (even); the J'^n_α are integer (half odd integer) if $L-(N_e-M'_n)$ is odd (even). According to ref. [14], we have the following inequalities:

$$|J^n_\alpha| \leqslant \tfrac{1}{2}\left(N_e - 2M' - \sum_{m=1}^{\infty} t_{nm}M_m - 1\right),$$
$$|J'^n_\alpha| \leqslant \tfrac{1}{2}\left(L - N_e + 2M' - \sum_{m=1}^{\infty} t_{nm}M'_m - 1\right),$$
$$0 < I_j \leqslant L. \tag{2.7}$$

where $t_{nm} = 2\min(n, m) - \delta_{nm}$.

We will now make the standard assumption that, in order to enumerate the different solutions of the system (2.5), it is sufficient to enumerate all possible sets of *non-repeating* (half odd) integers I_j, J^n_α and J'^n_α, satisfying eq. (2.7).

[This assumption mimics the similar assumption which is usually made for the spin-$\frac{1}{2}$ Heisenberg *XXX* model [15,16]. It is known, however *, that the actual distribution of the different types of solutions can be different from the one implied by this counting. We carefully studied this phenomenon in a recent paper [18], where we give a detailed discussion of the two-magnon sector of the *XXX* model. We explicitly show that the deviations from the above assumption can be viewed as a "redistribution phenomenon" which does not affect the total number of Bethe-ansatz states. For the Hubbard model we have a similar situation, which gets complicated further by the fact that there is a free coupling constant U in the model. In appendix B we analyze in detail the $N=1$, $M=1$ sector of the Hubbard model. Although we do find U-dependent redistributions among different types of solutions, we find agreement with the predictions based on the "ideal" string assumption for the total number of states in this sector, which is $\frac{1}{2}(L-1)(L+2)$ for the L-site model.]

From eq. (2.7) we read off that the numbers of allowed values for the (half odd)integers corresponding to each of the fundamental strings are

(1) L for a free k_i;
(2) $N_e - 2M' - \sum_{m=1}^{\infty} t_{nm}M_m$ for a Λ-string of length n;
(3) $L - N_e + 2M' - \sum_{m=1}^{\infty} t_{nm}M'_m$ for a $k-\Lambda$-string of length n.

The total number of ways to choose the (half odd)integers in a solution with multiplicities M_e, M_m and M'_m is therefore given by (remember that the integers are assumed to be non-repeating)

$$n(M_e, \{M_m\}, \{M'_m\}) = \binom{L}{M_e} \prod_{n=1}^{\infty} \binom{N_e - 2M' - \sum_{m=1}^{\infty} t_{nm}M_m}{M_n}$$
$$\times \prod_{n=1}^{\infty} \binom{L - N_e + 2M' - \sum_{m=1}^{\infty} t_{nm}M'_m}{M'_n}. \tag{2.8}$$

* This fact was actually already noticed in the original paper by Bethe [17].

The total number of solutions of eq. (1.11) with given numbers N and M is now obtained by summing $n(M_e, \{M_m\}, \{M'_m\})$ over all the M_e, M_m and M'_m, under the constraints (2.4).

Every solution to eq. (1.11) gives us a regular Bethe-ansatz state, which comes with an entire multiplet of eigenstates of the hamiltonian, the dimension $\dim_{M,N}$ of which is given in eq. (1.18). The full number of eigenstates that are obtained from the Bethe ansatz *and* the SO(4) symmetry is therefore given by

$$\#(\text{eigenstates}) = \sum_{\substack{M\geq 0\\ N-M\geq 0\\ N+M\leq L}} \sum_{N\geq 0} \left(\sum_{\substack{M_e=0\\ N+M=M_e+2\sum_{m=1}^{\infty} mM'_m\\ N-2M=M_e-2\sum_{m=1}^{\infty} mM_m}}^{\infty} \sum_{M_m=0}^{\infty} \sum_{M'_m=0}^{\infty} n(M_e, \{M_m\}, \{M'_m\}) \right) \dim_{M,N}. \tag{2.9}$$

The counting of the eigenstates that are obtained from the SO(4) extended nested Bethe ansatz has thus been reduced to a purely algebraic problem, which we will solve in sect. 3.

3. Counting eigenstates

In this section we will prove that for general even L the sum in eq. (2.9) equals 4^L. This will prove completeness. Before we come to that, we show the examples of the two-site and four-site models. The two-site model ($L = 2$) was discussed in ref. [7], where we presented the explicit form of a complete set of $4^2 = 16$ eigenstates of the hamiltonian. In table 1 we show how the counting presented in sect. 2 work out in this case.

The total number of 16 states splits into two singlets, two triplets and two quadruplets of SO(4). The ground state is the singlet with $M_1 = 1$ for the case $U > 0$ and the singlet with $M'_1 = 1$ for the case $U < 0$. In both cases it is a bound

TABLE 1

$L = 2$. n denotes the number of regular Bethe-ansatz states of a given type. There are a total number of $16 = 4^2$ eigenstates of the hamiltonian.

M_e	M_1	M'_1	M	N	n	$\dim_{M,N}$	#(states)
0	0	0	0	0	1	3	3
1	0	0	0	1	2	4	8
2	0	0	0	2	1	3	3
2	1	0	1	1	1	1	1
0	0	1	1	1	1	1	1
							16

TABLE 2
$L=4$. There are 60 regular Bethe-ansatz states, which, when weighted with the correct SO(4) multiplicities, give a total of $256 = 4^4$ eigenstates of the hamiltonian.

M_c	M_1	M_2	M_1'	M_2'	M	N	n	$\dim_{M,N}$	#(states)
0	0	0	0	0	0	0	1	5	5
1	0	0	0	0	0	1	4	8	32
2	0	0	0	0	0	2	6	9	54
3	0	0	0	0	0	3	4	8	32
4	0	0	0	0	0	4	1	5	5
2	1	0	0	0	1	1	6	3	18
0	0	0	1	0	1	1	3	3	9
3	1	0	0	0	1	2	8	4	32
1	0	0	1	0	1	2	8	4	32
4	2	0	0	0	2	2	1	1	1
4	0	1	0	0	2	2	1	1	1
2	1	0	1	0	2	2	6	1	6
0	0	0	2	0	2	2	1	1	1
0	0	0	0	1	2	2	1	1	1
4	1	0	0	0	1	3	3	3	9
2	0	0	1	0	1	3	6	3	18
									256

state of one spin-up and one spin-down electron with energy $E_0 = -\sqrt{\frac{1}{4}U^2+16}$. The counting for the four-site model ($L=4$) is presented in table 2, where we show how the total number of $4^4 = 256$ is obtained. (Notice that the total number of regular Bethe-ansatz states is only 60.)

We now turn to the proof that for general (even) L the sum in eq. (2.9) equals 4^L. We will split this proof into two steps as follows. In the first step we will prove the following two identities,

$$\sum_{\substack{M_1, M_2, \ldots = 0 \\ \sum_{m=1}^{\infty} m M_m = M}}^{\infty} \prod_{n=1}^{\infty} \binom{N - \sum_m t_{nm} M_m}{M_n} = \binom{N}{M} - \binom{N}{M-1} \tag{3.1}$$

and

$$\sum_{M=0}^{[N/2]} \left(\binom{N}{M} - \binom{N}{M-1} \right) (N - 2M + 1) = 2^N. \tag{3.2}$$

For later convenience we define

$$P_n = N - \sum_{m=1}^{\infty} t_{nm} M_m, \tag{3.3}$$

and

$$n(\{M_m\}) = \prod_{n=1}^{\infty} \binom{P_n}{M_n}, \tag{3.4}$$

where $t_{nm} = 2\min(n, m) - \delta_{nm}$ as before. In the second step we will then use identities (3.1) and (3.2) to perform the summation in eq. (2.9).

The auxiliary identities (3.1) and (3.2) have a natural interpretation in the context of the spin-$\frac{1}{2}$ Heisenberg *XXX* model [16]. Eq. (3.1) gives the total number of regular Bethe-ansatz states (defined by $M \leqslant [N/2]$) with M overturned spins in the *XXX* model on a lattice of length N. The second formula shows that the total number of states obtained by combining the regular Bethe ansatz with the SU(2) structure equals 2^N, which is the dimension of the Hilbert space of the *XXX* model. These relations thus establish the completeness of the SU(2) extended Bethe ansatz for the *XXX* model.

The fact that identities that have their origin in the *XXX* model play a role here should not come as a surprise. Indeed, our method of solution of the Hubbard model is the *nested* Bethe ansatz. The solutions to the Bethe ansatz are specified by two sets $\{k_j\}$ and $\{\Lambda_\alpha\}$ of spectral parameters. The k_j's are momenta of charge density waves, whereas the Λ_α's, which describe the "nesting" of the Bethe ansatz, are rapidities of spin density waves of the type encountered in the Heisenberg *XXX* model. This should make clear that the second stage of the nested Bethe ansatz for the Hubbard model is really a spin problem, which is very similar to the Bethe-ansatz analysis of the Heisenberg *XXX* model. Our two-step procedure for performing the summation is natural from the point of view of the nesting: in the first step we sum over the spin degrees of freedom, and in the second step we then sum over the charge degrees of freedom as well.

3.1. STEP 1

Let us now explain how eq. (3.1) can be derived. In the first step, one simply solves for $M_1 = M - \Sigma_{m=2}^{\infty} mM_m$ and substitutes this back into the left-hand side of eq. (3.1). Using this value for M_1, the quantities P_n reduce to

$$P_1 = N - \sum_{m=1}^{\infty} t_{1m} M_m = N - M + \sum_{m=3}^{\infty} (m-2) M_m,$$

$$P_n = N - \sum_{m=1}^{\infty} t_{nm} M_m = N - 2M + M_n + 2 \sum_{m=n+1}^{\infty} (m-n) M_m. \tag{3.5}$$

Let us now consider the summation over M_2. Although our summand in the left-hand side of eq. (3.1) has the form of an infinite product, only two of the factors contain the variable M_2. Singling these out, one finds that the summation over M_2 is as follows,

$$\Omega_2 = \sum_{M_2=0}^{\infty} \binom{N - 2M + M_2 + 2\Sigma_{m=3}^{\infty} (m-2) M_m}{M_2} \times \binom{N - M + \Sigma_{m=3}^{\infty} (m-2) M_m}{M - \Sigma_{m=2}^{\infty} mM_m}. \tag{3.6}$$

In order to perform this summation we will make use of the identity

$$\sum_{\alpha=0}^{\infty} \binom{B+\alpha}{\alpha} x^{\alpha} = (1-x)^{-1-B}, \tag{3.7}$$

which can easily be proved by induction. As a simple consequence, we have

$$(1-x^2)^{-1-\omega}(1+x)^{\eta}\big|_{x^A} = \sum_{\alpha=0}^{\infty} \binom{\omega+\alpha}{\alpha}\binom{\eta}{A-2\alpha}, \tag{3.8}$$

where the notation $|_{x^A}$ in the left-hand side means that we single out the coefficient of the power x^A. The right-hand side of eq. (3.6) is of the same form as eq. (3.8) and we find

$$\begin{aligned}
\Omega_2 &= (1-x^2)^{-1-[N-2M+2(M_3+2M_4+\dots)]}(1+x)^{N-M+(M_3+2M_4+\dots)}\big|_{x^{M-3M_3-4M_4-\dots}} \\
&= (1+x)^{N-M}(1-x^2)^{-N+2M-1}\prod_{n=3}^{\infty}\left(\mathscr{Z}_n^{(0)}\right)^{M_n}\big|_{x^M} \\
&= \frac{1}{2\pi i}\oint \frac{dx}{x^{M+1}}(1+x)^{N-M}(1-x^2)^{-N+2M-1}\prod_{n=3}^{\infty}\left(\mathscr{Z}_n^{(0)}\right)^{M_n},
\end{aligned} \tag{3.9}$$

where

$$\mathscr{Z}_n^{(0)} = \frac{x^n}{(1-x)^{2(n-2)}(1+x)^{n-2}}. \tag{3.10}$$

In the last line of eq. (3.9) we extracted the coefficient at x^M by performing a contour integral around the origin $x=0$. After performing the M_2 summation (3.1) now reads

$$\sum_{\substack{M_1,M_2,\dots=0 \\ \sum_{m=1}^{\infty} mM_m = M}}^{\infty} n(\{M_m\}) = \frac{1}{2\pi i}\oint \frac{dx}{x^{M+1}} A(x), \tag{3.11}$$

where

$$A(x) = (1+x)^{N-M}(1-x^2)^{-1-N+2M} \times \sum_{M_3,M_4,\dots=0}^{\infty}\prod_{n=3}^{\infty}\binom{N-2M+M_n+2\sum_{m=n+1}^{\infty}(m-n)M_m}{M_n}\prod_{l=3}^{\infty}\left(\mathscr{Z}_l^{(0)}\right)^{M_l}. \tag{3.12}$$

The summation over M_3 is given by

$$\Omega_3 = \sum_{M_3=0}^{\infty} \binom{N-2M+M_3+2\sum_{n=4}^{\infty}(n-3)M_n}{M_3} \left(\mathscr{Z}_3^{(0)}\right)^{M_3}$$

$$= \left(1-\mathscr{Z}_3^{(0)}\right)^{-1-N+2M-2\sum_{n=4}^{\infty}(n-3)M_n}. \tag{3.13}$$

At this point, the full expression for $A(x)$ has been reduced to

$$A(x) = (1+x)^{N-M}(1-x^2)^{-1-N+2M}\left(1-\mathscr{Z}_3^{(0)}\right)^{-1-N+2M}$$

$$\times \sum_{M_4,M_5,\ldots=0}^{\infty} \prod_{n=4}^{\infty} \binom{N-2M+M_n+2\sum_{m=n+1}^{\infty}(m-n)M_m}{M_n} \prod_{l=4}^{\infty}\left(\mathscr{Z}_l^{(1)}\right)^{M_l}, \tag{3.14}$$

where $\mathscr{Z}_n^{(1)}$ are defined through

$$\mathscr{Z}_n^{(1)} = \frac{\mathscr{Z}_n^{(0)}}{\left(1-\mathscr{Z}_3^{(0)}\right)^{2(n-3)}}. \tag{3.15}$$

From the above it is now clear, that the sum with respect to M_4 and all M_n with $n>4$ has the same structure as the sum with respect to M_3. Thus the final result, after performing all summations, will look like

$$A(x) = (1+x)^{N-M}F(x)^{-1-N+2M}, \tag{3.16}$$

where

$$F(x) = (1-x^2)\prod_{m=3}^{\infty}\left(1-\mathscr{Z}_m^{(m-3)}\right), \tag{3.17}$$

and we have the iteration formula

$$\mathscr{Z}_n^{(m)} = \frac{\mathscr{Z}_n^{(m-1)}}{\left(1-\mathscr{Z}_{m+2}^{(m-1)}\right)^{2(n-m-2)}}. \tag{3.18}$$

Our task is now to find a closed expression for $F(x)$ by exploiting this relation. We define

$$U_2 = x^{-2}, \qquad U_m = \frac{1}{\mathscr{Z}_m^{(m-3)}}, \qquad (m \geq 3), \tag{3.19}$$

so that $F(x)$ can be written as

$$F(x) = \prod_{m=2}^{\infty} \left(1 + \frac{1}{U_m}\right). \tag{3.20}$$

We now claim that the functions $U_m(x)$ satisfy the following recursion relation, to be denoted by RR I_p,

$$\text{RR I}_p: \quad (U_{p+3} - 1)^2 = U_{p+4} U_{p+2} \qquad (p \geqslant 0). \tag{3.21}$$

Together with the initial conditions

$$U_2 = x^{-2}, \qquad U_3 = \frac{(1-x)^2(1+x)}{x^3}, \tag{3.22}$$

these relations completely fix the functions $U_m(x)$ and thereby the function $F(x)$.

In order to prove the recursion relation (3.21), we first give a second recursion relation, which involves some of the other $\mathscr{Z}$'s and which we shall denoted by RR II_p

$$\text{RR II}_p: \quad \frac{\mathscr{Z}_{n+1}^{(p)}}{\mathscr{Z}_n^{(p)}} = \frac{U_{p+2}}{U_{p+3}} \qquad (p \geqslant 0, \qquad n \geqslant p+3). \tag{3.23}$$

Let us now show that the validity of both recursion relations can be proved by induction. We start at the point where we have $\mathscr{Z}_n^{(0)}$, which is defined by eq. (3.10), and U_2 and U_3 as above in eq. (3.22). One easily checks that RR $\mathrm{II}_{p=0}$ is valid. Using eq. (3.18) for the definition of $U_4 = 1/\mathscr{Z}_4^{(1)}$ and RR $\mathrm{II}_{p=0}$, one proves RR $\mathrm{I}_{p=0}$. This establishes the validity of both RR $\mathrm{I}_{p=0}$ and RR $\mathrm{II}_{p=0}$, which is the starting point for the proof by induction.

Let us now assume that we have proved the validity of both RR I_p and RR II_p for some given p. By using this induction assumption and the definition (3.18), one then proves the relation RR II_{p+1} (3.23). After that, by using the definition (3.18) and RR II_{p+1}, one then proves RR I_{p+1}. This completes the induction step. We may thus conclude that the relations RR I_p and RR II_p are valid for all $p \geqslant 0$.

One easily checks that the expressions

$$U_j = \left(\frac{a(x)^{j+1} - a(x)^{-j-1}}{a(x) - a(x)^{-1}} \right)^2 \tag{3.24}$$

with

$$a(x) = \tfrac{1}{2}\left(\sqrt{\frac{1-3x}{x}} + \sqrt{\frac{1+x}{x}} \right) \tag{3.25}$$

satisfy the recursion relations (3.21) and the initial conditions (3.22). The function $F^2(x)$ is now expressed as a convergent product

$$F^2(x) = \prod_{m=2}^{\infty} \frac{(U_m - 1)^2}{U_m^2} = \prod_{m=2}^{\infty} \frac{U_{m+1}U_{m-1}}{U_m^2} = \lim_{l\to\infty} \frac{U_1}{U_2}\frac{U_{l+1}}{U_l} = a(x)^2 x(x+1), \tag{3.26}$$

where we defined $U_1(x) = (x+1)/x$, in accord with eq. (3.21), and where we used eq. (3.21) in the second equality. This brings us to the following representation of the number of regular Bethe-ansatz states with M overturned spins [using (3.1), (3.11), (3.12), (3.16), (3.20), (3.25) and (3.26)]

$$\sum_{\substack{M_m=0 \\ M=\sum_{m=1}^{\infty} mM_m}}^{\infty} n(\{M_m\})$$

$$= \frac{1}{2\pi i}\oint \frac{\mathrm{d}x}{x^{M+1}}(1+x)^{N-M}\left(\frac{(1+x)+\sqrt{(1+x)(1-3x)}}{2}\right)^{2M-N-1}$$

$$= \frac{1}{2\pi i}\oint \frac{2\,\mathrm{d}y}{y^{M+1}}[2(1+y)]^{N-M}\left(1+y+\sqrt{1-y^2}\right)^{2M-N-1}. \tag{3.27}$$

where the contour is a small circle around the origin and we used the substitution $y = 2x/(1-x)$. Calling $y^{-1} = \cosh\phi$, the integral reduces to $I_+ - I_-$, where

$$I_\pm = \frac{1}{2\pi}\int_0^{2\pi} \mathrm{d}\phi\ \mathrm{e}^{-\phi(N-M\mp 1)}(1+\mathrm{e}^{\phi})^{N-1} = \binom{N-1}{N-M\mp 1}, \tag{3.28}$$

where we wrote $\phi = \Lambda - i\varphi$, with $\Lambda \to \infty$. This finally establishes the result (3.1) since

$$\sum_{\substack{M_m=0 \\ M=\sum_{m=1}^{\infty} mM_m}}^{\infty} n(\{M_m\}) = \binom{N-1}{N-M-1} - \binom{N-1}{N-M+1}$$

$$= \binom{N}{M} - \binom{N}{M-1}. \tag{3.29}$$

We still have to prove eq. (3.2), which can be done as follows,

$$\sum_{M=0}^{[N/2]}\left(\binom{N}{M}-\binom{N}{M-1}\right)(N-2M+1)$$

$$=\sum_{M=0}^{[N/2]}\binom{N}{M}(N-2M+1)-\sum_{M=0}^{[N/2]-1}\binom{N}{M}(N-2M-1)$$

$$=2\sum_{M=0}^{[N/2]-1}\binom{N}{M}+\binom{N}{[N/2]}(1+N-2[N/2])$$

$$=\sum_{M=0}^{[N/2]-1}\binom{N}{M}+\binom{N}{[N/2]}(1+N-2[N/2])+\sum_{N-[N/2-1]}^{N}\binom{N}{M}$$

$$=\sum_{M=0}^{N}\binom{N}{M}=2^N. \tag{3.30}$$

This completes the proof of eq. (3.2).

3.2. STEP 2

The total number of states that are obtained from the SO(4) extended Bethe ansatz for the Hubbard model is given by eqs. (2.8) and (2.9). The summations over the multiplities M_m and over the difference $N-M$ in the summation (2.9) are precisely of the type (3.1) and (3.2), respectively, if we substitute $M\to\frac{1}{2}(M_e-N+M)$ and $N\to M_e$. (Under these summations the total number of electrons N_e is kept fixed.) The summation that remains after this "spin summation" is

$$\#(\text{eigenstates})$$

$$=\sum_{N_e=0}^{L}(L-N_e+1)$$

$$\times\left[\sum_{\substack{M_e=0\\ N_e=M_e+2\sum_{m=1}^{\infty}mM'_m}}^{N_e}\ \sum_{M'_m=0}^{\infty}2^{M_e}\binom{L}{M_e}\prod_{n=1}^{\infty}\binom{L-N_e+\sum_{m=1}^{\infty}(2m-t_{nm})M'_m}{M'_n}\right], \tag{3.31}$$

where as before $t_{nm}=2\,\mathrm{Min}(n,m)-\delta_{nm}$. In the next step we perform the summation with respect to all M_n''s, using a similar kind of "summation device" as subsect. 3.1. As a consequence of eq. (3.7) we have

$$(1-x^2)^{-1-B}(1+2x)^L = \sum_{M_1'=0}^{\infty}\sum_{p=0}^{L}\binom{B+M_1'}{M_1'}\binom{L}{p}2^p x^{2M_1'+p}, \quad (3.32)$$

and therefore

$$\frac{1}{2\pi i}\oint\frac{\mathrm{d}x}{x^{\gamma+1}}(1-x^2)^{-1-B}(1+2x)^L = \sum_{M_1'=0}^{\infty}\binom{B+M_1'}{M_1'}\binom{L}{\gamma-2M_1'}2^{\gamma-2M_1'}. \quad (3.33)$$

The integration is along a small contour around zero. Defining $E=L-N_e$, $\gamma=N_e-2\Sigma_{m=2}^{\infty}mM_m'$ and $B=E+2\Sigma_{m=2}^{\infty}(m-1)M_m'$, the r.h.s. of eq. (3.33) becomes the summation over M_1' in eq. (3.31), if we solve the constraint in the sum in eq. (3.31) for $M_e=N_e-2\Sigma_{m=1}^{\infty}mM_m'$. Using eq. (3.33) in eq. (3.31) we then obtain the following expression for the number of eigenstates,

$$\#(\text{eigenstates}) = \frac{1}{2\pi i}\oint\frac{\mathrm{d}x}{x^{L+1}(1-x^2)}(1+2x)^L\sum_{E=0}^{L}(E+1)\frac{x^E}{(1-x^2)^E}F(x), \quad (3.34)$$

where

$$F(x) = \sum_{\substack{M_m'=0\\ m\geq 2}}^{\infty}\prod_{n=2}^{\infty}\binom{E+M_n'+2\Sigma_{m=n+1}^{\infty}(m-n)M_m'}{M_n'}\prod_{m=2}^{\infty}\left(\mathscr{Z}_m^{(0)}\right)^{M_m'}, \quad (3.35)$$

and

$$\mathscr{Z}_m^{(0)} = \frac{x^{2m}}{(1-x^2)^{2(m-1)}}. \quad (3.36)$$

The summations over M_2', M_3', ... have precisely the form of the l.h.s. of eq. (3.7) and can thus be performed easily. The result is

$$F(x) = \prod_{m=2}^{\infty}\left(1-\mathscr{Z}_m^{(m-2)}\right)^{-1-E}, \quad (3.37)$$

where

$$\mathscr{Z}_m^{(p)} = \frac{\mathscr{Z}_m^{(p-1)}}{\left(1-\mathscr{Z}_{p+1}^{(p-1)}\right)^{2(m-p-1)}}. \quad (3.38)$$

It can be shown along the lines given in subsect. 3.1, that the quantities $U_m = 1/\mathscr{Z}_m^{(m-2)}$ obey the recursion relation

$$(U_{p+2}-1)^2 = U_{p+3}U_{p+1} \qquad (p \geqslant 0) \tag{3.39}$$

with initial conditions

$$U_1 = x^{-2}, \qquad U_2 = \frac{(1-x^2)^2}{x^4}. \tag{3.40}$$

Eq. (3.23) is replaced by

$$\frac{\mathscr{Z}_{n+1}^{(p)}}{\mathscr{Z}_n^{(p)}} = \frac{U_{p+1}}{U_{p+2}} \qquad (p \geqslant 0, \qquad n \geqslant p+2). \tag{3.41}$$

Eq. (3.34) now can be written as

$$\#(\text{eigenstates}) = \frac{1}{2\pi i}\oint \frac{\mathrm{d}x}{x^{L+1}}(1+2x)^L \sum_{E=0}^{L}(E+1)x^E[f(x)]^{-E-1}, \tag{3.42}$$

where

$$f(x) = \prod_{l=1}^{\infty}(1-U_l^{-1}). \tag{3.43}$$

The solution of the recursion relation (3.39) is again of the form (3.24), i.e.,

$$U_j = \left(\frac{a(x)^{j+1}-a(x)^{-j-1}}{a(x)-a(x)^{-1}}\right)^2,$$

where now

$$a(x) = \frac{1}{2x} + \sqrt{\frac{1}{4x^2}-1} \tag{3.44}$$

due to the new initial conditions (3.40). Insertion of the resulting expression for U_l into eq. (3.43) leads to the following result for the function $f(x)$,

$$2f(x) = 1+\sqrt{1-4x^2} = 2x\,a(x). \tag{3.45}$$

Eq. (3.42) can now be rewritten as

$$\#(\text{eigenstates}) = \sum_{E=0}^{L}(E+1)I(E), \tag{3.46}$$

where

$$I(E)=\frac{1}{2\pi i}\oint \mathrm{d}\left(-\frac{1}{x}\right)\left(\frac{1}{x}+2\right)^{L}[a(x)]^{-E-1}. \tag{3.47}$$

The contour integration can be worked out as in sect. 2. Defining $\alpha=\Lambda-i\varphi$ with $\Lambda\gg 1$ and substituting $x=1/(\mathrm{e}^{\alpha}+\mathrm{e}^{-\alpha})$ we obtain

$$I(E)=I_{+}(E)-I_{-}(E), \tag{3.48}$$

where

$$I_{\pm}(E)=\frac{1}{2\pi}\int_{0}^{2\pi}\mathrm{d}\phi\ \mathrm{e}^{\pm\alpha}\ \mathrm{e}^{-(1+E)\alpha}\left(\mathrm{e}^{\alpha/2}+\mathrm{e}^{-\alpha/2}\right)^{2L}. \tag{3.49}$$

Expanding

$$\left(\mathrm{e}^{\alpha/2}+\mathrm{e}^{-\alpha/2}\right)^{2L}=\sum_{p=0}^{2L}\binom{2L}{p}\mathrm{e}^{\alpha(L-p)} \tag{3.50}$$

and then using

$$\frac{1}{2\pi}\int_{0}^{2\pi}\mathrm{d}\phi\ \mathrm{e}^{\pm i(n\varphi)}=\delta_{n,0} \tag{3.51}$$

in the resulting expression, we find that

$$I_{+}(E)=\binom{2L}{L-E},\qquad I_{-}(E)=\binom{2L}{L-E-2}. \tag{3.52}$$

Plugging these results into eq. (3.48) and then eq. (3.46) we are left with only a single summation

$$\#(\text{eigenstates})=\sum_{E=0}^{L}(E+1)\left\{\binom{2L}{L-E}-\binom{2L}{L-E-2}\right\}. \tag{3.53}$$

This summation can be performed the same way as eq. (3.30) and we finally obtain the desired result

$$\#(\text{eigenstates})=4^{L}. \tag{3.54}$$

This concludes our two-step evaluation of the sum (2.9).

Using the above, we can obtain a closed expression for the number of regular Bethe-ansatz states for given numbers M and N of spin-down and spin-up electrons:

$$\sum_{\substack{M_c=0 \\ N_c=M_c+2\Sigma mM'_m \\ M=\Sigma m(M_m+M'_m)}}^{\infty} \sum_{M_m=0}^{\infty} \sum_{M'_m=0}^{\infty} n(M_c, \{M_m\}, \{M'_m\})$$

$$= \binom{L}{N}\left(\binom{L}{M} + \binom{L}{M-2}\right) - \left(\binom{L}{N+1} + \binom{L}{N-1}\right)\binom{L}{M-1}. \qquad (3.55)$$

This formula is the close analogue of the result (3.1) for the XXX Heisenberg model.

We repeat once more our conclusion, where is that the combination of the nested Bethe ansatz with the SO(4) symmetry of the one-dimensional Hubbard model leads to a complete set of 4^L independent eigenstates.

It is a pleasure to thank C.N. Yang for proposing the ideas worked out in this paper. We thank S. Dasmahapatra for stimulating discussions. This work was supported by NSF Grant PHY-9107261.

Appendix A

BOUND STATES IN THE ONE-DIMENSIONAL HUBBARD MODEL

In this appendix we investigate the nature of the Λ- and k–Λ-strings in the one-dimensional Hubbard model. We show for explicit examples, that both kinds of string lead to certain kinds of bound states (i.e., the wave function decays exponentially with respect to the differences of coordinates).

A.1. Λ-strings

Let us consider the example of N electrons with spin up and two electrons with spin down forming a Λ-string, i.e.,

$$\Lambda_2 = \Lambda_1^* \qquad (k_1, \ldots, k_{N+2} \text{ real}). \qquad (A.1)$$

the set $\{\Lambda_1, \Lambda_2 \,|\, k_1, \ldots, k_{N+2}\}$ must fulfill the periodic boundary conditions (1.11).

The wave function corresponding to this set of spectra parameters is given by

$$\psi_{-1,-1,1,\ldots,1}(x_1, x_2, \ldots, x_{N+2})$$
$$= \sum_{P \in S_{2+N}} \text{sgn}(Q)\ \text{sgn}(P)\ \exp\left(i\Sigma_{j=1}^{2+N} k_{P_j} x_{Q_j}\right) \phi(y_1, y_2 | P), \qquad (A.2)$$

with amplitudes

$$\varphi(y_1, y_2|\mathrm{P}) = A_{\mathrm{id}}F_{\mathrm{P}}(\Lambda_1, y_1)F_{\mathrm{P}}(\Lambda_2, y_2) + A_{21}F_{\mathrm{P}}(\Lambda_2, y_1)F_{\mathrm{P}}(\Lambda_1, y_2)$$

$$= A_{\mathrm{id}}\left(\prod_{i=1}^{y_1-1}\frac{e_-^{(1)}(\mathrm{P}_i)}{e_+^{(1)}(\mathrm{P}_i)}\right)\frac{1}{e_+^{(1)}(\mathrm{P}_{y_1})}\left(\prod_{i=1}^{y_2-1}\frac{e_-^{(2)}(\mathrm{P}_i)}{e_+^{(2)}(\mathrm{P}_i)}\right)\frac{1}{e_+^{(2)}(\mathrm{P}_{y_2})}$$

$$+ A_{21}\left(\prod_{i=1}^{y_2-1}\frac{e_-^{(1)}(\mathrm{P}_i)}{e_+^{(1)}(\mathrm{P}_i)}\right)\frac{1}{e_+^{(1)}(\mathrm{P}_{y_2})}\left(\prod_{i=1}^{y_1-1}\frac{e_-^{(2)}(\mathrm{P}_i)}{e_+^{(2)}(\mathrm{P}_i)}\right)\frac{1}{e_+^{(2)}(\mathrm{P}_{y_1})}. \quad (\mathrm{A.3})$$

We want to show that this wave function decays exponentially with respect to the difference of the coordinates y_1 and y_2. The only nontrivial y-dependent part of the wave function are the amplitudes φ. Therefore, it is sufficient to prove that they decay exponentially.

Taking exponentially small corrections Δ into account, the Λ-string is of the form

$$\Lambda_1 = \Lambda + \tfrac{1}{2}(u + \Delta),$$

$$\Lambda_2 = \Lambda - \tfrac{1}{2}(u + \Delta) \qquad (\text{with } \Delta^* = -\Delta). \quad (\mathrm{A.4})$$

As all momenta k_j are real this leads to the following inequalities,

$$\left|\frac{e_-^{(1)}(i)}{e_+^{(1)}(i)}\right| > 1, \qquad \left|\frac{e_-^{(2)}(i)}{e_+^{(2)}(i)}\right| < 1. \quad (\mathrm{A.5})$$

Using the periodic boundary conditions,

$$\prod_{i=1}^{N+2}\frac{e_-^{(1)}(i)}{e_+^{(1)}(i)} = \frac{\Lambda_2 - \Lambda_1 - u}{\Lambda_2 - \Lambda_1 + u} = \frac{A_{\mathrm{id}}}{A_{21}}, \quad (\mathrm{A.6})$$

we can express A_{21} in terms of A_{id}. Using the second set of periodic boundary conditions,

$$\exp(ik_jL) = \frac{e_-^{(1)}(j)}{e_+^{(1)}(j)}\frac{e_-^{(2)}(j)}{e_+^{(2)}(j)}, \quad (\mathrm{A.7})$$

we can express products over $e_\pm^{(\alpha)}$ in terms of exponential factors of magnitude 1.

Straightforward computations yield

$$\phi(y_1, y_2|\mathrm{P}) = A_{\mathrm{id}} \exp\left(i \sum_{l=1}^{y_1-1} k_{\mathrm{P}_l} L\right)\left(\prod_{i=y_1}^{y_2-1} \frac{e_-^{(2)}(\mathrm{P}_i)}{e_+^{(2)}(\mathrm{P}_i)}\right) \frac{1}{e_+^{(1)}(\mathrm{P}_{y_1})} \frac{1}{e_+^{(2)}(\mathrm{P}_{y_2})}$$
$$+ A_{\mathrm{id}} \exp\left(i \sum_{l=1}^{y_2-1} k_{\mathrm{P}_l} L\right)\left(\prod_{i=1}^{y_1-1} \frac{e_-^{(2)}(\mathrm{P}_i)}{e_+^{(2)}(\mathrm{P}_i)}\right)\left(\prod_{i=y_2}^{N+2} \frac{e_-^{(2)}(\mathrm{P}_i)}{e_+^{(2)}(\mathrm{P}_i)}\right)$$
$$\times \frac{1}{e_+^{(1)}(\mathrm{P}_{y_2})} \frac{1}{e_+^{(2)}(\mathrm{P}_{y_1})}. \tag{A.8}$$

By definition $y_2 > y_1$ and the inequalities (A.5) ensure that the product in parentheses in the first term (and thus the whole term) in (A.8) decays exponentially for $y_2 \gg y_1$. The second term can be dropped, because for spin waves N plays the role of the lattice length, and in order to investigate asymptotic properties of the wave function we should consider the infinite volume limit, i.e., $N \to \infty$. In this limit the second term can be set to zero as

$$\left(\prod_{i=y_2}^{N+2} \frac{e_-^{(2)}(\mathrm{P}_i)}{e_+^{(2)}(\mathrm{P}_i)}\right)$$

vanishes.

A.2. k–Λ-strings

We consider the example $N = M = 2$, i.e., two electrons with spin up and two with spin down. The periodic boundary conditions read

$$\exp(ik_j L) = \frac{e_-^{(1)}(j)}{e_+^{(1)}(j)} \frac{e_-^{(2)}(j)}{e_+^{(2)}(j)} \qquad (j = 1, \ldots, 4),$$
$$\prod_{j=1}^{4} \frac{e_-^{(\beta)}(j)}{e_+^{(\beta)}(j)} = \prod_{\alpha=1}^{2} \frac{\Lambda_\alpha - \Lambda_\beta - u}{\Lambda_\alpha - \Lambda_\beta + u} \qquad (\beta = 1, 2). \tag{A.9}$$

A k–Λ-string solution of these equations takes the following form in the $L \to \infty$ limit,

$$\Lambda_1 = \Lambda - \tfrac{1}{2}u = \Lambda_2^*,$$
$$k_1 = \pi - \arcsin(\Lambda - u),$$
$$k_2 = \arcsin(\Lambda) = \pi - k_3,$$
$$k_4 = \pi - \arcsin(\Lambda + u). \tag{A.10}$$

In the finite volume there exist two distinct configurations that both lead to k–Λ-strings in the limit $L \to \infty$, depending on whether k_2 and k_3 are real or complex for finite L.

Case (i): k_2 and k_3 are complex. In this case the k's can be rearranged such that $\mathrm{Im}(k_1) < \mathrm{Im}(k_2) < 0 < \mathrm{Im}(k_3) < \mathrm{Im}(k_4)$ and the invariance of the periodic boundary conditions under complex conjugation gives the additional constraints

$$k_3^* = k_2, \qquad k_4^* = k_1. \tag{A.11}$$

Taking this into account one obtains the following solution of (A.9) for finite but large L,

$$\begin{aligned} \Lambda_1 &= \Lambda - \tfrac{1}{2}u - \tfrac{1}{2}\delta = \Lambda_2^*, \\ \sin(k_1) &= \Lambda - u - \tfrac{1}{2}\delta + \epsilon_1 = \sin(k_4^*), \\ \sin(k_2) &= \Lambda + \tfrac{1}{2}\delta + \epsilon_2 = \sin(k_3^*), \end{aligned} \tag{A.12}$$

where, δ is purely imaginary and $\epsilon_{1,2}$ are complex. The exponentially small corrections are of the orders

$$\begin{aligned} &\delta = \mathrm{O}(2u \exp[-i(k_1 + k_2^*)L]), \qquad \epsilon_1 = \mathrm{O}(-2u \exp(-ik_1 L)), \\ &\epsilon_2 = \mathrm{O}(-2u \exp(-ik_1 L)). \end{aligned} \tag{A.13}$$

The wave function is given by eq. (A.2) with $N = 2$ and eq. (A.3). After inserting the values of the spectral parameters found in eq. (A.13) into eqs. (A.2) and (A.3) and re-normalising the resulting expression one finds

$$\psi_{-1,-1,1,1}(x_1, x_2, x_3, x_4) = \begin{cases} \mathrm{sgn}(\mathrm{Q})(-1)^{y_1+y_2}\left[\exp\left(i\sum_{j=1}^{4} k_j x_{\mathrm{Q}_j}\right) - \exp\left(i\sum_{j=1}^{4} k_{\mathrm{R}_j} x_{\mathrm{Q}_j}\right)\right] \\ \qquad \text{if } (y_1, y_2) \notin \{(1,2),(3,4)\} \\ 0 \qquad \text{else,} \end{cases} \tag{A.14}$$

where R is the permutation (1, 3, 2, 4).

Due to the ordering of the imaginary parts of the momenta k_j both terms on the l.h.s. of eq. (A.14) decay exponentially with respect to the magnitudes of differences of coordinates $|x_k - x_j|$ and thus the wave function describes a bound state.

Case (ii): k_2 and k_3 are real. In this case we must drop the constraint $k_2^* = k_3$. We then find the following solution to eq. (A.9),

$$\begin{aligned}
\Lambda_1 &= \Lambda - \tfrac{1}{2}u - \tfrac{1}{2}\delta = \Lambda_2^*,\\
\sin(k_1) &= \Lambda - u - \tfrac{1}{2}\delta + \epsilon_1 = \sin(k_4^*),\\
\sin(k_2) &= \Lambda + \epsilon_2,\\
\sin(k_3) &= \Lambda + \epsilon_3,
\end{aligned} \tag{A.15}$$

where ϵ_1 is complex and $\epsilon_{2,3}$ are real, while δ is again purely imaginary. The corrections are of the orders

$$\begin{aligned}
\epsilon_1 &= \mathrm{O}(-2u\exp(-ik_1L)),\\
\delta &= \mathrm{O}\left(\frac{4i\,\mathrm{Re}(\epsilon_1)}{\cot(\frac{1}{2}k_2L) + \cot(\frac{1}{2}k_3L)}\right),\\
\epsilon_2 &= \mathrm{O}(-\tfrac{1}{2}i\cot(\tfrac{1}{2}k_2L)\delta),\\
\epsilon_3 &= \mathrm{O}(-\tfrac{1}{2}i\cot(\tfrac{1}{2}k_3L)\delta).
\end{aligned} \tag{A.16}$$

The computation of the wave function is analogous to case (i), the only difference being a new renormalisation constant. The wave function is given by the same expression as in case (i). Again it describes a bound state although k_2 and k_3 are now real.

Appendix B

THE $M = N = 1$ SECTOR IN THE HUBBARD MODEL

In this appendix we further work out the structure of the Bethe-ansatz wave functions in the sector $M = N = 1$. In that sector the wavefunctions depend on parameters k_1, k_2 and Λ. Our general analysis in sect. 3 and 4 gives the following possibilities: (i) we can have $M_e = 2$, $M_1 = 1$, which gives real values for k_1, k_2 and Λ, or (ii) we can have $M_1' = 1$, which gives a $m = 1$ k–Λ-string with complex conjugate k_1, k_2 and real Λ. According to the counting of sect. 3, we expect to have $\frac{1}{2}L(L-1)$ real solutions (i) and $L - 1$ string solutions (ii), giving a total number of $\frac{1}{2}(L-1)(L+2)$ Bethe-ansatz states in this sector. However, we already mentioned that, in the context of the *XXX* Heisenberg model, there is a "redistribution phenomenon" between different types of solutions, which does not affect the total number of states within a sector, but which does affect the distribution of those states over various types of string solutions and real solutions [18]. In this appendix we will establish a very similar result for the Hubbard model: we will find that the numbers of the real solutions (i) and string solutions (ii) are not always given by the values quoted above, but that the expected total number $\frac{1}{2}(L-1)(L+2)$ of solutions in this sector can be rigorously established.

Let us consider the Bethe equations for periodic boundary conditions in the sector $M=N=1$,

$$\exp(ik_1L)=\frac{\sin k_1-\Lambda-\frac{1}{2}u}{\sin k_1-\Lambda+\frac{1}{2}u},\qquad \exp(ik_2L)=\frac{\sin k_2-\Lambda-\frac{1}{2}u}{\sin k_2-\Lambda+\frac{1}{2}u},$$

$$\prod_{i=1}^{2}\frac{\sin k_i-\Lambda-\frac{1}{2}u}{\sin k_i-\Lambda+\frac{1}{2}u}=1. \tag{B.1}$$

Since we assume Λ to be finite, we can solve for it and find $\Lambda=\frac{1}{2}(\sin k_1+\sin k_2)$. The equations then reduce to

$$\exp(ik_1L)=\frac{\sin k_1-\sin k_2-u}{\sin k_1-\sin k_2+u},\qquad \exp[i(k_1+k_2)L]=1. \tag{B.2}$$

We can solve the second equation by putting $k_1+k_2=2\pi m/L$, with $m=0,1,\dots,2L-1$ and write $k_1=\pi m/L+x$ and $k_2=\pi m/L-x$. The remaining equation reads

$$\exp[i(\pi m+Lx)]=\frac{-(4/U)\cos(\pi m/L)\sin x-i}{-(4/U)\cos(\pi m/L)\sin x+i}. \tag{B.3}$$

One easily checks that, if x_0 solves this equation for $m=m_0$, then $x=x_0+\pi$ solves the equation for $m=m_0+L$, and that the resulting wave functions are the same. We can thus restrict our attention to $m=0,1,\dots,L-1$ and $-\pi\leqslant x<\pi$.

Let us now try to find real solutions x for eq. (B.3) for given m. Taking a logarithm we have

$$\arctan[-\tfrac{1}{4}U\cos(\pi m/L)\sin x]=-\tfrac{1}{2}Lx-\pi n, \tag{B.4}$$

where n is an arbitrary integer for m odd and half odd integer for m even. It is rather straightforward to solve this equations by a graphical method: one plots both the l.h.s. and the r.h.s. (for various n) of these equations on the interval $-\pi\leqslant x<\pi$ and reads of the intersection points, which are then solutions of the equation. For large enough $|U|$ this procedure is easily carried out and one finds the following.

For m even there are solutions for $n = -\frac{1}{2}(L+1), \ldots \frac{1}{2}(L-1)$, which are L solutions in total. Since solutions x and $-x$ are equivalent (and $x=0$ is not among the solutions) we should divide this number by 2. Using that there are $\frac{1}{2}L$ possible even values for m we thus find $\frac{1}{4}L^2$ solutions. For m odd one finds non-equivalent solutions for $n = 1, 2, \ldots, \frac{1}{2}(L-1)$ for each m, which gives a total number of $\frac{1}{4}(L^2 - 2L)$. (The solutions $x=0$, $x = \pm\pi$, which exist for generic U, given vanishing wavefunctions in general.) adding up the contributions from odd and even m, we find $\frac{1}{2}L(L-1)$, which is indeed the number predicted by the counting in sect. 3.

However, let us now assume that m is odd and that U is close to a critical value U_m, which we define by

$$\tfrac{1}{4}U_m \cos(\pi m/L) = \tfrac{1}{2}L. \tag{B.5}$$

At the value $U = U_m$ the curve for the l.h.s. of eq. (B.4) has slope $\frac{1}{2}L$ at $x = \pm\pi$, and at the value $U = -U_m$ the curve for the l.h.s. of eq. (B.4) has slope $\frac{1}{2}L$ at $x=0$. Since the r.h.s. is given by straight lines of slope $\frac{1}{2}L$, and since both curves already had intersections at $x=0$ (for $n=0$) and $x = \pm\pi$ (for $n = \pm\frac{1}{2}L$) (which did however not give rise to non-trivial wave functions), it will be clear that the number of intersections changes when U reaches the critical values $\pm U_m$. In fact, one finds one extra real solution x (together with the equivalent solution $-x$) for a given odd m as soon as $U < |U_m|$. For example, if U is such that $U_1 > U > U_3 > \ldots > 0$ there will be one extra real solution to eq. (B.2).

The complex values for x which solve eq. (B.3) are of the form $x = iy$ or $x = \pi + iy$ and y real. In a way similar to what we showed above, one can analyse the equations for the real quantity y by a graphical method. If $|U|$ is sufficiently large, one finds precisely one complex solution for $m = 1, 2, \ldots, L$ with the exception of $m = \frac{1}{2}L$. In that case, there are thus $L-1$ complex solutions, which is in agreement with the counting of sect. 3.

However, from the graphical analysis one finds that the complex solution for a given odd m disappears as soon as $|U|$ is chosen to be smaller than U_m. Note that this happens precisely in the regime where we have found one extra real solution!

We thus find a redistribution phenomenon, where solutions change their nature as a function of U, in analogy to what we found for the *XXX* Heisenberg model in ref. [18]. In the Hubbard model the phenomenon is easily understood: if $|U|$ is made small enough, the interactions become so weak that some of the bound states (with complex x) decay into real solutions (with real x).

When $|U|$ is chosen to be equal to one of the critical values U_m, there do exist non-trivial wave functions with $x=0$ or $x = \pm\pi$, i.e., with coinciding k_1 and k_2. These wave functions can be seen to be non-vanishing by a renormalisation à la l'Hôpital.

In all cases, the total number of eigenfunctions of the hamiltonian in the sector

$M = N = 1$ is found to be $\frac{1}{2}(L-1)(L+2)$, which is the value predicted by the counting in sect. 2, and used for the proof of completeness in sect. 3.

References

[1] P.W. Anderson and R. Schrieffer, Phys. Today (June 1991) 54–61
[2] C.N. Yang, Phys. Rev. Lett. 63 (1989) 2144
[3] C.N. Yang and S. Zhang, Mod. Phys. Lett. B4 (1990) 759
[4] C.N. Yang, preprint ITP-SB-91-38
[5] E. Lieb and F.Y. Wu, Phys. Rev. Lett. 20 (1968) 1445
[6] C.N. Yang, Phys. Rev. Lett. 19 (1967) 1315
[7] F.H.L. Eßler, V.E. Korepin and K. Schoutens, Nucl. Phys. B372 (1992) 559
[8] O.J. Heilmann and E.H. Lieb, Ann. New York Acad. Sci. 172 (1971) 583;
E.H. Lieb, Phys. Rev. Lett. 62 (1989) 1201
[9] I. Affleck in talk given at the Nato Advanced Study Institute on Physics, Geometry and Topology, Banff. August 1989;
M. Pernici, Europhys. Lett. 12 (1990) 75
[10] F.H.L. Eßler, V.E. Korepin and K. Schoutens, Phys. Rev. Lett. 67 (1991) 3848
[11] F. Woynarovich, J. Phys. (Paris) C15 (1982) 85
[12] B. Sutherland, Lecture Notes in Physics, 242 (Springer, Berlin, 1985) p. 1
[13] S. Zhang, Phys. Rev. Lett. 65 (1990) 120
[14] M. Takahashi, Prog. Theor. Phys. 47 (1972) 69
[15] L.D. Faddeev and L. Taktajan, Phys. Lett. A85 (1981) 375; Zap. Nauch.-Semin. LOMI 109 (1981) 134; Lecture Notes in Physics, 242 (Springer, Berlin, 1985) pp. 158, 175
[16] M. Takahashi, Prog. Theor. Phys. 46 (1971) 401
[17] H. Bethe, Z. Phys. 71 (1931) 205
[18] F.H.L. Eßler, V.E. Korepin and K. Schoutens, J. Phys. A25 (1992) 4115

Journal of Statistical Physics, Vol. 50, Nos. 1/2, 1988

Decorated Star–Triangle Relations and Exact Integrability of the One-Dimensional Hubbard Model

B. Sriram Shastry[1]

Received July 16, 1987; revision received September 25, 1987

The exact integrability of the one-dimensional Hubbard model is demonstrated with the help of a novel set of triangle relations, the decorated star–triangle relations. The covering two-dimensional statistical mechanical model obeys the star–triangle or Yang–Baxter relation. A conjecture is presented for the eigenvalues of the transfer matrix.

KEY WORDS: One-dimensional Hubbard model; exactly integrable systems; star–triangle relations.

1. INTRODUCTION

We have recently shown[1] that the one-dimensional (1d) Hubbard model possesses an infinite number of conservation laws by identifying a 2d classical statistical model for which a one-parameter family of transfer matrices commutes with the Hamiltonian. Reference 2 contains a demonstration that the one-parameter family of transfer matrices commute mutually and hence we have a new completely integrable problem. This embedding of the 1d Hubbard model into a covering 2d statistical model parallels the well-known relationship between the 1d XYZ model and the 2d eight-vertex models established by Sutherland and Baxter.[3,4] In this paper we present some further results on the covering statistical model, and also recover the previous results of Refs. 1 and 2 through a promising new line of argument.

The eigenfunctions of the 1d Hubbard model were found by Lieb and Wu[5] using the Bethe–Yang or nested Bethe Ansatz technique. The

[1] Theoretical Physics Section, Tata Institute of Fundamental Research, Bombay-400005, India.

applicability of the latter suggests the existence of conservation laws. This was emphasized by Heilmann and Lieb,[6] who diagonalized a six-membered, half-filled Hubbard ring and found surprising instances of level crossings and degeneracies.

Barma and Shastry[7] proposed a 2d statistical model with the help of Trotters' formula, for which the 1d Hubbard model is the logarithmic derivative of a transfer matrix [in the sense of Eq. (2.8) to order u]. This model was diagonalized by Bariev[8] through a variant of the coordinate space Bethe ansatz, and elaborated upon by Schotte and Truong.[9,10] This transfer matrix, however, does not commute with the Hamiltonian for general values of Boltzmann weights. In order to find a commuting transfer matrix, one needs further information. In Section 2.1 we present a novel algorithm which yields one nontrivial "current" operator that commutes with the Hamiltonian. This information can be used within the rather tight framework of the transfer matrix formulation (see Luscher[11]) to guess the other members of a commuting family. This is done in Section 2.2, where the form of the transfer matrix is proposed.

Section 3 contains a discussion of a novel class of triangle relations, which we call the decorated star–triangle relations (DSTR). These are intimately related to the star–triangle or Yang–Baxter relations, and rest essentially on the same algebraic structure. The terminology is suggested by the fact that these triangle relations can be pictured as the usual triangle diagrams with additional (diagonal) operators residing on the intermediate lines. However, the DSTRs are an independent set of relations from the STR, and we indicate how one may combine the two in order to get a richer set of STRs. The examples provided in Section 3 yield STRs for the free Fermi vertex models in the presence of fields. These examples are presaged to some extent in the work of Bazhanov and Stroganov,[12] which came to our notice after the completion of this work.

In Section 5, we explore the problem of diagonalizing the transfer matrix. We have not succeeded in an explicit diagonalization, but present a conjecture for the general eigenvalue from which the results of Lieb and Wu follow.

Finally, we mention the review of integrable models by Kulish and Sklyanin,[13] which contains an exhaustive list, and also a discussion of the difficulties of the 1d Hubbard model. Also, the results of Refs. 1 and 2 have been recently verified through a different route by Wadati *et al.*[14,15]

In Section 4 we consider a pair of free Fermi six-vertex models and show that the DSTR together with the STR enable us to construct the R matrix of the covering model for the Hubbard problem rather easily. The R matrix is given explicitly in a compact form.

2. CONSERVED CURRENTS

2.1. Introduction

In this section we present a novel algorithm for identifying conserved currents, i.e., commuting operators with respect to a Hamiltonian expressible in the form

$$H = H_0 + UH_1 \tag{2.1}$$

where H_0 is a "free" Hamiltonian, typically bilinear in fermionic operators, and H_1 is the interaction term quartic in fermionic operators. For the 1d Hubbard model

$$H_0 = -t \sum_{n,\sigma} (C^+_{n+1\sigma} C_{n\sigma} + C^+_{n\sigma} C_{n+1\sigma}) = \sum \varepsilon_{k\sigma} C^+_{k\sigma} C_{k\sigma} \tag{2.2}$$

with

$$C_{k\sigma} = \frac{1}{\sqrt{N}} \sum \exp(ikn)\, C_{n\sigma}, \qquad \varepsilon_k = -2t \cos k$$

and

$$H_1 = \sum n_{m\uparrow} n_{m\downarrow} = \frac{1}{N} \sum C^+_{k+q\uparrow} C_{k\uparrow} C^+_{p-q\downarrow} C_{p\downarrow} \tag{2.3}$$

The summation over m runs from 1 to N, and σ represents the two components of fermions ($\uparrow$ and $\downarrow$). Periodic boundary conditions are assumed everywhere in this work. The XXZ model is in the form (2.1) with U replaced by Δ, and with a single species of fermions.

The free part H_0 commutes with all bilinears in fermions of the form $\sum W_k C^+_{k\sigma} C_{k\sigma}$, and we expect that the currents for H, if they exist, should go over continuously to those of H_0 as $U \to 0$. The simplest Ansatz for a current is

$$j = j_0 + Uj_1 \tag{2.4}$$

where $j_0 = \sum W_k C^+_{k\sigma} C_{k\sigma}$ with some as yet undetermined W_k and j_1. In principle one could go on and add terms to (2.4) of $O(U^2)$, etc. However, we shall truncate at order U, guided by the known results for the XXZ model, where *all* currents are in the above form. Requiring $[j, H] = 0$ and equating various orders of U to zero separately, we find the set of equations

$$[j_0, H_0] = 0 \tag{2.5a}$$

$$[j_0, H_1] = [H_0, j_1^\perp] \tag{2.5b}$$

$$[j_1^\perp, H_1] = -[j_1^\parallel, H_1] \tag{2.5c}$$

Here we have decomposed the operator $j_1 = j_1^{\parallel} + j_1^{\perp}$, where the parallel part is defined by $[j_1^{\parallel}, H_0] = 0$.

The algorithm works as follows: we pick some W_k defining j_0 and calculate $j_1^{\perp}$ from (2.5b). This is done most simply by sandwiching (2.5b) between a pair of common eigenstates of H_0 and j_0. The "particle content" of $j_1^{\perp}$ must clearly mimic that of H_1; if H_1 is a four-fermion term, so is $j_1^{\perp}$. The resulting $j_1^{\perp}$ is inserted into the lhs of (2.5c) and one checks if the commutator vanishes or can be simplified to the form of the rhs. It is not guaranteed that (2.5c) can be satisfied; in general, one would have to try various W_k. The working of this scheme is straightforward, if tedious, in momentum space.

We tried this scheme for the Hubbard model and found the following Hermitian currents corresponding to the simplest choices of $W_k = \sin k$ and $\sin 2k$:

$$\begin{aligned} j_A = {} & (it) \sum (C^+_{n+1\sigma} C_{n\sigma} - C^+_{n\sigma} C_{n+1\sigma}) \\ & + (iU) \sum_{r=1}^{N} \sum_{n=1}^{N-1} C^+_{r+n\uparrow} C^+_{r+n\downarrow} C_{r\downarrow} C_{r\uparrow} (-)^n \\ & (N = \text{odd integer}) \end{aligned} \tag{2.6}$$

$$\begin{aligned} j_B = {} & (it) \sum (C^+_{m+2\sigma} C_{m\sigma} - C^+_{m\sigma} C_{m+2\sigma}) + (iU) \sum (C^+_{m+1\sigma} C_{m\sigma} - C^+_{m\sigma} C_{m+1\sigma}) \\ & + (iU) \sum_{m\sigma} [C^+_{m\sigma}(C_{m+1\sigma} - C_{m-1\sigma}) - (C^+_{m+1\sigma} - C^+_{m-1\sigma}) C_{m\sigma}]\, n_{m-\sigma} \end{aligned} \tag{2.7}$$

The operator j_B contains a nonvanishing $j_1^{\parallel}$, whereas j_A does not. The current j_A is tantalizing. First, it makes sense only for N odd. The first term (j_0) has $W_k = \sin k$, the group velocity corresponding to $\varepsilon_k \approx \cos k$, and is in fact the current operator in the usual sense. The second term (j_1) corresponds to a kind of long-ranged backflow of doubly occupied sites. In the sector with no double occupation ($U = \infty$) the commutation of j_A with H was first noticed by Brinkman and Rice,[16] who pointed out that the dc conductivity diverges as a consequence. We expect that j_A should be useful in conductivity calculations for U finite; however, in the remainder of this paper we do not encounter it again. Also, we set $t = 1$ in the following.

2.2. Inferring the Transfer Matrix

In this section we outline the considerations used to guess a transfer matrix embedding of the Hamiltonian. The standard models of 2d classical statistical mechanics, such as the six-vertex model, have a rich algebraic structure, as is well known. The transfer matrix depends on a spectral parameter u, and an expansion in powers of u about a suitable value (say zero) generates an infinite number of conserved currents commuting with the Hamiltonian. Generically we write

$$T(u)=T(0)\left[1+uH+\frac{u^2}{2!}H^2+\frac{u^2}{2!}(-i)j+O(u^3)\right] \tag{2.8}$$

where $T(0)$ is the right shift operator. Here H is the Hamiltonian and j the first nontrivial current.

In the case of the 1d Hubbard model the Hamiltonian can be written in the form

$$H=\sum H_{n+1,n} \tag{2.9}$$

$$H_{n+1,n}=(\sigma_n^+\sigma_{n+1}^-+\sigma_{n+1}^+\sigma_n^-)+(\tau_n^+\tau_{n+1}^-+\tau_{n+1}^+\tau_n^-)+\tfrac{1}{4}U\sigma_n^z\tau_n^z \tag{2.10}$$

(the nonsymmetric definition is convenient in later usage). This form is obtained from (2.1)–(2.3) by subtracting a constant from the original Hamiltonian, corresponding to writing $H_1=\sum(n_{m\uparrow}-1/2)(n_{m\downarrow}-1/2)$, and using a Jordan–Wigner transformation

$$C_{m\uparrow}=(\sigma_1^z\cdots\cdot\sigma_{m-1}^z)\sigma_m^- \tag{2.11a}$$

$$C_{m\downarrow}=(\sigma_1^z\cdots\sigma_N^z)(\tau_1^z\cdots\tau_{m-1}^z)\tau_m^- \tag{2.11b}$$

to eliminate the fermions in favor of two species of Pauli matrices σ and τ. The noninteracting problem $U=0$ corresponds to a pair of uncoupled XY models. In this case we know that the (free Fermi) six-vertex model transfer matrix commutes with the Hamiltonian for a single species, and hence we expect that the relevant statistical model for the Hubbard problem should consist of two copies of the six-vertex model coupled appropriately. The precise nature of the coupling is the subject of investigation in this section. We will find that an explicit knowledge of j is of great help in this regard.

The transfer matrix is written in the standard form

$$T(u)=\operatorname*{tr}_g[L_{N,g}(u)\,L_{N-1,g}(u)\cdots L_{1,g}(u)] \tag{2.12}$$

where g is the auxiliary space variable, corresponding to the horizontal arrows in the row-to-row transfer matrix. The local scattering matrix $L_{n,g}(u)$ is as yet unspecified, apart from the requirement that when $U=0$, it must reduce to

$$L_{n,g}(u)\xrightarrow[U\to 0]{} l_{ng}^{(\sigma)}(u)\otimes l_{ng}^{(\tau)}(u)\equiv l_{n,g}(u) \tag{2.13}$$

with

$$l^{(\sigma)}_{n,g}(u)=\frac{a+b}{2}+\frac{a-b}{2}\sigma^z_n\sigma^z_g+c(\sigma^+_n\sigma^-_g+\sigma^+_g\sigma^-_n) \tag{2.14}$$

The weights a, b, and c are parametrized by u and obey the free Fermi condition $a^2+b^2=c^2$. In order to fix ideas, set $a=1$, $b=\sinh(u)$, $c=\cosh(u)$.

Consider an expansion of $T(u)$ through second order in u. First we assume an expansion for $L_{n,g}$,

$$L_{n,g}(u)=P_{n,g}(1+uH_{n,g}+\tfrac{1}{2}u^2B_{n,g}+O(u^3)) \tag{2.15}$$

where $P_{n,g}$ is the permutation operator $[L_{n,g}(u=0)]$. The coefficients of expansion $H_{n,g}$ and $B_{n,g}$ determine the expansion of T as

$$T(u)=T(0)\left[1+u\sum_n H_{n+1,n}+\frac{u^2}{2}\sum_n B_{n+1,n}\right.$$
$$\left.+u^2\sum_{n\geqslant m+1} H_{n,n-1}H_{m,m-1}+O(u^3)\right] \tag{2.16}$$

Using (2.9), we rewrite (2.16) as

$$T^{-1}(0)\,T(u)=1+uH+\frac{u^2}{2}H^2$$
$$+\frac{u^2}{2}\left\{\sum_m(B_{m,m-1}-H^2_{m,m-1})+\sum_m[H_{m,m-1},H_{m-1,m-2}]\right\}+O(u^3) \tag{2.17}$$

This is in the form of (2.8) with the explicit representation

$$(-i)j=\sum_m(B_{m,m-1}-H^2_{m,m-1})+\sum_m[H_{m,m-1},H_{m-1,m-2}] \tag{2.18}$$

To summarize the working so far, we see that if we demand that a transfer matrix exist such that its first two coefficients in an expansion in u give the Hamiltonian of the Hubbard model and a nontrivial current operator commuting with H, then the first two coefficients of expansion of the local $L_{n,g}$ operators are constrained by Eqs. (2.10), (2.15), (2.17), and (2.18). The current j used in (2.18) must be calculated separately (as we did in Section 2.1) and should go over in the $U=0$ limit to the first term of (2.7) [i.e., $W_k \propto \sin(2k)$], since the rhs of (2.18) does so by actual calculation.

Thus, we expect that the current j_B in (2.7) and the form of H should constrain the form of $L_{n,g}$ sufficiently to enable us to guess it. Toward this end, we write (2.7) in terms of the Pauli matrices

$$(-i)j_B=\sum_m [(\sigma^+_{m-1}\sigma^z_m\sigma^-_{m+1}-\text{h.c.}+(\sigma\leftrightarrow\tau)] + \frac{U}{2}\sum_m \{\tau^z_m[(\sigma^+_{m+1}-\sigma^+_{m-1})\sigma^-_m-(\sigma^-_{m+1}-\sigma^-_{m-1})\sigma^+_m]+(\sigma\leftrightarrow\tau)\} \tag{2.19}$$

Since we know $H_{n+1,n}$ [Eq. (2.10)] and j_B, we can find [redacted] through (2.18). By straightforward calculation we find

$$H^2_{m+1,m}=\frac{1}{2}(1-\sigma^z_m\sigma^z_{m+1})+(\sigma\leftrightarrow\tau)+2(\sigma^+_{m+1}\sigma^-_m+\sigma^+_m\sigma^-_{m+1}) \times(\tau^+_{m+1}\tau^-_m+\tau^+_m\tau^-_{m+1})+\frac{1}{16}U^2 \tag{2.20}$$

$$[H_{m+1,m}, H_{m,m-1}]=[\sigma^z_m(\sigma^+_{m-1}\sigma^-_{m+1}-\text{h.c.})+\frac{1}{2}U\sigma^z_m\tau^z_m \times(\sigma^+_m\sigma^-_{m-1}+\sigma^+_{m-1}\sigma^-_m)]+(\sigma\leftrightarrow\tau) \tag{2.21}$$

From (2.18)–(2.21) we find

$$B_{m,g}=\frac{1}{2}(1-\sigma^z_m\sigma^z_g)+\frac{1}{2}(1-\tau^z_m\tau^z_g)+2(\sigma^+_m\sigma^-_g+\text{h.c.})(\tau^+_m\tau^-_g+\text{h.c.}) +\frac{1}{2}U(\tau^+_m\tau^-_g+\tau^+_g\tau^-_m)\sigma^z_g\tau^z_g+\frac{1}{2}U(\sigma^+_m\sigma^-_g+\sigma^+_g\sigma^-_m)\sigma^z_g\tau^z_g+\frac{1}{16}U^2 \tag{2.22}$$

[we have equated the summands in (2.18) and replaced $m-1$ by g]. The form of the L operator is easy to guess at this stage from Eqs. (2.15), (2.10), and (2.22). In particular, (2.22) indicates that $L_{n,g}$ is probably $l_{n,g}$ postmultiplied by a function with the first derivative equal to $U/4\sigma^z_g\tau^z_g$ and the second derivative $U^2/16$. We therefore guess

$$L_{ng}(u)=l_{ng}(u)\exp(h\sigma^z_g\tau^z_g) \tag{2.23}$$

where $h=h(u)$, with $h(0)=0$, $h'(0)U/4$, $h''(0)=0$.

In Refs. 1 and 2 we showed that a transfer matrix (2.12) with L_{ng} chosen as in (2.11) indeed provides a covering model of the Hubbard model with a proper choice of h and u. In the following sections we provide an alternative and rather compact demonstration of the same results.

3. DECORATED STAR-TRIANGLE RELATIONS AND A FUSION PRINCIPLE

In this section we point out the existence of a modified triangle relation satisfied by the generic eight-vertex model, in addition to the usual star–triangle, or Yang–Baxter, relation. The "decorated" STR is an independent relation, which is in some sense a consequence of the structure of the STR, and can be used in conjunction with the latter through a kind of fusion principle to generate new models obeying the STR. As a prelude let us summarize the STR.[4] The relation is encountered when we consider the commutation of two transfer matrices in the form of Eq. (2.12) with different Boltzmann weights (vectors) W_1 and W_2,

$$TT' = \operatorname*{tr}_{g_1 g_2} \overleftarrow{\prod_n} \{L_{ng_1}(W_1)\, L_{ng_2}(W_2)\} \tag{3.1a}$$

$$T'T = \operatorname*{tr}_{g_1 g_2} \overleftarrow{\prod_n} \{L_{ng_1}(W_2)\, L_{ng_2}(W_1)\} \tag{3.1b}$$

The symbol $\prod_n^{\leftarrow}$ stands for an ordered product as in (2.12). The commutator $[T, T']$ vanishes, as first noted by Baxter, when an invertible operator R exists such that

$$L_{32}(W_1)\, L_{31}(W_2)\, R_{12}(W_3) = R_{12}(W_3)\, L_{32}(W_2)\, L_{31}(W_1) \tag{3.2}$$

(writing $n \to 3$, $g_1 \to 2$, $g_2 \to 1$). Writing $R_{12} = P_{12} S_{12}$, with P as the permutation operator, we find

$$L_{31}(W_1)\, L_{32}(W_2)\, S_{12}(W_3) = S_{12}(W_3)\, L_{32}(W_2)\, L_{31}(W_1) \tag{3.3}$$

The form of the operator S_{12} need not in general be the same as that of L; in fact, S may act upon a different kind of Hilbert space[13] from L. The parameters W_3 in general depend on W_1 and W_2 independently and may be indicated in the form $(W_2 | W_1)$. Considering the product $L_{0,1}(W_1)\, L_{0,2}(W_2)\, L_{0,3}(W_3)$, there are two distinct ways of rewriting this using (3.3) (corresponding to the two usual "braids"), which implies

$$[L_{03} L_{02} L_{01},\, S_{12}^{-1} S_{13}^{-1} S_{23}^{-1} S_{12} S_{13} S_{23}] = 0$$

Hence we expect

$$\begin{aligned} &S_{31}(W_1 | W_3)\, S_{32}(W_2 | W_3)\, S_{12}(W_2 | W_1) \\ &\quad = S_{12}(W_2 | W_1)\, S_{32}(W_2 | W_3)\, S_{31}(W_1 | W_3) \end{aligned} \tag{3.4}$$

This relation is not strictly a consequence of (3.3), but follows if the product $L_{01} L_{02} L_{03}$ is sufficiently nontrivial.[13] In any case it has to be checked independently.

In the case of the eight-vertex model, the famous result of Baxter[4] is in the form of (3.3) with

$$l_{31}^{8V}(W_1)=\frac{a_1+b_1}{2}+\frac{a_1-b_1}{2}\sigma_1^z\sigma_3^z+\frac{c_1+d_1}{2}\sigma_1^x\sigma_3^x+\frac{c_1-d_1}{2}\sigma_1^y\sigma_3^y \quad (3.5)$$

where the Boltzmann weight vector $W_1=(a_1, b_1, c_1, d_1)$, $l_{32}(W_2)$ is the same as above with $\sigma_1^\alpha \to \sigma_2^\alpha$, and $(a_1 b_1 c_1 d_1) \to (a_2 b_2 c_2 d_2)$. The S_{12} is also in the same form with $\sigma_3^\alpha \to \sigma_2^\alpha$ and $(a_1 \cdots) \to (a_3 \cdots)$. The consistency conditions for the Boltzmann weights are summarized in terms of the invariants

$$\Delta_n \equiv (a_n^2+b_n^2-c_n^2-d_n^2)/2a_n b_n \quad (3.6a)$$

$$\Gamma_n = c_n d_n / a_n b_n \quad (3.6b)$$

The consistency condition becomes

$$\Delta_1=\Delta_2=\Delta_3\equiv\Delta, \qquad \Gamma_1=\Gamma_2=\Gamma_3\equiv\Gamma \quad (3.7)$$

The weights a_3, b_3, c_3, d_3 can be computed explicitly in terms of $(a_2 \cdots)$ and $(a_1 \cdots)$ and are given by

$$\begin{aligned} a_3 &= a_1(c_1c_2-d_1d_2)(b_1^2c_2^2-c_1^2a_2^2)/c_1 \\ b_3 &= b_1(d_1c_2-c_1d_2)(a_1^2c_2^2-d_1^2a_2^2)/d_1 \\ c_3 &= c_1(b_1b_2-a_1a_2)(a_1^2c_2^2-d_1^2a_2^2)/a_1 \\ d_3 &= d_1(a_1b_2-b_1a_2)(b_1^2c_2^2-c_1^2a_2^2)/b_1 \end{aligned} \quad (3.8)$$

(The notation used here differs from that of Baxter[4] in that we use a_1, a_2, a_3, etc., to denote a, a', a'', etc.)

The decorated STRs are given by the relation

$$l_{31}(W_1)\, l_{32}(W_2)\, \sigma_2^z l_{12}(W_4) = l_{12}(W_4)\, \sigma_2^z l_{32}(W_2)\, l_{31}(W_1) \quad (3.9)$$

This is in the form of (3.3) with σ^z inserted in the places indicated. We can easily find the conditions on the Boltzmann weights necessary for (3.9) to hold by noting the identity

$$\begin{aligned} l_{31}(a_1, -b_1, c_1, -d_1) &= \sigma_3^z l_{31}(a_1, b_1, c_1, d_1)\sigma_1^z \\ &= \sigma_1^z l_{31}(a_1, b_1, c_1, d_1)\sigma_3^z \end{aligned} \quad (3.10)$$

Equation (3.10) follows from the definition (3.5) and the usual commutation relations of the Pauli matrices. The four weights are explicitly displayed in (3.10). We use the abbreviation $\bar{W}_n = (a_n, -b_n, c_n, -d_n)$ in the following. Multiplying (3.9) from the left by $\sigma_2^z \sigma_1^z$ and from the right by σ_3^z and using $[l_{32}, \sigma_2^z \sigma_3^z] = 0$, we find

$$l_{31}(\bar{W}_1)\, l_{32}(W_2)\, l_{12}(W_4) = l_{12}(W_4)\, l_{32}(W_2)\, l_{31}(\bar{W}_1) \tag{3.11}$$

This is just the STR with weights $\bar{W}$ in place of W in (3.3). We can thus borrow completely from the previously stated results for STR of Baxter and conclude that the decorated STR (3.9) holds if

$$-\Delta_1 = \Delta_2 = \Delta_4, \qquad \Gamma_1 = \Gamma_2 = \Gamma_4 \tag{3.12}$$

[the invariant Δ_1 changes sign from (3.6a), whereas Γ_1 does not]. The weights a_4, b_4, c_4, and d_4 can be found from (3.8) by merely negating b_1 and d_1 in the rhs.

Thus, the decorated STR connects models with Δ's negated as in (3.2) and does not appear to be very useful in the general case. For the free Fermi case, however, one has independently two sets of triangle relations, Eqs. (3.3) and (3.9), for the *same set* of scattering operators. This fact can be used to advantage, as we now demonstrate in two examples.

3.1. Free Fermi Eight-Vertex Model in a Horizontal Field

Consider the free Fermi case $\Delta_n = 0$, in which case we have both triangle relations (3.3) and (3.9) obeyed. We can add the two with (real) arbitrary coefficients and find the general relation

$$l_{31}(W_1)\, l_{32}(W_2)\, g_{12}^+ = g_{12} l_{32}(W_2)\, l_{31}(W_1) \tag{3.13}$$

where

$$g_{12} \equiv \alpha l_{12}(W_3) + \beta l_{12}(W_4)\sigma_2^z \tag{3.14}$$

The Hermitian conjugate on the lhs of (3.13) is given by

$$g_{12}^+ \equiv \alpha l_{12}(W_3) + \beta \sigma_2^z l_{12}(W_4) \tag{3.15}$$

We now observe that (3.13) could be used in the following decorated eight-vertex model, where $L_{31}(W_1) = l_{31}(W_1) I_1$ and $L_{32}(W_2) = l_{32}(W_2) I_2$, with I_1 and I_2 as "decoration" operators acting nontrivially only on the sites 1 and 2. We seek the S operator in (3.3) corresponding to the above L's. Since the (as yet unspecified) operators I_1 and I_2 can be pulled through operators independent of sites 1 and 2, Eq. (3.3) simplifies to

$$l_{31}(W_1)\, l_{32}(W_2)\, I_1 I_2 S_{12} (I_1 I_2)^{-1} = S_{12} l_{32}(W_2)\, l_{31}(W_1) \tag{3.16}$$

Comparing with (3.13)–(3.15), we infer

$$S_{12} = g_{12} \tag{3.17}$$

$$I_1 I_2 g_{12} = g_{12}^{+} I_1 I_2 \tag{3.18}$$

Therefore, if we can find decoration operators I_1 and I_2 and a pair α, β such that Eq. (3.18) is satisfied, then g_{12} is the S operator in the sense of the STR (3.3), with the L operators given by

$$L_{ng}(W_g) = l_{ng}(W_g) I_g \tag{3.19}$$

An inspection of Eq. (3.18) in fact suggests the form of the decoration operators

$$I_1 = \exp(h_1 \sigma_1^z); \qquad I_2 = \exp(h_2 \sigma_2^z) \tag{3.20}$$

The decorated eight-vertex model thus has nontrivial horizontal electric fields. The explicit solution of (3.18) is rather simple. Considering diagonal matrix elements where the spins 1 and 2 are not flipped, the equation is trivially satisfied. The off-diagonal elements $|\uparrow\downarrow\rangle \to |\downarrow\uparrow\rangle$ and $|\uparrow\uparrow\rangle \to |\downarrow\downarrow\rangle$ respectively yield the constraints

$$(\alpha c_3 - \beta c_4)\, e^{h_2 - h_1} = (\alpha c_3 + \beta c_4) e^{h_1 - h_2}$$

and

$$(\alpha d_3 + \beta d_4) e^{-h_1 - h_2} = (\alpha d_3 - \beta d_4) e^{h_1 + h_2} \tag{3.21}$$

Thus,

$$\frac{\alpha}{\beta}\frac{c_3}{c_4} = \frac{\cosh(h_2 - h_1)}{\sinh(h_2 - h_1)}; \qquad \frac{\alpha}{\beta}\frac{d_3}{d_4} = \frac{\cosh(h_2 + h_1)}{\sinh(h_2 + h_1)} \tag{3.22}$$

It is clear from (3.22) that h_1 and h_2 cannot be arbitrary; we eliminate α/β to find

$$\frac{c_3 d_4}{c_4 d_3} = \frac{\tanh(h_2 + h_1)}{\tanh(h_2 - h_1)}$$

Using (3.8) and analogous equations for a_4, b_4, c_4, and d_4 [obtained by ngating b_1 and d_1 in the rhs of (3.8)], we find after some elementary manipulations the constraint

$$\sinh(2h_1)\,\frac{a_1^2 - b_1^2}{a_1 b_1} = \sinh(2h_2)\,\frac{a_2^2 - b_2^2}{a_2 b_2} = g \tag{3.23}$$

where g is some fixed constant. To summarize, we have shown that the free Fermi eight-vertex model in a horizontal field with Boltzmann weights obeying $\Delta = 0$ and (3.23) constraining the parameters provides a one-parameter family of commuting transfer matrices. The R matrix (or S matrix) is given by Eq. (3.14), with

$$\beta/\alpha = c_3/c_4 \tanh(h_2 - h_1) = d_3/d_4 \tanh(h_2 + h_1)$$

In terms of the standard elliptic function parametrization of Baxter, we write

$$a_n : b_n : c_n : d_n = \mathrm{sn}(K - iu_n) : \mathrm{sn}(iu_n) : 1 : k\, \mathrm{sn}(iu_n)\, sn(K - iu_n) \tag{3.24}$$

with $u_3 = u_2 - u_1$ and $u_4 = u_2 + u_1$. Note that the R matrix is not a function of the spectral parameter difference u_3 alone, but depends also on $u_4 = u_2 + u_1$. This is a common feature to all the models discussed in this paper. We also note that the commutation of the transfer matrix of the eight-vertex model in a field with an appropriate XYZ model Hamiltonian was first discussed by Krinsky[17] and is an infinitesimal statement of the above result (corresponding to b_2 small).

3.2. Free Fermi Six-Vertex Model in Arbitrary Fields

Specializing to the six-vertex case $d_1 = d_2 = 0$, the elliptic parametrization degenerates into a trigonometric parametrization and we set

$$a_n = \cos(\theta_n); \qquad b_n = \sin(\theta_n); \qquad c_n = 1 \tag{3.25}$$

with $\theta_3 = \theta_2 - \theta_1$ and $\theta_4 = \theta_4 + \theta_1$. The entire argument leading to (3.20) is then common, and the only nontrivial matrix element of (3.18) is $|\uparrow\downarrow\rangle \to |\downarrow\uparrow\rangle$, leading to the constraint

$$\beta/\alpha = \tanh(h_2 - h_1) \tag{3.26}$$

Thus, the horizontal fields h_1 and h_2 are completely arbitrary and the resulting S matrix is (with ρ arbitrary)

$$S_{12}(\theta_2 \mid \theta_1) = \rho[\cosh(h_2 - h_1)\, l_{12}(\theta_2 - \theta_1) + \sinh(h_2 - h_1)\, l_{12}(\theta_2 + \theta_1)\, \sigma_2^z] \tag{3.27}$$

We note that vertical fields are easily included, since $[l_{12}, \sigma_1^z + \sigma_2^z] = 0$. To see this, write the expected relation

$$L_{31}(\theta_1)[\exp(b_1 \sigma_3^z)]\, L_{32}(\theta_2)[\exp(b_2 \sigma_3^z)]\, \hat{S}_{12} = \hat{S}_{12} L_{32}(\theta_2)[\exp(b_2 \sigma_3^z)]\, L_{31}(\theta_1) \exp(b_1 \sigma_3^z) \tag{3.28}$$

where $L_{31} = l_{31} \exp(h_1 \sigma_1^z)$, etc., and $\hat{S}$ is yet to be calculated [for $b_1 = b_2 = 0$, $\hat{S}$ is given by (3.27)]. Rewrite $b_2\sigma_3^z = b_2(\sigma_3^z + \sigma_1^z) - b_2\sigma_1^z$ in the rhs and $b_1\sigma_3^z = b_1(\sigma_3^z + \sigma_2^z) - b_1\sigma_2^z$ on the lhs; commuting factors through, we find a common factor $\exp(b_1 + b_2)\sigma_3^z$ on the extreme right of both sides. Canceling and rearranging, we find

$$L_{31} L_{32}[\exp(b_1\sigma_2^z)]\, \hat{S}_{12} \exp(-b_2\sigma_1^z)$$
$$= [\exp(b_1\sigma_2^z)]\, \hat{S}_{12}[\exp(-b_2\sigma_1^z)]\, L_{32} L_{31} \qquad (3.29)$$

Thus, we can choose

$$\hat{S}_{12} = [\exp(-b_1\sigma_2^z)]\, S_{12} \exp(b_2\sigma_1^z)$$

In the two examples that we have given here, the L operator is not Hermitian, since the decoration factor appears on one side of l only. However, we can trivially symmetrize using a "gauge transformation" $L' = QLQ^{-1}$ with an appropriate operator Q. Note that the inclusion of vertical fields has made no use of the free Fermi nature of l's.

Finally, in the case of the six-vertex model, I have checked that Eq. (3.4) is also valid in the form

$$S_{31}(\theta_1 \mid \theta_3)\, S_{32}(\theta_2 \mid \theta_3)\, S_{12}(\theta_2 \mid \theta_1)$$
$$= S_{12}(\theta_2 \mid \theta_1)\, S_{32}(\theta_2 \mid \theta_1)\, S_{31}(\theta_1 \mid \theta^3) \qquad (3.30)$$

by a brute force calculation. This implies that a more general inhomogeneous model is integrable with a transfer matrix

$$T(\theta \mid \{\theta_n\}) = \operatorname{tr}_g \overleftarrow{\prod_n} S_{ng}(\theta \mid \theta_n) \qquad (3.31)$$

such that $[T(\theta \mid \{\theta_n\}), T(\theta' \mid \{\theta_n\})] = 0$, with the R matrix again given by (3.27). The parameters $\{\theta_n\}$ are arbitrary constants.

4. INTEGRABILITY AND *R* MATRIX FOR THE COVERING MODEL FOR THE ONE-DIMENSIONAL HUBBARD MODEL

In Section 2 we introduced a model of a pair of six-vertex models coupled in a special way. The transfer matrix (2.12) is built out of local scattering operators, which were guessed to be in the form of Eq. (2.17),

$$L_{ng}(\theta) = l_{ng}(\theta) \exp(h\sigma_g^z \tau_g^z) \qquad (4.1)$$

with

$$l_{ng}(\theta) = l_{ng}^{(\sigma)}(\theta) \otimes l_{ng}^{(\tau)}(\theta) \qquad (4.2)$$

The operators $l^{(\sigma)}$ and $l^{(\tau)}$ are the usual free Fermi six-vertex scattering operators [Eq. (2.14)] parametrized by $a=\cos\theta$, $b=\sin\theta$, $c=1$. The constant h determines the "decoration" operator, i.e., the second factor in (4.1). We now show that the STR can be found for this model in a natural fashion, using the idea of fusion of decorated STRs explained in Section 3.

Let us write down the complete set of STRs and decorated STRs obeyed by the operators $l^{(\sigma)}$ and $l^{(\tau)}$:

$$l_{31}^{(\sigma)}(\theta_1)\, l_{32}^{(\sigma)}(\theta_2)\, l_{12}^{(\sigma)}(\theta_2-\theta_1) = l_{12}^{(\sigma)}(\theta_2-\theta_1)\, l_{32}^{(\sigma)}(\theta_2)\, l_{31}^{(\sigma)}(\theta_1) \tag{4.3}$$

$$l_{31}^{(\sigma)}(\theta_1)\, l_{32}^{(\sigma)}(\theta_2)\, \sigma_2^z l_{12}^{(\sigma)}(\theta_2+\theta_1) = l_{12}^{(\sigma)}(\theta_2+\theta_1)\, \sigma_2^z l_{32}^{(\sigma)}(\theta_2)\, l_{31}^{(\sigma)}(\theta_1) \tag{4.4}$$

We have two more equations of the same form as (4.3) and (4.4) with τ replacing σ. Taking direct products as in (4.2), we write down two resulting equations

$$l_{31}(\theta_1)\, l_{32}(\theta_2)\, l_{12}(\theta_2-\theta_1) = l_{12}(\theta_2-\theta_1)\, l_{32}(\theta_2)\, l_{31}(\theta_1) \tag{4.5}$$

$$l_{31}(\theta_1)\, l_{32}(\theta_2)\, \sigma_2^z\tau_2^z l_{12}(\theta_2+\theta_1) = l_{12}(\theta_2+\theta_1)\, \sigma_2^z\tau_2^z l_{32}(\theta_2)\, l_1(\theta_1) \tag{4.6}$$

Taking a linear combination, we find

$$l_{31}(\theta_1)\, l_{32}(\theta_2)\, g_{12}^{+} = g_{12} l_{32}(\theta_2)\, l_{31}(\theta_1) \tag{4.7}$$

with

$$g_{12} = \alpha l_{12}(\theta_2-\theta_1) + \beta l_{12}(\theta_2+\theta_1)\sigma_2^z\tau_2^z \tag{4.8}$$

[compare with Eqs. (3.13) and (3.14)]. We can now couple the σ and τ models through a "decoration" coupling given in (4.1). Thus, $L_{31} = l_{31}\exp(h_1\sigma_1^z\tau_1^z)$ and $L_{32} = l_{32}\exp(h_2\sigma_2^z\tau_2^z)$, and the expected STR (3.3) gives

$$l_{31}(\theta_1)\, l_{32}(\theta_2)[\exp(h_1\sigma_1^z\tau_1^z + h_2\sigma_2^z\tau_2^z)]\, S_{12} = S_{12} l_{32}(\theta_2)\, l_{31}(\theta_1)\exp(h_1\sigma_1^z\tau_1^z + h_2\sigma_2^z\tau_2^z) \tag{4.9}$$

Comparing with (4.7), we find

$$S_{12} = g_{12} \tag{4.10}$$

$$\exp[(h_1\sigma_1^z\tau_1^z + h_2\sigma_2^z\tau_2^z)]\, g_{12} = g_{12}^{+}\exp(h_1\sigma_1^z\tau_1^z + h_2\sigma_2^z\tau_2^z) \tag{4.11}$$

The condition (4.11) is trivially satisfied for two classes of terms: (1) diagonal in σ and in τ, (2) off-diagonal in both σ and in τ. Nontrivial constraints only arise when we consider a spin flip in one species and a diagonal term in the other. Using the equivalence of the two species, we need to consider two kinds of nontrivial terms $|\{\uparrow\downarrow\}\uparrow\uparrow\rangle \to |\{\downarrow\uparrow\}\uparrow\uparrow\rangle$ (i.e., $c\otimes a$) and $|\{\uparrow\downarrow\}\uparrow\downarrow\rangle \to |\{\downarrow\uparrow\}\uparrow\downarrow\rangle$ (i.e., $c\otimes b$), leading to the equations

$$\frac{\beta}{\alpha}\frac{a_4}{a_3} = \tanh(h_2 - h_1); \quad -\frac{\beta}{\alpha}\frac{b_4}{b_3} = \tanh(h_2 + h_1) \tag{4.12}$$

Eliminating β/α, we find the consistency condition

$$\frac{a_3}{b_3}\tanh(h_2 - h_1) = \frac{a_4}{b_4}\tanh(h_2 + h_1) \tag{4.13}$$

with

$$a_3 = \cos(\theta_2 - \theta_1), \qquad a_4 = \cos(\theta_2 + \theta_1)$$

$$b_3 = \sin(\theta_2 - \theta_1), \qquad b_4 = \sin(\theta_2 + \theta_1)$$

Simplifying further, we find

$$\frac{\sinh(2h_1)}{a_1 b_1} = \frac{\sinh(2h_2)}{a_2 b_2} = \frac{U}{2} \tag{4.14}$$

The constant on the rhs is chosen in conformity with Refs. 1 and 2. The S matrix follows from (4.10) (with ρ arbitrary)

$$\begin{aligned} S_{12}(\theta_2|\theta_1) = &= \rho[\cos(\theta_2+\theta_1)\cosh(h_2-h_1)\,l_{12}(\theta_2-\theta_1) \\ &+ \cos(\theta_2-\theta_1)\sinh(h_2-h_1)\,l_{12}(\theta_2+\theta_1)\,\sigma_2^z\tau_2^z] \end{aligned} \tag{4.15}$$

In summary, we have seen that the covering model of the 1d Hubbard model, defined by the transfer matrix (2.12) with the L operator given by (4.1), is an integrable system with the coupling h chosen according to (4.14). The S matrix (4.15) is essentially identical to the one found in Ref. 2, and differs only in that we have worked here with a non-Hermitian L operator, a simple "gauge transformation"

$$L' = [\exp(h\sigma_g^z\tau_g^z/2)]\,L\,\exp(-h\sigma_g^z\tau_g^z/2)$$

recovers completely the previous result.

I have checked some nontrivial matrix elements of (3.30) and believe it to be true in general; I am unable to give a tidy analytical proof of this result. This result implies that an inhomogeneous covering model with a transfer matrix (3.31) is also integrable, with $\{\theta_n\}$ arbitrary.

Some remarks concerning possible generalizations are in order at this stage. We have assumed that the spectral parameters θ_1 and θ_2 are the same for both $l^{(\sigma)}$ and $l^{(\tau)}$. The only other choice permitted is to negate one of the parameters, i.e., to consider $l'(\theta) = l^{(\sigma)}(\theta) \otimes l^{(\tau)}(-\theta)$, for otherwise we obtain too many consistency conditions in the sense of (4.12). The eight-vertex generalization of this scheme fails for the same reason; we end up with too many consistency conditions. One important question has been whether one could couple two general XXZ models in the same sense as the Hubbard model. We see that such a scheme is not possible, since the decorated STRs are useful in the above sense only for the free Fermi case. We have also verified that "natural" generalizations of the Hubbard problem to include more components in a *symmetric* fashion also fail because one obtains too many constraint equations. (The reader is urged to try the three-component problem independently.) Nonsymmetrical couplings might be allowed, althrough I have not checked these in detail.

We note that the model considered is intimately connected, in certain limits, to the isotropic Heisenberg antiferromagnet (XXX model). One correspondence follows from the degeneration to the Hubbard Hamiltonian to first order in θ [Eq. (2.8)], and the relationship of the latter to the XXX model in the limit of U large in the half-filled limit.[18] Another follows in the limit of h large and positive in the sector with all sites having either no particles (i.e., $\downarrow\downarrow$) or two particles ($\uparrow\uparrow$). Here the only allowed arrow configurations are six in number, with identical arrows on the σ and τ lattices, for which the invariant

$$\Delta = (a^4 + b^4 - c^4)/(2a^2b^2) = -1$$

(using $a^2 + b^2 = c^2$).

Let us remark that the STR (3.3) for the covering model has the feature that the infinitesimal limit of $S_{12}(w_3)$ as $w_1 \to w_2$ does not yield H_{12}, thereby sidestepping the difficulty mentioned in Ref. 13 [after Eq. (3.20)].

5. EIGENVALUES OF THE TRANSFER MATRIX

In this section we give a brief and regrettably incomplete account of the eigenvalues of the transfer matrix of the covering model (2.12). The problem is quite nontrivial, from either the coordinate-space Bethe Ansatz point of view or the algebraic Bethe Ansatz point of view.[19-21] One of the main difficulties is the absence of an obvious uniformizing parametrization of the S matrix (4.15), i.e., a parametrization in terms of which *all* the matrix elements are functions of the *difference* of appropriate spectral parameters.

From the algebraic point of view, a central role is ascribed to the global monodromy matrix

$$\Upsilon = \overleftarrow{\prod_n} \{l_{ng}(\theta) \exp(h\sigma_g^z \tau_g^z)\} \tag{5.1}$$

The four-dimensional auxiliary space g may be labeled in the form $|1\rangle = |\uparrow\uparrow\rangle$, $|2\rangle = |\downarrow\downarrow\rangle$, $|3\rangle = |\downarrow\uparrow\rangle$, and $|4\rangle = |\uparrow\downarrow\rangle$, and the matrix elements of Υ are denoted by T_{ij}. The pseudo commutation relations between $T_{ij}(\theta_1)$ and $T_{kl}(\theta_2)$ can be found from the STR (3.2) or (3.3) with the help of the explicit S matrix (4.15). We find a total of 256 relations, which may be written down with considerable labor. The state with all spins up $|\Omega\rangle$ is the vacuum state, and is an eigenfunction of T_{ii} for all i with ($a = \cos\theta$, $b = \sin\theta$)

$$\begin{aligned} T_{11}|\Omega\rangle &= a^{2N} e^{Nh} |\Omega\rangle; \qquad & T_{22}|\Omega\rangle &= b^{2N} e^{Nh} |\Omega\rangle \\ T_{33}|\Omega\rangle &= a^N b^N e^{-Nh} |\Omega\rangle; \qquad & T_{44}|\Omega\rangle &= a^N b^N e^{-Nh} |\Omega\rangle \end{aligned} \tag{5.2}$$

The difficulty of the problem arises from the proliferation of possible creation operators, T_{21}, T_{31}, T_{41}, T_{23}, T_{24} and composite operators T_{43}, T_{21}, etc. The only (rather trivial) case for which I have been able to construct eigenstates of T explicitly is the one with particles of one species only (say σ species), for which the state $T_{31}(\theta_1)\, T_{31}(\theta_2) \cdots T_{31}(\theta_n)|\Omega\rangle$, or $T_{24}(\theta_1)\, T_{24}(\theta_2) \cdots T_{24}(\theta_n)|\Omega\rangle$, is an exact eigenstate of T. The analysis of the commutation relations is sufficiently tedious and uninspiring as to prevent its inclusion here [the eigenvalues are consistent with Eq. (5.7) given below].

From the coordinate space point of view, the commutation of the transfer matrix with the Hubbard model implies that the eigenstates of the latter are also candidates for the former. Guided by the results of one- and two-spin deviation and the form of the results of the Bethe–Yang Ansatz for the Hubbard model, we conjecture the eigenvalue of the transfer matrix below. We show that the form of the eigenvalue, with the added constraint that poles in the eigenvalue on varying the Boltzmann weights in the finite part of the complex plane have vanishing residues, yields all the subsidiary conditions needed to fix the parameters.

We first parameterize the Boltzmann weights somewhat differently from before; let

$$a = 1/(e^{4x}+1)^{1/2}, \qquad b = e^{2x}/(e^{4x}+1)^{1/2}, \qquad c = 1 \tag{5.3}$$

whence Eq. (4.14) implies

$$\sinh(2h)\cosh(2x) = U/4 \tag{5.4}$$

Also define a "spin wave" function

$$\sigma_{\pm}(z) \equiv (e^{2x} + ze^{\pm 2h})/(1 - ze^{2x \pm 2h}) \tag{5.5}$$

The functions σ arise in the calculation of the wanted terms in the operation of T on the spin wave states $\sum z^n \sigma_n^- |\Omega\rangle$. In the sector with M particles with $M-K$ particles having spin up and K particles having spin down, the Bethe wave function is written in the form[22]

$$|M, K, \{z_n\}\rangle_Q = \sum_P A(Q|P)\, z_{P_1}^{n_1} z_{P_2}^{n_2} \cdots z_{P_M}^{n_M} \rho_{n_1}^{(Q\uparrow)} \cdots \rho_{n_{M-K}}^{(Q\uparrow)} \times \rho_{n_{M-K+1}}^{(Q\downarrow)} \cdots \rho_{M}^{(Q\downarrow)} |\Omega\rangle \tag{5.6}$$

where z_n are generalized momenta $z_n = e^{ik_n}$, P represents the permutation of the momentum set, and Q is a sector permutation label, $\rho_n^{(\uparrow)} \equiv \sigma_n^-$ and $\rho_n^{(\downarrow)} \equiv \tau_n^-$. For example, the state corresponding to the identity sector $Q = e$ has a string

$$\sigma_{n_1}^- \cdots \sigma_{n_{M-K}}^- \tau_{n_{M-K+1}}^- \cdots \tau_{n_M}^-$$

in (5.6). The amplitudes A are determined in the Hubbard model by requiring that (5.6) be an eigenfunction of H, and the eigenvalue condition determining Z_n requires the nested or Bethe–Yang Ansatz involving a new set of complex numbers $\{\lambda_m\}$, which are K in number.

The eigenvalue of T (actually the adjoint of T) on the state (5.6) is conjectured to be

$$\begin{aligned} &\Lambda_{M,K}(\theta, \{z_n\}, \{\lambda_m\}) \\ &= a^{2N} e^{Nh} \prod_{n=1}^{M} \sigma_-(z_n) + b^{2N} e^{Nh} (-1)^M \prod_{n=1}^{M} \sigma_+(z_n) \\ &+ a^N b^N e^{-Nh} (-1)^{M-K} \prod_{n=1}^{M} \sigma_-(z_n) \\ &\times \prod_{m=1}^{K} \left(\frac{e^{2h-2x} - e^{2x-2h} - \lambda_m + U/2}{e^{2h-2x} - e^{2x-2h} - \lambda_m - U/2} \right) + a^N b^N e^{-Nh} (-1)^K \\ &\times \prod_{n=1}^{M} \sigma_+(z_n) \prod_{m=1}^{K} \left(\frac{e^{-2h-2x} - e^{2x+2h} - \lambda_m - U/2}{e^{-2h-2x} - e^{2x+2h} - \lambda_m + U/2} \right) \end{aligned} \tag{5.7}$$

Some feeling for the numbers λ can be obtained from the results for $M = 2$, $K = 1$, where $\lambda \propto (z_1 - z_1^{-1} + z_2 - z_2^{-1})$. The conjecture (5.7) can be viewed as a kind of analytical Ansatz in the sense of Reshetekhin.[23] We now list a few important checks, which are fulfilled by (5.7) in support of the conjecture.

1. The limit $U=0$ is trivially satisfied; we merely multiply the eigenvalues of two free Fermi models.

2. The asymptotic behavior of the eigenvalue belonging to a given sector can be readily found for $\theta=-iu$, $u\to$ large and positive. Here

$$a=\cos\theta\to\tfrac{1}{2}e^{u},\qquad b=\sin\theta\to\tfrac{1}{2}e^{u-i\pi/2},\qquad c=1$$

The weights corresponding to the diagonal vertices a and b are dominant and hence spin deviations remain localized. We consider the two cases $U>0$ and $U<0$ separately.

(a) $U>0$. In this case $h>0$ and from $\sinh(2h)=U/2ab$, we find

$$e^{h}\to e^{u-i\pi/4+\phi/2},\qquad \text{where}\quad \phi=\ln(U/4)$$

The eigenvalue for $M-K$ spin-up particles and K spin-down particles follows from T_{11} and T_{22}, and is

$$\begin{aligned}\Lambda\to{}&2^{-2N}e^{3uN+N\phi/2-iN\pi}(e^{-iM\pi/2}+e^{-i\pi/2(2N-M)})\\&+O(e^{(3N-1)u})\end{aligned}\tag{5.8a}$$

(b) $U<0$. In this case $h<0$ and

$$e^{-h}\to e^{u-i\pi/4+\phi'/2},\qquad \phi'=\ln(|U|/4)$$

We find from T_{33} and T_{44}

$$\begin{aligned}\Lambda\to{}&2^{-2N}e^{3uN+N\phi'/2-3i\pi N/4}(e^{-i\pi(M-2K)/2}+e^{-i\pi(2K-M)/2})\\&+O(e^{(3N-1)u})\end{aligned}\tag{5.8b}$$

The eigenvalue (5.7) satisfies (5.8a) and (5.8b), as is readily seen.

3. Consider the adjoint of the transfer matrix T. From Eq. (5.1) the transfer matrix is obtained by taking the trace. We use the cyclic invariance of the trace to insert $1=(\sigma_g^x\tau_g^x)^2$ and write

$$T(\theta)=\operatorname*{tr}_{g}\,\sigma_g^x\tau_g^x\,\Upsilon(\theta)\,\sigma_g^x\tau_g^x$$

Using

$$[\sigma_g^x\tau_g^x,\sigma_g^z\tau_g^z]=0,\qquad \sigma_g^x\tau_g^x l_{ng}(\theta)\,\sigma_g^x\tau_g^x=l_{ng}^{*}(\pi/2-\theta)$$

(where the asterisk denotes Hermitian conjugation in the quantum variables n only), we find

$$T(\theta)=T^{+}(\pi/2-\theta)\qquad\text{or}\qquad T^{+}(\theta)=T(\pi/2-\theta)\tag{5.9}$$

This incidentally shows that the transfer matrix is normal, i.e., $[T(\theta), T^{+}(\theta')] = 0$, since $[T(\theta), T(\theta')] = 0$ for all θ, θ'. The eigenvalues therefore must satisfy the condition

$$\bar{\Lambda}(\theta) = \Lambda(\pi/2 - \theta) \tag{5.10}$$

where $\bar{\Lambda}$ is the eigenvalue of T^{+}. Equation (5.10) is in fact a constraint on the form of Λ,[23] since $\bar{\Lambda}$ should be alternately deducible from Λ by inverting $z_n \to z_n^{-1}$ (and $\lambda_m \to -\lambda_m$). Thus, we demand

$$\Lambda(\theta, \{z_n\}, \{\lambda_m\}) = \Lambda(\pi/2 - \theta, \{z_n^{-1}\}, \{-\lambda_m\}) \tag{5.11}$$

Equation (5.7) is readily seen to satisfy (5.11) on using (5.5), from which

$$\sigma_{\pm}(z) \xrightarrow[\theta \to \pi/2 - \theta]{z \to z^{-1}} (-1)\, \sigma_{\mp}(z) \tag{5.12}$$

4. We finally note an "inversion relation" satisfied by T. The S matrix (4.15) satisfies the condition

$$S_{12}(\theta_1 - \pi/2 \mid \theta_1) \propto p_{12}^{(-)} \tag{5.13}$$

where $p_{12}^{(-)}$ is the antisymmetrization operator [a direct product of the antisymmetrization operators $p_{12}^{\sigma(-)} \equiv \frac{1}{4}(1 - \sigma_1^z \sigma_2^z) - (\sigma_1^+ \sigma_2^+ + \text{h.c.})$, and a similar $p_{12}^{\tau(-)}$]. Therefore,

$$\begin{aligned} &L_{ng_1}(\theta_1)\, L_{ng_2}(\theta_1 - \pi/2)\, p_{g_1 g_2}^{(-)} \\ &\quad = p_{g_1 g_2}^{(-)} L_{ng_2}(\theta_1 - \pi/1)\, L_{ng_1}(\theta_1) \end{aligned} \tag{5.14}$$

Premultiplying by the symmetrization operator $p^{(+)}$, we find

$$p_{g_1 g_2}^{(+)} L_{ng_1}(\theta_1)\, L_{ng_2}(\theta_1 - \pi/2)\, p_{g_1 g_2}^{(-)} = 0 \tag{5.15}$$

This result has nontrivial consequences for the matrix product

$$T(\theta_1)\; T(\theta_1 - \pi/2) = \operatorname*{tr}_{g_1 g_2} \overleftarrow{\prod_n} \{L_{ng_1}(\theta_1)\, L_{ng_2}(\theta_1 - \pi/2)\} \tag{5.16}$$

We observe that the antisymmetric one-dimensional subspace $(\uparrow\downarrow - \downarrow\uparrow) \otimes (\uparrow\downarrow - \downarrow\uparrow)$ corresponding to a product of the singlets in g_1 and g_2 (for the σ and τ species) does not connect to the symmetric subspace and hence a block triangularity results. This argument is similar to that in Ref. 23 for the XXZ model. The matrix element within the 1d subspace is readily computed (using $h_1 = -h_2$) and we find

$$T(\theta_1)\, T(\theta_1 - \pi/2) = \cos^{4N}(\theta_1) + \tilde{T}(\theta_1) \tag{5.17}$$

The remainder $\tilde{T}$ contains powers of $\sin\theta_1$ and vanishes for $\theta_1=0$ [since $T(-\pi/2)$ is the left shift operator, i.e., $T(-\pi/2)=T(0)^{-1}$]. This equation implies a constraint on the eigenvalue Λ, namely the coefficient of a^{4N} in the product $\Lambda(\theta)\,\Lambda(\theta-\pi/2)$ should be unity. This is readily verifid for (5.7) using the result [remembering $h(\theta-\pi/2)=-h(\theta)$]

$$\sigma_\pm(z)\xrightarrow{\theta\to\theta-\pi/2}(-1)\frac{1}{\sigma_\pm(z)} \tag{5.18}$$

I have verified directly that (5.7) is true for $K=0$ (by the algebraic Ansatz), but have been unable to prove it in general. Next consider the singularities of Λ arising from fixing Z_n and λ_n and varying a and b, or equivalently x, in the finite part of the complex plane. As stressed by Baxter in his classical paper on the eight-vertex model,[4] such singularities must "go away" somehow, since the eigenfunction of the transfer matrix does not depend on the spectral parameter (θ or x), and hence singularities of the free energy on varying θ must be apparent only. In the case of poles of Λ, one simply demands that the residue should vanish.

The expression (5.7) has poles of two kinds, which we now discuss. The first and third factors have common simple poles corresponding to the vanishing of the denominator of (5.5). Consider one typical term $\sigma_-(z_n)$, which blows up for $e^{2x-2h}\to z_n^{-1}$. Equation the residue to zero, setting $(a/be^{2h})^N\to z_n^N$, and canceling common factors, we find the M relations

$$z_n^N=(-1)^{M-K-1}\prod_{m=1}^{K}\left(\frac{z_n-z_n^{-1}-\lambda_m+U/2}{z_n-z_n^{-1}-\lambda_m-U/2}\right) \tag{5.19}$$

The second and fourth terms have common poles from $\sigma_+(z_n)$, and equating the residues to zero, one again finds (5.19).

The second class of poles arise from the vanishing denominators of the third and fourth terms. Using (5.4), we see that poles are common, and a typical term has the pole condition $e^{2h-2x}-e^{2x-2h}\to\lambda_n+U/2$. Using the relation

$$\frac{\sigma_+(z)}{\sigma_-(z)}=\frac{e^{2h-2x}-e^{2x-2h}-(z-z^{-1})}{e^{2h-2x}-e^{2x-2h}-(z-z^{-1})-U} \tag{5.20}$$

we compute the residue and find the relations

$$\prod_{m\neq n}\left(\frac{\lambda_n-\lambda_m+U}{\lambda_n-\lambda_m-U}\right)=(-1)^M\prod_{m}\left(\frac{z_m-z_m^{-1}-\lambda_n-U/2}{z_m-z_m^{-1}-\lambda_n+U/2}\right) \tag{5.21}$$

In order to make contact with the results of Lieb and Wu, we recall

$$\frac{d}{d\theta}\ln T(\theta)\bigg|_{\theta=0}=H=\sum(\sigma_n^+\sigma_{n+1}^-+\text{h.c.})+(\sigma\leftrightarrow\tau)+\frac{U}{4}\sum\sigma_n^z\tau_n^z \quad (5.22)$$

Thus, the largest eigenvalue of T would give the highest energy state of H given above. In order to relate the *lowest energy* states of an appropriate Hamiltonian, we write

$$\bar{H}(|U|)=-\sum(\sigma_n^+\sigma_{n+1}^-+\text{h.c.})-\sum(\tau_n^+\tau_{n+1}^-+\text{h.c.})+\frac{|U|}{4}\sum\sigma_n^z\tau_n^z \quad (5.23)$$

We write $U=-|U|$ in (5.22) corresponding to $\sinh(2h)=-\frac{1}{2}|U|ab$ with $a, b>0$ (thus, $h<0$ in the principal domain)

$$\frac{d}{d\theta}\ln T(\theta)\bigg|_{\theta=0}=-\bar{H}(|U|) \quad (5.24)$$

The eigenvalue of $\bar{H}(|U|)$ can be read off from (5.7) easily by noting that as $\theta\to 0$ the first term dominates in the thermodynamic limit. Using $h\to|U|/4\theta+O(\theta^3)$, we readily find the eigenvalue of $\bar{H}$:

$$\bar{e}(M, K, \{z_n\}, \{\lambda_m\})=(N/4-M/2)|U|-\sum(z_n+z_n^{-1}) \quad (5.25)$$

The identification $Z_n\to e^{ik_n}$ recovers the results of Lieb and Wu provided we denote $\lambda_n\to 2i\Lambda_n$ and write $U=-|U|$ in (5.19) and (5.21).

NOTE ADDED IN PROOF

The article "Algebraic Geometry Methods in The Theory of Baxter-Yang Equations" by I. M. Krichever (Mathematical Physics Reviews Vol. 3, ed. S. P. Novikov (Harwood Academic Publisher) 1982), discusses a similar class of S matrices.

ACKNOWLEDGMENTS

It is a pleasure to thank Bill Sutherland and Leon Takhtajan for stimulating discussions, and E. H. Lieb, B. McCoy, and J. Perk for bringing relevant references to my attention.

REFERENCES

1. B. Sriram Shastry, *Phys. Rev. Lett.* **56**:1529 (1986); (E) **56**:2334 (1986).
2. B. Sriram Shastry, *Phys. Rev. Lett.* **56**:2453 (1986).
3. Bill Sutherland, *J. Math. Phys.* **11**:3138 (1970).
4. R. J. Baxter, *Ann. Phys. (N.Y.)* **70**:193 (1972); *Exactly Solved Models in Statistical Mechanics* (Academic, London, 1982).
5. E. H. Lieb and F. Y. Wu, *Phys. Rev. Lett.* **20**:1445 (1968).
6. O. J. Heilmann and E. H. Lieb, *Ann. N. Acad. Sci.* **172**:583 (1971).
7. M. Barma and B. S. Shastry, *Phys. Lett. A* **61**:15 (1977).
8. R. J. Bariev, *Teor. Mat. Fiz.* **49**:261 (1981).
9. T. T. Truong and K. D. Schotte, *Nucl. Phys. B* **230** [FS 10]:1 (1984).
10. K. D. Schotte, S. Iwabuchi, and T. T. Truong, *Z. Phys. B* **60**:255 (1985).
11. M. Luscher, *Nucl. Phys. B* **117**:475 (1976).
12. V. V. Bazhanov and Y. G. Stroganov, *Teor. Mat. Fiz.* **62**:377 (1985).
13. P. Kulish and E. Sklyanin, in *Integrable Quantum Field Theories*, J. Hietarinta and C. Montonen, eds. (Springer-Verlag, Berlin, 1981), p. 61.
14. M. Wadati, E. Olmedilla, and Y. Akutsu, *J. Phys. Soc. Jpn.* **56**:1340 (1987).
15. E. Olmedilla, M. Wadati, and Y. Akutsu, *J. Phys. Soc. Jpn.* **56**:2298 (1987).
16. W. Brinkman and T. M. Rice, *Phys. Rev. B* **2**:1324 (1970).
17. S. Krinsky, *Phys. Lett. A* **39**:169 (1972).
18. M. Takahashi, *J. Phys. C* **10**:1289 (1977).
19. L. D. Faddeev and L. A. Takhtajan, *Usp. Mat. Nauk* **34**:13 (1979).
20. E. K. Sklyanin, L. A. Takhtajan, and L. D. Faddeev, *Teor. Mat. Fiz.* **40**:194 (1979).
21. L. A. Takhtajan, in *Exactly Solvable Problems in Condensed Mater and Relativistic Field Theory*, B. S. Shastry, S. S. Jha, and V. Singh, eds. (Springer-Verlag, Berlin, 1986).
22. B. Sutherland, in *Exactly Solvable Problems in Condensed Matter and Relativistic Field Theory*, B. S. Shastry, S. S. Jha, and V. Singh, eds. (Springer-Verlag, Berlin, 1986).
23. N. Y. Reshetekhin, *Sov. Phys. JETP* **57**:691 (1983).

Prog. Theor. Phys. Vol. 83, No. 4, April 1990, Progress Letters

An Extension of the Thermal Bethe Ansatz

—— *One-Dimensional Hubbard Model* ——

Tohru KOMA

Department of Physics, Gakushuin University, Tokyo 171

(Received February 5, 1990)

The thermodynamics of the one-dimensional Hubbard model is reduced to a system of algebraic equations through an extension of our thermal Bethe ansatz. This new formalism is free from the string conjecture, or more basically from the unproven assumption of the completeness of the Bethe states.

In the previous papers,[1] we proposed a new method for the statistical mechanics of the spin-1/2 XXZ Heisenberg chain, which we call the thermal Bethe-ansatz method. It consists in (i) using the path integral idea to transform the partition function of the model into the one of a two-dimensional Ising system,[2] and (ii) applying the Bethe-ansatz method to find the maximum eigenvalue of the transfer matrix of the Ising system, which leads to the free energy in the thermodynamic limit.[1)~3)] This method is free from the string conjecture,[4] or more basically from the unproven assumption of the completeness of the Bethe states,[5),6)] and it gave the better results[1] than the previous methods did in the spin-1/2 XXZ Heisenberg chain.

In this paper, our thermal Bethe-ansatz method is extended to treat the one-dimensional Hubbard model. The thermodynamics of the model is thus reduced to a system of algebraic equations, as was the case of the Heisenberg model in the previous papers.[1]

Now, the Hamiltonian of the one-dimensional Hubbard model is given by

$$\begin{aligned}\hat{H}_{N_a}:=&\,t\sum_{\tau=\uparrow,\downarrow}\sum_{j=1}^{N_a-1}(\hat{a}_{j+1,\tau}^\dagger\hat{a}_{j,\tau}+\hat{a}_{j,\tau}^\dagger\hat{a}_{j+1,\tau})\\&+U\sum_{j=1}^{N_a}(\hat{a}_{j,\uparrow}^\dagger\hat{a}_{j,\uparrow}-1/2)(\hat{a}_{j,\downarrow}^\dagger\hat{a}_{j,\downarrow}-1/2)\\&-\mu\sum_{j=1}^{N_a}(\hat{a}_{j,\uparrow}^\dagger\hat{a}_{j,\uparrow}+\hat{a}_{j,\downarrow}^\dagger\hat{a}_{j,\downarrow})+N_a\mu\end{aligned}\tag{1}$$

(the symbol := signifies definition) with the free boundary condition, where t, U and μ are the hopping integral, the Coulomb interaction energy and chemical potential, respectively; $\hat{a}_{j,\tau}^\dagger$ and $\hat{a}_{j,\tau}$ are, respectively, the creation and annihilation operators of the electron with spin τ at the jth lattice site ($j=1,2,\cdots,N_a$; $\tau=\uparrow,\downarrow$).

We extend the thermal Bethe-ansatz method to calculate the free energy per electron in the thermodynamic limit,

$$f=-\lim_{N_a\uparrow\infty}(\beta N_a)^{-1}\log Z_{N_a}\,,\qquad Z_{N_a}=\mathrm{Tr}\exp[-\beta\hat{H}_{N_a}]\,,\tag{2}$$

for the sequence $N_a=2n$ ($n=1,2,\cdots$) in the following way, where β is the inverse temperature and Tr denotes the trace over the electron states.

We rewrite (1) into the form (3) below which is more convenient for the purpose of applying the path integral idea to the partition function Z_{2n} in (2). By using the Jordan-Wigner transformation,[7]

$$\hat{a}_{j,\uparrow}=(\hat{\sigma}^z_{1,\uparrow}\cdots\hat{\sigma}^z_{j-1,\uparrow})\hat{\sigma}^-_{j,\uparrow},$$

$$\hat{a}_{j,\downarrow}=(\prod_{l=1}^{N_a}\hat{\sigma}^z_{l,\uparrow})(\hat{\sigma}^z_{1,\downarrow}\cdots\hat{\sigma}^z_{j-1,\downarrow})\hat{\sigma}^-_{j,\downarrow}, \qquad (j=1,2,\cdots,2n)$$

we have

$$\hat{H}_{2n}=\sum_{\tau=\uparrow,\downarrow}\sum_{j=1}^{2n}\hat{H}_{j,j+1;\tau}+4^{-1}U\sum_{j=1}^{2n}\hat{\sigma}^z_{j,\uparrow}\hat{\sigma}^z_{j,\downarrow} \tag{3}$$

with

$$\hat{H}_{j,j+1;\tau}:=-2^{-1}t(\hat{\sigma}^x_{j,\tau}\hat{\sigma}^x_{j+1,\tau}+\hat{\sigma}^y_{j,\tau}\hat{\sigma}^y_{j+1,\tau})-4^{-1}\mu(\hat{\sigma}^z_{j,\tau}+\hat{\sigma}^z_{j+1,\tau}),$$

where we regard the lattice to be twofold as labeled by the spin $\tau=\uparrow,\downarrow$ and $\hat{\boldsymbol{\sigma}}_{j,\tau}$ is the Pauli spin operator for the jth site of the spin-τ lattice ($j=1,\cdots,2n$). We have changed the free boundary condition to the periodic one, $\hat{\boldsymbol{\sigma}}_{2n+1,\tau}=\hat{\boldsymbol{\sigma}}_{1,\tau}$ ($\tau=\uparrow,\downarrow$), because the thermodynamic limit in (2) is independent of the boundary condition. Here, we remark that in the case of (1) with the periodic boundary condition, the transformed Hamiltonian does not have a simple periodic structure.[7]

Following Suzuki,[2),8)] we divide the Hamiltonian (3) into three parts such that each one is a sum of operators all commuting with each other,

$$\hat{H}_A=\sum_{l=1}^{n}(\hat{H}_{2l-1,2l;\uparrow}+\hat{H}_{2l,2l+1;\downarrow}),$$

$$\hat{H}_B=\sum_{l=1}^{n}(\hat{H}_{2l,2l+1;\uparrow}+\hat{H}_{2l-1,2l;\downarrow})$$

and

$$2\hat{H}_I=4^{-1}U\sum_{j=1}^{2n}\hat{\sigma}^z_{j,\uparrow}\hat{\sigma}^z_{j,\downarrow}$$

to write the finite-M approximate partition function,

$$Z_{2n}^{(M)}=\mathrm{Tr}[e^{-\beta\hat{H}_A/M}e^{-\beta\hat{H}_I/M}e^{-\beta\hat{H}_B/M}e^{-\beta\hat{H}_I/M}]^M. \tag{4}$$

In the same way as in the previous papers,[1)] i.e., by using the path integral idea, (4) can be written as

$$Z_{2n}^{(M)}=\mathrm{Tr}(\hat{T}_{2M}^{-1}\hat{R}_{2M}\hat{T}_{2M}\hat{R}_{2M})^n,$$

where

$$\hat{R}_{2M}:=[\prod_{l=1}^{M}\hat{V}_{2l-1,2l;\uparrow}\hat{V}_{2l,2l+1;\downarrow}]\prod_{j=1}^{2M}\exp[(8M)^{-1}\beta U\hat{\sigma}^z_{j,\uparrow}\hat{\sigma}^z_{j,\downarrow}],$$

$$\hat{V}_{j,j+1;\tau} := \exp[-(8M)^{-1}\beta\mu(\hat{\sigma}^z_{j,\tau} - \hat{\sigma}^z_{j+1,\tau})]\,\hat{V}^{(0)}_{j,j+1;\tau}\exp[(8M)^{-1}\beta\mu(\hat{\sigma}^z_{j,\tau} - \hat{\sigma}^z_{j+1,\tau})]\,,$$

$$\hat{V}^{(0)}_{j,j+1;\tau} := 2^{-1}(\exp[\beta t/M] + \hat{\sigma}^x_{j,\tau}\hat{\sigma}^x_{j+1,\tau} + \hat{\sigma}^y_{j,\tau}\hat{\sigma}^y_{j+1,\tau} - \exp[-\beta t/M]\,\hat{\sigma}^z_{j,\tau}\hat{\sigma}^z_{j+1,\tau})$$

with the periodic boundary condition, $\hat{\boldsymbol{\sigma}}_{2M+1,\tau} = \hat{\boldsymbol{\sigma}}_{1,\tau}$ $(\tau = \uparrow, \downarrow)$, and the operator $\hat{T}_{2M}$ shifts any periodic array of the spin states by one lattice unit backwards. Further, it can be shown easily that for the n's restricted to multiples of M,

$$Z^{(M)}_{2ML} = \mathrm{Tr}(\hat{U}_{2M})^{2ML}\,, \qquad (L = 1, 2, \cdots)$$

where

$$\hat{U}_{2M} := \hat{T}^{-1}_{2M}\hat{R}_{2M}\,.$$

The operator $\hat{U}_{2M}$ plays the role of the transfer matrix in the "space" direction.

The free energy per electron in the thermodynamic limit is given by

$$f = -\beta^{-1}\lim_{N_a\uparrow\infty}\lim_{M\uparrow\infty} N_a^{-1}\log Z^{(M)}_{N_a}\,.$$

But, the order of limits can be interchanged by the theorem for the symmetrized decomposition.[1),3)] Therefore, we have

$$f = -\beta^{-1}\lim_{M\uparrow\infty}\lim_{L\uparrow\infty}(2ML)^{-1}\log Z^{(M)}_{2ML}$$

$$= -\beta^{-1}\lim_{M\uparrow\infty}\log\Lambda_M{}^{\max}\,,$$

where $\Lambda_M{}^{\max}$ is the maximum eigenvalue of the transfer matrix $\hat{U}_{2M}$.

Now, we write down the eigenstate $|B_{N\uparrow,N\downarrow}\rangle$ of $\hat{U}_{2M}$ having $N_\uparrow$ down spins for the spin-$\uparrow$ lattice and $N_\downarrow$ down spins for the spin-$\downarrow$ lattice by the Bethe ansatz. We can in fact prove that the maximum eigenvalue $\Lambda_M{}^{\max}$ is obtained when $N_\uparrow = N_\downarrow = M$, by the same argument as in the previous papers.[1)] Therefore, our method makes us free from the question whether or not the string conjucture[4)] provides the complete system of eigenstates. Thus, we write down the extended form of the thermal Bethe ansatz,[1),9)]

$$|B_{N\uparrow,N\downarrow}\rangle := \sum_{y_1,y_2,\cdots,y_{N_e}}\sum_Q\sum_P\Big[\prod_{i=1}^{N_e-1}\Delta(y_i, y_{i+1}; \tau_{Qi}, \tau_{Q(i+1)})\Big]$$

$$\times\langle\tau_{Q1}, \cdots, \tau_{QN_e}|A(P)\rangle F(z_{P1}, \tau_{Q1}, y_1)\cdots F(z_{PN_e}, \tau_{QN_e}, y_{N_e})$$

$$\times\Big[\prod_{j=1}^{N_e}\hat{\sigma}^-_{y_j,\tau_j}\Big]|\uparrow, \cdots, \uparrow\rangle$$

$$(N_e = N_\uparrow + N_\downarrow;\ N_\uparrow, N_\downarrow = 1, 2, \cdots, 2M)$$

for the eigenstates of $\hat{U}_{2M}$, with

$$\Delta(y_i, y_{i+1}; \tau_{Qi}, \tau_{Q(i+1)}) := \begin{cases} 0 \begin{pmatrix} y_i = y_{i+1} = \text{even and} \\ (\tau_{Qi}, \tau_{Q(i+1)}) = (\uparrow, \downarrow) \end{pmatrix} \\ 0 \begin{pmatrix} y_i = y_{i+1} = \text{odd and} \\ (\tau_{Qi} = \tau_{Q(i+1)}) = (\downarrow, \uparrow) \end{pmatrix} \\ 1 \ (\text{otherwise}), \end{cases}$$

$$F(z, \tau, y) := \begin{cases} a(z) z^{(y+1)/2} & (y = \text{odd} \quad \text{and} \quad \tau = \uparrow) \\ z^{y/2} & (y = \text{even} \quad \text{and} \quad \tau = \uparrow) \\ a(z) b(z) z^{y/2} & (y = \text{even} \quad \text{and} \quad \tau = \downarrow) \\ b(z) z^{(y-1)/2} & (y = \text{odd} \quad \text{and} \quad \tau = \downarrow) \end{cases}$$

and

$$(\tau_1, \cdots, \tau_{N_e}) = (\underbrace{\uparrow, \cdots, \uparrow}_{N_\uparrow \text{ spins}}, \underbrace{\downarrow, \cdots, \downarrow}_{N_\downarrow \text{ spins}}), \quad \hat{\sigma}_{j,\tau}^{-} = 2^{-1}(\hat{\sigma}_{j,\tau}^{x} - i\hat{\sigma}_{j,\tau}^{y}),$$

where $|\uparrow, \cdots, \uparrow\rangle$ is the state with all the spins up; the first summation runs over all integers satisfying $1 \leqq y_1 \leqq y_2 \leqq \cdots \leqq y_{N_e} \leqq 2M$, and the next two summations run over all the permutations P, Q of $(1, \cdots, N_e)$. We shall determine the complex numbers z_j $(j=1, \cdots, N_e)$, the functions a, b, and the set of ket vectors $\{|A(P)\rangle\}_P$ as follows. Namely, as in the original Bethe ansatz,[10] the eigenvalue equation for $\hat{U}_{2M}$ leads to the conditions,

$$\hat{Y}_{j,j+1}(\theta_{Pj} - \theta_{P(j+1)})|A(P1, \cdots, Pj, P(j+1), \cdots, PN_e)\rangle$$
$$= |A(P1, \cdots, P(j+1), Pj, \cdots, PN_e)\rangle,$$
$$|A(P1, \cdots, PN_e)\rangle = \hat{P}_{1,2}\hat{P}_{2,3}\cdots\hat{P}_{N_e-1,N_e}|A(P2, \cdots, PN_e, P1)\rangle z_{P1}^{M},$$
$$z_j = \frac{W_\mu \lambda_j + s^2}{\lambda_j(\lambda_j - W_\mu^{-1})}, \quad a(z_j) = c^{-1}(\lambda_j - W_\mu^{-1}),$$
$$b^2(z_j) = z_j, \quad \exp[i\theta_j] = a(z_j) b(z_j)$$

and to the eigenvalue,

$$\Lambda_{M,N_e} = \exp[2^{-1}\gamma(N_e - M)] s^{2M-N_e} \lambda_1 \cdots \lambda_{N_e}, \tag{5}$$

where

$$\hat{Y}_{k,l}(\theta_i - \theta_j)$$
$$:= -[2\sin(\theta_i - \theta_j - i\gamma)]^{-1}\{[\sin(\theta_i - \theta_j - i\gamma) + \sin(\theta_i - \theta_j) - \sin(i\gamma)]$$
$$+ [\sin(\theta_i - \theta_j - i\gamma) + \sin(\theta_i - \theta_j) + \sin(i\gamma)]\,\hat{\tau}_k^z \hat{\tau}_l^z - 2\sin(\theta_i - \theta_j)\hat{P}_{k,l}\},$$
$$c := \cosh(\beta t/M), \qquad s := \sinh(\beta t/M),$$
$$W_\mu := \exp[(2M)^{-1}\beta\mu], \qquad \gamma := -(2M)^{-1}\beta\mu,$$

$\hat{\tau}_k$ is the Pauli spin operator acting on the kth τ spin state and the operator $\hat{P}_{k,l}$

exchanges the kth τ spin state and the lth one. These equations can be solved by using the quantum inverse scattering method.[11] As a result, we obtain the system of the thermal Bethe-ansatz equations with unknowns $(\lambda_1, \cdots, \lambda_{N_e}, \Theta_1, \cdots, \Theta_{N\downarrow})$,

$$\left[\frac{W_\mu \lambda_j + s^2}{\lambda_j(\lambda_j - W_\mu^{-1})}\right]^M = (-1)^{N\uparrow - 1} \prod_{k=1}^{N\downarrow} \frac{\sin(\theta(\lambda_j) - \Theta_k - i\gamma/2)}{\sin(\theta(\lambda_j) - \Theta_k + i\gamma/2)}$$

$$(j = 1, \cdots, N_e) \tag{6}$$

and

$$\prod_{i=1}^{N_e} \frac{\sin(\theta(\lambda_j) - \Theta_l + i\gamma/2)}{\sin(\theta(\lambda_j) - \Theta_l - i\gamma/2)} = (-1)^{N_e - 1} \prod_{k=1}^{N\downarrow} \frac{\sin(\theta_l - \Theta_k - i\gamma)}{\sin(\theta_l - \Theta_k + i\gamma)},$$

$$(l = 1, \cdots, N_\downarrow) \tag{7}$$

where

$$\exp[2i\theta(\lambda_j)] = \frac{(\lambda_j - W_\mu^{-1})(W_\mu \lambda_j + s^2)}{c^2 \lambda_j}.$$

Thus, in our thermal Bethe-ansatz method, it is sufficient to find only one solution to the system of the Bethe-ansatz equations with $N_\uparrow = N_\downarrow = M$ for calculating the free energy in the thermodynamic limit. For solving them numerically, we remark that in the high temperature limit $\beta \downarrow 0$, the system of the Bethe-ansatz equations (6) and (7) with $N_\uparrow = N_\downarrow = M$ has a solution leading to the maximum eigenvalue in (5). Therefore, by stepping down from $\beta = 0$ a decreasing sequence of temperature as in the previous papers,[1] one can obtain numerical solutions to the system of the Bethe-ansatz equations for any temperature β.

A full account of this work and numerical results will be given elsewhere.

The author wishes to express his sincere gratitude to Professor H. Ezawa for his constant guidance. Thanks are also due to Professor M. Suzuki of the University of Tokyo and to Professor M. Takahashi of the Institute for Solid State Physics for many helpful suggestions.

---

1) T. Koma, Prog. Theor. Phys. **78** (1987), 1213; **81** (1989), 783.
2) M. Suzuki, J. Stat. Phys. **43** (1986), 883, and further references therein.
3) M. Suzuki and M. Inoue, Prog. Theor. Phys. **78** (1987), 787.
4) M. Takahashi and M. Suzuki, Prog. Theor. Phys. **48** (1972), 2187.
M. Takahashi, Prog. Theor. Phys. **47** (1972), 69.
5) A. A. Vladimirov, Phys. Lett. **105A** (1984), 418.
6) T. Koma and H. Ezawa, Prog. Theor. Phys. **78** (1987), 1009.
7) E. Lieb, T. Schultz and D. Mattis, Ann. of Phys. **16** (1961), 407.
E. Lieb and D. Mattis, Phys. Rev. **125** (1962), 164.
8) M. Barma and B. S. Shastry, Phys. Lett. **61A** (1977), 15.
9) R. Z. Bariev, Teoret. Mat. Fiz. **49** (1981), 261 [Theor. Math. Phys. **49** (1982), 1021].
10) H. A. Bethe, Z. Phys. **71** (1931), 205.
C. N. Yang, Phys. Rev. Lett. **19** (1967), 1312.
E. H. Lieb and F. Y. Wu, Phys. Rev. Lett. **20** (1968), 1445.
11) L. A. Takhtadzhan and L. D. Faddeev, Uspekhi Mat. Nauk **34** (1979), 13 [Russian Math. Surveys **34** (1979), 11].
A. A. Belavin, Phys. Lett. **87B** (1979), 117.

ERRATA

Prog. Theor. Phys. 83 (1990) 655–659

On p. 657, the second equation in the parenthesis above the bottom line of the page should read

$$N_\uparrow, N_\downarrow = 0, 1, \cdots, 2M$$

On p. 658, the second equation above the bottom line of the page should read

$$\gamma := -(2M)^{-1}\beta U$$

On p. 659, Equation (7) should read

$$\prod_{j=1}^{N_e} \frac{\sin(\theta(\lambda_j) - \Theta_\ell + i\gamma/2)}{\sin(\theta(\lambda_j) - \Theta_\ell - i\gamma/2)} = (-1)^{N_e - 1} \prod_{k=1}^{N_\downarrow} \frac{\sin(\Theta_\ell - \Theta_k - i\gamma)}{\sin(\Theta_\ell - \Theta_k + i\gamma)},$$

$$(\ell = 1, \cdots, N_\downarrow)$$

Journal of The Physical Society of Japan
Vol. 60, No. 5, May, 1991, pp. 1460–1463

LETTERS

Temperature Dependence of Spin Correlation Length of Half-Filled One-Dimensional Hubbard Model

Hirokazu TSUNETSUGU

Institute for Solid State Physics, University of Tokyo, Roppongi 7-22-1, Minato-ku, Tokyo 106

(Received February 23, 1991)

We have calculated the spin correlation length of the half-filled one-dimensional Hubbard model down to the temperature of $T=0.01\times$(transfer integral) by the Bethe-Ansatz method for the quantum transfer matrix. Its temperature dependence shows the crossover from $(2 \log \pi T)^{-1}$ at high temperatures to aT^{-1} with the log corrections at low temperatures, the low-T behavior being consistent with the prediction by the conformal field theory. When the Coulomb repulsion U is small, we have found at low temperatures another crossover in the proportional constant a from the value for the free electron system to that for the antiferromagnetic Heisenberg model. This crossover is due to the disappearance of charge fluctuation. A similar crossover has been observed for the linear coefficient of specific heat, C/T.

The one-dimensional Hubbard model (HM) has been of great interest in the condensed-matter physics as the simplest model of strongly correlated electron systems. This model can be exactly solved with the Bethe-Ansatz (BA) techniques as shown by Lieb and Wu,[1] and the BA solution was generalized by Takahashi to the finite-temperature case on the string hypothesis.[2] Several thermodynamic quantities commuting with the Hamiltonian were obtained by using this method.[3,4]

In contrast to the thermodynamic quantities, it has been believed to be difficult to calculate correlation functions. As well as many other one-dimensional (1D) quantum systems, its correlation functions show power-law decays in space at $T=0$ and exponential decays at $T>0$. Recently, some groups have succeeded in calculating power-law exponents η of the spin, charge, and superconductivity correlation functions, etc., at $T=0$ by applying the conformal field theory (CFT).[5-7] In the CFT, the η's are given by the conformal dimension of each relevant operator,[8] and they obtained the conformal dimensions by the BA techniques. The CFT predicts the temperature dependence of the correlation length as $\xi\sim T^{-1}$ at sufficiently low temperatures.[8] At high temperatures, on the other hand, it is expected that the correlation lengths are close to those of noninteracting systems, $\xi\sim(\log T)^{-1}$.

Another important progress was made for the exactly solvable 1D quantum spin systems; their correlation lengths ξ at finite temperatures have been obtained by the BA method.[9-11] The calculation of ξ can be reduced to diagonalization of their quantum transfer matrix (QTM) via the formula $e^{-1/\xi}=\lambda_2/\lambda_1$, where λ_1 and λ_2 are the largest and the second largest eigenvalues, respectively. The QTM is equivalent to the diagonal-to-diagonal transfer matrix of a 2D 6- or 8-vertex model, and it is known that this can be solved by the BA method. The correlation lengths of the $S=1/2$ ferro- and antiferromagnetic Heisenberg and XYZ spin models were obtained by using this scheme.[9-11] The free energy can also be obtained via the formula, $-\beta f=\log\lambda_1$.[12]

As for the HM, Barma and Shastry showed that this model could be mapped into a staggered 6-vertex model (S6VM),[13] and Bariev solved this S6VM as a 2D classical system.[14] Koma obtained an equivalent BA equation as the solution of the QTM of the HM.[15] In this letter, we shall study the temperature dependence of the correlation length of antiferromagnetic spin fluctuations by numerical computation of the BA equation given by Bariev and Koma. We restrict ourselves to the half-filled case because the BA equation is simple in this case, and leave the non-half-filled case to further study. Our main interest is *the quantitative analysis of the crossover* of the correlation length from the high temperature limit to the low temperature limit and the effect of the charge fluctuation. We note that the correlation length of the spin fluctuation in the repulsive HM with the interaction of U is equal to that of the s-wave superconductivity in the attractive HM with $-U$.[16]

The Hamiltonian of the HM is

$$H=-t\sum_{j\sigma}(a^\dagger_{j+1\sigma}a_{j\sigma}+\text{h.c.})+U\sum_j n_{j\uparrow}n_{j\downarrow}$$
$$-\mu\sum_{j\sigma}n_{j\sigma},\quad (t=1), \tag{1}$$

with standard notations, and the mapping into the S6VM is as follows. The first procedure is the Trotter decomposition of $e^{-H/T}$, transforming the thermodynamics of the HM at a temperature T into that of the S6VM with the width of $2N$ in the time direction, where N is the Trotter number.[13] The second is a particle-hole transformation to recover the particle number conservation of the QTM.[15] (In this letter, we call down and up arrows by particles and holes, respectively.) The QTM consequently becomes the transfer matrix of the S6VM with the Boltzmann weights shown in Fig. 1(a), and an example of arrow configuration is drawn in Fig. 1(b). The solid (dashed) lines represent the motion of up-spin (down-spin) electrons, and electrons interact via the on-site Coulomb repulsion at the intersections of the solid and dashed lines. The extrapolation to $N\to\infty$ should be taken to calculate physical quantities;[17] however, because the Boltzmann weights are dependent on N, we cannot transform the BA equation into an integral equation as in the Lieb-Wu solution. We shall numerically take that limit instead.

We summarize the results of Bariev and Koma (please refer to original papers[14,15] for details). The QTM ($\mathbf{R}_1$) for the sites with odd r are different from those ($\mathbf{R}_2$) for even r as shown in Fig. 1(b), but because $\mathbf{R}_2=\mathbf{T}\mathbf{R}_1\mathbf{T}^{-1}$ ($\mathbf{T}$: the shift operator in the time direction) and $[\mathbf{T}^2,\mathbf{R}_1]=0$, it is sufficient to diagonalize the QTM defined by $\mathbf{R}\equiv\mathbf{T}\mathbf{R}_1$. In the half-filled case (i.e., $\mu=U/2$), eigenvalues λ of $\mathbf{R}$ are determined by solutions of the following BA equations:

$$\lambda=a^{2N-n}\,e^{-n\beta\mu/(2N)}\prod_{j=1}^{n}e^{-ip_j}$$
$$\times[b\cos p_j+\sqrt{c^2-b^2\sin^2 p_j}], \tag{2}$$

$$2Np_j=2\pi I_j+\sum_{\beta=1}^{n_1}\varphi_u(M(p_j)-\Lambda_\beta), \tag{3a}$$

$$\sum_{j=1}^{n}\varphi_u(M(p_j)-\Lambda_\alpha)=2\pi J_\alpha+\sum_{\beta=1}^{n_1}\varphi_{2u}(\Lambda_\beta-\Lambda_\alpha), \tag{3b}$$

$$\varphi_u(x)\equiv -i\log\left[-\frac{\sin(x+iu)}{\sin(x-iu)}\right],$$
$$M(p)\equiv\sin^{-1}(bc^{-1}\sin p),\quad u\equiv\beta U/4N, \tag{4}$$

where a, b, and c are given in Fig. 1(a), N is the Trotter number, n is the number of particles, and n_1 is the particle number on the dashed-line sublattice. The quantum numbers I_j and J_α are half-integers when N is even, and we shall treat this case. It is easily verified that when all variables are real numbers, their range can be reduced to $-\pi<p_j\le\pi$ and $-\pi/2<\Lambda_\alpha\le\pi/2$.

The largest eigenvalue λ_1 is obtained by $n=2N$, $n_1=N$ and by choosing $\{I_j\}$ and $\{J_\alpha\}$

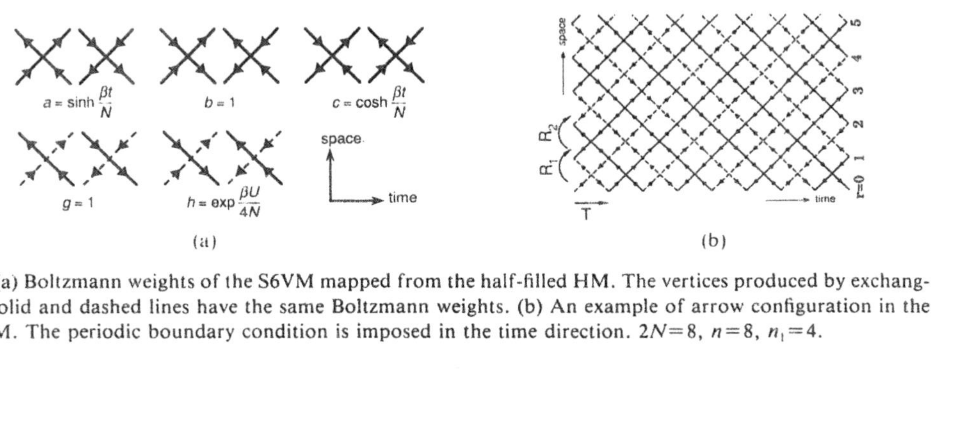

Fig. 1. (a) Boltzmann weights of the S6VM mapped from the half-filled HM. The vertices produced by exchanging solid and dashed lines have the same Boltzmann weights. (b) An example of arrow configuration in the S6VM. The periodic boundary condition is imposed in the time direction. $2N=8$, $n=8$, $n_1=4$.

consecutive numbers centered around zero.[14,15] We have found that the second largest eigenvalue λ_2 relevant to the spin correlation function is given by the same n, n_1, and the shift of $J_\alpha \to J_\alpha + 1$ for all α with $\{I_j\}$ unchanged, which we confirmed by direct diagonalization of the QTM up to $2N=8$. In these two cases, $\{p_j\}$ and $\{\Lambda_\alpha\}$ are all real numbers and have a characteristic distribution. In the case of λ_1, $\{p_j\}$ and $\{\Lambda_\alpha\}$ are symmetrically distributed around zero; in the case of λ_2, $p_{2N}=\pi$, $p_N=0$, $\Lambda_N=\pi/2$, $\Lambda_{N/2}=0$, and the other elements are symmetrically distributed around zero. Consequently, λ_1 and λ_2 are real numbers and $\lambda_1>0$, $\lambda_2<0$. This means that the most slowly decaying component is like $e^{-r/\xi}\cos\pi r$; i.e., the antiferromagnetic spin fluctuation is dominant.

We have numerically solved the BA eq. (3) and obtained λ_1, and λ_2 with decreasing temperature for various values of U. We have calculated the spin correlation length by the formula[9] of $\xi_s=[\log(\lambda_1/\lambda_2)]^{-1}$ and calculated several thermodynamic quantities, such as the free energy, the internal energy, the specific heat, etc., from λ_1 and its derivatives. Our results of the thermodynamic quantities coincide with those calculated by the conventional BA techniques.[1,4] We obtained these quantities by extrapolation to $N\to\infty$ from the data for several N, the largest system we used being $2N=4096$. Down to the temperature $T=0.01$ in units of t, our results are reliable within the error of 0.01% at worst except for the specific heat for large U (*e.g.*, $U=8$, the error is estimated to $\sim 0.5\%$ at $T=0.0143$).

We show the results of the temperature dependence of the spin correlation length ξ_s for some typical U in Fig. 2. Above $T\sim 1.0$, the effect of the Coulomb repulsion is small and the results are close to the value for the free electron system ($U=0$), $\xi_s^{\rm free}=\{2\log[\sqrt{(\pi T/2)^2+1}+\pi T/2]\}^{-1}$. The temperature dependence is, therefore, asymptotically $\sim(2\log\pi T)^{-1}$ in the high temperature region. The effect of the Coulomb repulsion develops below $T\sim 1.0$, and ξ_s shows the crossover from $(\log T)^{-1}$ to a power law. The CFT predicts the low-T asymptotic form as $\xi=v/\pi T$ for a wide class of 1D quantum systems with gapless excitations, where v is the velocity of the excitation. In our case, this is the spin wave velocity[18,19] $v_s=2I_1(2\pi/U)/I_0(2\pi/U)$ (I_0, I_1: the modified Bessel functions) because, in the half-filled case, the charge excitation has an energy gap called the Hubbard gap Δ no matter how small U is.[1] Owing to the Hubbard gap, when $T\ll\Delta$, the HM reduces to the universality class of the 1D antiferromagnetic Heisenberg model (AFHM) with the exchange integral of $J_{\rm eff}=2v_s/\pi$. However, because the gap is extremely small for small U,[18] $\Delta\sim 8\pi^{-1}\sqrt{U}\exp(-2\pi/U)$, in contrast to the large-U case ($\Delta\sim U-4$), the charge fluctuation should produce a considerable effect on ξ_s at the temperatures $\Delta<T$. Therefore, we expect that the critical temperature region is correspondingly narrow, particularly when U is small.

The crossover in Fig. 2 might not seem to be monotonous with respect to U, but we can clearly visualize the systematic behavior of the crossover by normalizing ξ_s and T by their characteristic values. The characteristic energy scale of temperature is only v_s owing to the Hubbard gap Δ at temperatures $T\ll\Delta$; this has the U-dependence like $v_s\sim 2-U/2\pi$ (for small U) and $v_s\sim 2\pi/U$ (for large U). Therefore, we plot $\pi\xi_s T/v_s$ as a function of T/v_s in Fig. 3, and the results of the AFHM[10] and the free electron system ($U=0$) are also shown for comparison. The value $\pi\xi_s T/v_s$ should converge to 1 in the limit of $T\to 0$ according to the CFT, while in the case of $U=0$, the limit is not 1 but 1/2 as seen from the analytical result $\xi_s^{\rm free}$. Figure 3 indicates that the temperature dependence shows a crossover from the weak interaction limit (the $U=0$ case) to the strong interaction limit (the AFHM) with decreasing temperature, and that the crossover temperature normalized by v_s is higher for large U. The crossover temperature is in the same order of Δ, and this behavior is consistent with the fact that the half-filled HM reduces to the universality class of the AFHM when $T<\Delta$. When U is so large that $v_s\ll\Delta$ (*e.g.*, $U=8$), the temperature region where the spin wave excitation is dominant is well separated from that for the charge excitation, and, therefore, the crossover to the $U\to\infty$ limit can be clearly found at temperatures $T<v_s$. On the other hand, when U is so

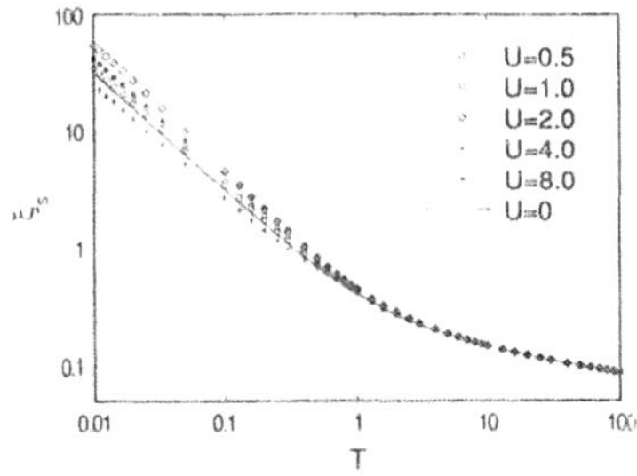

Fig. 2

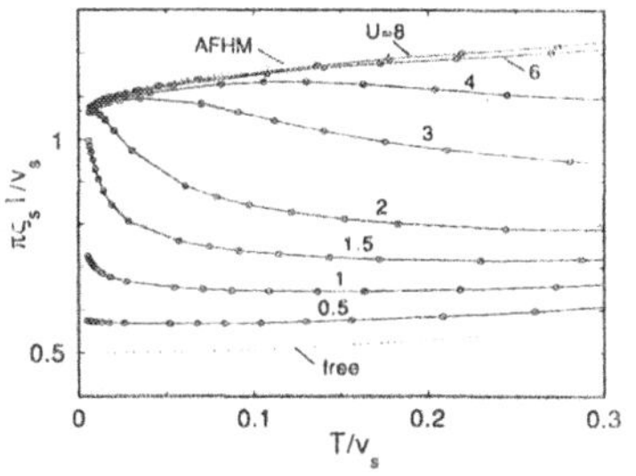

Fig. 3

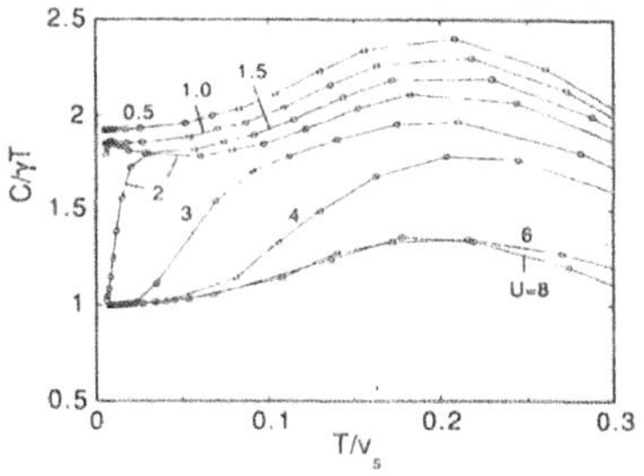

Fig. 4

Fig. 2. Temperature dependence of spin correlation length. Fig. 3. Effect of Coulomb repulsion on the crossover of spin correlation length. Fig. 4. Effect of Coulomb repulsion on the crossover of specific heat.

small that $\Delta < T$, the charge fluctuation has a considerable amplitude at this temperature and suppresses the growth of the spin correlation. The temperature dependence of ξ_s is consequently rather close to that of the free electron system down to the lowest temperature used here. Even in this case, when the temperature becomes $T \ll \Delta$, the curve should show the crossover to the AFHM.

The singular behavior near $T=0$ in Fig. 3 for large U is due to the log corrections, which were also observed in the AFHM.[10] Applying the CFT, we can estimate this from the log terms in the finite-size corrections of the excited state energies of the HM Hamiltonian given by Woynarovich and Eckle,[20] and the result is

$$\frac{1}{\xi_s}=\frac{\pi T}{v_s}\left[1-\frac{1}{2}\frac{1}{\log(T_0/T)}+(\text{higher terms})\right], \tag{5}$$

where T_0 is a characteristics energy scale dependent on U. However, because the log correction given by Woynarovich and Eckle was derived for the large systems where the charge fluctuation does not contribute due to the Hubbard gap, the temperature region where this correction (5) is dominant is again $T \ll \mathrm{Min}\{\Delta, v_s\}$. We have found that in the large-U case (*e.g.*, $U \geq 6$), the obtained curve is well fitted by the expression (5), but in the small-U case (*e.g.*, $U \leq 2$), as expected from the above argument, the curve has not yet reached the asymptotic behavior (5) down to the temperatures used here.

The amplitude of the charge fluctuation can be measured by the specific heat C. In Fig. 4, we plot C/T normalized by its limit value $\gamma \equiv \lim_{T\to 0} C/T = \pi/(3v_s)$. The free electron case is again an exception; $\lim_{T\to 0} C/T = \pi/3$, being twice $\pi/(3v_s)$.[3] The extra contribution is due to the charge fluctuation because this is also a gapless excitation when $U=0$. As well as ξ_s, $C/\gamma T$ also shows the crossover from the free electron limit to the AFHM limit with decreasing temperature, the decrease of which directly indicates the suppression of the charge fluctuation.

The author thanks M. Takahashi and J. Suzuki for valuable discussions and thanks K. Nomura and M. Yamada for providing the data on the AFHM.

References

1) E. Lieb and F. Y. Wu: Phys. Rev. Lett. **20** (1968) 1445.
2) M. Takahashi: Prog. Theor. Phys. **47** (1972) 69.
3) M. Takahashi: Prog. Theor. Phys. **52** (1974) 103.
4) N. Kawakami, T. Usuki and A. Okiji: Phys. Lett. A**137** (1989) 287.
5) H. Frahm and V. E. Korepin: Phys. Rev. B**42** (1990) 10553.
6) N. Kawakami and S.-K. Yang: Phys. Lett. A**148** (1990) 359.
7) H. J. Shulz: Phys Rev. Lett. **64** (1990) 2831.
8) For a review, see J. L. Cardy: *Phase Transitions and Critical Phenomena*, ed. C. Domb and J. L. Lebowitz (Academic, London, 1987) Vol. 11, p. 55.
9) M. Yamada: J. Phys. Soc. Jpn. **59** (1990) 848.
10) K. Nomura and M. Yamada: Phys. Rev. B**43** (1991) 8217.
11) M. Takahashi: Phys. Rev. B**43** (1991) 5788.
12) T. Koma: Prog. Theor. Phys. **81** (1989) 783.
13) M. Barma and S. Shastry: Phys. Lett. **61**A (1977) 15.
14) R. Z. Bariev: Teor. Mat. Fiz. **49** (1981) 261 [Theor. Math. Phys. **49** (1982) 1021].
15) T. Koma: Prog. Theor. Phys. **83** (1990) 655.
16) For example, see V. J. Emery: *Highly Conducting One-Dimensional Solids*, ed. J. T. Devreese *et al.* (Plenum, New York, 1979) p. 247.
17) M. Suzuki and M. Inoue: Prog. Theor. Phys. **78** (1987) 787.
18) A. A. Ovchinnikov: Zh. Eksp. Teor. Fiz. **57** (1969) 2137 [Sov. Phys. JETP **30** (1970) 1160].
19) M. Takahashi: Prog. Theor. Phys. **43** (1970) 1619.
20) F. Woynarovich and H-P. Eckle: J. Phys. A **20** (1987) L443.

Correlation functions in the one-dimensional Hubbard model

A. M. Finkel'shteǐn

L. D. Landau Institute of Theoretical Physics, USSR Academy of Sciences

(Submitted November 23, 1976)
Pis'ma Zh. Eksp. Teor. Fiz. **25**, No. 2, 83–86 (20 January 1977)

The asymptotic forms of the correlation functions in the one-dimensional Hubbard model, when umklapp processes are significant, are obtained. It is shown that two types of pairing in the ground state are simultaneously realized. The corresponding correlation functions fall off in power-law fashion at large distances.

PACS numbers: 71.10.+x

The asymptotic forms of the correlation functions were recently calculated[1] for a one-dimensional electron gas in the absence of umklapp processes (see also[2,3]). The kinetic energy of the free electrons was linearized near $\pm k_F$, and this made it possible to use the so-called "boson representation" for the field operators[4,5]

$$\psi_{is} = L^{-1/2}\Sigma\, e^{ikx} a_{is}(k) \to (2\pi a)^{-1/2}\exp\{(-1)^i[-ik_F x + \Sigma A(x,k)\rho_{is}(k)]\}. \tag{1}$$

The exponent $i=(1,2)$ corresponds here to electrons near $\pm k_F$, s is the spin index, $\rho_{is}(k)=\Sigma_p a^+_{is}(p+k)a_{is}(p)$, $A(x,k)=2\pi L^{-1}k^{-1}\exp(-\alpha\,|\,k\,|/2-ikx)$, and $v_F\alpha^{-1}$ is interpreted as the width of the conduction band.

It will be shown below that the boson representation technique makes it possible to find the asymptotic forms of the correlation functions also in those cases when the umklapp processes are significant. For simplicity we consider the one-dimensional Hubbard model with a half-filled band. The interaction constant of the electrons will be assumed to be small: $|g|\ll 2\pi v_F$. Such a system corresponds in the representation (1) to a Hamiltonian[6]

$$H = h\{\rho_i;-g\} + h\{\sigma_i;g\}, \tag{2}$$

where

$$h\{\rho_i;-g\} = 2\pi(v_F + g/2\pi)L^{-1}\sum_{k>0}[\rho_1(k)\rho_1(-k) + \rho_2(-k)\rho_2(k)]$$

$$+ gL^{-1}\Sigma\rho_1(k)\rho_2(-k) - g(2\pi a)^{-2}\int dx\{\exp[2^{1/2}\Sigma A(x,k)(\rho_1(k)+\rho_2(k))] + \text{c.c.}\}, \tag{3}$$

$h\{\sigma_i;g\}$ is determined by replacing ρ_i in (3) by σ_i and g by $-g$; $\rho_i=2^{-1/2}(\rho_{i\uparrow}+\rho_{i\downarrow})$, $\sigma_i=2^{-1/2}(\rho_{i\uparrow}-\rho_{i\downarrow})$, where

$$[\rho_i,\sigma_j]=0, \tag{4}$$

$$[\rho_i(k),\rho_j(-k')] = [\sigma_i(k),\sigma_j(-k')] = (-1)^i\delta_{ij}\frac{kL}{2\pi}\delta_{kk'}.$$

We note that in the Hubbard Hamiltonian (2) the umklapp processes correspond to the last term in $h\{\rho_i;-g\}$.

We are interested in the correlation functions that describe the fluctuations of singlet and triplet Cooper pairs (SCP and TCP), as well as fluctuations of the dielectric (CDW) and antiferromagnetic (SDW) type:

$$K_{CDW} = \langle \psi_{1\uparrow}(x,t)\,\psi^+_{2\downarrow}(x,t)\,\psi_{2\downarrow}(0.0)\,\psi^+_{1\uparrow}(0.0)\rangle = e^{2ik_Fx}K^+_\rho(-g)\,K^+_\sigma(g),$$

$$K_{SDW} = \langle \psi_{1\uparrow}(x,t)\,\psi^+_{2\downarrow}(x,t)\psi_{2\downarrow}(0.0)\psi^+_{1\uparrow}(0.0)\rangle = e^{2ik_Fx}K^+_\rho(-g)K^-_\sigma(g),$$

$$K_{SCP} = \langle \psi_{1\uparrow}(x,t)\,\psi_{2\downarrow}(x,t)\,\psi^+_{2\downarrow}(0.0)\,\psi^+_{1\uparrow}(0.0)\rangle = K^-_\rho(-g)K^+_\sigma(g), \qquad (5)$$

$$K_{TCP} = \langle\psi_{1\uparrow}(x,t)\psi_{2\uparrow}(x,t)\,\psi^+_{2\uparrow}(0.0)\psi^+_{1\uparrow}(0.0)\rangle = K^-_\rho(-g)K^-_\sigma(g),$$

where

$$K^\pm_\rho(-g) = (2\pi a)^{-1}\langle e^{ith}\exp[-2^{-1/2}\Sigma A(x,k)(\rho_1 \pm \rho_2)]\,e^{-ith}$$

$$\times \exp[2^{-1/2}\Sigma A(0,k)(\rho_1 \pm \rho_2)]\rangle_n, \qquad (6)$$

here $h = h\{\rho_i; -g\}$. $K^\pm_\sigma(g)$ is obtained from (6) by replacing ρ_i with σ_i and $h = h\{\sigma_i; g\}$.

The factorization in (5) was made possible by the fact that the field operators ψ_{is} are expressed in accordance with (1) in terms of an exponential of a function linear in ρ_{is}. Since the commutation relations of ρ_i and σ_i are the same, $K^\pm_\rho(-g) = K^\pm_\sigma(-g)$ and consequently the problem reduces to finding the four functions $K^\pm_\sigma(g \gtrless 0)$.

The $K^\pm_\sigma(g>0)$ can be calculated, inasmuch as a situation of the zero-charge type arises at $g>0$ following normalization in $h\{\sigma_i; g\}$. For the calculations we note that the parquet approximation was sufficient to find the correlation function in a gas with δ-function repulsion[7-9]

$$\widetilde{K}_{CDW}(g>0) \sim (\ln\omega)^{-3/2}\,\omega^{-g/2\pi v_F}, \qquad \widetilde{K}_{SDW}(g>0) \sim (\ln\omega)^{1/2}\,\omega^{-g/2\pi v_F} \qquad (7)$$

On the other hand, the factorization relations of the type (5) are valid also for these functions, the factors $K^\pm_\sigma(g)$ remaining unchanged, while $K^\pm_\rho(-g)$ are replaced by $\widetilde{K}^\pm_\rho(-g)$. The $\widetilde{K}^\pm_\rho(-g)$ are calculated from formula (6) with $h = \widetilde{h}\{\rho_i - g\}$; while $\widetilde{h}\{\rho_i; -g\}$ are determined by formula (3) but without the last term (this term describes umklapp processes, which do not exist in a gas with δ-function interaction). $\widetilde{h}\{\rho_i; -g\}$ is quadratic in ρ_i, and therefore the function $\widetilde{K}^+_\rho(-g)$ can be easily obtained[4]

$$\widetilde{K}^+_\rho(-g) \sim \omega^{-1,\,-g/2\pi v_F}. \qquad (8)$$

As a result we obtain from (7) and (8) the asymptotic forms of $K^\pm_\sigma(g>0)$

$$K^+_\sigma(g>0) = \widetilde{K}_{CDW}(g>0)/\widetilde{K}^+_\rho(-g) \sim \ln^{-3/2}[x^2-(v't)^2]/[x^2-(v't)^2]^{1/2}, \qquad (9)$$

$$K^-_\sigma(g>0) = \widetilde{K}_{CDW}(g>0)/\widetilde{K}^+_\rho(-g) \sim \ln^{1/2}[x^2-(v't)^2]/[x^2-(v't)^2]^{1/2}, \qquad (10)$$

where $v' = v_F - g/2\pi$.

Reducing the problem to a two-dimensional Coulomb gas, it was shown in[3] that at large distances we have the asymptotic behavior

$$K_\sigma^+(g < 0) \sim \text{const} . \tag{11}$$

It remains to explain the behavior of the function $K_\sigma^-(g<0)$. It was noted in[2] that at $g=-(6/5)\pi v_F$ it is possible to calculate $K_\sigma^-(x,t)$. This circumstance makes it possible to find the asymptotic form of $K_\sigma^-(g<0)$, inasmuch as at $g<0$, as the result of renormalization, the charge arrived at the point $\tilde{g}=-\frac{6}{5}\pi\tilde{v}_F$, ($\tilde{v}_F$ is the coefficient at

$$2\pi L^{-1} \sum_{k>0} [\sigma_1(k)\sigma_1(-k)+\sigma_2(-k)\sigma_2(k)]$$

in the renormalized $h\{\sigma_i;\tilde{g}\}$). Omitting certain details (we note only that $\tilde{v}_F = 1.25(v_F - g/2\pi) + O(g^2)$), we present the final result

$$K_\sigma^-(g < 0) \sim \frac{1}{(x/v'')^2 - t^2} \exp\{-2\Delta[(x/v'')^2 - t^2]^{1/2}\}, \tag{12}$$

where

$$v'' = v_F + |g|/2\pi \text{ and } \Delta \sim a^{-1}|g\,v_F|^{1/2}\exp(-\pi v_F/|g|)$$

is the gap in the spectrum of the fermion excitations $\epsilon(p) = [\Delta^2 + (v''p)^2]^{1/2}$.

It follows from (5) that $K_\sigma^-(g<0)$ enters in $K_{SCP}(g>0)$, $K_{SDW}(g<0)$, and $K_{TCP}(g \lessgtr 0)$. These functions fall off exponentially at large distances; in addition

$$\mathrm{Im}K^r_{SCP}(g > 0),\quad \mathrm{Im}K^r_{SDW}(g < 0),\quad \mathrm{Im}K^r_{TCP}(g \gtrless 0) \sim \theta(|\omega| - 2\Delta), \tag{13}$$

K^r is the susceptibility and describes the response of the system to the action of the corresponding external field.

In the remaining cases we obtain a power-law decrease of the correlation functions

$$K_{CDW}(g \gtrless 0) \sim e^{2ik_Fx}\ln^{-3/2}[x^2 - (v't)^2]/[x^2 - (v't)^2]^{1/2}, \tag{14}$$

$$K_{SDW}(g > 0) \sim e^{2ik_Fx}\ln^{1/2}[x^2 - (v't)^2]/[x^2 - (v't)^2]^{1/2}, \tag{15}$$

$$K_{SCP}(g < 0) \sim \ln^{1/2}[x^2 - (v't)^2]/[x^2 - (v't)^2]^{1/2}, \tag{16}$$

where $v_F' = v_F - |g|/2\pi$ is the velocity of the gapless excitations.

The slow (power-law) decrease of the correlation function at large distances means that a corresponding type of pairing is realized in the system, although no long range order is produced, owing to the strong quantum fluctuations inherent in a one-dimensional system. It follows from (14)—(16) that two types of pairing in the Hubbard model are realized simultaneously: SCP and CDW in the case of attraction, and SDW and CDW in the case of repulsion. To excite waves of another type it is necessary, according to (13), to expend an energy 2Δ to break the pair.

In conclusion, the author thanks A.I. Larkin, P.B. Vigman, and D.E. Khmel'nitskiĭ for useful discussions.

[1]A. Luther and V.J. Emery, Phys. Rev. Lett. 33, 589 (1974).
[2]P.A. Lee, Phys. Rev. Lett. 34, 1247 (1975).

[3]S. T. Chui and P. A. Lee, Phys. Rev. Lett. **35**, 315 (1975).
[4]A. Luther and I. Peshel, Phys. Rev. **B9**, 2911 (1974).
[5]D. Mattis, J. Math. Phys. **15**, 609 (1974).
[6]V. J. Emery, A. Luther, and I. Peshel, Phys. Rev. **B13**, 1272 (1976).
[7]Yu. A. Bychkov, L. P. Gor'kov, and I. E. Dzyaloshinskiĭ, Zh. Eksp. Teor. Fiz. **50**, 738 (1966) [Sov. Phys. JETP **23**, 489 (1966)].
[8]I. E. Dzyaloshinskiĭ and A. I. Larkin, Zh. Eksp. Teor. Fiz. **61**, 791 (1971) [Sov. Phys. JETP **34**, 422 (1972)].
[9]J. Solyom, J. Low Temp. Phys. **12**, 547 (1973).

International Journal of Modern Physics B Vol. 5, Nos. 1&2 (1991) 57–74

CORRELATED FERMIONS IN ONE DIMENSION

H.J. SCHULZ
Laboratoire de Physique des Solides
Université Paris-Sud
91405 Orsay, France

Abstract: A brief introduction to the bosonization method for interacting one-dimensional fermion systems is given. Using these results, the long-distance decay of correlation functions in the one-dimensional Hubbard model is determined exactly for arbitrary bandfilling and correlation strength, using the exact solution of Lieb and Wu. For infinite U the results are generalized to the case of nonzero nearest-neighbour interaction. The behaviour of thermodynamic quantities, of the frequency-dependent conductivity, and of the thermopower is also discussed, in particular in the proximity of the metal–insulator transitions occurring for half- and quarter-filling. The one-dimensional Luttinger liquid is shown to be unstable in the presence of interchain hopping. The results for the metal–insulator transition are compared with other scenarios developed in higher dimensions.

1 INTRODUCTION

A theoretical understanding of interacting fermion systems in one dimension is important for a number of reasons. On the one hand, in the physics of quasi-one-dimensional organic conductors [1] or of conducting polymers [2] interaction effects play a major role. On the other hand, one-dimensional models can be easier to understand than their higher-dimensional versions, or even exactly solvable, as is the case with the prototypical model of correlated fermions, the Hubbard model [3]. They therefore can provide valuable information on the role of correlation effects in higher dimension, e.g. on the physics of correlated fermions in two dimensions which is thought to be at the origin of the many interesting properties of high-temperature superconductors [4,5].

The theory of one-dimensional interacting fermions has made progress along two somewhat separate lines: (i) Perturbative renormalization group calculations have shown that different correlation functions (see e.g. eqs (2.13), (2.14) below) have a long-range power law behaviour, with interaction-dependent exponents [6,7]. These exponents in turn determine a number of physical properties: temperature dependence of the NMR relaxation rate [8] or X-ray scattering intensities [9], effect of impurities [10], or possible low-temperature ordered states in systems of coupled chains. (ii) Specific lattice models like the Hubbard model and its generalizations have been studied either by exact solutions [3] or numerically to obtain correlation functions [11] and the energetics of ground and excited states [12,13].

In the present paper, I shall show how results originally obtained in the weak-coupling limit (in particular using the "bosonization" method) can be used, for arbitrary interaction strength, to obtain precise information about low-temperature thermodynamics, transport properties, and correlation exponents from energies alone ***without the explicit calculation of correlation functions***. In the following section, the bosonization method is explained, the peculiar properties of a one-dimensional interacting fermion system are discussed ("Luttinger liquid"), and it is shown that the low-energy physical properties are determined by only three parameters: the velocities of collective charge- and spin-density oscillations ($u_{\rho,\sigma}$), and a coefficient K_ρ that determines the long-distance decay of correlation functions. In sec.3 the calculation of these parameters for specific lattice models is discussed [14]. The reasoning used is a generalization of arguments due to Haldane [15] to the case of spin-1/2 fermions. I will illustrate the method using the Hubbard model, where exact energies can be obtained even in the thermodynamic limit. Even then the eigenfunctions are so complicated that the direct calculation of correlation functions like (2.13), (2.14) is hard even for very small systems [16]. The present calculation then provides a rather detailed ***and exact*** description of the crossover between weak and strong correlation and of the metal-insulator transition occurring when the average particle number per site, n, approaches unity. Similar results have been reported recently by a number of authors for the one-dimensional Hubbard [17,18] or $t-J$ model [19]. Also, the thermopower in the vicinity of the metal-insulator transition is calculated to obtain the sign of the charge carriers. Sec. 4 is devoted to a discussion of the stability of the Luttinger liquid in a quasi-one-dimensional system, and in the concluding section the results are discussed and compared with scenarios for strongly correlated systems in higher dimensions.

2 BOSONIZATION, SPIN-CHARGE SEPARATION, LUTTINGER LIQUID

2.1 Bosonization Formalism

One of the important findings in the theory of one-dimensional interacting fermions is that fermion field operators can be expressed in terms of boson operators [20,21]. This equivalence then can be used to express the fermion Hamiltonian in a particularly simple form, in terms of boson fields only (see eq.(2.2) below). Consider for the moment spinless electrons, and define a boson field $\phi(x)$ by $\partial\phi/\partial x = \pi\rho(x)$, where ρ is the deviation from the average density. Then; creating a particle at site x means introducing a kink of height π in ϕ, i.e. at points on the left of x ϕ has to be shifted by π. Displacement operators are exponentials of momentum operators, and therefore a first guess would be $\psi^\dagger(x) \approx \exp(i\pi \int_{-\infty}^x \Pi(x')dx')$, where Π is the momentum density conjugate conjugate to ϕ: $[\phi(x), \Pi(y)] = i\delta(x-y)$. However, this operator commutes with itself, instead of satisfying canonical anticommutation relations. Anticommutation is achieved by multiplying with an operator, acting at site x, that changes sign each time a particle passes through x. Such an operator is $\exp(\pm i\phi(x))$. The final result then is

$$\psi_\pm^\dagger(x) = \lim_{\alpha\to 0} \frac{1}{\sqrt{2\pi\alpha}} \exp[\pm ik_F x \mp i\phi(x) + i\pi \int_{-\infty}^x \Pi(x')dx'] \ , \tag{2.1}$$

where the upper and lower sign refer to electrons near k_F and $-k_F$ respectively. A detailed derivation of this important relation as an operator identity is given in the literature [15,22]. One should also notice that (2.1) is strictly speaking valid for models with linear energy–momentum relation. If there is some curvature in the dispersion relation (as is necessarily the case in lattice models), all odd powers of $\exp(i\phi)$ appear [23].

For electrons with spin, one simply introduces one boson field for each spin orientation. Introducing charge and spin bosons via $\phi_{\rho,\sigma} = (\phi_\uparrow \pm \phi_\downarrow)/\sqrt{2}$, one then finds that the low-energy, large-distance behaviour of a one-dimensional fermion system with spin-independent interactions is described by the Hamiltonian [6,7]

$$H = H_\rho + H_\sigma + \frac{2g_1}{(2\pi\alpha)^2} \int dx \cos(\sqrt{8}\phi_\sigma) \ . \tag{2.2}$$

Here α is a short-distance cutoff, g_1 is the backward scattering amplitude, and for $\nu = \rho, \sigma$

$$H_\nu = \int dx [\frac{\pi u_\nu K_\nu}{2} \Pi_\nu^2 + \frac{u_\nu}{2\pi K_\nu} (\partial_x \phi_\nu)^2] \ . \tag{2.3}$$

The phase fields are [24]

$$\phi_\nu(x) = -\frac{i\pi}{L} \sum_{p\neq 0} \frac{1}{p} e^{-\alpha|p|x/2 - ipx} [\nu_+(p) + \nu_-(p)] - N_\nu \frac{\pi x}{L} \ , \tag{2.4}$$

$$\Pi_\nu(x) = \frac{1}{L} \sum_{p\neq 0} e^{-\alpha|p|x/2 - ipx} [\nu_+(p) - \nu_-(p)] + J_\nu/L \ . \tag{2.5}$$

Here $\rho_r(p)$ ($\sigma_r(p)$) are the Fourier components of the charge (spin) density operator for right–($r = +$) and left–($r = -$) going fermions. Introducing the *total* number operators (measured with respect to the ground state) $N_{r,s}$ for right– and left–going particles ($r = \pm$) of spin s, the (charge and spin) number and current operators are $N_\nu = [(N_{+\uparrow} + N_{-\uparrow}) \pm (N_{+\downarrow} + N_{-\downarrow})]/\sqrt{2}$, $J_\nu = [(N_{+\uparrow} - N_{-\uparrow}) \pm (N_{+\downarrow} - N_{-\downarrow})]/\sqrt{2}$, where the upper and lower sign refer to charge and spin, respectively.

The operators ϕ_ν and Π_ν in (2.2) obey Bose-like commutation relations: $[\phi_\nu(x), \Pi_\mu(y)] = i\delta_{\nu\mu}\delta(x-y)$, and consequently, at least for $g_1 = 0$, (2.2) describes independent long-wavelength oscillations of the charge and spin density, with linear dispersion relation

$\omega_\nu(k) = u_\nu|k|$, and the system is conducting [25]. The only nontrivial interaction effects in (2.2) come from the cosine term. However, for repulsive interactions ($g_1 > 0$), this term is renormalized to zero in the long-wavelength limit, and at the fixed point one has $K_\sigma^* = 1$. The three remaining parameters in (2.2) then completely determine the long-distance and low-energy properties of the system. It should be emphasized that (2.2) can be derived exactly for fermions with linear energy-momentum relation. For more general (e.g. lattice) models, (2.2) is still the correct effective Hamiltonian for low-energy excitations.

2.2 Spin–Charge Separation

One of the more spectacular consequences of the Hamiltonian (2.2) is the complete separation of the dynamics of the spin and charge degrees of freedom. For example, in general one has $u_\sigma \neq u_\rho$, i.e. the charge and spin oscillations propagate with different velocities. Only in a noninteracting system one has $u_\sigma = u_\rho = v_F$. To make the meaning of this fact more transparent, let us create an extra particle in the ground state, at $t = 0$ and spatial coordinate x_0. The charge and spin densities then are easily found, using $\rho(x) = -(\sqrt{2}/\pi)\partial\phi_\rho/\partial x$ (note that $\rho(x)$ is the deviation of the density from its average value) and $\sigma_z(x) = -(\sqrt{2}/\pi)\partial\phi_\sigma/\partial x$:

$$\begin{aligned} \langle 0|\psi_+(x_0)\rho(x)\psi_+^\dagger(x_0)|0\rangle &= \delta(x - x_0) \\ \langle 0|\psi_+(x_0)\sigma_z(x)\psi_+^\dagger(x_0)|0\rangle &= \delta(x - x_0) \ . \end{aligned} \tag{2.6}$$

Now, consider the time developement of the charge and spin distributions. The time-dependence of the charge and spin density operators is easily obtained from (2.2) (using the fixed point value $g_1 = 0$), and one obtains

$$\begin{aligned} \langle 0|\psi_+(x_0)\rho(x,t)\psi_+^\dagger(x_0)|0\rangle &= \delta(x - x_0 - u_\rho t) \\ \langle 0|\psi_+(x_0)\sigma_z(x,t)\psi_+^\dagger(x_0)|0\rangle &= \delta(x - x_0 - u_\sigma t) \ . \end{aligned} \tag{2.7}$$

Because in general $u_\sigma \neq u_\rho$, after some time charge and spin will be localized at completely different points in space, i.e. *charge and spin have separated completely.* A interpretation of this surprising phenomenon in terms of the Hubbard model will be given in sec.(3). Here a linear energy-momentum relation has been assumed for the electrons, and consequently the shape of the charge and spin distributions is time-independent. If the energy–momentum relation has some curvature (as is necessarily the case in lattice systems) the distributions will widen with time. However this widening is proportional to $\sqrt{t}$, and therefore much smaller than the distance between charge and spin. Thus, the qualitative picture of spin-charge separation is unchanged.

2.3 Luttinger Liquid

The simple form of the Hamiltonian (2.2) at the fixed point $g_1^* = 0$ makes the calculation of physical properties rather straightforward. First note that acoustic phonons in one dimension have a linear specific heat. Consequently, the low-temperature specific heat of interacting fermions is $C(T) = \gamma T$, with

$$\gamma/\gamma_0 = \frac{1}{2}(v_F/u_\rho + v_F/u_\sigma) \ . \tag{2.8}$$

Here γ_0 is the specific heat coefficient of noninteracting electrons of Fermi velocity v_F.

The spin susceptibility and the compressibility are equally easy to obtain. Note that in (2.2) the coefficient u_σ/K_σ determines the energy necessary to create a nonzero spin polarization, and u_ρ/K_ρ fixes the energy needed to change the particle density. Given that at the fixed point $K_\sigma^* = 1$, one finds

$$\chi/\chi_0 = v_F/u_\sigma \ , \kappa/\kappa_0 = v_F K_\rho/u_\rho \ , \tag{2.9}$$

where χ_0 and κ_0 are the susceptibility and compressibility of the noninteracting case. From eqs.(2.8) and (2.9) the Wilson ratio is

$$R_W = \frac{\chi}{\gamma}\frac{\gamma_0}{\chi_0} = \frac{2u_\rho}{u_\rho + u_\sigma} \ . \tag{2.10}$$

The quantity $\Pi_\rho(x)$ is proportional to the current density. Obviously, the Hamiltonian commutes with the total current, and therefore the frequency dependent conductivity is a delta function at $\omega = 0$. Using the Kubo formula, one straightforwardly finds

$$\sigma(\omega) = 2K_\rho u_\rho \delta(\omega) \ , \tag{2.11}$$

i.e. the product $K_\rho u_\rho$ determines the weight of the dc peak in the conductivity.

The above properties, linear specific heat, finite spin susceptibility, and dc conductivity are those of an ordinary Fermi liquid, the coefficients u_ρ, u_σ, and K_ρ determining renormalizations with respect to noninteracting quantities. We will now consider quantities which show that a one-dimensional interacting fermion system is *not* a Fermi liquid. Consider the single-particle Green's function which can be calculated using the representation (2.1) of fermion operators. One then finds for the momentum distribution function in the vicinity of k_F:

$$n_k \approx n_{k_F} - \beta \text{sign}(k - k_F)|k - k_F|^\alpha \ , \tag{2.12}$$

and for the single-particle density of states: $N(\omega) \approx |\omega|^\alpha$, with $\alpha = (K_\rho + 1/K_\rho - 2)/4$, and β is a model-dependent constant . Note that for any $K_\rho \neq 1$, i.e. ***for any nonvanishing interaction***, the momentum distribution function and the density of states have power-law singularities at the Fermi level, with a vanishing single particle density of states at E_F. This behaviour is obviously quite different from a standard Fermi liquid which would have a finite density of states and a step–like singularity in n_k. The unusual type of behaviour found here has been called **Luttinger liquid** by Haldane [15].

The coefficient K_ρ also determines the long-distance decay of all other correlation functions of the system: Using the representation (2.1) and the Hamiltonian (2.2) at the fixed point $g_1^* = 0$ one finds for example for the charge and spin correlation functions [26,27]

$$\begin{aligned} \langle n(x)n(0)\rangle &= K_\rho/(\pi x)^2 + A_1 \cos(2k_F x) x^{-1-K_\rho} \ln^{-3/2}(x) \\ &\quad + A_2 \cos(4k_F x) x^{-4K_\rho} + \ldots \qquad (2.13) \\ \langle \mathbf{S}(x)\cdot\mathbf{S}(0)\rangle &= 1/(\pi x)^2 + B_1 \cos(2k_F x) x^{-1-K_\rho} \ln^{1/2}(x) + \ldots \ , \qquad (2.14) \end{aligned}$$

with model dependent constants A_i, B_i. The ellipses in (2.13) and (2.14) indicate higher harmonics of $\cos(2k_F x)$ which are present but decay faster than the terms shown here. Similarly, correlation functions for singlet (SS) and triplet (TS) superconducting pairing decay like x^{-1-1/K_ρ}. The corresponding susceptibilities (i.e. the Fourier transforms of the above correlation functions) behave at low temperatures as

$$\chi_{CDW}(T) \approx T^{K_\rho - 1}|\ln(T)|^{-3/2} \ , \chi_{SDW}(T) \approx T^{K_\rho - 1}|\ln(T)|^{1/2} \ , \qquad (2.15)$$

$$\chi_{SS}(T) \approx T^{1/K_\rho - 1}|\ln(T)|^{-3/2} \ , \chi_{TS}(T) \approx T^{1/K_\rho - 1}|\ln(T)|^{1/2} \ , \qquad (2.16)$$

i.e. for $K_\rho < 1$ (spin or charge) density fluctuations at $2k_F$ are enhanced and diverge at low temperatures, whereas for $K_\rho > 1$ pairing fluctuations dominate.

3 HUBBARD MODEL

In a weakly interacting system the coefficients K_ρ and u_ν can be determined perturbatively. For example, for the Hubbard model, with Hamiltonian

$$H = -t\sum_{i,s}(a_{i,s}^\dagger a_{i+1,s} + a_{i+1,s}^\dagger a_{i,s}) + U\sum_i n_{i\uparrow}n_{i\downarrow} \ , \qquad (3.1)$$

one finds

$$K_\rho = 1 - U/(\pi v_F) + \ldots \ , \qquad (3.2)$$

where $v_F = 2t\sin(\pi n/2)$ is the Fermi velocity for n particles per site. For larger U higher operators appear in the continuum hamiltonian (2.2), e.g. higher derivatives of the fields or cosines of multiples of $\sqrt{8}\phi_\sigma$. These operators are irrelevant, i.e. they renormalize to zero and do not qualitatively change the long-distance properties, but they do lead to nontrivial corrections to the coefficients u_ν, K_ρ. In principle these corrections can be treated order-by order in perturbation theory. However, this approach is obviously unpractical for large U, and moreover it is likely that perturbation theory is not convergent. To obtain the physical properties for arbitrary U a different approach is therefore necessary.

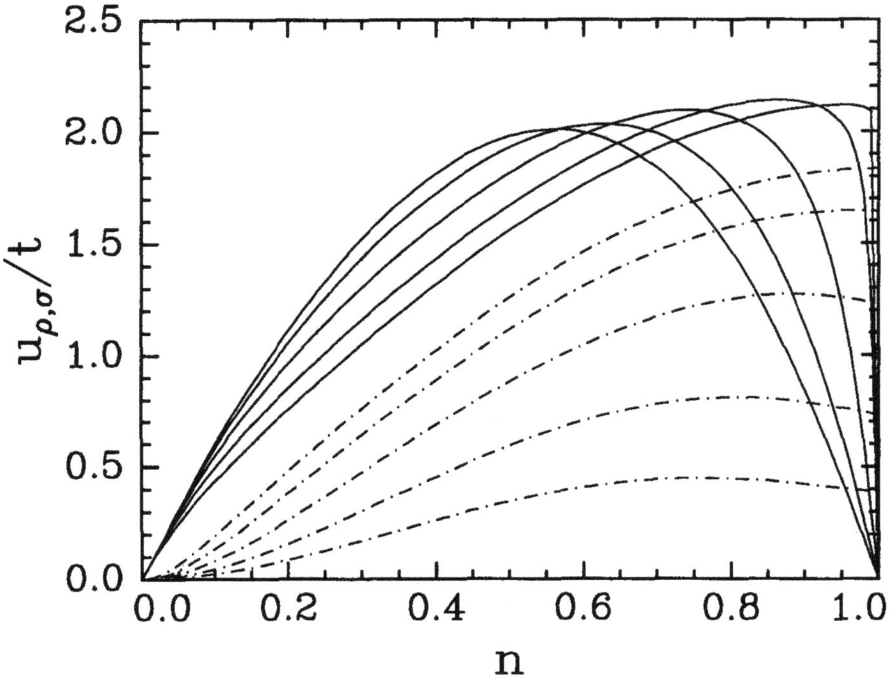

Figure 1: The charge and spin velocities u_ρ (full line) and u_σ (dash-dotted line) for the Hubbard model, as a function of the band filling for different values of U/t: for u_σ $U/t = 1, 2, 4, 8, 16$ from top to bottom, for u_ρ $U/t = 16, 8, 4, 2, 1$ from top to bottom in the left part of the figure.

I note two points: (i) in the small-U perturbative regime, interactions renormalize to the weak-coupling fixed point $g_1^* = 0, K_\sigma^* = 1$; (ii) the exact solution [3] does not show any singular behaviour at nonzero U, i.e. large U and small U are the same phase of the model, so that the long-range behaviour even of the large U case is determined by the fixed point $g_1^* = 0$. Thus, the low energy properties of the model are still determined by the three parameters $u_{\rho,\sigma}$ and K_ρ.

The velocity parameters are easily obtained from the exact solution. In fact, u_σ is the velocity of long-wavelength spinwaves, which has been calculated by Coll [28]. The identification of u_ρ is slightly more delicate: note that the $4k_F$-part of the density correlation function (2.13) is entirely determined by ϕ_ρ excitations. On the other hand, Coll's "particle-hole excitations" are gapless at $4k_F$, and consequently are expected to at the origin of the power-law in the $4k_F$ part of (2.13). Consequently, u_ρ is determined by the velocity of Coll's "electron-hole excitations" for $q \to 0$. Thus, u_ρ and u_σ can be found numerically from a straightforward solution of an integral equation. Results are shown in fig.1 for various values of U/t. Note that for $U = 0$ one has $u_\rho = u_\sigma =$

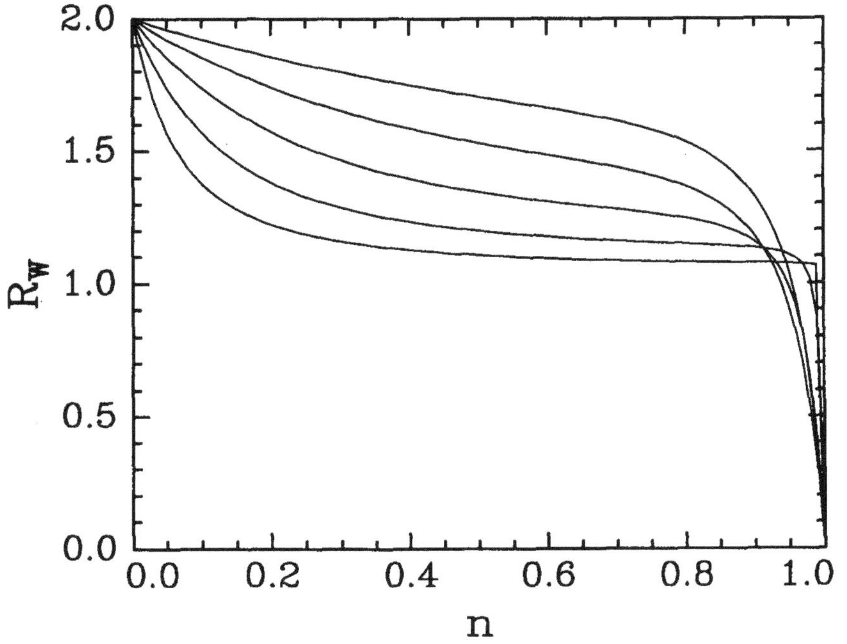

Figure 2: The Wilson ration R_W for the one-dimensional Hubbard model, as a function of the band filling for different values of U/t (U/t = 16, 8, 4, 2, 1 for the top to bottom curves).

$2t\sin(\pi n/2)$, whereas for $U \to \infty$ $u_\rho = 2t\sin(\pi n)$, $u_\sigma = (2\pi t^2/U)(1-\sin(2\pi n)/(2\pi n))$. In the noninteracting case $u_\sigma \propto n$ for small n, but for *any* positive U $u_\sigma \propto n^2$. The Wilson ratio, eq.(2.10) obtained from the velocities is shown in fig.2. For $U = 0$ one has $R_W = 1$, whereas for $U \to \infty$ $R_W = 2$ for $n \neq 1$.

To obtain the parameter K_ρ from the exact solution note that the gradient of the phase field ϕ_ρ is proportional to the particle density, and in particular a constant slope of ϕ_ρ represents a change of total particle number. Consequently, the coefficient u_ρ/K_ρ in eq. (2.3) is proportional to the variation of the ground state energy E_0 with particle number [29]:

$$\frac{1}{L}\frac{\partial^2 E_0(n)}{\partial n^2} = \frac{\pi}{2}\frac{u_\rho}{K_\rho} \quad . \tag{3.3}$$

Note that this quantity is the inverse of the compressibility. Equation (3.3) now allows the direct determination of K_ρ: $E_0(n)$ can be obtained solving (numerically) Lieb and Wu's [3] integral equation, and u_ρ is already known. The results for K_ρ as a function of particle density are shown in fig. 3 for different values of U/t. For small U one finds in all cases agreement with the perturbative expression, eq. (3.2), whereas for large U $K_\rho \to 1/2$. The limiting behaviour for large U can be understood noting that for $U = \infty$

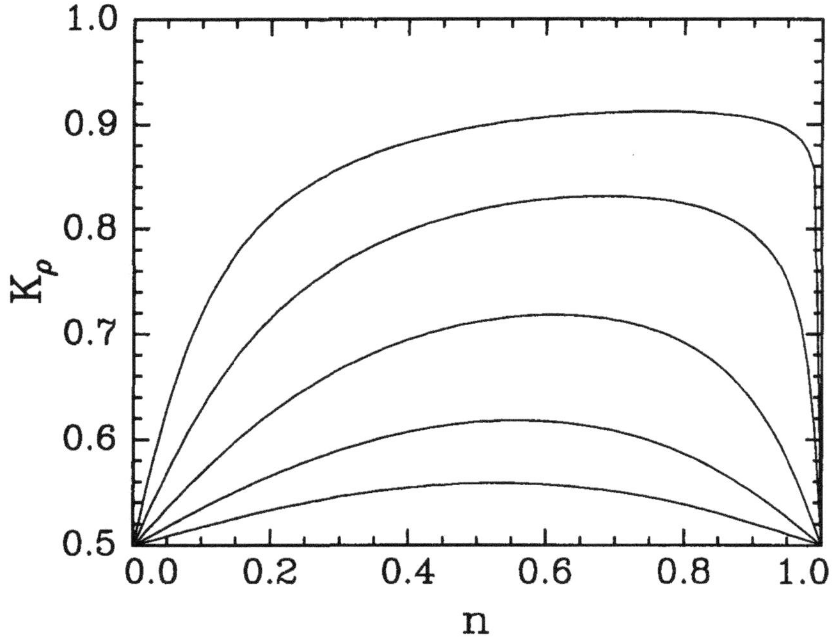

Figure 3: The correlation exponent K_ρ as a function of the bandfilling n for different values of U ($U/t = 1, 2, 4, 8, 16$ for the top to bottom curves). Note the rapid variation near $n = 1$ for small U.

the charge dynamics of the system can be described by noninteracting *spinless* fermions (the hard-core constraint then is satisfied by the Pauli principle) with k_F replaced by $2k_F$. Consequently one finds a contribution proportional to $\cos(4k_F x)x^{-2}$ in the density-density correlation function, which from eq. (2.13) implies $K_\rho = 1/2$. One then finds an asymptotic decay like $\cos(2k_F x)x^{-3/2}\ln^{1/2}(x)$ for the spin-spin correlations, eq.(2.14), and an exponent $\alpha = 1/8$ in the momentum distribution function [30]. Ogata and Shiba's numerical results [16] are quite close to these exact values.

As is apparent from fig. 3, the strong-coupling value $K_\rho = 1/2$ is also reached in the limits $n \to 0, 1$ *for any positive* U. For $n \to 0$ this behaviour is easily understood: the effective interaction parameter is U/v_F, but v_F goes to zero in the low-density limit (corresponding to the diverging density of states). The limit $n \to 1$ is more subtle: in this case nearly every site is singly occupied, with a very low density of holes. The only important interaction then is the short range repulsion between holes, which can be approximated by treating the *holes* as a gas of spinless noninteracting fermions. Using (3.3), one then again finds $K_\rho = 1/2$.

The exact solution of Lieb and Wu can also be combined with the long-wavelength effective hamiltonian (2.2) to obtain some information on the frequency-dependent con-

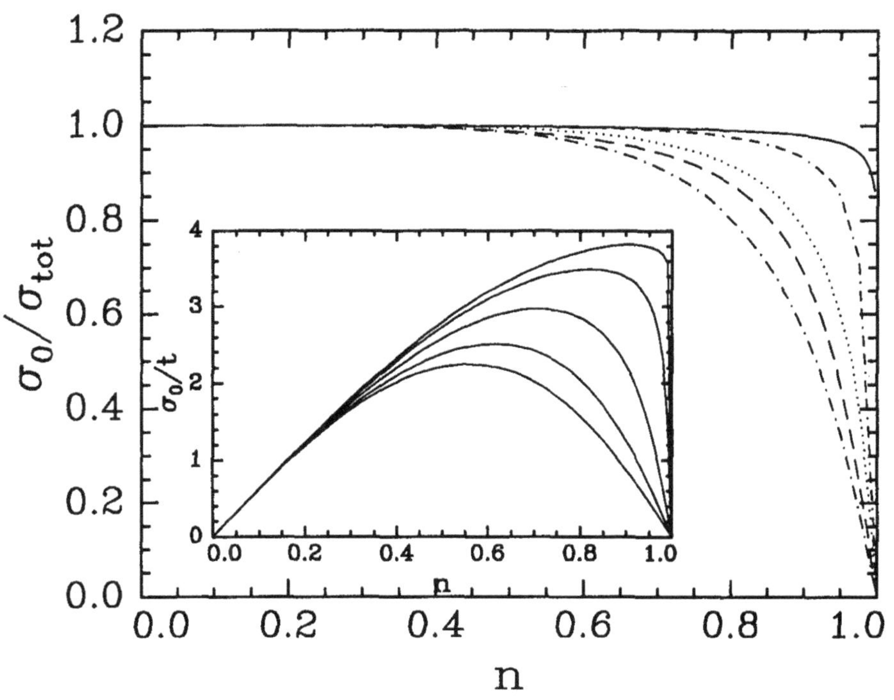

Figure 4: Variation of the relative weight of the dc peak in the total conductivity oscillator strength as a function of the bandfilling n for different values of U: $U/t = 1$ (full line), 4 (dashed), 16 (dash-dotted), 64 (dotted), and 256 (dash-double-dotted). The insert shows the weight of the dc peak in $\sigma(\omega)$ as a function of bandfilling for different values of U/t ($U/t = 1, 2, 4, 8, 16$ for the top to bottom curves).

ductivity $\sigma(\omega)$. On the one hand, from eq. (2.11) there is a delta function peak at zero frequency of weight $2K_\rho u_\rho$. On the other hand, the total oscillator strength is proportional to the kinetic energy [31]:

$$\sigma_{tot} = \int_{-\infty}^{\infty} \sigma(\omega) d\omega = -\pi \langle H_{kin} \rangle / L \ . \tag{3.4}$$

Thus, both the weight of the dc peak and the relative weight of the dc peak in the total conductivity can be obtained and are plotted in fig.4. As expected, far from half-filling, all the weight in σ_{tot} is in the dc peak. For exactly half-filling the dc conductivity vanishes, due to the existence of a gap for charge excitations Δ_c created by umklapp scattering, and all the weight is at $\omega > \Delta_c$. Fig.2 then shows that as $n \to 1$ umklapp scattering progressively transfers weight from zero to high frequency. The crossover is very sharp for small or large U, but rather smooth in intermediate cases ($U/t \approx 16$). This nonmonotonic behaviour as a function of U can be understood noting that initially with increasing U umklapp scattering plays an increasingly important role. Beyond $U/t \approx 16$, however, the spinless-fermion picture becomes more and more appropriate, and at $U = \infty$ one again has all the weight in the dc peak. The linear vanishing of σ_0 as $n \longrightarrow 1$ implies a linear variation of the ratio n/m^* with "doping".

An interesting question is the sign of the charge carriers, especially close to the metal–insulator transition. The standard way to determine this, the sign of the Hall constant, is useless in a one-dimensional system. As an alternative, the thermopower can be used which is negative (positive) for electron (hole) conduction. In general, calculation of the thermopower is a nontrivial task, as the curvature of the bands plays an important role, and the approximate form of the Hamiltonian (2.2) is therefore insufficient. Moreover, both charge ans spin entropies can play a role. However, close to the metal–insulator transition $u_\rho \ll u_\sigma$, and therefore the entropy of the charge degrees of freedom is much bigger than the spin entropy. In the presence of umklapp scattering, which becomes important close to half-filling, the charge part of the Hamiltonian can be transformed into a model of massive fermions, with energy–momentum relation $\varepsilon_k = \pm(v^2k^2 + \Delta^2)^{1/2}$ [32]. Δ is the charge excitation gap created by umklapp scattering. In general, the quasi-particles interact, however close to half-filling this interaction can be eliminated [33]. At half-filling all negative energy states are filled, all positive energy states are empty. Doping with a concentration n^* of holes, some of the negative energy states become empty and only states with $|k| > k_F^* \propto n^*$ are filled. Because of the vanishing interaction, a standard formula for the thermopower can be used [34] and gives

$$S = \frac{\pi^2 k_B^2 T}{6|e|} \frac{\Delta^2}{v^2(k_F^*)^2(v^2(k_F^*)^2 + \Delta^2)^{1/2}} \ , \tag{3.5}$$

i.e. *approaching the metal-insulator transition from $n < 1$, the thermopower is hole-like*, whereas obviously far from the transition ($n \ll 1$) it is electron-like. The exactly opposite behaviour can be found for $n > 1$.

For more complicated models, e.g. the "extended Hubbard model"

$$H = -t \sum_{i,s} (a_{i,s}^\dagger a_{i+1,s} + a_{i+1,s}^\dagger a_{i,s}) + U \sum_i n_{i\uparrow} n_{i\downarrow} + V \sum_i n_i n_{i+1} \ , \tag{3.6}$$

exact eigenvalues can not be obtained in the thermodynamic limit. The parameters in eq. (3.3) can however be calculated reliably for finite systems, and this gives rather good results, as shown in ref. 14 .

Exact exponents can be obtained for the model (3.6) in the limit $U \longrightarrow \infty$: then one has effectively spinless fermions (with $k_F \longrightarrow 2k_F$) with nearest neighbour interaction, a model which can be exactly solved using the Jordan–Wigner transformation into the XXZ spin chain. In particular, the $4k_F$-component of (2.13) is related to the correlation function of S_z. From the known results [35] one obtains, for a quarter-filled band ($n = 1/2$), $K_\rho = 1/(2 + (4/\pi)\sin^{-1}(v))$, $u_\rho = \pi t\sqrt{1 - v^2}/\cos^{-1}(v)$, with $v = V/2|t|$. Now $K_\rho < 1/2$ is possible. For $v > 1$ the system is in a dimerized insulating state. Approaching the insulating state from $v < 1$ both K_ρ and u_ρ remain finite, i.e. σ_0 jumps to zero at $v = 1$.

For $n \neq 1/2$ the parameters u_ρ, K_ρ can be obtained from numerical results [36]. Quite generally, one has $K_\rho > 1/8$, but $K_\rho = 1/2$ for $n \to 0, 1$, independent of v. On the other hand, $u_\rho \to 0$ as $n \to 1/2$ for $v > 1$, i.e. in that case the weight of the dc conductivity goes to zero continuously, the point $(v, n) = (1, 1/2)$ is thus highly singular. The same type of singularity also occurs at $U = 0, n = 1$ in the Hubbard model. Interestingly enough, one has $K_\rho > 1$ if $V < -\sqrt{2}|t|$, i.e. a finite amount of nearest-neighbor attraction is sufficient to lead to divergent superconducting fluctuations even for infinite on-site repulsion. Also note that the singularities in u_ρ and K_ρ at $v = -1$ (attractive interaction) represent a point of phase separation.

The Hubbard model also provides a rather straightforward interpretation of the spin–charge separation discussed above. Consider a piece of a Hubbard chain with a half-filled band. Then for strong U there will be no doubly-occupied sites, and because of the strong short-range antiferromagnetic order the typical local configuration will be

$$\cdots \uparrow\downarrow\uparrow\downarrow\uparrow\downarrow\uparrow\downarrow\uparrow\downarrow\uparrow\downarrow \cdots$$

Introducing a hole will lead to

$$\cdots \uparrow\downarrow\uparrow\downarrow\uparrow \; O \; \uparrow\downarrow\uparrow\downarrow\uparrow\downarrow \cdots$$

and after moving the hole one has (note that the kinetic term in the hamiltonian does not flip spins)

$$\cdots \uparrow\downarrow \; O \; \uparrow\downarrow\uparrow\uparrow\downarrow\uparrow\downarrow\uparrow\downarrow \cdots$$

Now the hole is surrounded by one up and one down spin, whereas somewhere else there are two adjacent up spins. Finally, a few exchange spin processes lead to

$$\cdots \uparrow\downarrow \; O \; \uparrow\downarrow\uparrow\downarrow\uparrow\downarrow\uparrow\uparrow\downarrow \cdots$$

Note that the original configuration, a hole surrounded by *two* up spins has split into a hole surrounded by antiferromagnetically aligned spins ("holon") and a domain-wall like configuration, two adjacent up spins, which contain an excess spin 1/2 with respect to the initial antiferromagnet ("spinon"). The exact solution by Lieb and Wu contains two types of quantum numbers which can be associated with the dynamics of the spinons and holons, respectively. We thus notice that spinons and holons [37,38] have a well-defined meaning in the present one-dimensional case. Also note that the ground state away from half-filling can be considered as a state with a finite concentration of holons, but no spinons: on the average, adjacent spins then are antiparallel, which leads to oscillations in the spin density of period $2/n$ (n is the particle density), i.e. of wavenumber $\pi n = 2k_F$, as in eq. (2.14). The exponent in (2.14) then can be understood as one contribution from the spin dynamics (for localized spins, correlations decay like $x^{-1}\ln^{1/2}(x)$) and another contribution, K_ρ, coming from the motion of the spins.

4 (IN)STABILITY OF THE LUTTINGER LIQUID

The question of the stability of the Luttinger liquid with spin–charge separation in higher dimension is clearly of importance and has been studied extensively in the context of quasi–one–dimensional conductors [8,39,40]. I here give a simple scaling argument for the effect of a nonzero but small interchain hopping term on an assembly of parallel chains. Interchain hopping is described by an extra term in the Hamiltonian:

$$H_{\perp} = t_{\perp} \sum_{(i,j)} \int dx [\psi^{\dagger}_{\pm,i}(x)\psi_{\pm,j}(x) + h.c.] \ . \tag{4.1}$$

Here i, j are chain indices, and the sum is over nearest neighbour pairs. Now notice that at zero temperature all excitations of the strictly one–dimensional system are massless and correlation functions decay as power laws, as is typical of a critical point. Consequently, crossover scaling arguments can be used to determine the relevance or irrelevance of the perturbation $H_{\perp}$. In particular, consider the lowest order correction to the free energy:

$$\delta F^{(2)} \approx t_{\perp}^2 \int dx d\tau G_{\pm}^2(x,\tau) \ , \tag{4.2}$$

where τ is the Matsubara imaginary time and $G_{\pm}$ is the single–particle Green's function for right or left going particles. Now

$$G_{\pm}(x_1 - x_2, \tau_1 - \tau_2) \approx |1-2|^{-(2+K_\rho+1/K_\rho)/4} \ , \tag{4.3}$$

where $|1-2|$ is an abbreviation:

$$|1-2| = \frac{v_F}{2\pi T}[\cosh(2\pi T(x_1 - x_2)/v_F) - \cos(2\pi T(\tau_1 - \tau_2))]^{1/2} \ . \tag{4.4}$$

For simplicity I have set $u_\sigma = u_\rho = v_F$, and a factor depending on the angle of the vector $(x_1 - x_2, \tau_1 - \tau_2)$ with the x-axis has been omitted. Note that the prefactor $1/T$ in (4.4) is important to reproduce the correct zero-temperature limit. After the rescaling $Tx_i/v_F \to z_i$, $T\tau_i \to y_i$, the integral in (4.2) becomes dimensionless, with the result

$$\delta F^{(2)} \propto t_{\perp}^2 T^{-1+(K_\rho+1/K_\rho)/2} \ . \tag{4.5}$$

Compared to the zeroth order term $F^{(0)} \propto T^2$ this becomes dominant at sufficiently low temperatures provided the exponent in (4.5) is smaller than 2, i.e. if

$$3 - \sqrt{8} < K_\rho < 3 + \sqrt{8} \ . \tag{4.6}$$

Then the interchain coupling is a relevant perturbation, which would imply that the low temperature properties are not determined by the one–dimensional Luttinger liquid fixed point, but by some other type of behaviour, possibly Fermi–liquid like. The condition (4.6) is always satisfied for the one–dimensional Hubbard model, but for more general models, e.g. the extended model, eq.(3.6) one can have $K_\rho < 3 - \sqrt{8}$. One would then conclude that interchain hopping is irrelevant, i.e. the Luttinger liquid would even exist in a quasi–one–dimensional system [41].

Consider now the next order:

$$\delta F^{(4)} \approx t_\perp^4 \int d1d2d3d4 \langle T_\tau \psi_+^\dagger(1)\psi_-^\dagger(2)\psi_-(3)\psi_+(4)\rangle^2 \ , \tag{4.7}$$

where, with the same approximations as before

$$\begin{aligned}&\langle T_\tau \psi_+^\dagger(1)\psi_-^\dagger(2)\psi_-(3)\psi_+(4)\rangle \\ &\quad\approx \left[\frac{|1-3||2-4|}{|1-2||3-4|}\right]^{(K_\rho-1/K_\rho)/4} [|1-4||2-3|]^{-(2+K_\rho+1/K_\rho)/4} \ .\end{aligned} \tag{4.8}$$

If one ignores the first term in square brackets, one simply recovers the square of $\delta F^{(2)}$. However, this is incorrect: for example, configurations with $1 \approx 3, 2 \approx 4$ in (4.7) give a contribution

$$\delta F^{(4,1)} \approx t_\perp^4 \int d1d2|1-2|^{-2-2K_\rho} \propto T^{2K_\rho} \ . \tag{4.9}$$

This now is a relevant perturbation *for any* $K_\rho < 1$. Similarly, configurations with $1 \approx 2, 3 \approx 4$ lead to a contribution proportonal to T^{2/K_ρ}, which is relevant for any $K_\rho > 1$, i.e. we conclude ***that for any nonzero interchain coupling the one-dimensional Luttinger liquid is unstable***. One should however notice that in the parameter region where $\delta F^{(2)}$ is irrelevant, the instability comes from (electron–hole or electron–electron) pair hopping processes in (4.7), which almost certainly lead to a broken symmetry groundstate (SDW or superconducting). Above the critical temperature of that symmetry breaking one then would still expect one–dimensional Luttinger liquid behaviour.

5 DISCUSSION AND CONCLUSION

In the present paper I have shown how the boson representation of one–dimensional fermions can be combined with the Lieb–Wu exact solution of the one-dimensional Hubbard model to obtain exact results on ground state correlations and the low–temperature thermodynamics. In particular, the behaviour in the vicinity of the metal-insulator transition can be studied in detail.

It seems worthwhile here to compare the metal–insulator transition in the one–dimensional Hubbard model with other scenarios for strongly correlated fermion systems in higher dimension (see the review by Vollhardt for details [42]). In the "nearly localized" picture, effective mass effects predominate and enhance both the specific heat and the spin susceptibility. Consequently, the Wilson ratio ($1/(1+F_0^a)$ in Fermi liquid language) remains nonzero as the metal–insulator is approached. On the other hand, in the "nearly ferromagnetic" (or paramagnon) picture, only the spin susceptibility is enhanced significantly, and therefore R_W can be much larger than unity. The behaviour found here in the one–dimensional case is quite different from both these scenarios: generally $R_W < 2$, and approaching the metal–insulator transition $R_W \to 0$. This occurs because generally an enhancement of the mass of the *charge* carriers (i.e. a decrease of u_ρ) has no influence on the spin degrees of freedom (see fig.1). This is rather straightforwardly understood in terms of ***spin–charge decoupling***, as explained at the end of sec.3: charge and spin excitations move nearly independently of each other, and in particular the spin dynamics is determined by nearest–neighbor exchange. This exchange energy is antiferromagnetic, whether there is a hole between two spins or not, and consequently the low–energy spin dynamics always is that of an antiferromagnetic chain of localized spins. In particular the spin susceptibility remains finite even when the mass of the charge carrier approaches infinity.

Let us discuss the metal–insulator transition in more detail. The fact that u_ρ and σ_0 vanish linearly as $n \to 1$ seems to be consistent with a divergent effective mass at constant carrier density because $u_\rho \approx 1/m^*$, $\sigma_0 \approx n/m^*$. A constant carrier density is also consistent with the fact that $k_F = \pi n/2$ is independent of U. It is ***not consistent*** with the hole–like sign of the thermopower as $n \to 1$ from below, nor with the electron–like sign as $n \to 1$ from above: if the carriers are holes, the carrier density is the density of holes: $n^* = 1 - n$. Treating the holes as spinless fermions, as already mentioned in sec.3, $\sigma_0 \to 0$ because $n^* \to 0$, and $\gamma \to \infty$ because the density of states of a one–dimensional band diverges at the band edges. What is not so easily understood in this picture is the fact that k_F (i.e. the location of the singularity of n_k) is given by its free–electron value $\pi n/2$, rather then being proportional to n^*. One should however notice that n_k is given by the single–particle Green's function, which contains both charge and spin degrees of freedom. The location of k_F then may possibly be explained by phase shifts due to holon–spinon interaction. This is in fact suggested by the structure of the wavefunction of the exact solution [3,16].

Anderson [43] has argued that two–dimensional correlated fermion systems can exhibit spin–charge separation very similar to the one–dimensional Luttinger liquid considered here. We here simply notice that the two scenarios discussed in the preceding paragraph would have drastically different consequences in two dimensions: in the first case ($m^* \to \infty, n \approx \mathit{const.}$) one expects $\gamma \to \infty$, whereas the density of states at the edges of a two–dimensional band is finite, and therefore in the second case $\gamma \to \mathit{const.}$ ($\sigma_0 \to 0$ in both cases).

It is interesting to notice that most of the results found here would be expected if the system had commensurate antiferromagnetic long–range order (which, because of zero–point fluctuations, is of course not the case in one dimension). One then would have two branches of quasi–particle excitations, with energies $\varepsilon_k = \pm\sqrt{v_F^2(k - \pi/2)^2 + \Delta^2}$. At half–filling the lower band is filled, the upper one is empty, i.e. the system is an insulator, but retains the finite spin susceptibility of an antiferromagnet. Doping now with electrons or holes one of the bands get partially occupied, leading to a conductivity proportional to dopant concentration, a specific heat coefficient inversely proportional to dopant concentration, and hole (electron) like thermopower for hole (electron) doping. All this is in agreement with the findings of sec.3. One would also expect a singularity in n_k at k_F, but due to the broken long-range order and the associated halving of the Brillouin zone, another singularity would appear at $\pi - k_F$. This last effect is obviously an artefact due to the (false) assumption of long-range order, however, interestingly enough all properties which are not directly affected by the change of structure of k-space are found to have qualitatively the correct behaviour.

Finally, I notice that the present results place some constraints on the way experimental systems can be modelled. For example, in the quasi-one-dimensional organic compound TTF-TCNQ one observes strong diffuse X-ray scattering at $4k_F$ [9]. From eq.(2.13) one then concludes that $K_\rho < 1/2$, and therefore the Hubbard model alone cannot be sufficient to describe correlation effects in this compound. The experimentally determined exponents in a number of other compounds [8,9] also imply $K_\rho < 1/2$, i.e. finite-range interactions seem to be rather important in many cases.

Acknowledgement - I thank D. Jérome and J.P. Pouget for discussions of their experimental results. This work was in part supported by ESPRIT basic research contract no. 3121.

References

[1] D. Jérome and H.J. Schulz, Adv. Phys. **31**, 299 (1982).

[2] A.J. Heeger, S. Kivelson, J.R. Schrieffer, and W.P. Su, Rev. Mod. Phys. **60**, 781 (1988).

[3] E.H. Lieb and F.Y. Wu, Phys. Rev. Lett. **20**, 1445 (1968).

[4] P. W. Anderson, Science **235**, 1196 (1987).

[5] P. W. Anderson, in *Frontiers and Borderlines in Many Particle physics* (North Holland, Amsterdam, 1988).

[6] J. Sólyom, Adv. Phys. **28**, 209 (1979).

[7] V.J. Emery, in *Highly Conducting One-Dimensional Solids* (Plenum, New York, 1979), p.327.

[8] C. Bourbonnais, F. Creuzet, D. Jérome, K. Bechgaard, and A. Moradpour, J. Phys. (Paris) Lett. **45**, L755 (1984).

[9] J. P. Pouget, in *Highly Conducting Quasi-One-Dimensional Crystal Semiconductors and Semimetals*, vol. 27 (Pergamon, New York, 1988), p.87.

[10] T. Giamarchi and H.J. Schulz, Phys. Rev. B **37**, 325 (1988).

[11] J.E. Hirsch and D.J. Scalapino, Phys. Rev. B **29**, 5554 (1984).

[12] Z.G. Soos and S. Ramasesha, Phys. Rev. Lett. **51**, 2374 (1983).

[13] D.K. Campbell, T.A. DeGrand, and S. Mazumdar, Phys. Rev. Lett. **52**, 1717 (1984).

[14] H.J. Schulz, Phys. Rev. Lett. **64**, 2831 (1990).

[15] F.D.M. Haldane, J. Phys. C **14**, 2585 (1981).

[16] M. Ogata and H. Shiba, Phys. Rev. B **41**, 2326 (1990).

[17] N. Kawakami and S.K. Yang, Phys. Lett. A **148**, 359 (1990).

[18] H. Frahm and V.E. Korepin, Stony Brook preprint ITP–SB–90–33.

[19] N. Kawakami and S.K. Yang, Phys. Rev. Lett. **65**, 2309 (1990).

[20] A. Luther and I. Peschel, Phys. Rev. B **9**, 2911 (1974).

[21] D.C. Mattis, J. Math. Phys. **15**, 609 (1974).

[22] R. Heidenreich, R.Seiler, and D.A. Uhlenbrock, J. Stat. Phys. **22**, 27 (1980).

[23] F.D.M. Haldane, Phys. Rev. Lett. **47**, 1840 (1981).

[24] The N_ν- and J_ν-terms are discussed in ref. 15 .

[25] I here consider incommensurate band filling ($n \neq 1$). For $n = 1$ umklapp scattering, represented by an additional operator $\cos(\sqrt{8}\phi_\rho)$, leads to an insulating groundstate.

[26] The time- and temperature dependence is also easily obtained, see ref. 7 .

[27] The logarithmic factors have been derived by T. Giamarchi and H.J. Schulz, Phys. Rev. B **39**, 4620 (1989).

[28] C.F. Coll, Phys. Rev. B **9**, 2150 (1974).

[29] This formula can also be obtained using the conformal invariance of H_ρ.

[30] The result $\alpha = 1/8$ has also been found by P.W. Anderson and Y. Ren (preprint), and by A. Parola and S. Sorella, Phys. Rev. Lett. **64**, 1831 (1990).

[31] D. Baeriswyl, J. Carmelo, and A. Luther, Phys. Rev. B **33**, 7247 (1986).

[32] V.J. Emery, A. Luther, and I. Peschel, Phys. Rev. B **13**, 1272 (1976).

[33] H.J. Schulz, Phys. Rev. B **22**, 5274 (1980).

[34] P.M. Chaikin, R.L. Greene, S. Etemad, and E. Engler, Phys. Rev. B **13**, 1627 (1976).

[35] A. Luther and I. Peschel, Phys. Rev. B **12**, 3908 (1975).

[36] F.D.M. Haldane, Phys. Rev. Lett. **45**, 1358 (1980).

[37] S. Kivelson, D. Rokhsar, and J. Sethna, Phys. Rev. B **35**, 8865 (1987).

[38] Z. Zou and P.W. Anderson, Phys. Rev. B **37**, 627 (1988).

[39] C. Bourbonnais and L.G. Caron, Physica **143B**, 450 (1986).

[40] S. Brazovskii and V. Yakovenko, J. Phys. (Paris) Lett. **46**, L111 (1985).

[41] X.G. Wen, preprint.

[42] D. Vollhardt, Rev. Mod. Phys. **56**, 99 (1984).

[43] P. W. Anderson, Phys. Rev. Lett. **64**, 1839 (1990).

J. Phys.: Condens. Matter 3 (1991) 757–765. Printed in the UK

LETTER TO THE EDITOR

Generalization of the Landau liquid concept: example of the Luttinger liquids

J Carmelo† and A A Ovchinnikov‡

† Max-Planck-Institut für Festkörperforschung, D-7000 Stuttgart 80, Federal Republic of Germany

‡ Institute of Chemical Physics, Kosigin str. 4, 117334, Moscow, USSR

Received 12 November 1990

Abstract. We introduce a new representation of the Bethe *ansatz* solutions for 1D Luttinger liquids which describes the spectral properties and asymptotic behaviour of the correlation functions of the 1D Hubbard model in terms of the renormalized interaction of charge and spin pseudoparticles. The study of the low-lying eigenstates is reduced to the familiar languages of band theory and of the Fermi liquid. Our results provide a better understanding of the decoupling of charge and spin degrees of freedom in 1D interacting systems, generalize the concept of a Landau liquid and may be relevant to the physics of higher dimensional systems.

The purpose of this letter is to provide evidence for the fact that the usual Landau–Fermi liquids constitute an example of a much wider class of non-trivial many-body fermionic systems (probably of all dimensions). The present study is restricted to one-dimensional (1D) Luttinger liquids [1] that are soluble by the Bethe *ansatz*, and in particular to the (1D repulsive Hubbard model. Nonetheless, we believe that many of the features found in this letter have a universal character and are also present in the low-energy physics of nearly all non-trivial many-body fermionic liquids. These common features follow essentially from the fact that the low-energy properties are fully controlled by the departure of the pseudo-momentum distribution(s) of the pseudoparticles (often many-body collective modes specific to each system) from its (their) value(s) in the interacting ground state. Moreover, after renormalization the generalized Landau liquids have only forward (or exchange) scattering. As in the Fermi liquid, the two-pseudoparticle f-functions (second functional derivatives of the energy with respect to the fluctuations) regulate the forward scattering renormalized interaction of the pseudoparticles in the low-energy regime. Our results are fully consistent with the original idea of Anderson [2] that the Luttinger liquid is a fixed point of the same renormalization group which, in some three-dimensional systems, leads to the Landau–Fermi liquid as a unique fixed point.

Although we concentrate our investigation on the case of the 1D Hubbard model, the generalization of the results to the other 1D Luttinger liquids is straightforward and will be presented elsewhere. A full understanding of the one-dimensional Hubbard model solution [3] is of interest in its own right, and may provide clues to the understanding of higher dimensional systems [2]. We consider the 1D Hubbard model

0953-8984/91/060757+09$03.50

at arbitrary magnetic field

$$\hat{H} = -t\sum_{j,\sigma}\left(c^{+}_{j\sigma}c_{j+1\sigma} + c^{+}_{j+1\sigma}c_{j}\right) + U\sum_{j} c^{+}_{j\uparrow}c_{j\uparrow}c^{+}_{j\downarrow}c_{j\downarrow} - \mu_0 H\sum_{j,\sigma}\sigma c^{+}_{j\sigma}c_{j\sigma} \tag{1}$$

where $c^{+}_{j\sigma}$ $(c_{j\sigma})$ is the creation (annihilation) operator for an electron with spin σ at site j. The system consists of N electrons on N_a sites. We make use of the following notation: the dimensionless on-site repulsion $u = U/4t$, the density $n = N/N_a$ ($k_F = \pi n/2$), the spin density $s = (k_{F\uparrow} - k_{F\downarrow})/2\pi$, where $k_{F\sigma} = \pi N_\sigma/N_a$, and $N_\uparrow = M'$ and $N_\downarrow = M$ are the number of up and down spins in the system.

Lieb and Wu [3] used the Bethe-*ansatz* technique to reduce the eigenvalue problem for (1) to that of solving a set of coupled algebraic equations. The crystal momentum P and the energy E are given by

$$P = \sum_{j=1}^{N} q_j + \sum_{\alpha=1}^{M} p_\alpha \tag{2}$$

$$E = \sum_{j=1}^{N}(-2t\cos K_j) - 2\mu_0 N_a H s \tag{3}$$

where we make use of the notation $q_j = (2\pi/N_a)\,I_j$ and $p_\alpha = (2\pi/N_a)\,J_\alpha$. An eigenstate of the many-body system is uniquely specified by a particular choice of the quantum numbers $\{I_j\}$, $\{J_\alpha\}$ (or pseudo-momentum distributions $\{q_j\}, \{p_\alpha\}$). For instance, in the ground state I_j and J_α are consecutive integers (or half-odd integers) centred around the origin [3]. After choosing the set $\{q_j\}$, $\{p_\alpha\}$, the Lieb and Wu algebraic equations determine the charge and spin rapidities $K_j = K_j(q_j)$ and $S_\alpha = S_\alpha(p_\alpha)$, respectively ($S_\alpha = \Lambda_\alpha/u$) [3]. Although our formulation can be extended to excited states described by complex roots [4], we restrict the present study mainly to excitations involving only real rapidities.

In the thermodynamic limit ($N_a \to \infty$, n fixed) the roots $K_j = K_j(q_j)$ and $S_\alpha = S_\alpha(p_\alpha)$ proliferate on the real axis and the Lieb and Wu equations lead to [4]

$$K(q) = q + \frac{1}{\pi}\int_{-k_{F\uparrow}}^{k_{F\uparrow}} \mathrm{d}p'\, N_\downarrow(p')\tan^{-1}[S(p') - (1/u)\sin K(q)] \tag{4}$$

$$p = \frac{1}{\pi}\int_{-\pi}^{\pi} \mathrm{d}q'\, M_c(q')\tan^{-1}[S(p) - (1/u)\sin K(q')]$$
$$\qquad - \frac{1}{\pi}\int_{-k_{F\uparrow}}^{k_{F\uparrow}} \mathrm{d}p'\, N_\downarrow(p')\tan^{-1}[\tfrac{1}{2}(S(p) - S(p'))]\,. \tag{5}$$

($K(q)$, $S(p)$ are simply related to the distributions of [3].) Moreover, the energy (3) yields

$$E = \frac{N_a}{2\pi}\int_{-\pi}^{\pi} \mathrm{d}q'\, M_c(q')[-2t\cos K(q')] - 2\mu_0 N_a H s \tag{6}$$

where

$$s = \frac{1}{2\pi}\left(\frac{1}{2}\int_{-\pi}^{\pi} \mathrm{d}q'\, M_c(q') - \int_{-k_{F\uparrow}}^{k_{F\uparrow}} \mathrm{d}p'\, N_\downarrow(p')\right). \tag{7}$$

In the right-hand sides of equations (4)–(7) $M_c(q)$ and $N_\downarrow(p)$ may be interpreted as pseudo-momentum distributions of charge and spin pseudoparticles, respectively. For eigenstates involving only real rapidities, we always have: $K(\pm\pi) = \pm\pi$ and $S(\pm k_{F\uparrow}) = \pm\infty$. Moreover, for the ground state at fixed magnetization $K(q)$ and $S(p)$ are odd functions such that $K(2k_F) = Q$ and $K(k_{F\downarrow}) = B/u$, where Q and B are the usual cut offs of the Lieb and Wu equations [3]. In this case the distributions $M_c(q)$ and $N_\downarrow(p)$ read

$$M_c^0(q) = \Theta(2k_F - |q|) \qquad N_\downarrow^0(p) = \Theta(k_{F\downarrow} - |p|) \,. \tag{8}$$

As in Fermi liquid theory these distributions do not depend on the interaction. In the present two-fluid Landau liquid the charge and spin pseudo-Fermi surfaces are defined as the set of points $\{q = \pm 2k_F,\ p = \pm k_{F\downarrow}\}$ separating the occupied from the unoccupied region. The limits of the pseudo-Brillouin zones of the charge and spin pseudoparticles are $\{q = \pm\pi\}$, $\{p = \pm k_{F\uparrow}\}$, respectively. The restrictions on the numbers I_j and J_α [3] imply that each pseudo-momentum value cannot be occupied by more than one pseudoparticle, i.e. the pseudoparticles have fermionic character [4]. These can be identified with the 'pseudo-fermions' considered in [5] and are related to the holons and spinons [2, 4]. The spin pseudoparticles are of the same kind of the ones of the Heisenberg chain [4], being closely related to the spin-$\frac{1}{2}$ 'spin waves' introduced in [6]. They are many-body collective modes and in contrast to the quasi-particles of the Fermi liquid theory, which in the limit of vanishing interaction map onto real particles (electrons), the present class of pseudoparticles cannot exist outside the many-body system for any value of the bare interaction (including vanishing interaction). This feature of the Landau–Luttinger pseudoparticles [4] is related to the 'infrared catastrophe' of [2].

We consider small pseudo-momentum fluctuations around the ground-state distributions, equations (8)

$$M_c(q) = M_c^0(q) + \delta_c(q) \qquad N_\downarrow(p) = N_\downarrow^0(p) + \delta_\downarrow(p) \,. \tag{9}$$

As in Fermi liquid theory, the departure of the pseudo-momentum distribution functions of the pseudoparticles from their values in the ground state $\delta_c(q), \delta_\downarrow(p)$, equations (9), fully controls the low-energy physics of the model: This is true both for elementary excitations involving real and complex rapidities [4]. The main point in our approach is to consider $K(q)$, $S(p)$ and E (see equations (4)–(6)) as functionals of the pseudoparticle distributions. Provided that these involve a small number of pseudoparticles, an expansion of the energy $E = E_0 + E_1 + E_2 + \cdots$ can be performed to arbitrary order in the fluctuations. In the case of excitations described only by real rapidities, the leading order corrections read

$$E_1 = \frac{N_a}{2\pi}\int_{-\pi}^{\pi} dq\,\delta_c(q)\epsilon_c(q) + \frac{N_a}{2\pi}\int_{-k_{F\uparrow}}^{k_{F\uparrow}} dp\,\delta_\downarrow(p)\epsilon_s(p) \tag{10}$$

$$\begin{aligned} E_2 = {} & \frac{N_a}{4\pi^2}\int_{-\pi}^{\pi} dq \int_{-\pi}^{\pi} dq'\,\delta_c(q)\delta_c(q')\tfrac{1}{2}f_{cc}(q,q') \\ & + \frac{N_a}{4\pi^2}\int_{-k_{F\uparrow}}^{k_{F\uparrow}} dp \int_{-k_{F\uparrow}}^{k_{F\uparrow}} dp'\,\delta_\downarrow(p)\delta_\downarrow(p')\tfrac{1}{2}f_{ss}(p,p') \\ & + \frac{N_a}{4\pi^2}\int_{-\pi}^{\pi} dq \int_{-k_{F\uparrow}}^{k_{F\uparrow}} dq\,\delta_c(q)\delta_\downarrow(p)f_{cs}(q,p) \,. \end{aligned} \tag{11}$$

As in Fermi liquid theory, the first and second functional derivatives of energy with respect to the fluctuations define the bands $\epsilon_c(q)$ and $\epsilon_s(p)$ of the pseudoparticles and their interactions, the f-functions $f_{cc}(q,q')$, $f_{ss}(p,p')$ and $f_{cs}(q,p)$, respectively. The latter are related to the pseudoparticle zero-momentum transfer forward scattering amplitudes [4]. Although the fluctuations of the right-hand sides of equations (10) and (11) are arbitrary in the sense that the expressions for the bands and f-functions are independent of them, only appropriate choices of $\delta_c(q)$ and $\delta_\downarrow(p)$ describe true eigenstates of the many-body system [4, 7]. The charge and spin pseudoparticle bands in the presence of a magnetic field can be expressed as [4]

$$\epsilon_c(q) = \int_{Q}^{K(q)} \mathrm{d}k'\, 2t\eta_c(k') \tag{12}$$

$$\epsilon_s(p) = \int_{B/u}^{S(p)} \mathrm{d}v'\, 2t\eta_s(v') \tag{13}$$

where $K(q)$ and $S(p)$ are the ground-state solutions of equations (4) and (5). We note that $\mu_c = \epsilon_c(2k_{\mathrm{F}}) = 0$, $\mu_s = \epsilon_s(k_{\mathrm{F}\downarrow}) = 0$. In the right-hand sides of equations (12) and (13) the distributions $2t\eta_c(k)$ and $2t\eta_s(v)$ are solutions of coupled integral equations of the same form as the Lieb and Wu equations except that the inhomogeneous term of the first of these equations is replaced by $2t\sin k$ (and $2\pi\sigma(\Lambda)$ by $(1/u)2t\eta_s(v)$, $v = \Lambda/u$) [3, 4]. The pseudoparticle velocities are defined as $v_c(q) = \mathrm{d}\epsilon_c(q)/\mathrm{d}q$ and $v_s(p) = \mathrm{d}\epsilon_s(p)/\mathrm{d}p$. In particular $v_c(2k_{\mathrm{F}})$ and $v_s(k_{\mathrm{F}\downarrow})$ are the same velocities as the ones of [8, 9] (i.e. obey the same integral equations). A crucial advantage of our choice of variables q, p is that, given an eigenstate described by distributions (9), the corresponding crystal momentum P is always additive in the pseudoparticle pseudomomenta (see equation (2)). In fact, in contrast to the usual representation of Bethe *ansatz*, the back-flow effect only affects the energy bands $\epsilon_c(q)$ and $\epsilon_s(p)$ [4]. The charge band, equation (12), is such that its bandwidth $4t$, pseudo-Brillouin zone width 2π and pseudo-Fermi surface points $\pm 2k_{\mathrm{F}}$ remain unaltered when varying H and u, which slightly change $\epsilon_c(q)$. In contrast to $\epsilon_c(q)$, the spin pseudoparticle band (13) is clearly affected by the magnetic field. In fact, the pseudo-Brillouin zone limits and Fermi surface points are given by $p = \pm k_{\mathrm{F}\uparrow}$ and $p = \pm k_{\mathrm{F}\downarrow}$, respectively. The effect of u (H) on the band $\epsilon_s(p)$, equation (13), is essentially to modify its bandwith (pseudo-Brillouin zone). On the other hand n affects both the bandwidth and the pseudo-Brillouin zone.

Except for spin singlet excitations involving complex roots (anti-bound states of down-spin pseudoparticles) [4], the elementary excitations can be described in terms of pseudoparticle–hole processes in the charge and spin bands. The full description of the low-lying excitations in the absolute ground state ($H = 0$ and $k_{\mathrm{F}\downarrow} = k_{\mathrm{F}}$) involves a 'frozen' up-spin pseudoparticle band which is always filled ('holes' are not allowed in the frozen band [4, 7]) and an upper 'conduction' charge band, $\epsilon_c^h(q) = U - \epsilon_c(q)$ [4, 7]. The spectra of the charge gapless [10] and across-gap [11] excitations, as well as of the triplet two-parametric excitations [10], can be written simply as

$$E_c = \epsilon_c(q_1) - \epsilon_c(q_0) \qquad P = q_1 - q_0 \qquad |q_0| < 2k_{\mathrm{F}} \quad |q_1| > 2k_{\mathrm{F}} \tag{14}$$

$$E_c^{\Delta} = \epsilon_c^h(q_1) - \epsilon_c(q_0) \qquad P = q_1 - q_0 \qquad |q_0| < 2k_{\mathrm{F}} \tag{15}$$

$$E_t = -\epsilon_s(p_1) - \epsilon_s(p_0) \qquad P = 2k_{\mathrm{F}} - p_1 - p_0 \qquad |p_0|, |p_1| < k_{\mathrm{F}} \tag{16}$$

respectively, where the bands involved in the pseudoparticle–hole processes of the right-hand sides of equations (14)–(16) are given by equations (12) and (13) for the particular case of the absolute ground state ($B = \infty$). The excitations in the presence of a magnetic field are studied in [7].

As well as reducing the study of low lowing excitations to the usual language of band theory, the Landau–Luttinger approach allows the straightforward evaluation of the magnetic susceptibility (which involves the bands and f-functions) and low temperature specific heat: These can be readily obtained by replacing in the right-hand sides of equations (10) and (11) fluctuations $\delta_c(q)$ and $\delta_\downarrow(p)$ of appropriate form [4, 7].

The f-functions of the right-hand side of equation (11), $f_{cc}(q,q')$, $f_{ss}(p,p')$, $f_{cs}(q,p)$ read [4]

$$\begin{aligned} f_{cc}(q,q') &= 2\pi v_c(q)\Phi_{cc}(q,q') + 2\pi v_c(q')\Phi_{cc}(q',q) \\ &\quad + [2\pi v_c(2k_F)] \sum_{j=\pm1} \Phi_{cc}(2k_F j,q)\,\Phi_{cc}(2k_F j,q') \\ &\quad + [2\pi v_s(k_{F\downarrow})] \sum_{j=\pm1} \Phi_{sc}(k_{F\downarrow} j,q)\,\Phi_{sc}(k_{F\downarrow} j,q') \end{aligned} \tag{17}$$

$$\begin{aligned} f_{ss}(p,p') &= 2\pi v_s(p)\Phi_{ss}(p,p') + 2\pi v_s(p')\Phi_{ss}(p',p) \\ &\quad + [2\pi v_s(k_{F\downarrow})] \sum_{j=\pm1} \Phi_{ss}(k_{F\downarrow} j,p)\,\Phi_{ss}(k_{F\downarrow} j,p') \\ &\quad + [2\pi v_c(2k_F)] \sum_{j=\pm1} \Phi_{cs}(2k_F j,p)\,\Phi_{cs}(2k_F j,p') \end{aligned} \tag{18}$$

$$\begin{aligned} f_{cs}(q,p) &= 2\pi v_c(q)\Phi_{cs}(q,p) + 2\pi v_s(p)\Phi_{sc}(p,q) \\ &\quad + [2\pi v_c(2k_F)] \sum_{j=\pm1} \Phi_{cc}(2k_F j,q)\,\Phi_{cs}(2k_F j,p) \\ &\quad + [2\pi v_s(k_{F\downarrow})] \sum_{j=\pm1} \Phi_{ss}(k_{F\downarrow} j,p)\,\Phi_{sc}(k_{F\downarrow} j,q)\,. \end{aligned} \tag{19}$$

In order to define the functions $\Phi_{cc}(q,q'), \Phi_{cs}(q,p), \Phi_{ss}(p,p'), \Phi_{sc}(p,q)$ appearing in the right-hand sides of equations (17)–(19), it is useful to introduce the auxiliary functions $\tilde{\Phi}_{cc}(k,k')$, $\tilde{\Phi}_{cs}(k,v)$, $\tilde{\Phi}_{ss}(v,v')$, $\tilde{\Phi}_{sc}(v,k)$ such that $\tilde{\Phi}_{cc}(K(q),K(q')) = \Phi_{cc}(q,q')$, $\tilde{\Phi}_{cs}(K(q),S(p)) = \Phi_{cs}(q,p)$, $\tilde{\Phi}_{ss}(S(p),S(p')) = \Phi_{ss}(p,p')$, $\tilde{\Phi}_{sc}(S(p),K(q)) = \Phi_{sc}(p,q)$, where $K(q)$, $S(p)$ are the ground-state solutions of equations (4) and (5). The auxiliary functions obey the following system of coupled integral equations [4]:

$$\tilde{\Phi}_{cc}(k,k') = \int_{-B/u}^{B/u} dv'\, A_1^{(v')}(v',k)\tilde{\Phi}_{sc}(v',k') \tag{20}$$

$$\tilde{\Phi}_{cs}(k,v) = A_1(v,k) + \int_{-B/u}^{B/u} dv'\, A_1^{(v')}(v',k)\tilde{\Phi}_{ss}(v',v) \tag{21}$$

$$\widetilde{\Phi}_{ss}(v,v') = A_2(v,v') - \int_{-Q}^{Q} dk'\, A_1^{(k')}(v,k')\widetilde{\Phi}_{cs}(k',v') - \int_{-B/u}^{B/u} dv''\, A_2^{(v'')}(v'',v)\widetilde{\Phi}_{ss}(v'',v') \tag{22}$$

$$\widetilde{\Phi}_{sc}(v,k) = -A_1(v,k) - \int_{-Q}^{Q} dk' A_1^{(k')}(v,k')\widetilde{\Phi}_{cc}(k',k) - \int_{-B/u}^{B/u} dv'\, A_2^{(v')}(v',v)\widetilde{\Phi}_{sc}(v',k) \tag{23}$$

where $A_1(v,k) = (1/\pi)\tan^{-1}(v-(1/u)\sin k)$, $A_2(v,v') = (1/\pi)\tan^{-1}(\frac{1}{2}(v-v'))$ and $A_1^{(k)}(v,k) = dA_1(v,k)/dk$, $A_1^{(v)}(v,k) = dA_1(v,k)/dv$, $A_2^{(v)}(v,v') = dA_2(v,v')/dv$.

The functions $\Phi_{cc}(q,q')$, $\Phi_{cs}(q,p)$, $\Phi_{ss}(p,p')$, $\Phi_{sc}(p,q)$ are the pseudoparticle renormalized scattering phase shifts [4]. In fact, the usual Bethe *ansatz* phase shifts, which for the 1D Hubbard model are four in number and can be evaluated by the method introduced by Korepin [12] for the massive Thirring model, may be written as a sum of two terms [4]. The first of these terms can be considered to be the scattering part of the phase shift. For the present model the scattering part of the four phase shifts are given by equations (20)–(23) [4]. The nature of the second term is discussed in [4].

The f-functions (17)–(19) regulate the electronic spectral properties of the model. In fact, the 'single-particle excitations' are described by fluctuations involving pairs of charge and spin pseudoparticles [4]. Each point (k,ω) of the two-dimensional space where the electronic spectral weight function is defined can be associated to one (or two) pair(s) of pseudoparticles [4]. On the other hand, there is a clear connection between the value of the electronic spectral function at a point (k,ω) and the interaction of the pair (or pairs) of pseudoparticles associated with that point [4]. Moreover, the interaction of the pseudoparticles determines the form of the electronic correlation functions. Particularly, the non-classical critical exponents which characterize the power law anomalies of the electronic momentum and the asymptotic behaviour of the correlation functions, are determined by the interaction of pseudoparticles with pseudo-momenta in the neighbourhood of the pseudo-Fermi points $q = \pm 2k_F$, $p = \pm k_{F\downarrow}$ [4]. These exponents can be derived by the conformal field approach [8, 9]. To illustrate the general character of the Landau–Luttinger liquid theory, we show that the finite size energy corrections and expressions for conformal dimensions of the fields in the former theory can be obtained by choosing particular forms for the fluctuations of the energy functional, equations (10) and (11), in the latter theory. For simplicity we restrict our considerations to the case when the number of electrons N and down-spin electrons M in the system remain unaltered [4]. We introduce the matrices $\mathbf{R}^{+}(q,p)$ and $\mathbf{R}^{-}(q,p)$ given by

$$\mathbf{R}^{\pm}(q,p) = \begin{pmatrix} \Phi_{cc}(q,\pm 2k_F) & \Phi_{cs}(q,\pm k_{F\downarrow}) \\ \Phi_{sc}(p,\pm 2k_F) & \Phi_{ss}(p,\pm k_{F\downarrow}) \end{pmatrix}. \tag{24}$$

These matrices describe the scattering of charge and spin pseudoparticles of arbitrary pseudo-momenta q, p, respectively, with right (+) and left (−) moving pseudoparticles of momenta at the pseudo-Fermi points.

It follows from equations (20) and (23) that the dressed charge matrix of [8, 9] (here we use the definition of Waynarovich [9], which is the transpose of that of [8]) can be rewritten as

$$\mathbf{Z} = \mathbf{1} + \mathbf{R}^{+}(2k_{\mathrm{F}}, k_{\mathrm{F}\downarrow}) - \mathbf{R}^{-}(2k_{\mathrm{F}}, k_{\mathrm{F}\downarrow}) . \tag{25}$$

The form of equation (25) evidences that the matrix elements of $\mathbf{Z}$ are combinations of phase shifts associated with the scattering of pseudoparticles with pseudo-momenta at the pseudo-Fermi surfaces.

We consider now fluctuations $\delta_c(q) = M_c(q) - M_c^0(q)$, $\delta_\downarrow(p) = N_\downarrow(p) - N_\downarrow^0(p)$, such that

$$\begin{aligned} M_c(q) = \Theta\left(2k_{\mathrm{F}} + (\operatorname{sgn} q)q_c - |q|\right) + \frac{2\pi}{N_a}\Big\{ \sum_p [\delta(q - q_p^+) + \delta(q - q_p^-)] \\ - \sum_h [\delta(q - q_h^+) + \delta(q - q_h^-)] \Big\} \end{aligned} \tag{26}$$

$$\begin{aligned} N_\downarrow(p) = \Theta\left(k_{\mathrm{F}\downarrow} + (\operatorname{sgn} p)p_s - |p|\right) + \frac{2\pi}{N_a}\Big\{ \sum_p [\delta(p - p_p^+) + \delta(p - p_p^-)] \\ - \sum_h [\delta(p - p_h^+) + \delta(p - p_h^-)] \Big\} \end{aligned} \tag{27}$$

where $|q_c|$, $|2k_{\mathrm{F}} \mp q_p^{\pm}|$, , $|2k_{\mathrm{F}} \mp q_h^{\pm}| \ll 2k_{\mathrm{F}}$ and $|p_s|$, $|k_{\mathrm{F}\downarrow} \mp p_p^{\pm}|$, $|k_{\mathrm{F}\downarrow} \mp p_h^{\pm}| \ll k_{\mathrm{F}\downarrow}$. The first term of the right-hand side of equation (26) ((27)) includes charge (spin) pseudoparticle–hole processes from pseudo-momenta close to $-2k_{\mathrm{F}}(-k_{\mathrm{F}\downarrow})$ to pseudo-momenta in the neighbourhood of $2k_{\mathrm{F}}(k_{\mathrm{F}\downarrow})$. $D_c = (N_a/2\pi)\,q_c$ ($D_s = (N_a/2\pi)\,q_s$) gives the number of pseudoparticles transferred ($D_c \ll N$, $D_s \ll M$). On the other hand, the second term of the right-hand side of equation (26) ((27)) describes charge (spin) pseudoparticle–hole processes around the points $\pm 2k_{\mathrm{F}}$ ($\pm k_{\mathrm{F}\downarrow}$). The indices p and h refer to particle and hole summations. + and − refer to right- and left-moving pseudoparticles. The asymptotic behaviour of the correlation functions is determined by these pseudoparticle–hole processes which involve exclusively pseudo-momenta in the neighbourhood of the pseudo-Fermi points.

We define the numbers

$$\begin{aligned} N_c^{\pm} &= \pm\frac{N_a}{2\pi}\left(\sum_p q_p^{\pm} - \sum_h q_h^{\pm}\right) \\ N_s^{\pm} &= \pm\frac{N_a}{2\pi}\left(\sum_p p_p^{\pm} - \sum_h p_h^{\pm}\right) . \end{aligned} \tag{28}$$

To evaluate the integrals of the right-hand sides of the energies (10) and (11) we expand $\epsilon_c(q)$ and $\epsilon_\downarrow(p)$ around the pseudo-Fermi points. Moreover to the two leading orders only the f-functions connecting pseudomomenta at the pseudo-Fermi points give contributions to the energy corrections. The energy and momentum associated with the distributions (26) and (27) are given by [4]

$$\begin{aligned} EN_a = E_0 N_a &+ \Big(2\pi v_c(2k_F) + \sum_{j=\pm1}(j) f_{cc}(2k_F, 2k_F j)\Big) D_c^2 \\ &+ \Big(2\pi v_s(k_{F\downarrow}) + \sum_{j=\pm1}(j) f_{ss}(k_{F\downarrow}, k_{F\downarrow} j)\Big) D_s^2 \\ &+ \Big(\sum_{j=\pm1}(j) f_{cs}(2k_F, k_{F\downarrow} j)\Big) 2D_c D_s \\ &+ 2\pi v_c(2k_F)[N_c^+ + N_c^-] + 2\pi v_s(k_{F\downarrow})[N_s^+ + N_s^-] \end{aligned} \tag{29}$$

$$P = \frac{2\pi}{N_a}[N D_c + M D_s + N_c^+ - N_c^- + N_s^+ - N_s^-]\,. \tag{30}$$

The use of equations (17)–(19), (24) and (25) allows us to rewrite the energy (29) as follows:

$$E = E_0 + \frac{2\pi}{N_a}[v_c(2k_F)(\Delta_c^+ + \Delta_c^-) + v_s(k_{F\downarrow})(\Delta_s^+ + \Delta_s^-)] \tag{31}$$

where

$$\Delta_c^{\pm} = \tfrac{1}{2}(\xi_{cc} D_c + \xi_{cs} D_s)^2 + N_c^{\pm} \qquad \Delta_s^{\pm} = \tfrac{1}{2}(\xi_{sc} D_c + \xi_{ss} D_s)^2 + N_s^{\pm} \tag{32}$$

and ξ_{cc}, ξ_{cs}, ξ_{sc} and ξ_{ss} are the elements of matrix (25). As our fluctuations refer to the case when $\Delta N = \Delta M = 0$ (N and M remain unaltered), equations (30) and (31) and (32) are precisely the momentum, finite-size energy correction and conformal field dimensions $\Delta_c^{\pm}$, $\Delta_s^{\pm}$, respectively, of [8, 9]. The leading term in the asymptotic expansion of the correlation functions decays with critical exponents obtained from (32) by minimizing with respect to D_c, D_s (i.e., by minimizing with respect to $\delta_c(q)$, $\delta_\downarrow(p)$) [4, 8, 9]. When $\Delta N \neq 0$ or $\Delta M \neq 0$, extra energy boundary terms appear in the right-hand sides of equations (29) and (32) [4, 8, 9]. These terms are functions purely of ΔN, ΔM and of the renormalized pseudoparticle phase shifts.

It follows from the present results that the critical exponents of the model are fully determined by the pseudoparticle renormalized interactions [4]. In fact, these exponents are exclusively functions of renormalized scattering phase shifts associated to the pseudoparticle forward scattering processes such that both pseudo-momenta are pseudo-Fermi points. On the other hand, the Landau–Luttinger liquid formulation introduces a more general framework: it contains full information about the pseudoparticle renormalized interactions for any pair of pseudo-momenta [4].

In this letter we have introduced the concept of a Landau–Luttinger liquid. In addition to clarifying the physics by reducing the study of the low-lying excitations to the familiar language of band theory, the formulation used here allows explicit calculation of the f-functions which, as in Fermi liquid theory, are related to the forward scattering amplitudes of the pseudoparticles. Moreover, our results show that the renormalized Landau–Luttinger theory has only forward scattering. Although the formal similarities with the Fermi liquid theory are striking, we would like to stress the crucial differences with the latter. It is important to realize that in the former the pseudoparticles involved in the description of the low-energy properties refer to exact eigenstates of the many-body system. This is in contrast to Fermi liquid theory where the quasiparticles describe approximate eigenstates of finite lifetime near the Fermi surface. Finally, there is a second important difference which we believe to be common to all non-trivial higher dimensional fermionic liquids for which the overlap integral of [2] vanishes ('infrared catastrophe'): there is no one-to-one correspondence between the pseudoparticles of such non-trivial liquids (including the present Landau–Luttinger liquids) and the real particles (electrons) upon turning off adiabatically the bare interaction. This is obviously due to the fact that in these non-trivial liquids the usual Fermi liquid fixed point is excluded.

The present results may offer insight into the physics of higher dimensional systems [2] where, in contrast to the 1D case, the renormalized interaction of the new Landau liquid pseudoparticles could eventually produce bound states, providing a mechanism for high-T_c superconductivity. Moreover, we believe they are relevant to quasi-one-dimensional materials. For example coupling the charge and spin pseudoparticles to $4k_F$ and $2k_F$ phonon modes [4], respectively, results in $4k_F$ (charge) and $2k_F$ (spin) instabilities for large and intermediate on-site U, in agreement with experiment [13].

This work has been initiated while we were attending the high-T_c Program at the ISI, Torino. JC was supported by the Alexander von Humboldt-Stiftung, Un Évora and CFMC-INIC (Lisbon). We thank P W Anderson, N Andrei, D Baeriswyl, P A Bares, P Fulde, P Horsch, M Ogata and T M Rice for stimulating discussions.

References

[1] Haldane F D M 1981 *J. Phys. C: Solid State Phys.* **14** 2585
[2] Anderson P W 1990 *Phys. Rev. Lett.* **64** 1839
Anderson P W and Ren Y 1990 *Preprint* Princeton University
[3] Lieb E H and Wu F Y 1968 *Phys. Rev. Lett.* **20** 1445
[4] Carmelo J, Ovchinnikov A A, Horsch P, Bares P A 1990 in preparation
[5] Carmelo J, Baeriswyl D 1988 *Int. J. Mod. Phys.* **1** 1013
1988 *Phys. Rev.* B **37** 7541
[6] Faddeev L D, Takhtajan L A 1981 *Phys. Lett.* **85A** 375
[7] Carmelo J, Horsch P, Bares P A, Ovchinnikov A A 1990 in preparation
[8] Frahm H, Korepin V E 1990 *Preprint*
[9] Woynarovich F 1989 *J. Phys. A: Math. Gen.* **22** 4243
[10] Ovchinnikov A A 1969 *Zh. Eksp. Teor. Fiz.* **57** 2137 (Engl. transl. 1970 *Sov. Phys.-JETP* **30** 1160)
Coll C F III 1974 *Phys. Rev.* B **9** 2150
Choy T C, Young W 1982 *J. Phys. C: Solid State Phys.* **15** 521
[11] Woynarovich F 1982 *J. Phys. C: Solid State Phys.* **15** 85, 97
[12] Korepin V E 1979 *Teor. Mat. Fiz.* **41** 169
[13] Pouget J P, Khanna S K, Denoyer F, Comès R 1976 *Phys. Rev. Lett.* **37** 437

J. Phys. A: Math. Gen. **20** (1987) L443–L449. Printed in the UK

LETTER TO THE EDITOR

Finite-size corrections for the low lying states of a half-filled Hubbard chain

F Woynarovich† and H-P Eckle

Fachbereich Physik, WE 5, Freie Universität Berlin, Arnimallee 14, D-1000 Berlin 33, West Germany

Received 2 February 1987

Abstract. The finite-size corrections to the ground state and the energy of the low magnetisation ($S \ll N$) states as a function of the size N are calculated analytically for the one-dimensional half-filled Hubbard model with on-site repulsion ($U > 0$). It is found that the contribution of the charge degrees of freedom is negligible, while the contribution of the spin degrees is the same as that in the one-dimensional isotropic Heisenberg model. The analytical results are compared to numerical ones obtained for the chain lengths up to $N = 512$.

As is well known, several strictly one-dimensional quantum systems are in critical phases at zero temperature. These systems—similarly to those higher-dimensional ones which exhibit real phase transitions at finite temperatures—are believed to form universality classes. Within these classes the microscopic details do not play an important role, and the critical exponents are common. Due to recent developments in studying conformal invariance (Cardy 1984, 1986a, b, Blöte *et al* 1986, Affleck 1986)—a symmetry widely accepted to be present in critical systems—it is known that the dependence of the ground-state energy and low lying part of the spectrum on the size of these systems is also universal:

$$E_0 = AL - \pi c/6L \qquad E_n - E_0 = 2\pi x_n/L \tag{1}$$

where the E_n are the energy eigenvalues, x_n are the scaling dimensions of the scaling operators and L is the size of the system. The conformal anomaly number c classifies the system. Systems for which c and x_n coincide are expected to show identical critical behaviour.

In the present letter we report on analytical and numerical studies on the one-dimensional half-filled Hubbard model with on-site repulsion, which is known to be critical. We have calculated analytically the finite-size corrections to the ground-state energy and the size dependence of the mass gap. We have found that both quantities follow the rule (1) with $c = 1$ and $x_S = S^2/2$, just as the one-dimensional isotropic Heisenberg chain does (Avdeev and Dörfel 1986, Hamer 1985, 1986, Woynarovich and Eckle 1987). It is also found that the next corrections are also the same in the two models. The one-dimensional Hubbard model exhibits two kinds of excitations, one connected with the charge, the other with the spin degrees of freedom (Woynarovich 1982a, b, 1983). The latter ones are gapless, and in the infinite repulsion limit they

† Permanent address: Central Research Institute for Physics, POB 49, H-1525 Budapest 114, Hungary.

0305-4470/87/070443+07$02.50

coincide with the excitations of the isotropic antiferromagnetic Heisenberg chain. Thus the critical behaviour should coincide in that limit. Since $c=1$ would allow for a coupling dependence of the critical behaviour (as in the anisotropic Heisenberg model) it is remarkable that the spin part of the Hubbard model shows the same critical behaviour as the isotropic Heisenberg model for all non-zero values of the on-site repulsion.

The one-dimensional Hubbard model described by the Hamiltonian

$$H=-\sum_{i=1}^{N}\sum_{\sigma}(c_{i+1\sigma}^{+}c_{i\sigma}+\mathrm{HC})+U\sum_{i=1}^{N}c_{i\uparrow}^{+}c_{i\uparrow}c_{i\downarrow}^{+}c_{i\downarrow} \qquad N+1\equiv 1 \tag{2}$$

(where $c_{i\sigma}$ are electron creation and destruction operators) can be diagonalised by solving the set of equations (Lieb and Wu 1968)

$$Nk_j=2\pi I_j-\sum_{\beta=1}^{M}2\tan^{-1}\frac{\sin k_j-\lambda_\beta}{U/4} \tag{3}$$

$$\sum_{j=1}^{N_e}2\tan^{-1}\frac{\lambda_\alpha-\sin k_j}{U/4}=2\pi J_\alpha+\sum_{\beta=1}^{M}2\tan^{-1}\frac{\lambda_\alpha-\lambda_\beta}{U/2}. \tag{4}$$

Here k_j are the momenta of the electrons and λ_α are connected with the spin distribution. I_j and J_α are the actual quantum numbers. The magnetisation and the energy per site of the N_e electrons described by a solution of (3) and (4) are given by

$$S=\tfrac{1}{2}N_e-M \tag{5}$$

$$\varepsilon=-\frac{1}{N}\sum_{j=1}^{N_e}2\cos k_j. \tag{6}$$

In order to obtain the lowest energy state of the half-filled band ($N_e=N(=\text{even})$) with a given magnetisation one has to choose the I_j and J_α sets as

$$I_{j+1}=I_j+1 \qquad I_1=-N/2+\begin{cases}0 & (N/2-S=\text{even})\\ \tfrac{1}{2} & (N/2-S=\text{odd})\end{cases} \qquad j=1,2,\ldots,N \tag{7}$$

$$J_{\alpha+1}=J_\alpha+1 \qquad J_1=-[N/2-(S+1)]/2 \qquad \alpha=1,2,\ldots,N/2-S. \tag{8}$$

The ground and first excited states are characterised by (7) and (8) with $S=0$ and $S=1$, respectively.

In calculating the finite-size effects we closely follow the method given by de Vega and Woynarovich (1985) and further developed by Woynarovich and Eckle (1987). We introduce the functions

$$w(k)=\frac{1}{2\pi}\left(k+\frac{1}{N}\sum_{\beta}2\tan^{-1}\frac{\sin k-\lambda_\beta}{U/4}\right) \qquad \frac{\mathrm{d}w(k)}{\mathrm{d}k}=\rho_N(k) \tag{9}$$

$$z(\lambda)=\frac{1}{2\pi}\left(\frac{1}{N}\sum_{j}2\tan^{-1}\frac{\lambda-\sin k_j}{U/4}-\frac{1}{N}\sum_{\beta}2\tan^{-1}\frac{\lambda-\lambda_\beta}{U/2}\right) \qquad \frac{\mathrm{d}z(\lambda)}{\mathrm{d}\lambda}=\sigma_N(\lambda). \tag{10}$$

With these definitions (3) and (4) take the form

$$w(k_j)=I_j/N \tag{11}$$

$$z(\lambda_\alpha)=J_\alpha/N \tag{12}$$

and a straightforward manipulation leads to the energy per site

$$\varepsilon=\varepsilon_\infty^{(0)}-\int_{-\pi}^{\pi}\varepsilon_c(k)R(k)-\int_{-\infty}^{\infty}\varepsilon_s(\lambda)S(\lambda). \tag{13}$$

Here $\varepsilon_\infty^{(0)}$ is the ground-state energy per site for an infinite system

$$\varepsilon_\infty^{(0)}=-4\int_0^{\infty}J_0(\omega)J_1(\omega)\frac{\exp(-\omega U/2)}{1+\exp(-\omega U/2)}\frac{d\omega}{\omega} \tag{14}$$

and $\varepsilon_c(k)$ and $\varepsilon_s(\lambda)$ are the same as the excitation energies connected with the charge and spin excitations (holes in the k and λ distributions, respectively) (Woynarovich 1983):

$$\varepsilon_c(k)=2\cos k+4\int_0^{\infty}J_1(\omega)\frac{\exp(-\omega U/2)}{1+\exp(-\omega U/2)}\cos(\omega\sin k)\frac{d\omega}{\omega} \tag{15}$$

$$\varepsilon_s(\lambda)=2\int_0^{\infty}\frac{J_1(\omega)\cos\omega\lambda}{\cosh(\omega U/4)}\frac{d\omega}{\omega}. \tag{16}$$

The $R(k)$ and $S(\lambda)$ are shorthand notations for

$$R(k)=\frac{1}{N}\sum_j\delta(k-k_j)-\rho_N(k)\qquad S(\lambda)=\frac{1}{N}\sum_\beta\delta(\lambda-\lambda_\beta)-\sigma_N(\lambda) \tag{17}$$

while $J_0(\omega)$ and $J_1(\omega)$ are Bessel functions.

We note that, similarly to the excitation energies, the finite-size corrections to the ground state also split up into two contributions, one coming from the charge, the other from the spin degrees of freedom. Now we show that for a state characterised by an I_j set given by (7) the charge contribution is negligible in the sense that as $N\to\infty$ it disappears faster than any power of $1/N$. For this we use the formula

$$\frac{1}{N}\sum_{I=I_4}^{I_N}f\left(\frac{I}{N}\right)-\int_{(I_1-1/2)/N}^{(I_N+1/2)/N}f(w)\,dw$$
$$=\sum_{m=1}^{\mu}\frac{A_m}{N^{2m}}\left(f^{(2m-1)}\left(\frac{I_N+1/2}{N}\right)-f^{(2m+1)}\left(\frac{I_1-1/2}{N}\right)\right)+\frac{A_\mu(f,N)}{N^{2\mu+1}}. \tag{18}$$

Here A_m are f and N independent constants and $A_\mu(f,N)$ depends on both f and N, but if the $(2\mu+1)$th derivative of f is finite then it has an N-independent upper bound. If f is a smooth periodic function with a period 1, all the terms on the RHS are zero except the last one. Since μ can be any large value, the LHS disappears faster than any power of $1/N$ as $N\to\infty$.

Changing the variables in the contribution of the charge part from k to w (k_j to I_j/N) transforms this contribution into the form of the LHS of (18), with $f=\varepsilon_c(k(w))$, which is, for any $u>0$, a smooth periodic function with a period of 1. Thus according to the above paragraph this contribution is negligible. This is not a surprising result: the charge excitation spectrum possesses a gap, so its contribution is expected to be exponentially small (de Vega and Woynarovich 1985).

According to the previous paragraph, the contribution of $R(k)$ can be neglected. Moreover, also in (10), we may replace $(1/N)\sum_j$ by $\int dk\,\rho_N(k)$ without introducing a significant error. The remaining equation

$$\sigma_N(\lambda)=\frac{1}{2\pi}\left(\frac{1}{2\pi}\int_{-\pi}^{\pi}2\frac{U/4}{(U/4)^2+(\lambda-\sin k)^2}dk\right.$$
$$\left.-\int_{-\infty}^{\infty}2\frac{U/2}{(U/2)^2+(\lambda-\lambda')^2}(\sigma_N(\lambda)+S(\lambda))\,d\lambda\right) \tag{19}$$

is extremely similar to the analogous equation of the isotropic Heisenberg model. Actually it can be treated in the same way (Woynarovich and Eckle 1987). Using the formula

$$\frac{1}{2N}\left(g\left(\frac{J_1}{N}\right)+2\sum_{\alpha=2}^{N/2-S-1} g\left(\frac{J_\alpha}{N}\right)+g\left(\frac{J_{N/2-S}}{N}\right)\right)$$
$$=\frac{1}{12N^2}\left(g'\left(\frac{J_{N/2-S}}{N}\right)-g'\left(\frac{J_1}{N}\right)\right)+\mathrm{O}\left(\frac{\max g'''}{N^3}\right) \qquad (20)$$

to calculate the contributions of $S(\lambda)$, denoting the largest λ_α ($\lambda_{N/2-S}$) by Λ, introducing the functions

$$\sigma_N^+(\lambda)=\begin{cases}\sigma_N(\lambda+\Lambda) & \text{if } \lambda>0\\ 0 & \text{if } \lambda<0\end{cases}$$
$$\sigma_N^-(\lambda)=\begin{cases}0 & \text{if } \lambda>0\\ \sigma_N(\lambda+\Lambda) & \text{if } \lambda<0\end{cases} \qquad (21)$$

and using Fourier transforms, (19), (12) and (13) can be transformed into the set of equations

$$\tilde\sigma_N^-(\omega)+\frac{\tilde\sigma_N^+(\omega)}{1+\exp(-|\omega|U/2)}$$
$$=\frac{1}{2\pi}\frac{1}{2\cosh(\omega U/4)}\exp(\mathrm{i}\omega\Lambda)J_0(\omega)-\frac{1}{2\pi}\left[\left(\frac{1}{2N}-\frac{\mathrm{i}\omega}{12N^2\sigma(\Lambda)}\right)\right.$$
$$\left.+\left(\frac{1}{2N}+\frac{\mathrm{i}\omega}{12N^2\sigma(\Lambda)}-2\pi\tilde\sigma_N^+(-\omega)\right)\exp(\mathrm{i}\omega2\Lambda)\right]\frac{\exp(-|\omega|U/2)}{1+\exp(-|\omega|U/2)} \qquad (22)$$

$$2\pi\tilde\sigma_N^+(0)=(S+\tfrac{1}{2})/N \qquad (23)$$

$$\sigma_N(\Lambda)=\int_{-\infty}^{\infty}(\tilde\sigma_N^+(\omega)+\tilde\sigma_N^-(\omega))\,\mathrm{d}\omega \qquad (24)$$

$$\varepsilon_N^{(S)}-\varepsilon_\infty^{(0)}=16\pi\exp(-2\pi\Lambda/U)I_1\left(\frac{2\pi}{U}\right)\left[\tilde\sigma+\left(-\frac{\mathrm{i}2\pi}{U}\right)-\frac{1}{2\pi}\frac{1}{2N}+\frac{1}{12N^2\sigma_N(\Lambda)U}\right]$$
$$+\mathrm{O}\left[I_1\left(\frac{6\pi}{U}\right)\exp(-6\pi\Lambda/U)\tilde\sigma_N^+\left(-\frac{\mathrm{i}6\pi}{U}\right)\right] \qquad (25)$$

with $J_0(\omega)$ and $I_1(x)$ being Bessel functions. Solving (22) by the method given by Yang and Yang (1966) (23)-(25) can be calculated giving

$$\varepsilon_N^{(0)}-\varepsilon_\infty^{(0)}=-\frac{\pi}{3}\frac{I_1(2\pi/U)}{I_0(2\pi/U)}\frac{1}{N^2}\left[1+0.3433\frac{1}{\ln[NI_0(2\pi/U)]^3}+\mathrm{O}\left(\frac{\ln(\ln N)}{(\ln N)^2}\right)\right] \qquad (26)$$

$$\varepsilon_N^{(S)}-\varepsilon_\infty^{(0)}=2\pi\frac{I_1(2\pi/U)}{I_0(2\pi/U)}\frac{S^2}{N^2}\left[1-\frac{1}{2}\frac{1}{\ln[NI_0(2\pi/U)]}+\mathrm{O}\left(\frac{\ln(\ln N)}{(\ln N)^2}\right)\right]. \qquad (27)$$

Since the dispersion of the spin excitations is given by (Woynarovich 1983)

$$p_s(\lambda)=\frac{\pi}{2}-\int_0^\infty\frac{J_0(\omega)\sin(\omega\lambda)}{\omega\cosh(\omega U/4)}\mathrm{d}\omega \qquad (\sim\exp(-2\pi\lambda/U)I_0(2\pi/U)\text{ for }\lambda\gg1)$$
$$\varepsilon_s(\lambda)=2\int_0^\infty\frac{J_1(\omega)\cos(\omega\lambda)}{\omega\cosh(\omega U/4)}\frac{\mathrm{d}\omega}{\omega} \qquad (\sim4\exp(-2\pi\lambda/U)I_1(2\pi/U)\text{ for }\lambda\gg1) \qquad (28)$$

after properly normalising the Hamiltonian (von Gehlen *et al* 1986) (i.e. to have instead of (28) the simple relation $\varepsilon_s(\lambda)=p_s(\lambda)$ for small momenta) and returning from the energy per site to the energy, (26) and (27) yield (1) with the same c and x_S as the isotropic Heisenberg model, i.e. with $c=1$ and $x_S=S^2/2$. Moreover, the powers and the coefficients of the first logarithmic corrections are also the same in the two models:

$$E_N^{(0)}=N\varepsilon_\infty^{(0)}-\frac{\pi}{6N}\left(1+0.3433\frac{1}{\{\ln[NI_0(2\pi/U)]\}^3}+\ldots\right) \tag{29}$$

$$E_N^{(S)}-E_N^{(0)}=2\pi\frac{S^2}{2}\left(1-\frac{1}{2}\frac{1}{\ln[NI_0(2\pi/U)]}+\ldots\right). \tag{30}$$

Based on the above results one may expect the half-filled Hubbard model to exhibit the same critical behaviour as the isotropic Heisenberg model. We have to stress, however, that this analogy holds only for $U>0$ for two reasons. One is that in the $U\to 0$ limit the corrections due to the charge degrees of freedom grow up: for small U the derivatives of $\varepsilon_c(k(w))$ grow up, and in the $U=0$ limit $\varepsilon_c(k(w))$ is not a smooth (although it is still a periodic) function, and the argument presented just after (18) does not hold. Actually, in this limit the contribution of the charge degrees of freedom is just as large as that of the spins. The other reason is that the spin part of the corrections itself cannot be continued down to $U=0$ due to the essential singularity of the model at this point. In our calculation $\Lambda\gg 1$ has been supposed, but for $U=0$ for all λ_α, and so for Λ too, $|\lambda|<1$. Actually, at the point $U=0$ one can solve the equations which describe the spin contribution only (equations (31) and (32)) exactly, and one finds that the result for the correction to the ground state does not coincide with (26).

We also carried out numerical calculations for the finite-size corrections. We have solved numerically by iteration for $N=8, 16, 32, \ldots, 512$ and several U values that form of (4) in which the Σ_i is replaced by the $N\int\rho_N(k)$, i.e.

$$\frac{1}{2\pi}\int_{-\pi}^{\pi}2\tan^{-1}\frac{\lambda_\alpha-\sin k}{U/4}\,\mathrm{d}k=2\pi J_\alpha+\sum_{\beta=1}^{M}2\tan^{-1}\frac{\lambda_\alpha-\lambda_\beta}{U/2}. \tag{31}$$

For the energy per site we used the expression

$$\varepsilon=-\int_{-\pi}^{\pi}2\cos k\rho_N(k)\,\mathrm{d}k=-\frac{4}{N}\sum_\alpha\{[(U/4)^2+\cos^2(x_\alpha/2)]^{1/2}-U/4\}$$
$$x_\alpha=2\sin^{-1}\left(\frac{[(U/4)^2+(\lambda_\alpha+1)^2]^{1/2}-[(U/4)^2+(\lambda-1)^2]^{1/2}}{2}\right) \tag{32}$$

and for $\varepsilon_\infty^{(0)}$ we used (14).

Note that (31) and (32) do not contain the finite-size corrections due to the charge degrees of freedom. Our findings are plotted in figures 1 and 2. Both the correction to the ground state and the mass gap are normalised to their values for $N\to\infty$. The individual curves are labelled by the value of U and ∞ indicates the Heisenberg limit. On the both sets of curves one can observe the tendency that, for fixed but large enough N with decreasing U, the points approach the $N\to\infty$ values. This is due to the fact that in the argument of the logarithm N enters together with $I_0(2\pi/U)$ which increases with decreasing U. It is striking, however, that for small U and not large enough N the corrections to the ground-state energy are very far away from their $N\to\infty$ values. This is due to the power law type corrections (indicated in (25) but not in (26)). The

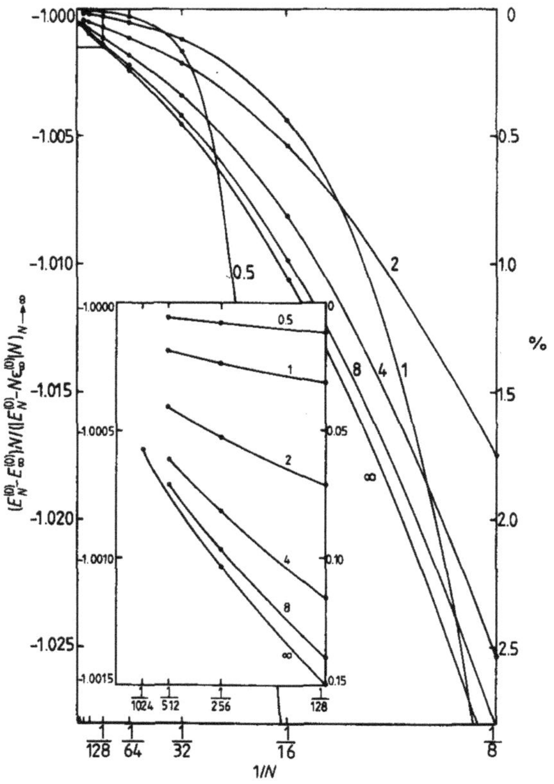

Figure 1. Corrections to the ground-state energy normalised to -1 in the $N\to\infty$ limit plotted against $1/N$. The individual curves are labelled by the value of U.

terms yielding power type corrections are in general of the form

$$\left[I_1\left(\frac{(2n+1)2\pi}{U}\right)\exp\{-[(2n+1)2\pi\Lambda]/U\}\right]$$
$$\times\left[I_0\left(\frac{(2m+1)2\pi}{U}\right)\exp\{-[(2m+1)2\pi\Lambda]/U\}\right]. \tag{33}$$

Since

$$\exp(-2\pi\Lambda/U)/I_0(2\pi/U)\sim N \tag{34}$$

for small U the terms of (33) take the form

$$\left[I_1\left(\frac{(2n+1)2\pi}{U}\right)I_0\left(\frac{(2m+1)2\pi}{U}\right)\{[I_0(2\pi/U)]^{2(m+n)+2}\}^{-1}\right]$$
$$\times\frac{1}{N^{2(m+n)+2}}\sim\frac{1}{N^2}\frac{1}{(\sqrt{U}N)^{2(m+n)}} \tag{35}$$

i.e. their significance is enhanced when U is small. Nevertheless they decay faster than the logarithmic terms. The form (35) is an indication that our calculation breaks down in the $U\to 0$ limit, as discussed earlier.

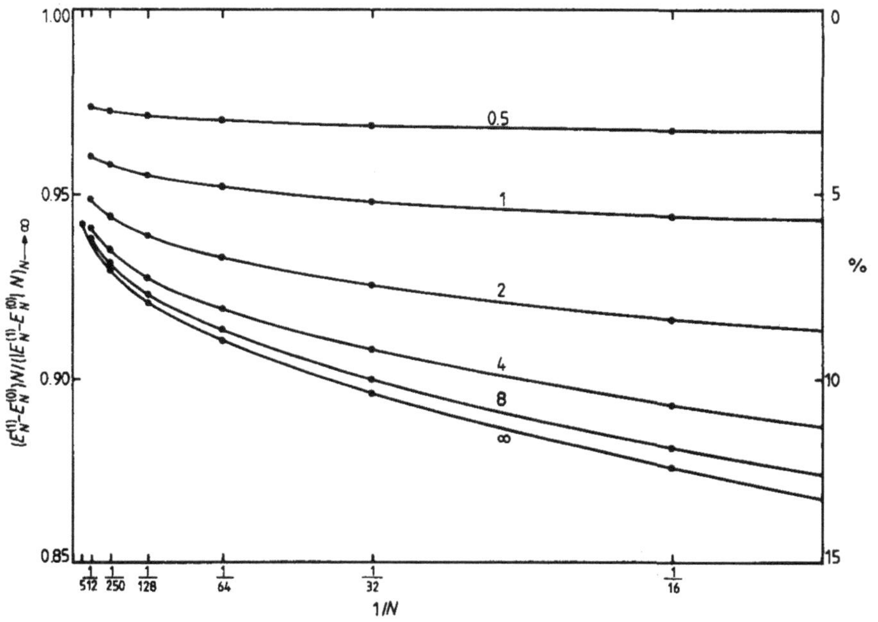

Figure 2. The gap between the first excited and ground states normalised to 1 in the $N \to \infty$ limit plotted against $1/N$. The individual curves are labelled by the value of U.

We are very grateful to Professor I Peschel for many inspiring discussions. We also would like to acknowledge the advice and interest of Professor K-D Schotte and Dr T Truong. This research has been supported by DFG under contract Scho 158/7-2.

References

Affleck I 1986 *Phys. Rev. Lett.* **56** 746
Avdeev L V and Dörfel B-D 1986 *J. Phys. A: Math. Gen.* **19** L13
Blöte H W J, Cardy J L and Nightingale M P 1986 *Phys. Rev. Lett.* **56** 742
Cardy J L 1984 *J. Phys. A: Math. Gen.* **17** L385
—— 1986a *Nucl. Phys.* B **270** 1986
—— 1986b *Preprint University of California at Santa Barbara* 79-1986
de Vega H J and Woynarovich F 1985 *Nucl. Phys.* B **251** [FS13] 439
Hamer C J 1985 *J. Phys. A: Math. Gen.* **18** L1133
—— 1986 *J. Phys. A: Math. Gen.* **19** 3335
Lieb E H and Wu F Y 1968 *Phys. Rev. Lett.* **20** 1445
von Gehlen G, Rittenberg V and Ruegg H 1986 *J. Phys. A: Math. Gen.* **19** 107
Woynarovich F 1982a *J. Phys. B: At. Mol. Phys.* **15** 85
—— 1982b *J. Phys. B: At. Mol. Phys.* **15** 97
—— 1983 *J. Phys. B: At. Mol. Phys.* **16** 5293
Woynarovich F and Eckle H-P 1987 *J. Phys. A: Math. Gen.* **20** L97
Yang C N and Yang C P 1966 *Phys. Rev.* **150** 327

International Journal of Modern Physics B Vol. 3, No. 3 (1989) 427–439

THE ROLE OF QUASI-ONE-DIMENSIONAL STRUCTURES IN HIGH-T_c SUPERCONDUCTIVITY

N. M. BOGOLIUBOV and V. E. KOREPIN
Leningrad Branch of Steklov Mathematical Institute, Fontanka 27, Leningrad D-11, USSR 191011

Received 23 September 1988

The critical exponents describing the decrease of correlation functions on long distances for the one-dimensional Hubbard model is obtained. The behaviour of correlators shows that Cooper pairs of electrons are formed. The electron tunneling between the chains leads to the existence of the anomalous mean values and to the superconductive current. The anisotropy of the quasi-one-dimensional system leads to the rise of critical temperature T_c.

1. Introduction

The discovery of high temperature superconductivity[1] has excited research activity. Various theoretical proposals have been made to interpret the origin of this phenomenon.[2] We shall consider the one-dimensional attractive Hubbard model. In Ref. 3 the arguments were given in favour of this model. The superconductors of $YBa_2Cu_3O_7$ type (with transition temperature $T_c \simeq 90$ K) may be considered as quasi-one-dimensional structures consisting of Cu-O chains. Taking into account the virtual transitions among the electronic shells of oxygen O^{2-} and O^0 the one-dimensional Hubbard model with attraction was obtained in Ref. 3.

The Hamiltonian of the model is

$$H_0 = -\sum_{j=1}^{N}\sum_{s}\{\Psi^+_{j+1,s}\Psi_{j,s} + \Psi^+_{j-1,s}\Psi_{j,s}\} - 4U\sum_{j=1}^{N}\Psi^+_{j,\uparrow}\Psi_{j,\uparrow}\Psi^+_{j,\downarrow}\Psi_{j,\downarrow}\,. \tag{1}$$

Here N is the number of sites on the one-dimensional lattice. Canonical Fermi field $\Psi_{j,s}$ describes the electron with spin 1/2 ($s = \uparrow,\downarrow$) on the site j ($j = 1, \ldots, N$), $\{\Psi^+_{j,s}\Psi_{i,t}\} = \delta^i_j\delta^s_t$. The value $U > 0$ is the coupling constant.

The Hamiltonian commutes with the operators of a number of up and down spins

$$M_s = \sum_{j=1}^{N}\Psi^+_{j,s}\Psi_{j,s}\,.$$

The operator of number of electrons is equal to

$$M = M_\uparrow + M_\downarrow \tag{2}$$

and

$$2S = M_\uparrow - M_\downarrow \tag{3}$$

is the total spin operator.

Considering the system in arbitrary external fields it is convenient to pass to the Hamiltonian:

$$H = H_0 - AM - hS \tag{4}$$

where $A < 0$ is the chemical potential and h is the magnetic field. In the thermodynamical limit $N, M \to \infty$ the electronic density $D = N/M$ is finite. Density is connected with doping in the following way, $\delta = 1 - D$. Limiting cases $\delta = 0$ and $\delta = 1$ correspond to half-filled band and empty band respectively.

We start with the detailed investigation of the one dimensional model at the zero temperature, especially concentrating on the asymptotics of correlators, then we shall study the model at nonzero temperature. For the quasi-one-dimensional model the temperature of the superconductive transition will be obtained.

For the one-dimensional quantum models the anomalous amplitude $\langle \Psi_{n\uparrow} \Psi_{j\downarrow} \rangle$ is equal to zero, it means the absence of superconductivity in the literal sense. As pointed out by different authors[4,5] there is the analogy to the long range order in one dimension — the power decay of the correlator of pairs of fields in the singlet state (SCP) regarded as the existence of the superconducting state

$$\langle \Psi^+_{n+j,\uparrow} \Psi^+_{n,\downarrow} \Psi_{i,\uparrow} \Psi_{1,\downarrow} \rangle \underset{n\to\infty}{\to} \frac{1}{n^\gamma}; \qquad \gamma > 0 \tag{5}$$

(i,j fixed).

Let us consider also the correlator of Fermi fields (FF) $\langle \Psi^+_{n,s} \Psi_{1,s} \rangle$, if it decays exponentially:

$$\langle \Psi^+_{n,s} \Psi_{1,s} \rangle \underset{n\to\infty}{\to} e^{-n/\xi}; \qquad \xi > 0\,, \tag{6}$$

then we can say that the electrons are confined in the Cooper pairs. These asymptotics mean that the average distance between the pairs are large, but the average distance between electrons in the pair ξ is small. If both correlators decay as a power of the distance — the average distance between pairs of electrons is the same as between single electrons. So the pairs have no remarkable properties and the electrons are more or less independent. The important properties of the critical behaviour are described by the charge-density correlation function (CDW)

$$\langle \Psi_{n,s}^{+} \Psi_{n,s} \Psi_{1,t}^{+} \Psi_{1,t} \rangle .$$

In this paper we shall show, that for the attractive Hubbard model ($U > 0$) in the range of the magnetic field $h < h_c$ (the value of the critical magnetic field h_c is connected with the gap in the model spectrum) the asymptotics of the FF and SCP correlation functions is of the required form (5) and (6), while the asymptotics of the CDW correlator is:

$$\langle\langle \Psi_{n,s}^{+} \Psi_{n,s} \Psi_{1,t}^{+} \Psi_{1,t} \rangle\rangle = \langle \Psi_{n,s}^{+} \Psi_{n,s} \Psi_{1,t}^{+} \Psi_{1,t} \rangle$$
$$- \langle \Psi_{1,t}^{+} \Psi_{1,t} \rangle \underset{n \to \infty}{\to} \cos(\pi D n) \frac{1}{n^{1/\gamma}}; \qquad \gamma > 0 . \tag{7}$$

The critical exponent ξ is positive and inversely proportional to the gap in the electron excitation spectrum: $\xi = v\Delta^{-1}$. The critical exponent γ is the function of the Fermi velocity v and compressibility:

$$\gamma^{-1} = \frac{\pi}{2} v \frac{\partial D}{\partial A} . \tag{8}$$

The complete integrability of the model gives the opportunity to express this critical exponent in terms of the solution of the "dressing" integral equation and to prove that γ changes monotonically in the range $1/2 \leq \gamma < 1$. It is equal to unity ($\gamma = 1$) only for the half-filled band $D = 1$ ($M_\uparrow = M_\downarrow$). Comparing asymptotics (5) and (7) we see that for the nonzero doping δ the CDW correlator decays faster than the SCP correlator. If the magnetic field $h > h_c = \Delta$ then all the correlations mentioned above will decay as a power of distance. The investigation of this case as well as the case of the Hubbard model with repulsion ($U < 0$) will be published later.

At the finite temperatures all the correlators decay exponentially in one-dimension. For the quasi-one-dimensional systems the electron tunneling from one chain into another leads to the existence of the ordered state at finite temperatures, and this state will be superconductive, as is clear from the above.

In the following, Sec. 2 is concerned with the nature of the ground state and the energy of the excited states, Sec. 3 with calculation of the correlation function asymptotics both for $T = 0$ and for nonzero temperature in one-dimension. In Sec 4 we consider the quasi-one-dimensional model and give the expression for the critical temperature of the superconductive transition.

These results were announced in Ref. 6.

2. Ground State

The Hubbard model was exactly solved by means of the Bethe Ansatz[7] and was imbedded into the quantum inverse scattering method[8] in Ref. 9. The case of the attractive model was considered in Refs. 10 and 11 and the thermodynamics was constructed in Ref. 12. The excitation spectrum was treated in Refs. 13 and 14.

The ground state of the model with attraction at the zero magnetic field is constructed by filling the Fermi sphere with two-particle bound singlets. The distribution functions of pairs are defined by the equation:

$$\sigma(\lambda) + \int_{-\Lambda}^{\Lambda} \mathcal{K}(\lambda,\mu)\sigma(\mu)d\mu = \frac{1}{2\pi}\bar{P}'(\lambda). \tag{9}$$

Here $P'(\lambda)$ is the derivative of the bare momentum of the pair

$$p(\lambda) = 2\mathrm{Re}\arcsin(\lambda - iU);$$
$$p'(\lambda) = 2\mathrm{Re}[1 - (\lambda - iU)]^{-1/2}. \tag{10}$$

The integral kernel is equal to

$$\mathcal{K}(\lambda,\mu) = \frac{1}{2\pi}\frac{\partial}{\partial\lambda}\theta(\lambda,\mu) = \frac{2U}{\pi}\frac{1}{4U^2 + (\lambda-\mu)^2};$$
$$\theta(\lambda,\mu) = 2\,\mathrm{arctg}\frac{\lambda-\mu}{2U}. \tag{11}$$

The momentum of the pair is given by

$$P(\lambda) = p(\lambda) - \int_{-\Lambda}^{\Lambda}\theta(\lambda,\mu)\sigma(\mu)d\mu. \tag{12}$$

The parameter Λ is determined by the condition:

$$D = 2\int_{-\Lambda}^{\Lambda}\sigma(\lambda)d\lambda. \tag{13}$$

where D is the density of the electron liquid ($\pi D = 2P_F$, $P(\Lambda) = P_F$ is the Fermi momentum).

The energy of the elementary excitations over the ground state is defined by the integral equations

$$\epsilon(\lambda) + \int_{-\Lambda}^{\Lambda} \mathcal{K}(\lambda,\mu)\epsilon(\mu)d\mu = E(\lambda) - 2A\,, \tag{14}$$

$$e(k) = -2\cos k - A - 2\int_{-\Lambda}^{\Lambda} \mathcal{K}(2\sin k, 2\mu)\epsilon(\mu)d\mu\,,$$

$$-\infty \le \lambda \le \infty; \qquad -\pi \le k \le \pi \tag{15}$$

demanding that

$$\epsilon(\pm\Lambda) = 0\,. \tag{16}$$

Here $\epsilon(\lambda)$ is the excitation energy of pairs, and $e(k)$ of unpaired electrons, respectively. The bare energy of the pair is:

$$E(\lambda) = -4\mathrm{Re}[1 - (\lambda - iU)^2]^{1/2}\,. \tag{17}$$

The ground state energy is

$$\mathcal{F}(D) = \int_{-\Lambda}^{\Lambda} E(\lambda)\sigma(\lambda)d\lambda\,,$$

$$\frac{\partial\mathcal{F}(D)}{\partial D} = A\,. \tag{18}$$

It follows from (14) and (16) that the spectrum of the pair excitation is gapless with the linear dispersion law in the vicinity of Fermi level:

$$\epsilon(\lambda) \simeq v(P(\Lambda) - P_F)\,. \tag{19}$$

Here v is the Fermi velocity

$$v = \frac{\epsilon'(\Lambda)}{P(\Lambda)} = \frac{\epsilon'(\Lambda)}{2\pi\sigma(\Lambda)}\,. \tag{20}$$

The gap in the spectrum of unpaired electrons is equal to

$$\Delta = e(0) = -2 - A - \frac{U}{\pi}\int_{-\Lambda}^{\Lambda} \frac{\epsilon(\mu)}{U^2 + \mu^2}d\mu\,. \tag{21}$$

The problem of the critical exponent γ calculation is reduced to the solution of the "dressing" linear integral equation[16,17]:

$$Z(\lambda) + \int_{-\Lambda}^{\Lambda} \mathcal{K}(\lambda,\mu) Z(\mu) d\mu = 1 . \tag{22}$$

The critical exponent γ is defined by the equality

$$\gamma = 2Z^2(\Lambda) . \tag{23}$$

The solutions of Eqs. (22) and (14) are connected:

$$2Z(\lambda) = -\partial\epsilon(\lambda)/\partial A . \tag{24}$$

One can prove that the gap Δ is positive for arbitrary Λ, D and $U > 0$. The solution of Eq. (22) is a monotonically decreasing function of Λ. Moreover

$$\frac{1}{2} \leq Z(\lambda) \leq 1 ; \qquad \frac{\partial Z(\lambda)}{\partial \Lambda} < 0; \qquad \frac{\partial Z(\lambda)}{\partial D} < 0; \qquad \frac{\partial Z(\lambda)}{\partial U} > 0 . \tag{25}$$

We state that the critical behaviour of the attractive Hubbard model at the zero magnetic field is defined by Eqs. (14), (22) and by the equality (21). Considering some limiting cases we can solve these equations. The simplest one is the case of the nearly half-filled ($\Lambda \to \infty$) $D = 1 - \delta$. For the small doping ($\delta \to 0$, U-fixed) one gets, applying the Winer-Hopf method

$$\gamma = 1 - \frac{1}{2\ln(\mathbb{C}/\delta)} . \tag{26}$$

where $\mathbb{C} = \sqrt{8/\pi e}\, I_0(\pi/2U)$, I_j is the Bessel function. For the half-filled band ($\delta = 0$)

$$\gamma = 1 . \tag{27}$$

To the second order of δ the gap is:

$$\Delta_\Lambda = \Delta + \pi\delta \frac{I_1(\pi/2U)}{I_0(\pi/2U)} + \frac{\pi}{2}\delta^2 \frac{I_1(\pi/2U)}{I_0^2(\pi/2U)} + O(\delta^3) . \tag{28}$$

Here Δ is the gap for the half-filled band[13]:

$$\Delta = \frac{2}{U} \int_1^\infty \frac{\sqrt{y^2 - 1}}{\sinh\left(\frac{\pi}{2U} y\right)} dy . \tag{29}$$

Treating the equations by the perturbation method in the small density limit ($D \to 0$, U fixed, $D/U \to 0$) we obtain a power series expansion for γ as follows:

$$\gamma = \frac{1}{2}\left(1 + \frac{D}{2}\sqrt{1 + \frac{1}{U^2}}\right) + O(D^2) \tag{30}$$

and for Δ_Λ

$$\Delta_\Lambda = 2\left(\sqrt{1 + U^2} - 1\right) - D^2\frac{\pi^2}{16\sqrt{1 + U^2}} + D^3\frac{\pi^2}{24U} + O(D^4)\,. \tag{31}$$

The quantity $2(\sqrt{1 + U^2} - 1)$ is the binding energy of the pair for $D = 0$.

The results of this section are valid for the magnetic fields $h < h_c$, where

$$h_c = \Delta\,. \tag{32}$$

Here Δ is the gap (29).

3. Critical Behavior

The long distance asymptotics of the correlation functions for the integrable models in the external fields was studied in Refs. 16, 17 and 18. The explicit formulae for the critical exponents were obtained.

To study correlators it is convenient to insert the complete set of eigenstates of Hamiltonian between fields in the first and n-th sites of the lattice. Asymptotics of correlators of fields ϕ are defined by change of energy of the ground state under the excitation caused by the field ϕ.

Critical behaviour of the system with a gapless excitation spectrum having a linear dispersion law in the vicinity of the Fermi level is described by conformal field theory.[19] To obtain the complete information about infrared asymptotics one has to determine the central charge and conformal dimensions of the fields. It is known, that if central charge $C \geq 1$ then the critical exponents may continuously depend on the parameters of the model. The central charge of the Hubbard Hamiltonian in the sector under consideration is $C = 1$. Now we can apply the method of Ref. 18, based on the finite-size corrections,[20] to obtain the spectrum of conformal dimensions:

$$2\Delta_\pm = \frac{m^2}{2\gamma} + \frac{q^2\gamma}{2} \pm qm + 2I_\pm\,. \tag{33}$$

Here m, q, $I_\pm$ are integer positive numbers; the critical exponent γ is proportional to the difference between the ground state energies of the model with $M + 2$ particles and with M particles respectively,

$$\frac{\pi v}{N}\gamma = N\left\{\mathscr{F}\left(D+\frac{2}{N}\right)-\frac{2A}{N}-\mathscr{F}(D)\right\}=\frac{2}{N}\frac{\partial A}{\partial D}. \tag{34}$$

Thus

$$\gamma^{-1}=\frac{\pi}{2}v\frac{\partial D}{\partial A}. \tag{35}$$

Here $\mathscr{F}$ is the energy of the ground state (18), A the chemical potential, D the density (13), $\partial D/\partial A$ is the compressibility, and the Fermi velocity is defined by (20). Finally, calculating the derivative in (35) using definition (13) and Eq. (14) we obtain[17,18] expression (23) for the critical exponent. The equality (23) and Eq. (22) gives us the opportunity to express γ in terms of microscopic quantities. It follows from the properties of $Z(\lambda)$ function (25) and equalities (27) and (30) that the exponent γ is the monotonic function of density D. Moreover

$$\frac{1}{2}\leq\gamma\leq 1, \qquad \frac{\partial\gamma}{\partial D}>0. \tag{36}$$

For the half-filled band $\gamma = 1$.[21]

The long distance asymptotics of the correlation function can be written in the form

$$\langle\phi_q(W,\bar{W})\phi_{-q}(0,0)\rangle-\langle\phi_q\rangle^2\rightarrow\sum_{m,I_+,I_-}B(m,I_+,I_-)\frac{\cos(2P_F mn)}{W^{2\Delta_+}\bar{W}^{2\Delta_-}}. \tag{37}$$

Here q is the "change" of the field $\phi_q(W,\bar{W})$ ($W = in + v\tau$, τ is the Euclidean time, v the Fermi velocity (20)). In the case under consideration the uncharged field $q = 0$ is the operator of the conserved current $\phi_0(W,\bar{W}) = \Psi^+_{n,s}(\tau)\Psi_{n,s}(\tau)$; the field with the unit change $q = 1$ is the creation operator of Cooper pair $\phi_1(W,\bar{W}) = \Psi^+_{n,\uparrow}(\tau)\Psi^+_{n,\downarrow}(\tau)$; the example of the charged field with $q > 1$ is the product

$$\phi_q(W,\bar{W})=\prod_{j=1}^{q}\Psi^+_{n+j,\uparrow}(\tau)\Psi^+_{n+j,\downarrow}(\tau);\qquad \phi_{-q}(W,\bar{W})=\phi_q^+(W,\bar{W}).$$

Since the ground state is uncharged $\langle\phi_q\rangle = 0$ for $q\geq 1$. The summation is performed over integers (33), B is the amplitude independent of distance n and Euclidian time τ and P_F is the Fermi momentum.

Using (33) we can write the leading terms of the asymptotics

$$\langle \phi_q(W,\bar{W})\phi_{-q}(0,0)\rangle \underset{|W|\to\infty}{\to} \frac{1}{|W|^{q^2\gamma}}\left\{a + b\left[\frac{1}{W^2}+\frac{1}{\bar{W}^2}\right] + c\frac{\cos(2P_F n)}{|W|^{1/\gamma}}\left[\left(\frac{W}{\bar{W}}\right)^q + \left(\frac{\bar{W}}{W}\right)^q\right]\right\}. \tag{38}$$

The SCP correlator (5) is the special case of this expression. For time dependent CDW correlator we obtain

$$\langle\langle \phi_0(W,\bar{W})\phi_0(W,\bar{W})\rangle\rangle \underset{|W|\to\infty}{\to} c\frac{\cos(2P_F n)}{|W|^{1/\gamma}} + b\left[\frac{1}{W^2}+\frac{1}{\bar{W}^2}\right]. \tag{39}$$

The asymptotics of many point time-dependent correlation function is of the following form:

$$\left\langle \prod_{n=1}^{S} \phi_{\nu_n}(W_n,\bar{W}_n)\right\rangle \to \prod_{n>k} |W_n - W_k|^{\nu_n\nu_k\gamma}, \qquad |W_n - W_k| \to \infty, \quad \nu_n = \pm 1, \quad \sum \nu_n = 0. \tag{40}$$

Here we use notations: $W_n = iX_n + v\tau_n$, $\phi_1(W_n,\bar{W}_n) = \Psi^+_{x_n,\uparrow}(\tau_n)\Psi^+_{x_n,\downarrow}(\tau_n)$ and $\phi_{-1} = \phi_1^+$.

To compute the correlator of the fields (6) we must remove one electron from the Fermi sphere, i.e. to destroy the bound pair. This is the reason for the gap Δ (21) in the spectrum of Hamiltonian in this sector which leads to the exponential decay of FF correlator

$$\langle \Psi^+_{n,s}\Psi_{1,s}\rangle \underset{n\to\infty}{\to} C_1 e^{-n/\xi}, \tag{41}$$

where $\xi = v/\Delta$. Similarly, we calculate the spin wave correlation function:

$$\langle \Psi_{n,\uparrow}\Psi^+_{n,\downarrow}\Psi^+_{1,\uparrow}\Psi_{1,\downarrow}\rangle \underset{n\to\infty}{\to} C_2 e^{-2n/\xi}. \tag{42}$$

So far we discussed only the zero temperature case. At nonzero temperatures all the correlators decay exponentially. The explicit expression for this asymptotics is necessary when the quasi-one-dimension limit is considered. We shall find the asymptotics ($n\to\infty$, $T\to 0$) of the Matsubara's correlation functions (function with the period $1/T$ along the imaginary time axis) using the conformal invariance of the gapless spectrum. The correlation functions are to be calculated in the strip geometry and are determined in terms of the infinite plane correlation functions at $T = 0$ by using the conformal mapping $U = (v/2\pi T)\ \ln W$; $W = n + iv\tau$. For the "charged" correlation function (38) we obtain

$$\langle \phi_q(W,\bar{W})\phi_{-q}(W,\bar{W})\rangle$$

$$\to \left|\frac{(\pi/v)T}{\sinh\left(\frac{\pi T}{v}W\right)}\right|^{q^2\gamma}\left\{a + b\left(\frac{\pi}{v}T\right)^2\left[\frac{1}{\left(\sinh\left(\frac{\pi T}{v}W\right)\right)^2} + \frac{1}{\left(\sinh\left(\frac{\pi T}{v}\bar{W}\right)\right)^2}\right]\right.$$

$$\left. + C\cos(2P_F n)\left|\frac{(\pi/v)T}{\sinh\left(\frac{\pi T}{v}W\right)}\right|^{1/\gamma}\left[\left(\frac{\sinh\left(\frac{\pi T}{v}W\right)}{\sinh\left(\frac{\pi T}{v}\bar{W}\right)}\right)^q + \left(\frac{\sinh\left(\frac{\pi T}{v}\bar{W}\right)}{\sinh\left(\frac{\pi T}{v}W\right)}\right)^q\right]\right\}. \tag{43}$$

The CDW correlator is written as

$$\langle\langle \phi_0(W,\bar{W})\phi_0(W,\bar{W})\rangle\rangle \to C\cos(2P_F n)\left|\frac{(\pi/v)T}{\sinh\left(\frac{\pi T}{v}W\right)}\right|^{1/\gamma}$$

$$+ b\left[\frac{1}{\left(\sinh\left(\frac{\pi T}{v}W\right)\right)^2} + \frac{1}{\left(\sinh\left(\frac{\pi T}{v}\bar{W}\right)\right)^2}\right]\left(\frac{\pi}{\gamma}T\right)^2. \tag{44}$$

While deriving this expression we used the transformation law of the primary fields.[19] Comparing this result we note that the correlation length of CDW function is smaller than that of SCP function: $R_{CDW}/R_{SCP} = \gamma^2 < 1$.

It must be mentioned here that the critical exponent γ does not depend on the value of the magnetic field h, when $h < h_c$, while $\xi = v/(\Delta - h)$.

4. Quasi-One-Dimensional Model

Let us treat quasi-one-dimensional model consisting of the attractive Hubbard chain and interchain tunneling. By $t_\perp$ we shall denote an amplitude of tunneling $t_\perp \ll \Delta$.

If $\delta > 0$, the exponent $\gamma < 1$ and, as is seen from (5) and (7) CDW correlator decays faster than SCP correlator. In that case the CDW phase is suppressed and the system is in the SCP phase, by means of techniques similar to Refs. 22, 23 one can show that the anomalous average appears $\langle \Psi_{n,\uparrow} \Psi_{1,\downarrow} \rangle \neq 0$ and superconductive current exists. It must be mentioned here that magnetic flux quantisation in a high temperature superconductor implies that the charge carries of superconductivity are Cooper pairs.

Let us calculate quasi-one-dimensional correlator. First let us denote the one-dimensional SCP correlator by $G_0(x,\tau|x',\tau')$

$$\langle \Psi_\uparrow^+(x,\tau)\Psi_\downarrow^+(x,\tau)\Psi_\uparrow(x',\tau')\Psi_\downarrow(x',\tau')\rangle \equiv G_0(x,\tau|x',\tau') = a\left|\frac{(\pi/v)T}{\sinh\left(\frac{\pi T}{v}W\right)}\right|^\gamma . \quad (45)$$

Here we replace the lattice variable n by continuous space variable x. (We study the vicinity of the phase transition point.) Quasi-one-dimensional correlator of SCP is denoted by $G_{ij}(x,\tau|x',\tau')$. Here i,j are the numbers of the chains. We suppose that chains form two-dimensional lattice. $\mathbf{a}_{ij}$ is denoted as a two-dimensional vector, orthogonal to the chain, and pointing from the j-th chain to the i-th chain. Following the perturbation theory of Refs. 22 and 23 we obtain Dyson equations for G_{ij}:

$$G_{ij}(x,\tau|x',\tau') = \delta_j^i G_0(x,\tau|x',\tau') + \left(\frac{t_\perp}{\Delta}\right)^2 \sum_{\langle i,k\rangle} \int_0^{1/T} dt \int_{-\infty}^{\infty} dy G_0(x,\tau|y,t)G_{kj}(y,t|x',\tau') . \quad (46)$$

Here $\langle i,k\rangle$ denotes the nearest neighbour chains. All correlators depend only on the difference of space (time) variables. This fact permits to solve the equation by means of Fourier transformation

$$G(K_\parallel,K_\perp,\omega) = \int_{-\infty}^{\infty} dx e^{iK_\parallel x} \int_0^{1/T} d\tau e^{i\omega\tau} \sum_j e^{i\mathbf{K}_\perp \mathbf{a}_{j,0}} G_{j,0}(x,\tau|0,0) . \quad (47)$$

We obtain an equation

$$\Psi(K_\perp) = \sum_{\langle j,0\rangle} e^{i\mathbf{K}_\perp \mathbf{a}_{j,0}} , \quad (48)$$

$$G(K_{\parallel},K_{\perp},\omega) = G_0(K_{\parallel},\omega) + \left(\frac{t_{\perp}}{\Delta}\right)^2 \Psi(K_{\perp}) G_0(K_{\parallel},\omega) G(K_{\parallel},K_{\perp},\omega) .$$

The solution of this equation is

$$G(K_{\parallel},K_{\perp},\omega) = \frac{G_0(K_{\parallel},\omega)}{1 - \left(\frac{t_{\perp}}{\Delta}\right)^2 \Psi(K_{\perp}) G_0(K_{\parallel},\omega)} . \tag{49}$$

This formula describes Fourier transformation of quasi-one-dimensional correlator in the normal state (above T_c). The pole appears first at $K_{\perp} = K_{\parallel} = \omega = 0$ (as temperature goes down). This pole corresponds to the instability of the normal state and to SC transition.[22-24] So the equation for T_c has the form

$$1 = PG_0(0,0) . \tag{50}$$

Here $P = q(t_{\perp}/\Delta)^2$ (q is a number of nearest neibours chains) is a probability of interchain tunneling. The dependence of $G_0(0,0)$ on the tempeature can be estimated from the formula (45)

$$G_0(0,0) = \int_{x>\xi} dx \int_0^{1T} d\tau\, G_0(x,\tau|0,0) \sim T^{\gamma-2} . \tag{51}$$

So T_c can be estimated as follows

$$T_c \sim \Delta P^{1/(2-\gamma)} .$$

The anisotropy of the quasi-one-dimensional type generally leads to the increase of critical temperature T_c.[23] To estimate the value of T_c we must replace Hamiltonian (1) by the "physical" one

$$H_{\text{phys}} = -t \sum_{j=1}^{N} \sum_{s} \{\Psi^{+}_{j+1,s}\Psi_{j,s} + \Psi^{+}_{j-1,s}\Psi_{j,s}\} - 4U \sum_{j=1}^{N} \Psi^{+}_{j,\uparrow}\Psi_{j,\uparrow}\Psi^{+}_{j,\downarrow}\Psi_{j,\downarrow}$$

and substitute the coupling constant U and the width of the band t from the chemical reactions ($\sim 1/10$ eV), we can then obtain $T_c \simeq 100$ K.

When $T \gtrsim T_c$ the specific heat is the linear function of temperature

$$C = \frac{\pi}{6v} T ,$$

the consequence of the gapless excitation spectrum of the Cooper pairs.

5. Conclusion

We summarise the results of the previous sections. The ground state of the Hubbard model with attraction is constructed of bounded electrons in the singlet state, the charge of such a pair is equal to $2e$ and the spin is equal to zero. The analysis of the correlation function behaviour has shown that the Cooper pairing hold in the system. A small magnetic field does not change asymptotics: but there exists the critical value of magnetic field which destroys Cooper pairs. It means that both correlators FF and SCP decay as a power of the distance. Considering the quasi-one-dimensional system with the interchain tunneling (the three-dimensional system with strong anisotropy) at the non zero doping we have obtained the superconductive phase with a rather high critical temperature.

References

1. J. G. Bednorz and K. A. Müller, *Z. Phys.* **B64** (1986) 189.
2. P. W. Anderson, *Science* **235** (1987) 1196.
3. A. K. Zvezdin and D. I. Khomsky, *Solid State Commun.* **66** (1987) 651.
4. V. N. Popov, *Functional Integrals in Quantum Field Theory and Statistical Physics*, (D. Reidel Dordrecht, 1983), p. 183.
5. J. M. Kosterlitz and D. Thouless, *J. Phys.* **C6** (1973) 1691.
6. N. M. Bogoliubov and V. E. Korepin, *Mod. Phys. Lett.* **B1** (1988) 349.
7. E. H. Lieb and F. Y. Wu, *Phys. Rev. Lett.* **20** (1968) 1445.
8. L. D. Faddeev, *Sov. Sci. Rev., Math. Phys.* **C1** (1981) 107.
9. B. S. Shastry, *Phys. Rev. Lett.* **56** (1986) 2453.
10. M. Takahashi, *Progr. Theor. Phys.* **42** (1969) 1098.
11. B. Sutherland, *Lect. Notes Phys.* **242** (1985) 32.
12. M. Takahashi, *Prog. Theor. Phys.* **47** (1972) 69.
13. A. A. Ovchinikov, *Zh. Eksp. Teor. Fiz.* **57** (1969) 2137.
14. F. Woynarovich, *J. Phys.* **C15** (1982) 85.
15. A. G. Izergin and V. E. Korepin, *Pis'ma Zh. Eksp. Teor. Fiz.* **42** (1985) 414.
16. N. M. Bogoliubov, A. G. Izergin and V. E. Korepin, *Nucl. Phys.* **B275** (1986) 687.
17. N. M. Bogoliubov, A. G. Izergin and N. Yu. Reshetikhin, *J. Phys.* **A20** (1987) 5361.
18. A. A. Belavin, A. M. Polyakov and A. B. Zamolodchikov, *Nucl. Phys.* **B241** (1984) 333.
19. J. L. Cardy, *J. Phys.* **A17** (1984) L385.
20. A. M. Finkelstein, *Pis'ma Zh. Eksp. Teor. Fiz.* **25** (1977) 83.
21. J. Solyom, *Adv. Phys.* **28** (1979) 201.
22. Yu. A. Firsov, V. N. Prigodin and Chr. Seidel, *Phys. Reports* **126** (1985) 245.
23. K. B. Yefetov and A. I. Larkin, *Zh. Eksp. Teor. Fiz.* **66** (1974) 2290.

J. Phys. A: Math. Gen. **22** (1989) 4243–4256. Printed in the UK

Finite-size effects in a non-half-filled Hubbard chain

F Woynarovich

Central Research Institute for Physics, POB 49, H-1525 Budapest 114, Hungary

Received 14 February 1989

Abstract. The finite-size effects in the spectrum of a Hubbard chain are obtained for both the repulsive and attractive cases. It is shown that the finite-size corrections—similar to the case of a Heisenberg chain or Bose gas—are non-analytic unless some conditions are imposed on the chemical potential, magnetic field and chain length. If these conditions are met, the spectrum shows a similar tower structure as expected in conformal theories, although the model in general is not conformally invariant. In the special case when the two Fermi velocities are equal, the model is conformally invariant with $c = 2$, the indices are similar to the Gaussian form and there are four marginal operators.

1. Introduction

In the understanding of the critical two-dimensional classical and $(1+1)$-dimensional quantum systems the concept of conformal symmetry put forward by Belavin *et al* (1984) has proven to be a very fruitful one. This symmetry provides an abstract classification according to the central charge (c) of the Virasoro algebra describing the conformal symmetry of the system (Friedan *et al* 1984). The conformal anomaly c and the scaling dimensions of the primary-order parameters are directly accessible through the finite-size effects in an affiliated system defined on an infinitely long but finitely wide strip (Blöte *et al* 1986, Affleck 1986). These results have prompted several groups to study the finite-size effects both numerically and analytically in different critical and conformally invariant systems.

A condition for a critical system to be also conformally invariant is that the group velocity be the same for all elementary excitations. If this holds, the spectrum of the Hamiltonian for a chain of length N (in 2D statistical systems, the spectrum of the logarithm of the transfer matrix acting along the infinitely long strip of width N) should have the so-called tower structure (Cardy 1986a, b) which in the most general case (Bogoliubov *et al* 1988, Berkovich and Murthy 1988) means

$$E(n, N^+, N^-) - E_0 = \frac{2\pi v_F}{N}(x_n + N^+ + N^-) \tag{1.1}$$

$$E_0 = N\varepsilon_{\rm inf} - \frac{\pi v_F}{6N} c \tag{1.2}$$

$$P(n, N^+, N^-) - P_0 = \frac{2\pi}{N}(s_n + N^+ + N^-) + D2k_F. \tag{1.3}$$

Here $\varepsilon_{\rm inf}$ is the ground-state energy density of the infinite system and E_0 is the ground-state energy of the finite one, x_n and s_n are the scaling dimensions and spins of the primary scaling operators, N^+ and N^- are non-negative integers, D is the number of particles moved from one Fermi point to the other, P is the momentum of the system, $v_{\rm F}$ is the Fermi velocity and c is the central charge (conformal anomaly).

There are several examples for systems which are critical, but do not show conformal symmetry. These are systems in which there are several kinds of excitations, all posessing linear dispersion, but the different excitations have different velocities. Prime examples for such systems are the 1D spin-$\frac{1}{2}$ Fermi gas with δ interaction or its lattice version, the Hubbard chain. In these systems there are two kinds of excitations, one connected with the charge degrees of freedom, the other with the spins. Both are fermion-like but they have different Fermi velocities. These systems are not conformally invariant but are expected to be treatable in terms of two conformal fields (Korepin *et al* 1988).

In the present work we study the spectrum of the Hubbard chain. It is known that in the half-filled case the charge excitations possess a gap (Woynarovich 1982a, b, 1983a, b), only the spin excitations are critical and they have a spectrum of the form (1.1)–(1.3) with $c=1$ and x_n, s_n of the Gaussian form (Woynarovich and Eckle 1987b). Now, to have both degrees of freedom critical, we study the non-half-filled band, and to have the possible most general case we introduce also a magnetic field. Thus the Hamiltonian is

$$\hat{H}=-\sum_{i=1}^{N}\sum_{\sigma}(c^{+}_{i+1,\sigma}c_{i,\sigma}+c^{+}_{i,\sigma}c_{i+1,\sigma})+U\sum_{i=1}^{N}n_{i\uparrow}n_{i\downarrow}$$

$$+\mu\sum_{i=1}^{N}(n_{i\uparrow}+n_{i\downarrow})-\tfrac{1}{2}h\sum_{i=1}^{N}(n_{i\uparrow}-n_{i\downarrow}) \tag{1.4}$$

where $c_{i,\sigma}$ ($\sigma=\downarrow$ or $\uparrow$) are the spin-$\frac{1}{2}$ fermion operators at site i, $n_{i\uparrow}$ and $n_{i\downarrow}$ are the numbers of up and down spin particles at site i, and μ and h are the chemical potential and magnetic field, respectively.

As is well known, this system is exactly treatable by the Bethe ansatz (Lieb and Wu 1968), and by now there is also a well established method to calculate the finite-size corrections in Bethe ansatz systems (de Vega and Woynarovich 1985, Woynarovich and Eckle 1987a). Treating the model this way we have determined the low-lying part (scaling with $1/N$) of the spectrum for the Hamiltonian (1.4) for both positive and negative U. We have found the following.

(i) This spectrum ((2.44) for $U>0$ and (2.50) for $U<0$) is not analytic in N unless extra conditions imposed on μ and h are also satisfied (similar to the case of the Heisenberg chain in magnetic field or the 1D Bose gas (Woynarovich *et al* 1989)).

(ii) If the extra conditions are satisfied, the spectrum has a structure (3.7) which can be considered as the generalisation of (1.2) and (1.3): it looks as the finite-size corrections would come from two independent $c=1$ fields, but this independence is not true, however, since both of the x depend on the state of both Fermi seas.

(iii) There are special values of U, μ and h where the two Fermi velocities coincide. In these points the system is conformally invariant with $c=2$. The scaling indices are of generalised Gaussian form (3.21) (similar to that found in nested Bethe ansatz systems by Suzuki (1988)), and there are four marginal operators.

In the next section we give a detailed derivation of the energy and in § 3 we discuss our results in detail.

2. The energy of a finite chain

The Bethe ansatz equations for the Hubbard chain are

$$Nk_j = 2\pi I_j + \sum_{\beta=1}^{N_s} 2\tan^{-1}\left(\frac{\sin k_j - \lambda_\beta}{u}\right) \tag{2.1a}$$

$$\sum_{j=1}^{N_c} 2\tan^{-1}\left(\frac{\lambda_\alpha - \sin k_j}{u}\right) = 2\pi J_\alpha + \sum_{\beta=1}^{N_s} 2\tan^{-1}\left(\frac{\lambda_\alpha - \lambda_\beta}{2u}\right). \tag{2.1b}$$

Here N_c is the total number of particles and N_s is the number of down spins, u is the interaction strength in units of the bandwidth $u = U/4$, and the quantum numbers I_j and J_α are integers or half-odd-integers, depending on the parities of the numbers N_c and N_s:

$$I_j = N_s/2 \,(\text{mod } 1) \qquad J_\alpha = (N_c + N_s + 1)/2 \,(\text{mod } 1). \tag{2.2}$$

Once these equations are solved, the energy and the momentum of the system are given by

$$E = -2\sum_{j=1}^{N_c} \cos k_j + \mu N_c + h(N_s - N_c/2) \tag{2.3}$$

$$P = \sum_{j=1}^{N_c} k_j = \frac{2\pi}{N}\left(\sum_j I_j + \sum_\alpha J_\alpha\right). \tag{2.4}$$

We solve these equations at $U > 0$ for those states which have a spectrum scaling like $1/N$. (The $U < 0$ case can be obtained from the $U > 0$ one through a transformation (Woynarovich 1983b).) For this we choose the I_j and J_α sets as follows: we choose $I^\pm = (N_s + 1)/2$ (mod 1) and $J^\pm = (N_c + N_s)/2$ (mod 1) so that

$$\begin{aligned} I^+ - I^- &= N_c \qquad & I^+ + I^- &= 2D_c \\ J^+ - J^- &= N_s \qquad & J^+ + J^- &= 2D_s \end{aligned} \tag{2.5}$$

with $D_{c(s)} \ll N$. The I_j are all the numbers equal to $N_s/2$ (mod 1) between I^+ and I^- while the J_α are all the numbers equal to $(N_c + N_s + 1)/2$ (mod 1) between J^+ and J^-. This corresponds to two Fermi seas with D_c and D_s particles moved from the left Fermi points to the right ones. (Later on particle-hole pairs can also be introduced but care must be taken that the number of holes and particles must be the same around all four Fermi points separately (i.e. not to change N_c, N_s, D_c and D_s). Excitations with complex k and λ will not be considered as they have a gap.)

Now we define

$$\begin{aligned} z_c(k) &= \frac{1}{2\pi}\left[k + \frac{1}{N}\sum_\beta 2\tan^{-1}\left(\frac{\sin k - \lambda_\beta}{u}\right)\right] \\ z_s(\lambda) &= \frac{1}{2\pi}\left[\frac{1}{N}\sum_j 2\tan^{-1}\left(\frac{\lambda\text{-}\sin k_j}{u}\right) - \frac{1}{N}\sum_\beta 2\tan^{-1}\left(\frac{\lambda - \lambda_\beta}{2u}\right)\right] \end{aligned} \tag{2.6}$$

and

$$\rho_c(k) = \partial z_c(k)/\partial k \qquad \rho_s(\lambda) = \partial z_s(\lambda)/\partial\lambda. \tag{2.7}$$

With this notation

$$z_c(k_j) = I_j/N \qquad z_s(\lambda_\alpha) = J_\alpha/N. \tag{2.8}$$

Using the formula

$$\frac{1}{N}\sum_{n=n_1}^{n_2} f\left(\frac{n}{N}\right) \simeq \int_{(n_1-1/2)/N}^{(n_2+1/2)/N} f(x)\,\mathrm{d}x + \frac{1}{24N^2}\left(f'\left(\frac{n_1-1/2}{N}\right) - f'\left(\frac{n_2+1/2}{N}\right)\right) \tag{2.9}$$

(2.7) can be written in the form

$$\rho_c(k) = \frac{1}{2\pi}\left(1 - \frac{1}{24N^2\rho_s(\lambda^-)}\cos(k)K_1'(\sin k-\lambda^-) + \frac{1}{24N^2\rho_s(\lambda^+)}\cos(k)K_1'(\sin k-\lambda^+)\right.$$
$$\left. + \cos(k)\int_{\lambda^-}^{\lambda^+} K_1(\sin k-\lambda')\rho_s(\lambda')\,\mathrm{d}\lambda'\right) \tag{2.10a}$$

$$\rho_s(\lambda) = \frac{1}{2\pi}\left(-\frac{1}{24N^2\rho_c(k^-)}\cos(k^-)K_1'(\lambda-\sin k^-)\right.$$
$$+\frac{1}{24N^2\rho_c(k^+)}\cos(k^+)K_1'(\lambda-\sin k^+)$$
$$+\frac{1}{24N^2\rho_s(\lambda^-)}K_2'(\lambda-\lambda^-) - \frac{1}{24N^2\rho_s(\lambda^+)}K_2'(\lambda-\lambda^+)$$
$$\left.+\int_{k^-}^{k^+} K_1(\lambda-\sin k')\rho_c(k')\,\mathrm{d}k' - \int_{\lambda^-}^{\lambda^+} K_2(\lambda-\lambda')\rho_s(\lambda')\,\mathrm{d}\lambda'\right). \tag{2.10b}$$

Here

$$K_1(x) = 2\frac{u}{u^2+x^2} \qquad K_2(x) = 2\frac{2u}{(2u)^2+x^2}. \tag{2.11}$$

and $K_{1,2}'$ are the derivatives of $K_{1,2}$ and $k^\pm$ and $\lambda^\pm$ satisfy the equations

$$z_c(k^\pm) = I^\pm/N \qquad z_s(\lambda^\pm) = J^\pm/N. \tag{2.12}$$

These four equations (2.12), together with the definitions of ρ and (2.5), are equivalent to

$$\int_{k^+}^{k^+} \rho_c(k) = \frac{N_c}{N} \qquad -\frac{1}{2}\left(\int_{k^+}^{\pi}\rho_c(k) - \int_{-\pi}^{k^-}\rho_c(k)\right) - \frac{1}{2\pi}\int_{\lambda^-}^{\lambda^+} 2\tan^{-1}(\lambda/u)\rho_s(\lambda) = \frac{D_c}{N} \tag{2.13}$$

$$\int_{\lambda^-}^{\lambda^+} \rho_s(\lambda) = \frac{N_s}{N} \qquad -\frac{1}{2}\left(\int_{\lambda^+}^{\infty}\rho_s(\lambda) - \int_{-\infty}^{\lambda^-}\rho_s(\lambda)\right) = \frac{D_s}{N}. \tag{2.14}$$

In the following a central role will be played by the solutions of equations of the type

$$x_c(k\,|\,k^\pm,\lambda^\pm) = x_{0c}(k) + \frac{\cos k}{2\pi}\int_{\lambda^-}^{\lambda^+} K_1(\sin k-\lambda')x_s(\lambda'\,|\,k^\pm,\lambda^\pm)\,\mathrm{d}\lambda'$$

$$x_s(\lambda\,|\,k^\pm,\lambda^\pm) = x_{0s}(\lambda) + \frac{1}{2\pi}\int_{k^-}^{k^+} K_1(\lambda-\sin k')x_c(k'\,|\,k^\pm,\lambda^\pm)\,\mathrm{d}k' \tag{2.15}$$
$$-\frac{1}{2\pi}\int_{\lambda^-}^{\lambda^+} K_2(\lambda-\lambda')x_s(\lambda'\,|\,k^\pm,\lambda^\pm)\,\mathrm{d}\lambda'.$$

This system of equations we shall write in the symbolic form

$$\boldsymbol{x}(k, \lambda \,|\, k^{\pm}, \lambda^{\pm}) = \boldsymbol{x}_0(k, \lambda) + \mathbf{K}(k, \lambda \,|\, k', \lambda' | \, k^{\pm}, \lambda^{\pm}) \otimes \boldsymbol{x}(k', \lambda' | \, k^{\pm}, \lambda^{\pm}). \tag{2.16}$$

Here $\boldsymbol{x}(k, \lambda)$ is a column vector with upper and lower elements $x_c(k)$ and $x_s(\lambda)$, respectively, and $\mathbf{K}$ is a 2×2 matrix with integral-operator elements which can be read out from (2.15). We shall also use the equation

$$\boldsymbol{y}(k, \lambda \,|\, k^{\pm}, \lambda^{\pm}) = \boldsymbol{y}_0(k, \lambda) + \mathbf{K}^{\mathrm{T}}(k, \lambda \,|\, k', \lambda' | \, k^{\pm}, \lambda^{\pm}) \otimes \boldsymbol{y}(k', \lambda' | \, k^{\pm}, \lambda^{\pm}) \tag{2.17}$$

which is analogous to (2.15) but the integral-operator matrix $\mathbf{K}^{\mathrm{T}}$ is the transpose of that in (2.15):

$$\mathbf{K}^{\mathrm{T}}(k, \lambda \,|\, k', \lambda' | \, k^{\pm}, \lambda^{\pm}) = \frac{1}{2\pi} \begin{pmatrix} 0 & \displaystyle\int_{\lambda^-}^{\lambda^+} \mathrm{d}\lambda' \, K_1(\sin k - \lambda') \ldots \\ \displaystyle\int_{k^-}^{k^+} \mathrm{d}k' \, K_1(\lambda - \sin k') \cos k' \ldots & \displaystyle\int_{\lambda^-}^{\lambda^+} \mathrm{d}\lambda' \, K_2(\lambda - \lambda') \ldots \end{pmatrix}. \tag{2.18}$$

It is clear that

$$\begin{aligned} \boldsymbol{\rho}(k, \lambda) = \boldsymbol{\rho}_{\infty}(k, \lambda) + \frac{1}{24N^2} \bigg(& \frac{1}{\rho_c(k^+)} \boldsymbol{\rho}_1(k, \lambda \,|\, k^{\pm}, \lambda^{\pm}) \\ & + \frac{1}{\rho_c(k^-)} \boldsymbol{\rho}_1(-k, -\lambda \,|\, -k^{\mp}, -\lambda^{\mp}) + \frac{1}{\rho_s(\lambda^+)} \boldsymbol{\rho}_2(k, \lambda \,|\, k^{\pm}, \lambda^{\pm}) \\ & + \frac{1}{\rho_s(\lambda^-)} \boldsymbol{\rho}_2(-k, -\lambda \,|\, -k^{\mp}, -\lambda^{\mp}) \bigg) \end{aligned} \tag{2.19}$$

with $\boldsymbol{\rho}_{\infty}$, $\boldsymbol{\rho}_1$ and $\boldsymbol{\rho}_2$ determined by (2.16) with the inhomogeneous part $\boldsymbol{x}_0$ replaced by

$$\begin{pmatrix} \dfrac{1}{2\pi} \\ 0 \end{pmatrix} \qquad \begin{pmatrix} 0 \\ \dfrac{1}{2\pi} \cos(k^+) K_1'(\lambda - \sin k^+) \end{pmatrix} \qquad \begin{pmatrix} \dfrac{1}{2\pi} \cos(k) K_1'(\sin k - \lambda^+) \\ -\dfrac{1}{2\pi} K_2'(\lambda - \lambda^+) \end{pmatrix} \tag{2.20}$$

respectively.

The energy according to (2.3) and (2.9) is

$$\begin{aligned} E = N \int_{k^-}^{k^+} (h_c - 2\cos k) \rho_c(k) + N h_s \int_{\lambda^-}^{\lambda^+} \rho_s(\lambda) \\ + \frac{1}{24N\rho_c(k^+)} 2\sin k^+ - \frac{1}{24N\rho_c(k^-)} 2 \sin k^- \end{aligned} \tag{2.21}$$

with $h_c = \mu - h/2$ and $h_s = h$. This, using (2.19), can be written in the form

$$\begin{aligned} E = N\varepsilon_{\infty}(k^+, k^-, \lambda^+, \lambda^-) + \frac{1}{24N} (& \varepsilon_1(k^+, k^-, \lambda^+, \lambda^-) \\ & + \varepsilon_1(-k^-, -k^+, -\lambda^-, -\lambda^+) + \varepsilon_2(k^+, k^-, \lambda^+, \lambda^-) \\ & + \varepsilon_2(-k^-, -k^+, -\lambda^-, -\lambda^+)) \end{aligned} \tag{2.22}$$

with

$$\varepsilon_\infty = \int_{k^-}^{k^+} (h_c - 2\cos k)\rho_{\infty_c}(k) + \int_{\lambda^-}^{\lambda^+} h_s\rho_{\infty_s}(\lambda) \tag{2.23}$$

$$\varepsilon_1 = \frac{1}{\rho_c(k^+)}\left(2\sin k^+ - \int_{k^-}^{k^+} (h_c - 2\cos k)\rho_{1c}(k) - \int_{\lambda^-}^{\lambda^+} h_s\rho_{1s}(\lambda)\right) \tag{2.24}$$

$$\varepsilon_2 = \frac{-1}{\rho_s(\lambda^+)}\left(\int_{k^-}^{k^+} (h_c - 2\cos k)\rho_{2c}(k) + \int_{\lambda^-}^{\lambda^+} h_s\rho_{2s}(\lambda)\right). \tag{2.25}$$

ε_∞ is actually the energy density of an infinite system at the given $k^\pm$ and $\lambda^\pm$. It is instructive to write it in another form. Since the formal solution of an equation of the type (2.16) is

$$\boldsymbol{x}(k,\lambda) = \sum_{n=1}^{\infty} (\mathbf{K}(k,\lambda\,|\,k',\lambda'\,|\,k^\pm,\lambda^\pm)\otimes)^n \boldsymbol{x}_0(k',\lambda') \tag{2.26}$$

ρ_∞ will be given by (2.26) with $\boldsymbol{x}_0$ replaced by the first column vector in (2.20). Substituting this into (2.23) one obtains that

$$\varepsilon_\infty = \frac{1}{2\pi}\int_{k^-}^{k^+} \varepsilon_c(k\,|\,k^\pm,\lambda^\pm)\,dk \tag{2.27}$$

where

$$\boldsymbol{\varepsilon}(k,\lambda\,|\,k^\pm,\lambda^\pm) = \sum_{n=1}^{\infty} (\mathbf{K}^{\mathrm{T}}(k,\lambda\,|\,k',\lambda'\,|\,k^\pm,\lambda^\pm)\otimes)^n \boldsymbol{\varepsilon}_0(k',\lambda') \tag{2.28}$$

with

$$\boldsymbol{\varepsilon}_0 = \begin{pmatrix} h_c - 2\cos k \\ h_s \end{pmatrix} \tag{2.29}$$

i.e. $\boldsymbol{\varepsilon}(k,\lambda)$ satisfies (2.17) with y_0 replaced by $\boldsymbol{\varepsilon}_0$. $\varepsilon_c(k)$ and $\varepsilon_s(\lambda)$ can be considered as the dressed energies.

The infinite chain is in the ground state at the given μ and h if ε_∞ is minimal with respect to $k^\pm$ and $\lambda^\pm$. This condition, using (2.27) and the integral equations determining ε_c, lead to the conditions

$$\varepsilon_c(k^+|k^\pm,\lambda^\pm) = 0 \qquad \varepsilon_c(k^-|k^\pm,\lambda^\pm) = 0 \tag{2.30}$$

$$\varepsilon_s(\lambda^+|k^\pm,\lambda^\pm) = 0 \qquad \varepsilon_s(\lambda^-|k^\pm,\lambda^\pm) = 0 \tag{2.31}$$

i.e. in the ground state the dressed energies are zero at the Fermi points. From symmetry it is clear that in the ground state $k^- = -k^+$ and $\lambda^- = -\lambda^+$. Let us denote the ground-state values of k^+ and λ^+ by k_0 and λ_0, respectively. Now we can expand ε_∞ up to second order in $(k^\pm \mp k_0)$ and $(\lambda^\pm \mp \lambda_0)$. Since the conditions (2.30) and (2.31) are satisfied at $k^\pm = \pm k_0$, $\lambda^\pm = \pm\lambda_0$ there are no cross derivatives and we find

$$\begin{aligned}
\varepsilon_\infty(k^+,k^-,\lambda^+,\lambda^-) = {} & \varepsilon_\infty(k_0,-k_0,\lambda_0,-\lambda_0) + \frac{1}{\rho_{\infty c}(k_0)}\frac{\partial}{\partial k}\,\varepsilon c(k\,|\pm k_0,\pm\lambda_0)\Big|_{k=k_0} \\
& \times\tfrac{1}{2}[(\rho_{\infty c}(k_0)(k^+ - k^0))^2 + (\rho_{\infty c}(k_0)(k^- + k^0))^2] \\
& + \frac{1}{\rho_{\infty s}(\lambda_0)}\frac{\partial}{\partial\lambda}\,\varepsilon_s(\lambda\,|\pm k_0,\pm\lambda_0)\Big|_{\lambda=\lambda_0} \\
& \times\tfrac{1}{2}[(\rho_{\infty s}(\lambda_0)(\lambda^+ - \lambda_0))^2 + (\rho_{\infty s}(\lambda_0)(\lambda^- + \lambda_0))^2].
\end{aligned} \tag{2.32}$$

Utilising the equation for ε and those for $\boldsymbol{\rho}_1$ and $\boldsymbol{\rho}_2$ it is not hard to see that with an accuracy of $1/N^2$ (the error coming from the $O(1/N^2)$ difference between $\boldsymbol{\rho}$ and $\boldsymbol{\rho}_\infty$)

$$\begin{aligned}\frac{1}{\rho_{\infty c}(k_0)}\frac{\partial}{\partial k}\varepsilon_c(k|\pm k_0,\pm\lambda_0)\bigg|_{k=k_0} &= \varepsilon_1(k_0,-k_0,\lambda_0,-\lambda_0)\\ \frac{1}{\rho_{\infty s}(\lambda_0)}\frac{\partial}{\partial \lambda}\varepsilon_s(\lambda\,|\pm k_0,\pm\lambda_0)\bigg|_{\lambda=\lambda_0} &= \varepsilon_2(k_0,-k_0,\lambda_0,-\lambda_0).\end{aligned} \tag{2.33}$$

As a next step we express $(k^{\pm}\mp k_0)$ and $(\lambda^{\pm}\mp\lambda_0)$ by the deviation of N_c, N_s, D_c and D_s from their ground-state values. For this we can use (2.13*a*, *b*). First we note that, since the energy contains the square of these deviations only, it is enough to calculate them up to $O(1/N)$, i.e. in (2.13*a*, *b*) ρ_c and ρ_s can be replaced by $\rho_{\infty c}$ and $\rho_{\infty s}$, respectively. Denoting $N_c/N=\nu_c$, $N_s/n=\nu_s$, $D_c/N=\delta_c$, $D_s/N=\delta_s$, from (2.13*a*, *b*) we have

$$\frac{\partial\nu_c}{\partial k^+}=-\frac{\partial\nu_c}{\partial k^-}=\rho_{\infty c}(k_0)\left(1+\int_{-k_0}^{k_0}\sigma_{1c}(k)\right)=\rho_{\infty c}(k_0)\xi_{11} \tag{2.34a}$$

$$\frac{\partial\nu_s}{\partial k^+}=-\frac{\partial\nu_s}{\partial k^-}=\rho_{\infty c}(k_0)\int_{-\lambda_0}^{\lambda_0}\sigma_{1s}(\lambda)=\rho_{\infty c}(k_0)\xi_{12} \tag{2.34b}$$

$$\frac{\partial\nu_c}{\partial \lambda^+}=-\frac{\partial\nu_c}{\partial \lambda^-}=\rho_{\infty s}(\lambda_0)\int_{-k_0}^{k_0}\sigma_{2c}(k)=\rho_{\infty s}(\lambda_0)\xi_{21} \tag{2.34c}$$

$$\frac{\partial\nu_s}{\partial \lambda^+}=-\frac{\partial\nu_s}{\partial \lambda^-}=\rho_{\infty s}(\lambda_0)\left(1+\int_{-\lambda_0}^{\lambda_0}\sigma_{2s}(\lambda)\right)=\rho_{\infty s}(\lambda_0)\xi_{22} \tag{2.34d}$$

$$\frac{\partial\delta_c}{\partial k^+}=\frac{\partial\delta_c}{\partial k^-}=\rho_{\infty c}(k_0)\left[\frac{1}{2}\left(1-\int_{k_0}^{\pi}\sigma_{1c}+\int_{-\pi}^{-k_0}\sigma_{1c}\right)\right.$$
$$\left.-\frac{1}{\pi}\int_{-\lambda_0}^{\lambda_0}\tan^{-1}(\lambda/u)\sigma_{1s}\right]=\rho_{\infty c}(k_0)z_{11} \tag{2.35a}$$

$$\frac{\partial\delta_s}{\partial k^+}=\frac{\partial\delta_s}{\partial k^-}=-\rho_{\infty c}(k_0)\frac{1}{2}\left(\int_{\lambda_0}^{\infty}\sigma_{1s}-\int_{-\infty}^{-\lambda_0}\sigma_{1s}\right)=\rho_{\infty c}(k_0)z_{12} \tag{2.35b}$$

$$\frac{\partial\delta_c}{\partial \lambda^+}=\frac{\partial\delta_c}{\partial \lambda^-}=-\rho_{\infty s}(\lambda_0)\left[\frac{1}{2}\left(\int_{k_0}^{\pi}\sigma_{2c}-\int_{-\pi}^{-k_0}\sigma_{2c}\right)\right.$$
$$\left.+\frac{1}{\pi}\tan^{-1}(\lambda_0/u)+\frac{1}{\pi}\int_{-\lambda_0}^{\lambda_0}\tan^{-1}(\lambda/u)\sigma_{2s}\right]=\rho_{\infty_s}(\lambda_0)z_{21} \tag{2.35c}$$

$$\frac{\partial\delta_s}{\partial \lambda^+}=\frac{\partial\delta_s}{\partial \lambda^-}=\rho_{\infty s}(\lambda_0)\frac{1}{2}\left(1-\int_{\lambda_0}^{\infty}\sigma_{2s}+\int_{-\infty}^{-\lambda_0}\sigma_{2s}\right)=\rho_{\infty s}(\lambda_0)z_{22} \tag{2.35d}$$

where $\boldsymbol{\sigma}_1$ and $\boldsymbol{\sigma}_2$ are defined by (2.16) with the inhomogeneous part $\boldsymbol{x}_0$ replaced by

$$\begin{pmatrix} 0 \\ \dfrac{1}{2\pi}K_1(\lambda-\sin k_0)\end{pmatrix}\qquad\begin{pmatrix}\dfrac{1}{2\pi}\cos(k)K_1(\sin k-\lambda_0)\\ -\dfrac{1}{2\pi}K_2(\lambda-\lambda_0)\end{pmatrix} \tag{2.36}$$

respectively. Using the formal solution (2.26) of (2.16) for $\boldsymbol{\sigma}_1$ one can convince oneself that the matrix $\boldsymbol{\xi}$ in (2.34) is

$$\boldsymbol{\xi} = \begin{pmatrix} \xi_{11} & \xi_{12} \\ \xi_{21} & \xi_{22} \end{pmatrix} = \begin{pmatrix} \xi_{11}(k_0) & \xi_{12}(k_0) \\ \xi_{21}(\lambda_0) & \xi_{22}(\lambda_0) \end{pmatrix} \tag{2.37}$$

where the $\boldsymbol{\xi}(k, \lambda)$ matrix is defined through the equation

$$\boldsymbol{\xi}(k, \lambda) = \mathbf{I} + \mathbf{K}^{\mathrm{T}}(k, \lambda \,|\, k', \lambda' \,|\, \pm k_0, \pm\lambda_0) \otimes \boldsymbol{\xi}(k', \lambda') \tag{2.38}$$

with $\mathbf{I}$ being the 2×2 identity matrix. This $\boldsymbol{\xi}$ matrix can be considered as a generalisation of the dressed charge (Korepin 1979). Taking the derivatives of the elements of $\boldsymbol{\xi}$ and reintegrating the first row from k_0 to π and the second row from λ_0 to infinity one finds that the $\mathbf{z}$ matrix in (2.35) is

$$\mathbf{z} = \tfrac{1}{2}(\boldsymbol{\xi}^{\mathrm{T}})^{-1} \tag{2.39}$$

with the upper index T corresponding to transposition. Compiling (2.22), (2.32), (2.33)-(2.37) and (2.39) one arrives at the energy expression correct to $\mathrm{O}(1/N^2)$

$$\begin{aligned} E = {} & N\varepsilon_\infty(k_0, -k_0, \lambda_0, -\lambda_0) \\ & + \frac{1}{N}\varepsilon_1\left(\frac{[\xi_{22}(N_c - \nu_c N) - \xi_{21}(N_s - \nu_s N)]^2}{4(\det \xi)^2} + (\xi_{11}D_c + \xi_{12}D_s)^2 - \frac{1}{12}\right) \\ & + \frac{1}{N}\varepsilon_2\left(\frac{[\xi_{12}(N_c - \nu_c N) - \xi_{11}(N_s - \nu_s N)]^2}{4(\det \xi)^2} + (\xi_{21}D_c + \xi_{22}D_s)^2 - \frac{1}{12}\right) \end{aligned} \tag{2.40}$$

where ε_1 and ε_2 are given by (2.24) and (2.25) with $k^\pm = \pm k_0$ and $\lambda^\pm = \pm\lambda_0$, and ν_c and ν_s are the densities of the particles and down spins, respectively, in the ground state of an infinite system at μ and h.

The total momentum of the system according to (2.4) and (2.5) is given by

$$P = \frac{1}{2\pi}(N_c D_c + N_s D_s). \tag{2.41}$$

As a final step we have to deal with the excitations. As an example we treat a particle-hole pair around the right Fermi point of the k sea. Suppose that the particle and hole are characterised by the quantum numbers I_{p}^+ and I_{h}^+, respectively. This defines their positions in the k space:

$$z_c(k_{\mathrm{p}}^+) = I_{\mathrm{p}}^+/N \qquad z_c(k_{\mathrm{h}}^+) = I_{\mathrm{h}}^+/N. \tag{2.42}$$

The presence of this particle-hole pair modifies $\boldsymbol{\rho}(k, \lambda)$ by $-\boldsymbol{\rho}_1(k, \lambda \,|\, k^\pm, \lambda^\pm)(k_{\mathrm{p}}^+ - k_{\mathrm{h}}^+)/N$ and gives a contribution to the energy $\varepsilon_1 \rho_c(k^+)(k_{\mathrm{p}}^+ - k_{\mathrm{h}}^+)/N$. According to (2.42) and the definition of ρ, $\rho_c(k^+)(k_{\mathrm{p}}^+ - k_{\mathrm{h}}^+) = (I_{\mathrm{p}}^+ - I_{\mathrm{h}}^+)/N$. The momentum of such a particle-hole pair is $2\pi(I_{\mathrm{p}}^+ - I_{\mathrm{h}}^+)/N$ and thus, due to this excitation we have contributions $\varepsilon_1(I_{\mathrm{p}}^+ - I_{\mathrm{h}}^+)/N$ and $2\pi(I_{\mathrm{p}}^+ - I_{\mathrm{h}}^+)/N$ to the energy (2.40) and momentum (2.41), respectively. This and the analogue calculations for the particle-hole excitations in the λ sea justify the notation

$$\varepsilon_1 = 2\pi v_c \qquad \varepsilon_2 = 2\pi v_s \tag{2.43}$$

with v_c and v_s being the Fermi velocities for the two Fermi seas. It is not hard to see that the energy and the momentum of a state with all four possible particle-hole excitations are

$$E = N\varepsilon_\infty(k_0, -k_0, \lambda_0, -\lambda_0)$$
$$+\frac{2\pi v_c}{N}\left(\frac{[\xi_{22}(N_c-\nu_c N)-\xi_{21}(N_s-\nu_s N)]^2}{4(\det\xi)^2}\right.$$
$$\left.+(\xi_{11}D_c+\xi_{12}D_s)^2-\tfrac{1}{12}+n_c^+ + n_c^-\right)$$
$$+\frac{2\pi v_s}{N}\left(\frac{[\xi_{12}(N_c-\nu_c N)-\xi_{11}(N_s-\nu_s N)]^2}{4(\det\xi)^2}\right.$$
$$\left.+(\xi_{21}D_c+\xi_{22}D_s)^2-\tfrac{1}{12}+n_s^+ + n_s^-\right) \tag{2.44}$$

$$P=\frac{2\pi}{N}(N_c D_c+n_c^+ - n_c^- + N_s D_s + n_s^+ - n_s^-). \tag{2.45}$$

Here

$$n_c^+ = \sum_{\mathrm{p}} I_{\mathrm{p}}^+ - \sum_{\mathrm{h}} I_{\mathrm{h}}^+ \qquad n_c^- = \sum_{\mathrm{h}} I_{\mathrm{h}}^- - \sum_{\mathrm{p}} I_{\mathrm{p}}^-$$
$$n_s^+ = \sum_{\mathrm{p}} J_{\mathrm{p}}^+ - \sum_{\mathrm{h}} J_{\mathrm{h}}^+ \qquad n_s^- = \sum_{\mathrm{h}} J_{\mathrm{h}}^- - \sum_{\mathrm{p}} J_{\mathrm{p}}^- \tag{2.46}$$

With I_{p}^+ (I_{h}^+), I_{p}^- (I_{h}^-), J_{p}^+ (J_{h}^+) and J_{p}^- (J_{h}^-) being the quantum numbers of the particles (holes) near to the right $(+)$ and left $(-)$ Fermi points of the k and λ seas.

The above results are valid for the case $U>0$ but can be easily translated for negative U. A way to do this is provided by the 'complementer solutions' of the Lieb–Wu equations (2.1*a*, *b*) (Woynarovich 1983b). Suppose that in a $\{k_j, \lambda_\alpha\}$ solution of these equations the k_j are distributed in (k^+, k^-) according to $\rho_c(k)$ with holes at k_{h} and particles at k_{p}. In the complementer solution the λ_α are unchanged, but the k_j set is replaced by a k_g set in which there are real k and also complex k pairs. The real k_g are distributed in $(-\pi, k^-)$ and (k^+, π) according the corresponding part of $\rho_c(k)$ ($\rho_c(k)$ is defined by (2.7) for the whole $(-\pi, \pi)$ interval) with holes at k_{p} and particles at k_{h}, while the complex k_g pairs are determined by the λ:

$$\sin k_\alpha^\pm = \lambda_\alpha \mp \mathrm{i}u \qquad \pm\mathrm{Im}\, k_\alpha^\pm > 0. \tag{2.47}$$

The total number of k_g is $N+2N_s-N_c$. Also this $\{k_g, \lambda_\alpha\}$ set will solve (2.1*a*, *b*) of course with an $\{I_g, J'_\alpha\}$ set different from $\{I_j, J_\alpha\}$ (but $I_j - I_g =$ integer and $J_\alpha - J'_\alpha =$ integer). The sum of the energies and momenta of these complementer solutions is

$$\sum -2\cos k_j + \sum -2\cos k_g = N_s U$$
$$\sum k_j + \sum k_g = \pi(N+N_s+1). \tag{2.48}$$

Utilising all this one can see that the set $\{k_g+\pi, -\lambda_\alpha\}$ will satisfy (2.1*a*, *b*) with u replaced by $-u$ and $\{I_j, J_\alpha\}$ replaced by $\{I_g+N/2, J'_\alpha\}$. This solution describes an eigenstate of the attractive chain (interaction: $-U$) in which there are $N_{\mathrm{f}} = N - N_c$ 'free' particles and $N_b = N_s$ bound pairs, and the total spin is $S = N_{\mathrm{f}}/2$. (The parities of the numbers $2I_g + N$ and $2J'_\alpha$ will correspond to the periodic boundary condition (2.2), if in the positive u equations the quantum numbers are chosen as $I_j = (N+N_{\mathrm{b}})/2$ (mod 1); $J_\alpha = (N_{\mathrm{b}}+N_{\mathrm{f}}+1)/2$ (mod 1) (accordingly $I^\pm = (N+N_{\mathrm{b}}+1)/2$ (mod 1) and $J^\pm = (N_{\mathrm{f}}+N_{\mathrm{b}})/2$ (mod 1)).) Through (2.48) one can see that, if for the chain with $-U$

we chose $\mu'=(U+h)/2$ and $h'=U+2\mu$, the eigenstate for this attractive chain described by $\{k_g+\pi, -\lambda_\alpha\}$ will have the same energy as the state $\{k_j, \lambda_\alpha\}$ in the repulsive chain (1.4) (apart from a macroscopic constant). Thus, after substituting

$$\begin{array}{llll} N_c=N-N_f & D_c=D_f & \nu_c=1-\nu_f & n_c^\pm=n_f^\mp \\ N_s=N_b & D_s=-D_b & \nu_s=\nu_b & n_s^\pm=n_b^\mp \end{array} \tag{2.49}$$

into (2.44) we have

$$\begin{aligned} E(-U)=&N(\varepsilon_\infty(k_0,-k_0,\lambda_0,-\lambda_0)-\mu+h/2)\\ &+\frac{2\pi v_c}{N}\left(\frac{[\xi_{22}(N_f-\nu_f N)+\xi_{21}(N_b-\nu_b N)]^2}{4(\det\xi)^2}\right.\\ &\left.+(\xi_{11}D_f-\xi_{12}D_b)^2-\tfrac{1}{12}+n_f^++n_f^-\right)\\ &+\frac{2\pi v_s}{N}\left(\frac{[\xi_{12}(N_f-\nu_f N)+\xi_{11}(N_b-\nu_b N)]^2}{4(\det\xi)^2}\right.\\ &\left.+(\xi_{21}D_f-\xi_{22}D_b)^2-\tfrac{1}{12}+n_b^++n_b^-\right) \end{aligned} \tag{2.50}$$

and (2.45), (2.48) and (2.49) gives

$$P=\pi N+\frac{2\pi}{N}(N_f D_f+n_f^+-n_f^-+N_b D_b+n_b^+-n_b^-). \tag{2.51}$$

Since $2\nu_s \leqslant \nu_c<1$, $\nu_f+2\nu_b\leqslant 1$.

3. Discussion of the energy spectrum and finite-size effects

First we have to notice that the spectrum (2.44) is not analytic in N: for the finite system in the ground state N_c and N_s should minimise (2.44), but since N_c, N_s and N are integers, the optimal values of N_c and N_s are not analytic functions of N. This phenomenon is known already for other systems (Woynarovich *et al* 1989) and is thought to be connected with the possibility of a consistent definition of the continuum limit for the system. In the present case the spectrum will be analytic if ν_c and ν_s are rational and only special values of N are allowed. To be definite, if

$$\nu_c=p_c/q_c \qquad \nu_s=p_s/q_s \tag{3.1}$$

with p_c and q_c (p_s and q_s) being relative prime integers, then only

$$N=qN' \tag{3.2}$$

values are allowed for N, where N' is an integer and q is the least integer dividable by both q_c and q_s. If (3.1) and (3.2) are met, the ground-state values N_c and N_s are

$$N_{c0}=p_c(q/q_c)N' \qquad N_{s0}=p_s(q/q_s)N' \tag{3.3}$$

and in the excited states

$$\Delta N_c=N_c-\nu_c N \qquad \Delta N_s=N_s-\nu_s N \tag{3.4}$$

are integers. Whether D_c and D_s are integers or half-odd integers depends on the parity of the numbers N_c and N_s. Due to the restrictions (2.2)

$$\begin{aligned} D_c &= (N_c + N_s + 1)/2 \qquad (\text{mod } 1) \\ D_s &= N_c/2 \qquad (\text{mod } 1). \end{aligned} \tag{3.5}$$

In the ground state both D are zero only if N_{c0} is even and N_{s0} is odd (otherwise at least one of the two D is $\pm\frac{1}{2}$, i.e. the ground state is degenerate). This imposes a restriction on the numbers p_c, p_s, q/q_c, q/q_s and N': p_s, q/q_s and N' should be odd, while one of p_c and q/q_c should be even. If these requirements are met, the finite-size corrections to the ground-state energy are

$$E_0 - N\varepsilon_\infty = -\frac{\pi v_c}{6N} - \frac{\pi v_s}{6N} \tag{3.6}$$

just as it would be in the case of two independent conformal fields. These fields are, however, not independent as all ΔN_c, ΔN_s, D_c and D_s appear multiplied by both Fermi velocities:

$$\begin{aligned} E - E_0 = {} & \frac{2\pi v_c}{N}\left(\frac{(\xi_{22}\Delta N_c - \xi_{21}\Delta N_s)^2}{4(\det \xi)^2} + (\xi_{11}D_c + \xi_{12}D_s)^2 + n_c^+ + n_c^-\right) \\ & + \frac{2\pi v_s}{N}\left(\frac{(\xi_{12}\Delta N_c - \xi_{11}\Delta N_s)^2}{4(\det \xi)^2} + (\xi_{21}D_c + \xi_{22}D_s)^2 + n_s^+ + n_s^-\right). \end{aligned} \tag{3.7}$$

Another interesting feature is that D_c and D_s are not independent of ΔN_c and ΔN_s, as even if the parameters are such that in the ground state $\Delta N_c = \Delta N_s = D_c = D_s = 0$, in the excited states due to (3.5)

$$\begin{aligned} D_c &= (\Delta N_c + \Delta N_s)/2 \qquad (\text{mod } 1) \\ D_s &= \Delta N_c/2 \qquad (\text{mod } 1). \end{aligned} \tag{3.8}$$

Examining (2.50) analogous conclusions can be drawn for the case of the attractive chain.

An important special case is when $U > 0$, $h = 0$, i.e. the ground state is non-magnetic ($\nu_s = \nu_c/2$). From our formulae we can get this case by taking the $\lambda_0 \to \infty$ limit. This can be done by solving the equations for the λ-dependent quantities at $\lambda \gg 1$ with Wiener–Hopf techniques and then taking the $\lambda_0 \to \infty$ limit. As a result one obtains that

$$\lim_{\lambda_0 \to \infty} \xi = \begin{pmatrix} \xi & \xi/2 \\ 0 & 1/\sqrt{2} \end{pmatrix} \tag{3.9}$$

where

$$\xi = \xi(\sin k_0) \tag{3.10}$$

with $\xi(x)$ solving the equation

$$\begin{aligned} \xi(x) &= 1 + \frac{1}{2\pi}\int_{-\sin k_0}^{\sin k_0} \bar{K}(x - x')\xi(x')\, dx' \\ \bar{K}(x) &= \int_{-\infty}^{\infty} \frac{\exp(-|\omega| u)}{2\cosh \omega u} \exp(i\omega x)\, d\omega. \end{aligned} \tag{3.11}$$

In this case (2.44) and (2.45) simplify to

$$E = N\varepsilon_\infty + \frac{2\pi v_c}{N}\left(\frac{(\Delta N_c)^2}{4\xi^2} + \xi^2(D_c + D_s/2)^2 - \tfrac{1}{12} + n_c^+ + n_c^-\right)$$
$$+\frac{2\pi v_s}{N}(\tfrac{1}{2}S^2 + \tfrac{1}{2}D_s^2 - \tfrac{1}{12} + n_s^+ + n_s^-) \quad (3.12)$$

and

$$P = \frac{2\pi}{N}N_{c0}(D_c + D_s/2) + \frac{2\pi}{N}[\Delta N_c(D_c + D_s/2) + n_c^+ + n_c^-]$$
$$+\frac{2\pi}{N}(-SD_s + n_s^+ - n_s^-) \quad (3.13)$$

with

$$S = N_c/2 - N_s. \quad (3.14)$$

It is also interesting to give the Fermi velocities in this limit:

$$2\pi v_c = \varepsilon_c'(k_0)/\rho_c(k_0)$$
$$2\pi v_s = \left(\int_{-k_0}^{k_0} \exp\left(\frac{\pi}{2u}\sin k\right)\varepsilon_c'(k)\right)\left(\int_{-k_0}^{k_0} \exp\left(\frac{\pi}{2u}\sin k\right)\rho_c(k)\right)^{-1}. \quad (3.15)$$

Here $\rho_c(k)$ and $\varepsilon_c'(k)$ satisfy the equations

$$\rho_c(k) = \frac{1}{2\pi} + \frac{1}{2\pi}\cos(k)\int_{-k_0}^{k_0} \bar{K}(\sin k - \sin k')\rho_c(k') \quad (3.16)$$

and

$$\varepsilon_c'(k) = 2\sin k + \frac{1}{2\pi}\cos(k)\int_{-k_0}^{k_0} \bar{K}(\sin k - \sin k')\varepsilon_c'(k'). \quad (3.17)$$

For the negative U case the $\lambda_0 \to \infty$ limit corresponds to the half-filled band $\nu_f + 2\nu_b = 1$.

We have to comment also on the $k_0 \to \pi$ limit. For the positive U case this corresponds to the half-filled band. In a strictly half-filled band $N_c = N$, and there are no such charge excitations as described in this work since there is no place for the particles (for $N_c = N$ the only possible charge excitations are those with complex k, but those have a gap). It is possible to create such particle-hole excitations only if first N_c is decreased, i.e. a $\Delta N_c < 0$ is introduced. Even in this case, however, we do not get a contribution to the energy as $v_c = 0$ if $k_0 = \pi$ (indicating that the spectrum is quadratic, i.e. the excitation energy $O(1/N^2)$. If U is negative, the $k_0 = \pi$ limit corresponds to the zero magnetisation, i.e. to states with no 'free' particles, only bound pairs. It is possible to break up pairs to create free particles and this does not cost large amounts of energy since the magnetic field and chemical potential are such that both the bottom of the band for the free particles and the Fermi points of the sea of bound pairs are at zero. The fact that $v_c = 0$ indicates that the spectrum of the free particles is quadratic, as it should be at the bottom of a band.

An important case is when $v_c = v_s$: in this point the model is conformally invariant with $c = 2$. In the Hubbard model (1.4) there are three parameters U, μ and h, or equivalently U, ν_c and ν_s. In principle one can define through the equation

$$v_c(U, \nu_c, \nu_s) = v_s(U, \nu_c, \nu_c) \quad (3.18)$$

a function $U(\nu_c, \nu_s)$ which gives the value of U where (3.18) is satisfied at the given ν_c and ν_s. Not all (ν_c, ν_s) pairs define such a U, but for a certain part of the $0<2\nu_s\leq\nu_c<1$ parameter space such a U can be found. (An example for this is the $\nu_s=\nu_c/2$ case, where for $\nu_c\to 0$ $(k_0\to 0)$, $v_s/v_c\to 0$, while for $\nu_c\to 1$ $(k_0\to\pi)$, $v_s/v_c\to\infty$ at any U (see (3.15)), so there is a ν_c to any U where $v_c=v_s$. This means that there is a whole range in $0<\nu_c<1$ where, with $\nu_s=\nu_c/2$, (3.18) can be satisfied.) In those points, where ν_c and ν_s are rational and satisfy the requirements discussed in the first paragraphs of this section, the model has a conformally invariant continuum limit with $c=2$, provided U is chosen according to (3.18). In this case the scaling indices of the primary operators are

$$
\begin{aligned}
x(\Delta N_c, \Delta N_s, D_c, D_s) &= \left(\frac{(\xi_{22}\Delta N_c-\xi_{21}\Delta N_s)^2}{4(\det\xi)^2}+(\xi_{11}D_c+\xi_{12}D_s)^2\right)\\
&\quad+\left(\frac{(\xi_{12}\Delta N_c-\xi_{11}\Delta N_s)^2}{4(\det\xi)^2}+(\xi_{21}D_c+\xi_{22}D_s)^2\right) \qquad (3.19)\\
s(\Delta N_c, \Delta N_s, D_c, D_s) &= \Delta N_c D_c+\Delta N_s D_s
\end{aligned}
$$

with ΔN_c, ΔN_s being integers, and D_c, D_s satisfying (3.8). It is worth noting that (3.19) is a generalisation of the Gaussian form: with the notation

$$
\begin{pmatrix}\Delta N_c\\ \Delta N_s\end{pmatrix}=\boldsymbol{\Delta N} \qquad \begin{pmatrix}D_c\\ D_s\end{pmatrix}=\boldsymbol{D} \qquad \xi^{\mathrm{T}}\xi=\mathbf{X} \qquad (3.20)
$$

we have

$$
\begin{aligned}
x(\boldsymbol{\Delta N}, \boldsymbol{D}) &= \tfrac{1}{4}\boldsymbol{\Delta N}^{\mathrm{T}}\mathbf{X}^{-1}\boldsymbol{\Delta N}+\boldsymbol{D}^{\mathrm{T}}\mathbf{X}\boldsymbol{D}\\
s(\boldsymbol{\Delta N}, \boldsymbol{D}) &= \boldsymbol{\Delta N}^{\mathrm{T}}\boldsymbol{D}.
\end{aligned} \qquad (3.21)
$$

In addition to the above operators, there is a class of operators with $\boldsymbol{\Delta N}=\boldsymbol{D}=0$ and

$$
x=n_c^++n_c^-+n_s^++n_s^- \qquad s=n_c^+-n_c^-+n_s^+-n_s^-. \qquad (3.22)
$$

It is remarkable that four of them are marginal ($x=2$; $s=0$). Although we cannot read out the scaling indices from the spectrum directly if $v_c\neq v_s$, we expect that some of these operators are marginal even if $v_c\neq v_s$. The reason for this is that in the most general 1D model of spin-$\frac{1}{2}$ fermions there are several coupling constants and the model is critical in a whole region of a four-dimensional parameter space (for a review see Sólyom 1979). Thus there must be a set of marginal operators which govern the motion of the Hamiltonian in this parameter space. The Hubbard model is one special line parametrised by U in the critical region of the more general model, and the operator which, by adding it to the Hamiltonian, changes the value of U can be constructed readily:

$$
\sum_i\left(n_{i\uparrow}n_{i\downarrow}+\frac{\partial\mu}{\partial u}(n_{i\uparrow}+n_{i\downarrow})-\frac{1}{2}\frac{\partial h}{\partial u}(n_{i\uparrow}-n_{i\downarrow})\right) \qquad (3.23)
$$

where $\partial\mu/\partial u$ and $\partial h/\partial u$ are partial derivatives at fixed ν_c and ν_s. This operator does not change the complete integrability of the model. Nevertheless—if U was such that $v_c=v_s$—it drives out the system from the conformally invariant point. The operators driving the system off the Hubbard line are not present explicitly in the Hubbard Hamiltonian. Nevertheless they should also be marginal.

Acknowledgments

I am very grateful to Drs V E Korepin, N M Bogoliubov and N Yu Reshetikhin and to Professor J Sólyom for extremely interesting and valuable discussions. This work was partially supported by DFG under contract Seho-158/8-1.

References

Affleck I 1986 *Phys. Rev. Lett.* **56** 746
Belavin A A, Polyakov A M and Zamolodchikov A B 1984 *J. Stat. Phys.* **34** 763
Berkovich A and Murthy G 1988 *J. Phys. A: Math. Gen.* **21** L381
Blöte H W J, Cardy J L and Nightingale M P 1986 *Phys. Rev. Lett.* **56** 742
Bogoliubov N M, Izergin A G and Reshetikhin N Yu 1987 *J. Phys. A: Math. Gen.* **20** 5631
Cardy J L 1986a *Nucl. Phys.* B **270** [FS16] 186
—— 1986b *Phase Transitions and Critical Phenomena* vol 11, ed C Domb and J L Lebowitz (New York: Academic)
de Vega H J and Woynarovich F 1985 *Nucl. Phys.* B **251** [FS13] 439
Friedan D, Qiu Z and Shenker S 1984 *Phys. Rev. Lett.* **52** 1575
Korepin V E 1979 *Teor. Mat. Fiz.* **41** 169
Korepin V E, Bogoliubov N M and Reshetikhin N Yu 1988 private communication
Lieb E M and Wu F Y 1968 *Phys. Rev. Lett.* **20** 1445
Sólyom J 1979 *Adv. Phys.* **28** 201
Suzuki J 1988 *Preprint* Tokyo
Woynarovich F 1982a *J. Phys. C: Solid State Phys.* **15** 85
—— 1982b *J. Phys. C: Solid State Phys.* **15** 97
—— 1983a *J. Phys. C: Solid State Phys.* **16** 5293
—— 1983b *J. Phys. C: Solid State Phys.* **16** 6593
Woynarovich F and Eckle H-P 1987a *J. Phys. A: Math. Gen.* **20** L97
—— 1987b *J. Phys. A: Math. Gen.* **20** L443
Woynarovich F, Eckle H-P and Truong T T 1989 *J. Phys. A: Math. Gen.* **22** 4027

PHYSICAL REVIEW B VOLUME 43, NUMBER 7 1 MARCH 1991

Correlation functions of the one-dimensional Hubbard model in a magnetic field

Holger Frahm*
Department of Physics, University of Virginia, Charlottesville, Virginia 22901

V. E. Korepin
Institute for Theoretical Physics, State University of New York at Stony Brook, Stony Brook, New York 11794-3840

(Received 24 August 1990)

We present a general method for the calculation of correlation functions in the repulsive one-dimensional Hubbard model at less than half-filling in a magnetic field h. We describe the dependence of the critical exponents that drive their long-distance asymptotics on the Coulomb coupling, the density, and h. This dependence can be described in terms of a set of coupled Bethe-*Ansatz* integral equations. It simplifies significantly in the strong-coupling limit, where we give explicit formulas for the dependence of the critical exponents on the magnetic field. In particular, we find that at small field the functional dependence of the critical exponents on h can be algebraic or logarithmic—depending on the operators involved. In addition, we evaluate the singularities of the Fourier images of the correlation functions. It turns out that switching on a magnetic field gives rise to singularities in the dynamic field-field correlation functions that are absent at $h=0$.

I. INTRODUCTION

Recently, the relevance of correlation effects in interacting electron systems to high-T_c superconductivity has led to growing interest in the computation of correlation functions for the one-dimensional Hubbard model[1-7] since it combines the essentials of correlated electrons on one hand with the attractive feature of complete integrability on the other.

As a number of other systems in one spatial dimension, the Hubbard model has a critical point at zero temperature. For these models correlation functions decay as powers of the distance asymptotically. The calculation of these powers—the critical exponents—is of major interest. For some of these systems, for example, spinless fermions and the spin-$\frac{1}{2}$ Heisenberg chain, this can be done within Haldane's *Luttinger liquid* approach,[8-11] based on the fact that these models belong to the same universality class as the Gaussian model.[12] Due to conformal invariance,[13] the possible universality classes of critical theories are related to a single dimensionless number: the central charge c of the underlying Virasoro algebra ($c=1$ for the Gaussian model). Furthermore, the finite-size corrections in the spectra of these models are closely related to the scaling dimensions of the fields present in the theory,[14,15] which in turn determine the critical exponents. Common to all of these systems is the property that they have a single critical degree of freedom only. As is known from Lieb and Wu's[16] exact solution the Hubbard model does not belong to this class: in general, both charge- and spin-density waves are critical. Only in a sufficiently strong magnetic field—where the spin-density waves have a gap—or at half-filling—where the Hubbard model is insulating and the charge excitations are massive—the remaining critical degree of freedom can be described within the scheme outlined above.

Based on the Bethe-*Ansatz* solution the finite-size corrections for the half-filled Hubbard model at zero magnetic field have been calculated analytically and the critical theory has been found to be described by a single Virasoro algebra with central charge $c=1$.[17] The Hubbard model with attractive interaction at arbitrary filling also has central charge $c=1$; the critical exponents depend on the density of electrons.[18] The magnetic field dependence of the critical behavior in a theory of this type has first been investigated in the Heisenberg spin chain.[19] The critical theory still corresponds to $c=1$. The scaling dimensions, however, are found to depend on the magnetic field.

For systems with more than one critical degree of freedom—the situation is more complicated: the understanding of this class of systems is based on exact finite-size calculations[20,21] in models that are soluble by a hierarchy[22] of Bethe *Ansätze* (the Hubbard model belongs to this class). It has been found that they can be understood as a direct product of Virasoro algebras each having central charge $c=1$. As in the one-component case this continues to hold in a generic situation, i.e., with external fields coupled to the critical degrees of freedom.[21]

Woynarovich has calculated the finite-size corrections in the spectrum of the Hubbard model at less than half-filling analytically.[23] His results can be interpreted in the framework of conformal quantum field theory following the lines mentioned above: the critical theory describing both charge- and spin-density waves is based on a product of two Virasoro algebras each having central charge $c_i=1$. In a previous paper we have used these results to calculate the critical exponents for the Hubbard model.[5] There we have shown that while the central charges are universal the critical exponents depend on all the system parameters, i.e., density (or, alternatively, the chemical potential), coupling constant, and the magnetic field. In the present paper we concentrate on the magnetic field dependence. It is found to be logarithmic in the generic case—very similar to the situation in the spin-$\frac{1}{2}$ Heisenberg model.[19] In fact, this similarity has been exploited before to show that the small field magnetic susceptibility χ exhibits logarithmic singularities as $h \to 0$ and as a further consequence the specific heat coefficient is singular.[24] Since the critical exponents are closely related to thermodynamic quantities such as χ it is not surprising that they show the same singular field dependence.[5] For the full Hubbard model (that is, including charge degrees of freedom) we find an additional linear field dependence in the corresponding contributions to the critical exponents. This is much weaker than the logarithmic one and shows up in certain correlators only.

At this point we would like to emphasize the difference between our approach[5] and that of several other authors:[3,6,7] the exact solution of the model and, in particular, the spectrum of low-lying excitations[23] show that the contributions of charge- and spin-density waves cannot be described by two independent effective Hamiltonians—charge- and spin-density waves do interact. This means that the spectrum of conformal operator dimensions in this model is determined by a 2×2 matrix (the so-called dressed charge matrix, see below) rather than two scalar coupling constants. At zero magnetic field this fact is obscured to some extent since the dressed charge matrix is found to be triangular and the nonzero off-diagonal element is just one-half of one of the diagonal ones.[23] Due to this fact, the critical exponents obtained when starting from two independent critical theories[3,6,7] coincide with the ones found on the basis of the exact finite-size corrections.[5]

In the presence of a magnetic field, however, this situation changes: no simple relation between different elements of the dressed charge matrix holds and, as we show in this paper, the functional dependence of the various matrix elements on the field is significantly different. An important consequence of this is the appearance of an additional singularity of the electronic field correlation function in momentum space if the magnetic field is switched on. We do not think that this effect can be obtained within an approach that starts from separate effective theories for charge- and spin-density waves, respectively.

The Hubbard model describes spin-$\frac{1}{2}$ electrons on the lattice. The electrons are described by canonical Fermi fields ψ_σ, $\psi_\sigma^\dagger$ and the Hamiltonian is given by the following expression:

$$\begin{aligned}\mathcal{H} = &-\sum_{j=1}^{N}\sum_{\sigma}\left(\psi_{j+1,\sigma}^\dagger\psi_{j,\sigma}+\psi_{j,\sigma}^\dagger\psi_{j+1,\sigma}\right)\\ &+4u\sum_{j=1}^{N}n_{j\uparrow}n_{j\downarrow}+\mu\sum_{j=1}^{N}(n_{j\uparrow}+n_{j\downarrow})\\ &-\frac{h}{2}\sum_{j=1}^{N}(n_{j\uparrow}-n_{j\downarrow})\,. \end{aligned} \tag{1.1}$$

Here $n_{j,\sigma} = \psi_{j,\sigma}^\dagger\psi_{j,\sigma}$ is the number of spin σ electrons at site j, $4u > 0$ is the on-site Coulomb repulsion, μ is the chemical potential, and h is an external magnetic field.

Our paper is organized as follows: In the following section we shall review the Bethe-*Ansatz* solution of the model (1.1) in the aspects relevant to the present work. In Sec. III we consider the strong-coupling limit $u \to \infty$. In this limit the dependence of the critical exponents on the magnetic field h simplifies essentially. In particular, we investigate this dependence for small fields and close to the critical field h_c where all the spins are aligned and the model becomes ferromagnetic. In Sec. IV we apply these results to compute the critical exponents for a few interesting correlation functions. In the final section we calculate the singularities in the Fourier images of these correlation functions.

II. THE BETHE-*ANSATZ* SOLUTION FOR THE HUBBARD MODEL

Lieb and Wu[16] have constructed a complete set of eigenfunctions of the Hubbard Hamiltonian (1.1). In the thermodynamic limit the zero-temperature ground state of the model consists of two Fermi seas, characterized by distribution functions $\rho_c(k)$ of charges with "holon" momentum k and $\rho_s(\lambda)$ of down spins with spin-wave ("spinon") rapidity λ. Lieb and Wu have written down integral equations describing this configuration:

$$\rho_c(k) = \frac{1}{2\pi} + \frac{\cos k}{2\pi}\int_{-\lambda_0}^{\lambda_0} d\lambda K(\sin k - \lambda; u)\rho_s(\lambda)\,, \tag{2.1}$$

$$\begin{aligned}\rho_s(\lambda) = &\frac{1}{2\pi}\int_{-k_0}^{k_0} dk K(\lambda - \sin k; u)\rho_c(k)\\ &-\frac{1}{2\pi}\int_{-\lambda_0}^{\lambda_0} d\mu K(\lambda-\mu; 2u)\rho_s(\mu)\,.\end{aligned}$$

The kernels of these equations are given by

$$K(x;\alpha) = \frac{2\alpha}{\alpha^2 + x^2}\,. \tag{2.2}$$

The values of λ_0 and k_0 are related to the number of electrons per lattice site $n_c = N_c/N$ and the magnetization $\mathcal{M} = (N_\uparrow - N_\downarrow)/2N$ (N_σ is the number of spin-σ

electrons) by the following equations:

$$\int_{-k_0}^{k_0} dk\rho_c(k) = \frac{N_c}{N} \equiv n_c \ ,$$
$$\int_{-\lambda_0}^{\lambda_0} d\lambda\rho_s(\lambda) = \frac{N_s}{N} \equiv \frac{1}{2}n_c - \mathcal{M} \ . \tag{2.3}$$

Another way to describe the system is in terms of integral equations for excitation energies:

$$\varepsilon_c(k) = \varepsilon_c^{(0)}(k) + \frac{1}{2\pi}\int_{-\lambda_0}^{\lambda_0} d\lambda K(\sin k - \lambda; u)\varepsilon_s(\lambda) \ ,$$
$$\varepsilon_s(\lambda) = \varepsilon_s^{(0)}(\lambda) + \frac{1}{2\pi}\int_{-k_0}^{k_0} dk \cos k K(\lambda - \sin k; u)\varepsilon_c(k) - \frac{1}{2\pi}\int_{-\lambda_0}^{\lambda_0} d\mu K(\lambda - \mu; 2u)\varepsilon_s(\mu) \ . \tag{2.4}$$

Here $\varepsilon_c(k)$ is the energy of a charge-density excitation with momentum k, $\varepsilon_s(\lambda)$ that of a spin-density wave with rapidity λ. The bare energies $\varepsilon_{c,s}^{(0)}$ are

$$\varepsilon_c^{(0)}(k) = \mu - \frac{h}{2} - 2\cos k, \qquad \varepsilon_s^{(0)}(\lambda) = h \ . \tag{2.5}$$

The solutions of Eqs. (2.4) define the energy bands. The ground-state configuration corresponds to the filling of all states with $\varepsilon_c(k) < 0$ and $\varepsilon_s(\lambda) < 0$. Consequently, the conditions

$$\varepsilon_c(k_0) = 0, \qquad \varepsilon_s(\lambda_0) = 0 \tag{2.6}$$

provide another way to define the values of k_0 and λ_0 for the ground state for given magnetic field h and chemical potential μ. At zero field $h = 0$ one finds $\lambda_0 = \infty$, the critical field $h = h_c$ where all the spins are pointing up, corresponds to $\lambda_0 = 0$. From (2.6) it is found to be

$$h_c = \frac{2u}{\pi}\int_{-\pi n_c}^{\pi n_c} dk \cos k \frac{\cos k - \cos \pi n_c}{u^2 + \sin^2 k} \ . \tag{2.7}$$

At large coupling u the value of h_c scales like $1/u$:

$$h_c \sim \frac{2}{u}\left(n_c - \frac{1}{2\pi}\sin 2\pi n_c\right) + O\left(\frac{1}{u^3}\right), \quad \text{for } u \gg 1 \ . \tag{2.8}$$

The quantity that determines the critical exponents is the dressed charge matrix:[5]

$$Z \equiv \begin{pmatrix} Z_{cc} & Z_{cs} \\ Z_{sc} & Z_{ss} \end{pmatrix} = \begin{pmatrix} \xi_{cc}(k_0) & \xi_{cs}(\lambda_0) \\ \xi_{sc}(k_0) & \xi_{ss}(\lambda_0) \end{pmatrix} \ . \tag{2.9}$$

Here the matrix ξ_{ij} is defined as the solution of the following integral equations:[21,23,25]

$$\xi_{cc}(k) = 1 + \frac{1}{2\pi}\int_{-\lambda_0}^{\lambda_0} d\lambda \xi_{cs}(\lambda) K(\lambda - \sin k; u) \ ,$$
$$\xi_{cs}(\lambda) = \frac{1}{2\pi}\int_{-k_0}^{k_0} dk \cos k \xi_{cc}(k) K(\sin k - \lambda; u) - \frac{1}{2\pi}\int_{-\lambda_0}^{\lambda_0} d\mu \xi_{cs}(\mu) K(\mu - \lambda; 2u) \ ,$$
$$\xi_{sc}(k) = \frac{1}{2\pi}\int_{-\lambda_0}^{\lambda_0} d\lambda \xi_{ss}(\lambda) K(\lambda - \sin k; u) \ ,$$
$$\xi_{ss}(\lambda) = 1 + \frac{1}{2\pi}\int_{-k_0}^{k_0} dk \cos k \xi_{sc}(k) K(\sin k - \lambda; u) - \frac{1}{2\pi}\int_{-\lambda_0}^{\lambda_0} d\mu \xi_{ss}(\mu) K(\mu - \lambda; 2u) \ . \tag{2.10}$$

Other quantities of relevance are the Fermi velocities v_c and v_s of charge- and spin-density waves:

$$v_c = \frac{1}{2\pi\rho_c(k_0)}\varepsilon_c'(k_0) > 0 \ ,$$
$$v_s = \frac{1}{2\pi\rho_s(\lambda_0)}\varepsilon_s'(\lambda_0) > 0 \ . \tag{2.11}$$

and the Fermi momenta $\mathcal{P}_{F,\uparrow}$ and $\mathcal{P}_{F,\downarrow}$ for electrons with spin up and down, respectively, i.e.,

$$\mathcal{P}_{F,\uparrow(\downarrow)} = \tfrac{1}{2}(\pi n_c \pm 2\pi\mathcal{M}) \ . \tag{2.12}$$

We already mentioned in the Introduction that the critical behavior of the repulsive Hubbard model at less than filling ($n_c < 1$) is described by the direct product of two Virasoro algebras, each having central charge c=1. The corresponding conformal dimensions of the primary fields are given in terms of the elements of (2.9) as[5]

$$2\Delta_c^{\pm}(\Delta\mathbf{N}, \mathbf{D}) = \left(Z_{cc}D_c + Z_{sc}D_s \pm \frac{Z_{ss}\Delta N_c - Z_{cs}\Delta N_s}{2\det Z}\right)^2 ,$$
$$2\Delta_s^{\pm}(\Delta\mathbf{N}, \mathbf{D}) = \left(Z_{cs}D_c + Z_{ss}D_s \pm \frac{Z_{cc}\Delta N_s - Z_{sc}\Delta N_c}{2\det Z}\right)^2 . \tag{2.13}$$

Here ΔN_c and ΔN_s are integers which describe the quantum numbers of the operators involved in the correlation functions considered, i.e., the change in the number of charges and down spins with respect to the ground state. D_c and D_s are integer or half integer depending on the parities of ΔN_c and ΔN_s:

$$D_c = \frac{\Delta N_c + \Delta N_s}{2} \bmod 1\,, \qquad D_s = \frac{\Delta N_c}{2} \bmod 1\,. \tag{2.14}$$

Knowing the conformal dimensions (2.13) the long-distance asymptotics of zero-temperature correlation functions is then given as a sum of terms

$$\frac{\exp(-2iD_c\mathcal{P}_{F,\uparrow}x)\exp[-2i(D_c+D_s)\mathcal{P}_{F,\downarrow}x]}{(x-iv_c\tau)^{2\Delta_c^+}(x+iv_c\tau)^{2\Delta_c^-}(x-iv_s\tau)^{2\Delta_s^+}(x+iv_s\tau)^{2\Delta_s^-}} \tag{2.15}$$

(we use Euclidean time τ). As discussed above, ΔN_c and ΔN_s are fixed by the correlator one considers while summation has to be performed with respect to D_c and D_s. The leading term in the asymptotics is given by the term in this sum which minimizes $\Delta^{\pm}$.

Conformal quantum field theory also describes the correlation functions at small nonzero temperatures. Their exponential decay is given by the same dimensions (2.13) (see Ref. 5).

III. THE STRONG-COUPLING LIMIT

To illustrate the dependence of the critical exponents on the magnetic field let us consider now the limit $u \to \infty$. This simplifies the integral equations (2.1), (2.4), and (2.10) significantly. After proper rescaling they can be rewritten in the following way:

$$\rho_s(\lambda) = \frac{n_c}{\pi}\frac{1}{1+\lambda^2} - \frac{1}{2\pi}\int_{-\lambda_0}^{\lambda_0} d\mu K(\lambda-\mu;2)\rho_s(\mu) + O\left(\frac{1}{u^2}\right)\,,$$

$$\varepsilon_s(\lambda) = h - \frac{h_c}{1+\lambda^2} - \frac{1}{2\pi}\int_{-\lambda_0}^{\lambda_0} d\mu K(\lambda-\mu;2)\varepsilon_s(\mu) + O\left(\frac{1}{u^2}\right)\,, \tag{3.1}$$

$$\xi_{ss}(\lambda) = 1 + \frac{\sin k_0}{\pi u}\frac{2Z_{sc}}{1+\lambda^2} - \frac{1}{2\pi}\int_{-\lambda_0}^{\lambda_0} d\mu K(\lambda-\mu;2)\xi_{ss}(\mu) + O\left(\frac{1}{u^2}\right)\,.$$

(λ_0 differs from its value used in the previous section by a factor of u.) At $u = \infty$ the value of the critical field h_c (2.8) vanishes—an infinitesimal magnetic field is sufficient to magnetize the system completely. From (3.1) we see, however, that at first order in $1/u$ we already obtain nontrivial behavior.

The density and energy of the charge-density waves are given by the following expressions:

$$\rho_c(k) = \frac{1}{2\pi} + \frac{\cos k}{2\pi u}\int_{-\lambda_0}^{\lambda_0} d\lambda \frac{2\rho_s(\lambda)}{1+\lambda^2} + O\left(\frac{1}{u^2}\right) \quad \text{for } k \le k_0\,,$$

$$\varepsilon_c(k) = 2(\cos k_0 - \cos k) + O\left(\frac{1}{u^2}\right)\,. \tag{3.2}$$

For the calculation of the elements of the dressed charge matrix (2.9) we shall neglect the corrections in $1/u$. These contributions can, however, be obtained from (3.1) and similar expressions for the other elements. (For h=0 they have been written down in Ref. 5.) In this approximation they are given by

$$Z_{cc} = 1, \quad Z_{cs} = 0,$$

$$Z_{sc} = \frac{1}{2\pi}\int_{-\lambda_0}^{\lambda_0} d\lambda \xi_{ss}(\lambda)K(\lambda;1)\,. \tag{3.3}$$

For vanishing magnetic field $\lambda_0 = \infty$ and Eqs. (3.1) can be solved by Fourier transformation. This case has been discussed in great detail in our previous paper.[5] The dressed charge matrix has been found to be

$$Z = \frac{1}{2}\begin{pmatrix} 2 & 0 \\ 1 & \sqrt{2} \end{pmatrix}\,. \tag{3.4}$$

For small magnetic field (and λ_0 large but finite) one can use the Wiener-Hopf (WH) method as outlined in the Appendix together with the condition

$$\varepsilon_s(\lambda_0) = 0 \tag{3.5}$$

to find the field dependence of λ_0:

$$\lambda_0 = \frac{2}{\pi}\ln\left(\frac{h_0}{h}\right)\,, \quad h_0 = \sqrt{\frac{\pi^3}{2e}}\,h_c\,. \tag{3.6}$$

(A similar dependence has been found for the isotropic spin-$\frac{1}{2}$ Heisenberg magnetic chain.[19]) For the magnetization $\mathcal{M}$ we obtain in an analogous calculation

$$\mathcal{M} = \frac{2n_c}{\pi^2 h_c}\,h \tag{3.7}$$

and for the elements of the dressed charge matrix in a small magnetic field:

$$Z_{ss} = \frac{1}{\sqrt{2}}\left(1 + \frac{1}{4\ln(h_0/h)}\right) + O\left(\frac{1}{[\ln(h_0/h)]^2}\right)\,. \tag{3.8}$$

To calculate the leading correction to Z_{sc} for finite λ_0 we make use of the fact that Eq. (3.3) can be written alternatively as

$$Z_{sc} = \frac{1}{n_c}\int_{-\lambda_0}^{\lambda_0} d\lambda \rho_s(\lambda) = \tfrac{1}{2} - \frac{\mathcal{M}}{n_c} \tag{3.9}$$

[this is possible, since the kernel of the integral equations (3.1) is symmetric]. Hence, we obtain with (3.7)

$$Z_{sc} = \tfrac{1}{2} - \frac{2}{\pi^2}\frac{h}{h_c} + O\left(\frac{h}{h_c \ln(h_0/h)}\right) . \tag{3.10}$$

Note the different functional dependence on h of Z_{ss} and Z_{sc} near h=0. This has an interesting consequence on the conformal dimensions (2.13) entering the expressions for the correlation functions:

$$\begin{aligned} 2\Delta_c^{\pm}(\Delta\mathbf{N},\mathbf{D}) &= (D_c + \tfrac{1}{2}D_s \pm \tfrac{1}{2}\Delta N_c)^2 \\ &\quad - \frac{4h}{\pi^2 h_c}(D_c + \tfrac{1}{2}D_s \pm \tfrac{1}{2}\Delta N_c)D_s , \end{aligned} \tag{3.11}$$

$$\begin{aligned} 2\Delta_s^{\pm}(\Delta\mathbf{N},\mathbf{D}) &= \tfrac{1}{2}[D_s \pm (\Delta N_s - \tfrac{1}{2}\Delta N_c)]^2 \\ &\quad + \frac{1}{4\ln(h_0/h)}[D_s^2 - (\Delta N_s - \tfrac{1}{2}\Delta N_c)^2] . \end{aligned}$$

The magnetic field dependence of the critical dimensions for the charge excitations is much weaker than that of the spin excitations. This is not surprising since the magnetic field couples directly to the spin degree of freedom. In general, the exponents of equal time correlators (where only the sum of Δ_c and Δ_s enters) will be dominated by the latter; in time-dependent quantities, however, this effect should become observable.

As h approaches the critical field (2.8) from below, i.e., near the ferromagnetic state, λ_0 vanishes like

$$\lambda_0 = \left(\frac{h_c - h}{h_c}\right)^{1/2} . \tag{3.12}$$

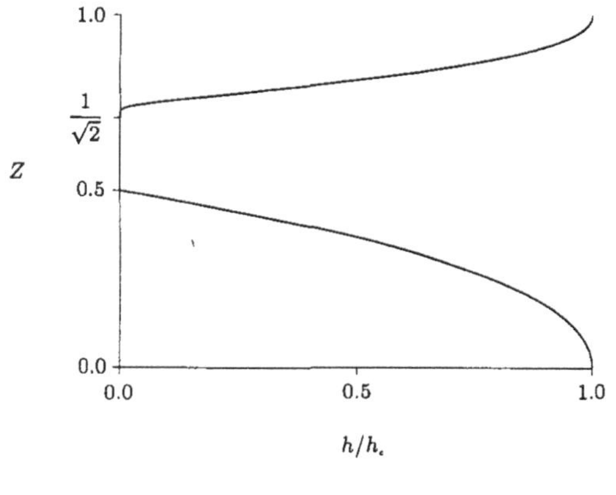

FIG. 1. Magnetic field dependence of the elements Z_{ss} (upper curve) and Z_{sc} (lower curve) of the dressed charge matrix in the strong-coupling limit. Note the drastically different functional behavior as $h \to 0$.

The dressed charge matrix in this regime is given by

$$Z = \begin{pmatrix} 1 & 0 \\ 0 & 1 \end{pmatrix} + \frac{1}{\pi}\left(\frac{h_c - h}{h_c}\right)^{1/2} \begin{pmatrix} 0 & 0 \\ 2 & -1 \end{pmatrix} \tag{3.13}$$

(this agrees with the limiting case $h = h_c$ considered in Ref. 5). The field dependence of the conformal dimensions to leading order is given by

$$\begin{aligned} 2\Delta_c^{\pm}(\Delta\mathbf{N},\mathbf{D}) &= (D_c \pm \tfrac{1}{2}\Delta N_c)^2 + \frac{4}{\pi}\left(1 - \frac{h}{h_c}\right)^{1/2}(D_c \pm \tfrac{1}{2}\Delta N_c)D_s , \\ 2\Delta_s^{\pm}(\Delta\mathbf{N},\mathbf{D}) &= (D_s \pm \tfrac{1}{2}\Delta N_s)^2 - \frac{2}{\pi}\left(1 - \frac{h}{h_c}\right)^{1/2}(D_s \pm \tfrac{1}{2}\Delta N_s)[D_s \pm (\Delta N_c - \tfrac{1}{2}\Delta N_s)] . \end{aligned} \tag{3.14}$$

At $h \geq h_c$ a phase transition similar to the one found at half-filling occurs: excitations with spin develop a gap and the corresponding contributions to the correlation functions decay like exponentials asymptotically.

In Fig. 1 we present numerical results based on Eqs. (3.1) and (3.3) for the magnetic field dependence of Z_{ss} and Z_{sc} for the entire region $0 \leq h \leq h_c$.

IV. CORRELATION FUNCTIONS IN THE STRONG-COUPLING LIMIT

The results obtained in the last section can now be used to obtain the magnetic field dependence of the critical exponents of certain correlation functions. The basic procedure for this has been discussed in our earlier paper.[5]

First we consider the field-field correlation functions

$$G_{\psi\psi}^{(\sigma)}(x,t) = \langle \psi_\sigma(x,t)\psi_\sigma^\dagger(0,0)\rangle , \qquad \sigma = \uparrow, \downarrow . \tag{4.1}$$

For $\sigma = \uparrow$ the quantum numbers of this operator are ΔN_c=1 and ΔN_s=0; from (2.14) we find that D_c and D_s both take half odd integer values. Hence the leading contribution with wave number $\mathcal{P}_{F,\uparrow}$ to $G_{\psi\psi}^{(\uparrow)}$ is

$$\frac{\exp(i\mathcal{P}_{F,\uparrow}x)}{(x - iv_c\tau)^{2\Delta_c^+}(x + iv_c\tau)^{2\Delta_c^-}(x - iv_s\tau)^{2\Delta_s^+}(x + iv_s\tau)^{2\Delta_s^-}} , \tag{4.2}$$

where the expressions for the $\Delta_{c,s}^{\pm}$ are given by Eq. (2.13) with $\Delta N_c = 1$, $\Delta N_s = 0$, $D_c = -D_s = -\frac{1}{2}$. Near $h = 0$ these expressions simplify considerably due to the simpler structure of the Z matrix. We obtain

$$2\Delta_c^+ = \tfrac{1}{16} - \frac{h}{2\pi^2 h_c}, \quad 2\Delta_c^- = \tfrac{9}{16} + \frac{3h}{2\pi^2 h_c},$$
$$2\Delta_s^+ = \frac{1}{2}\left(\frac{1}{4\ln(h_0/h)}\right)^2, \tag{4.3}$$
$$2\Delta_s^- = \tfrac{1}{2} + \frac{1}{2}\left(\frac{1}{4\ln(h_0/h)}\right)^2.$$

The logarithmic field dependence of $\Delta_s^\pm$ cancels to first order. Fortunately, the next order is completely fixed by the leading correction to Z_{ss} (3.8).

As h approaches h_c one obtains for this contribution

$$2\Delta_c^+ = 0, \quad 2\Delta_c^- = 1 - \frac{2}{\pi}\left(1 - \frac{h}{h_c}\right)^{1/2},$$
$$2\Delta_s^+ = \tfrac{1}{4} - \frac{3}{2\pi}\left(1 - \frac{h}{h_c}\right)^{1/2}, \tag{4.4}$$
$$2\Delta_s^- = \tfrac{1}{4} + \frac{1}{2\pi}\left(1 - \frac{h}{h_c}\right)^{1/2}.$$

[Here we neglect contributions of order $O(1 - h/h_c)$.]

There is also a contribution with wave number $\mathcal{P}_{F,\uparrow} + 2\mathcal{P}_{F,\downarrow}$ (corresponding to $D_c = D_s = -\frac{1}{2}$) with

$$2\Delta_c^+ = \tfrac{1}{16} - \frac{h}{2\pi^2 h_c}, \quad 2\Delta_c^- = \tfrac{25}{16} - \frac{5h}{2\pi^2 h_c},$$
$$2\Delta_s^+ = \tfrac{1}{2} + \frac{1}{2}\left(\frac{1}{4\ln(h_0/h)}\right)^2, \tag{4.5}$$
$$2\Delta_s^- = \frac{1}{2}\left(\frac{1}{4\ln(h_0/h)}\right)^2.$$

for small magnetic field $h \ll h_c$ (again the corrections of order $[\ln(h_0/h)]^{-1}$ cancel). As h approaches h_c one obtains for this contribution

$$2\Delta_c^+ = 0, \quad 2\Delta_c^- = 1 + \frac{2}{\pi}\left(1 - \frac{h}{h_c}\right)^{1/2},$$
$$2\Delta_s^+ = \tfrac{1}{4} + \frac{1}{2\pi}\left(1 - \frac{h}{h_c}\right)^{1/2}, \tag{4.6}$$
$$2\Delta_s^- = \tfrac{1}{4} - \frac{3}{2\pi}\left(1 - \frac{h}{h_c}\right)^{1/2}.$$

For the spin-down field correlator we have $\Delta N_c = 1$, $\Delta N_s = 1$ and D_c now runs through all integer, D_s through all half odd integer numbers. The leading terms in the correlation functions are the ones given above for the spin-up correlators with $\mathcal{P}_{F,\uparrow}$ and $\mathcal{P}_{F,\downarrow}$ interchanged. For small magnetic field the dimensions are the ones given in Eqs. (4.3) and (4.5) with h replaced by $-h$ in the expressions for $\Delta_c^\pm$. For $h \to h_c$ the exponents for the contribution with wave number $\mathcal{P}_{F,\downarrow}$ ($D_c = 0$, $D_s = -\frac{1}{2}$) are

$$2\Delta_c^\pm = \tfrac{1}{4} \mp \frac{1}{\pi}\left(1 - \frac{h}{h_c}\right)^{1/2}, \quad 2\Delta_s^+ = 0,$$
$$2\Delta_s^- = 1 - \frac{2}{\pi}\left(1 - \frac{h}{h_c}\right)^{1/2}, \tag{4.7}$$

and for the component with wave number $2\mathcal{P}_{F,\uparrow} + \mathcal{P}_{F,\downarrow}$ ($D_c = -1$, $D_s = \frac{1}{2}$) we find

$$2\Delta_c^+ = \tfrac{1}{4} - \frac{1}{\pi}\left(1 - \frac{h}{h_c}\right)^{1/2},$$
$$2\Delta_c^- = \tfrac{9}{4} - \frac{3}{\pi}\left(1 - \frac{h}{h_c}\right)^{1/2}, \tag{4.8}$$
$$2\Delta_s^+ = 1 - \frac{2}{\pi}\left(1 - \frac{h}{h_c}\right)^{1/2}, 2\Delta_s^- = 0.$$

Another correlation function of interest is the density-density correlation function

$$G_{nn}(x,t) = \langle n(x,t)n(0,0)\rangle,$$
$$n(x,t) = n_\uparrow(x,t) + n_\downarrow(x,t). \tag{4.9}$$

This operator has quantum numbers $\Delta N_c = \Delta N_s = 0$ and D_c, D_s take integer values. The leading contributions to the asymptotics apart from the constant are found to have wave number $2\mathcal{P}_{F,\uparrow}$ (corresponding to $D_c = -D_s = -1$). For small magnetic field the corresponding critical dimensions are

$$2\Delta_c^\pm = \tfrac{1}{4} + \frac{2h}{\pi^2 h_c}, \quad 2\Delta_s^\pm = \tfrac{1}{2} + \frac{1}{4\ln(h_0/h)}, \tag{4.10}$$

The contribution with wave number $2\mathcal{P}_{F,\downarrow}$ ($D_c = 0$ and $D_s = -1$) has the same dimensions with h in the expression for $\Delta_c^\pm$ replaced by $-h$. At $h = 0$ the amplitudes of these contributions are known to be vanishing in the strong-coupling limit.[1]

As $h \to h_c$ the leading terms beyond the constant in the asymptotics of the density-density correlation function are found at zero wave number:

$$\frac{1}{(x - iv_c\tau)^2} + \frac{1}{(x + iv_c\tau)^2} + \frac{1}{(x - iv_s\tau)^2} + \frac{1}{(x + iv_s\tau)^2}, \tag{4.11}$$

and at wave number $2(\mathcal{P}_{F,\uparrow} + \mathcal{P}_{F,\downarrow})$ ($D_c = -1$, $D_s = 0$): the dimensions do not depend on the magnetic field to leading order. At infinite coupling u they are known to be $2\Delta_c^\pm = 1$ and $2\Delta_s^\pm = 0$.[5]

The longitudinal spin-spin correlation function

$$G^z_{\sigma\sigma}(x,t) = \langle S^z(x,t)S^z(0,0)\rangle \tag{4.12}$$

(where $S^z(x,t) = [n_\uparrow(x,t) - n_\downarrow(x,t)]/2$) has the same quantum numbers as the density-density correlator. Hence, the leading term beyond the constant is the one with wave number $2\mathcal{P}_{F,\uparrow}$ and dimensions (4.10). The contribution of the spin-density waves $\Delta_s^\pm$ is the same as the one found in the isotropic Heisenberg chain.[19] Finally we want to consider the transversal spin-spin correlation function

$$G^{\perp}_{\sigma\sigma}(x,t) = \langle S^-(x,t)S^+(0,0)\rangle \ , \tag{4.13}$$

with $S^+(x,t) = \psi^\dagger_\uparrow(x,t)\psi_\downarrow(x,t)$. The leading term in the asymptotics has wave number $\mathcal{P}_{F,\uparrow} + \mathcal{P}_{F,\downarrow}$ and with the corresponding quantum numbers $\Delta N_c = 0$, $\Delta N_s = 1$, $D_c = -\frac{1}{2}$ and $D_s = 0$ we find for the dimensions in the strong-coupling limit

$$2\Delta_c^\pm = \tfrac{1}{4}, \quad 2\Delta_s^\pm = \tfrac{1}{2} - \frac{1}{4\ln(h_0/h)} \quad \text{for } h \to 0,$$
$$2\Delta_c^\pm = \tfrac{1}{4}, \quad 2\Delta_s^\pm = \tfrac{1}{4} + \frac{1}{2\pi}\left(1 - \frac{h}{h_c}\right)^{1/2} \quad \text{for } h \to h_c \ . \tag{4.14}$$

In our previous paper[5] we have evaluated the long distance asymptotics of correlators for singlet and triplet pairs in addition to the ones presented here. The calculation of the corresponding critical exponents is straightforward but will not be presented here, since they are larger than 2 for all values of h.

V. CORRELATION FUNCTIONS IN MOMENTUM SPACE

The long-distance asymptotics of zero-temperature correlation functions consists of terms of the form (2.15), i.e.,

$$g(x,\tau) = \frac{\exp(ik_0x)}{(x - iv_c\tau)^{2\Delta_c^+}(x + iv_c\tau)^{2\Delta_c^-}(x - iv_s\tau)^{2\Delta_s^+}(x + iv_s\tau)^{2\Delta_s^-}} \ . \tag{5.1}$$

To compute the Fourier transforms

$$\tilde{g}(k,\omega) = \int dx \int dt e^{-i(kx-\omega t)} g(x,t) \ , \tag{5.2}$$

we first have to rotate the time coordinate from Euclidean time τ to real time t. Analyticity requires

$$\tau = it + \text{sgn}(t) \ . \tag{5.3}$$

Equivalently, the correct regularization of the cuts in the complex time coordinate can be obtained by replacing $\tau = it$ in (5.1) and giving an infinitesimal imaginary part to the velocity:

$$v_i \to v_i e^{-i0} \ . \tag{5.4}$$

All the following equations are to be understood that way. This gives

$$g(x,t) = \frac{\exp(ik_0x)}{(x + v_ct)^{2\Delta_c^+}(x - v_ct)^{2\Delta_c^-}(x + v_st)^{2\Delta_s^+}(x - v_st)^{2\Delta_s^-}} \ . \tag{5.5}$$

As mentioned above, this expression holds asymptotically only. Nevertheless, it allows us to calculate the behavior of the Fourier transforms $\tilde{g}(k,\omega)$ near the singularities

$$\omega = \pm v_{c,s}(k - k_0) \ . \tag{5.6}$$

Standard methods yield

$$\tilde{g}(k,\omega) \sim \begin{cases} \text{const}[\omega \mp v_c(k-k_0)]^{2(\Delta_s^+ + \Delta_s^- + \Delta_c^\pm)-1} & \text{for } \omega \approx \pm v_c(k-k_0) \\ \text{const}[\omega \mp v_s(k-k_0)]^{2(\Delta_c^+ + \Delta_c^- + \Delta_s^\pm)-1} & \text{for } \omega \approx \pm v_s(k-k_0) \ . \end{cases} \tag{5.7}$$

This formula is applicable only if all the dimensions $\Delta_i^\pm$ are nonzero. If one of them vanishes (as is the case for the field-field correlator at zero magnetic field) the corresponding singularity disappears, i.e., const=0.

Note that the integral in (5.2) is not absolutely convergent in general. To prove (5.7) mathematically one should consider the case where $\sum \Delta_a^\pm > 1$ and the sum of three of the Δ's less than $\frac{1}{2}$ and then continue analytically to obtain this expression.

For the Fourier transform of equal time correlators, i.e., terms like

$$\tilde{g}(k) = \int dx e^{-ikx} g(x, t = 0^+) = \int dx \frac{e^{-i(k-k_0)x}}{(x - i0)^{2\Delta^+}(x + i0)^{2\Delta^-}} \tag{5.8}$$

(here $\Delta^\pm = \Delta_c^\pm + \Delta_s^\pm$), one has to consider $k > k_0$ and

$k < k_0$ separately. Contour integration yields

$$\frac{\tilde{g}(k_0+p)}{\tilde{g}(k_0-p)} = \frac{\sin 2\pi\Delta^-}{\sin 2\pi\Delta^+} = (-1)^{2s}, \quad p>0 \tag{5.9}$$

where $s = \Delta_+ - \Delta_-$ is the conformal spin of the operator under consideration. From (2.13) it is clear that $2s$ is always an integer. Using this we find that

$$\tilde{g}(k \approx k_0) \sim [\mathrm{sgn}(k-k_0)]^{2s}|k-k_0|^{\nu}, \qquad \nu = 2(\Delta_c^+ + \Delta_c^- + \Delta_s^+ + \Delta_s^-) - 1 . \tag{5.10}$$

The extra sign will appear in correlation functions of Fermi fields, e.g., the field-field correlator (4.1).

The results (5.10) and (5.7) are very general and can be applied to any correlation function.[5] To illustrate this let us consider here a few examples in the strong-coupling limit.

The static field-field correlator has a singularity at $\mathcal{P}_{F,\uparrow}$

$$\tilde{G}^{(\uparrow)}_{\psi\psi}(k \approx \mathcal{P}_{F,\uparrow}) \sim \mathrm{sgn}(k-\mathcal{P}_{F,\uparrow})|k-\mathcal{P}_{F,\uparrow}|^{\nu},$$
$$\nu = \tfrac{1}{8} + \frac{1}{[4\ln(h_0/h)]^2} \quad \text{for } h \to 0, \tag{5.11}$$
$$\nu = \tfrac{1}{2} - \frac{3}{\pi}\left(1-\frac{h}{h_c}\right)^{1/2} \quad \text{for } h \to h_c .$$

The h=0 exponent $\nu = \frac{1}{8}$ for the strong-coupling limit has been obtained before.[3-7] At zero magnetic field ν is a monotonous function of the coupling constant u and goes to 0 as $u \to 0$. For large but finite u the leading correction is[5]

$$\nu(h=0) = \tfrac{1}{8} - \frac{3\ln 2}{4\pi u}\sin \pi n_c . \tag{5.12}$$

Another singularity is at $\mathcal{P}_{F,\uparrow} + 2\mathcal{P}_{F,\downarrow}$ (b is a positive constant):

$$\tilde{G}^{(\uparrow)}_{\psi\psi}(k \approx \mathcal{P}_{F,\uparrow}+2\mathcal{P}_{F,\downarrow}) \sim \mathrm{sgn}(k-\mathcal{P}_{F,\uparrow}-2\mathcal{P}_{F,\downarrow}) \times |k-\mathcal{P}_{F,\uparrow}-2\mathcal{P}_{F,\downarrow}|^{\nu}, \tag{5.13}$$

$$\nu = \begin{cases} \frac{9}{8} + \dfrac{1}{[4\ln(h_0/h)]^2} & \text{for } h \to 0, \\ \frac{1}{2} + \dfrac{1}{\pi}\left(1-\dfrac{h}{h_c}\right)^{1/2} & \text{for } h \to h_c . \end{cases}$$

As mentioned above, the corresponding singularities for the spin-down correlation functions at small magnetic fields are obtained by replacing $\mathcal{P}_{F,\uparrow}$ with $\mathcal{P}_{F,\downarrow}$ and vice versa. Near $h = h_c$ one obtains

$$\tilde{G}^{(\downarrow)}_{\psi\psi}(k \approx \mathcal{P}_{F,\downarrow}) \sim \mathrm{sgn}(k-\mathcal{P}_{F,\downarrow})|k-\mathcal{P}_{F,\downarrow}|^{\nu},$$
$$\nu = \tfrac{1}{2} - \frac{2}{\pi}\left(1-\frac{h}{h_c}\right)^{1/2}, \tag{5.14}$$

$$\tilde{G}^{(\downarrow)}_{\psi\psi}(k \approx 2\mathcal{P}_{F,\uparrow}+\mathcal{P}_{F,\downarrow}) \sim \mathrm{sgn}(k-2\mathcal{P}_{F,\uparrow}-\mathcal{P}_{F,\downarrow}) \times|k-2\mathcal{P}_{F,\uparrow}-\mathcal{P}_{F,\downarrow}|^{\nu},$$
$$\nu = \tfrac{5}{2} - \frac{6}{\pi}\left(1-\frac{h}{h_c}\right)^{1/2} .$$

The singularities of the correlation function $G^{(\uparrow)}_{\psi\psi}(\omega,k)$ can be obtained from (5.7). For the contributions with wave number $k \approx \mathcal{P}_{F,\uparrow}$ we find

$$\tilde{G}^{(\uparrow)}_{\psi\psi}(\omega,k) \sim [\omega - v_c(k-\mathcal{P}_{F,\uparrow})]^{\nu}, \quad \text{for } \omega \approx v_c(k-\mathcal{P}_{F,\uparrow}),$$

with (5.15)

$$\nu = \begin{cases} -\frac{7}{16} + \dfrac{1}{[4\ln(h_0/h)]^2} & \text{as } h \to 0, \\ -\frac{1}{2} - \dfrac{1}{\pi}\left(1-\dfrac{h}{h_c}\right)^{1/2} & \text{as } h \to h_c, \end{cases}$$

$$\tilde{G}^{(\uparrow)}_{\psi\psi}(\omega,k) \sim [\omega + v_c(k-\mathcal{P}_{F,\uparrow})]^{\nu}, \quad \text{for } \omega \approx -v_c(k-\mathcal{P}_{F,\uparrow}),$$

with (5.16)

$$\nu = \begin{cases} \frac{1}{16} + \dfrac{1}{[4\ln(h_0/h)]^2} & \text{as } h \to 0, \\ \frac{1}{2} - \dfrac{3}{\pi}\left(1-\frac{h}{h_c}\right)^{1/2} & \text{as } h \to h_c, \end{cases}$$

$$\tilde{G}^{(\uparrow)}_{\psi\psi}(\omega,k) \sim [\omega - v_s(k-\mathcal{P}_{F,\uparrow})]^{\nu}, \quad \text{for } \omega \approx v_s(k-\mathcal{P}_{F,\uparrow}),$$

with (5.17)

$$\nu = \begin{cases} -\frac{3}{8} + \dfrac{1}{2[4\ln(h_0/h)]^2} & \text{as } h \to 0, \\ \frac{1}{4} - \dfrac{7}{2\pi}\left(1-\dfrac{h}{h_c}\right)^{1/2} & \text{as } h \to h_c . \end{cases}$$

For the Hubbard model without a magnetic field these singularities have also been found by Ren and Anderson.[7] For finite h there appears an additional singularity at $\omega + v_s(k-\mathcal{P}_{F,\uparrow}) \approx 0$ (this effect exists at any value of the coupling u):

$$\tilde{G}^{(\uparrow)}_{\psi\psi}(\omega,k) \sim [\omega + v_s(k-\mathcal{P}_{F,\uparrow})]^{\nu} \ , \quad \text{for } \omega \approx -v_s(k-\mathcal{P}_{F,\uparrow}) \ ,$$

with (5.18)

$$\nu = \begin{cases} \frac{1}{8} + \dfrac{1}{2[4\ln(h_0/h)]^2} & \text{as } h \to 0 \ , \\ \frac{1}{4} - \dfrac{3}{2\pi}\left(1-\dfrac{h}{h_c}\right)^{1/2} & \text{as } h \to h_c \ . \end{cases}$$

At $h=0$ the amplitude of this contribution to $\tilde{G}^{(\uparrow)}_{\psi\psi}(k,\omega)$ vanishes since $\Delta_s^+ \to 0$ in this limit. Similarly, there will appear an additional singularity at $\omega - v_s(k - \mathcal{P}_{F,\uparrow} - 2\mathcal{P}_{F,\downarrow}) \approx 0$ in the presence of a magnetic field.

For the static spin-spin correlator (4.13) we find for the singularity near wave number $\mathcal{P}_{F,\uparrow} + \mathcal{P}_{F,\downarrow}$

$$\tilde{G}^{\perp}_{\sigma\sigma}(k \approx \mathcal{P}_{F,\uparrow} + \mathcal{P}_{F,\downarrow}) \sim |k - \mathcal{P}_{F,\uparrow} - \mathcal{P}_{F,\downarrow}|^{\nu} \ , \tag{5.19}$$

where

$$\nu = \begin{cases} \frac{1}{2} - \dfrac{1}{2\ln(h_0/h)} & \text{for } h \to 0 \\ \dfrac{1}{\pi}\left(1-\dfrac{h}{h_c}\right)^{1/2} & \text{for } h \to h_c. \end{cases} \tag{5.20}$$

VI. CONCLUSION

In Ref. 5 and the current paper we give a complete description of correlation functions in the one-dimensional Hubbard model. Based on exact finite size calculations and the principles of conformal quantum field theory we find expressions for the critical exponents that describe the long-distance asymptotics of the correlation functions in coordinate space or, equivalently, the singularities of their Fourier transforms in momentum space. The values of the critical exponents are given in terms of the elements of the dressed charge matrix, which in turn is defined as the solution of a set of Bethe-*Ansatz* integral equations (2.10). The critical behavior depends on all the system parameters, i.e., the density of electrons, the coupling constant, and the magnetic field. In Ref. 5 we have concentrated on the dependence on the density and the coupling constant at zero field and at the critical field where the ground state becomes ferromagnetic. In the present work we have computed the magnetic field dependence of the critical exponents in the limit of strong coupling.

To conclude, we would like to emphasize again that the integral equations (2.10) fix the critical behavior for arbitrary values of the system parameters. The reason for studying limiting cases is to simplify the expressions for the critical exponents. Furthermore, the analysis of the critical behavior is not restricted to the correlation functions considered here. The method presented can be applied to arbitrary operators.

ACKNOWLEDGMENTS

V.E.K. gratefully acknowledges the hospitality of M. Fowler and the Physics Department (University of Virginia), where much of this work was performed. This work has been supported by the National Science Foundation (NSF) under Grant Nos. DMR-8810541 and PHY-89-08495.

APPENDIX: WIENER-HOPF METHOD FOR EQS. (3.1)

In this appendix we review the Wiener-Hopf method used above to solve integral equations of the type

$$f(\lambda) = f^{(0)}(\lambda) - \mathcal{K}_{\lambda_0} * f(\lambda) \ , \tag{A1}$$

where $\mathcal{K}$ is an integral operator whose action is defined by

$$\mathcal{K}_{\lambda_0} * f(\lambda) = \frac{1}{2\pi}\int_{-\lambda_0}^{\lambda_0} d\mu K(\lambda-\mu;2) f(\mu) \tag{A2}$$

for large but finite λ_0.

First we extend the definition of the bare function $f^{(0)}$ and of the kernel K to the entire real axis. Following Yang and Yang[26] we rewrite (A1) as

$$(1+\mathcal{K}_\infty) * f(\lambda) = f^{(0)}(\lambda) + \frac{1}{2\pi}\left(\int_{-\infty}^{-\lambda_0} + \int_{\lambda_0}^{\infty}\right) \times d\mu K(\lambda-\mu;2) f(\mu) \ . \tag{A3}$$

The operator $(1+\mathcal{K}_\infty)$ can be inverted by Fourier transform. Denoting

$$(1+\mathcal{K}_\infty)^{-1} \equiv 1 - \bar{\mathcal{K}} \tag{A4}$$

the integral equation (A3) can be written as

$$f(\lambda) = f_\infty(\lambda) + \frac{1}{2\pi}\left(\int_{-\infty}^{-\lambda_0} + \int_{\lambda_0}^{\infty}\right) d\mu \bar{K}(\lambda-\mu) f(\mu) \ , \tag{A5}$$

where f_∞ is the solution of (A1) for $\lambda_0 = \infty$ and $\bar{K}$ is the kernel of the integral operator $\bar{\mathcal{K}}$ introduced in Eq. (A4). In the system considered here it is given in terms of its Fourier transform by

$$\bar{K}(\omega) = \frac{1}{1+\exp(2|\omega|)} \ . \tag{A6}$$

Using that $f(\lambda)$ is an even function of λ and shifting the variables in (A5) $\lambda = \lambda_0 + x$ we obtain finally

$$f(x) = f_\infty(\lambda_0 + x) + \int_0^\infty dy \bar{K}(x-y) f(y) + \int_0^\infty dy \bar{K}(2\lambda_0 + x + y) f(y) \ . \tag{A7}$$

Note that $\bar{K}(2\lambda_0 + x)$ is $O(\lambda_0^{-2})$ for large λ_0 and positive x. This fact allows us to treat the last integral in Eq. (A7) as a perturbation. To every order the resulting equations are of Wiener-Hopf type:

$$g(x) - \int_0^\infty dy \bar{K}(x-y) g(y) = g^{(0)}(x) \ . \tag{A8}$$

Fourier transformation yields

$$[1 - \bar{K}(\omega)] g^+(\omega) + g^-(\omega) = g^{(0)}(\omega) \ , \tag{A9}$$

where $g^\pm(\omega) = \int dx \Theta(\pm x) g(x) \exp(i\omega x)$ are the parts of $g(\omega)$ that are analytic for $\pm\mathrm{Im}(\omega) > 0$ [$\Theta(x)$ is the step function]. The key to the solution of this equation is to find a decomposition of the kernel into factors $G^\pm$ that are analytic in the upper and lower complex ω plane, respectively:

$$1 - \bar{K}(\omega) = [G^+(\omega) G^-(\omega)]^{-1} \ , \qquad \lim_{\omega \to \infty} G^\pm(\omega) = 1 \ . \tag{A10}$$

For the present problem that is straightforward:

$$[G^-(\pi x)]^{-1} = [G^+(-\pi x)]^{-1} = \frac{1}{\sqrt{2\pi}} \Gamma(\tfrac{1}{2} + ix)(ix)^{-ix} e^{ix} \ . \tag{A11}$$

Using this factorization Eq. (A9) becomes

$$[G^+(\omega)]^{-1} g^+(\omega) + G^-(\omega) g^-(\omega) = Q^+(\omega) + Q^-(\omega) \ , \tag{A12}$$

where $Q^\pm(\omega)$ are analytic for $\pm\mathrm{Im}(\omega) > 0$,

$$Q^+(\omega) + Q^-(\omega) = G^-(\omega) g^{(0)}(\omega) \ . \tag{A13}$$

The analytic properties of the functions involved allow for the solution of Eq. (A12):

$$g^+(\omega) = G^+(\omega) Q^+(\omega) \ . \tag{A14}$$

In terms of this solution we have

$$\int_0^\infty dx g(x) = g^+(\omega = 0), \quad g(x=0) = -i \lim_{\omega \to \infty} \omega g^+(\omega) \ . \tag{A15}$$

To illustrate the scheme outlined above we apply it to the calculation of the magnetic field dependence of λ_0 (3.6): starting from Eq. (3.1) for ε_s we find

$$Q^+(\omega) + Q^-(\omega) = G^-(\omega) \left(\pi h \delta(\omega) - e^{-i\omega\lambda_0} \frac{\pi h_c}{2\cosh\omega} \right) \ . \tag{A16}$$

From this we obtain

$$Q^+(\omega) = -\frac{h}{2i} \frac{G^-(0)}{\omega + i0} + e^{-\pi\lambda_0/2} \frac{\pi h_c G^-(-i\pi/2)}{2i(\omega + i\pi/2)} + O(e^{-3\pi\lambda_0/2}) \ . \tag{A17}$$

The second equation in (A15) gives

$$\varepsilon_s(\lambda_0) = \frac{hG^-(0)}{2} - e^{-\pi\lambda_0/2} \frac{\pi h_c G^-(-i\pi/2)}{2} \ . \tag{A18}$$

Finally using (A11) and (3.5) we have the result (3.6).

---

*Address after April 1, 1991: Institut für Theoretische Physik, Universität Hannover, D-3000 Hannover 1, Federal Republic of Germany.

[1] M. Ogata and H. Shiba, Phys. Rev. B **41**, 2326 (1990).

[2] A. Parola and S. Sorella, Phys. Rev. Lett. **64**, 1831 (1990).

[3] H. J. Schulz, Phys. Rev. Lett. **64**, 2831 (1990).

[4] N. Kawakami and S.-K. Yang, Phys. Lett. A **148**, 359 (1990).

[5] H. Frahm and V. E. Korepin, Phys. Rev. B **42**, 10 553 (1990).

[6] A. M. Tsvelick (unpublished).

[7] Y. Ren and P. W. Anderson (unpublished).

[8] F. D. M. Haldane, Phys. Rev. Lett. **45**, 1358 (1980); J. Phys. C **14**, 2589 (1981).

[9] F. D. M. Haldane, Phys. Lett. **81A**, 153 (1981).

[10] F. D. M. Haldane, Phys. Rev. Lett. **47**, 1840 (1981).

[11] F. D. M. Haldane, in *Electron Correlation and Magnetism in Narrow-Band Systems*, edited by T. Moriya (Springer-Verlag, Berlin, 1981), p. 150.

[12] L. Kadanoff and A. C. Brown, Ann. Phys. (N.Y.) **121**, 318 (1979).

[13] A. A. Belavin, A. M. Polyakov, and A. B. Zamolodchikov, Nucl. Phys. **B241**, 333 (1984).

[14] J. L. Cardy, Nucl. Phys. **B270** [FS16], 186 (1986).

[15] H. W. J. Blöte, J. L. Cardy, and M. P. Nightingale, Phys. Rev. Lett. **56**, 742 (1986); I. Affleck, *ibid.* **56**, 746 (1986).

[16] E. H. Lieb and F. Y. Wu, Phys. Rev. Lett. **20**, 1445 (1968).

[17] F. Woynarovich and H.-P. Eckle, J. Phys. A **20**, L443 (1987).

[18] N. M. Bogoliubov and V. E. Korepin, Mod. Phys. Lett. B **1**, 349 (1988); Int. J. Mod. Phys. B **3**, 427 (1989).

[19] A. G. Izergin and V. E. Korepin, Pis'ma Zh. Eksp. Teor. Fiz. **42**, 414 (1985) [JETP Lett. **42**, 512 (1985)]; N. M. Bogoliubov, A. G. Izergin, and V. E. Korepin, Nucl. Phys. **B275** [FS17], 687 (1986).

[20] H. J. de Vega, J. Phys. A **21**, L1089 (1988); J. Suzuki, *ibid.* **21**, L1175 (1988).

[21] A. G. Izergin, V. E. Korepin, and N. Yu. Reshetikhin, J. Phys. A **22**, 2615 (1989).

[22] C. N. Yang, Phys. Rev. Lett. **19**, 1312 (1967).

[23] F. Woynarovich, J. Phys. A **22**, 4243 (1989).

[24] K. Lee and P. Schlottmann, Phys. Rev. Lett. **63**, 2299 (1989).

[25] H. Frahm and N.-C. Yu, J. Phys. A **23**, 2115 (1990).

[26] C. N. Yang and C. P. Yang, Phys. Rev. **150**, 327 (1966).

VOLUME 65, NUMBER 2 PHYSICAL REVIEW LETTERS 9 JULY 1990

Twisted Boundary Conditions and Effective Mass in Heisenberg-Ising and Hubbard Rings

B. Sriram Shastry
AT&T Bell Laboratories, Murray Hill, New Jersey 07974

Bill Sutherland
Department of Physics, University of Utah, Salt Lake City, Utah 84112
(Received 16 March 1990)

We identify the boundary energy of a many-body system of fermions on a lattice under twisted boundary conditions as the inverse of the effective charge-carrying mass, or the stiffness, renormalizing nontrivially under interactions due to the absence of Galilean invariance. We point out that this quantity is a sensitive and direct probe of the metal-insulator transitions possible in these systems, i.e., the Mott-Hubbard transition or density-wave formation. We calculate exactly the stiffness, or the effective mass, in the 1D Heisenberg-Ising ring and the 1D Hubbard model by using the ansatz of Bethe. For the Hubbard ring we also calculate a spin stiffness by extending the nested ansatz of Bethe-Yang to this case.

PACS numbers: 71.30.+h, 75.10.Jm

It is intuitively clear that one can distinguish between a metal and an insulator by studying the variation of eigenvalues under changes in boundary conditions (BC's). This was proposed by Kohn[1,2] as a means of studying the Mott transition—a metal-insulator transition that requires the combination of strong correlations and a single-band model of fermions. We present here what we believe is the first application of this idea, in two nontrivial many-body problems in 1D exhibiting the metal-insulator transition. These are the Heisenberg-Ising model undergoing a CDW (charge-density-wave) transition, and the Hubbard model undergoing a Mott transition.

We first sharpen the arguments of Kohn, specializing to a one-band d-dimensional-lattice Fermi system, and deduce the main implications—some of which seem to be insufficiently appreciated in literature. Consider a d-dimensional hypercubic lattice of linear dimension L, with spinless fermions having a nearest-neighbor hopping matrix element t and arbitrary density-dependent interactions that are lattice-translation invariant, and assume periodic BC's. We now introduce a uniform vector potential $A_x\hat{\mathbf{x}}$, which modifies the hopping in $\hat{\mathbf{x}}$-directed bonds by the usual Peierls phase factor, $t \rightarrow t\exp(\pm i\Phi/L)$, where $\Phi = LA_x e/\hbar c$ and the lattice constant $a_0 = 1$. Expanding the exponential we find the perturbed Hamiltonian $H' = H - \Phi j_x/L - \frac{1}{2}\Phi^2 T_x/L^2 + O(\Phi^3)$, where $j_x = 2t\sum \sin k_x C_k^\dagger C_k$, $T_x = -2t\sum \cos k_x C_k^\dagger C_k$, and H is the Hamiltonian for the interacting Fermi system. The energy shift of the ground state (g.s.) in the presence of the field is $E_0(\Phi) - E_0(0) \equiv D\Phi^2/L^{2-d} + O(\Phi^4)$, with the stiffness constant D given by second-order perturbation theory as

$$D = \frac{1}{L^d}\left[\frac{1}{2d}\langle -T\rangle - \sum_{\nu\neq 0}\frac{|\langle 0|j_x|\nu\rangle|^2}{E_\nu - E_0}\right], \tag{1}$$

where $\langle T\rangle$ is the kinetic-energy expectation in the g.s. and $E_0(0) \equiv E_0$. We have assumed that $\langle j_x\rangle$ is zero. Higher-order (nonquadratic) terms in the energy-shift formula are important when the energy shift is comparable to the energy gaps in the spectrum of H. The latter are $O(1/L)$ in metals and so in this case corrections arise when Φ is $O(1/L^{(d-1)/2})$. Level crossings would occur and perturbation theory would break down for Φ of order π.

We next specialize to $A_x \rightarrow A_x^0 \exp(-i\omega t)$ leading to an electric field $\mathbf{E}_x = A_x(i\omega/c)\hat{\mathbf{x}}$, from which the usual linear-response formula[3] gives the imaginary part of the ac conductivity,

$$\mathcal{I}\sigma_{xx}(\omega) = \frac{2e^2}{L^d\hbar^2\omega}\left[\frac{1}{2d}\langle -T\rangle - \mathcal{P}\sum_{\nu\neq 0}\frac{|\langle 0|j_x|\nu\rangle|^2(E_\nu - E_0)}{(E_\nu - E_0)^2 - \hbar^2\omega^2}\right]. \tag{2}$$

From (1) and (2) we see that $\lim_{\omega\to 0}\omega\mathcal{I}\sigma_{xx}(\omega) = (2e^2/\hbar^2)D$ and $\lim_{\omega\to\infty}\omega\mathcal{I}\sigma_{xx}(\omega) = (e^2/d\hbar^2L^d)\langle -T\rangle$. The high-frequency behavior of the imaginary part of the conductivity implies for the real part, through the usual dispersion relations, the well-known[3] f-sum rule:

$$\int_{-\infty}^{\infty}\mathcal{R}\sigma_{xx}(\omega)d\omega = \frac{\pi e^2}{d\hbar^2L^d}\langle -T\rangle.$$

More interesting is the small-ω behavior, implying that

$$\mathcal{R}\sigma_{xx}(\omega) = \frac{2\pi e^2}{\hbar}\left[D\delta(\hbar\omega) + \frac{1}{L^d}\sum_{\nu\neq 0}|\langle 0|j_x|\nu\rangle|^2\delta((E_\nu - E_0)^2 - \hbar^2\omega^2)\right]. \tag{3}$$

The coefficient of $\delta(\hbar\omega)$, if nonzero, implies free acceleration or infinite dc conductivity, which is reasonable here since there is no dissipative mechanism in the model at $T=0$. The coefficient is essentially the inverse of the effective current-carrying mass (for free electrons it is $\pi\rho e^2/m$). Therefore the f-sum rule is satisfied by the *sum of two terms of the same order*, the stiffness D and the "intraband dipole matrix elements." A method to calculate $E_0(\Phi)$ is to study different BC's—we can absorb the Peierls phases by a pseudo-gauge-transformation[4] and shift the effect of Φ into twisted BC's for the wave functions:

$$\Psi(\ldots,\mathbf{r}+L\mathbf{x},\ldots)=\exp\{i\Phi\}\Psi(\ldots,\mathbf{r},\ldots). \quad (4)$$

A crucial point (familiar from Landau's Fermi-liquid theory) is that for a Galilean-invariant interacting system, an analogous calculation would give the coefficient of $\delta(\hbar\omega)$ in (3) unrenormalized by interactions since $[j_x,H]=0$ [the first term in (1) becomes the particle density]. For lattice fermions the operator j_x commutes with the hopping part of H, but not with the interaction piece in general[5] and hence for *interacting lattice fermions* there is the possibility that the two terms in (1) cancel as some parameter is varied, signaling a metal-insulator transition. The absence of Galilean invariance thus allows the charge-carrying effective mass to vary with interactions, and in fact to diverge.[6]

We now consider the 1D Heisenberg-Ising (H-I) model of spinless fermions, with twisted BC's on a ring of length L described by

$$H=-\sum(C_n^\dagger C_{n+1}+\text{H.c.})-2\Delta\sum(\rho_n-\tfrac{1}{2})(\rho_{n+1}-\tfrac{1}{2}),$$

with $\rho_n=C_n^\dagger C_n$. Much is known about the model without the twist, and we merely note here that it has a gapless phase for $-1\leq\Delta\leq+1$ which is the conducting phase, and an ordered state with a gap for $-1>\Delta$, the insulating state. Bethe's ansatz is readily generalized to the case of twisted BC's (4) and the g.s. energy is known[7,8] for all Φ for the repulsive case $(0\geq\Delta\geq-1)$. The angle Φ has the physical interpretation of a magnetic flux through the ring in units of $\hbar c/e$. In brief, the Bethe equations generalize to $Lk_n=2\pi I_n+\Phi-\sum_{m\neq n}\Theta(k_n,k_m)$ with the usual phase shift[9] Θ. In the sector with $M=L/2$ particles the g.s. quantum numbers are $I_n=-(L+1)/2+n$ for $1\leq n\leq L$; this is the half-filled band corresponding to $S^z_{\text{tot}}\equiv L/2-M=0$ in the spin representation. In general, a calculation of the stiffness D requires a precision in total energy of order $1/L$ in 1D. In this problem, however, it is possible to obtain D through a thermodynamic calculation[8] using a remarkable property of the generalized Bethe equations, and the result (with $\Delta=-\cos\mu$) is

$$D=\frac{\pi}{4}\frac{\sin\mu}{\mu(\pi-\mu)}.$$

As $\Delta\to-1$, $\mu\to0$ and D approaches a nonzero value $\frac{1}{4}$. For $\Delta<-1$ there is a gap in the spectrum and D is zero—thus the stiffness and the effective mass have a jump discontinuity.

This transition is tracked by the interesting variation of certain eigenvalues of the H-I model. The state of the H-I model obtained[8] from the g.s. by adiabatically increasing Φ from 0 to 2π is one with a total momentum π, and can be found from the set of generalized Bethe equations by shifting all g.s. integers by unity; the energy above the ground state is $\Delta E_1=4D\pi^2/L=\pi^3(\sin\mu)/\mu(\pi-\mu)L$. A third state of relevance is the g.s. in the sector $S^z=1$ (corresponding to removing a particle) with an energy (above the absolute g.s.) given by Yang and Yang[9] as $\Delta E_2=\pi(\pi-\mu)(\sin\mu)/\mu L$ in the entire gapless range, $-1\leq\Delta\leq+1$. These levels cross at the critical point where $\Delta\to-1$. This degeneracy is accounted for by the rotational invariance of the H-I model at $\Delta=-1$.

In the ordered state $\Delta<-1$, the second state above (with quantum numbers leading to ΔE_1) is asymptotically degenerate with the g.s. (the splitting vanishing more rapidly than $1/L^n$ for any n). Its energy splitting from the g.s. is fortunately available from the work of Baxter who calculated the interfacial tension of the six- and eight-vertex models.[10] Baxter's beautiful result translates into $D\sim\exp(-L/\xi)$, where the correlation length

$$\xi=1\bigg/\ln\left[\frac{1}{2x^{1/2}}\left(\prod_{m=1}^{\infty}\frac{1+x^{4m-2}}{1+x^{4m}}\right)^2\right],$$

with $\Delta=-\cosh\lambda$, $x=\exp(-\lambda)$. This phase is therefore insulating in the thermodynamic limit. The third state above corresponds to removing a particle, and develops an energy gap in this region,[9] $\Delta<-1$, with

$$\Delta E_2=2(\sinh\lambda)\sum_{n=-\infty}^{\infty}\frac{(-1)^n}{\cosh(n\lambda)}.$$

The above behavior of D implies that although the system is insulating in the infinite-lattice limit, for a finite system, provided L/ξ is not too large, we should see a small "free acceleration" response arising from adiabatic sliding between the almost degenerate "ground" states.

We next consider the repulsive $U\geq0$ Hubbard model in 1D containing two species of particles, spin up and spin down. The boundary angles for the two are treated as independent parameters $\Phi_\uparrow$ and $\Phi_\downarrow$. There are two independent stiffnesses that we may calculate. Setting $\Phi_\uparrow=\Phi_\downarrow=\Phi$ the energy shift gives the "charge stiffness" D_c and setting $\Phi_\downarrow=-\Phi_\uparrow=\Phi$ gives the "spin stiffness" D_s. These have expressions identical to Eq. (1), with D_c involving the *sum* of the up and down currents and D_s the *difference* in the matrix element, and both containing the total kinetic-energy expectation. This general case requires a generalization of the Bethe-Yang ansatz that was employed by Lieb and Wu[11] for the solution of the model with periodic BC's. We present here, in brief, the analysis necessary to ensure that the model remains solvable with the twisted BC's.

The model is described by the usual Hamiltonian and we denote by L, N, and M the number of sites, particles, and down-spin particles. The Bethe wave function is written in the form

$$|\Psi\rangle = \sum_{\substack{1\le x_1\le\cdots\le x_N\le L\\ 1\le j_1\le\cdots\le j_M\le N}} \sum_P \left[\exp\left(i\sum_n k_{P_n}x_n\right)A(\{j_n\}\,|\,P)\right] \times C^\dagger_{x_1\uparrow}\cdots C^\dagger_{x_{j_1}\downarrow}\cdots C^\dagger_{x_{j_M}\downarrow}\cdots C^\dagger_{x_N\uparrow}|0\rangle,$$

where P is a sum over the $N!$ permutations and $A(\{j_n\}|P)$ is the amplitude. The wave function satisfies the difference equations that follow from the Hubbard model in the interior of the chain as usual with energy $E=-2\sum\cos k_n$ (setting $t\to 1$), provided the amplitudes satisfy the usual consistency conditions.[11,12] We impose the boundary conditions Eq. (4) with different boundary angles for the two spin species by transporting the particle at x_1 to x_1+L, and this gives

$$\exp(ik_{P_1}L)A(\{j_n'\}\,|\,P') = \{\exp(i\Phi_\downarrow\delta_{j_1,1})+\exp[i\Phi_\uparrow(1-\delta_{j_1,1})]\}A(\{j_n\}\,|\,P). \quad (5)$$

Here P' is obtained from P by a cyclic permutation and $j_n'=j_n-1(\mathrm{mod}N)$. It is convenient to write these in vector form by introducing $|A(P)\rangle=\sum|\{j_n\}\rangle A(\{j_n\}|P)$ with the vector $|\{j_n\}\rangle$ denoting the basis state with overturned "spins" located at the "sites" $j_1,\ldots,j_n$. The BC's Eq. (5) translate into the following N eigenvalue conditions that must be simultaneously satisfied: $\exp(ik_jL)|A_0\rangle=L_j|A_0\rangle$, where $|A_0\rangle$ is the vector for the identity permutation, the N-string operators are

$$L_j=X_{j+1,j}\cdots X_{N,j}D_jX_{1,j}\cdots X_{j-1,j},$$

the operators $X_{i,j}=(y_{i,j}-P_{i,j})/(y_{i,j}-1)$, with $y_{i,j}=2i\times(\sin k_i-\sin k_j)/U$ and $P_{i,j}$ the usual permutation operator, and the new operator is

$$D_j=\exp(i\Phi_\uparrow)(1+\sigma_j^z)/2+\exp(i\Phi_\downarrow)(1-\sigma_j^z)/2.$$

We must now verify that the N operators L_j commute, and then diagonalize these. This task is neatly performed with a generating (monodromy) operator Υ_g acting on a $(N+1)$-site spin chain, $\Upsilon_g(\lambda)=D_g l_{N,g}(\lambda)\cdots l_{1,g}(\lambda)$, where g is the extra $(N+1)$th site and the scattering operator

$$l_{n,g}(\lambda)=[i(\sin k_n-\lambda)-U/2P_{n,g}]/[i(\sin k_n-\lambda)-U/2].$$

The N-string operators L_j can be obtained from the generating operator by using that $\mathrm{Tr}_g\Upsilon_g(\lambda=\sin k_j)=L_j^\dagger(-[k_n])$. The commutation relations between L_j are guaranteed if $\mathrm{Tr}_g\Upsilon_g(\lambda)$ commutes with similar operators differing in the spectral parameter λ. This is in turn true[10] if an operator $R_{g,g'}$ exists such that $\Upsilon_g(\lambda)\Upsilon_{g'}(\mu)R_{g,g'}(\lambda,\mu)=R_{g,g'}(\lambda,\mu)\Upsilon_g(\mu)\Upsilon_{g'}(\lambda)$. In the present problem the Υ operator differs from the zero-flux case through the D_j operators with the property that $D_gD_{g'}=c\exp[d(\sigma_g^z+\sigma_{g'}^z)]$. Noting that the $R_{g,g'}$ for the Heisenberg spin chain fulfills the commutation rules in the zero-flux case and further commutes with $\sigma_g^z+\sigma_{g'}^z$, we conclude that the twisted-BC case is also satisfied by the same R operator. The diagonalization of the L_j operators was done by a variant of the nested Bethe-Yang ansatz and the resulting transcendental equations are

$$Lk_n=2\pi I_n+\Phi_\uparrow+2\sum_{j=1}^{M}\arctan[4(\Lambda_j-\sin k_n)/U], \quad (6)$$

$$2\sum_{n=1}^{N}\arctan[4(\Lambda_j-\sin k_n)/U]=2\pi J_j+\Phi_\downarrow-\Phi_\uparrow+2\sum_{i\neq j}^{M}\arctan[2(\Lambda_j-\Lambda_i)/U], \quad (7)$$

with I_n,J_j as the usual quantum numbers (integer or half odd integer).

In order to study D_c, the charge stiffness, we set $\Phi_\uparrow=\Phi_\downarrow=2\pi$, and argue that the excitation energy is $(4\pi^2/L)D_c$. The underlying assumption here and in the next section (justified for the H-I model in Ref. 7) is that the energy $E_0(\Phi)$ remains quadratic in Φ out to this value, in spite of a level crossing that occurs prior to it. With this assumption, we can calculate D_c, in the general case, from the excitation energy of the state, obtained by adding unity to the g.s. quantum numbers I_n. An evaluation of D_c requires a detailed study of the finite-size effects. Here we are content to observe that the general structure of the equations forces D_c to vanish as $N\to L$, i.e., as we approach half filling for any nonzero value of U. This follows from the fact I_n+L and I_n lead to the same solution, and further, at half filling, the set of $N=L$ g.s. integers I_n exhaust all the allowed distinct values $-(L-1)/2,\ldots,(L-1)/2$, whereby $D_c=0$. To leading order in $1/U$ we can see this explicitly. Here Λ_j are of $O(U)$, and hence the two sets of equations decouple. It is readily seen that the charge stiffness is identical to that of spinless fermions with a density $\delta=(L-N)/L$, and hence $D_c\to 0$ linearly as $\delta\to 0$. It is also worth remarking that this vanishing stiffness can be equally well interpreted as a vanishing of the density of the effective carriers of charge, the "holons" of Anderson.[13]

The spin stiffness can, however, be related with the help of a remarkable identity to the bulk spin susceptibility, which in turn can be calculated readily by the method of integral equations for relevant densities. Consider the state for even N, with $M=N/2$ and $\Phi_\uparrow=\Phi_\downarrow$

VOLUME 65, NUMBER 2 PHYSICAL REVIEW LETTERS 9 JULY 1990

$=0$; this is the g.s. for this filling and has the quantum numbers taking on values $I_n = -(N+1)/2+n$ for $1 \le n \le N$ and $J_j = -(N/2+1)/2+j$ for $1 \le j \le N/2$. Suppose that we have found the solutions for k_n and Λ_j. We now turn on Φ so that $\Phi_\uparrow = -\pi$ and $\Phi_\downarrow = \pi$, thereby "deforming" the previous solution. This case, however, can be solved by a neat observation (analogous to the H-I model in Ref. 8). The self-consistent solution is that $\Lambda_{max} = +\infty$. Equation (7) for $j = j_{max}$ is identically satisfied, and the remaining $N/2-1$ equations for $1 \le j \le N/2-1$ can be written as

$$2\sum_{n=1}^{N} \arctan[4(\Lambda_j - \sin k_n)/U] = 2\pi J_j' + 2\sum_{l \ne j}^{N/2-1} \arctan[2(\Lambda_j - \Lambda_l)/U],$$

with $J_j' = J_j + \frac{1}{2}$; in Eq. (6) we drop the Φ and sum j over the $N/2-1$ finite Λ's. This set is recognizable as the g.s. equations in the sector $M = N/2-1$. Hence, the spin stiffness $D_s = (L/4\pi^2)[E_0(N,N/2) - E_0(N,N/2-1)]$. With the magnetization variable $y = 1-2M/N$, the g.s. energy in a sector with fixed M is $E_0(N,M) = Ne_0(y) + \frac{1}{2}N\chi^{-1}y^2 + O(y^4)$, defining the susceptibility. From the smallest allowed value of $y = 2/N$ we conclude that $D_s = (1/2\pi^2)(L/N)\chi^{-1}$. This identity is true at all U and can be used to extract D_s from the calculation of the susceptibility. The latter has been calculated numerically[14] at several fillings and U. Qualitatively it is nonzero at all fillings, and resembles the Pauli susceptibility renormalized by U. Physically a nonvanishing D_s implies that the model has long-ranged (presumably power-law) spin correlations at all fillings.

The origin of the above relationship, between D_s and χ, is in the rotation invariance of the model for any U or filling. It follows from the degeneracy of the lowest excitations of $S_z = 0$ with those of $S_z = 1$ (with appropriate momenta). Apart from the normalization factor of L/N this is the same relation as in the Heisenberg-Ising model at $\Delta = -1$, i.e., the isotropic point. In general, the relation between the two for the H-I model is $D = [1/2(\pi-\mu)^2]\chi^{-1}$.

In conclusion, we have given two nontrivial examples where a metal-insulator transition occurs due to interactions and is reflected directly in the effective mass obtained by twisting the BC's. For the 1D Hubbard model the spin stiffness has been related to the bulk susceptibility through an interesting identity. It is clear that the ideas explored here have possible applications in higher-dimensional models, where numerical investigations with twisted BC's are possible for small systems.

The authors thank Professor C. N. Yang for his hospitality at the Institute for Theoretical Physics, State University of New York at Stony Brook, where this work was initiated. B.S.S. would like to thank Professor P. W. Anderson and Dr. A. J. Millis for helpful discussions.

[1] W. Kohn, Phys. Rev. 133, A171 (1964).

[2] A similar idea for Anderson localization is due to D. J. Thouless, Phys. Rep. 13C, 94 (1974). This notion has found extensive application in disordered systems. See also M. E. Fisher, M. N. Barber, and D. Jasnow, Phys. Rev. A 8, 1111 (1973).

[3] See, for example, D. Pines and P. Nozieres, *Theory of Quantum Liquids* (Benjamin, New York, 1966); R. Bari, D. Adler, and R. V. Lang, Phys. Rev. B 2, 2898 (1970); I. Sadakata and E. Hanamura, J. Phys. Soc. Jpn. 34, 882 (1973); P. F. Maldague, Phys. Rev. B 16, 2437 (1977).

[4] N. Byers and C. N. Yang, Phys. Rev. Lett. 7, 46 (1961).

[5] Interestingly enough, in both the examples considered here, one can add a suitable four-fermionic term to the current operator j_x so that the resulting "generalized current" commutes with the total Hamiltonian. In the case of the 1D Hubbard model, this is discussed in B. S. Shastry, J. Stat. Phys. 50, 57 (1988), and a similar argument works for the Heisenberg-Ising model as well.

[6] Similar arguments clearly apply for the same reasons in other *lattice* many-body problems involving superfluidity and superconductivity. Indeed, the Heisenberg-Ising model discussed here may alternatively be viewed as a hard-core Bose gas undergoing a (quasi) superfluid-solid transition accompanied by a vanishing superfluid stiffness.

[7] C. J. Hamer, G. R. W. Quispel, and M. T. Batchelor, J. Phys. A 20, 5677 (1987). These authors note the connection between the finite-size corrections in the twisted 1D Heisenberg-Ising model and the dependence of the "central charge" on the value of q in the 2D "q-state" Potts model. See also F. C. Alcaraz, M. N. Barker, and M. T. Batchelor, Phys. Rev. Lett. 58, 771 (1987).

[8] B. Sutherland and B. S. Shastry, "Adiabatic Transport in an Exactly Soluble 1D Quantum Many-Body Problem," 1990 (to be published).

[9] C. N. Yang and C. P. Yang, Phys. Rev. 147, 303 (1966); 150, 321 (1966); 150, 327 (1966).

[10] R. J. Baxter, *Exactly Solved Models in Statistical Mechanics* (Academic, New York, 1982).

[11] E. H. Lieb and F. Y. Wu, Phys. Rev. Lett. 20, 1445 (1968).

[12] C. N. Yang, Phys. Rev. Lett. 19, 1312 (1967).

[13] P. W. Anderson, Phys. Rep. 184, 195 (1989); P. W. Anderson and Y. Ren (to be published).

[14] H. Shiba, Phys. Rev. B 6, 930 (1972).

PHYSICAL REVIEW B VOLUME 45, NUMBER 20 15 MAY 1992-II

Persistent current of a Hubbard ring threaded with a magnetic flux

Naichang Yu and Michael Fowler
Department of Physics, University of Virginia, Charlottesville, Virginia 22901
(Received 15 August 1991; revised manuscript received 13 December 1991)

The persistent current in microscopic and mesoscopic Hubbard rings threaded with magnetic flux is studied as a function of the flux and the Coulomb repulsion parameter U. For microscopic rings having very large U, we find that as the flux is increased by one quantum, a magnon hole traverses the magnon sea, generating a periodicity of $1/N_c$ in the persistent current, but with no changes in the overall spin magnetization, in contrast to the earlier suggestion of Kusmartsev. For non-half-filled mesoscopic rings, we use methods developed by Woynarovich for the zero-flux case to build a rather complete picture of the variation of persistent current with magnetic flux. We find the periodicity to be a half flux quantum or a whole one, and show how the critical flux values at which the current reverses vary with system parameters. We show how the behavior characteristic of microscopic rings goes over to that of mesoscopic rings as N_c/U increases.

I. INTRODUCTION

Recently, several authors have examined the behavior of microscopic and mesoscopic Hubbard rings threaded by magnetic flux and, in particular, have examined the persistent currents generated in such rings by the flux.[1-5] The Hubbard ring is interesting as an example of a strongly-correlated-electron system that is solvable using the Bethe ansatz.[6] It has been argued to be a good model for the optical properties of such quasi-one-dimensional conductors as TTF-TCNQ,[7] various aromatic molecules, and systems of connected quantum dots.[4] It is also possible that understanding its properties might be helpful in analyzing the physics of the Aharonov-Bohm effect in mesoscopic metal rings.[8]

In this paper we map the persistent current as a function of magnetic flux for Hubbard rings over a wide range of ring sizes and values of the Hubbard Coulomb-repulsion parameter U. For convenience we usually plot the ground-state energy as a function of the threading magnetic flux; the persistent current is then just minus the slope of this curve. For a Hubbard ring, the effect of a small magnetic flux is to twist the boundary conditions on electronic eigenstates through an angle Φ proportional to the flux. A flux quantum corresponds to a twisting angle of 2π, and so $\Phi/2\pi$ measures the threading magnetic field in units of the flux quantum hc/e. Evidently, the ground-state energy $E_0(\Phi)$ of the electronic system is periodic in Φ with period 2π. (By "ground state" we mean the lowest-energy state of the system for a given angle of twist, which is in general not the state given by following adiabatically the quantum state lowest for zero twist as the angle is increased. A trivial counterexample is provided by a single noninteracting electron.)

Very recently, Kusmartsev[4] has pointed out that for microscopic Hubbard rings with very strong Coulomb repulsion U between electrons, that is, $U/N_c \gg 1$, where N_c is the number of electrons in the system, the ground-state energy varies with a surprisingly short period, $\Delta\Phi = 2\pi/N_c$. He states that this oscillation is driven by spin-flip processes, the number of spin-up electrons jumping by 1 for each period, so that the spin magnetization of the system varies over its maximum possible range as the flux increases by one flux unit. We examine these small Hubbard systems in Sec. III and confirm Kusmartsev's result that the ground-state energy (and hence, of course, the persistent current) does oscillate with the shorter period as Φ increases. However, we do not agree with his assertion that the spin magnetization varies over a wide range as the flux changes by one quantum. It is true that if U is actually infinite, the energy levels, including the ground state, become degenerate with respect to spin magnetization, and so Kusmartsev's choice of the successive ground states is as good as any other. Nevertheless, in the physically interesting situation of large but not infinite U, the degeneracy is broken, and we find that over the whole range of Φ, states having total spin 0, $\frac{1}{2}$, or 1 (depending on the number of particles present) have lower energies than the spin-magnetized states discussed by Kusmartsev.

To see what happens when an increasing magnetic flux threads these small Hubbard rings with large U, consider the Bethe-ansatz equations (2.3) (given below) for U going to infinity, giving (3.1) (following Kusmartsev). The point to note is that in this limit, the magnon sea contributes the same constant phase shift to each of the charge (or holon) momenta, so that its effect is the same as that of an added fractional flux proportional to the total momentum of the magnon sea. Now, on increasing the external twist angle Φ on the system (i.e., the magnetic flux) from zero, the lowest-energy state of the system for a particular Φ is given at large U by generating a compensating momentum in the magnon sea, that is to say, a momentum which will counterbalance as much as possible the extra holon phase shift Φ and, hence, minimize the increase in energy of the holon distribution. Of course, for finite U, there is some energy cost associated with creating momentum in the magnon sea. However, for large U, the spin degrees of freedom are equivalent to those of a Heisenberg antiferromagnetic chain with coupling of order N_c/NU. Thus, for large enough U, it is always worth creating momentum in the magnon sea, which costs energy of order N_c/NU, because the consequent lowering of energy of the charge Fermi sea is independent of U in leading order. It is of order $1/N$. As discussed in Sec. III, the energetically most economical way to create a given magnon momentum less than π is to create a single des Cloizeaux–Pearson excitation, that is, a single hole in the sea. As Φ is increased, this hole moves from one Fermi point to the other.

It should be noted that since the magnon momentum is itself quantized, it cannot perfectly compensate for the effect of a smoothly increasing twist angle Φ, and from (3.3) it is evident that this remaining imbalance leads to the energy oscillating with the short period $2\pi/N_c$. The picture is somewhat complicated by the parity requirements (integer or half odd integer) on the numbers of charges and spins. It is necessary to deal with the four possibilities separately, as they lead to different sets of ground-state configurations. A full analysis is presented in Sec. III.

Thus, for large enough U in a microscopic ring, the function $E_0(\Phi)$ has a sequence of N_c parabolic segments between zero and 2π. The analysis outlined above enables us to identify the states of the system corresponding to these curves However, it also suggests how $E_0(\Phi)$ will change as U becomes smaller or the size of the system increases. It is a good initial approximation just to add the appropriate magnon energy to each parabolic segment. This clearly means that, as U decreases, segments corresponding to higher magnon energies are raised and their share of the range in Φ shrinks and disappears. Beyond a certain point, only two segments remain: those corresponding to magnon momenta zero and π. By numerically solving the Bethe-ansatz equations, we have confirmed that this is indeed what happens, even when higher-order terms are included. For example, we find that for 16 electrons on a chain having 32 sites, there are only two segments remaining if U is less than 100. This means that for U below this value, there are *no holes* in the sequence of magnon quantum numbers in the lowest-energy state.

Since the relevant parameter in determining the number of segments is N_c/U from the discussion above, it certainly seems safe to conclude that for *mesoscopic* rings having U in the range of physical interest ($U \le 100$, say), there will be no holes in the magnon distribution in the ground state. This implies that we can construct $E_0(\Phi)$ for mesoscopic rings by extending some of the work of Woynarovich on the finite-size corrections to the ground-state energy of a Hubbard ring with the usual periodic-boundary conditions (no flux).[9] Among other things, Woynarovich found that the energy changes in these rings when the particle sea and/or the magnon sea is shifted over by discrete amounts (i.e., particles or spin quantum numbers are moved from one Fermi point to the other). But this is precisely what happens (for particles) when a magnetic flux is introduced through the ring, except that the flux gives a continuous shift rather than a discrete one. Woynarovich found the energy of the set of states generated in this way to vary as the square of the displacement, with a coefficient that could be expressed in terms of the Fermi velocities and the dressed charge. (The dressed charge, a renormalization factor, can be calculated from the Bethe-ansatz equations.[10]) Using his results, we can map out exactly the lowest-energy state of the system as a function of the magnetic flux. The graph is a sequence of parabolic segments, as it would be for noninteracting fermions, but the interaction changes both the curvature of the parabolas and their relative minima, so the segments are in general of different lengths. In fact, we find that the parabolic segment centered at the origin for a ring of $4n$ electrons disappears entirely near half filling. Another important difference from the noninteracting case is that the allowed quantum numbers for the particles can be either integer or half odd integer, depending on the numbers of excitations present, and this affects the ordering of energy levels at a given flux, as discussed in detail in Sec. IV.

It is instructive to compare the $E_0(\Phi)$ derived for mesoscopic rings using Woynarovich's approach with that following from our approximate analysis, and exact numerical work, on microscopic rings. The connection is clear. The mesoscopic $E_0(\Phi)$ given by (4.14) has two parabolic segments centered at zero and π. That at π is raised by $\pi v_s/N$, the magnon energy of a π magnon in a finite system. This is just what we found above on extrapolating the microscopic analysis. (It should be noted that in both microscopic and mesoscopic cases, what we refer to as the magnon energy here means the energy in the *spin* degrees of freedom. Introducing that same magnon into a system constrained by *periodic*-boundary conditions would cost more energy, because of the holon-phase shifting. That is the same effect, with opposite twist, as increasing Φ, which is what we have introduced the magnon to compensate.) However, the change in curvature of the parabolic segments, in other words, the dressed charge, is *not* given by our simple approximation of just adding the magnon energy, although it is a $1/U$ effect. This point is discussed in more detail in Sec. III.

Finally, we consider the exactly half-filled Hubbard ring. The curvature of the energy as a function of flux near the origin, which is essentially the Drude weight measuring the low-frequency optical response of the system, has been analyzed by Fye *et al.*[2] They find it to be negative (paramagnetic) for rings with $4n$ sites, positive for $4n+2$ sites, and exponentially vanishing with increasing system size. Stafford, Millis, and Shastry[3] emphasize what they term a rather peculiar property, namely, that for N electrons on a ring of $N=4n$ sites, the distribution of charge quantum numbers, which must be integers in this case, is necessarily not quite symmetric, going from $-N/2$ to $N/2-1$ or $-N/2+1$ to $N/2$, giving a ground-state momentum of $\pm\pi$. Unfortunately, our methods based on Woynarovich's results do not work for the half-filled case. This is because there is no headroom at the Fermi surface to insert extra particles—that is why the charge excitations have a gap. We did, however, note one unexpected point concerning the distribution of momenta in the ground state. Even though the distribution of quantum numbers may not be symmetric in the $4n$ ground state, as discussed above, the distribution of momenta is. The reason is that one of the momenta is exactly at π (for zero flux), well away from the others in a finite system. It is immediately apparent, on inspecting the Bethe-ansatz equations (2.3) below, that this is the momentum value corresponding to the quantum number $N/2$. When the flux moves away from zero, this particle rapidly descends, and the overall energy drops, although by a much smaller amount than the single-particle contribution as a result of backflow in the rest of the Fermi sea. For this system no level crossings take place as the flux varies. The $(4n+2)$-particle systems vary in a similar way, except that the maximum energy as a function of flux is now at π rather than at 0. This rapid movement of an isolated root as the twist angle varies through a symmetry point is closely related to that found by Sutherland and Shastry for the Heisenberg-Ising chain.[11]

II. BETHE-ANSATZ FORMALISM

We first review the standard Bethe-ansatz solution to the Hubbard ring. The Hamiltonian of the model is

$$\mathcal{H}=-\sum_{j=1}^{N}\sum_{\sigma}(e^{-ieA}\psi^{\dagger}_{j+1,\sigma}\psi_{j,\sigma}+e^{ieA}\psi^{\dagger}_{j,\sigma}\psi_{j+1,\sigma})+U\sum_{j=1}^{N}n_{j\uparrow}n_{j\downarrow}\ . \tag{2.1}$$

Here N is the number of lattice sites of the ring, $n_{j,\sigma}=\psi^{\dagger}_{j,\sigma}\psi_{j,\sigma}$ is the number of spin-σ electrons at site j, and $U>0$ is the on-site Coulomb repulsion. $A=\Phi/N$ is the vector potential for the magnetic flux. We have neglected the interaction of the spin of the electrons with the magnetic field. We will study rings with a fixed number of electrons. The current in the ring at zero temperature is given by[12]

$$j=-\frac{\partial E_0(\Phi)}{\partial\Phi}\ , \tag{2.2}$$

where we used units $\hbar=1$ and $e=1$.

The eigenstates of a chain with $N_c=N_\uparrow+N_\downarrow$ electrons and $N_s=N_\downarrow$ down spins are characterized by the momenta k_j of charges and the rapidities λ_α of spin waves. For a chain with twisted-boundary conditions, the Bethe-ansatz equations are[6,1]

$$\begin{aligned}Nkj&=2\pi I_j+\Phi-\sum_{\beta=1}^{N_s}2\arctan\left[\frac{4(\sin k_j-\lambda_\beta)}{U}\right],\\ \sum_{j=1}^{N_c}2\arctan\left[\frac{4(\lambda_\alpha-\sin k_j)}{U}\right]&=2\pi J_\alpha+\sum_{\beta=1}^{N_s}2\arctan\left[\frac{2(\lambda_\alpha-\lambda_\beta)}{U}\right].\end{aligned} \tag{2.3}$$

Here N is the number of sites in the ring. The quantum numbers I_j and J_α are either integers or half odd integers, depending on the parities of the numbers of down- and up-spin electrons, respectively:

$$I_j=\frac{N_s}{2}\ (\mathrm{mod}\ 1),\quad J_\alpha=\frac{N_c-N_s+1}{2}\ (\mathrm{mod}\ 1)\ . \tag{2.4}$$

The energy and momentum of the system in a state corresponding to a solution of (2.3) are given by

$$\begin{aligned}E&=-2\sum_{j=1}^{N_c}\cos k_j\ ,\\ P&=\sum_{j=1}^{N_c}\left[k_j-\frac{\Phi}{N}\right]=\frac{2\pi}{N}\left[\sum_j I_j+\sum_\alpha J_\alpha\right].\end{aligned} \tag{2.5}$$

III. MICROSCOPIC RINGS

We study in this section a small chain with very strong on-site repulsion, $U/N_c\gg 1$. As discussed by Kusmartsev,[4] in the limit $U/N_c\to\infty$, the $\sin k_j$ terms in (2.3) can be neglected, leading to

$$Nk_j=2\pi\left[I_j+\frac{1}{N_c}\sum_\alpha J_\alpha\right]+\Phi\ . \tag{3.1}$$

Equation (3.1) is identical to that describing a set of noninteracting spinless fermions on a ring threaded with a flux:

$$\Phi/2\pi+\frac{1}{N_c}\sum_\alpha J_\alpha\ . \tag{3.2}$$

If the I_j's are consecutive quantum numbers, the energy of the state is

$$E_0(\Phi)=-E_m\cos\left[\frac{2\pi}{N}\left[\frac{\Phi}{2\pi}+\frac{1}{N_c}\sum_\alpha^{N_s}J_\alpha+D_c\right]\right]\ , \tag{3.3}$$

where D_c is defined by $D_c=(I_{\max}+I_{\min})/2$, and E_m is a positive constant:

$$E_m=2\frac{\sin(\pi N_c/N)}{\sin(\pi/N)}\ . \tag{3.4}$$

We assume in the following that the system is not exactly half filled, for if it were, the above expression for E_m would be zero. The energy $E_0(\Phi)$ can be minimized for Φ by choosing the set J_α such that

$$\sum_{\alpha=1}^{N_s}J_\alpha=-p\quad\text{for}\quad\frac{2p-1}{2N_c}<\frac{\Phi}{2\pi}+D_c<\frac{2p+1}{2N_c}\ . \tag{3.5}$$

It is evident from (3.3) that, with these J_α, the graph of $E_0(\Phi)$ as a function of Φ is a sequence of identical parabolic segments (strictly, parabolic in the limit of large N), giving a function with period $1/N_c$ of a flux quantum. The problem is that this infinite-U system is highly degenerate. There are many ways of choosing sets of J_α's to give the same sum—one can adjust the total number of down spins, following Kusmartsev. There are, however, other possibilities. For example, gaps can be introduced in the magnon quantum-number distribution. The question to resolve is what states are the lowest-energy states when U is large but finite and the degeneracy is lifted. To find out we examine the leading-order $1/U$ corrections to the Bethe-ansatz equations for infinite U.

This $1/U$ expansion is actually quite tricky, as we shall discuss later. On examining the Bethe-ansatz equation (2.3), we note that, for large U, the λ_α's will be of order U, whereas the $\sin k_j$'s are, of course, of order unity; for a small system, the λ_α's will be widely scattered and, with one possible exception, will have $|\lambda| \gg 1$. With this picture in mind, we expand the arctangent functions to leading order in $\sin(k_j)/U$ and define scaled variables x_α by

$$x_\alpha = \lim_{U\to\infty} (2\lambda_\alpha/U) . \tag{3.6}$$

For large but finite U, the k_j's in (2.3) have leading $1/U$ corrections:

$$\delta k_j = -\frac{2\sin(k_j)}{NU}\sum_\alpha \frac{1}{\frac{1}{4}+x_\alpha^2} , \tag{3.7}$$

where the x_α satisfy the equations

$$2N_c \arctan(2x_\alpha) = 2\pi J_\alpha + \sum_{\beta=1}^{N_s} 2\arctan(x_\alpha - x_\beta) . \tag{3.8}$$

The corresponding correction to the energy is easily obtained from (2.5):

$$E_{\rm spin} = -\frac{4}{NU}\left[\sum_{j=1}^{N_c}\sin^2 k_j\right]\sum_{\alpha=1}^{N_s}\frac{1}{\frac{1}{4}+x_\alpha^2}$$

$$= -J\sum_\alpha \frac{1}{\frac{1}{4}+x_\alpha^2} . \tag{3.9}$$

We note that the energy $E_{\rm spin}$ and Eq. (3.8) are just the energy and Bethe-ansatz equations of an antiferromagnetic Heisenberg spin chain of exchange coupling J, with N_c sites and with N_s spins down.

Finding the lowest-energy state of a Hubbard chain with sufficiently large U in a magnetic field is thus reduced to finding the state of lowest energy of a Heisenberg spin chain with a certain momentum $q = 2\pi\sum J_\alpha/N_c$. This problem has been studied by des Cloizeaux and Pearson.[13] For zero momentum the state of lowest energy of the Heisenberg chain is the singlet state. However, for a given nonzero momentum, the lowest-energy state is the des Cloizeaux–Pearson spin-wave excitation, a triplet state described by real $\{x_\alpha\}$, with a hole in the distribution of the quantum numbers $\{J_\alpha\}$. Creating a single-hole excitation of this kind is the most energy-efficient way of generating a given total magnon momentum, because the spin-wave energy plotted as a function of its momentum curves downward below its low-energy linear form, so it would cost more energy to produce the same momentum using several excitations. In particular, we find that for finite U, the states in Ref. 4, where the total spin magnetization undergoes large fluctuations as Φ increases, have higher energies than the single-hole states with the same magnon momentum. States with complex λ_α are also found to have higher energies than these states (compare the remarks of Woynarovich in Ref. 9). For large U this follows from a consideration of the corresponding states of the equivalent finite antiferromagnetic Heisenberg chain. Translating the quantum numbers back to the Hubbard model, we find that, for $U \gg 1$, the states minimizing the energy for nonzero flux depend on the values of N_c, N_s (mod 4).

A. $N_c = 4n+2$, $N_s = 2n+1$

In this case, from (2.4), the I_j's are half odd integers and the J_α's are integers. For zero flux, the quantum numbers I_j's and J_α's are both distributed symmetrically about the origin. For nonzero flux, the new ground state has a hole in the J_α distribution. The ground state for a chain with flux

$$(2p-1)/2N_c < \Phi/2\pi < (2p+1)/2N_c$$

is

$$\begin{aligned} &I_1,\ldots,I_{N_c} = -(N_c-1)/2,\ldots,(N_c-1)/2 ,\\ &J_1,\ldots,J_{N_s} = -(N_s+1)/2, -(N_s-1)/2,\ldots,-(N_s-2p+1)/2, -(N_s-2p-3)/2,\ldots,(N_s-1)/2 . \end{aligned} \tag{3.10}$$

The J_α from the hole at $-(N_s-2p-1)/2$ has been moved to the left Fermi surface. This is a state with a magnon excitation of momentum $-2\pi p/N$ and is the lowest-energy state from approximately $\Phi/2\pi = (2p-1)/2N_c$ to $\Phi/2\pi = (2p+1)/2N_c$. We say "approximately," because different parabolic segments have had their bottoms raised by different amounts of order $1/U$, so the points of intersection will have shifted to this order.

B. $N_c = 4n$, $N_s = 2n$

The I_j's must be integers, and all the J_α's must be half odd integers. The energy at zero flux is minimized by taking

$$I_j = -N_c/2, -(N_c-2)/2,\ldots,(N_c-2)/2 , \tag{3.11}$$

$$J_\alpha = -(N_s-3)/2, -(N_s-5)/2,\ldots,(N_s+1)/2 . \tag{3.12}$$

We note that one of the J_α's has been moved from the left Fermi surface to the right one and the state has momentum $P = 2\pi N_s/N$. For nonzero flux, $(2p-1)/2N_c < \Phi/2\pi < (2p+1)/2N_c$ (approximately, as discussed above), the ground state is the one with the pth J_α from the left moved to the left Fermi surface.

C. $N_c = 4n+1$, $N_s = 2n$

For this case the I_j's and J_α's all have to be integers. At zero flux the quantum numbers are

$$I_j = -(N_c-1)/2, \ldots, (N_c-1)/2 ,$$
$$J_\alpha = -N_s/2, \ldots, -1, 1, \ldots, N_s/2 . \quad (3.13)$$

There is a hole at $J_\alpha = 0$. This state has momentum zero. For $(2p-1)/2N_c < \Phi/2\pi < (2p+1)/2N_c$, the ground state will be the one with $J_\alpha = p$ moved to 0. We note in particular that for $\Phi = \pi/2$, the J_α's are consecutive integers.

D. $N_c = 4n-1, N_s = 2n-1$

In this case all the I_j's and J_α's must be half odd integers. For zero flux the ground-state quantum numbers are

$$I_j = -N_c/2, \ldots, (N_c-2)/2 ,$$
$$J_\alpha = -(N_s-2)/2, \ldots, -\tfrac{1}{2}, \tfrac{3}{2}, \ldots, (N_s+2)/2 . \quad (3.14)$$

We note that there is a hole between $-\frac{1}{2}$ and $\frac{3}{2}$. As the flux increases, the hole will move to the right. For

$$(2p-1)/2N_c < \Phi/2\pi < (2p+1)/2N_c ,$$

the hole is at $J_\alpha = (2p+1)/2$. At Φ equal to half a flux quantum, the J_α quantum numbers all consecutive half odd integers. Beyond that the I_k quantum numbers will all shift to the right by 1, and the hole in the J_α distribution will move to the negative side.

What we have done in the above analysis is to find, by evaluating the leading term in $1/U$ for a small system, just which of the many degenerate (at infinite U) states has lowest energy at finite but large U. At the same time, we have determined approximately what the energy splitting is and how the parabolic segments are moved up and down relative to each other by amounts equal to the appropriate magnon-energy differences. We wish to examine the range of validity of this picture as we go to smaller U or to larger systems. From (3.9) we see that the energy cost of creating the magnon has order of magnitude N_c/NU. From (3.3) the untwisting of the boundary angle made possible by creating the magnon lowers the energy of the k distribution by an amount of order $1/N$. Thus the relevant parameter in assessing the reliability of our picture is N_c/U. It is also clear from our remarks before Eq. (3.7) why this is so. For mesoscopic systems, $N_c \gg U$, and many λ_α's are of order unity. In this case, taking only the leading-order term in the Taylor expansion for each of them and adding is clearly not a reliable approximation.

Despite these limitations the analysis gives a picture of $E_0(\Phi)$ as a function of Φ, in good agreement with numerical results from infinite U down to U of order 50. The main point is that the sequence of parabola bottoms—and for infinite U there are N_c of them per period—are raised by amounts of order $1/U$, reflecting the magnon energy at the appropriate momentum. Those parabolic sectors raised least therefore become the lowest-energy states over larger and larger intervals in Φ. Thus sectors corresponding to large spin-wave energies disappear from the $E_0(\Phi)$ curve, until finally only two sectors remain: those corresponding to magnon momentum zero and π.

The above picture can be verified for small chains by direct diagonalization of the Hamiltonian for various flux values. Figure 1 gives the ground-state energy of a ring of eight sites and four electrons, calculated by both direct diagonalization and minimization of energy of states within the sector of real λ using the Bethe ansatz, with the same result. As can be seen, for very large U ($U=200$ in the figure), there are four cusps and four parabolic segments in the energy-versus-flux curve in one flux quantum, as many as the number of electrons in the ring. However, for smaller U, the width of some of the segments gets smaller and smaller, until at some U they are taken over by others.

The assumption that only real λ appear in the ground state for any flux thus proves to be correct for large U, where a map to the Heisenberg model is possible, and for small chains, where direct diagonalization of the Hamiltonian is possible. As will be seen in the next section, the assumption is also correct for large systems, in which case the energy of various states can be analytically calculated up to $1/N$. We therefore assume it to be true for any chain size and U and minimize the energy of the system by using only real λ for various sizes of the chain. This is, of course, much more feasible than allowing general complex λ. Figure 2 is the ground-state energy $E(\Phi)$ calculated for chains with $U=100$ and density $n_c = 0.5$. The energy for the chain with four sites has four pronounced cusps. For the chain with eight electrons, the energy has eight parabolic segments in a period, but some become very narrow, while the segments around $\Phi=0$ and π widen. For 16 electrons, these two branches take over the whole period.

As mentioned above, these two sectors correspond to the singlet ground state of the spin chain and the spin-wave state with momentum $q=\pi$, respectively. These two states have the same energy in the thermodynamic limit, as indicated by the spin-wave dispersion relation. However, for finite chains, the spin-wave state has an excitation energy proportional to $1/N_c$. For large yet finite chains, this spin-wave excitation energy scales as a function of the chain size N the same way as the energy

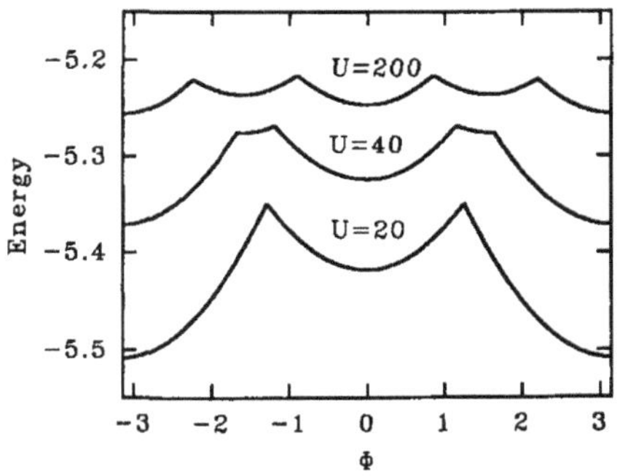

FIG. 1. Ground-state energy $E(\Phi)$ for a chain of eight sites and two spin-up and two spin-down electrons for the Hubbard repulsion $U=20$, 40, and 200.

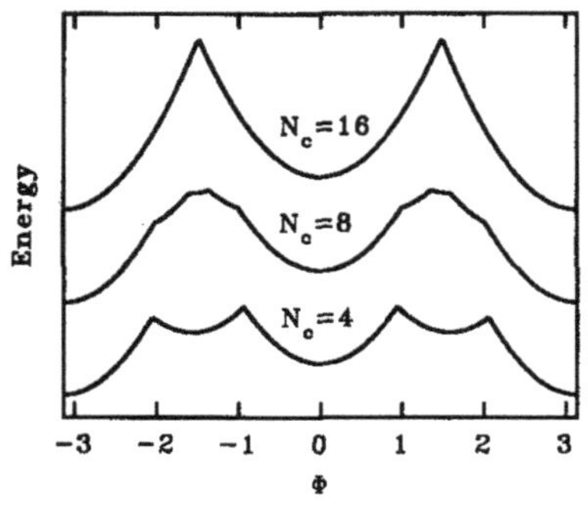

FIG. 2. Ground-state energy $E(\Phi)$ for chains with $U=100$ and density $n_c=0.5$. For the chain with 4 electrons, the energy has 4 pronounced cusps in one period. For 8 electrons there are 8 parabolic segments, but some are very narrow. For 16 electrons there are only two segments. Larger chains also have two level crossings. Note that the energies have been scaled.

reduction achieved in E_0 by the partial flux cancellation. As a result, for large chains, the widths of the two sectors at $\Phi=0$ and π approach values independent of the size of the chain. However, the width of the two sectors is dependent on the on-site repulsion U and the density of electrons.

Again, for chains with $U \gg 1$, this relative width is easy to find, at least in certain limits. For very large U, the magnon energy is negligible, and the two sectors should have almost the same width. However, for chains with density close to half filling, E_m in Eq. (3.4) is proportional to the hole density and is very small. The magnon energy in this case becomes dominant, and the sector with a magnon of $q=\pi$ will not be the ground state for any flux value. In this case the ground-state energy will have only one parabolic sector in the whole period.

In the next section, we will discuss mesoscopic chains with $N \gg 1$ and $N/U \gg 1$. In this case the k_j and λ_α become continuously distributed on the real axis. We can analytically discuss the flux dependence of the ground-state energy without resorting to the $1/U$ expansion. We will find that any chain with $N/U \gg 1$ has qualitatively the same behavior as chains with $N/U \gg 1$ and $U \gg 1$, independent of the value of U.

IV. MESOSCOPIC RINGS

In the thermodynamic limit, the momenta k_j of the electrons and rapidities λ_α of the magnons are continuously distributed on the real axis, and the Bethe-ansatz equations can be reduced to a set of integral equations, making them much easier to solve. This is the standard approach for finding the ground-state energy of a large system, for example. However, it has to be used carefully for the problem we are considering. The persistent current has an order of magnitude equivalent to that generated by a single electron at the Fermi level. In other words, for a noninteracting gas, this would be the current resulting from moving all the electrons to the right in momentum space by just one quantum state—a $1/N$ effect. In fact, the change in the ground-state energy arising from such a shift is down by a factor $1/N^2$ from the total ground-state energy, since to leading order the energy gained in the shift at one Fermi point is lost at the other. This implies that simply replacing the sum over momenta in the Bethe-ansatz equations by an integral will miss the effect we are looking for, since that introduces an error of order $1/N^2$.

Fortunately, a complete analysis of these finite-size corrections for the usual periodic-boundary conditions has been carried out by Woynarovich,[9] and it is not difficult to adapt his work to the case of twisted-boundary conditions. Woynarovich succeeded in calculating the energies of those states which resemble the ground state in that both momentum and spin quantum numbers form consecutive sets; that is to say, there are no holes in the distributions. (He later went on to add low-energy excitations, but these are not relevant to our present considerations.) These hole-free states differ from the standard half-filled nonmagnetic Hubbard ground state in that they may have different total numbers of particles, N_c, and of down spins, N_s, and they also may be shifted off center, by displacements D_c and D_s, respectively. It is easy to see that for the states under consideration these quantum numbers can be written in terms of the maximum and minimum occupied single-particle and -magnon quantum numbers $I_{\max}$, $I_{\min}$, $J_{\max}$, and $J_{\min}$ as follows;[9]

$$\begin{aligned} &I_{\max}-I_{\min}+1=N_c, \quad I_{\max}+I_{\min}=2D_c, \\ &J_{\max}-J_{\min}+1=N_s, \quad J_{\max}+J_{\min}=2D_s. \end{aligned} \tag{4.1}$$

Woynarovich found that, to order $1/N$, the energy of a hole-free distribution defined as above is

$$\begin{aligned} E=N\varepsilon_\infty &+\frac{2\pi v_c}{N}\left[\frac{(N_c-n_cN)^2}{4\xi^2}+\xi^2\left[D_c+\frac{D_s}{2}\right]^2-\frac{1}{12}\right] \\ &+\frac{2\pi v_s}{N}\left[\frac{1}{2}\left[\frac{N_c}{2}-N_s\right]^2+\tfrac{1}{2}D_s^2-\frac{1}{12}\right]. \end{aligned} \tag{4.2}$$

Here v_c and v_s are the so-called holon and spinon velocities, that is to say, the speeds of excitations near the Fermi points in the charge and spin distributions, ξ is the dressed charge, and n_c is the ground-state density in the thermodynamic limit. These parameters can all be found as functions of the electron density and on-site repulsion strength by solving some integral equations, as has been shown by Woynarovich.

The important thing to note in Eq. (4.2) is that the energy has a simple quadratic dependence on the four parameters N_c, D_c, N_s, and D_s.

Now adding a phase twist Φ to the boundary conditions in the Bethe-ansatz equations is equivalent to a uniform shift in the quantum numbers $I_j \to I_j+\Phi/2\pi$. Evidently, then, the generalization of (4.2) to twisted-boundary conditions is just to replace D_c by $D_c+\Phi/2\pi$, giving

$$E=N\varepsilon_\infty+\frac{2\pi v_c}{N}\left[\frac{(N_c-n_cN)^2}{4\xi^2}+\xi^2\left[D_c+\frac{\Phi}{2\pi}+\frac{D_s}{2}\right]^2-\frac{1}{12}\right]+\frac{2\pi v_s}{N}\left[\frac{1}{2}\left[\frac{N_c}{2}-N_s\right]^2+\tfrac{1}{2}D_s^2-\frac{1}{12}\right]. \tag{4.3}$$

It is clear from this equation that the curvature constant, the coefficient of Φ^2, is just the charge-stiffness constant discussed in Refs. 1–3. That is, we define the charge-stiffness constant by

$$\mathcal{D}_c=\frac{v_c\xi^2}{2\pi}. \tag{4.4}$$

The analogous spin-stiffness constant $\mathcal{D}_s$ is defined by

$$\mathcal{D}_s=\frac{v_s}{2\pi}, \tag{4.5}$$

for the systems that we consider. (For systems having finite spin magnetization, the dressing factors are more complicated.) The spin stiffness becomes relevant when the lowest-energy state for a given flux has a shifted magnon sea.

It is now straightforward to find how the ground-state energy of the Hubbard ring varies with the magnetic flux enclosed. For a given total number of electrons, N_c, and given magnetic flux Φ threading the system, the expression for the energy in (4.3) is minimized by appropriate choice of the other quantum numbers. Obviously, as Φ is increased from zero, at certain values the best choice of these other quantum numbers will change, and so the graph of ground-state energy as a function of Φ will be a sequence of parabolic segments.

The only other complication in this analysis is a bookkeeping one—the quantum numbers I_j and J_α, and hence the displacements D_c and D_s, are integer or half odd integer, depending on the parities of the numbers of electrons and of down spins, so one must consider separately the different possible total numbers of electrons *modulo 4*.

The physics of the problem is contained in the curvature of the parabolic segments, that is to say, the second derivative of total energy as a function of Φ, and the switch points, or level crossings, the values of Φ at which the lowest-energy state moves from one parabolic segment to another one defined by a different set of quantum numbers. These determine the periodicity of the energy, and hence of the persistent current, as a function of the threading magnetic field. It is evident from (4.3) above that these points are determined by the Fermi-point velocities and the dressed charge, which in turn can be calculated from the total electron density and on-site Coulomb-repulsion parameter U.

For the convenience of the reader, we summarize here the equations derived by Woynarovich for computing the Fermi-point velocities and dressed charge (for the case of no net macroscopic spin magnetization) and solve them analytically in some simple limits.

The dressed charge ξ is given by

$$\xi=\xi(k_0), \tag{4.6}$$

where $\xi(x)$ satisfies the integral equation

$$\xi(x)=1+\frac{1}{2\pi}\int_{-\sin k_0}^{\sin k_0}\bar{K}(x-x')\xi(x')dx',\qquad \bar{K}(x)=\int_{-\infty}^{\infty}\frac{\exp(-|\omega|U/4)}{2\cosh(\omega U/4)}\exp(i\omega x)d\omega. \tag{4.7}$$

The holon and spinon velocities are given by

$$2\pi v_c=\varepsilon_c'(k_0)/\rho_c(k_0),\qquad 2\pi v_s=\left[\int_{-k_0}^{k_0}\exp\left[\frac{2\pi\sin k}{U}\right]\varepsilon_c'(k)dk\right]\times\left[\int_{-k_0}^{k_0}\exp\left[\frac{2\pi\sin k}{U}\right]\rho_c(k)dk\right]^{-1}. \tag{4.8}$$

Here $\rho_c(k)$ and $\varepsilon_c'(k)$ satisfy the integral equations

$$\rho_c(k)=\frac{1}{2\pi}+\frac{1}{2\pi}\cos k\int_{-k_0}^{k_0}\bar{K}(\sin k-\sin k')\rho_c(k')dk' \tag{4.9}$$

and

$$\varepsilon_c'(k)=2\sin k+\frac{1}{2\pi}\cos k\int_{-k_0}^{k_0}\bar{K}(\sin k-\sin k')\varepsilon_c'(k')dk'. \tag{4.10}$$

These equations can be solved analytically in the following limits.

A. Strong coupling limit

In the limit $U\gg 1$, the kernel of the integral equations $\bar{K}\to 4\pi\ln 2/U$. The Fermi velocities and ξ can be explicitly obtained:

$$v_c=2\sin(\pi n_c)\left[1-\frac{4\pi\ln 2}{U}n_c[2+\cos(\pi n_c)]\right],\qquad v_s=\frac{4}{\pi U}[2\pi n_c-\sin(2\pi n_c)],\qquad \xi=1+\frac{4\ln 2}{\pi U}\sin(\pi n_c). \tag{4.11}$$

We note that the charge-stiffness constant approaches the finite value corresponding to free spinless fermions, while the spin-stiffness constant goes to zero.

B. Close to half-filled chain

As is well known, a half-filled repulsive chain is an insulator, and the charge stiffness $\mathcal{D}_c$ is zero. Close to the half-filled limit, the charge stiffness is proportional to the hole density:[14]

$$\mathcal{D}_c=\frac{v_c\xi^2}{2\pi}=4n_c(1-n_c)b/a^2, \tag{4.12}$$

where b and a are two functions of U introduced in Ref. 14. On the other hand, the spin-stiffness constant is related to the magnetic susceptibility, which has been obtained by Shiba in Ref. 15:

$$2\pi\mathcal{D}_s = v_s = \frac{1}{2\pi\chi} = 2\frac{I_0(2\pi/U)}{I_1(2\pi/U)} , \tag{4.13}$$

where I_ν is the Bessel function of imaginary argument of order ν. We see that in this limit the spin stiffness is a nonzero constant, while the charge-stiffness constant goes to zero as the density approaches the half-filled limit. In general, the two stiffness constants are of comparable value and have to be calculated numerically.

We now discuss the possible level crossings as the flux through the ring changes. As always, the ground state of an individual ring depends crucially on the parities of the numbers of up- and down-spin electrons. Below we give the ground state and its energy as a function of flux for each of the four possible cases. Note that because $E(-\Phi)=E(\Phi)$, we only give the ground-state energy for $0<\Phi<\pi$.

C. $N_c = 4n+2$

In this case, $N_s = 2n+1$. There are odd numbers of spin-up and spin-down electrons in the ground state. The quantum numbers I_j must be half odd integers, and the J_α are integers. From (2.4) this gives $D_c=0$ (mod 1) and $D_s=0$ (mod 1). For Φ close to zero, the ground state is the state with $D_c=D_s=0$. We will call this state (0,0). However, for Φ close to π, the state $(0,-1)$ with $D_c=0$, $D_s=-1$ may have lower energy. The energies of the two states are

$$\begin{aligned} E_{0,0}(\Phi)-E(0) &= \frac{v_c}{2\pi N}\xi^2\Phi^2 , \\ E_{0,-1}(\Phi)-E(0) &= \frac{v_c}{2\pi N}\xi^2(\pi-\Phi)^2 + \frac{\pi v_s}{N} . \end{aligned} \tag{4.14}$$

Here $E(0)$ is the ground-state energy at $\Phi=0$. These two energy levels will cross at

$$\Phi_c = \frac{\pi}{2} + \frac{\pi v_s}{v_c\xi^2} . \tag{4.15}$$

We first discuss the current for Φ close to the origin. The current is diamagnetic and is proportional to the flux. Its magnitude is particularly easy to find in the two limits discussed above. In the strongly repulsive case, differentiating (4.14) with respect to Φ and using (4.11), we find

$$j = \frac{2}{N}\sin(\pi n_c)\left[1-\frac{4\ln 2}{\pi U}[2n_c\pi^2+n_c\pi^2\cos(\pi n_c) + 2\sin(\pi n_c)]\right]\Phi . \tag{4.16}$$

The first-order term is actually the current of a spinless fermion ring with density n_c. We see that for low density the current is proportional to the electron density, while close to half filling the current is proportional to $(1-n_c)$. In fact, close to half filling, the current is always proportional to $(1-n_c)$ for any nonzero U, as can be seen from (4.12). The current is maximum for quarter-filled rings. Now we discuss where these levels cross.

In the strong-coupling limit, the spin stiffness is very small, and these two levels cross at the point $\Phi_c=\pi/2$. The energy is close to periodic in the flux with period half a flux quantum. At the level crossing, the current changes from a diamagnetic one to an paramagnetic one. We emphasize that here the period halving is caused by level crossing; none of the usual averaging has been introduced.

Near half filling, on the other hand, from (4.12) and (4.13), the charge-stiffness constant is very small and the spin stiffness is not small. It follows that, close enough to half filling, the two levels in (4.14) will cross at some point *beyond* $\Phi=\pi$, and in this case the state (0,0) will first intersect $(-1,0)$, which then becomes the lowest-energy state, and $(0,-1)$ is never the lowest-energy state. In this case the period of the current is one flux quantum.

For general U and filling, we have numerically calculated the level-crossing point Φ_c. Figure 3 is a contour graph of the width of the branch of the parabola centered at $\Phi=\pi$ as a function of U and the filling. We note that, close to half filling, there is a region in the (U,n_c) plane where the width is zero; in other words, the level crossing mentioned above does not happen at all, whereas for large U and away from half filling the level crossing occurs somewhere near $\frac{1}{4}$ of the flux quantum.

D. $N_c = 4n$ with n integer

The ground state should have $N_s=2n$. According to (2.4), all the I_j's must be integers and all the J_α's must be half odd integers. We have

$$D_c = \tfrac{1}{2} \ (\mathrm{mod}\ 1) \quad \text{and} \quad D_s = 0 \ (\mathrm{mod}\ 1) . \tag{4.17}$$

For Φ around 0, the ground state should have $D_c=-\frac{1}{2}$ and $D_s=1$. The energy of this state is

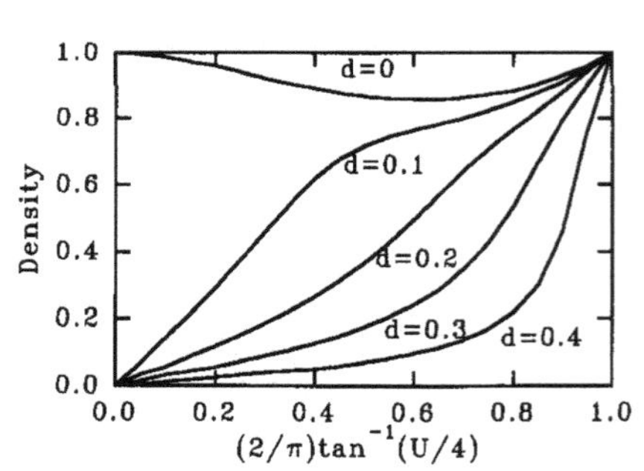

FIG. 3. Width d of the local minimum centered at $\Phi=0$ for a chain with $4n$ electrons. Close to the top of the graph (close to half filling) is a region where the width is zero; this minimum is never the real ground state, and the current is paramagnetic. As the density is decreased and/or the repulsion gets stronger, the width of that local minimum gets bigger. On the lines shown in the figure, the width of the minimum is constant. From top to bottom: $d=0$, 0.1, 0.2, 0.3, and 0.4 flux quantum.

$$E(\Phi)-E(0)=\frac{v_c}{2\pi N}\xi^2\Phi^2+\frac{\pi v_s}{N}\ , \quad (4.18)$$

while for Φ around π, the ground state should have $D_c=-\frac{1}{2}$ and $D_s=0$. The energy of this state is

$$E(\Phi)-E(0)=\frac{v_c}{2\pi N}\xi^2(\pi-\Phi)^2\ . \quad (4.19)$$

These two energy levels cross at

$$\Phi=\pi-\Phi_c\ , \quad (4.20)$$

where Φ_c is given by (4.15). We note that the spectrum is same as in Sec. IV C shifted by half a flux quantum.

E. $N_c=4n+1$

In this case we assumed $N_s=2n$ ($N_s=2n+1$ will give the same result). Then we have $D_c=0$ (mod 1) and $D_s=\frac{1}{2}$ (mod 1). The ground state for Φ between zero and π is the state with $D_c=0$ and $D_s=-\frac{1}{2}$. The energy is

$$E(\Phi)-E(0)=\frac{v_c}{2\pi N}\xi^2(\Phi-\pi/2)^2+\frac{\pi v_s}{4N}\ . \quad (4.21)$$

Extending the above formula to $\Phi<0$, we find that level crossing occurs at $\Phi=0$ and $\Phi=\pm\pi$. Unlike the level crossings for chains with even numbers of electrons, these level crossings are caused by the electron statistics and also occur for free electrons.

F. $N_c=4n+3$

The ground state in this case is $D_c=-\frac{1}{2}$ and $D_s=\frac{1}{2}$. The ground-state energy is the same as in Sec. IV E.

Let us now summarize our findings concerning the persistent current in mesoscopic Hubbard rings. For rings with an *odd* number of electrons, the period of the current is a half a flux quantum, independent of the interaction strength. The current is *paramagnetic* around $\Phi=0$:

$$j=B(\pi/2-|\Phi|)\mathrm{sgn}(\Phi),\quad B=v_c\xi^2/\pi N\ , \quad (4.22)$$

where sgn(x) is the signum function of x. There is a renormalization of the magnitude of the current due to the interaction, but the periodicity and sign of the current are the same as those of rings of odd numbers of free electrons.

For rings with an *even* numbers of electrons, the current is *diamagnetic* around $\Phi=0$ [except rings whose density is very close to half filling, in which case the current may be paramagnetic for a ring whose electron number is 0 (mod 4)]. For a chain whose numbers of spin-up and -down electrons are both odd, the current will become paramagnetic at $\Phi_c, \pi/2<\Phi_c<\pi$,

$$j=\begin{cases}-B\Phi, & |\Phi|<\Phi_c\\ -B(\pi-|\Phi|), & \Phi_c<|\Phi|<\pi\ ,\end{cases} \quad (4.23)$$

whereas for a chain in which the numbers of spin-up and -down electrons are both even, the switch occurs at $\pi-\Phi_c$,

$$j=\begin{cases}-B\Phi, & |\Phi|<\pi-\Phi_c\\ -B(\pi-|\Phi|), & \pi-\Phi_c<|\Phi|<\pi\ .\end{cases} \quad (4.24)$$

An interesting observation is that for chains with strong repulsion $U\gg1$, the switch occurs at $\Phi_c\simeq\pi/2$. For such chains the period of the persistent current of each individual ring is half a flux quantum. A comparison with the persistent current of free-electron rings[16] shows that in this case the current is drastically different from the free-electron case. The current for a ring with $4n$ electrons for small Φ is, e.g., changed from a paramagnetic one into a diamagnetic one.

For rings with not very strong interactions or with a density close to half filling, the period of the persistent current of an individual ring is still one flux quantum. However, just as demonstrated by Loss and Goldbart, and by Kusmartsev[16] for free-electron rings, the average current of a collection of rings with random numbers of electrons has a period of half a flux quantum. Assuming there are equal numbers of rings with even and odd numbers of electrons, a simple average of the current yields

$$j_{\mathrm{av}}=\begin{cases}B(\mathrm{sgn}\Phi-4\Phi), & |\Phi|<\pi-\Phi_c\\ B(2\pi\,\mathrm{sgn}\Phi-4\Phi), & \pi-\Phi_c<|\Phi|<\Phi_c\\ B(3\pi\,\mathrm{sgn}\Phi-4\Phi), & \Phi_c<|\Phi|<\pi\ .\end{cases} \quad (4.25)$$

The period of the average current is half a flux quantum, and the current is paramagnetic. The periodicity and sign of the average current are the same as those of free-electron rings.

ACKNOWLEDGMENTS

The authors thank A. Dorsey for his interest in this work and a careful reading of the manuscript. We thank H. Frahm, P. Goldbart, and Dr. S. Ullah for their helpful discussions. This work is supported by the National Science Foundation (NSF) through Grant No. DMR-8810541.

[1] B. S. Shastry and B. Sutherland, Phys. Rev. Lett. **65**, 243 (1990).

[2] R. M. Fye, M. J. Martins, D. J. Scalapino, J. Wagner, and H. Hanke, Phys. Rev. B **44**, 6909 (1991).

[3] C. A. Stafford, A. J. Millis, and B. S. Shastry, Phys. Rev. B **43**, 13 660 (1991).

[4] F. V. Kusmartsev, J. Phys. Condens. Matter **3**, 3199 (1991).

[5] A. A. Zvyagin, Fiz. Nizk. Temp. **16**, 1299 (1990) [Sov. J. Low Temp. Phys. **16**, 745 (1990)].

[6] E. H. Lieb and F. Y. Wu, Phys. Rev. Lett. **20**, 1445 (1968).

[7] P. F. Maldague, Phys. Rev. B **16**, 2437 (1977).

[8] L. P. Levy, G. Dolan, J. Dunsmuir, and H. Bouchiat, Phys. Rev. Lett. **64**, 2074 (1990).

[9] F. Woynarovich, J. Phys. A **22**, 4243 (1989).

[10] H. Frahm and V. E. Korepin, Phys. Rev. B **42**, 10 533 (1990).

[11] B. Sutherland and B. S. Shastry, Phys. Rev. Lett. **65**, 1833

(1990).

[12]M. Büttiker, Y. Imry, and R. Landauer, Phys. Lett. **96A**, 365 (1983); R. Landauer and M. Büttiker, Phys. Rev. Lett. **54**, 2049 (1985).

[13]J. des Cloizeaux and J. J. Pearson, Phys. Rev. **128**, 2131 (1962). See also J. C. Bonner, B. Sutherland, and P. Richards, in *Magnetism and Magnetic Materials—1974 (San Francisco)*, Proceedings of the 20th Annual Conference on Magnetism and Magnetic Materials, edited by C. D. Graham, G. H. Lander, and J. J. Rhyne, AIP Conf. Proc. No. 24 (AIP, New York, 1975), p. 335.

[14]N. Kawakami and S. K. Yang, Phys. Rev. Lett. **65**, 3063 (1990).

[15]H. Shiba, Phys. Rev. B **6**, 930 (1972).

[16]D. Loss and P. Goldbart, Phys. Rev. B **43**, 13 972 (1991). Also Kusmartsev has discussed this point for spinless-free-fermion chains (see F. Kusmartsev, Pis'ma Zh. Eksp. Teor. Fiz. **53**, 27 (1991) [JETP Lett. **53**, 28 (1991)]).

II

The supersymmetric t-J model

II. THE SUPERSYMMETRIC t-J MODEL

The (two-dimensional) t-J model was invented by F. C. Zhang and T. M. Rice as an effective model for describing the copper-oxide planes in high-T_c superconductors[126]. The model describes electrons on a lattice with a hamiltonian that includes nearest neighbour hopping (denoted t) and nearest neighbour spin exchange and charge interactions (J). The Hilbert space of the model is constrained to exclude double occupancy of any single site, which corresponds to an infinite on-site repulsion. In the limit $J \ll t$ the t-J model can be obtained from a large-U Hubbard model[127]; for other values of the couplings the t-J model seems to exhibit quite different phenomena like phase separation (for $J \gg 2t$)[128–130].

A particularly interesting feature of the t-J model in arbitrary dimensions is the supersymmetry[131–134] of the model for the special case $J = \pm 2t$ (Ref. 135). Here supersymmetry is to be understood as invariance under a graded Lie algebra, that does not contain the Poincaré algebra, and is thus different from the supersymmetry encountered in relativistic quantum field theories. Supersymmetry in Solid State Physics was first discussed in Ref. 136.

The history of the exact solution of the supersymmetry t-J model is quite complicated and we will attempt to give a chronological account of its evolution below.

Surprisingly the exact solution of the t-J model at the integrable point $J = \pm 2t$ in one dimension pre-dates the paper of Zhang and Rice by 14 years. This apparent contradiction is due to the equivalence (up to a trivial term in the hamiltonian) of the supersymmetric t-J model to a quantum lattice gas of hard-core bosons and fermions, which was introduced and solved by means of a nested Bethe *Ansatz* by C. K. Lai in 1974[137] (the t-J model is equivalent to the case of one species of bosons and two species of fermions). In his seminal 1975 paper B. Sutherland[138] pointed out a mistake in Lai's results and introduced a more general multicomponent lattice gas, which he then solved by a Bethe *Ansatz* [repr.II.1] (the derivation of the Bethe *Ansatz* Equations (BAE) is based on Sutherland's paper Ref. 139). He also determined the ground state and low-lying excitations for a gas of N species of fermions. The hamiltonians of these lattice gas models can be written as a sum over permutation operators. Due to their different choices of reference states for the Bethe *Ansatz* the form of the Bethe *Ansatz* equations derived by Sutherland and by Lai are quite different. In the literature these two different forms of BAE are referred to as Sutherland and Lai representations respectively. A third form of the BAE was discovered by F. H. L. Eßler and V. E. Korepin in Ref. 140.

In 1987 P. Schlottmann[141] [repr.II.2] investigated the question of integrability of the model introduced by Zhang and Rice in one dimension. He discovered the solution at $J = \pm 2t$ and derived the Bethe *Ansatz* equations (in Lai's representation). He discussed the integral equations for the ground state in a magnetic field as well as the thermodynamics. He also found the connection with the integrable quantum lattice gas of Lai and Sutherland.

In 1990 and 1991 P. A. Bares, G. Blatter and M. Ogata obtained the Bethe *Ansatz* solution of the t-J hamiltonian in both Sutherland's and Lai's representation and gave a detailed account of the excitation spectrum in both representations[142–144] [repr.II.3]. At the same time as Bares *et al.*, S. Sarkar independently discovered the supersymmetry of the model and constructed the Bethe *Ansatz* solution in Sutherland's form. Sarkar derived the integral equations for the ground state in a magnetic field. He discussed the ground state and low-lying excitations close to half-filling[145] [repr.II.4]. He also constructed a basis of supercoherent states for the model[146]. The magnetization curves for the model were determined by M. Quaisser, A. Schadschneider and J. Zittartz[147].

The first ones to study the asymptotic behaviour of correlation functions were N. Kawakami and S.-K. Yang[148,116] [repr.II.5]. Applying the methods of conformal field theory for Bethe *Ansatz* solvable models, developed by H. Frahm and V. E. Korepin for the Hubbard model (see [repr.I.23]), they evaluated the finite-size corrections and critical exponents for the supersymmetric t-J model and compared their results with the ones for the repulsive Hubbard model. Shortly after the work of Kawakami and Yang, who performed their calculations in the Lai representation, Bares, Blatter and Ogata performed studies of the conformal properties in Sutherland's representation[143]. A detailed account of the derivation of many of the results mentioned above can be found in Bares' thesis[144]. Exact results for the t-J model can be obtained by Bethe *Ansatz* only at the supersymmetric point $J = \pm 2t$. To obtain exact results for other values of the couplings, one therefore has to use other techniques. Pursuing this path, P. Schlottmann succeeded in obtaining an asymptotically exact solution for regions away from the supersymmetric point $J = \pm 2t$ for small band fillings[149].

The integrability of the model in the sense of the existence of a family of commuting transfer matrices and an infinite set of conservation laws was established by F. H. L. Eßler and V. E. Korepin[140], who studied the model in the framework of the (Graded) Quantum Inverse Scattering Method (QISM)[150,151]. The first to apply the QISM to the permutation-type models was P. P. Kulish, who re-derived Sutherland's form of the BAE[150] by a nested Algebraic Bethe Ansatz[152–154].

The question of completeness of the Bethe Ansatz states was resolved by A. Förster and M. Karowski[155,156] [repr.II.6]. They proved a highest weight theorem for the Bethe *Ansatz* states with respect to the $su(1|2)$ supersymmetry of the t-J model and showed completeness of the set of states obtained by acting with the $su(1|2)$ raising operators on the Bethe *Ansatz* states. Quantum deformations of the supersymmetric t-J model were found by A. Klümper, A. Schadschneider and J. Zittartz[157], R. Z. Bariev[158,159], and A. Foerster and M. Karowski[160]. R. Z. Bariev also studied the correlation functions of such an anisotropic t-J model[161].

The transport properties of the model in an external magnetic field were studied in the framework of the Landau-Luttinger approach by P. A. Bares, J. M. P. Carmelo, J. Ferrer and P. Horsch[162].

PHYSICAL REVIEW B VOLUME 12, NUMBER 9 1 NOVEMBER 1975

Model for a multicomponent quantum system*

Bill Sutherland†

The Institute for Theoretical Physics, State University of New York, Stony Brook, New York 11794

(Received 3 March 1975)

In a recent paper, Lai introduced a lattice-gas model. In this paper we generalize Lai's model, making application to various systems such as dilute Heisenberg magnets, higher-spin systems, and a lattice of SU(3) triplets. By a careful consideration of general thermodynamic stability, and by variational arguments, we demonstrate Lai's solution to be incorrect, and in turn produce the correct solution in this case and in other cases including higher-dimensional models. The remaining cases we treat in one dimension by Bethe's ansatz, reducing the problem to coupled integral equations. We locate the singularities of the ground-state energy in the phase plane; and we explicitly calculate the absolute-ground-state energy, excitations above the absolute ground state, and the first correction to the absolute ground state for small concentrations of impurities.

I. INTRODUCTION

In a recent paper,[1] Lai introduced a model for a quantum lattice gas and presented a solution in the case of one dimension. However, his solution is incorrect for his most interesting example $\Delta=-1$ (our example B^2F), and thus his conclusion that two phases will coexist for some fixed concentration is invalid. This can be seen by noting that upon expanding the ground-state energy E/L as a function of M/L for fixed N/L, both of Lai's equations (15) and (20) give a positive correction. However, E/L for fixed N/L must be symmetric in M/L about $N/2L$ and by the requirement of thermodynamic stability, must be concave upward. Thus the first correction must be negative or zero.

The reason for Lai not obtaining the true ground state appears to be that neither of his assumptions for the distribution of quasimomentum in Bethe's ansatz corresponds to the ground-state distribution. In fact, it is unclear in what sense Bethe's ansatz is a solution in this case.

We have avoided this problem in the present paper by instead using general arguments of thermodynamic stability coupled with lower bounds on the ground-state energy from the variational principle. This method has a definite advantage over an explicit solution for the wave function and, in fact, gives results in higher dimension as well.

The other situation that Lai considered, his case $\Delta=+1$, corresponds to our BF^2, and in this instance our equations agree.

Although our primary purpose is to discuss Lai's solution, we have in addition considerably generalized and rephrased the original problem. For instance, our Eq. (1) would be a natural choice for a prototypical Hamiltonian to exhibit the phenomena of mixing in a multicomponent system. In this formulation, we no longer have the arbitrary restriction of a coupling constant to particular values, i.e., $\Delta=\pm 1$. Finally, numerous applications are made to various branches of physics in which multicomponent systems are of interest.

The final portion of our paper is devoted to explicit calculation of properties of interest and the demonstration of their dependence on the number of components.

II. STATEMENT OF THE PROBLEM

Consider a periodic one-dimensional lattice of N sites. Place N objects, or particles, on this lattice exactly one to a site. Let the operator $P_{j,k}$ permute whatever objects occupy sites j and k. We then wish to compute the eigenvalues and eigenfunctions of the Hamiltonian H,

$$H=\pm\sum_{j=1}^{N} P_{j,j+1} \quad (N+1=1). \tag{1}$$

The problem is obviously invariant under the permutation group S_N We shall also have occasion to consider the corresponding problem on a square or cubic lattice.

We must further specify the nature of the N objects. We assume P species, or components, denoted $A, B, \ldots, P$. Objects of a given species A are identical, but may either be bosons $A=+1$, or fermions $A=-1$. This specification of species statistics we designate as type $\{P\}=\{A, B, \ldots, P\}$. Further, if x of the species are bosons and y are fermions, $x+y=P$, then we shall often write the type $\{P\}$ as B^xF^y.

Let us define N_A as the number of objects of species A. The N_A are constants of motion, for the permutation operators do not create or annihilate particles. If we further define N_{AB} as the number of nearest-neighbor pairs of species A and B, then we have the following relationships:

$$\sum_A N_A=N, \tag{2a}$$

$$\sum_{A\geq B}\sum N_{AB}=N, \tag{2b}$$

$$N_A = N_{AA} + \frac{1}{2}\sum_{B(\neq A)} N_{AB}$$

$$= \frac{1}{2}\left(N_{AA} + \sum_B N_{AB}\right). \quad (2c)$$

Corresponding quantities normalized to the total site number N we denote by lower case n, i.e., $n_A = N_A/N$, $n=1$.

It is useful to define a restricted Hamiltonian $H(\{P\})$ which will be equivalent to H if we only consider states of type $\{P\}$,

$$H(\{P\}) = \pm\left(\sum_A A N_{AA} + \sum_{A>B}\sum \phi_{AB}\right). \quad (3)$$

The operator ϕ_{AB} permutes only nearest neighbors of species A and B, otherwise it gives zero. This restricted Hamiltonian is invariant under $S_{N_A}\otimes\cdots\otimes S_{N_P}$, not S_N.

If two operators are related by a unitary transformation, they have the same spectrum. In preparation for later discussion, we catalogue here the following unitary transformations, and their effect on H and $\{P\}$:

(a) Let us multiply all wave functions by the wave function completely antisymmetric in all objects. This operation we denote by L, and

$$LHL = -H, \quad L\{P\} = \{-P\}. \quad (4)$$

We see that the $\pm$ sign in the original expression for H is unnecessary, if we vary the type. We henceforth choose the minus sign.

(b) Consider the following operations, denoted by J_A: Proceed around the lattice (N even) and at each even numbered site, multiply the wave function by $+1$ if no particle of species A is present, -1 if there is a particle of species A present. This operator then transforms H by

$$J_A H J_A = H + 2\sum_{B(\neq A)} \phi_{AB}, \quad J_A\{P\} = \{P\}. \quad (5)$$

Thus J_A on H changes the sign of ϕ_{AB}, all $B\neq A$.

(c) The previous operations may be applied to higher dimension, as well as to the one-dimensional problem. However, there exists a transformation—the Jordan-Wigner transformation—applicable only to one dimension. This transformation may be used to change bosons into fermions and vice versa.

We first note the well-known lattice-gas analogy between hard-core bosons on a lattice, and two-component spins on a lattice: If $S_z = -\frac{1}{2}$, there is no boson; if $S_z = +\frac{1}{2}$, there is one boson. Then, in terms of the Pauli spin operators, the Jordan-Wigner transformation to fermion creation and annihilation operators $C^\dagger$, C is

$$C_j = \sigma_j^- \prod_{k=1}^{j-1} \sigma_k^z. \quad (6)$$

We denote such a Jordan-Wigner transformation on the species A by K_A. Then the effect of K_A is

$$K_A H K_A = H + 2\phi_{AA}, \quad K_A\{A, B, \ldots, P\} = \{-A, B, \ldots, P\}. \quad (7)$$

That is, the transformation changes the sign of ϕ_{AA}. On the other hand, if we vary the signs of all the terms ϕ_{AA} in H, we need not vary the type $\{P\}$ at all.

The restricted Hamiltonian is invariant under K_A, but since the type of any wave function does change, we conclude that the eigenvalues of the restricted Hamiltonian are independent of the type.

(d) We define $K \equiv \Pi_A^P K_A$, and thus

$$KHK = \sum_A \phi_{AA} - \sum_{A>B}\sum \phi_{AB}, \quad K\{P\} = \{-P\}. \quad (8)$$

(e) Finally, if we multiply K and L to give a transformation $J \equiv KL$, then

$$JHJ = -\sum_A \phi_{AA} + \sum_{A>B}\sum \phi_{AB}, \quad J\{P\} = \{P\}. \quad (9)$$

III. GENERALITIES

In this paper we shall consider the ground-state energy E_0, or $\epsilon \equiv E_0/N$, for the Hamiltonian of Eq. (1) with the minus sign.[2] First, we shall derive some general properties of $\epsilon(\{P\}; n_A, \cdots, n_P)$.

It is easy to verify the stability property that $\epsilon(\{P\}; n_A)$ considered as a function of the concentrations n_A is concave upwards: To derive an upper bound on ϵ at a concentration

$$n_A = a n_A^1 + (1-a) n_A^2 \quad (0 \le a \le 1), \quad (10)$$

we divide the system into two fractions; aN and $(1-a)N$. In the first portion, we take as a trial wave function the ground state at a concentration n_A^1, and in the second, the ground state at a concentration n_A^2. Then by the variational principle,

$$\epsilon(n_A) \le a\epsilon(n_A^1) + (1-a)\epsilon(n_A^2). \quad (11)$$

This is precisely the statement that $\epsilon(n_A)$ concaves upwards as a function of n_A.

Suppose we have two species A and B which are both bosons; i.e., $A = B = +1$. Let us consider the reduced Hamiltonian,

$$H(\{P\}) = -(N_{AA} + N_{BB} + \phi_{AB}) - \sum_{C(\neq A,B)} (\phi_{AC} + \phi_{BC}) - \sum_{\substack{C>D \\ (\neq A,B)}}\sum \phi_{CD}. \quad (12)$$

The minimum of the operator $-\phi_{AB}$ is equal to $-N_{AB}$. Thus, if we let the label S represent either A or B,

$$H(\{P\}) \ge -N_{SS} - \sum_{C(\neq S)} N_{CS} - \sum_{\substack{C>D \\ (\neq S)}}\sum \phi_{CD}. \quad (13)$$

If we consider eigenstates of $H(\{P'\}$ of type $\{P'\}$, where

$$\{P'\}=\{S=1,\ C, D, \ldots, P\}, \tag{14}$$

then these states will serve as acceptable states for $H(\{P\})$. We then choose the ground state $\psi_0(\{P'\})$ to minimize the right-hand side of Eq. (13) with the ground-state energy $\epsilon(\{P'\};\ n_S,\ n_C, \ldots, n_P)$. We finally conclude

$$\epsilon(\{1,1,C,\ldots,P\};\ an,\ (1-a)n,\ n_C,\ldots,n_p) \geq \epsilon(\{1,C,\ldots,P\};\ n,\ n_C,\ldots,n_P)\quad (0\leq a\leq 1). \tag{15}$$

But by the concavity property, the reverse must also be true, and hence we have an equality,

$$\epsilon(\{1,1,C,\ldots,P\};\ an,\ (1-a)n,\ n_C,\ldots,n_P) = \epsilon(\{1,C,\ldots,P\};\ n,\ n_C,\ldots,n_P)\quad (0\leq a\leq 1). \tag{16}$$

In words: If two components of a P-component system are bosons, then the ground-state energy is identical to a $(P-1)$-component system.

Before restricting ourselves entirely to one dimension, we briefly summarize the few solutions to Eq. (1) which are known exactly in three dimensions. First, the one-component systems are obviously trivial. Second, by Sec. II, the ground-state energy for B^xF^y is equivalent to BF^y. Thus for B^x, $\epsilon=-1$.

Third, let us consider BF. We use a representation where bosons are represented by vacancies on the lattice and the fermions hop about. Then the corresponding reduced Hamiltonian is

$$H=N_{FF}-N_{BB}-\phi_{FB} = N_F-N_B-\phi_{FB}. \tag{17}$$

Thus all eigenfunctions are given by the familiar expression

$$\Psi(\vec{X}_1,\ldots,\vec{X}_{N_F})=\det[\exp(i\vec{K}_j\cdot\vec{X}_l)], \tag{18}$$

where

$$K_{j_\alpha}=2\pi n_{j_\alpha}/N^{1/3},\quad n_{j_\alpha}\ \text{are integers},$$
$$j=1,\ldots,N_F;\quad \alpha=1,2,3\ . \tag{19}$$

Then the energies are

$$\epsilon=n_F-n_B-\frac{2}{N}\sum\cos K_{j_\alpha}. \tag{20}$$

For the ground state in one dimension,

$$\epsilon_0=2n_F-1-(2/\pi)\sin(\pi n_F). \tag{21}$$

Lastly, Eq. (20) or (21) also gives the ground-state energy for B^xF.

IV. EXAMPLES

In the remainder of this paper, we will consider only the case of one dimension. Before presenting the exact solution, we will first give a few examples of systems to which the Hamiltonian of Eq. (1) would apply. For the case of two species $P=2$, it is natural to use the language of spin-1/2 systems. The reduced Hamiltonian is

$$H=-(AN_{AA}+BN_{BB})-\phi_{AB}. \tag{22}$$

We consider each site as a spin $\frac{1}{2}$, with species A given by spin-up, species B given by spin-down. Then, in terms of the Pauli spin matrices σ_x, σ_y, σ_z, we have a representation of

$$\phi_{AB}=\sum_{nn}\tfrac{1}{2}(\sigma_x\sigma_x'+\sigma_y\sigma_y'), \tag{23a}$$

and

$$AN_{AA}+BN_{BB}=\begin{cases}+\sum_{nn}\frac{1}{2}(1+\sigma_z\sigma_z'), & BB\\ -\sum_{nn}\frac{1}{2}(1+\sigma_z\sigma_z'), & FF\\ \pm\sum\sigma_z, & FB\end{cases}. \tag{23b}$$

The expression $\sum_{nn}$ represents a summation over nearest-neighbor pairs.

Since $\sum\sigma_z$ is a constant of the motion, equal to N_A-N_B, which may be fixed at will, we find the following cases:

(a) BB,

$$H=-\frac{1}{2}\sum_{nn}(1+\vec{\sigma}\cdot\vec{\sigma}'). \tag{24}$$

Thus this type corresponds to a Heisenberg ferromagnet.

(b) FF,

$$H=+\frac{1}{2}\sum_{nn}(1+\sigma_z\sigma_z'-\sigma_x\sigma_x'-\sigma_y\sigma_y'). \tag{25}$$

Let us transform this Hamiltonian by the operator J of Eq. (9), which changes the sign of ϕ_{AB}. Then the transformed H is

$$H=+\frac{1}{2}\sum_{nn}(1+\vec{\sigma}\cdot\vec{\sigma}'). \tag{26}$$

This type corresponds to the Heisenberg antiferromagnet.

(c) FB,

$$H=-\frac{1}{2}\sum_{nn}(\sigma_x\sigma_x'+\sigma_y\sigma_y')\mp\sum\sigma_z. \tag{27}$$

This type corresponds to the $X-Y$ model.

Exact solutions have been found for all of these cases, and we will make use of these results in our later analysis of the three-component system. Lattice-gas interpretations for the above systems are also familiar in the literature.

We now consider the case of three components $P=3$. There are four distinct choices for the particle statistics. We discuss each in turn, again using the language of spin $\frac{1}{2}$: Place M spins on a lattice of N sites, one to a site—thus $M \leq N$. We then call an empty site, or vacancy, species A; spin-up species B; spin-down species C. In an obvious notation, $A=0$, $B=\uparrow$, $C=\downarrow$.

(a) BBB, $A=B=C=1$.

We write the Hamiltonian as

$$H = -(\phi_{0\uparrow}+\phi_{0\downarrow}) - N_{00} - (N_{\uparrow\uparrow}+N_{\downarrow\downarrow}+\phi_{\uparrow\downarrow}). \tag{28}$$

The first-term "hops" spins to nearby vacancies and thus represents the kinetic energy of the spins. We denote it by T. The last term may be written with Pauli matrices as

$$-(N_{\uparrow\uparrow}+N_{\downarrow\downarrow}+N_{\uparrow\downarrow}) = -\sum_{\text{s.p.}} \tfrac{1}{2}(1+\vec{\sigma}\cdot\vec{\sigma}\,'). \tag{29}$$

The symbol $\sum_{\text{s.p.}}$ represents a summation over nearest-neighbor spin pairs only.

Using the relationships between the various N's given in Eq. (2), we may rewrite N_{00} as

$$N_{00} = N_0 - N_\uparrow - N_\downarrow + N_{\uparrow\uparrow} + N_{\downarrow\downarrow} + N_{\uparrow\downarrow}$$

$$= N - 2M + \sum_{\text{s.p.}} 1\,. \tag{30}$$

The first two terms are constants of motion, which may be fixed. The last term is the number of spin pairs. If we combine this with Eq. (29), we obtain an interaction V between nearest-neighbor spins of the form

$$V = -\sum_{\text{s.p.}} \tfrac{1}{2}(3+\vec{\sigma}\cdot\vec{\sigma}\,') \tag{31}$$

Thus, if energies are measured with respect to separated spins, we find that spin pairs in a singlet state have zero energy, while spin pairs in a triplet state have energy -2.

The final form for our Hamiltonian is

$$H = T + V - (N-2M)\,. \tag{32}$$

The obvious interpretation is as a dilute Heisenberg ferromagnet which prefers to form bound pairs.

The discussion of the other three types proceeds similarly, and we will present them with less detail.

(b) FBB, $-A=B=C=1$.

We find the Hamiltonian to be

$$H = T + V + N - 2M, \tag{33}$$

where T is the kinetic energy as before, but now the interaction energy V between spins is

$$V = \sum_{\text{s.p.}} \tfrac{1}{2}(1-\vec{\sigma}\cdot\vec{\sigma}\,')\,. \tag{34}$$

Again the interpretation is as a dilute Heisenberg ferromagnet, but this time there is no preference to form bound pairs over separated spins.

(c) BFF, $A=-B=-C=1$.

We find the Hamiltonian to be

$$H = T - (N-2M)$$
$$-\frac{1}{2}\sum_{\text{s.p.}} (1-\sigma_z\sigma_z' + \sigma_x\sigma_x' + \sigma_y\sigma_y')\,. \tag{35}$$

We now apply a unitary transformation JJ_A from Eqs. (5) and (9) which has the effect of changing the sign of $\phi_{\uparrow\downarrow}$. Thus the final form for the transformed H is

$$H = T + V - (N-2M)\,, \tag{36}$$

with

$$V = -\frac{1}{2}\sum_{\text{s.p.}} (1-\vec{\sigma}\cdot\vec{\sigma}\,'). \tag{37}$$

The system is now a dilute Heisenberg antiferromagnet, inclined to form bound pairs of spins.

(d) FFF, $A=B=C=-1$.

Again we apply the transformation JJ_A to give a Hamiltonian

$$H = T + V + N - 2M, \tag{38}$$

with

$$V = \frac{1}{2}\sum_{\text{s.p.}} (3+\vec{\sigma}\cdot\vec{\sigma}\,'). \tag{39}$$

This is the case of a dilute Heisenberg antiferromagnet, now with no preference for bound spin pairs.

Thus our original Hamiltonian of Eq. (1) contains a wealth of particular models. For the examples just given, that of dilute magnets, we may expect both magnetic and electric behavior, with the possibility of singular behavior in the ground state.

Other interpretations are possible for three-component systems. The first that comes to mind might be a spin-1 system, with $S_z=+1, 0, -1$. But this is not the most natural, and we delay discussion for the moment. Instead, we return to the two-component system and note the Heisenberg magnets,

$$H = \pm\sum_{\text{nn}} \vec{\sigma}\cdot\vec{\sigma}\,'. \tag{40}$$

We may consider the three Pauli spin matrices σ_α ($\alpha=1,2,3$) as the generators of SU(2); H is similar in form to the Casimir operator. From this viewpoint, a natural generalization of Eq. 40 to the three-component case would be a Hamiltonian

$$H = \pm 2\sum_{\text{nn}} \sum_{\alpha=1}^{8} F_\alpha F_\alpha'\,, \tag{41}$$

where F_α ($\alpha=1, \ldots, 8$) are the generators of SU(3).

The equivalence of Eq. (1) and (42) is most easily seen by using instead of F_α's, the nine traceless 3×3 matrices A_ν^μ.

$$(A_\nu{}^\mu)_{jK} = \delta_{\mu K}\delta_{\nu j} - \tfrac{1}{3}\delta_{\mu\nu}\delta_{jK}. \tag{42}$$

Then the Casimir operator is

$$2F^2 \equiv \sum A_\mu{}^\nu A_\nu{}^\mu. \tag{43}$$

We note the relationships

$$\sum_{\text{nn}} (A_\nu{}^\mu A_\mu'{}^\nu + A_\mu{}^\nu A_\nu'{}^\mu) = \phi_{\mu\nu} \tag{44}$$

($\mu \neq \nu$; no sum over μ, ν), and

$$\sum_{\text{nn}} A_\mu{}^\mu A_\mu'{}^\mu = \phi_{\mu\mu} - \tfrac{1}{3}2N_\mu + \tfrac{1}{9}N. \tag{45}$$

Thus,

$$\sum_{\text{nn}} A_\nu{}^\mu A_\mu'{}^\nu = \sum_{\mu,\nu} \phi_{\mu\nu} - \tfrac{1}{3}N. \tag{46}$$

In fact, due to the extensive symmetry of the original problem, we may actually solve the more general problem of either a chain

$$\cdots \{3\}\{3\}\{3\}\{3\}\cdots, \tag{47a}$$

or

$$\cdots\{3\}\{\overline{3}\}\{3\}\{\overline{3}\}\cdots. \tag{47b}$$

$\{3\}$ and $\{\overline{3}\}$ are the two nonequivalent contragradient triplet representations of SU(3), often called quark and antiquark.

In the same way that Eq. (40) is invariant under the total SU(2) group, Eq. (41) is invariant under the total SU(3) group. The conserved quantities—total third component of isospin and total hypercharge—are linear combinations of the particle numbers N_A.

Finally, we remark that a similar correspondence exists between the P-component system and an SU(P)-invariant interaction.

We now return to consider a spin-1 representation for our system. This is most easily done by writing the SU(3) generators in terms of the three spin operators S_μ ($\mu = 1, 2, 3$) and the six tensor operators $T_{\mu\nu} = S_\mu S_\nu + S_\nu S_\mu$. The six tensor operators are not independent, however, for

$$T_{\mu\mu} = 2S_\mu S_\mu = 4I. \tag{48}$$

Then we have the identity,

$$\sum_{\text{nn}} A_\nu{}^\mu A_\mu'{}^\nu = 2\sum_{\text{nn}} (S_\mu S_\mu' + \tfrac{1}{2}T_{\mu\nu}T'_{\mu\nu}) - \tfrac{1}{3}16N. \tag{49}$$

Thus

$$\sum_{\mu,\nu} \phi_{\mu\nu} = \sum_{\text{nn}} (2S_\mu S_\mu' + T_{\mu\nu}T'_{\mu\nu}) - 5N. \tag{50}$$

Considered as a model of a spin-1 magnet, we have not simply dipole interactions but also tensor interactions. Again, there are some nonequivalent choices of sign for the various terms in Eq. (50). The total spin components are individually conserved.

V. SOLUTION OF THE PROBLEM

We now present the exact solution of the one-dimensional problem. First, we restrict ourselves to types of either F^P or BF^P. If we have these solutions, then by the arguments of Sec. VII, we have the ground-state energy for all types. This is not to say the other types are not interesting, for we do not have the excited states. However, until the excited states of the Heisenberg ferromagnet are understood, we can not expect to proceed with the multicomponent cases.

We now order the species so that the first P species are fermions with $N_1 > N_2 > \cdots > N_P$. We define the partial sums,

$$M_l = \sum_{j>l} N_j = N - \sum_{j\le l} N_j. \tag{51}$$

We now consider the fermions of species 1 as a background through which the other M_1 objects move. We write the Hamiltonian as

$$H = N_{11} + T - \sum_{A \geq B}{}' \phi_{AB}. \tag{52}$$

However, we may use the relation

$$N_{11} = N - 2M_1 + \sum_{A \geq B} N_{AB} \tag{53}$$

to write

$$H = N - 2M_1 + T - \sum_{A>B} (\phi_{AB} - N_{AB}). \tag{54}$$

We seek solutions to this Hamiltonian as a wave function $\Psi(X_1, \cdots, X_M)$, where $X_1, \ldots, X_M$ are the locations of the $M = M_1$ objects, of the form known as Bethe's ansatz,

$$\Psi_Q = \sum_P A(Q, P) \exp\, i\sum_{j=1}^{M} X_j K_{Pj}. \tag{55}$$

Here P is one of the $M!$ permutations of the K's (not to be confused with the number of species of fermions), Q is one of the $M!$ permutations of the objects, and the X's are ordered so that $X_1 < X_2 < \cdots < X_M$. We may easily determine the eigenvalues of H, provided that such a wave function exists, by considering the particles to be separated. Then the final term of Eq. (54) gives zero, and we have

$$E = N - 2M - 2\sum_{j=1}^{M} \cos K_j. \tag{56}$$

As is familiar in such verifications of Bethe's ansatz, we arrange the $M!$ coefficients $A(Q, P)$, for fixed P, as a column vector ξ_P. Then we find that the boundary conditions at $X_4 = X_5 - 1$ can be satisfied provided

$$\xi \cdots ij \cdots = Y_{ji}^{45}\, \xi \cdots ji \ldots, \tag{57}$$

where

$$Y_{ij}^{\alpha\beta} = (X_{ij} + P_{\alpha\beta})/(1 - X_{ij}) . \tag{58}$$

Here $P_{\alpha\beta}$ interchanges Q_α and Q_β, and

$$\frac{1}{X_{ij}} = \frac{e^{iK_j} - e^{iK_i}}{(1+e^{iK_i})(1+e^{iK_j})} . \tag{59}$$

We now make a change of variables,

$$\alpha_j(K_j) = \tfrac{1}{2}\tan(\tfrac{1}{2}K_j) , \tag{60}$$

so that

$$X_{ij} = i/(\alpha_i - \alpha_j) . \tag{61}$$

However, this is precisely the form treated by Yang[3] and more generally by Sutherland.[4] Thus we immediately have that the Eqs. (57) are consistent, and that requiring the problem to be periodic imposes the following coupled algebraic equations on the K's:

$$e^{iK(\alpha)N} = \prod_\beta \left(\frac{\alpha-\beta+\frac{1}{2}i}{\alpha-\beta-\frac{1}{2}i}\right) \prod_\alpha \left(\frac{\alpha-\alpha'-i}{\alpha-\alpha'+i}\right) , \tag{62α}$$

$$\prod_{\beta'}\left(\frac{\beta-\beta'-i}{\beta-\beta'+i}\right) = \prod_\alpha\left(\frac{\beta-\alpha-i/2}{\beta-\alpha+i/2}\right)\prod_\gamma\left(\frac{\beta-\gamma-\frac{1}{2}i}{\beta-\gamma+\frac{1}{2}i}\right) \tag{62β}$$

$$\vdots$$

$$\prod_{\zeta'}\left(\frac{\zeta-\zeta'-i}{\zeta-\zeta'+i}\right) = \prod_\delta\left(\frac{\zeta-\delta-\frac{1}{2}i}{\zeta-\delta+\frac{1}{2}i}\right) , \quad F^P$$

or (62ζ)

$$\prod_\delta\left(\frac{\zeta-\delta-\frac{1}{2}i}{\zeta-\delta+\frac{1}{2}i}\right) = 1, \quad BF^P .$$

Equation (62ζ) presents the two alternate choices for the final equation, the first corresponding to F^P and the second to BF^P. All intermediate equations have the form of Eq. (62β). In all, there are $P-1$ equations for M_1 variables α, M_2 variables $\beta, \ldots, M_{P-1}$ variables ζ if F^P; or P equations for M_1 variables $\alpha, \ldots, M_P$ variables ζ if BF^P.

We now take the logarithm of these equations,

$$NK(\alpha) = 2\pi J_\alpha - \sum_{\alpha'} \theta(\alpha - \alpha) + \sum_\beta \theta(2\alpha - 2\beta), \tag{63α}$$

$$\sum_{\beta'} \theta(\beta - \beta') = 2\pi J_\beta + \sum_\alpha \theta(2\beta - 2\alpha) + \sum_\gamma \theta(2\beta - 2\gamma) , \tag{63β}$$

$$\vdots$$

$$\sum_{\zeta'} \theta(\zeta - \zeta') = 2\pi J_\zeta + \sum_\delta \theta(2\zeta - 2\delta), \quad F^P$$

or (63ζ)

$$0 = 2\pi J_\zeta + \sum_\delta \theta(2\zeta - 2\delta), \quad BF^P .$$

Here $\theta(x) = -2\tan^{-1}(x)$, and $J_\alpha, J_\beta, \ldots, J_\zeta$ are integers (half odd integers) which arise from the logarithm of $+1(-1)$ and serve as quantum numbers.

Finally, we consider the ground state where the the J's are dense about the origin and the variables are smoothly distributed with densities $R_j(\alpha)$ between limits $\pm B_j$. These densities are normalized so that

$$\int_{-B_j}^{+B_j} R_j(\alpha)\, d\alpha = 2\pi \frac{M_j}{N} \equiv 2\pi m_j . \tag{64}$$

Then the Eqs. (63) become integral equations for the densities $R_j(\alpha)$. If we arrange the densities $R_j(\alpha)$ as the elements of a column vector $\underline{R}$, then the integral equations may be put in the concise matrix form

$$\underline{\xi} = \underline{R} + \underline{K}\,\underline{B}\,\underline{R} . \tag{65}$$

Here,

$$[\underline{\xi}]_j = \delta_{j1}\frac{dK}{d\alpha} = \frac{4}{1+4\alpha^2}\delta_{j1} , \tag{66}$$

$$[\underline{B}\,\underline{R}]_j = \begin{cases} R_j(\alpha), & |\alpha| < B_j \\ 0, & |\alpha| > B_j \end{cases} , \tag{67}$$

and $\underline{K}$ is a matrix whose elements $[\underline{K}]_{ij}$ are integral operators. Let us define K_n as the integral operator

$$K_n \psi(\alpha) \equiv \frac{1}{2\pi}\int_{-\infty}^{+\infty} \frac{2n\, d\beta}{1+n^2(\alpha-\beta)^2}\, \psi(\beta) . \tag{68}$$

It has eigenvalues given by the Fourier transform $\tilde{K}_n(k)$

$$\tilde{K}_n(k) = e^{-|k|/n} . \tag{69}$$

We note the relation

$$K_n K_m = K_{nm/(n+m)} \tag{70}$$

We may now write out the matrix $\underline{K}$ for the two cases: If F^P, $\underline{R}$ is of dimension $P-1$, and

$$[K]_{ij} = \delta_{ij}K_1 - (\delta_{i,j+1} + \delta_{i,j-1})K_2 . \tag{71}$$

If BF^P, R is of dimension P, and

$$[K]_{ij} = (\delta_{i,j} - \delta_{iP}\delta_{jP})K_1 - (\delta_{i,j+1} + \delta_{i,j-1})K_2 . \tag{72}$$

We may write the ground-state energy per particle ϵ also in matrix form as

$$\epsilon = 1 - \frac{1}{2\pi}\underline{\xi}^\dagger \underline{B}\,\underline{R}$$

$$\equiv 1 - \frac{1}{2\pi}\int_{-B_1}^{B_1} R_1(\alpha)\frac{dK}{d\alpha}\, d\alpha . \tag{56'}$$

Finally, let us rewrite Eqs. (64) as matrix equations.

$$\underline{m} = \frac{1}{2\pi}\underline{\eta}^\dagger \underline{B}\,\underline{R} , \tag{64'}$$

where

$$[\underline{\eta}^\dagger \underline{B}\,\underline{R}]_j \equiv \int_{-B_j}^{+B_j} d\alpha\, R_j(\alpha) . \tag{73}$$

To summarize, we collect in the very concise matrix form our basic Eqs. (65), (56′), and (64′):

$$\underline{\xi} = \underline{R} + \underline{K}\,\underline{B}\,\underline{R}\,, \tag{65}$$

$$2\pi \underline{m} = \underline{\eta}^\dagger \underline{B}\,\underline{R}\,, \tag{56'}$$

$$\epsilon = 1 - \underline{\xi}^\dagger \underline{B}\,\underline{R}/2\pi\,. \tag{64'}$$

We are to solve for $\underline{R}$ in terms of $\underline{B}$, calculate ϵ, $\underline{m}$ as functions of $\underline{B}$, and then eliminate $\underline{B}$ to find $\epsilon(\underline{m})$.

If all limits are finite, then Eq. (65) is a nonsingular matrix Fredholm equation, and the solution is an analytic function of the limits B_j. Thus, in turn, the ground-state energy must be an analytic function of the concentrations. Thus the region $0< B_j < +\infty$ we call the fundamental domain, and we now determine the region of concentrations to which it corresponds.

First, if $B_j = 0$, then $m_j = 0$, and we conclude

$$B_j = 0 \rightarrow n_k = 0, \qquad k > j. \tag{74}$$

Second, suppose $B_j = \infty$; then we integrate the equation for R_j from $-\infty$ to $+\infty$ and find

$$\delta_{j1} = 2m_j - m_{j+1} - m_{j-1}, \qquad j < P$$

or (75)

$$B_j = +\infty \rightarrow n_{j+1} = n_{j-1}, \qquad j < P\,.$$

The P equation, however, gives

$$B_P = +\infty \rightarrow n_{P-1} = 0\,. \tag{76}$$

In this case, the problem is reduced from $F^P B$ to $F^{P-1}B$.

We finally conclude that the fundamental domain corresponds to

$$n_1 > n_2 > \ldots > n_P\,, \tag{77}$$

and singularities occur only at the boundaries. By permuting the fermion concentrations, the fundamental domain is mapped onto the entire physical region of concentrations, and the ground-state energy obeys this permutation symmetry S_P. Figure 1 illustrates the surfaces of singularities in the phase plane for the two three-component systems F^3 and BF^2; the shaded region is the fundamental domain.

We note also, that if we are constrained to a surface of singularities, $B_j = +\infty$, then the corresponding R_j may be eliminated so that the resulting equations are again a nonsingular matrix Friedholm equation. We then conclude that restricted to a singular surface, singularities occur only at the intersection with another singular surface.

VI. ABSOLUTE GROUND STATE OF THE F^P PROBLEM

As an example, we may explicitly calculate the energy per particle $\epsilon(P)$ for the absolute ground

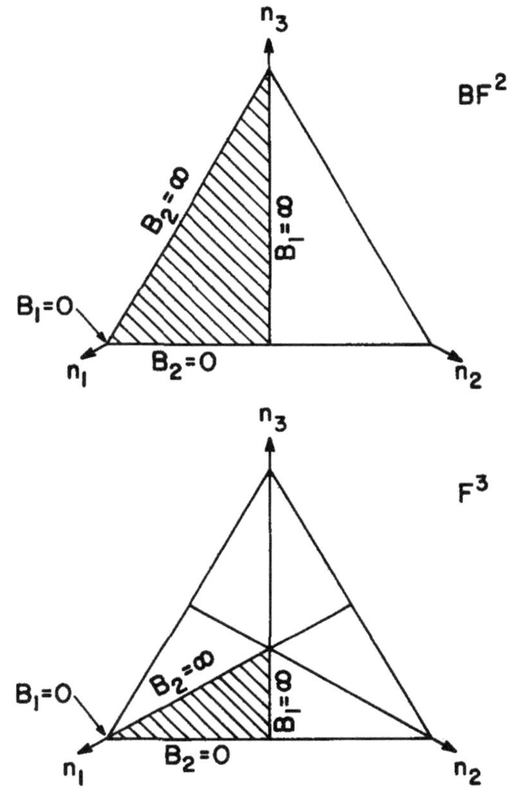

FIG. 1. We show the lines of singularities in the phase plane for the two three-component systems BF^2 and F^3. The shaded region is the fundamental domain, and the ground-state energy is analytic within this region. The limits of the integral equations are here correlated with the boundaries of the fundamental domain.

state of the F^P problem. The limits are all $+\infty$, the concentrations are all equal, and thus if we Fourier transform, we obtain a matrix equation. The Fourier transform of $\underline{K}$ is a $(P-1)\times(P-1)$ matrix,

$$[\tilde{I} + \tilde{K}]_{ij} = \delta_{ij}(1+X^2) - (\delta_{i,j+1} + \delta_{i,j-1})X \tag{78}$$

where $X \equiv e^{-|k|/2}$. The Fourier transform of the resolvent $(\underline{\tilde{I}} + \underline{\tilde{J}})$ is the inverse of this matrix and may be calculated by first finding the eigenvectors,

$$\psi_n(j) = (2/P)^{1/2}\sin(\pi n j/P) \qquad (n = 1, \ldots, P-1), \tag{79}$$

and eigenvalues

$$\lambda_n = 2X\cosh\tfrac{1}{2}k - \cos(\pi n/P)\,. \tag{80}$$

Then we have

$$[\underline{\tilde{I}} + \underline{\tilde{J}}]_{jk} = \sum_{n=1}^{P-1} \lambda_n^{-1}\psi_n(j)\psi_n^\dagger(k)\,. \tag{81}$$

We then may use this expression to solve for the Fourier transform of the densities,

$$\tilde{R}_j(k) = \frac{2\pi}{P} \sum_{n=1}^{P-1} \frac{\sin(\pi jn/P)\sin(\pi n/P)}{\cosh(\frac{1}{2}k) - \cos(\pi n/P)} \,. \tag{82}$$

To sum this series, we Fourier transform back and obtain

$$R_j(\alpha) = \frac{4\pi}{P} \frac{1}{\sinh 2\pi\alpha} \times \sum_{n=1}^{P-1} \sin(\pi jn/P)\sinh[2\pi\alpha(1 - n/P)] \,. \tag{83}$$

The summation now is simply four finite geometric series, and we have

$$R_j(\alpha) = \frac{2\pi}{P} \frac{\sin(\pi j/P)}{\cosh(2\pi\alpha/P) - \cos(\pi j/P)} \,. \tag{84}$$

The Fourier transform of this expression is

$$\tilde{R}_j(k) = 2\pi \frac{\sinh\frac{1}{2}k(P-j)}{\sinh\frac{1}{2}Pk} \,. \tag{85}$$

A simple check is to calculate the concentrations m_j by

$$m_j = (1/2\pi)R_j(0) = (P-j)/P \,. \tag{86}$$

Thus all $n_j = 1/P$.

Using the evaluation of Eq. (85), we may return and evaluate the Fourier transform of the resolvent, a symmetric matrix,

$$[\underline{\tilde{I}} + \underline{\tilde{J}}]_{jl} = e^{|K|/2} \frac{\sinh\frac{1}{2}K(P-j)\sinh\frac{1}{2}Kl}{\sinh\frac{1}{2}PK\sinh\frac{1}{2}K} \quad (j \geq l) \,. \tag{87}$$

The energy is given as

$$\begin{aligned} \epsilon(P) &= 1 - \frac{1}{2\pi}\int_{-\infty}^{+\infty} \frac{4d\alpha}{1+4\alpha^2} R_1(\alpha) \\ &= 1 - \int_{-\infty}^{+\infty} dK\, e^{-|K|/2} \frac{\sinh\frac{1}{2}k(P-1)}{\sinh\frac{1}{2}kP} \,. \end{aligned} \tag{88}$$

Let us define a new variable,

$$y \equiv e^{-Pk} \,. \tag{89}$$

We may rewrite $\epsilon(P)$ as

$$\epsilon(P) = 1 - \frac{2}{P}\int_0^1 dy \frac{y^{1/P-1} - 1}{1-y} \,. \tag{90}$$

Such an integral may be rewritten in terms of Euler's digamma function,

$$\psi(x) \equiv \frac{d}{dx} \ln\Gamma(x) \,. \tag{91}$$

We find

$$\begin{aligned} \epsilon(P) &= 1 - (2/P)[\psi(1) - \psi(1/P)] \\ &= -1 + (2/P)[\psi(1+1/P) - \psi(1)] \,. \end{aligned} \tag{92}$$

Typical values are

$$\psi(1) = -C = -0.577215\ldots, \tag{93}$$

where C is Euler's constant, which cancels out in the explicit expressions

$$\begin{aligned} &\psi(1) - \psi(\tfrac{1}{2}) = 2\ln 2, \\ &\psi(1) - \psi(\tfrac{1}{3}) = \pi/2\sqrt{3} + 3\ln\tfrac{3}{2}, \\ &\psi(1) - \psi(\tfrac{1}{4}) = \pi/2 + 3\ln 2. \end{aligned} \tag{93a}$$

Finally, an expression useful for large P is

$$\epsilon(P) = -1 + 2\sum_{k=2}^{\infty} \frac{(-1)^k \zeta(k)}{P^k} \,. \tag{94}$$

Here $\zeta(k)$ is Riemann's zeta function,

$$\zeta(k) = \sum_{n=1}^{\infty} \frac{1}{n^k} \,. \tag{95}$$

The behavior of $\epsilon(P)$ is shown in Fig. 2.

As for any problem which is solved by Bethe's ansatz, the low-lying excited states are obtained from the original algebraic equations by making alternate choices for the quantum numbers. Thus, we select one quantum number from any of the $P-1$ equations, and change it; the removal creates a hole and the new choice creates a particle. The necessary manipulations of the integral equations are by now familiar and lead to the following expressions for the energy $\Delta\epsilon$ and momentum Δk:

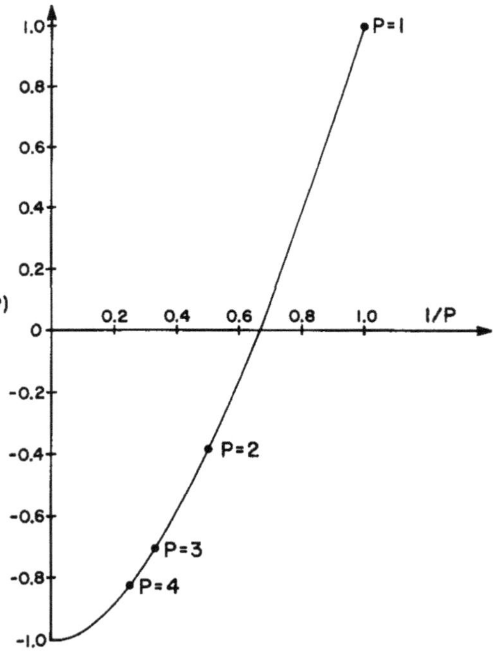

FIG. 2. Absolute ground-state energy per particle for the P-component system F^P is shown as a function of the number of components.

$$\Delta\epsilon = \sum R_i(B_i),$$
$$\Delta k = \sum \int^{B_i} d\alpha_i R_i(\alpha_i). \tag{96}$$

The variables B_i are to be eliminated between the two equations.

Let us consider the low-lying excited states about the absolute ground state. Then $R_i(\alpha)$ are given by Eq. (96). We decompose $\Delta\epsilon$ and Δk as a sum over single excitation dispersion curves and calculate each to be

$$\epsilon_j = \frac{2\pi}{P}\frac{\sin(\pi j/P)}{\cosh(2\pi B/P) - \cos(\pi j/P)},$$
$$k_j = 2\tan^{-1}[\cot(\pi j/2P)\tanh(2\pi B/P)] - (\pi - \pi j/P). \tag{97}$$

Eliminating B, we obtain

$$\epsilon_j(k) = \frac{2\pi}{\sin(\pi j/P)}[\cos(\pi j/P - |k|) - \cos(\pi j/P)] \tag{98}$$

for $|k| < 2\pi j/P$. The function is periodic with period $2\pi j/P$.

We see that, in general, there are $P-1$ distinct branches. However, at $k=0$, all branches have a common slope, with

$$\epsilon_j(k) \approx \frac{2\pi}{P}|k|, \tag{99}$$

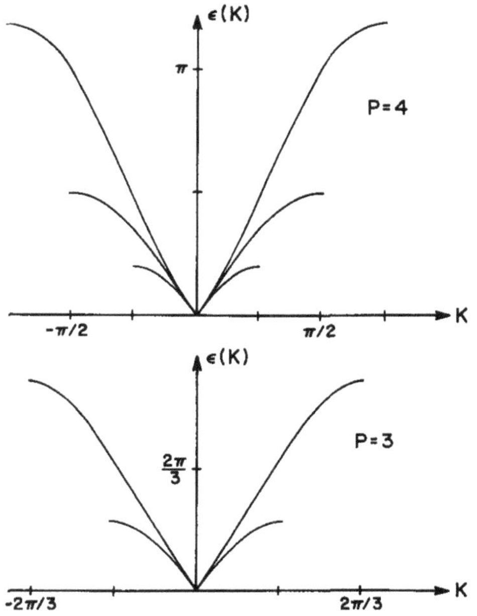

FIG. 3. $P-1$ dispersion curves for the low-lying excitations above the absolute ground state of the F^P system are shown for the cases $P=4$ and $P=3$.

and thus a common velocity of sound. In Fig. 3 we show the cases of $P=3$ and $P=4$.

Suppose we have a system F^P which is at the absolute ground state and replace some of the particles with $\delta n \ll 1$ impurities. Then the equations for the $P-1$ original densities become

$$\underline{\xi} + \underline{\xi}' = (\underline{I} + \underline{K})\underline{R}, \tag{100}$$

where

$$[\underline{\xi}']_j = \delta_{j\,P-1}\,\delta n\; 4/(1+4\alpha^2)\ . \tag{101}$$

Therefore our solution is given by

$$\underline{R} = \underline{R}^0 + \underline{R}', \tag{102}$$

where $\underline{R}^0$ is the previous solution, and the Fourier transform of $\underline{R}'$ is

$$\tilde{R}'_j(k) = 2\pi\frac{\sinh\frac{1}{2}jk}{\sinh\frac{1}{2}Pk}\,\delta n\,. \tag{103}$$

Then we verify that

$$m_j = (P-j)/P + \delta n j/P,$$

or

$$n_j = (1-\delta n)/P \quad (j = 1, \ldots, P)\,. \tag{104}$$

We then calculate the energy as

$$\epsilon = \epsilon(P) - \frac{\delta n}{2\pi}\int_{-\infty}^{+\infty}\frac{4\,d\alpha R'_1(\alpha)}{1+4\alpha^2} \tag{105}$$

or

$$\Delta\epsilon = -\delta n\int_{-\infty}^{+\infty} dk\, e^{-|k|/2}\frac{\sinh\frac{1}{2}k}{\sinh\frac{1}{2}Pk}. \tag{106}$$

Again, we define a variable as in Eq. (89) and write

$$\Delta\epsilon = -\frac{2\delta n}{P}\int_0^1 \frac{y^{-1/2} - y^{-1/2+1/P}}{1-y}\ . \tag{107}$$

This may be written in terms of the digamma function as

$$\Delta\epsilon = -\frac{2\delta n}{P}\left[\psi\left(\frac{1}{2}+\frac{1}{P}\right) - \psi\left(\frac{1}{2}\right)\right]. \tag{108}$$

We have the special values

$$\frac{d\epsilon}{dn} = \begin{cases} -4, & P=1 \\ -2\ln 2, & P=2 \\ -\frac{1}{4}\pi + \frac{1}{2}\ln 2, & P=4. \end{cases} \tag{109}$$

In Fig. 4 we show the general curve. We note the limiting form

$$\frac{d\epsilon}{dn} \sim -\left(\frac{\pi}{P}\right)^2. \tag{110}$$

We remark that this is the correct form no matter what the nature of the impurities—fermions, bosons, or mixed.

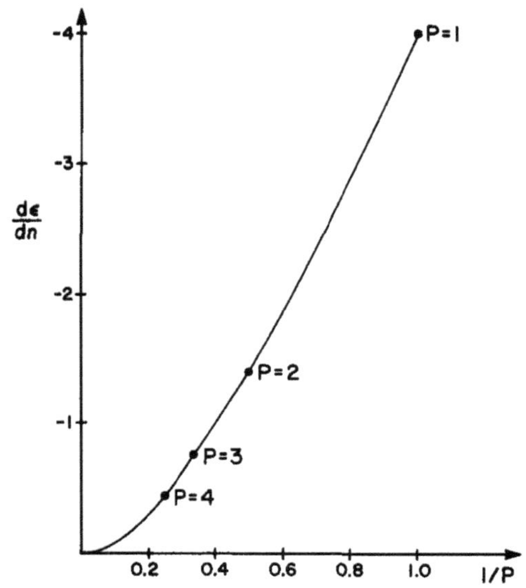

FIG. 4. We show the first correction to the absolute ground-state energy of the F^P system as a small number of particles are replaced by impurities. The correction is shown by the first derivative of the ground-state energy with respect to impurity concentration at zero-impurity concentration, as a function of the number of components P.

In the previous discussion, we have assumed that the point $B_j = \infty$ or $n_j = 1/P$ is the absolute ground state of the F^P system. This is certainly reasonable, and we may indeed verify the claim by examining the correction to the ground-state energy in the vicinity of $B_j = \infty$. We first transform the integral equations over the domain $\underline{B}$ to equations over the domain $(\underline{I} - \underline{B})$ by multiplying our basic Eq. (65) by the resolvent of Eq. (87) to give

$$\underline{R}^0 \equiv (\underline{I} + \underline{J})\underline{\xi} = \underline{R} + \underline{J}(\underline{I} - \underline{B})\underline{R}. \tag{111}$$

We may then write Eqs. (74) and (73) as

$$\Delta\epsilon \equiv \epsilon - (1 - \underline{\xi}^\dagger \underline{R}^0/2\pi) = (1/2\pi)\underline{R}^{0\dagger}(\underline{I} - \underline{B})\underline{R}, \tag{112}$$

and

$$-[(\underline{\tilde{I}} + \underline{\tilde{K}})](0)\underline{\Delta m} = -[(\underline{\tilde{I}} + \underline{\tilde{K}})_{(0)}](\underline{\gamma} - \underline{\eta}^\dagger \underline{R}^0/2\pi) = \underline{\eta}^\dagger(\underline{I} - \underline{B})\underline{R}/2\pi. \tag{113}$$

We note that the components on the left-hand side are

$$2\pi(n_j - n_{j+1}) = [\underline{\eta}^\dagger(\underline{I} - \underline{B})\underline{R}]_j = (\Delta n_j - \Delta n_{j+1})2\pi \geq 0. \tag{114}$$

If all B's are large—that is, we are in the vicinity of $n_j = 1/P$—then we may approximate the inhomogeneous term $\underline{R}^0$ of Eq. (111), as given in Eq. (84), by expanding

$$R_j^0(\alpha) \approx 4\pi \sin(\pi j/P)\, e^{-2\pi\alpha/P}/P. \tag{115}$$

We see that to this order,

$$R_j(\alpha) = R_{P-j}(\alpha). \tag{116}$$

Further, we see by looking at the explicit formula for the resolvent $\underline{I} + \underline{J}$ given in Eq. (87) that

$$[\underline{I} + \underline{J}]_{j,l} = [\underline{I} + \underline{J}]_{P-j,P-l}. \tag{117}$$

Thus to the lowest order in Δn_j,

$$\epsilon(1/P + \Delta n_1, 1/P + \Delta n_2, \ldots, 1/P + \Delta n_P) = \epsilon(1/P - \Delta n_P, \ldots, 1/P - \Delta n_2, 1/P - \Delta n_1). \tag{118}$$

But since the energy is symmetric in permutations of the P particles, we have

$$\epsilon(1/P + \Delta n_1, 1/P + \Delta n_2, \ldots, 1/P + \Delta n_P) = \epsilon(1/P - \Delta n_1, 1/P - \Delta n_2, \ldots, 1/P - \Delta n_P). \tag{119}$$

Thus we conclude that the ground-state energy is, to the lowest order in n_j, symmetric about the point $n_j = 1/P$.

However, by the basic concavity property of Eq. (11), we conclude then that $\Delta n_j = 0$ is a local minimum of the ground-state energy, and hence, the absolute ground state (with no restriction on the concentrations).

We note that this symmetry requires that for low-lying states, conjugate representations of SU(P) must be degenerate.

We have been unable to calculate explicitly the first-order correction to the absolute ground-state energy for general variations of concentration n_j. However, in two instances we may make a calculation:

(a) If all B_j's are equal,

$$R_j^0 = 2\pi(2/P)^{1/2} e^{-2\pi B/P} e^{-2\pi\sigma/P} \psi_1(j). \tag{120}$$

We have defined a new variable σ by $\alpha = B + \sigma$. If we also define

$$\zeta = (2/P)^{1/2} e^{-2\pi B/P}, \tag{121}$$

then we approximate Eq. (111) by

$$2\pi\zeta\, e^{-2\pi\sigma/P} \underline{\psi}_1 = \underline{S}(\sigma) + \int_0^\infty d\tau\, \underline{J}(\sigma - \tau)\underline{S}(\tau). \tag{122}$$

Let us make an eigenvector decomposition of $\underline{S}$,

$$\underline{S} = \sum S_n(\sigma)\underline{\psi}_n \tag{123}$$

We then find $S_n = 0$, $n \neq 1$, and

$$S_1(\sigma)+\int_0^\infty d\tau\, J_1(\sigma-\tau)\, S_1(\tau)=2\pi\zeta\, e^{-2\pi\sigma/P}, \qquad (124)$$

with

$$\bar{I}+\bar{J}_N+\frac{1}{\lambda_n}=\frac{\frac{1}{2}e^{|k|}}{2[\cosh\frac{1}{2}k-\cos\pi n/P]}\ . \qquad (125)$$

This is a Wiener-Hopf equation, and as in Ref. 2, we have

$$\Delta\epsilon=[(2\pi/P)\,\Delta q]^2\,(1-\cos\pi/P), \qquad (126)$$

with

$$\Delta q=-\sum_{j=1}^{P-1}\sin\frac{\pi j}{P}\,\Delta m_j\ . \qquad (127)$$

We have the restriction on Δm_j of

$$\Delta m_j=-\sin\left(\frac{\pi j}{P}\right)\frac{2\Delta q}{P}\ . \qquad (128)$$

(b) In the second case, suppose that all $B_l=\infty$, except $B_j\equiv B$. In this case we define

$$\zeta_j=(2/P)\sin(\pi j/P)\, e^{-2\pi B/P} \qquad (129)$$

The equation we are to solve is

$$2\pi\zeta_j\, e^{-2\pi\sigma/P}=S_j(\sigma)+\int_0^\infty d\tau\, J_{jj}(\sigma-\tau)\, S_j(\tau). \qquad (130)$$

Again, it is a Wiener-Hopf equation, and we find

$$\Delta\epsilon=(\pi\Delta n)^2\, j(P-j)/P^2, \qquad (131)$$

where

$$\Delta n_l=\Delta n_{l+1},\qquad l\neq j$$
$$\Delta n_j-\Delta n_{j+1}\equiv\Delta n\ . \qquad (132)$$

We finally remark that all of these examples are consistent with

$$\Delta\epsilon=\frac{\pi^2}{P}\sum_{j=1}^{P}(\Delta n_j)^2, \qquad (133)$$

and conjecture that this is in fact the exact form. And thus, as a consequence, we conjecture that all second derivatives of the ground-state energy with respect to the concentrations are continuous within the whole phase space.

ACKNOWLEDGMENTS

We would like to thank Professor C. N. Yang for suggesting the original problem, and for the hospitality of the Institute for Theoretical Physics, where this paper was in large part written. Also, we very much benefited from conversations with Y. Dothan.

---

*Work supported in part by National Science Foundation Grant No. CP 38905.

†Permanent address: Physics Department, University of Utah, Salt Lake City, Utah 84112.

[1] C. K. Lai, J. Math. Phys. 15, 1675 (1974).

[2] Our treatment in this section, and in all later sections, owes much to the treatment of the Heisenberg-Ising problem by C. N. Yang and C. P. Yang [Phys. Rev. 147, 303 (1966); 150, 321 (1966); 151, 258 (1966)].

[3] C. N. Yang, Phys. Rev. Lett. 19, 1312 (1967).

[4] B. Sutherland, Phys. Rev. Lett. 20, 98 (1968).

PHYSICAL REVIEW B VOLUME 36, NUMBER 10 1 OCTOBER 1987

Integrable narrow-band model with possible relevance to heavy-fermion systems

P. Schlottmann
Department of Physics, Temple University, Philadelphia, Pennsylvania 19122
(Received 16 March 1987)

A lattice model consisting of a single narrow band is introduced to describe some aspects of heavy electrons. The model excludes double occupancy of the sites and electrons on nearest-neighbor sites interact via a charge interaction and spin exchange. The model is integrable in one dimension for some special values of the coupling constants. These cases are related to the SU(3) invariance. The Bethe-ansatz equations are obtained and ground-state and thermodynamic properties are discussed and solved in some limiting cases.

I. INTRODUCTION

Heavy-electron metals[1-3] have received a large amount of attention in recent years, in particular because of their unusual low-temperature properties. Characteristic to heavy fermions is a very large electronic specific heat at low temperatures, $C=\gamma T$, where γ corresponds to a very high density of states at the Fermi level, or, equivalently, to an effective electron mass of 10^2-10^3 times that of the free-electron mass. As a consequence of the large density of states, these systems typically have a large Pauli susceptibility or order antiferromagnetically. The temperature dependence of C and χ can be explained in terms of narrow resonant levels or a narrow band with a typical width of a few meV or less. The narrow peak in the density of states has been attributed to a Kondo resonance[2,3] arising from the screening of the magnetic moment of the quasilocalized f electrons by the conduction electrons. Heavy-fermion behavior occurs in a variety of Ce-, U-, and Yb-based alloys and compounds.

The resistivity of stoichiometric heavy-fermion compounds initially increases as one lowers T (which can be attributed to the Kondo effect), then goes through a large maximum and shows a sharp decrease at very low T. Both features, the existence of a maximum and the high resistivity at the maximum, are uncommon to normal metals. The rapid decrease of $\rho(T)$ at low T is caused by a transition from incoherent to coherent scattering of the conduction electrons by the rare-earth (actinide) ions.[1-3]

Compounds may become antiferromagnetic or superconducting at low T as a consequence of the coherence of the low-energy excitation spectrum. Anomalous superconductivity has been discovered in some U compounds with highly unusual properties which may be due to triplet pairing.[4,5] The parameters of interactions determining the low-T phase, i.e., superconducting, magnetically ordered, or a Pauli paramagnet, are still to be understood.

The systems are usually described within the framework of the Kondo and Anderson lattices. Numerous approximate treatments have been applied to these models, which have been extensively reviewed in Refs. 2, 3, 6, and 7. In particular, the $1/N$ approaches (diagrammatic[8,9] and functional integral method[10-13]), variational methods,[14-16] and local Fermi-liquid theories[17-19] gave rise to important results and contributed to a preliminary understanding of heavy-fermion compounds. In a stoichiometric compound at low T, the Kondo resonances of the different rare-earth (actinide) sites superimpose coherently and form a narrow band at the Fermi level of width T_K. The low-temperature properties of the system, and in particular the coherence effects, are governed by the low-energy excitations of this narrow band.

Exact results, even for a simplified model which does not have all the features of the Anderson on Kondo lattices, are always useful and provide a testing ground for approaches intended for the full problem. The condition of exact diagonalization imposes limitations on the choice of the Hamiltonian. First, the integrability by means of a Bethe ansatz, i.e., the existence of an exact solution, restricts the model to one space dimension. The integrability requires further that in a scattering process the momenta of the outgoing particles are the same as those of the incoming particles. This restricts the model to have only one bandwidth or Fermi velocity. Second, since the Kondo resonance in a lattice consists of a narrow band at the Fermi level and its width is a fundamental energy scale, we describe the dynamics of the electrons by a nearest-neighbor tight-binding hopping model. A continuum model with a parabolic or linear dispersion would not provide a natural bandwidth parameter. Third, since f electrons are highly correlated, and, in particular, Ce compounds have only one f electron per Ce ion, it is reasonable to exclude the multiple occupancy of the sites. Fourth, since the f electrons are spin compensated at low temperatures, in part by the conduction electrons but possibly mainly by antiferromagnetic correlations among the rare-earth moments themselves, it is necessary to assume that the scattering is different if the two electrons involved form a singlet or a triplet state. The model is then considerably different from the traditional Hubbard model.[20]

The rest of the paper is organized as follows. In Sec. II we explicitly introduce the model, derive the two-particle scattering matrix, obtain the conditions for the integrability, and state the discrete Bethe-ansatz equations. In Sec. III we obtain the ground-state integral equations for the most important case, i.e., when two electrons in the singlet state are scattered, but they are not scattered if in a triplet state. Some ground-state properties are derived in Sec. IV. The thermodynamics of the model is obtained in Sec. V. A summary and discussion follows in Sec. VI.

II. MODEL, SCATTERING MATRIX, AND CONDITIONS FOR INTEGRABILITY

A. The model

We consider electrons with spin $\frac{1}{2}$ on a one-dimensional lattice with nearest-neighbor hopping. We assume a large on-site Coulomb repulsion that *excludes* the *double-occupancy* of every site. In other words each lattice site is constrained to have either one electron (with spin up or down) or none. Two types of interactions are considered between electrons on nearest-neighbor sites: A charge interaction independent of the spin of strength V and a spin exchange interaction J. The Hamiltonian is then given by

$$H=-\sum_{i,\sigma}(c_{i\sigma}^{\dagger}c_{i+1\sigma}+c_{i+1\sigma}^{\dagger}c_{i\sigma})+J\sum_{i,\sigma,\sigma'}c_{i\sigma}^{\dagger}\mathbf{S}_{\sigma\sigma'}c_{i\sigma'}\cdot c_{i+1\sigma'}^{\dagger}\mathbf{S}_{\sigma'\sigma}c_{i+1\sigma}+V\sum_{i,\sigma,\sigma'}c_{i\sigma}^{\dagger}c_{i\sigma}c_{i+1\sigma'}^{\dagger}c_{i+1\sigma'}\ , \tag{2.1}$$

where $c_{i\sigma}$ is the annihilation operator for an electron with spin σ at the site i and $\mathbf{S}_{\sigma\sigma'}$ represents the $S=\frac{1}{2}$ spin matrices. The hopping matrix element has been set equal to -1, such that the bandwidth is 2.

The above model is not integrable for arbitrary values of J and V. We next derive the two-electron scattering matrix to obtain the conditions for the integrability of the model.

B. Two-electron scattering matrix

Consider the linear chain described by (2.1) with only two electrons. Let $|0\rangle$ be the empty-lattice state, i.e., without electrons, then the two-particle wave function can be written in the following form:

$$\Psi_{\sigma_1\sigma_2}=\sum_{n_1,n_2}[a_{\sigma_1\sigma_2}(n_1,n_2)c_{n_1\sigma_1}^{\dagger}c_{n_2\sigma_2}^{\dagger}|0\rangle+a_{\sigma_2\sigma_1}(n_1,n_2)c_{n_1\sigma_2}^{\dagger}c_{n_2\sigma_1}^{\dagger}|0\rangle]\ . \tag{2.2}$$

If $\sigma_1=\sigma_2$ the two terms are identical. The wave function obeys Schrödinger's equation $H\Psi_{\sigma_1\sigma_2}=E\Psi_{\sigma_1\sigma_2}$, giving rise to the following relations for the coefficients $a_{\sigma_1\sigma_2}(n_1,n_2)$:

$$-a_{\sigma_1\sigma_2}(n_1+1,n_2)-a_{\sigma_1\sigma_2}(n_1-1,n_2)-a_{\sigma_1\sigma_2}(n_1,n_2+1)-a_{\sigma_1\sigma_2}(n_1,n_2-1)+\left[\left(V-\frac{J}{4}\right)a_{\sigma_1\sigma_2}(n_1,n_2)+\frac{J}{2}a_{\sigma_2\sigma_1}(n_1,n_2)\right]\delta_{n_1+1,n_2}=Ea_{\sigma_1\sigma_2}(n_1,n_2)\ . \tag{2.3}$$

The solution of this recursion relation is of the form

$$a_{\sigma_1\sigma_2}(n_1,n_2)=C_{\sigma_1\sigma_2}^{(1)}\exp(ik_1n_1+ik_2n_2)+C_{\sigma_1\sigma_2}^{(2)}\exp(ik_2n_1+ik_1n_2) \tag{2.4}$$

with

$$E=-2\cos k_1-2\cos k_2\ , \tag{2.5}$$

where k_1 and k_2 are the momenta of the particles. The coefficients $C_{\sigma_1\sigma_2}^{(l)}$ are not all independent but related by

$$(C_{\sigma_1\sigma_2}^{(2)}+C_{\sigma_2\sigma_1}^{(2)})/(C_{\sigma_1\sigma_2}^{(1)}+C_{\sigma_2\sigma_1}^{(1)})=\exp(-2i\Psi_{k_1k_2}^{t})\ , \tag{2.6a}$$

$$(C_{\sigma_1\sigma_2}^{(2)}-C_{\sigma_2\sigma_1}^{(2)})/(C_{\sigma_1\sigma_2}^{(1)}-C_{\sigma_2\sigma_1}^{(1)})=\exp(-2i\Psi_{k_1k_2}^{s})\ , \tag{2.6b}$$

where $\Psi_{k_1k_2}^{t,s}$ is given by

$$\cot\Psi_{k_1k_2}^{t,s}=x^{t,s}\frac{\cot\frac{1}{2}k_2-\cot\frac{1}{2}k_1}{(1-x^{t,s})\cot\frac{1}{2}k_1\cot\frac{1}{2}k_2-(1+x^{t,s})} \tag{2.7}$$

with $x^t=(V/2)+J/8$ and $x^s=(V/2)-3J/8$. If the pair of electrons forms a singlet state, then $\Psi^t=0$ and conversely if they are in a triplet state, $\Psi^s=0$. The expression (2.7) for the phases Ψ is analogous to the one for the anisotropic spin-$\frac{1}{2}$ Heisenberg chain,[21] where x is the anisotropy parameter. Taking into account the antisymmetry of a fermion wave function we obtain the following two-particle scattering matrix

$$\hat{S}(k_1,k_2)=-\tfrac{1}{2}[\exp(-2i\Psi_{k_1k_2}^{t})-\exp(-2i\Psi_{k_1k_2}^{s})]\hat{I}-\tfrac{1}{2}[\exp(-2i\Psi_{k_1k_2}^{t})+\exp(-2i\Psi_{k_1k_2}^{s})]\hat{P}\ , \tag{2.8}$$

where $\hat{I}$ is the identity matrix and $\hat{P}$ permutes the spin indices.

C. Condition for integrability

Let us now consider N electrons in the lattice arranged such that $n_1<n_2<\cdots<n_N$. The wave function can then be written in the form

$$\Psi_{\sigma_1,\sigma_2,\ldots,\sigma_N}=\sum_{n_1<n_2<n_3<\cdots<n_N}\sum_{P}a_{\sigma_{P1},\sigma_{P2},\ldots,\sigma_{PN}}(n_1,n_2,\ldots,n_N)c_{n_1\sigma_{P1}}^{\dagger}c_{n_2\sigma_{P2}}^{\dagger}\cdots c_{n_N\sigma_{PN}}^{\dagger}|0\rangle\ , \tag{2.9}$$

where $P=(P1,\ldots,PN)$ is a permutation of the integers $1,\ldots,N$. There are $N!$ such permutations. Other arrangements than $n_1<n_2<\cdots<n_N$ follow from the antisymmetry of the wave function with respect to the permutation of two particles.

In analogy to the two-particle problem, the coefficients in (2.9) are not all independent, but related by a scattering matrix. The condition for the applicability of Bethe's hypothesis is that the scattering matrix can be factorized into a product of two-particle scattering matrices. The single-valuedness of the wave function is ensured by Yang's triangular relation,[22,23] which is a sufficient condition for the integrability of the model,

$$S_{\sigma_2\sigma_2'}^{\sigma_1\sigma_1'}(k_1,k_2)S_{\sigma_3\sigma_3'}^{\sigma_1'\sigma_1''}(k_1,k_3)S_{\sigma_3'\sigma_3''}^{\sigma_2'\sigma_2''}(k_2,k_3)=S_{\sigma_3\sigma_3'}^{\sigma_2\sigma_2'}(k_2,k_3)S_{\sigma_3'\sigma_3''}^{\sigma_1\sigma_1'}(k_1,k_3)S_{\sigma_2'\sigma_2''}^{\sigma_1'\sigma_1''}(k_1,k_2)\ . \qquad (2.10)$$

The two-particle scattering matrix, (2.8), however, does not satisfy the triangular relation but for special values of the couplings J and V. These special cases are as follows:

(a) $x^t=0$, $x^s=\pm 1$ or equivalently $V=\pm\frac{1}{2}$ and $J=\mp 2$,

(b) $x^s=0$, $x^t=\pm 1$ or equivalently $V=\pm\frac{3}{2}$ and $J=\pm 2$,

and

(c) $x^s=x^t$, i.e., $J=0$ and V is arbitrary (or $J\to\pm\infty$).

Below we discuss the scattering matrices for the special cases (a) and (b) and by imposing periodic boundary conditions to the system we obtain the discrete Bethe-ansatz equations. Case (c) can be mapped onto the XXZ Heisenberg chain.

D. Two-particle scattering matrix for the integrable cases

In the special case (a), electrons forming a triplet state are not scattered. On the other hand, if their spins are in a singlet state the scattering depends on their crystal momenta k_1 and k_2. The scattering matrix (2.8) reduces to

$$\hat{S}(k_1,k_2)=\frac{p_1-p_2}{p_1-p_2\pm i}\hat{I}+\frac{\pm i}{p_1-p_2\pm i}\hat{P}\ , \qquad (2.11)$$

where $\pm$ refers to the sign of x^s and $p=\frac{1}{2}\cot\frac{1}{2}k$ if $x^s=+1$ and $p=\frac{1}{2}\tan\frac{1}{2}k$ if $x^s=-1$.

In the special case (b), on the other hand, electrons in a triplet state scatter, while they do not scatter if their spins form a singlet state. The scattering matrix for this case is

$$\hat{S}(k_1,k_2)=-\frac{p_1-p_2}{p_1-p_2\pm i}\hat{I}+\frac{\pm i}{p_1+p_2\pm i}\hat{P}\ , \qquad (2.12)$$

where $\pm$ refers to the sign of x^t and $p=\frac{1}{2}\cot\frac{1}{2}k$ if $x^t=+1$ and $p=\frac{1}{2}\tan\frac{1}{2}k$ if $x^t=-1$.

In case (c) when $J=0$, the scattering is independent of the spin. The model then basically reduces to the anisotropic $S=\frac{1}{2}$ (XXZ) Heisenberg chain,[21] where the total number of electrons plays the role of the magnetization and the chemical potential is the magnetic field. The spin degree of freedom of the electrons introduces an additional constant term to the entropy given by the number of electrons times ln2.

E. Bethe-Ansatz equations

The integrable cases (a) and (b) discussed above can now be solved by a standard procedure.[20,22–24] Imposing periodic boundary conditions the problem reduces to the simultaneous solution of N eigenvalue equations. For N-M electrons with spin up, and M electrons with spin down we obtain by means of a second Bethe Ansatz the following set of coupled algebraic equations (N_a is the number of lattice sites and assumed to be even). For case (a),

$$\left[\frac{p_j+i/2}{p_j-i/2}\right]^{N_a}=\prod_{\beta=1}^{M}\frac{p_j-\Lambda_\beta+i/2}{p_j-\Lambda_\beta-i/2}\ ,\quad j=1,\ldots,N$$

$$\prod_{j=1}^{N}\frac{\Lambda_\alpha-p_j+i/2}{\Lambda_\alpha-p_j-i/2}=-\prod_{\beta=1}^{M}\frac{\Lambda_\alpha-\Lambda_\beta+i}{\Lambda_\alpha-\Lambda_\beta-i},\quad \alpha=1,\ldots,M\ . \qquad (2.13)$$

For case (b),

$$\left[\frac{p_j+i/2}{p_j-i/2}\right]^{N_a}=(-1)^N\prod_{l=1}^{N}\frac{p_j-p_l+i}{p_j-p_l-i}\prod_{\beta=1}^{M}\frac{p_j-\Lambda_\beta-i/2}{p_j-\Lambda_\beta+i/2}\ ,\quad j=1,\ldots,N \qquad (2.14)$$

$$\prod_{j=1}^{N}\frac{\Lambda_\alpha-p_j+i/2}{\Lambda_\alpha-p_j-i/2}=-\prod_{\beta=1}^{M}\frac{\Lambda_\alpha-\Lambda_\beta+i}{\Lambda_\alpha-\Lambda_\beta-i},\quad \alpha=1,\ldots,M\ .$$

Here Λ_α are rapidities related to the spin degrees of freedom. The total energy of the system is, in both cases, given by

$$E=\mp 2N\pm 2\sum_{i=1}^{N}\frac{\frac{1}{2}}{p_i^2+\frac{1}{4}}\ , \qquad (2.15)$$

where the $\pm$ refers to the sign of x^s or x^t, respectively.

Since in a Kondo system the spins are compensated into a singlet state, case (a), i.e., when electrons paired in a singlet state are scattered, is the physically more relevant situation. In the remaining sections we restrict ourselves to derive the properties of case (a). Note that the Bethe-ansatz equations for this case, (2.13), are closely related to those of the SU(3)-invariant $S=1$ Heisenberg chain with ferromagnetic and antiferromagnetic coupling, respectively (see Sec. VI).

III. GROUND-STATE INTEGRAL EQUATIONS

In order to obtain the ground-state properties in the case of singlet scattering only [case (a)] we have to find the solution to Eqs. (2.13). The sets of rapidities $\{p_j\}$ and $\{\Lambda_\alpha\}$ have real and complex solutions. Complex solutions for the Λ_α corresponds to excited states and are discussed in Sec. V. The rapidities p_j may be real or complex.

Consider (2.13) for large N_a. If p_j has a positive imaginary part, there exists Λ_β such that to order $\exp(-N_a)$

$$p_j=\Lambda_\beta+i/2\ . \qquad (3.1a)$$

Similarly, if p_j has a negative imaginary part, there exists Λ'_β such that to order $\exp(-N_a)$

$$p_j' = \Lambda'_\beta - i/2 \; . \tag{3.1b}$$

Since the energy must be real, it follows that the set $\{p_j\}$ consists of real p_j values (unpaired charge modes) and pairs of complex conjugated p_j values (spin-paired electrons). Depending on the sign of x^s spin-paired electrons are energetically favorable compared to unpaired particles or vice versa.

The set $\{p_j\}$ consists of N-$2M$ real rapidities and M pairs of complex conjugated p_j-values related to the real Λ_α by

$$p_\alpha^{\pm} = \Lambda_\alpha \pm i/2 \; . \tag{3.2}$$

Inserting (2.2) into (2.13) we obtain

$$\left[\frac{p_j+i/2}{p_j-i/2}\right]^{N_a} = \prod_{\beta=1}^{M} \frac{p_j-\Lambda_\beta+i/2}{p_j-\Lambda_\beta-i/2}, \quad j=1,\ldots,N\text{-}2M$$

$$\left[\frac{\Lambda_\alpha+i}{\Lambda_\alpha-i}\right]^{N_a} = -\prod_{j=1}^{N-2M} \frac{\Lambda_\alpha-p_j+i/2}{\Lambda_\alpha-p_j-i/2} \times \prod_{\beta=1}^{M} \frac{\Lambda_\alpha-\Lambda_\beta+i}{\Lambda_\alpha-\Lambda_\beta-i}, \quad \alpha=1,\ldots,M \tag{3.3}$$

and the energy and the spin are given by

$$E = \mp 2N \pm 2 \sum_{j=1}^{N-2M} \frac{\frac{1}{2}}{p_j^2+\frac{1}{4}} \pm 2 \sum_{\alpha=1}^{M} \frac{1}{\Lambda_\alpha^2+1} \; , \tag{3.4}$$

$$S_z = \frac{N}{2} - M \; . \tag{3.5}$$

The sign of the energy refers to $x^s = \pm 1$.

Logarithmizing Eqs. (3.3) and in the thermodynamic limit ($N_a \to \infty$ with M/N_a and N/N_a being fixed) we obtain the following integral equations

$$\frac{1}{\pi}\frac{1}{\Lambda^2+1} = \sigma_h'(\Lambda)+\sigma'(\Lambda) + \frac{1}{\pi}\int d\Lambda' \sigma'(\Lambda') \frac{1}{(\Lambda-\Lambda')^2+1} + \frac{1}{\pi}\int dp\, \rho(p) \frac{\frac{1}{2}}{(\Lambda-p)^2+\frac{1}{4}} \; , \tag{3.6}$$

$$\frac{1}{\pi}\frac{\frac{1}{2}}{p^2+\frac{1}{4}} = \rho_h(p)+\rho(p)+\frac{1}{\pi}\int d\Lambda\, \sigma'(\Lambda) \frac{\frac{1}{2}}{(p-\Lambda)^2+\frac{1}{4}} \; , \tag{3.7}$$

where $\rho(p)$ and $\sigma'(\Lambda)$ are the distribution density functions for the p and Λ rapidities and $\rho_h(p)$ and $\sigma_h'(\Lambda)$ are the respective hole distribution functions. The intervals in which ρ and σ are nonvanishing depend on the total number of particles, the total spin and the energy

$$S_z/N_a = \tfrac{1}{2}\int dp\, \rho(p) \; , \tag{3.8}$$

$$N/N_a = \int dp\, \rho(p) + 2\int d\Lambda\, \sigma'(\Lambda) \; , \tag{3.9}$$

$$E/N_a = \mp 2N/N_a \pm 2\int dp\, \rho(p) \frac{\frac{1}{2}}{p^2+\frac{1}{4}} \pm 2\int d\Lambda\, \sigma'(\Lambda) \frac{1}{\Lambda^2+1} \; . \tag{3.10}$$

These equations are solved in Sec. IV for some situations.

IV. GROUND-STATE PROPERTIES

A. Filled-band solution

We assume that there is one electron per lattice site, i.e., $N/N_a = 1$. Fourier transforming (3.6) and using (3.9) it follows that the hole-distribution function $\sigma_h'(\Lambda)$ vanishes identically. Equation (3.7) then becomes

$$\rho_h(\xi)+\rho(\xi) - \int d\xi' \rho(\xi') G_1(\xi-\xi') = G_0(\xi) \; , \tag{4.1}$$

where

$$G_l(\xi) = \frac{1}{2\pi}\int_{-\infty}^{\infty} d\omega\, e^{-i\omega\xi} \frac{e^{-(l/2)|\omega|}}{2\cosh(\omega/2)} \; . \tag{4.2}$$

The solution of this integral equation depends on the magnetization of the system. In the absence of an external magnetic field two solutions are of interest: the nonmagnetic and the ferromagnetic states.

If the state is nonmagnetic $\rho(\xi) \equiv 0$ and the solution of (3.6) is $\sigma'(\Lambda) = G_1(\Lambda)$. The energy of the system is straightforwardly obtained, $E/N_a = \mp 2\ln 2$. For the ferromagnetic state, on the other hand, we have $\sigma'(\Lambda) \equiv 0$ and $\rho_h(\xi) \equiv 0$, such that

$$\rho(\xi) = \frac{1}{\pi}\frac{\frac{1}{2}}{\xi^2+\frac{1}{4}} \; . \tag{4.3}$$

In this case the spins of all the electrons are parallel and the total energy of the system is zero.

Hence, if $x^s = +1$ the ground state is nonmagnetic, while if $x^s = -1$ the ground state is the ferromagnetic state. For $x^s = +1$ the states with lowest energy correspond to large $|\Lambda|$ and $|p|$ values (long-wavelength states), such that $\sigma'(\Lambda)$ is nonzero only in the intervals $|\Lambda| > Q$ and $\rho(p)$ in the intervals $|p| > B$. $\sigma_h'(\Lambda)$ and $\rho_h(p)$ are complementary functions. For $x^s = -1$ we have that states with small $|\Lambda|$ and $|p|$ values have lowest energy (again long-wavelength states), such that $\sigma'(\Lambda)$ and $\rho(p)$ are nonvanishing in the intervals $|\Lambda| < Q$ and $|p| < B$. The interval limits are determined from the total number of particles and the magnetization.

B. Magnetic susceptibility ($x^s = +1$)

For $x^s = +1$ the magnetization vanishes in the absence of an external magnetic field. If we apply an arbitrarily small magnetic field the integration limit B can be made much larger than any given Q. By Fourier transformation of Eqs. (3.6) and (3.7) we obtain the following integral equation for ρ:

$$\rho_h(\xi)+\rho(\xi) - \left[\int_{-\infty}^{-B} + \int_{B}^{\infty}\right] d\xi' \rho(\xi') G_1(\xi-\xi') = G_0(\xi) + \int_{-Q}^{Q} d\xi' \sigma_h'(\xi') G_0(\xi-\xi') \; . \tag{4.4}$$

Since $B \gg Q$ it is convenient to define $y(\xi) = \rho(\xi+B)$, such that

$$y_h(\xi)+y(\xi)-\int_0^\infty d\xi' y(\xi')G_1(\xi-\xi')=G_0(\xi+B)+\int_{-Q}^{Q} d\xi'\sigma_h'(\xi')G_0(\xi+B-\xi')+\int_0^\infty d\xi' y(\xi')G_1(\xi+\xi'+2B)\ , \tag{4.5}$$

where for the last term we have assumed that $\rho(\xi)$ is an even function. This integral equation can be solved by iteration,[25] by writing $y=y_1+y_2+\dots$ with y_1 and y_2 satisfying

$$y_{1h}(\xi)+y_1(\xi)-\int_0^\infty d\xi' y_1(\xi')G_1(\xi-\xi')=RG_0(\xi+B)\ , \tag{4.6a}$$

$$y_{2h}(\xi)+y_2(\xi)-\int_0^\infty d\xi' y_2(\xi')G_1(\xi-\xi') =\int_0^\infty d\xi' y_1(\xi')G_1(\xi+\xi'+2B)\ . \tag{4.6b}$$

Here we used $B \gg Q$ such that for $\xi>0$

$$\int_{-Q}^{Q} d\xi'\sigma_h'(\xi')G_0(\xi+B-\xi') \simeq G_0(\xi+B)\int_{-\infty}^{\infty} d\xi'\sigma_h'(\xi')\exp(\pi\xi')$$

and

$$R=1+\int_{-Q}^{Q} d\xi'\sigma_h'(\xi')\exp(\pi\xi')\ . \tag{4.7}$$

In (4.6), $y_1(\xi)$ is the leading contribution to $y(\xi)$ if the field is small. Since $G_1(\xi)$ falls off like $1/\xi^2$ for large ξ, $y_2(\xi)$ is the next leading correction. Higher-order contributions can be obtained in a similar way. Both, (4.6a) and (4.6b), are standard integral equations of the Wiener-Hopf type;[25,26] the solution for $y_1(\xi)$ is

$$y_1(\xi)=\frac{1}{2\pi i}\int_{-\infty}^{\infty} dx\frac{e^{-i\xi x}}{g^+(x)}\int_{-\infty}^{\infty}\frac{dy}{2\pi}\frac{Re^{-iyB}}{y-x-io} \times\left[g^-(y)2\cosh\frac{y}{2}\right]^{-1} \tag{4.8}$$

and the leading contribution to the magnetization is given by

$$\int d\xi\, y_1(\xi)=[2R/(2\pi e)^{1/2}]\exp(-\pi B) +O(\exp(-2\pi B))\ . \tag{4.9}$$

Here the magnetic field is proportional to $\exp(-\pi B)$ and

$$g^+(x)=g^-(-x)=(2\pi)^{-1/2}\left[\frac{-ix+o}{2\pi e}\right]^{i(x/2\pi)} \times\Gamma\left[\frac{1}{2}-i\frac{x}{2\pi}\right] \tag{4.10}$$

with o being a positive infinitesimal and Γ the gamma function.

In order to solve (4.6b) we insert $y_1(\xi)$, (4.8), into the driving term. After some algebra we obtain to leading order in the field

$$\int d\xi\, y_2(\xi)=\frac{1}{\pi}\frac{R}{(2\pi e)^{1/2}}\frac{\exp(-\pi B)}{2B} \times\left[1-\frac{\ln 2B}{2\pi B}+\cdots\right], \tag{4.11}$$

such that the low-field magnetization is given by

$$S_z=\frac{2R}{(2\pi e)^{1/2}}e^{-\pi B}\left[1+\frac{1}{4\pi B}-2\left[\frac{1}{4\pi B}\right]^2\ln 2B+\cdots\right]. \tag{4.12}$$

Since H is proportional to $\exp(-\pi B)$ we have that the susceptibility has logarithmic corrections, in analogy to the isotropic SU(2) Heisenberg antiferromagnet.[25,27] The leading-order relation between the field H and B will be given at the end of Sec. V.

C. Charge fluctuations ($x^s=+1$)

Since the magnetization vanishes in the absence of an external magnetic field, the integral equation to be solved in this case is

$$\sigma_h'(\xi)+\sigma'(\xi)+\frac{1}{\pi}\int_{-\infty}^{\infty} d\xi'\sigma'(\xi')\frac{1}{(\xi-\xi')^2+1} =\frac{1}{\pi}\frac{1}{\xi^2+1}\ , \tag{4.13}$$

where $\sigma'(\xi)$ vanishes for $|\xi|<Q$ and $\sigma'_h(\xi)$ is the complementary function. This integral equation has a simple solution only if the band is almost half-filled or almost empty.

If the band is almost half-filled Q is small and (4.13) is more conveniently written as

$$\sigma'(\xi)+\sigma_h'(\xi)-\int_{-Q}^{Q} d\xi'\sigma_h'(\xi')G_1(\xi-\xi')=G_1(\xi)\ . \tag{4.14}$$

This equation can now be solved by iteration by writing $\sigma'(\xi)=\sigma_0'(\xi)+\sigma_1'(\xi)+\cdots$, where $\sigma_0'(\xi)$ and $\sigma_1'(\xi)$ satisfy

$$\sigma_0'(\xi)+\sigma_{0h}'(\xi)=G_1(\xi)=\frac{1}{2\pi}\mathrm{Re}\left[\Psi\left[1+i\frac{\xi}{2}\right] -\Psi\left[\frac{1}{2}+i\frac{\xi}{2}\right]\right], \tag{4.15}$$

$$\sigma_1'(\xi)+\sigma_{1h}'(\xi)=\int_{-Q}^{Q} d\xi'\sigma_{0h}'(\xi')G_1(\xi-\xi')\ . \tag{4.16}$$

Here ψ is the digamma function. To first order in Q we obtain

$$\sigma'(\xi)=\begin{cases}\left[1+2Q\frac{\ln 2}{\pi}\right]G_1(\xi), & |\xi|>Q\\ 0, & |\xi|<Q\end{cases} \tag{4.17}$$

and the occupation of the band is given by

$$N/N_a=2\int_{-\infty}^{\infty}d\xi\sigma'(\xi)=1-2Q\frac{\ln 2}{\pi}\ . \tag{4.18}$$

On the other hand, if the band is almost empty Q is very large and a similar procedure as for the spin susceptibility can be used. Since $\sigma'(\xi)$ is nonvanishing only for $|\xi|>Q$, we define $y(\xi)=\sigma'(\xi+Q)$, such that (assuming that $\sigma'(\xi)$ is an even function)

$$y_h(\xi)+y(\xi)+\frac{1}{\pi}\int_0^{\infty}d\xi' y(\xi')\frac{1}{(\xi-\xi')^2+1}$$
$$=\frac{1}{\pi}\frac{1}{(\xi+Q)^2+1}$$
$$-\frac{1}{\pi}\int_0^{\infty}d\xi' y(\xi')\frac{1}{(\xi+\xi'+2Q)^2+1}\ . \tag{4.19}$$

This equation is solved by iteration; we write $y(\xi)=y_1(\xi)+y_2(\xi)+\cdots$, where $y_1(\xi)$ and $y_2(\xi)$ obey

$$y_{1h}(\xi)+y_1(\xi)+\frac{1}{\pi}\int_0^{\infty}d\xi' y_1(\xi')\frac{1}{(\xi-\xi')^2+1}$$
$$=\frac{1}{\pi}\frac{1}{(\xi+Q^2)+1}\ , \tag{4.20a}$$

$$y_{2h}(\xi)+y_2(\xi)+\frac{1}{\pi}\int_0^{\infty}d\xi' y_2(\xi')\frac{1}{(\xi-\xi')^2+1}$$
$$=-\frac{1}{\pi}\int_0^{\infty}d\xi' y_1(\xi')\frac{1}{(\xi+\xi'+2Q)^2+1}\ . \tag{4.20b}$$

Both equations are of the Wiener-Hopf type[26] and can straightforwardly by solved. We obtain for $y_1(\xi)$

$$y_1(\xi)=\frac{1}{2\pi i}\int_{-\infty}^{\infty}dx\, e^{-i\xi x}g^+(x)$$
$$\times\int_{-\infty}^{\infty}\frac{dy}{2\pi}\frac{e^{-iyQ-|y|}}{y-x-i0}g^-(y) \tag{4.21}$$

and the leading contribution to the occupation number is given by

$$N/N_a=2\int_0^{\infty}d\xi y_1(\xi)=\frac{1}{\pi Q}+\frac{1}{2}\left[\frac{1}{\pi Q}\right]^2\ln Q+\cdots\ . \tag{4.22}$$

Note that the leading contribution of $y_2(\xi)$ is of the order of $(1/\pi Q)^2$. The relation between Q and the Fermi energy is obtained at the end of Sec. V. Similarly, it is possible to obtain the number of particles for the ferromagnetic case, i.e., for $x^s=-1$, if the band is nearly empty or nearly full.

V. THERMODYNAMIC EQUATIONS

Below we derive the thermodynamic Bethe-ansatz equations for case (a) ($x^s=\pm 1$), i.e., when electrons in a singlet state are scattered.

A. Excitations

The excitations of the system are given by the solutions of (2.13). The structure of the solutions is similar as for the fermion gas with attractive δ-function potential, solved by Lai[28] and Takahashi,[29] and for the $j=\frac{1}{2}$ Anderson impurity in the $U\to\infty$ limit.[30] The rapidities can be classified according to (i) (N-$2M'$) real charge rapidities, which correspond to unpaired propagating electrons; (ii) $2M'$ complex charge rapidities, which correspond to bound or paired electron states, of the form

$$p_\alpha^{\pm}=\Lambda'_\alpha\pm i/2\ , \tag{5.1}$$

where Λ'_α is a real spin rapidity, $\alpha=1,\ldots,M'$; (iii) M_n strings of complex spin rapidities of length n, $n=1,\ldots,\infty$ which correspond to bound spin states and are of the form

$$\Lambda_{\alpha,n}^{\mu}=\Lambda_{\alpha,n}+i\frac{\mu}{2}\ ,$$
$$\mu=-(n-1),-(n-3),\ldots,(n-3),(n-1)\ , \tag{5.2}$$

where $\Lambda_{\alpha,n}$ is a real parameter.

The integers M' and M_n satisfy the relation

$$M'+\sum_{n=1}^{\infty}nM_n=M\ . \tag{5.3}$$

The magnetization and the energy are given by

$$S_z=\frac{N}{2}-M'-\sum_{n=1}^{\infty}nM_n\ , \tag{5.4}$$

$$E=\mp 2N\pm 2\sum_{i=1}^{N-2M'}\frac{\frac{1}{2}}{p_i^2+\frac{1}{4}}\pm 2\sum_{\alpha=1}^{M'}\frac{1}{\Lambda_\alpha'^2+1}\ . \tag{5.5}$$

The above rapidities are inserted into Eqs. (2.13) and the resulting coupled equations for p_j, Λ'_α, and $\Lambda_{\alpha,n}$ are logarithmized. We define the usual distribution functions for the rapidities as $\rho(p)$ for the real p's, $\sigma'(\Lambda)$ for the Λ'_α and $\sigma_n(\Lambda)$ for the $\Lambda_{\alpha,n}$. In the thermodynamic limit we obtain a set of linearly coupled integral equations for the distribution functions. Introducing the corresponding "hole"-distribution functions and Fourier-transforming the equations, we have, after some algebra,

$$\hat\sigma_{m+1,h}(\omega)+\hat\sigma_{m-1,h}(\omega)$$
$$=2\cosh\frac{\omega}{2}[\hat\sigma_m(\omega)+\hat\sigma_{m,h}(\omega)],\quad m\geq 2$$

$$\hat\sigma_{2,h}(\omega)+\hat\rho(\omega)=2\cosh\frac{\omega}{2}[\hat\sigma_1(\omega)+\hat\sigma_{1,h}(\omega)]\ ,$$

$$\hat\sigma_{1,h}(\omega)+\hat\sigma'_h(\omega)+1=2\cosh\frac{\omega}{2}[\hat\rho(\omega)+\hat\rho_h(\omega)]\ , \tag{5.6}$$

$$e^{-(1/2)|\omega|}\hat\sigma'_h(\omega)-\hat\rho(\omega)+e^{-(1/2)|\omega|}$$
$$=2\cosh\frac{\omega}{2}[\hat\sigma'(\omega)+\hat\sigma'_h(\omega)]\ ,$$

where the caret denotes a Fourier transform. These equations differ only by the driving terms (independent terms) from the corresponding ones for the one-dimensional fermion gas with attractive δ-function interaction[28,29] and the $j=\frac{1}{2}$ Anderson impurity in the $U\to\infty$ limit.[30]

B. Minimization of the free energy

The distribution functions, ρ, σ', and σ_n are actually determined by minimizing the free-energy functional, $F=E-TS$, where

$$E/N_a=\mp 2\int_{-\infty}^{\infty} dp\,\rho(p)\mp 4\int_{-\infty}^{\infty} d\Lambda\,\sigma'(\Lambda) \pm 2\int_{-\infty}^{\infty} dp\,\rho(p)\frac{\frac{1}{2}}{p^2+\frac{1}{4}} \pm 2\int_{-\infty}^{\infty} d\Lambda\,\sigma'(\Lambda)\frac{1}{\Lambda^2+1}\,, \tag{5.7}$$

T is the temperature, and S is the distribution entropy, which, e.g., for $\rho(p)$ is given by

$$S_\rho=\int dp[(\rho+\rho_h)\ln(\rho+\rho_h)-\rho\ln\rho-\rho_h\ln\rho_h]\,. \tag{5.8}$$

The minimization must be carried out subject to the constraints (5.6) and the conservation of electrons and the total spin, i.e.,

$$N/N_a=\int_{-\infty}^{\infty} dp\,\rho(p)+2\int_{-\infty}^{\infty} d\Lambda\sigma'(\Lambda)\,, \tag{5.9}$$

$$S_z/N_a=\tfrac{1}{2}\int_{-\infty}^{\infty} dp\,\rho(p)-\sum_{n=1}^{\infty} n\int_{-\infty}^{\infty} d\Lambda\sigma_n(\Lambda)\,. \tag{5.10}$$

The corresponding Lagrange multipliers are the chemical potential (Fermi-energy) and the magnetic field.

Introducing the following functions

$$\rho_h/\rho=\exp(\varepsilon/T),\quad \sigma_h'/\sigma'=\exp(\Psi/T)\,,\qquad \sigma_{n,h}/\sigma_n=\exp(\varphi_n/T)\,, \tag{5.11}$$

we obtain, if we consider $\sigma_{n,h}$, ρ, and σ_h' as independent functions,

$$\begin{aligned}
\varepsilon&=\pm 2\pi G_0(\xi)+TG_0*\ln[(1+e^{\Psi/T})/(1+e^{\varphi_1/T})]\,,\\
\Psi&=\mp 2-A\pm 2\pi G_1(\xi)+TG_0*\ln(1+e^{-\varepsilon/T})\\
&\quad+TG_1*\ln(1+e^{\Psi/T})\,,\\
\varphi_1&=TG_0*\ln[(1+e^{-\varepsilon/T})(1+e^{\varphi_2/T})]\,,\\
\varphi_n&=TG_0*\ln[(1+e^{\varphi_{n-1}/T})(1+e^{\varphi_{n+1}/T})],\quad n\geq 2
\end{aligned} \tag{5.12}$$

where G_1 is defined by (4.2) and $*$ denotes convolution. Here A is the chemical potential and $A'=A\pm 2$ is the energy from the bottom and top of the band, respectively. The field boundary condition is given by

$$\lim_{n\to\infty}\frac{\varphi_n(\xi)}{n}=H\,. \tag{5.13}$$

It is convenient to replace $\pm 2\pi G_l(\xi)$ in (5.12) by $2\pi x^s G_l(\xi)$. Differentiating (5.12) with respect to x^s, we obtain from similarity with Eqs. (5.6) that

$$\begin{aligned}
\sigma_{n,h}&=-\frac{1}{2\pi}\frac{\partial\varphi_n}{\partial x^s}/(1+e^{-\varphi_n/T})\,,\\
\rho&=\frac{1}{2\pi}\frac{\partial\varepsilon}{\partial x^s}/(1+e^{\varepsilon/T})\,,\\
\sigma_h'&=\frac{1}{2\pi}\frac{\partial\Psi}{\partial x^s}/(1+e^{-\psi/T})\,.
\end{aligned} \tag{5.14}$$

and similarly the complementary functions.

The free energy of the system is given by

$$F=-\Psi(0)-2A\mp 2\,. \tag{5.15}$$

If, on the other hand, we consider σ_n, ρ, and σ' as independent functions and eliminate all the "hole"-distribution functions by means of (5.6), we obtain another set of integral equations, which is equivalent to (5.12),

$$\begin{aligned}
\varepsilon&=\mp 2\pm 2\frac{\frac{1}{2}}{\xi^2+\frac{1}{4}}-\frac{H}{2}-A+T\frac{1}{\pi}\frac{\frac{1}{2}}{\xi^2+\frac{1}{4}}*\ln(1+e^{-\Psi/T})-T\sum_{n=1}^{\infty}\frac{1}{\pi}\frac{n/2}{\xi^2+(n/2)^2}*\ln(1+e^{-\varphi_n/T})\,,\\
\Psi&=\mp 4\pm 2\frac{1}{\xi^2+1}-2A+T\frac{1}{\pi}\frac{\frac{1}{2}}{\xi^2+\frac{1}{4}}*\ln(1+e^{-\varepsilon/T})+T\frac{1}{\pi}\frac{1}{\xi^2+1}*\ln(1+e^{-\Psi/T})\,,\\
\varphi_n&=nH-T\ln(1+e^{-\varphi_n/T})+T\frac{1}{\pi}\frac{n/2}{\xi^2+(n/2)^2}*\ln(1+e^{-\varepsilon/T})+T\sum_{m=1}^{\infty}\Theta_{m,n}(\xi)*\ln(1+e^{-\varphi_m/T})\,,
\end{aligned} \tag{5.16}$$

where

$$\Theta_{m,n}(\xi)=\int\frac{d\omega}{2\pi}e^{-i\xi\omega}\coth\frac{|\omega|}{2}\times\left[e^{-(|\omega|/2)|m-n|}-e^{-(|\omega|/2)(m+n)}\right]. \tag{5.17}$$

Rewriting

$$\pm 2\frac{l/2}{\xi^2+(l/2)^2}\to 2x^s\frac{l/2}{\xi^2+(l/2)^2}$$

in the first two equations of (5.16) for $l=1$ and 2, respectively, and differentiating with respect to x^s, we obtain, after some algebra and making use of (5.14), the original set of relations (5.6).

An alternative expression for the free energy is

$$F/N_a=-\frac{T}{\pi}\int_{-\infty}^{\infty} d\xi\ln(1+e^{-\varepsilon/T})\frac{\frac{1}{2}}{\xi^2+\frac{1}{4}} -\frac{T}{\pi}\int_{-\infty}^{\infty} d\xi\ln(1+e^{-\Psi/T})\frac{1}{\xi^2+1}\,. \tag{5.18}$$

The equivalence between (5.15) and (5.18) is straightforwardly shown via the second equation of (5.16).

C. Special limits

In this subsection we discuss the high temperature and $T\to 0$ limits of the thermodynamic equations. If the temperature is much larger than the bandwidth, i.e., $T\gg 2$, we can neglect the independent terms in (5.12). In the absence of driving terms the potentials ε, Ψ, and φ_n do not depend on ξ and (5.23) can be reduced to an algebraic system of equations, similar to the one discussed by Takakashi[29] for the fermion gas with attractive δ-function potential,

$$\begin{aligned} e^{2\varepsilon/T}&=(1+e^{\Psi/T})/(1+e^{\varphi_1/T}) ,\\ e^{2(\Psi+A)/T}&=(1+e^{-\varepsilon/T})(1+e^{\Psi/T}) ,\\ e^{2\varphi_1/T}&=(1+e^{-\varepsilon/T})(1+e^{\varphi_2/T}) ,\\ e^{2\varphi_n/T}&=(1+e^{\varphi_{n+1}/T})(1+e^{\varphi_{n-1}/T}),\quad n\geq 2 : \end{aligned} \tag{5.19}$$

The general solution is given by

$$1+e^{\varphi_n/T}=\left[\sinh\left[\frac{nH}{2T}+\mu\right]/\sinh\left[\frac{H}{2T}\right]\right]^2 ,$$

where

$$e^{\mu}=e^{H/T}\left\{\left[1+\exp\left[-\frac{H}{2T}-\frac{A}{T}\right]\right]\Big/\left[1+\exp\left[\frac{H}{2T}-\frac{A}{T}\right]\right]\right\}^{1/2}$$

and the free energy is the one corresponding to three degrees of freedom per site

$$F/N_a=-T\ln\left[e^{A/T}2\cosh\frac{H}{2T}+1\right] . \tag{5.20}$$

Let us now analyze the thermodynamic equations in the limit $T\to 0$. From the last two equations it follows that $\varphi_n>0$ for $n=1,\ldots,\infty$, such that $\sigma_n\equiv 0$ for all n as $T\to 0$. The functions ε and Ψ, on the other hand, change sign as a function of ξ. As $T\to 0$ the first two equations of (5.16) yield

$$\begin{aligned} \varepsilon(\xi)&=\mp 2+2x^s\frac{\frac{1}{2}}{\xi^2+\frac{1}{4}}-\frac{H}{2}-A\\ &\quad-\frac{1}{\pi}\int_{-\infty}^{\infty}d\xi'\frac{\frac{1}{2}}{(\xi-\xi')^2+\frac{1}{4}}\Psi^-(\xi') ,\\ \Psi(\xi)&=\mp 4+2x^s\frac{1}{\xi^2+1}-2A\\ &\quad-\frac{1}{\pi}\int_{-\infty}^{\infty}d\xi'\frac{1}{(\xi-\xi')^2+1}\Psi^-(\xi')\\ &\quad-\frac{1}{\pi}\int_{-\infty}^{\infty}d\xi'\frac{\frac{1}{2}}{(\xi-\xi')^2+\frac{1}{4}}\varepsilon^-(\xi') , \end{aligned} \tag{5.21}$$

where $\Psi=\Psi^++\Psi^-$ and $\varepsilon=\varepsilon^++\varepsilon^-$ with Ψ^+, $\varepsilon^+>0$, and Ψ^-, $\varepsilon^-<0$. Differentiating with respect to x^s and using (5.14) we recover the ground-state equations.

D. Integration limits $B(H)$ and $Q(A)$.

Equations (5.21) can be used to determine the integration limits $B(H)$ and $Q(A)$ introduced in Sec. IV A. In (4.12) we expressed the magnetization for small fields as a function of B and in (4.22) the number of particles as a function of Q for $x^s=+1$.

We Fourier transform (5.21) and express Ψ^- as a function of ε^- and Ψ^+ in the second equation and insert it into the first equation. We obtain after Fourier transforming back

$$\begin{aligned} \varepsilon^+(\xi)+\varepsilon^-(\xi)-\left[\int_{-\infty}^{-B}+\int_{B}^{\infty}\right]d\xi'\varepsilon^-(\xi')G_1(\xi-\xi')\\ =\int_{-Q}^{Q}d\xi'\Psi^+(\xi')G_0(\xi-\xi')+2\pi G_0(\xi)-\frac{H}{2} . \end{aligned} \tag{5.22}$$

Equation (5.22) is a linear integral equation for ε. It is convenient to separate $\varepsilon^{\pm}=\varepsilon_a^{\pm}+\varepsilon_b^{\pm}$ with

$$\begin{aligned} \varepsilon_a^+(\xi)+\varepsilon_a^-(\xi)-\int_{-\infty}^{\infty}d\xi'\varepsilon_a^-(\xi')G_1(\xi-\xi')\\ =\int_{-Q}^{Q}d\xi'\Psi^+(\xi')G_0(\xi-\xi')+2\pi G_0(\xi) ,\\ \varepsilon_b^+(\xi)+\varepsilon_b^-(\xi)-\int_{-\infty}^{\infty}d\xi'\varepsilon_b^-(\xi')G_1(\xi-\xi')=-\frac{H}{2} . \end{aligned}$$

Comparing with (4.4) we have that $\varepsilon_a^+(\xi)=2\pi\rho_h(\xi)$ and $\varepsilon_a^-(\xi)=2\pi\rho(\xi)$, as well as $\Psi^+(\xi)=2\pi\sigma_h(\xi)$. The procedure to obtain ε_b is to rewrite the integral equation in the Wiener-Hopf from and solve it iteratively for large B, in analogy to Sec. IV B. To leading order in the field we obtain

$$\varepsilon_b^-(\xi+B)=-i\frac{H}{\sqrt{2}}\int_{-\infty}^{\infty}\frac{d\omega}{2\pi}\frac{e^{-i\xi\omega}}{\omega+i0}[g^+(\omega)]^{-1} . \tag{5.23}$$

The condition $\varepsilon(B)=0$ determines B. Using that

$$\varepsilon^-(B)=\lim_{\xi\to+0}\varepsilon^-(\xi+B)=-\lim_{\omega\to\infty}i\omega\hat{\varepsilon}^-(\omega)=0$$

we obtain

$$B=-\frac{1}{\pi}\ln\left[H/2\pi R\left[\frac{2\pi}{e}\right]^{1/2}\right] . \tag{5.24}$$

Using (5.24) in (4.12) the susceptibility is straightforwardly obtained.

Similarly, $Q(A)$ can be obtained. In the absence of a field $\varepsilon^-\equiv 0$, such that

$$\hat{\Psi}^+(\omega)+\hat{\Psi}^-(\omega)(1+e^{-|\omega|})=2\pi e^{-|\omega|}-4\pi(A+2)\delta(\omega) . \tag{5.25}$$

We assume the band is almost empty (Q is very large) and write $\Psi=\Psi_1+\Psi_2$ with

$$\hat{\Psi}_1^+(\omega)+\hat{\Psi}_1^-(\omega)(1+e^{-|\omega|})=2\pi e^{-|\omega|} ,$$

$$\hat{\Psi}_2^+(\omega)+\hat{\Psi}_2^-(\omega)(1+e^{-|\omega|})=-4\pi(A+2)\delta(\omega) .$$

In comparison with (4.13) we have that $\Psi_1^-(\xi)=2\pi\sigma'(\xi)$. The integral equation for Ψ_2 is put into the Wiener-Hopf form and solved iteratively. Q is then obtained from $\Psi(Q)=0$

$$Q=(A+2)^{-1/2} . \tag{5.26}$$

Hence $A\geq-2$, i.e., the Fermi level must be above the bottom of the band, and Q diverges with a square root singularity as expected for a one-dimensional system.

VI. CONCLUDING REMARKS

We have introduced a narrow-band model which includes some of the main features of heavy-fermion compounds. We considered a one-dimensional lattice of spin-$\frac{1}{2}$ electrons hopping between nearest-neighbor sites. Double occupancy of every site has been excluded and nearest-neighbor electrons interact via a charge interaction and a spin exchange. The integrability of the model imposes restrictions on the dimensionality and the interaction parameters. We obtain the Bethe-ansatz solution for the integrable cases. Of particular interest is the case of singlet scattering, since, in heavy-fermion systems, the f-electron spins are compensated in part by the conduction electrons and in part by antiferromagnetic spin fluctuations. We treated this case explicitly, obtained the ground-state and thermodynamic integral equations, and solved them for several special limits.

Although the model for general parameters seems to be nonintegrable and approximate methods have to be employed to solve it, the exact solution of special cases is of interest and provides an important testing ground for the approximations used. The model can be generalized to more internal degrees of freedom of the electrons, e.g., N components. These components may arise from combined spin and orbital degrees of freedom of the electrons. The model seems to be integrable for arbitrary N under similarly restrictive conditions as for $N=2$. The $1/N$ expansions applied previously to the Kondo and Anderson lattices, as well as to the Hubbard model, are adequate approximation schemes to solve the model for general parameters.

The integrable cases of our model can be mapped onto the multicomponent quantum lattice gas introduced by Sutherland.[31,32] Each lattice site can be in three possible states, namely empty or occupied with an electron with spin up or down. The problem can then be rewritten in terms of spin operators corresponding to $S=1$. Our four cases $x^s=\pm1$ with $x^t=0$ and $x^t=\pm1$ with $x^s=0$ correspond then to the examples discussed in Ref. 31. The logarithmic dependences obtained for the magnetization and the occupation number in Sec. IV are possibly a consequence of the SU(3) invariance of the model. If the coupling is weaker, e.g., $|x^s|<1$, we expect an analytic behavior, while if $|x^s|>1$, we expect an exponential activation in analogy to the anisotropic Heisenberg chain.

[1] G. R. Stewart, Rev. Mod. Phys. **56**, 755 (1984).
[2] P. A. Lee, T. M. Rice, J. W. Serene, L. J. Sham, and J. W. Wilkins, Comments Conden. Matt. Phys. **12**, 98 (1986).
[3] *Theory of Heavy Fermions and Valence Fluctuations,* Vol. 62 of *Solid State Sciences,* edited by T. Kasuya and T. Saso (Springer-Verlag, Heidelberg, 1985).
[4] H. R. Ott, H. Rudigier, T. M. Rice, K. Ueda, Z. Fisk, and J. L. Smith, Phys. Rev. Lett. **52**, 1915 (1984).
[5] P. W. Anderson, Phys. Rev. B **30**, 1549 (1984).
[6] H. Keiter and G. Morandi, Phys. Rep. **109**, 227 (1984).
[7] G. Czycholl, Phys. Rep. **143**, 277 (1986).
[8] N. Grewe, Solid State Commun. **50**, 19 (1984).
[9] Y. Kuramoto, in *Theory of Heavy Fermions and Valence Fluctuations,* Ref. 3, p. 152.
[10] N. Read and D. M. Newns, Solid State Commun. **52**, 993 (1984).
[11] P. Coleman, J. Magn. Magn. Mater. **47&48**, 323 (1985).
[12] A. Auerbach and K. Levin, Phys. Rev. Lett. **57**, 877 (1986).
[13] M. Lavagna, A. J. Millis, and P. Lee, Phys. Rev. Lett. **58**, 266 (1987).
[14] B. H. Brandow, Phys. Rev. B **33**, 215 (1986).
[15] T. M. Rice and K. Ueda, Phys. Rev. Lett. **55**, 995 (1985).
[16] C. M. Varma, W. Weber, and L. J. Randall, Phys. Rev. B **33**, 1015 (1986).
[17] F. J. Ohkawa, J. Phys. Soc. Jpn. **52**, 3886 (1983); **53**, 2697 (1984).
[18] M. Tachiki, S. Maekawa, and S. Takahashi, Phys. Rev. B **31**, 228 (1985).
[19] H. Jichu, T. Matsuura, and Y. Kuroda, Prog. Theor. Phys. **72**, 366 (1984); K. Miyake, T. Matsuura, H. Jichu, and Y. Nagaoka, *ibid.* **72**, 1063 (1984).
[20] E. H. Lieb and F. Y. Wu, Phys. Rev. Lett. **20**, 1445 (1968).
[21] R. Orbach, Phys. Rev. **112**, 309 (1958); J. des Cloizeaux and M. Gaudin, J. Math. Phys. **7**, 1384 (1966).
[22] C. N. Yang, Phys. Rev. Lett. **19**, 1312 (1967).
[23] A. M. Tsvelick and P. B. Wiegmann, Adv. Phys. **32**, 453 (1983).
[24] M. Gaudin, Phys. Lett. **24A**, 55 (1967).
[25] H. M. Babujian, Nucl. Phys. **B215**, [FS7], 317 (1983).
[26] P. M. Morse and H. Feshbach, *Methods of Theoretical Physics,* (McGraw-Hill, New York, 1953).
[27] C. N. Yang and C. P. Yang, Phys. Rev. **150**, 327 (1966).
[28] C. K. Lai, Phys. Rev. Lett. **26**, 1472 (1971).
[29] M. Takahashi, Prog. Theor. Phys. **46**, 1388 (1971).
[30] P. Schlottmann, Z. Phys. B **52**, 127 (1983).
[31] B. Sutherland, Phys. Rev. B **12**, 3795 (1975).
[32] C. K. Lai, J. Math. Phys. **15**, 1675 (1974).

VOLUME 64, NUMBER 21 PHYSICAL REVIEW LETTERS 21 MAY 1990

Supersymmetric *t-J* Model in One Dimension: Separation of Spin and Charge

P. A. Bares and G. Blatter

Theoretische Physik, Eidgenössische Technische Hochschule–Hönggerberg, CH-8093 Zürich, Switzerland

(Received 5 December 1989)

Using the Bethe-*Ansatz* technique, we diagonalize exactly the one-dimensional *t-J* Hamiltonian for the supersymmetric case $T=J$. In this limit it is identical with models considered previously by Sutherland and by Schlottmann. The ground state is a liquid of singlet pairs of varying spatial separation for all band fillings. We find two types of gapless excitations with effective Fermi surfaces at $2k_F$ and k_F which we identify with the holon and the spinon excitations near half filling.

PACS numbers: 75.10.Jm, 05.50.+q, 71.28.+d, 74.65.+n

The discovery of high-temperature superconductors has greatly stimulated the interest in strongly correlated systems. In particular, Anderson has suggested that the *t-J* model is an appropriate starting model.[1,2] The *t-J* model is characterized by a lattice Hamiltonian $\mathcal{H}$ which describes fermions with hard-core repulsion, nearest-neighbor hopping (t), and spin exchange (J). Considered in its own right, the model can be studied for any dimension and for all values of the ratio J/t. In this Letter, we consider one dimension and study the exact solution at $t=J>0$. We emphasize that the model we solve is not identical to the large-U limit of the repulsive Hubbard model,[3,4] which maps onto the limit $J\ll t$.

Using the Bethe-*Ansatz* method, this model was first solved by Sutherland[5] in a study of a multicomponent lattice gas. In particular, he derived the Bethe-*Ansatz* equations for the case of two fermions and one boson which reduces to the *t-J* model. A different form of the Bethe-*Ansatz* equations was discovered by Schlottmann,[6] who solved them for the ground state and discussed the thermodynamic properties of the model, applying the results to heavy-fermion systems. In addition, numerical calculations on finite clusters have been recently performed by Imada and Hatsugai[7] and by von Szczepanski *et al.*[8] In this Letter, we present for the first time a detailed analysis of the ground state and of the elementary-excitation spectrum at arbitrary filling.[9] We interpret the spectra in terms of solitonlike excitations which we identify as holons and spinons[10] near half filling. We show that doping the system with holes naturally leads to the separation of the spin and charge degrees of freedom.[10]

In the following we solve the Bethe equations for the ground state by means of a two-string *Ansatz* for the electron rapidities. The solution can be interpreted as a liquid of bound singlet pairs of varying spatial separation and binding energy. We solve for the elementary excitations of the model and show how the two branches involving charge and spin excitations can be interpreted as holons and spinons, respectively.

Consider a one-dimensional lattice of N_a sites with N electrons where each site is capable of accommodating at most one fermion. The dynamics is described by the *t-J* Hamiltonian

$$\mathcal{H}=-t\mathcal{P}\sum_{\langle i,j\rangle,\sigma}c_{i\sigma}^{\dagger}c_{j\sigma}\mathcal{P}+J\sum_{\langle i,j\rangle}(\mathbf{S}_i\cdot\mathbf{S}_j-n_in_j/4)\,,$$

where the projector $\mathcal{P}=\prod_i(1-n_{i\uparrow}n_{i\downarrow})$ restricts the Hilbert space by the constraint of no double occupancy. The symmetries of this Hamiltonian are U(1) gauge, SU(2) spin, and lattice translational invariance. In addition, the model becomes supersymmetric at $t=J$.[11,12] The short-range nature of the interaction motivates the following *Ansatz*[13] for the amplitudes

$$\psi_{\sigma_1\cdots\sigma_N}(x_1,\ldots,x_N)$$

in the sector $x_{Q_1}<x_{Q_2}<\cdots<x_{Q_N}$:

$$\psi_{\sigma_1\cdots\sigma_N}(x_1,\ldots,x_N)=\sum_P(-1)^PA_{\sigma_{Q_1}\cdots\sigma_{Q_N}}(QP)\times\exp\left(i\sum_{j=1}^{N}k_{P_j}x_j\right).$$

P and Q denote permutations of $1,\ldots,N$, $(-1)^P$ is the sign of the permutation P, and we choose $x_i\neq x_j$ whenever $i\neq j$. The condition that $|\Psi\rangle$ be an eigenstate of $\mathcal{H}$ establishes a linear relation between the amplitudes $A_{\sigma_{Q_1}\cdots\sigma_{Q_N}}(QP)$. The multiparticle scattering matrix defined by these relations factorizes into a product of two-particle scattering matrices provided the Yang-Baxter equations are fulfilled.[13] Also, the Yang-Baxter equations represent the conditions for the consistency of the Bethe *Ansatz*[14] and require that $t=|J|$. Applying the quantum inverse-scattering method[15] we obtain a set of coupled algebraic equations[6,16] for the electron rapidities $\{v_j\}$, $j=1,\ldots,N$, and the spin rapidities $\{\Lambda_\alpha\}$, $\alpha=1,\ldots,M$ (M is the "number of down spins" in the solution):

$$\prod_{j=1}^{N}\frac{v_j-\Lambda_\alpha+i/2}{v_j-\Lambda_\alpha-i/2}=-\prod_{\beta=1}^{M}\frac{\Lambda_\beta-\Lambda_\alpha+i}{\Lambda_\beta-\Lambda_\alpha-i}\,,$$
$$\left(\frac{v_j+i/2}{v_j-i/2}\right)^{N_a}=\prod_{\beta=1}^{M}\frac{v_j-\Lambda_\beta+i/2}{v_j-\Lambda_\beta-i/2}\,, \qquad (1)$$

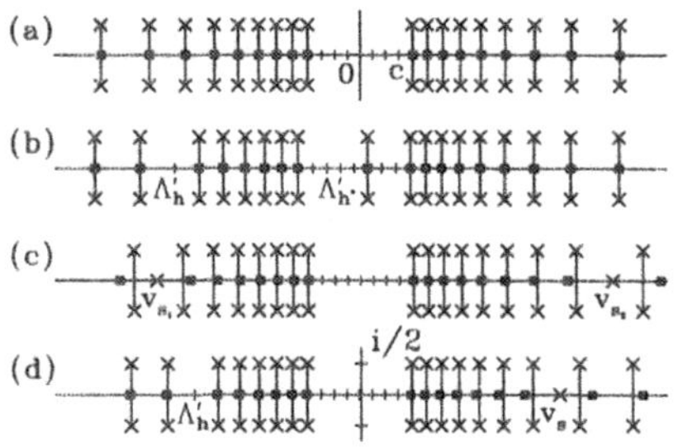

FIG. 1. Electron rapidities in the complex plane: Crosses denote the quantum numbers v_a' describing kinetic degrees of freedom, and solid squares denote the quantum numbers Λ_a' associated with spin degrees of freedom. (a) Ground state: Electron rapidities occur in complex pairs, $v_a' = \Lambda_a' \pm i/2$, describing singlet pairs of range $2/\ln(1+\Lambda_a'^{-2})$. The parameter c determines the filling factor N/N_a. (b) Holon-antiholon (h-h^*) excitation: A string at Λ_h' is transferred to a higher-energy state at Λ_{h^*}'. (c) Triplet (s-s) excitation: A string is broken up into two real rapidities v_{s_1} and v_{s_2}, each of which is describing a spinon. The two spinons combine into a triplet excitation as one of the spin rapidities Λ_a' has been removed. (d) Real-particle (s-h) excitation: Removing one electron leaves the system in an excited state with one holon at Λ_h' and one spinon at v_s. The holon and the spinon account for the charge and spin degrees of freedom, respectively, of the many-body state.

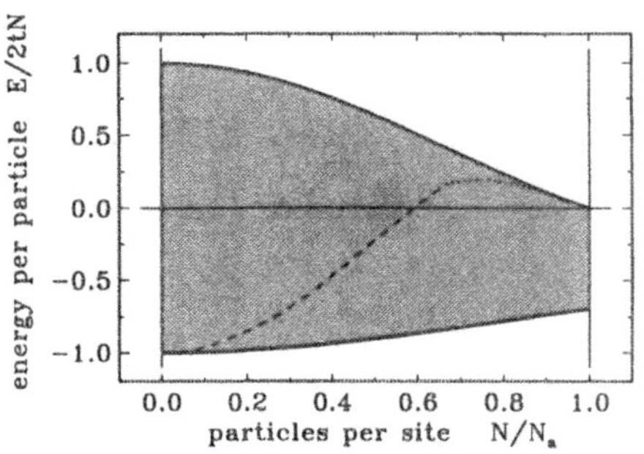

FIG. 2. Energy per particle $E/2tN$ vs filling factor N/N_a. The ground state is a liquid of singlet pairs of varying range described by complex pairs of rapidities; see Fig. 1(a). The highest accessible state is the ferromagnetic state with real rapidities v_a' and no spin rapidities Λ_a'. The dashed line denotes the lowest singlet state with real rapidities v_a'. This state is forced into a state with finite magnetization for a filling $N/N_a > 2/3$ (dotted line). The overall width in energy decreases from the free-electron value $4t$ per particle for $N/N_a \to 0$ to $2t\ln 2$ in the Heisenberg limit $N/N_a \to 1$.

where $2v_j = \cot(k_j/2)$ for $t = J$. For even N the low-energy states are parametrized by a sea of two-strings in the complex plane with $v_a' = \Lambda_a' \pm i/2 + O(e^{-N_a})$. Taking the logarithm of Eq. (1) we introduce the bare quantum numbers I_a', $a = 1, \dots, M = N/2$, which specify the roots of the equation. The I_a' are integers or half-odd integers and restricted to the interval $|I_a'| \le I_{\max}' = (N_a - M - 1)/2$. For arbitrary filling the number of available quantum numbers I_a' exceeds the number of actual two-strings, so there is freedom in the choice of the set $\{I_a'\}$ to be occupied.

For the ground state, the $I_a' = I_a^{0\prime}$ are chosen symmetrically with respect to zero, $0 \le I_{\min}' \le |I_a^{0\prime}| \le I_{\max}'$. The corresponding distribution of two-strings in the complex plane is shown in Fig. 1(a). In the thermodynamic limit ($N_a \to \infty$, $N/N_a = \text{const}$) we obtain an integral equation of Fredholm type for the distribution of the roots Λ_a'. At half filling, this integral can be solved in closed form and the ground-state energy of the Heisenberg chain, $E/N = -2t\ln 2$, is recovered.[6,17] Away from half filling, the integral equation has to be solved numerically: The corresponding ground-state energy is shown in Fig. 2. For all fillings N/N_a the ground state is a liquid of singlet bound pairs, where each singlet can be associated with a two-string at Λ_a'. The coherence length ξ of a particular pair depends on the position Λ_a' of the associated rapidity, $\xi = 2/\ln(1+\Lambda_a'^{-2})$. In particular, the ground state involves pairs of arbitrarily weak binding energy ($|\Lambda_a'| \to \infty$) resulting in a gapless excitation spectrum. As in the *attractive* Hubbard model[18,19] the ground state of the t-J model is parametrized by complex pairs of rapidities. However, the physics is more like that of the *repulsive* Hubbard model. There is no jump in the chemical potential for adding one or two particles and we do not observe any transition as a function of filling N/N_a. Note that the repulsive Hubbard model involves real electron and spin rapidities. For the t-J model, real rapidities lead to an excited state as shown in Fig. 2.

There are two types of elementary excitations, which do not change particle number, involving (i) charge and (ii) spin degrees of freedom: (i) Charge excitations away from half filling involve the transfer of a particular bare quantum number $I_h' \in \{I_a^{0\prime}\}$ to a previously unoccupied state I_{h^*}' above the pseudo Fermi surface at $I_{\min}'$. This excitation transfers a charge e (not $2e$) into a higher-energy state but differs from the usual particle-hole excitation in a Fermi liquid as no spin is involved. According to Anderson's terminology for the strongly correlated Hubbard model,[10] we identify this excitation with a holon (kink of charge 1) and the corresponding spectrum with the holon-antiholon branch. The spectrum is obtained by solving the Bethe equations for the rapidities Λ_a' to order $1/N_a$. A shown in Fig. 1(b) the electron rapidities v_a' and the spin rapidities Λ_a' remain aligned. The holon-antiholon spectrum is shown in Fig. 3. $N/N_a = 2/3$ marks the special filling (of high symmetry) above which a gap in momentum occurs, excluding excitations with momenta between $2\pi - 3k_F$ and $3k_F$ ($k_F = \pi N/2N_a$). Keeping the antiholon fixed (e.g., $\Lambda_{h^*}' = c$) and moving the holon over the allowed parameter range, we obtain the holon excitation spectrum with an effective Fermi surface spanning $4k_F$. With respect

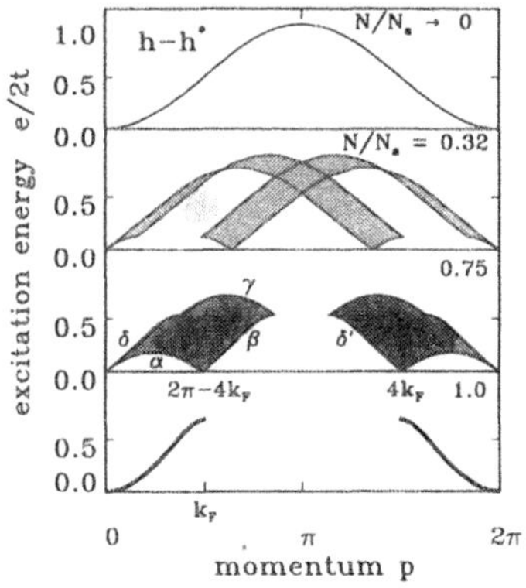

FIG. 3. Holon-antiholon (h-h^*) excitation spectrum for several values of the filling factor N/N_a. Starting with $\Lambda_h' = \Lambda_{h^*}' = c$ [see Fig. 1(b)], branch α is obtained by moving Λ_{h^*}' to $-c$, branch β is due to moving Λ_h' out to ∞, γ corresponds to moving Λ_{h^*}' back to c, and δ completes the loop as Λ_h' is moved back to c. The branches δ and δ' make up the holon excitation spectrum spanning a Fermi surface of range $4k_F$ ($k_F = \pi N/2N_a$). Note that a gap appears in momentum for filling $N/N_a < 2/3$, where the Hilbert space starts to shrink rapidly due to the constraint of no double occupancy.

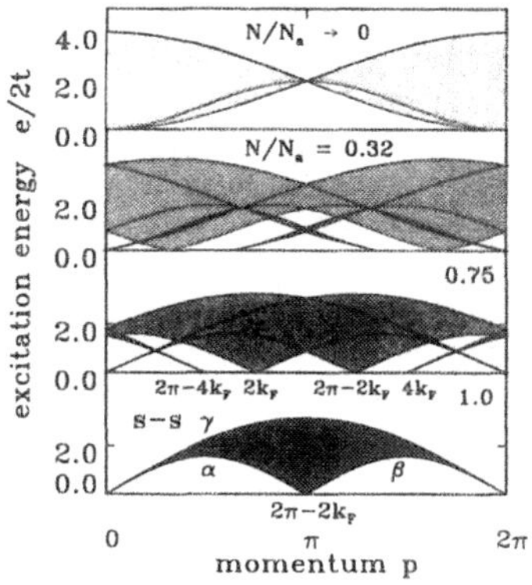

FIG. 4. Triplet (spinon-spinon, s-s) excitation spectrum for several values of the filling factor N/N_a. According to Ref. 16 a triplet excitation is two-parametric, one parameter describing a kink of spin $\frac{1}{2}$ or spinon. For $N/N_a \to 1$ the spectrum of the Heisenberg model is recovered: The branches α, β, and γ are obtained by starting with $v_{s_1} = v_{s_2} = \infty$ and moving $v_{s_1} \to -\infty$ (α), $v_{s_2} \to -\infty$ (β), and finally taking $v_{s_1} = v_{s_2}$ together back to ∞ (γ); see Fig. 1(c). The branch α is the spinon excitation spectrum spanning a Fermi surface of $2k_F$. The lowest (gapless) excitation is obtained by breaking a singlet pair at $\Lambda_a' = \pm\infty$, where the binding energy goes to zero. As $N/N_a \to 0$ the free-particle triplet excitation spectrum is recovered.

to Fig. 1(b) we identify the boundaries $\pm c$ with the pseudo Fermi surface for the holons.

(ii) The spin excitations, on the other hand, involve the breaking of a pair with (triplet) or without (singlet) spin flip. Here we restrict ourselves to the triplet excitations for the case of an even number of particles. The excitation consists in transferring a pair of complex roots onto the real axis and simultaneously removing the spin rapidity associated with the pair. As a consequence the remaining spin rapidities Λ_a' shift with respect to the electron rapidities v_a' as shown in Fig. 1(c). Again the excitation is two-parametric. At half filling the real rapidities v_{s_1} and v_{s_2} describe kinks of spin $\frac{1}{2}$ which combine into a triplet (or singlet) excitation as shown by Faddeev and Takhtajan[20] for the Heisenberg chain. The excitation spectrum shown in Fig. 4 is gapless for all fillings. This is due to the presence of asymptotically unbound pairs, i.e., pairs of arbitrarily weak binding energy. Real rapidities $|v_s| > c$ embedded in the sea of singlet pairs are identified with spinons as the corresponding excitation carries spin and no charge. On the other hand, isolated rapidities with $|v_s| < c$ are associated with real-particle excitations (carrying both spin and charge). Upon decreasing the filling N/N_a, the spinon excitation spectrum gradually transforms into a real-particle excitation spectrum as shown in Fig. 4. We find that the effective Fermi surface for the spinons is at k_F [corresponding to the points $v_s = \pm\infty$ in Fig. 1(c)].

Finally, the real-particle excitation spectrum involving a change in particle number is shown in Fig. 5. Removing a real particle near half filling leaves the system in an excited state [see Fig. 1(c)]: The hole splits into two solitonlike excitations, a spin-$\frac{1}{2}$ kink (spinon) and a spinless kink of charge e (holon). As in the strongly correlated Hubbard model, the excitation spectrum goes linearly to zero at $|k| = k_F$ and at $|k| = 3k_F$.

In conclusion, we have determined the ground state

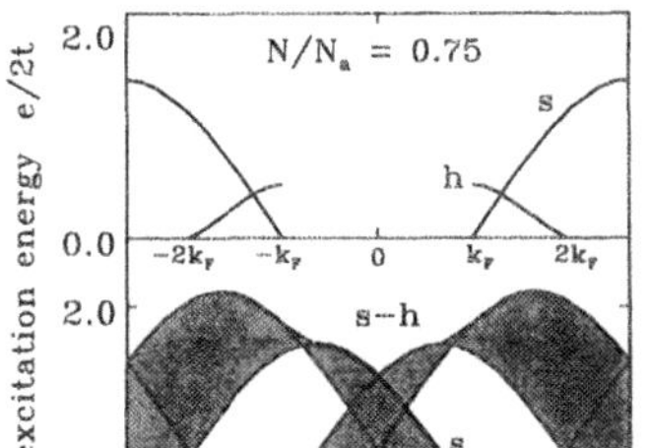

FIG. 5. Single-particle (s-h) excitation spectrum. Removing a particle leaves the system back in an excited state characterized by a spinon s and a holon h; see Fig. 1(d). Top: holon and spinon excitation spectra with Fermi surfaces at $2k_F$ and at k_F, respectively. Bottom: combination of the s and h excitation spectra into a real-particle (s-h) excitation spectrum. The state at k_F ($3k_F$) is a combination of a $2k_F$ holon and a $-k_F$ (k_F) spinon. The spectrum has been folded back into the first Brillouin zone.

and the elementary-excitation spectrum of the t-J model for arbitrary filling $N/N_a \leq 1$. We find that the repulsive on-site interaction dominates the attractive spin interaction J. We believe that the model belongs to the same universality class as the repulsive Hubbard model. Therefore we do not expect a phase transition in the interval $0 < J/t \leq 1$. This is consistent with the renormalized mean-field theory in one dimension by Zhang *et al.*[21] The gapless excitations at $2k_F$ (spin) and $4k_F$ (charge) produce long-range incommensurate spin and charge correlations in the t-J model.[22] Schulz has recently shown how to determine the asymptotic form of the correlation functions using results of the Bethe-*Ansatz* solution.[23]

We are very much indebted to T. M. Rice for many illuminating discussions and for his continuous encouragement during this work. We wish to thank G. Felder, M. Karowski, M. Luchini, M. Ogata, W. Puttika, J. Rhyner, P. Schlottmann, and P. Wiegmann for helpful discussions and correspondence.

[1]P. W. Anderson, Science **235**, 1196 (1987).

[2]F. C. Zhang and T. M. Rice, Phys. Rev. B **37**, 3759 (1988).

[3]E. H. Lieb and F. Y. Wu, Phys. Rev. Lett. **20**, 1445 (1968).

[4]C. F. Coll, Phys. Rev. B **9**, 2150 (1974).

[5]B. Sutherland, Phys. Rev. B **12**, 3795 (1975).

[6]P. Schlottmann, Phys. Rev. B **36**, 5177 (1987).

[7]M. Imada and Y. Hatsugai, J. Phys. Soc. Jpn. **58**, 3752 (1989).

[8]K. J. von Szczepanski, P. Horsch, W. Stephen, and M. Ziegler, Phys. Rev. B **41**, 2017 (1990).

[9]For details, see P. A. Bares and G. Blatter (to be published).

[10]P. W. Anderson, in Proceedings of the Kathmandu Summer School Lectures [Princeton University report (to be published)]; P. W. Anderson and Y. Ren, Princeton University report (to be published).

[11]The t-J model is supersymmetric up to a shift in the chemical potential. Indeed, there exist two supercharge operators (connecting the Fermi and Bose sectors) $Q_\sigma^{(\dagger)}$, $\sigma = \uparrow, \downarrow$, such that $[\mathcal{H} + 2\mathcal{N} - N_a, Q_\sigma^{(\dagger)}] = 0$, where $\mathcal{N}$ is the number operator.

[12]P. B. Wiegmann, Phys. Rev. Lett. **60**, 821 (1988).

[13]C. N. Yang, Phys. Rev. Lett. **19**, 1312 (1967).

[14]N. Andrei, K. Furuya, and J. H. Lowenstein, Rev. Mod. Phys. **55**, 331 (1983).

[15]L. A. Takhtajan, in *Early Solvable Problems in Condensed Matter and Relativistic Field Theory,* edited by B. S. Shastry *et al.* (Springer-Verlag, Berlin, 1985), p. 175.

[16]These equations have been derived also by A. Tsvelick [P. B. Wiegmann (private communication)].

[17]H. Bethe, Z. Phys. **71**, 205 (1931).

[18]F. Woynarovich, J. Phys. C **16**, 6593 (1983).

[19]Kong-Ju-Bock Lee and P. Schlottmann, Phys. Rev. B **38**, 11 566 (1988).

[20]L. D. Faddeev and L. A. Takhtajan, Phys. Lett. **85A**, 375 (1981).

[21]F. C. Zhang, C. Gros, T. M. Rice, and H. Shiba, Supercond. Sci. Technol. **1**, 36 (1988).

[22]M. Ogata and H. Shiba, Phys. Rev. B **41**, 2326 (1990).

[23]H. J. Schulz, Laboratoire de Physique des Solides, Université Paris-Sud, report, 1990 (to be published).

J. Phys. A: Math. Gen. **24** (1991) 1137–1151. Printed in the UK

The supersymmetric $t-J$ model in one dimension

Sarben Sarkar†

Centre for Theoretical Studies, Royal Signals and Radar Establishment, Great Malvern WR14 3PS, UK

Received 22 March 1990, in final form 5 November 1990

Abstract. The $t-J$ model (related to the strong-correlation limit of the Hubbard model) is shown to be soluble in one dimension using the Bethe ansatz. The solution holds only when the Hamiltonian is supersymmetric. The ground state in the presence of holes is found to be gapless, and to have no magnetization.

1. Introduction

There is a strong belief (Anderson 1988, Fukuyama *et al* 1989) that electron correlations are important in distinguishing the new high-temperature superconductors from conventional ones. The existence of antiferromagnetism in the absence of doping for the new materials is evidence for this. Hubbard (1963) was very influential in the study of such correlations. He proposed a lattice Hamiltonian,

$$H=\sum_{\substack{\langle ij\rangle\\ \sigma}} tc^{\dagger}_{i\sigma}c_{j\sigma}+\frac{U}{2}\sum_{i,\sigma} n_{i,\sigma}\sigma_{i,-\sigma} \tag{1}$$

as an example of a system which clearly accommodates the atomic ($t/U\to 0$) and band theory ($U/t\to 0$) limits. i and j are nearest-neighbour sites and $c_{j\sigma}$ destroys an electron with z-component of spin σ at site j. $n_{i,\sigma}$ is $c^{\dagger}_{i\sigma}c_{i\sigma}$, the number operator. Only the low-energy states can have any possible relevance to superconductivity. Since we are interested in the strong correlation ($U/t\gg 1$) limit we may take

$$H_0=\frac{U}{2}\sum_{i,\sigma} n_{i,\sigma}n_{i,-\sigma} \tag{2}$$

for the unperturbed Hamiltonian. For a lattice with N sites and $(N-n)$ electrons the ground state of H_0 cannot have more than one electron per site. There is a set of

$$\binom{N}{N-n}2^{N-n}$$

degenerate ground states of H_0. From degenerate perturbation theory (Lindgren and Morrison 1986, Pike *et al* 1991) we can construct an effective Hamiltonian $H_{\rm eff}$ which operates on this set but has the same low energy spectrum as H. Since the Hilbert

† Present address: Department of Physics, King's College London, Strand, London WC2R 2LS, UK.

0305-4470/91/051137+15$03.50

space of $H_{\rm eff}$ is much smaller than that for H this 'effective' description is very economical. The resultant $H_{\rm eff}$ is

$$H_{\rm eff}=P\Bigg(t\sum_{\substack{\langle ij\rangle\\ \sigma}} c^{\dagger}_{i\sigma}c_{j\sigma}+J\sum_{\langle ij\rangle}(\boldsymbol{S}_i\cdot\boldsymbol{S}_j-\tfrac{1}{4})n_in_j$$

$$-\frac{J}{2}\sum_{\substack{\langle ikj\rangle\\ \sigma}}(c^{\dagger}_{i\sigma}n_{k,-\sigma}c_{j\sigma}+c^{\dagger}_{i,-\sigma}c_{k,-\sigma}c^{\dagger}_{k\sigma}c_{j\sigma})\Bigg)P \tag{3}$$

where P is the projection operator onto the set of ground states of H_0, $J=2t^2/U(\ll t)$, $\boldsymbol{S}_i$ is a spin operator and site k is a nearest neighbour to sites i and j. The connection of H and $H_{\rm eff}$ with the copper oxide-based high-temperature superconductors is not obvious. Indeed it is generally accepted that there is substantial overlap of the electron orbitals on copper and oxygen. This would naturally lead to an H also involving creation and annihilation operators for oxygen orbitals. Zhang and Rice (1988) nonetheless showed that an effective model similar to (3) can emerge where $J\leqslant t$. Further support for the validity of this model with parameter values as large as $J\sim 1.53t$ has been given recently (Jefferson 1990).

For $n/N\ll 1$ the three-site terms in $H_{\rm eff}$ are small compared with the other terms and are often ignored. The resulting $H_{\rm eff}$ is called for obvious reasons the $t-J$ Hamiltonian (H_{t-J}). In a recent letter (Sarkar 1990a) the method of solution for this model in one dimension where $J/2t=1$ has been indicated. At this point in parameter space $H_{\rm eff}$ is invariant under the group of transformations of a supergroup U(1/2) (Wiegmann 1988, Cornwell 1989, Sarkar 1990b). Details of this solution which uses the Bethe ansatz (Bethe 1931) will now be given. In particular we will find the ground state and excited state energies as a function of concentration near half filling. This solution is not a simple consequence of the Bethe ansatz solution of the Hubbard model (Lieb and Wu 1968) for two reasons. Firstly the large U/t limit of the Hubbard model has three-site terms and secondly J is not very much less than t.

2. Supersymmetry

H_{t-J} operates on a Hilbert space which is spanned by states of the form

$$\bigotimes_i |\alpha_i\rangle$$

where $\otimes$ denotes a direct product and $|\alpha_i\rangle$ is $|0\rangle$, $|\downarrow\rangle$ or $|\uparrow\rangle$. $|\downarrow\rangle$ and $|\uparrow\rangle$ represent Wannier states with spin down and spin up respectively. $|0\rangle$ is a hole state. With the basis $|\alpha_i\rangle$ it is natural to associate operators

$$X_i^{\alpha\beta}=|\alpha_i\rangle\langle\beta_i|. \tag{4}$$

Now $X_i^{0\uparrow}$ has a fermionic nature since it destroys an up-spin electron whereas $X_i^{\uparrow\downarrow}$ is bosonic. This leads naturally to an operator algebra involving both commutators and anticommutators, the latter occurring only if both operators are fermionic. The resulting so-called superalgebra is

$$[X_i^{\alpha\beta},X_j^{\alpha'\beta'}]_{\pm}=\delta_{ij}(X_i^{\alpha\beta'}\delta_{\beta\alpha'}\pm X_i^{\alpha'\beta}\delta_{\beta'\alpha}) \tag{5}$$

where $[\,,\,]_+$ is an anticommutator and $[\,,\,]_-$ is a commutator. It is standard (Bars and Günaydin 1983) to represent these operators in terms of a bosonic and two fermionic harmonic oscillators, e.g.

$$X_i^{0\sigma}=f_i^{\sigma}b_i^{\dagger} \qquad X_i^{\sigma\sigma'}=f_i^{\sigma\dagger}f_i^{\sigma'} \qquad X_i^{00}=b_i^{\dagger}b_i \tag{6}$$

where

$$[f_i^{\sigma},f_j^{\dagger\sigma'}]_+=\delta_{\sigma\sigma'}\delta_{ij} \qquad [f_i^{\sigma},f_j^{\sigma'}]_+=[f_i^{\sigma\dagger},f_j^{\sigma'\dagger}]_+=0 \tag{7}$$

and

$$[b_i,b_j^{\dagger}]_-=\delta_{ij} \qquad [b_i,b_j]_-=[b_i^{\dagger},b_j^{\dagger}]_-=0. \tag{8}$$

This representation will be useful later.

Using the X-operators it is possible to rewrite H_{t-J} without the formal use of projection operators P. Since we are dealing with spin-$\frac{1}{2}$ we can write

$$J(\boldsymbol{S}_i\cdot\boldsymbol{S}_j-\tfrac{1}{4})=\frac{J}{4}(\boldsymbol{\sigma}_i\cdot\boldsymbol{\sigma}_j-1) \tag{9}$$

where $\boldsymbol{\sigma}_i$ are the Pauli spin matrices at site i. This can be further rewritten as

$$\frac{J}{2}(P_{ij}-1)$$

where

$$P_{ij}=\tfrac{1}{2}(\boldsymbol{\sigma}_i\cdot\boldsymbol{\sigma}_j+1). \tag{10}$$

It is easy to verify that

$$P_{ij}|\sigma\rangle_i|\sigma'\rangle_j=|\sigma'\rangle_i|\sigma\rangle_j \tag{11}$$

so that P_{ij} is a permutation on spin labels. In terms of the X-operators:

$$P_{i\,i+1}=\sum_{\sigma,\sigma'}X_i^{\sigma\sigma'}X_{i+1}^{\sigma'\sigma}. \tag{12}$$

We need to write $\delta_{n_i1}\delta_{n_{i+1}1}$ (where $n_i=n_{i\uparrow}+n_{i\downarrow}$) in terms of X-operators as well. We first note that

$$\sum_i\delta_{n_i0}=\sum_i\delta_{n_i0}\delta_{n_{i+1}0}+\tfrac{1}{2}\sum_i(\delta_{n_i0}\delta_{n_{i+1}1}+\delta_{n_i1}\delta_{n_{i+1}0}) \tag{13}$$

$$\sum_i(\delta_{n_i0}+\delta_{n_i1})=N \tag{14}$$

and

$$\sum_i(\delta_{n_i0}\delta_{n_{i+1}0}+\delta_{n_i1}\delta_{n_{i+1}1}+\delta_{n_i0}\delta_{n_{i+1}1}+\delta_{n_i1}\delta_{n_{i+1}0})=N \tag{15}$$

where N is the number of lattice sites. The validity of (13) and (15) is best established by examining examples. From (13) and (15) we have

$$2\sum_i(\delta_{n_i0}-\delta_{n_i0}\delta_{n_{i+1}0})+\sum_i\delta_{n_i1}\delta_{n_{i+1}1}+\sum_i\delta_{n_i0}\delta_{n_{i+1}0}=N \tag{16}$$

and on using (14) we deduce

$$\sum_i\delta_{n_i1}\delta_{n_{i+1}1}=\sum_i\delta_{n_i0}\delta_{n_{i+1}0}-\sum_i\delta_{n_i0}+\sum_i\delta_{n_i1}. \tag{17}$$

Since

$$\sum_i \delta_{n_i 0}\delta_{n_{i+1} 0} = \sum_i X_i^{00} X_{i+1}^{00} \tag{18}$$

(and $\Sigma_i\, \delta_{n_i 0}$ and $\Sigma_i\, \delta_{n_i 1}$ are constants) we can write

$$\frac{J}{2}\sum_i (P_{i\,i+1}-1) = \frac{J}{2}\sum_i X_i^{\sigma\sigma'} X_{i+1}^{\sigma'\sigma} - \frac{J}{2}\sum_i X_i^{00} X_{i+1}^{00} \tag{19}$$

up to an additive constant. (A summation convention is understood.) A Hamiltonian H' which is more general than H_{t-J} may then be written as

$$H' = g\sum_i (X_i^{\sigma 0} X_{i+1}^{0\sigma} + X_{i+1}^{\sigma 0} X_i^{0\sigma}) + g'\sum_i X_i^{\sigma\sigma'} X_{i+1}^{\sigma'\sigma} + g''\sum_i X_i^{00} X_{i+1}^{00} \tag{20}$$

g, g' and g'' being constants.

H_{t-J} is obtained when

$$g = -t \qquad g' = \frac{J}{2} = -g'' \tag{21}$$

(provided we make the canonical transformation $b_i \to (-1)^i b_i$ in (6)).

The generators $X^{0\sigma}$ $(=\Sigma\, X_i^{0\sigma})$, $X^{\sigma 0}$ $(=\Sigma\, X_i^{\sigma 0})$ and $X^{\sigma\sigma'}$ $(=\Sigma\, X_i^{\sigma\sigma'})$ form a superalgebra isomorphic to the single site superalgebra. The bosonic generators $X^{\sigma\sigma'}$ commute with the Hamiltonian for arbitrary g, g' and g''. For H' to be supersymmetric (with respect to this superalgebra), it remains to check that the condition

$$\left[\sum_j X_j^{0\sigma}, H'\right]_- = 0 \tag{22}$$

is satisfied. Now

$$\left[\sum_j X_j^{0\sigma}, g\sum_i (X_i^{\sigma' 0} X_{i+1}^{0\sigma'} + X_{i+1}^{\sigma' 0} X_i^{0\sigma'})\right]_-$$

$$= g\sum_i (X_i^{00} X_{i+1}^{0\sigma} + X_i^{\sigma'\sigma} X_{i+1}^{0\sigma'} + X_{i+1}^{00} X_i^{0\sigma} + X_{i+1}^{\sigma'\sigma} X_i^{0\sigma'}) \tag{23}$$

$$\left[\sum_j X_j^{0\sigma}, g'\sum_i X_i^{\sigma'\sigma''} X_{i+1}^{\sigma''\sigma'}\right]_- = g'\sum_i (X_i^{0\sigma'} X_{i+1}^{\sigma'\sigma} + X_i^{\sigma'\sigma} X_{i+1}^{0\sigma'}) \tag{24}$$

and

$$\left[\sum_j X_j^{0\sigma}, g''\sum_i X_i^{00} X_{i+1}^{00}\right]_- = -g''\sum_i (X_i^{0\sigma} X_{i+1}^{00} + X_i^{00} X_{i+1}^{0\sigma}) \tag{25}$$

and so

$$\left[\sum_j X_j^{0\sigma}, H'\right]_- = (g-g'')\sum_i (X_i^{00} X_{i+1}^{0\sigma} + X_{i+1}^{00} X_i^{0\sigma}) + (g+g')\sum_i (X_i^{\sigma'\sigma} X_{i+1}^{0\sigma'} + X_i^{0\sigma'} X_{i+1}^{\sigma'\sigma}). \tag{26}$$

For (22) to hold we require

$$g = g'' = -g' \tag{27}$$

which we will call the supersymmetric $t-J$ model. In the next section we will show how the resulting Hamiltonian can be interpreted as a generalized permutation operator.

Owing to the predominance of supersymmetry applications in a particle physics context it is often assumed that the supersymmetric algebra contains the Poincaré algebra as a subalgebra (Cornwell 1989, p 79). Consequently the Hamiltonians can be expressed in terms of bilinears in the fermionic generators (e.g. see de Crombrugghe and Rittenberg 1983). The $t-J$ Hamiltonian is not of this kind and the supersymmetry is of a kinematic type. Our use of the term 'supersymmetry' is similar to that adopted quite commonly in, for example, nuclear physics (Iachello 1985).

3. Generalized permutation operator

In terms of the harmonic oscillator representation the hopping term in H_{t-J} is

$$t\sum_i (b_i b_{i+1}^\dagger f_i^{\sigma\dagger} f_{i+1}^{\sigma} + b_{i+1} b_i^\dagger f_{i+1}^{\sigma\dagger} f_i^{\sigma}).$$

As a consequence each site of the lattice is occupied by the boson or a fermion (either an up or down fermion). The constraint on the Hilbert space of no double occupancy is

$$b_i^\dagger b_i + f_{i\sigma}^\dagger f_{i\sigma} = 0. \tag{28}$$

A generalized permutation operator will interchange fermions and bosons (as well as interchange just fermions) with the same amplitude. The relevance of such operators will now be discussed. The permutation aspect of the Heisenberg term has already been discussed and was in fact noticed by Bethe (1931). The hopping term is also a permutation operator but now between bosons and fermions. We will examine this aspect through an example. A lattice with $(N-4)$ up spins, two down spins and two hole has a state $|\psi\rangle$ of the form

$$|\psi\rangle = \sum_{i,j,k,l} a(i,j,k,l)\dots b_i^\dagger \dots b_j^\dagger \dots f_k^{\downarrow\dagger} \dots f_l^{\downarrow\dagger} \dots |0\rangle \tag{29}$$

where ... denote creation operators for up spins. For definiteness we consider the part of $|\psi\rangle$

$$a(i, i+1, k, k+1)\dots b_i^\dagger b_{i+1}^\dagger f_{i+2}^{\uparrow\dagger} \dots f_{k-1}^{\uparrow\dagger} f_k^{\downarrow\dagger} f_{k+1}^{\downarrow\dagger} f_{k+2}^{\uparrow\dagger} \dots |0\rangle$$

and operate on it with

$$t(b_{i+1} b_{i+2}^\dagger f_{i+1}^{\sigma\dagger} f_{i+2}^{\sigma} + b_{i+2} b_{i+1}^\dagger f_{i+2}^{\sigma\dagger} f_{i+1}^{\sigma})$$

a part of H_{t-J}. Now

$$\dots b_{i+1} b_{i+2}^\dagger f_{i+1}^{\uparrow\dagger} f_{i+2}^{\uparrow} b_i^\dagger b_{i+1}^\dagger f_{i+2}^{\uparrow\dagger} \dots |0\rangle = \dots b_{i+1} b_{i+2}^\dagger b_i^\dagger b_{i+1}^\dagger f_{i+1}^{\uparrow\dagger} f_{i+2}^{\uparrow} f_{i+2}^{\uparrow\dagger} \dots |0\rangle \tag{30}$$

and

$$f_{i+1}^{\uparrow\dagger} f_{i+2}^{\uparrow} f_{i+2}^{\uparrow\dagger} = f_{i+1}^{\uparrow\dagger}(1 - f_{i+2}^{\uparrow\dagger} f_{i+2}^{\uparrow}). \tag{31}$$

The last term in (31) when pulled through in (30) gives zero. Consequently (30) becomes

$$\dots b_{i+1} b_{i+2}^\dagger b_i^\dagger b_{i+1}^\dagger f_{i+1}^{\uparrow\dagger} \dots |0\rangle = \dots b_{i+2}^\dagger b_i^\dagger f_{i+1}^{\uparrow\dagger} \dots |0\rangle = \dots b_i^\dagger f_{i+1}^{\uparrow\dagger} b_{i+2}^\dagger \dots |0\rangle. \tag{32}$$

The hole at $(i+1)$ and the up spin at $(i+2)$ have thus been swapped. This is just the effect of a permutation operator which will be denoted by $P_{i+1,i+2}^{(0,\uparrow)}$. The same result is found by examining other cases. H_{t-J} can then be written as

$$H_{t-J} = t\sum_{i,\sigma} P_{i\,i+1}^{(0,\sigma)} + \frac{J}{2}\sum_i (P_{i\,i+1} - P_{i\,i+1}^{(0,0)}) \tag{33}$$

with

$$P^{(0,0)}_{i\,i+1}(\ldots b_i^{\dagger} b_{i+1}^{\dagger} \ldots)|0\rangle = (\ldots b_{i+1}^{\dagger} b_i^{\dagger} \ldots)|0\rangle. \tag{34}$$

The arguments above easily generalize to a situation when there are more 'flavours' of fermions.

The permutation symmetry of the ground state can be determined without a detailed calculation of energies for different Young tableau representations. The argument has been essentially given by Lai and Yang (1971) who restricted themselves to two flavours but their reasoning applies also to the case of more flavours. If there is an odd number N^i of fermions of flavour i ($i = 1, \ldots, m$) and N^0 bosons (with $\Sigma_i N^i + N^0 = N$) then the permutation symmetry of the ground state is given by the Young tableau

$$(m + N^0, m^{N_m - 1}, (m-1)^{N_{m-1} - N_m}, (m-2)^{N_{m-2} - N_{m-1}}, \ldots, 1^{(N_1 - N_2)}). \tag{35}$$

The precise nature of the Hamiltonian played no role in the discussion of Lai and Yang. They dealt with a continuum and allowed double occupation at a site. The continuum aspect was not relevant to their argument while the amplitude for double occupation can be made arbitrarily small by adding an energy penalty term to their Hamiltonian. Consequently the result of (35) is implied by Lai and Yang also for our case. It will be convenient to work with the conjugate Young tableau representation (Andrei *et al* 1983) which is equivalent to a canonical transformation on the variables (Sarkar 1990a).

4. Bethe ansatz

As an example of a simple case away from half-filling let us consider a lattice with $(N-2)$ up spins, one down spin and one hole. Any state $|\psi\rangle$ of this lattice has the form

$$|\psi\rangle = \sum_{x_1, x_2} \alpha(x_1, x_2) f_1^{\uparrow\dagger} f_2^{\uparrow\dagger} \ldots f_{x_1-1}^{\uparrow\dagger} f_{x_1}^{\downarrow\dagger} f_{x_1+1}^{\uparrow\dagger} \ldots b_{x_2}^{\dagger} \ldots f_N^{\uparrow\dagger}|0\rangle. \tag{36}$$

The down spin and hole are located at x_1 and x_2 respectively. We now demand that $|\psi\rangle$ is an energy eigenstate and so

$$H|\psi\rangle = E|\psi\rangle. \tag{37}$$

For x_1 and x_2 far apart

$$E\alpha(x_1, x_2) = \frac{-J}{2}(\alpha(x_1+1, x_2) + \alpha(x_1-1, x_2)) + t(\alpha(x_1, x_2-1) + \alpha(x_1, x_2+1)). \tag{38}$$

The Bethe ansatz is

$$\begin{aligned} \alpha(x_1, x_2) &= \alpha_Q(x_1, x_2) \\ &= A_1(Q)\exp(\mathrm{i}(k_1 x_{Q1} + k_2 x_{Q2})) + A_2(Q)\exp(\mathrm{i}(k_2 x_{Q1} + k_1 x_{Q2})) \end{aligned} \tag{39}$$

where Q is an element of S_2 the permutation group on two objects and defines a sector $x_{Q1} < x_{Q2}$. For brevity

$$Q = \begin{pmatrix} 1 & 2 \\ 1 & 2 \end{pmatrix}$$

will be denoted by 1 and

$$Q=\begin{pmatrix}1 & 2\\ 2 & 1\end{pmatrix}$$

by 2. $A_1(Q)$, $A_2(Q)$, k_1, and k_2 are constants which need to be determined.

Equation (37) implies that

$$\begin{aligned}
&E(A_1(1)\exp[i(k_1x_1+k_2x_2)]+A_2(1)\exp[i(k_2x_1+k_1x_2)])\\
&\quad=-\frac{J}{2}(A_1(1)\exp\{i[k_1(x_1+1)+k_2x_2]\}+A_2(1)\exp\{i[k_2(x_1+1)+k_1x_2]\}\\
&\qquad+A_1(1)\exp\{i[k_1(x_1-1)+k_2x_2]\}+A_2(1)\exp\{i[k_2(x_1-1)+k_1x_2]\})\\
&\qquad+t(A_1(1)\exp\{i[k_1x_1+k_2(x_2-1)]\}+A_2(1)\exp\{i[k_2x_1+k_1(x_2-1)]\}\\
&\qquad+A_1(1)\exp\{i[k_1x_1+k_2(x_2+1)]\}+A_2(1)\exp\{i[k_2x_1+k_1(x_2+1)]\})\\
&\quad=2\left(-\frac{J}{2}\cos k_1+t\cos k_2\right)A_1(1)\exp[i(k_1x_1+k_2x_2)]\\
&\qquad+2\left(-\frac{J}{2}\cos k_2+t\cos k_1\right)A_2(1)\exp[i(k_2x_1+k_1x_2)].
\end{aligned}\tag{40}$$

(In (40) we have ignored an overall constant energy shift.) Hence a necessary condition for an energy eigenstate is

$$-\frac{J}{2t}=1.\tag{41}$$

This is also the supersymmetric condition (27). Henceforth we will choose units and phases so that $t=-1$ and also require (41) to hold. For the term in (36) proportional to

$$f_1^{\uparrow\dagger}f_2^{\uparrow\dagger}b_{x_1}^{\dagger}f_{x_1+1}^{\downarrow\dagger}f_{x_1+2}^{\uparrow\dagger}\cdots f_N^{\uparrow\dagger}|0\rangle$$

the energy eigenstate condition gives

$$\begin{aligned}
&(2(\cos k_1+\cos k_2)+1)\alpha(x_1,x_1+1)\\
&\quad=\alpha(x_1+1,x_1)+\alpha(x_1,x_1+2)+\alpha(x_1-1,x_1+1)
\end{aligned}\tag{42}$$

and

$$\begin{aligned}
&(2(\cos k_1+\cos k_2)+1)\alpha(x_1+1,x_1)\\
&\quad=\alpha(x_1,x_1+1)+\alpha(x_1+1,x_1-1)+\alpha(x_1+2,x_1).
\end{aligned}\tag{43}$$

These two equations lead to

$$A_1(2)=u^{12}A_2(2)+v^{12}A_2(1)\qquad A_1(1)=u^{12}A_2(1)+v^{12}A_2(2)\tag{44}$$

where

$$u^{12}=-\frac{(1+e^{ik_1})(1+e^{ik_2})}{e^{i(k_1+k_2)}+2\,e^{ik_2}+1}\tag{45}$$

and

$$v^{12}=\frac{e^{ik_1}-e^{ik_2}}{e^{i(k_1+k_2)}+2\,e^{ik_2}+1}.\tag{46}$$

We need, of course, to be also able to consider situations other than that of (29). If there are $N^{\downarrow}$ down spins at $x_1, \ldots, x_{N^{\downarrow}}$ and N^0 holes at $x_{N^{\downarrow}+1}, \ldots, x_{N^{\downarrow}+N^0}$ then the Bethe ansatz is

$$\alpha(x_1, \ldots, x_{N^{\downarrow}+N^0}) = \sum_{\substack{P,Q \\ \in S_{N^{\downarrow}+N^0}}}' A_P(Q) \exp\left(i \sum_{j=1}^{N^{\downarrow}+N^0} k_{Pj} x_{Qj}\right) \theta(x_Q) \tag{47}$$

where $\theta(x_Q)$ denotes the region $x_{Q1} < x_{Q2} < \ldots < X_{Q(N^{\downarrow}+N^0)}$.

For the Bethe ansatz solution to work, entities such as u^{12} and v^{12} have to satisfy some identities. However we will not explicitly check these consistency conditions since we have shown elsewhere (Sarkar 1990a) that the supersymmetric $t-J$ model can be mapped onto a model of Lai (1974). The latter model has been shown by Lai to be soluble by the Bethe ansatz.

In order to proceed further, periodic boundary conditions have to be imposed on the wavefunction. This is by now a standard although somewhat complicated procedure (Yang 1967, Sutherland 1975, Lai and Yang 1971) and is known as the generalized Bethe hypothesis. It is discussed in detail by Andrei *et al* (1983) and so we will just give the results of the procedure. Apart from the ks involved in the Bethe ansatz, some auxiliary variables Λ appear which are related to a proper description of the permutation symmetry of $A_P(Q)$. The equations that emerge are

$$e^{ik_jN} = \prod_{\gamma=1}^{N^0} \left(\frac{i(\Lambda_\gamma - \alpha_j) + \frac{1}{2}}{i(\Lambda_\gamma - \alpha_j) - \frac{1}{2}}\right) \prod_{k=1}^{N^{\downarrow}+N^0} \frac{i(\alpha_j - \alpha_k) + 1}{i(\alpha_j - \alpha_k) - 1} \tag{48}$$

and

$$\prod_{j=1}^{N^{\downarrow}+N^0} \left(\frac{i(\Lambda_\delta - \alpha_j) + \frac{1}{2}}{i(\Lambda_\delta - \alpha_j) - \frac{1}{2}}\right) = 1 \tag{49}$$

where $\alpha_j = \frac{1}{2} \tan \frac{1}{2} k_j$. It is customary to take the logarithm of these equations. If

$$\theta(x) = -2 \tan^{-1} x \tag{50}$$

then

$$e^{i\theta} = \frac{1 - ix}{1 + ix}. \tag{51}$$

On taking logarithms of both sides of (51) we get

$$\theta = -i \log\left(\frac{1 - ix}{1 + ix}\right) + 2\pi J \tag{52}$$

where J is an integer. Using these elementary facts, (48) and (49) give

$$Nk_j = 2\pi J_j - \sum_{k=1}^{N^{\downarrow}+N^0} \theta(\alpha_j - \alpha_k) + \sum_{\gamma=1}^{N^0} \theta(2\alpha_j - 2\Lambda_\gamma) \tag{53}$$

$$\sum_{j=1}^{N^{\downarrow}+N^0} \theta(2\Lambda_\gamma^1 - 2\alpha_j) + 2\pi J_\gamma^1 = 0. \tag{54}$$

These equations readily generalize to the m fermion flavour case alluded to earlier. In the generalized Bethe hypothesis if there are m fermion flavours then there are $(m-1)$

flavours of Λ. The resulting equations are

$$Nk_j = 2\pi J_j - \sum_{k=1}^{N-N^1} \theta(\alpha_j - \alpha_k) + \sum_{\gamma=1}^{N-N^1-N^2} \theta(2\alpha_j - 2\Lambda^1_\gamma)$$

$$\sum_{\gamma'=1}^{N-\Sigma^r_{j=1} N^j} \theta(\Lambda^{r-1}_\gamma - \Lambda^{r-1}_{\gamma'})$$

$$= 2\pi J^{r-1}_\gamma + \sum_{\gamma'=1}^{N-\Sigma^{r-1}_{j=1} N^j} \theta(2\Lambda^{r-1}_\gamma - 2\Lambda^{r-2}_{\gamma'})$$

$$+ \sum_{\epsilon=1}^{N-\Sigma^{r+1}_{j=1} N^j} \theta(2\Lambda^{r-1}_\gamma - 2\Lambda^r_\epsilon) \qquad (2 \le r \le m-1) \tag{55}$$

and

$$\sum_{\gamma=1}^{N-\Sigma^m_{j=1} N^j} \theta(2\Lambda^{m-1}_\epsilon - 2\Lambda^{m-2}_\gamma) + 2\pi J^{m-1}_\epsilon = 0.$$

The r on Λ^r is the flavour index and the range of γ is $1 \le \gamma \le (N - \Sigma^{r+1}_{j=1} N^j)$.

We will leave further discussion of the m flavour case for elsewhere. For any lattice of macroscopic size, (53) and (54) are too complicated. Consequently we will follow the customary practice and convert to a set of coupled integral equations. The reasonable assumption is made that

$$\alpha_{j+1} - \alpha_j \sim O\left(\frac{1}{N}\right) \tag{56}$$

and

$$\Lambda_{\gamma+1} - \Lambda_\gamma \sim O\left(\frac{1}{N}\right).$$

From (53)

$$2\pi(J_{j+1} - J_j) = \sum_{k=1}^{N-N^1} (\theta(\alpha_{j+1} - \alpha_k) - \theta(\alpha_j - \alpha_k))$$

$$- \sum_{\gamma=1}^{N-N^1-N^2} (\theta(2\alpha_{j+1} - 2\Lambda^1_\gamma) - \theta(2\alpha_j - 2\Lambda^1_\gamma))$$

$$+ 2N[\tan^{-1}(2\alpha_{j+1}) - \tan^{-1}(2\alpha_j)]. \tag{57}$$

The general experience with the Bethe ansatz shows that for the ground state $J_{j+1} - J_j = 1$. Excited states appear when there are j' such that $J_{j'+1} - J_{j'} = 2$. J' is called a hole. Hence

$$2\pi\left(1 - \sum_{j'} \delta_{jj'}\right) + 2\pi 2 \sum_{j'} \delta_{jj'}$$

$$= \sum_{k=1}^{N-N^1} \theta'(\alpha_j - \alpha_k)(\alpha_{j+1} - \alpha_j) \sum_{\gamma=1}^{N-N^1-N^2} \theta'(2\alpha_j - 2\Lambda^1_\gamma) 2(\alpha_{j+1} - \alpha_j)$$

$$+ 2N \frac{1}{1 + 4\alpha_j^2} 2(\alpha_{j+1} - \alpha_j). \tag{58}$$

If N_h is the number of holes then

$$\sum_{j=1}^{N_h} \delta(\alpha - \alpha^h_j) + \rho_1(\alpha) = \frac{1}{N} \sum_{k=1}^{N-N_1} \theta'(\alpha - \alpha_k) - \frac{2}{N} \sum_{\gamma=1}^{N-N^1-N^2} \theta'(2\alpha - 2\Lambda^1_\gamma) + \frac{4}{1 + 4\alpha^2} \tag{59}$$

where

$$\rho_1(\alpha_j) = \frac{2\pi}{N} \frac{1}{(\alpha_{j+1} - \alpha_j)}. \tag{60}$$

In the $N \to \infty$ limit ρ_1 is a distribution and the sums on the right-hand side of (59) become integrals, and consequently

$$\rho_1(\alpha)+\frac{2\pi}{N}\sum_{j=1}^{N_h}\delta(\alpha-\alpha_j^h)$$
$$=-\frac{1}{\pi}\int\frac{1}{1+(\alpha-\alpha')^2}\rho_1(\alpha')\,d\alpha'$$
$$+\frac{2}{\pi}\int\frac{1}{1+4(\alpha-\Lambda^1)^2}\rho_2(\Lambda^1)\,d\Lambda^1+\frac{4}{1+4\alpha^2}. \tag{61}$$

These integrals will have limits which we will take to be $[-\alpha_0, \alpha_0]$ and $[-\Lambda_0^1, \Lambda_0^1]$ and we will see how these are determined by the concentrations of down spins and holes. Similarly from (54)

$$2\pi(J_{\gamma+1}^1-J_\gamma^1)=-\sum_{j=1}^{N-N^1}(\theta(2\Lambda_{\gamma+1}^1-2\alpha_j)-\theta(2\Lambda_\gamma^1-2\alpha_j)). \tag{62}$$

If Λ_γ^{1h} are the hole values of Λ^1, then

$$\rho_2(\Lambda^1)+\frac{2\pi}{N}\sum_{\gamma=1}^{N_h^1}\delta(\Lambda^1-\Lambda_\gamma^{1h})=\frac{2}{\pi}\int_{-\alpha_0}^{\alpha_0}d\alpha\frac{\rho_1(\alpha)}{1+4(\Lambda^1-\alpha)^2} \tag{63}$$

where $\rho_2(\Lambda^1)$ is the continuum limit of $(2\pi/N)[1/(\Lambda_{\gamma+1}^1-\Lambda_\gamma^1)]$.

We recall that the index j in α_j lies in the interval $[1, N-N^1]$, and the index γ in Λ_γ^1 lies in the interval $[1, N-N^1-N^2]$. Hence

$$\int_{-\alpha_0}^{\alpha_0}d\alpha\,\rho_1(\alpha)=\lim_{N\to\infty}\sum_j(\alpha_{j+1}-\alpha_j)\frac{2\pi}{N}\frac{1}{(\alpha_{j+1}-\alpha_j)}=\frac{2\pi(N-N^1)}{N} \tag{64}$$

and

$$\int_{-\Lambda_0^1}^{\Lambda_0^1}d\Lambda^1\,\rho_2(\Lambda^1)=\lim_{N\to\infty}\sum_\gamma(\Lambda_{\gamma+1}^1-\Lambda_\gamma^1)\frac{2\pi}{N}\frac{1}{\Lambda_{\gamma+1}^1-\Lambda_\gamma^1}=\frac{2\pi(N-N^1-N^2)}{N}. \tag{65}$$

The energy E is

$$E=2(2N^1+N^2)-2N-2\sum_j\cos k_j. \tag{66}$$

Since

$$\cos k_j=\frac{1-\tan^2\frac{1}{2}k_j}{1+\tan^2\frac{1}{2}k_j}=\frac{1-4\alpha_j^2}{1+4\alpha_j^2} \tag{67}$$

then

$$\frac{E}{N}=2-\frac{1}{\pi}\int_{-\Lambda_0^1}^{\Lambda_0^1}d\Lambda^1\,\rho_2(\Lambda^1)-\frac{1}{2\pi}\int_{-\alpha_0}^{\alpha_0}d\alpha\frac{4\rho_1(\alpha)}{1+4\alpha^2}. \tag{68}$$

From (64) and (65) it is clear that

$$\Lambda_0^1=0 \qquad \text{and} \qquad \alpha_0=\infty \tag{69}$$

corresponds to the half-filling case where the model reduces to the Heisenberg model. In general the integral equations (61) and (63) need to be solved numerically or through approximate application of Wiener–Hopf techniques (Andrei *et al* 1983). We will give an analytic treatment valid near half-filling, i.e. α_0 very large and Λ_0^1 very small. This will enable us to obtain limited information such as the gaplessness of the ground state. Let us first check what sort of half-filling is implied by (69). Equation (61) becomes

$$\frac{4}{1+4\alpha^2}=\rho_1(\alpha)+\frac{1}{2\pi}\int_{-\infty}^{\infty}\frac{2\rho_1(\alpha')}{1+(\alpha-\alpha')^2}\,d\alpha'. \tag{70}$$

On writing

$$\rho_1(\alpha)=\int_{-\infty}^{\infty}\frac{dp}{2\pi}\,e^{ip\alpha}\,\tilde{\rho}_1(p) \tag{71}$$

and on noting that

$$\frac{1}{\alpha^2+1}=\pi\int\frac{dp}{2\pi}\,e^{ip\alpha}\,e^{-|p|} \tag{72}$$

it is found that

$$\tilde{\rho}_1(p)=2\pi\frac{e^{-(1/2)|p|}}{1+e^{-|p|}}=\frac{\pi}{\cosh\frac{1}{2}|p|}. \tag{73}$$

Since

$$2\pi\frac{(N-N^1)}{N}=\int_{-\infty}^{\infty}\rho_1(\alpha)\,d\alpha=\tilde{\rho}_1(0)=\pi \tag{74}$$

$$N^1/N=\tfrac{1}{2} \tag{75}$$

and so there are an equal number of up and down spins.

We shall consider the effect of introducing a small macroscopic number of real holes (as opposed to Bethe ansatz holes). Consequently Λ_0^1 will be small. Let us write

$$\rho_1(\alpha)=\rho_1^{(0)}(\alpha)+\rho_1^{(1)}(\alpha)+\ldots \tag{76}$$

and

$$\rho_2(\Lambda^1)=\rho_2^{(0)}(\Lambda^1)+\rho_2^{(1)}(\Lambda^1)+\ldots \tag{77}$$

$$\rho_2^{(0)}(\Lambda^1)=\frac{2}{\pi}\int_{-\infty}^{\infty}d\alpha'\,\frac{\rho_1^{(0)}(\alpha')}{1+4(\Lambda^1-\alpha')^2} \tag{78}$$

where $\rho_1^{(1)}(\alpha)$ and $\rho_2^{(1)}(\Lambda^1)$ are small corrections due to doping.

From (61) in the absence of Bethe ansatz holes we obtain

$$\rho_1^{(1)}(\alpha)\simeq\frac{1}{\pi}\int_{\alpha_0}^{\infty}d\alpha'\,\frac{1}{1+(\alpha-\alpha')^2}\,\rho_1^{(0)}(\alpha')+\frac{1}{\pi}\int_{-\infty}^{-\alpha_0}d\alpha'\,\frac{1}{1+(\alpha-\alpha')^2}\,\rho_1^{(0)}(\alpha')$$
$$-\frac{1}{\pi}\int_{-\alpha_0}^{\alpha_0}d\alpha'\,\frac{1}{1+(\alpha-\alpha')^2}\,\rho_1^{(1)}(\alpha')+\frac{4}{\pi}\,\Lambda_0^1\rho_2^{(0)}(0)\,\frac{1}{1+4\alpha^2}. \tag{79}$$

($\rho_1^{(1)}(\alpha)$ actually also depends implicitly on α_0 and Λ_0^1 and more properly should be written as $\rho_1^{(1)}(\alpha,\alpha_0,\Lambda_0^1)$.)

Since $\rho_1^{(1)}(\alpha')$ is small it is a good approximation to write

$$\int_{-\alpha_0}^{\alpha_0}d\alpha'\,\frac{1}{1+(\alpha-\alpha')^2}\,\rho_1^{(1)}(\alpha')\sim\int_{-\infty}^{\infty}d\alpha'\,\frac{1}{1+(\alpha-\alpha')^2}\,\rho_1^{(1)}(\alpha'). \tag{80}$$

(Similarly from (63) for completeness we note

$$\rho_2^{(1)}(\Lambda^1) \simeq -\frac{2}{\pi}\int_{\alpha_0}^{\infty} d\alpha \frac{\rho_1^{(0)}(\alpha)}{1+4(\Lambda^1-\alpha)^2} - \frac{2}{\pi}\int_{-\infty}^{-\alpha_0} d\alpha' \frac{1}{1+4(\Lambda^1-\alpha)^2} + \frac{2}{\pi}\int_{-\infty}^{\infty} d\alpha \frac{\rho_1^{(1)}(\alpha)}{1+4(\Lambda^1-\alpha)^2} \tag{81}$$

although we will not need the explicit form of $\rho_2^{(1)}(\Lambda^1)$ for our first-order calculation.)

On solving (79) by Fourier transformation we find

$$\rho_1^{(1)}(\alpha) = \frac{2}{\pi} e^{-2\pi\alpha_0}\left(\frac{1}{1+(\alpha_0-\alpha)^2} + \frac{1}{1+(\alpha_0+\alpha)^2}\right) + \frac{4\Lambda_0^1}{\pi} \rho_2^{(0)}(0) \frac{1}{1+4\alpha^2}. \tag{82}$$

The magnetization M is

$$M = \tfrac{1}{2}(N^1 - N^2) = \tfrac{1}{2}(N - 2(N - N^1) + (N - N^1 - N^2)) \tag{83}$$

and on using (82) we have

$$\frac{M}{N} = \frac{1}{2}\left(\frac{e^{-2\pi\alpha_0}}{\pi}\left(3 + \frac{1}{\pi\alpha_0}\right) + \frac{4}{\pi^2} e^{-\pi\alpha_0} \Lambda_0^1 \rho_2^{(0)}(0)\right). \tag{84}$$

However α_0 is a function of Λ_0, i.e. given a certain doping level the spins align themselves in such a way so as to minimize the energy. We therefore need to calculate the energy. From (68) we have

$$\varepsilon \equiv \frac{E}{N} \simeq 2 - \frac{2}{\pi}\rho_2^{(0)}(0)\Lambda_0^1 - \frac{1}{2\pi}\int_{-\alpha_0}^{\alpha_0} d\alpha \frac{4\rho_1^{(0)}(\alpha)}{1+4\alpha^2} - \frac{1}{2\pi}\int_{-\alpha_0}^{\alpha_0} d\alpha \frac{4\rho_1^{(1)}(\alpha)}{1+4\alpha^2} \tag{85}$$

and so

$$\left.\frac{\partial\varepsilon}{\partial\alpha_0}\right|_{\Lambda_0^1} = -\frac{2}{\pi}\left(\frac{\rho_1^{(0)}(\alpha_0) + \rho_1^{(0)}(-\alpha_0) + \rho_1^{(1)}(\alpha_0, \alpha_0, \Lambda_0^1) + \rho_1^{(1)}(-\alpha_0, \alpha_0, \Lambda_0^1)}{1+4\alpha_0^2}\right) - \frac{2}{\pi}\int_{-\alpha_0}^{\alpha_0} d\alpha \frac{1}{1+4\alpha^2}\frac{\partial}{\partial\alpha_0}\rho_1^{(1)}(\alpha, \alpha_0, \Lambda_0^1). \tag{86}$$

After a certain amount of analysis it is possible to show that

$$\frac{\partial\varepsilon}{\partial\alpha_0} = -\frac{1}{\pi\alpha^2}\left(\frac{2\pi}{\cosh(2\pi\alpha_0)} + \frac{4}{\pi} e^{-2\pi\alpha_0}(f(0) + f(2\alpha_0)) + \Lambda_0^1\rho_2^{(0)}(0)\right) - \frac{8}{\pi} e^{-2\pi\alpha_0}(-2\pi g(\alpha_0) + g'(\alpha_0)) \tag{87}$$

where

$$f(\alpha) = \tfrac{1}{2}\sum_{r=1}^{\infty}(-1)^{r+1}\frac{r}{r^2+\alpha^2} \tag{88}$$

and

$$g(\alpha) = \tfrac{1}{2}\sum_{r=1}^{\infty}(-1)^{r+1}\frac{r+\frac{1}{2}}{(r+\frac{1}{2})^2+\alpha^2}. \tag{89}$$

Using asymptotic estimates for f and g we can deduce

$$\frac{\partial\varepsilon}{\partial\alpha_0} < 0 \tag{90}$$

and so the minimum of energy is found for $\alpha_0=\infty$. Equation (84) then implies that

$$M=0 \tag{91}$$

in the ground state.

In order to consider excited states we have to examine the effect of Bethe ansatz holes (Andrei *et al* 1983). Now we let

$$\rho_1(\alpha)\rightarrow\rho_1(\alpha)+\Delta\rho_1(\alpha) \tag{92}$$

and

$$\rho_2(\Lambda^1)\rightarrow\rho_2(\Lambda^1)+\Delta\rho_2(\Lambda^1) \tag{93}$$

where $\Delta\rho_1(\alpha)$ and $\Delta\rho_2(\Lambda^1)$ are changes in ρ_1 and ρ_2 due to the presence of Bethe ansatz holes. Clearly from (61) and (63)

$$\Delta\rho_1(\alpha)+\frac{2\pi}{N}\sum_{j=1}^{N_h}\delta(\alpha-\alpha_j^h)$$
$$=-\frac{1}{\pi}\int_{-\infty}^{\infty}\frac{\Delta\rho_1(\alpha')}{1+(\alpha-\alpha')^2}\,d\alpha'+\frac{2}{\pi}\int_{-\Lambda_0^1}^{\Lambda_0^1}\frac{\Delta\rho_2(\Lambda^1)}{1+4(\alpha-\Lambda^1)^2}\,d\Lambda^1 \tag{94}$$

and

$$\Delta\rho_2(\Lambda^1)+\frac{2\pi}{N}\sum_{\gamma=1}^{N_h^1}\delta(\Lambda^1-\Lambda_\gamma^{1h})=\frac{2}{\pi}\int_{-\infty}^{\infty}d\alpha\,\frac{\Delta\rho_1(\alpha)}{1+4(\Lambda^1-\alpha)^2}. \tag{95}$$

These equations can be solved by Fourier transforming. We obtain

$$\Delta\tilde{\rho}_1(p)=\frac{\Lambda_0^1\Delta\rho_2(0)}{\cosh\frac{1}{2}p}-\frac{2\pi}{N}\frac{\sum_{j=1}^{N_h}e^{ip\alpha_j^h}}{1+e^{-|p|}} \tag{96}$$

where $\Delta\tilde{\rho}_1$ is the Fourier transform of $\Delta\rho_1$. Similarly

$$\Delta\tilde{\rho}_2(p)=e^{-(1/2)|p|}\Delta\tilde{\rho}_1(p)-\frac{2\pi}{N}\sum_{\gamma=1}^{N_h^1}e^{-ip\Lambda_\gamma^{1h}}. \tag{97}$$

Consequently, since

$$\Delta\rho_2(0)=\int\frac{dp}{2\pi}\Delta\tilde{\rho}_2(p) \tag{98}$$

we have

$$\Delta\rho_2(0)=-\frac{2\pi}{N}\sum_{j=1}^{N_h}\frac{1}{\cosh(2\pi\alpha_j^h)}-\frac{2\pi}{N}\sum_{\gamma=1}^{N_h^1}\delta(\Lambda_\gamma^{1h})\left(1-\frac{2}{\pi}(\log 2)\Lambda_0^1\right)^{-1}. \tag{99}$$

We can now calculate the change in energy ΔE due to the Bethe ansatz holes. From (68), (85) and (96) we have

$$\frac{\Delta E}{N}\simeq\frac{2\pi}{N}\left[\left(\frac{2\log 2}{\pi}\Lambda_0^1+1\right)\sum_{j=1}^{N_h}\frac{1}{\cosh(2\pi\alpha_j^h)}+\frac{2\log 2}{\pi}\Lambda_0^1\sum_{\gamma=1}^{N_h^1}\frac{\delta(\Lambda_\gamma^{1h})}{N}\right]-\frac{2}{\pi}\Lambda_0^1\Delta\rho_2(0). \tag{100}$$

The change in the magnetization ΔM by definition is

$$\Delta M=\tfrac{1}{2}(-2\Delta(N-N^1)+\Delta(N-N^1-N^2)) \tag{101}$$

which on using

$$\frac{2\pi}{N}\Delta(N-N^1)=\int_{-\infty}^{\infty}\Delta\rho_1(\alpha)\,\mathrm{d}\alpha \tag{102}$$

and

$$\frac{2\pi}{N}\Delta(N-N^1-N^2)\simeq 2\Lambda_0^1\Delta\rho_2(0) \tag{103}$$

gives

$$\Delta M=\tfrac{1}{2}N_{\mathrm{h}} \tag{104}$$

and

$$\Delta(N-N^1)=-\tfrac{1}{2}N_{\mathrm{h}}-\Lambda_0^1\left(\sum_{j=1}^{N_{\mathrm{h}}}\frac{1}{\cosh(2\pi\alpha_j^{\mathrm{h}})}+\sum_{\gamma=1}^{N_{\mathrm{h}}^1}\delta(\Lambda_\gamma^{1\mathrm{h}})\right). \tag{105}$$

The gap above the ground state in (100) is zero since $[\cosh(2\pi\alpha_j^{\mathrm{h}})]^{-1}$ can be chosen to be arbitrarily small (or α_j^{h} arbitrarily large) and Λ_{h}^1 taken to be non-zero. For these same conditions $\Delta(N-N^1)$ is $-\frac{1}{2}N_{\mathrm{h}}$ which has to be an integer. Consequently the least complicated zero-energy excitation that has been constructed has angular momentum 1. We have thus obtained valuable information from (61) and (63) with our simple approximation. Our analysis of the $t-J$ model bears throughout a strong resemblance to that for the Heisenberg model. Many generalizations of the latter are possible but for both physical and mathematical reasons the supersymmetric generalization that we have considered is a particularly non-trivial one.

Acknowledgments

I would like to thank A C Hewson, J H Jefferson and N Andrei for helpful discussions, and NATO for a collaborative research grant.

References

Anderson P W 1988 *Frontiers and Borderlines in Many Particle Physics* ed R A Broglia and J R Schrieffer (Amsterdam: North-Holland)
Andrei N, Furuya K and Lowenstein J H 1983 *Rev. Mod. Phys.* **55** 331
Bars I and Günaydin M 1983 *Commun. Math. Phys.* **91** 31
Bethe H 1931 *Z. Phys.* **71** 205
Cornwell J F 1989 Supersymmetries and infinite dimensional algebras *Group Theory in Physics* vol 3 (New York: Academic)
de Crombrugghe M and Rittenberg V 1983 *Ann. Phys.* **151** 99
Fukuyama H, Maekawa S and Malozemoff A P (ed) 1989 *Strong Correlation and Superconductivity* (Berlin: Springer)
Hubbard J 1963 *Proc. R. Soc.* A **276** 238
Iachello F 1985 *Physica* **15D** 85
Jefferson J H 1990 Derivation of the $t-J$ model for high temperature superconductivity *Preprint* RSRE
Lai C K 1974 *J. Math. Phys.* **15** 1675
Lai C K and Yang C N 1971 *Phys. Rev.* A **3** 393
Lieb E H and Wu F Y 1968 *Phys. Rev. Lett.* **20** 1445
Lindgren I and Morrison J 1986 *Atomic Many-Body Theory* (Berlin: Springer)

Pike E R, Jefferson J H and Sarkar S 1991 *An Introduction to Electron Correlations and High Temperature Superconductivity* (Bristol: Hilger) to be published
Sarkar S 1990a *J. Phys. A: Math. Gen.* **23** L409
—— 1990b Supercoherent states for the $t-J$ model *Preprint* RSRE
Sutherland B 1975 *Phys. Rev.* B **12** 3795
Wiegmann P B 1988 *Phys. Rev. Lett.* **60** 821
Yang C N 1967 *Phys. Rev. Lett.* **19** 1312
Zhang F C and Rice T M 1988 *Phys. Rev.* B **37** 3759

J. Phys.: Condens. Matter 3 (1991) 5983–6008. Printed in the UK

Luttinger liquid properties of highly correlated electron systems in one dimension

Norio Kawakami and Sung-Kil Yang†
Yukawa Institute for Theoretical Physics, Kyoto University, Kyoto 606, Japan

Received 20 August 1990, in final form 28 January 1991

Abstract. An exact description is given of the long-distance behaviour of the one-dimensional t–J model at $t = J$. We employ the Bethe *ansatz* method and the finite-size scaling technique in conformal field theory. The charge and spin degrees of freedom are separated, and described by two independent $c = 1$ conformal theories. The critical exponents for the charge, spin, electron and superconducting correlation functions are obtained for arbitrary band filling. We then make detailed comparison of the t–J model with the repulsive Hubbard model with emphasis on their Luttinger liquid properties. Analysing the electron filling dependence we observe the enhancement of the superconducting correlations compared with the highly correlated Hubbard model. The effect of the external magnetic field at and near half-filling is also discussed.

1. Introduction

Almost ten years ago Haldane introduced the concept of Luttinger liquids that is valid in understanding the low-energy behaviour of a large class of one-dimensional (1D) conducting fermion systems [1,2]. The universal role of Fermi liquids in higher dimensions is thus replaced by Luttinger liquids in one dimension. These two types of quantum fluids have quite distinct features. Among others, in ordinary Fermi liquid theory, well-defined propagation of electron quasiparticles implies a finite jump discontinuity in the momentum distribution function at the Fermi momentum. This should be contrasted with the power-law singularity near the Fermi point in Luttinger liquids, which corresponds to the soliton-like excitation instead of the quasiparticle excitation [3].

Recent studies of high-T_c superconductivity have renewed interest in low-dimensional electron systems. In superconducting compounds the quantum fluctuation inherent in low dimensions is believed to play a crucial role in addition to the strong correlation effect near the insulating phase [4]. To find an appropriate model of high-T_c superconductors it is of particular importance to clarify if non-Fermi liquid behaviour appears in the normal state of low-dimensional highly correlated systems. A fundamental model Hamiltonian to study such correlated systems may be provided by the Hubbard model, or more simplified t–J model. In one dimension these systems are the simplest examples which have been expected to possess a non-Fermi liquid nature. Very recently numerical computations to resolve this issue in the highly correlated Hubbard chain have been done by Sorella *et al* [5,6], Imada

† Present address: National Laboratory for High Energy Physics (KEK), Ibaraki 305, Japan.

0953-8984/91/325983+26$03.50

and Hatsugai [7], and Ogata and Shiba [8]. Their works motivated us to find exact correlation exponents in 1D correlated systems and to clarify their Luttinger liquid nature.

In this paper we will describe exactly the long-distance behaviour of various correlation functions in the 1D t–J model at $t = J$. In addition to a full exposition of the results announced in a previous communication [9] an analysis of the magnetic field effect is also reported. The results will be compared in detail with the properties of the 1D Hubbard model whose correlation exponents have also been obtained quite recently by Schulz [10], Kawakami and Yang [11], and Frahm and Korepin [12]. bosonization formulae for the Hubbard model are given by Affleck [13, 14]. These models have been known to be exactly solved by the Bethe ansatz for arbitrary electron filling [15, 16]. In the Bethe *ansatz* approach it is a formidable task to deal with correlation functions. However, recent developments in two-dimensional conformal field theory have made it possible to calculate the correlation exponents [17]. The point is that under a conformal mapping the scaling operators and the eigenstates of the transfer matrix on a finite periodic strip have a one-to-one correspondence [18]. Consequently the critical exponents are obtained if one knows the gap due to the finite-size effect in the spectrum of the Hamiltonian at criticality. On the other hand, computation of the energy gap is the most tractable problem in the Bethe ansatz, and hence we are able to compute exactly various correlation exponents based on the finite-size scaling analysis [19–23].

Interacting 1D quantum systems may carry several low-energy excitations with linear dispersion relations, but with different Fermi velocities. Hence the systems will not be Lorentz invariant. When the motions of these excitations are decoupled, however, we can still apply the conformal theory technique [22, 23]. This is indeed the case for the 1D t–J model and the Hubbard model, where the charge and spin degrees of freedom are separated in the continuum limit, as will be seen. Consequently the charge fluctuation is described by a $c = 1$ conformal theory with continuously varying exponents as functions of the electron filling. Here c is the central charge of the Virasoro algebra. The spin fluctuation belongs to the universality class of the antiferromagnetic spin-$\frac{1}{2}$ Heisenberg chain irrespective of the electron filling. This class is a well known $c = 1$ SU(2) Kac–Moody theory.

In section 2 we recapitulate the Bethe *ansatz* solutions to the 1D t–J model at $t = J$ and compute the finite-size corrections in the energy spectrum. The long-distance properties of the charge, spin, electron and superconducting correlation functions for arbitrary band filling are described in section 3. The magnetic field dependence of correlation exponents at and near half-filling is also studied. In section 4 we first review the properties of Luttinger liquids in the light of our result for the t–J model, and then make the comparison with the Hubbard model. The relationship between the critical exponents and the bulk quantities is also discussed. The final section is devoted to our conclusions. In appendices A and B we summarize some technical details.

2. Finite-size scaling behaviour of the energy spectrum

The 1D t–J model consists of spin-$\frac{1}{2}$ electrons hopping around nearest-neighbour lattice sites with the hopping matrix element $-t < 0$. We assume there is no double-occupancy of every site, reflecting a large on-site Coulomb repulsion. Furthermore the motion of highly correlated electrons (or holes) is supposed to be strongly affected by the spin fluctuation through the antiferromagnetic coupling $J > 0$. The Hamiltonian is then given by [4]

$$\mathcal{H} = -t\sum_{i,\sigma}(c^\dagger_{i\sigma}c_{i+1\sigma} + c^\dagger_{i+1\sigma}c_{i\sigma}) + 2J\sum_i(S_i \cdot S_{i+1} - \tfrac{1}{4}n_i n_{i+1})$$
$$-\mu\sum_i n_i - \frac{1}{2}H\sum_i(n_{i\uparrow} - n_{i\downarrow}) \tag{2.1}$$

where $c_{i\sigma}$ ($\sigma =\uparrow$ or $\downarrow$) is the spin-σ electron annihilation operator at the ith site, $S_i = c^\dagger_{i\sigma}S_{\sigma\sigma'}c_{i\sigma'}$ with the spin-$\frac{1}{2}$ matrix $\mathbf{S}$, the number operator $n_{i\sigma} = c^\dagger_{i\sigma}c_{i\sigma}$, $n_i = n_{i\uparrow} + n_{i\downarrow}$, and μ and H are the chemical potential and the external magnetic field, respectively.

As is well known, this Hamiltonian is formally obtained by the canonical transformation from the Hubbard model, but with the limitation $J \ll t$. In (2.1), however, one can regard t and J as free parameters. Therefore the model we shall treat here ($t = J$) has an extremely large exchange coupling compared with the strong correlation limit of the Hubbard model. The relevance of such a model to the high-T_c superconductivity was first demonstrated by Zhang and Rice [24]. Subsequently many attempts have been made to clarify the nature of the t–J model, in particular laying stress upon the competition between magnetism and superconductivity.

Schlottmann found that the 1D t–J model (2.1) can be solved by the Bethe ansatz for the special case of $t = J$ [15]. At this integrable point the model is mapped onto the multicomponent quantum lattice gas whose exact solution was obtained by Sutherland [25]. The diagonalization is performed in two steps. First we seek for the wave function as a superposition of the plane waves characterized by the electron momenta p_j ($j = 1 \sim N_c$). Here we consider a 1D lattice of even N sites with N_c electrons among which M electrons are spin down. The complete integrability is then ensured by the factorization of the multiparticle scattering matrix (Yang–Baxter relation). On applying periodic boundary conditions we reduce the problem to the ancillary one in spin space. This problem can be solved by the generalized Bethe ansatz by introducing the spin rapidity Λ_α ($\alpha = 1 \sim M$) related to the internal degrees of freedom. The resulting Bethe–Yang transcendental equations are written in terms of the rapidities $k_j = \frac{1}{2}\cot(p_j/2)$ and Λ_α [25, 15]

$$\left(\frac{k_j + \mathrm{i}/2}{k_j - \mathrm{i}/2}\right)^N = \prod_{\beta=1}^{M}\frac{k_j - \Lambda_\beta + \mathrm{i}/2}{k_j - \Lambda_\beta - \mathrm{i}/2} \qquad j = 1, \ldots, N_c$$
$$\prod_{j=1}^{N_c}\frac{\Lambda_\alpha - k_j + \mathrm{i}/2}{\Lambda_\alpha - k_j - \mathrm{i}/2} = -\prod_{\beta=1}^{M}\frac{\Lambda_\alpha - \Lambda_\beta + \mathrm{i}}{\Lambda_\alpha - \Lambda_\beta - \mathrm{i}} \qquad \alpha = 1, \ldots, M. \tag{2.2}$$

For convenience we will set $t = J = 1$ from this point on.

The set of rapidities $\{k_j\}$ contains complex $k^\pm_\alpha$ of spin paired electrons ($\alpha = 1 \sim M$), where $k^\pm_\alpha$ are determined by real (down-) spin rapidities Λ_α through $k^\pm_\alpha = \Lambda_\alpha \pm \mathrm{i}/2$ [15]. At first sight the complex solutions $k^\pm_\alpha$ seem to generate the charge excitation gap as in the attractive Hubbard model. It turns out, however, that they describe the massless charge excitation except for the half-filled band, where

$N_c = N$ [26]. We note that similar observation was first made in the singlet ground state of the Anderson model for the Kondo problem [27]. The real solutions k_j describe the spin excitation at zero temperature. Other string solutions for excited states are not necessary for the present investigation.

Substituting the complex solution into (2.2) and taking the logarithm we obtain [15]

$$2N\tan^{-1}(2k_j) = 2\pi I_j + 2\sum_{\beta=1}^{M}\tan^{-1}(2(k_j - \Lambda_\beta)) \qquad j = 1,\ldots,N_c - 2M \tag{2.3}$$

$$2N\tan^{-1}(\Lambda_\alpha) = 2\pi J_\alpha + 2\sum_{j=1}^{N_c-2M}\tan^{-1}(2(\Lambda_\alpha - k_j)) + 2\sum_{\beta=1}^{M}\tan^{-1}(\Lambda_\alpha - \Lambda_\beta) \qquad \alpha = 1,\ldots,M \tag{2.4}$$

where

$$I_j = \frac{M}{2} \bmod 1 \qquad J_\alpha = \frac{N_c + M + 1}{2} \bmod 1\,. \tag{2.5}$$

The energy and the momentum are given by

$$\begin{aligned} E &= -2\sum_{j=1}^{N_c}\cos p_j - \mu N_c + H\left(M - \frac{N_c}{2}\right) \\ &= -2N_c + 2\sum_{j=1}^{N_c-2M}\frac{1/2}{k_j^2 + 1/4} \\ &\quad + 2\sum_{\alpha=1}^{M}\frac{1}{\Lambda_\alpha^2 + 1} - \mu N_c + H\left(M - \frac{N_c}{2}\right) \end{aligned} \tag{2.6}$$

$$P = \sum_{j=1}^{N_c} p_j = -\frac{2\pi}{N}\left(\sum_{j=1}^{N_c-2M} I_j + \sum_{\alpha=1}^{M} J_\alpha\right). \tag{2.7}$$

It is convenient to introduce

$$z_{s,N}(k) = \frac{1}{2\pi}\left(2\tan^{-1}(2k) - \frac{1}{N}\sum_{\beta=1}^{M} 2\tan^{-1}(2(k - \Lambda_\beta))\right) \tag{2.8}$$

$$z_{c,N}(\Lambda) = \frac{1}{2\pi}\left(2\tan^{-1}(\Lambda) - \frac{1}{N}\sum_{j=1}^{N_c-2M} 2\tan^{-1}(2(\Lambda - k_j)) - \frac{1}{N}\sum_{\beta=1}^{M} 2\tan^{-1}(\Lambda - \Lambda_\beta)\right) \tag{2.9}$$

$$\rho_{s,N}(k) = \frac{\partial z_{s,N}(k)}{\partial k} \qquad \rho_{c,N}(\Lambda) = \frac{\partial z_{c,N}(\Lambda)}{\partial \Lambda} \tag{2.10}$$

so that

$$z_{s,N}(k_j) = \frac{I_j}{N} \qquad z_{c,N}(\Lambda_\alpha) = \frac{J_\alpha}{N}\,. \tag{2.11}$$

Inspecting (2.11) one finds that the real solutions $\{k_j\}$ and $\{\Lambda_\alpha\}$ distribute over the regions $k < B^-$ and $k > B^+$, $\Lambda < Q^-$ and $\Lambda > Q^+$, respectively. Correspondingly the distributions of the quantum numbers I_j and J_α become $I < I^-$ and $I > I^+$, $J < J^-$ and $J > J^+$, where

$$z_{s,N}(B^\pm) = \frac{I^\pm}{N} \qquad z_{c,N}(Q^\pm) = \frac{J^\pm}{N} \tag{2.12}$$

and

$$\begin{aligned} I^+ - I^- &= N - N_s \qquad & I^+ + I^- &= 2D_s \\ J^+ - J^- &= N - N_c \qquad & J^+ + J^- &= 2D_c\,. \end{aligned} \tag{2.13}$$

Here $N_s = N_c - M$ is the number of up spins and D_s (or D_c) denotes the number of particles which transfer from a Fermi level of the spinon (or holon) to the other Fermi level.

2.1. *Corrections to the ground-state energy*

We now take the large-N limit while keeping the terms which scale as $1/N$ in the energy spectrum [19]. First from (2.10) we get

$$\rho_{s,N}(k) = \frac{1}{2\pi}\Big(T_{cs}(k) - \frac{1}{N}\sum_{\beta=1}^{M} T_{sc}(k-\Lambda_\beta)\Big) \tag{2.14}$$

$$\rho_{c,N}(\Lambda) = \frac{1}{2\pi}\Big(T_{cc}(\Lambda) - \frac{1}{N}\sum_{j=1}^{N_c-2M} T_{cs}(\Lambda - k_j) - \frac{1}{N}\sum_{\beta=1}^{M} T_{cc}(\Lambda-\Lambda_\beta)\Big) \tag{2.15}$$

where

$$T_{sc}(x) = T_{cs}(x) = \frac{1}{x^2+1/4} \qquad T_{cc}(x) = \frac{2}{x^2+1} \qquad T_{ss}(x) \equiv 0\,. \tag{2.16}$$

Using the Euler–Maclaurin formula

$$\frac{1}{N}\sum_{n=n_1}^{n_2} f\Big(\frac{n}{N}\Big) = \int_{(n_1-1/2)/N}^{(n_2+1/2)/N} f(x)\,\mathrm{d}x - \frac{1}{24N^2}\Big(f'(\frac{n_2+1/2}{N}) - f'(\frac{n_1-1/2}{N})\Big) \tag{2.17}$$

we obtain from (2.14) and (2.15)

$$\begin{aligned} \rho_{\alpha,N}(\lambda_\alpha) = {} & \frac{1}{2\pi}\tilde{a}_\alpha(\lambda_\alpha) + \frac{1}{24N^2}\sum_\beta\Big(\frac{T'_{\alpha\beta}(\lambda_\alpha - q_\beta^+)}{2\pi\rho_{\beta,N}(q_\beta^+)} - \frac{T'_{\alpha\beta}(\lambda_\alpha - q_\beta^-)}{2\pi\rho_{\beta,N}(q_\beta^-)}\Big) \\ & - \sum_{\beta=c,s}\int_{\pm\beta}\frac{\mathrm{d}\lambda'}{2\pi}T_{\alpha\beta}(\lambda_\alpha - \lambda')\rho_{\beta,N}(\lambda') \qquad \alpha = c,s \end{aligned} \tag{2.18}$$

where we have introduced the notation $\lambda_c = \Lambda$, $\lambda_s = k$, $q_c^\pm = Q^\pm$, $q_s^\pm = B^\pm$, $\tilde{a}_s(x) = T_{cs}(x)$, $\tilde{a}_c(x) = T_{cc}(x)$ and the integral

$$\int_{\pm\beta} = \int_{q_\beta^+}^{+\infty} + \int_{-\infty}^{q_\beta^-} .$$

The solution to (2.18) may be written as

$$\rho_{\alpha,N}(\lambda_\alpha) = \rho_\alpha(\lambda_\alpha|q^\pm) + \frac{1}{24N^2}\sum_\beta\left(\frac{f_{\alpha\beta}^+(\lambda_\alpha|q^\pm)}{2\pi\rho_{\beta,N}(q_\beta^+)} - \frac{f_{\alpha\beta}^-(\lambda_\alpha|q^\pm)}{2\pi\rho_{\beta,N}(q_\beta^-)}\right) \tag{2.19}$$

where

$$\rho_\alpha(\lambda_\alpha|q^\pm) = \frac{1}{2\pi}\tilde{a}_\alpha(\lambda_\alpha) - \sum_\beta\int_{\pm\beta}\frac{d\lambda'}{2\pi}T_{\alpha\beta}(\lambda_\alpha - \lambda')\rho_\beta(\lambda'|q^\pm) \tag{2.20}$$

$$f_{\alpha\beta}^\pm(\lambda_\alpha|q^\pm) = T'_{\alpha\beta}(\lambda_\alpha - q_\beta^\pm) - \sum_\gamma\int_{\pm\gamma}\frac{d\lambda'}{2\pi}T_{\alpha\gamma}(\lambda_\alpha - \lambda')f_{\gamma\beta}^\pm(\lambda'|q^\pm) . \tag{2.21}$$

Notice that $\rho_\alpha(\lambda_\alpha|q^\pm)$ in (2.20) are the rapidity distribution functions in the thermodynamic limit $N \to \infty$ with $N_c/N = n_c$ and $N_s/N = n_s$ being kept fixed. In this limit the electron density n_c and the magnetization $\mathcal{M}$ are obviously given by

$$\begin{aligned} n_c &= \int_{\pm s} dk\, \rho_s(k) + 2\int_{\pm c} d\Lambda\, \rho_c(\Lambda) \\ \mathcal{M} &= n_s - \frac{n_c}{2} = \frac{1}{2}\int_{\pm s} dk\, \rho_s(k) . \end{aligned} \tag{2.22}$$

For the ground state the rapidity distribution is symmetric, $q_\alpha^\pm = \pm q_\alpha$. In the absence of the external magnetic field the ground state turns out to be singlet, $\mathcal{M} = 0$ [15].

To calculate the energy we first apply the formula (2.17) to (2.6). Then, using (2.19) and (2.21), we have

$$E = N\epsilon(q^\pm) - \frac{\pi}{6N}\sum_\alpha\frac{1}{2\pi\rho_\alpha(q_\alpha)}\left(-\tilde{\epsilon}_\alpha^{0\prime}(q_\alpha) - \sum_\beta\int_{|\lambda|\geqslant q_\beta}\frac{d\lambda}{2\pi}\tilde{\epsilon}_\beta^0(\lambda)f_{\beta\alpha}^+(\lambda)\right) + O(N^{-2}) \tag{2.23}$$

where

$$\epsilon(q^\pm) = \sum_\alpha\int_{\pm\alpha} d\lambda\, \tilde{\epsilon}_\alpha^0(\lambda)\rho_\alpha(\lambda|q^\pm) \tag{2.24}$$

$$\tilde{\epsilon}_s^0(\lambda) = -(2+\mu) - \frac{H}{2} + T_{cs}(\lambda) \qquad \tilde{\epsilon}_c^0(\lambda) = -2(2+\mu) + T_{cc}(\lambda) . \tag{2.25}$$

Since the second term in (2.23) is of order N^{-1} we have replaced $\rho_\alpha(\lambda|q^\pm)$ by $\rho_\alpha(\lambda) = \lim_{q_\beta^\pm \to \pm q_\beta}\rho_\alpha(\lambda|q^\pm)$ which is the solution to (2.20) with $q_\beta^\pm$ being replaced by $\pm q_\beta$. Likewise $f_{\alpha\beta}^+(\lambda_\alpha) = \lim_{q_\gamma^\pm \to \pm q_\gamma} f_{\alpha\beta}^+(\lambda_\alpha|q^\pm)$.

Let us introduce the dressed energy functions

$$\epsilon_\alpha(\lambda|q^\pm) = -\tilde{\epsilon}^0_\alpha(\lambda) - \sum_\beta \int_{\pm\beta} \frac{d\lambda'}{2\pi} \epsilon_\beta(\lambda'|q^\pm) T_{\beta\alpha}(\lambda' - \lambda) \tag{2.26}$$

with the condition

$$\epsilon_\alpha(q^\pm_\alpha|q^\pm) = 0\,. \tag{2.27}$$

We iterate (2.26) and take the derivative. The result is compared with the expression obtained by inserting the iteration solution $f^+_{\alpha\beta}$ of (2.21) into (2.23). One finds

$$\epsilon'_\alpha(q_\alpha) \equiv \frac{\partial}{\partial\lambda}\epsilon_\alpha(\lambda)\bigg|_{\lambda=q_\alpha} = -\tilde{\epsilon}^{0\prime}_\alpha(q_\alpha) - \sum_\beta \int_{|\lambda|\geqslant q_\beta} \frac{d\lambda}{2\pi} \tilde{\epsilon}^0_\beta(\lambda) f^+_{\beta\alpha}(\lambda) \tag{2.28}$$

where $\epsilon_\alpha(\lambda) = \lim_{q^\pm_\beta \to \pm q_\beta} \epsilon_\alpha(\lambda|q^\pm)$. On the other hand, the Fermi velocities of the low-lying excitations are determined from

$$v_\alpha = \frac{1}{2\pi\rho_\alpha(q_\alpha)} \epsilon'_\alpha(q_\alpha)\,. \tag{2.29}$$

Therefore it can be seen that

$$E = N\varepsilon_\alpha(q^\pm) - \frac{\pi v_s}{6N} - \frac{\pi v_c}{6N} + O(N^{-2})\,. \tag{2.30}$$

Consequently we find that the ground-state energy scales as

$$E_0 = N\varepsilon_0 - \frac{\pi v_s}{6N} - \frac{\pi v_c}{6N} + O(N^{-2}) \tag{2.31}$$

where the bulk energy density $\varepsilon_0 = \varepsilon(q^\pm)|_{q^\pm_\alpha = \pm q_\alpha} \equiv \varepsilon(\pm q)$.

2.2. Corrections due to the excitations

Our next task is to compute the energy gap $E - E_0$ due to the elementary excitations. There exist two types of excitations: the excitations which cause the change of the symmetric Fermi level $\pm B$ and $\pm Q$ of the ground state to the asymmetric ones $B^\pm$ and $Q^\pm$, thereby with large momentum transfer, and the particle–hole excitations with small momentum transfer near the Fermi levels.

In order to calculate the contribution of the excitations with large momentum transfer, it is convenient to convert the integrals $\int_{\pm\gamma}$ into $\int_{q^-_\gamma}^{q^+_\gamma}$. This can be performed by Fourier transform. The integral equation (2.20) for the rapidity distribution then turns out to be

$$\rho_\alpha(\lambda|q^\pm) = \frac{1}{2\pi} a_\alpha(\lambda) + \sum_\beta \int_{q^-_\beta}^{q^+_\beta} \frac{d\lambda'}{2\pi} K_{\alpha\beta}(\lambda - \lambda') \rho_\beta(\lambda'|q^\pm) \tag{2.32}$$

where $K_{cs}(x) = K_{sc}(x) = T_{sc}(x)$, $K_{ss}(x) = -T_{cc}(x)$, $K_{cc}(x) \equiv 0$, and $a_s(x) = K_{sc}(x)$, $a_c(x) = 0$. For the dressed energy (2.26) we obtain

$$\epsilon_\alpha(\lambda|q^\pm) = \epsilon^0_\alpha(\lambda) + \sum_\beta \int_{q_\beta^-}^{q_\beta^+} \frac{d\lambda'}{2\pi} \epsilon_\beta(\lambda'|q^\pm) K_{\beta\alpha}(\lambda' - \lambda) \tag{2.33}$$

where $\epsilon^0_s(x) = H - K_{cs}(x)$, $\epsilon^0_c(x) = 2 + \mu - H/2$. The energy takes the form

$$\varepsilon(q^\pm) = \mu - \frac{H}{2} + \sum_\alpha \int_{q_\alpha^-}^{q_\alpha^+} d\lambda\, \epsilon^0_\alpha(\lambda) \rho_\alpha(\lambda|q^\pm)\,. \tag{2.34}$$

The integrations of the rapidity distributions over the closed intervals yield

$$\int_{Q^-}^{Q^+} d\Lambda\, \rho_c(\Lambda) = 1 - n_c \qquad \int_{B^-}^{B^+} dk\, \rho_s(k) = 1 - n_s \tag{2.35}$$

which are the number of holes and the number of 'holes' with respect to up spins, respectively. Notice that these are quite consistent with (2.13).

We turn now to the derivation of the explicit form of $\varepsilon(q^\pm)$. We first minimize $\varepsilon(q^\pm)$ with respect to the electron number and the magnetization. This condition is equivalent to demanding $\partial\varepsilon(q^\pm)/\partial q^\pm_\alpha = 0$, which is realized by virtue of the condition (2.27) for the dressed energy; $\epsilon_\alpha(q^\pm_\alpha|q^\pm) = 0$. See appendix A. Let us next expand $\varepsilon(q^\pm)$ about the ground-state energy density $\varepsilon_0 = \varepsilon(\pm q)$

$$\varepsilon(q^\pm) = \varepsilon_0 + \frac{1}{2}\sum_\alpha \left\{ \left(\frac{\partial}{\partial q^+_\alpha}\right)^2 \varepsilon\big|(q^+_\alpha - q_\alpha)^2 + \left(\frac{\partial}{\partial q^-_\alpha}\right)^2 \varepsilon\big|(q^-_\alpha + q_\alpha)^2 \right\} \tag{2.36}$$

where the vertical bar stands for setting $q^\pm_\alpha = \pm q_\alpha$. There is no cross derivative due to (2.27). We now wish to express the variations $dq^\pm_\alpha = q^\pm_\alpha \mp q_\alpha$ in terms of the change of the numbers of electrons and up spins. The details of this calculation are left to appendix A. The result reads

$$\varepsilon(q^\pm) - \varepsilon_0 = \frac{2\pi}{N^2}\left(\frac{1}{4} I^t (\mathbf{Z}^{-1})^t \mathbf{V} \mathbf{Z}^{-1} I + D^t \mathbf{Z} \mathbf{V} \mathbf{Z}^t D\right) + O(N^{-2}) \tag{2.37}$$

where $\mathbf{V} = \mathrm{diag}(v_c, v_s)$. Here we have introduced the 2×2 dressed charge matrix $\mathbf{Z}$ [22, 23, 28] whose elements $Z_{\alpha\beta} = \xi_{\alpha\beta}(q_\beta)$ are given by the solutions to the integral equations

$$\xi_{\alpha\beta}(\lambda_\beta) = \delta_{\alpha\beta} + \sum_{\gamma=c,s} \int_{-q_\gamma}^{q_\gamma} \frac{d\lambda}{2\pi} \xi_{\alpha\gamma}(\lambda) K_{\gamma\beta}(\lambda - \lambda_\beta)\,. \tag{2.38}$$

Here

$$D = \begin{pmatrix} D_c \\ D_s \end{pmatrix} \qquad I = \begin{pmatrix} I_c \\ I_s \end{pmatrix} = \begin{pmatrix} -(N_c - n^0_c N) \\ -(N_s - n^0_s N) \end{pmatrix} \tag{2.39}$$

with n^0_α being the ground-state value of $n_\alpha, \alpha = c, s$. As pointed out in [28, 29] n^0_c, n^0_s and N should meet certain commensuration conditions to be consistent with the conformal limit. I_c and I_s are then non-negative integers, implying that we are counting the hole number and the number of 'holes' with respect to the up spins, respectively.

It is straightforward to include the particle–hole excitations. Their contributions are specified by the set of non-negative integers $N_c^{\pm}$ and $N_s^{\pm}$. The final expression for the energy gap is thus obtained as

$$E - E_0 = \frac{2\pi v_c}{N} x_c + \frac{2\pi v_s}{N} x_s + O(N^{-2}) \tag{2.40}$$

$$\begin{aligned} x_c &= \left(\frac{Z_{ss}I_c - Z_{cs}I_s}{2\det \mathbf{Z}}\right)^2 + (Z_{cc}D_c + Z_{sc}D_s)^2 + N_c^+ + N_c^- \\ x_s &= \left(\frac{Z_{sc}I_c - Z_{cc}I_s}{2\det \mathbf{Z}}\right)^2 + (Z_{cs}D_c + Z_{ss}D_s)^2 + N_s^+ + N_s^- . \end{aligned} \tag{2.41}$$

The momentum takes the form

$$P - P_0 = (2\pi - 2k_{F\uparrow} - 2k_{F\downarrow})D_c + (2\pi - 2k_{F\uparrow})D_s + \frac{2\pi}{N} \sum_{\alpha=c,s} (I_\alpha D_\alpha + N_\alpha^+ - N_\alpha^-) \tag{2.42}$$

where P_0 is the ground-state momentum and the Fermi momentum $k_{F\uparrow}$ ($k_{F\downarrow}$) for the up- (down)-spin electrons is given by

$$k_{F\uparrow(\downarrow)} = \tfrac{1}{2}\pi(n_c \pm 2\mathcal{M}) . \tag{2.43}$$

Equation (2.42) is easily checked if one notes to rewrite (2.7) as

$$P = \frac{2\pi}{N}\left(\sum I_j + \sum J_\alpha\right) + (I^{\pm}, J^{\pm} \text{ independent term}) \tag{2.44}$$

where the sums are taken over $I_j \in [I^-, I^+]$ and $J_\alpha \in [J^-, J^+]$.

This completes our derivation of the finite-size corrections in the energy spectrum. Now, conformal invariance of 1D quantum critical systems dictates that the ground-state energy scales like [30]

$$E_0 = \varepsilon_0 N - \frac{\pi c}{6N} v + O(N^{-1}) \tag{2.45}$$

where v is the Fermi velocity and c is the central charge of the Virasoro algebra. The energy gaps of the excited states are related to the scaling dimensions x_n of the scaling operators of the theory [18]

$$E_n - E_0 = \frac{2\pi v}{N} x_n + O(N^{-1}) . \tag{2.46}$$

Thus our expressions (2.31) and (2.40) indicate that the critical behaviour of the t–J model is described by two independent $c = 1$ conformal theories. They are associated with the massless excitations, the holon and spinon, which are characterized by the Fermi velocities v_c and v_s.

3. Correlation functions

One of the remarkable observations in two-dimensional conformal field theory is that the critical exponents of the scaling operators are read off from the energy gaps as described in (2.46). To write down explicitly the correlation functions at long distance let us rewrite (2.40) and (2.42) as

$$E(I,D) - E_0 = \frac{2\pi}{N} \sum_{\alpha=c,s} v_\alpha(\Delta_\alpha^+ + \Delta_\alpha^-) + O(N^{-1}) \tag{3.1}$$

$$P(I,D) - P_0 = (2\pi - 2k_{F\uparrow} - 2k_{F\downarrow})D_c + (2\pi - 2k_{F\uparrow})D_s + \frac{2\pi}{N} \sum_{\alpha=c,s} (\Delta_\alpha^+ - \Delta_\alpha^-) \tag{3.2}$$

where $\Delta_\alpha^\pm$ are the left and right conformal weights in the sector α; $\alpha = c$ (holon), $\alpha = s$ (spinon). Here $x_\alpha = \Delta_\alpha^+ + \Delta_\alpha^-$ and we have

$$\Delta_c^\pm(I,D) = \frac{1}{2}\left(Z_{cc}D_c + Z_{sc}D_s \pm \frac{Z_{ss}I_c - Z_{cs}I_s}{2\det \mathbf{Z}}\right)^2 + N_c^\pm \tag{3.3}$$

$$\Delta_s^\pm(I,D) = \frac{1}{2}\left(Z_{cs}D_c + Z_{ss}D_s \pm \frac{Z_{cc}I_s - Z_{sc}I_c}{2\det \mathbf{Z}}\right)^2 + N_s^\pm . \tag{3.4}$$

The two-point correlation functions of the scaling fields $\phi_{\Delta^\pm}(x,t)$ with conformal weights $\Delta^\pm$ then take the form

$$\begin{aligned}\langle \phi_{\Delta^\pm}(x,t)\phi_{\Delta^\pm}(0,0)\rangle &\equiv G(\Delta^\pm(I,D)|x,t) \\ &= \frac{\exp(\mathrm{i}(2\pi - 2k_{F\uparrow} - 2k_{F\downarrow})D_c x)\exp(\mathrm{i}(2\pi - 2k_{F\uparrow})D_s x)}{(x - \mathrm{i}v_c t)^{2\Delta_c^+}(x + \mathrm{i}v_c t)^{2\Delta_c^-}(x - \mathrm{i}v_s t)^{2\Delta_s^+}(x + \mathrm{i}v_s t)^{2\Delta_s^-}} .\end{aligned} \tag{3.5}$$

We consider the following correlation functions.

(i) Electron correlator

$$G_\sigma(x,t) = \langle c_\sigma^\dagger(x,t)c_\sigma(0,0)\rangle \qquad \sigma = \uparrow \text{ or } \downarrow . \tag{3.6}$$

(ii) Charge density correlator

$$N(x,t) = \langle n(x,t)n(0,0)\rangle \qquad n(x,t) = n_\uparrow(x,t) + n_\downarrow(x,t) . \tag{3.7}$$

(iii) Spin correlator

$$\chi(x,t) = \langle S_z(x,t)S_z(0,0)\rangle \qquad S_z(x,t) = \tfrac{1}{2}(n_\uparrow(x,t) - n_\downarrow(x,t)) . \tag{3.8}$$

(iv) Singlet and triplet pair superconducting correlators

$$\begin{aligned}P_s(x,t) &= \langle c_\uparrow^\dagger(x+1,t)c_\downarrow^\dagger(x,t)c_\uparrow(1,0)c_\downarrow(0,0)\rangle \\ P_t(x,t) &= \langle c_\uparrow^\dagger(x+1,t)c_\uparrow^\dagger(x,t)c_\uparrow(1,0)c_\uparrow(0,0)\rangle .\end{aligned} \tag{3.9}$$

Generically the field operators we have introduced will renormalize to a certain linear combination of the scaling operators at long distance. The correlation functions (3.6)–(3.9) are thus expressed as

$$\sum A(\boldsymbol{I},\boldsymbol{D},\boldsymbol{N}^{\pm})G(\Delta^{\pm}(\boldsymbol{I},\boldsymbol{D})|x,t) \tag{3.10}$$

where $A(\boldsymbol{I},\boldsymbol{D},\boldsymbol{N}^{\pm})$ are constant coefficients and we have neglected possible logarithmic corrections.

In order to determine the scaling dimensions we now have to assign the quantum numbers $(\boldsymbol{I},\boldsymbol{D},\boldsymbol{N}^{\pm})$ to the field operators, as has been done for the Hubbard mode [12]. Notice that these quantum numbers are subject to the restrictions

$$D_c = \frac{I_c + I_s}{2} \bmod 1 \qquad D_s = \frac{I_c}{2} \bmod 1 \tag{3.11}$$

which can be checked from (2.5). Upon inspecting the explicit form of the field operators in (3.6)–(3.9) one finds the assignment

$$\begin{aligned}
&G_\uparrow(x,t) : (I_c = 1, I_s = 1, D_c \in \mathbb{Z}, D_s \in \mathbb{Z} + \tfrac{1}{2})\\
&G_\downarrow(x,t) : (I_c = 1, I_s = 0, D_c \in \mathbb{Z} + \tfrac{1}{2}, D_s \in \mathbb{Z} + \tfrac{1}{2})\\
&N(x,t) : (I_c = 0, I_s = 0, D_c \in \mathbb{Z}, D_s \in \mathbb{Z})\\
&\chi(x,t) : (I_c = 0, I_s = 0, D_c \in \mathbb{Z}, D_s \in \mathbb{Z})\\
&P_s(x,t) : (I_c = 2, I_s = 1, D_c \in \mathbb{Z} + \tfrac{1}{2}, D_s \in \mathbb{Z})\\
&P_t(x,t) : (I_c = 2, I_s = 2, D_c \in \mathbb{Z}, D_s \in \mathbb{Z})\,.
\end{aligned} \tag{3.12}$$

In the following we first study the correlation functions for zero magnetic field and then the effect of the external magnetic field at and near half-filling is discussed.

3.1. *Zero magnetic field*

It is readily seen that $B \to +\infty$ for zero magnetic field. Using the Fourier transform technique we obtain the simple form of the dressed charge matrix **Z**

$$\begin{pmatrix} Z_{cc} & Z_{cs} \\ Z_{sc} & Z_{ss} \end{pmatrix} = \begin{pmatrix} \xi_c(Q) & 0 \\ \xi_c(Q)/2 & 1/\sqrt{2} \end{pmatrix} \tag{3.13}$$

where $Z_{ss} = 1/\sqrt{2}$ is derived with the aid of the Wiener–Hopf method [28]. Here $\xi_c(\Lambda)$ is the solution to the equation

$$\xi_c(\Lambda) = 1 + \int_{-Q}^{Q} \mathrm{d}\Lambda'\, R(\Lambda - \Lambda')\xi_c(\Lambda') \tag{3.14}$$

with the kernel being

$$R(x) = \int_{-\infty}^{\infty} \frac{\mathrm{d}\omega}{2\pi} \frac{\exp(-\mathrm{i}\omega x)}{1 + \exp|\omega|}\,. \tag{3.15}$$

The conformal weights (3.3) and (3.4) are reduced to

$$\Delta_c^{\pm}(I,D) = \frac{1}{2}\left(\frac{I_c}{2\xi_c(Q)} \pm \xi_c(Q)(D_c + \frac{D_s}{2})\right)^2 + N_c^{\pm}$$
$$\Delta_s^{\pm}(I,D) = \frac{1}{4}\left(I_s - \frac{I_c}{2} \mp D_s\right)^2 + N_s^{\pm}\,. \tag{3.16}$$

Let us first consider the charge density correlation function. From (3.12) we write down the asymptotic form of the equal-time correlator

$$N(r,0) \sim \text{constant} + A_0 r^{-2} + A_2 r^{-\alpha_s}\cos(2k_F r) + A_4 r^{-\alpha_c}\cos(4k_F r) \tag{3.17}$$

where $k_{F\uparrow} = k_{F\downarrow} \equiv k_F$ since $\mathcal{M} = 0$ for zero field. The $4k_F$ piece arises from the excitation of $(I_c, I_s, D_c, D_s) = (0,0,\pm1,0)$, while the $2k_F$ piece from $(I_c, I_s, D_c, D_s) = (0,0,\pm1,\mp1)$ and $(0,0,0,\pm1)$. The non-oscillating part is due to the lowest particle–hole excitation. We thus find

$$\alpha_c = 2\xi_c(Q)^2 \qquad \alpha_s = 1 + \alpha_c/4\,. \tag{3.18}$$

Notice that both the holon and spinon excitations are responsible for the $2k_F$ oscillation part. On the other hand the $4k_F$ piece is dominated by the holon excitation alone. The same observation holds for the Hubbard model [10–12] and the Tomonaga–Luttinger model [3]. The spin correlation function $\chi(r,0)$ has the same form as (3.17) except that the $4k_F$ part is absent. The critical exponent for the $2k_F$ part is equal to α_s of the charge density correlation.

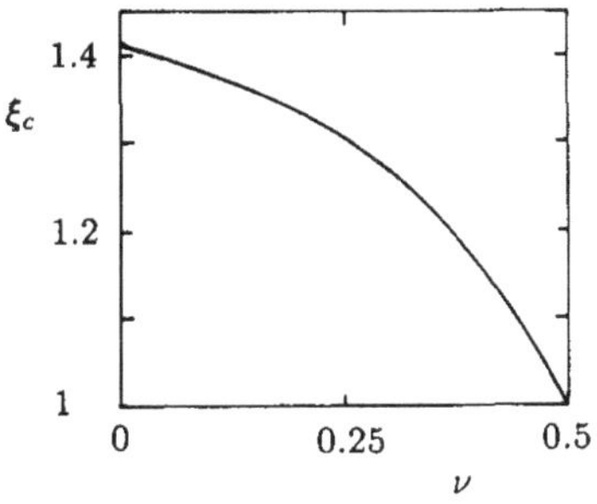

Figure 1. The dressed charge $\xi_c(Q)$ of the holon as a function of the electron concentration ν ($\nu = \frac{1}{2}$ for half-filling).

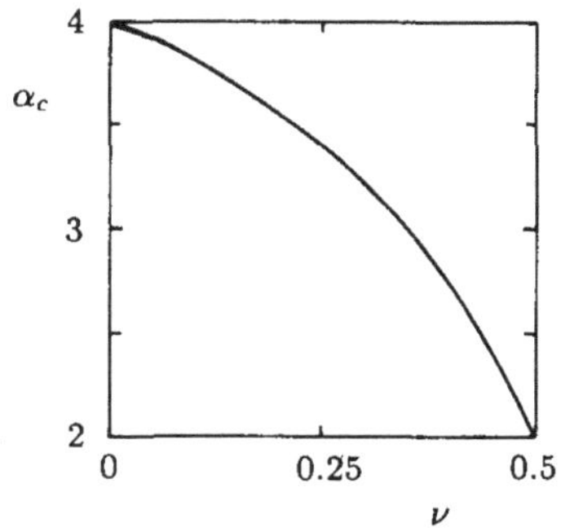

Figure 2. The charge density $4k_F$ exponent α_c as a function of ν.

The dressed charge $\xi_c(Q)$ of the holon is shown in figure 1, where $\nu = n_c^0/2$ and $\nu = \frac{1}{2}$ corresponds to the half-filled band. The $4k_F$ exponent α_c then behaves as depicted in figure 2. Near half-filling we obtain $\alpha_c \sim 2 + 8(\frac{1}{2} - \nu)$ as shown in appendix B. Note that in the low-density limit $\alpha_c = 4$, i.e. the value for the non-interacting model.

The long-distance behaviour of the electron correlation function is governed by the excitation specified by $(I_c, I_s, D_c, D_s) = (1,1,0,\pm\frac{1}{2})$. We thus obtain

$$G_\uparrow(r,0) \sim r^{-\eta}\cos(k_F r) \qquad \eta = (\alpha_c + 4)^2/(16\alpha_c)\,. \tag{3.19}$$

$G_{\downarrow}(r,0)$ follows the same behaviour, but with the excitation $(1,0,\pm\frac{1}{2},\mp\frac{1}{2})$. Consequently the momentum distribution function close to $k_{\rm F}$ has the form

$$\langle n_k\rangle = \langle n_{k_{\rm F}}\rangle - \text{constant}\,|k-k_{\rm F}|^{\theta}\text{sgn}(k-k_{\rm F}) \tag{3.20}$$

which is the typical power-law singularity of the Luttinger liquid [3] and we find

$$\theta = \eta - 1 = (\alpha_c - 4)^2/(16\alpha_c)\,. \tag{3.21}$$

From figure 3 we see that as ν deviates from half-filling θ decreases monotonically from $\frac{1}{8}$ to zero, and hence the momentum distribution in the low-density regime exhibits an abrupt change around $k_{\rm F}$.

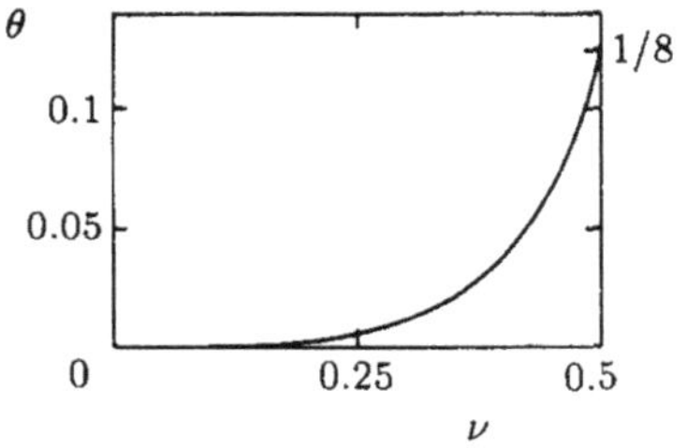

Figure 3. The exponent θ for the momentum distribution as a function of ν.

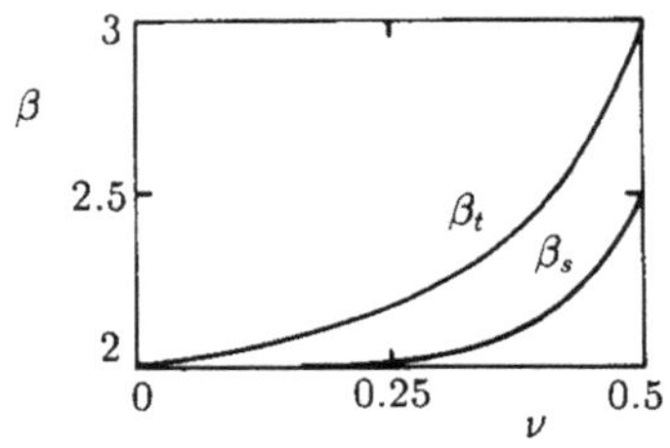

Figure 4. The superconducting correlation exponents as a function of ν. β_s and β_t are for the singlet and triplet pair, respectively.

We now turn to the superconducting correlation functions. The excitations relevant to the singlet and triplet pair correlations are specified by $(I_c, I_s, D_c, D_s) = (2,1,\pm\frac{1}{2},0)$ and $(2,2,0,0)$, respectively. We then obtain for the singlet pair

$$P_s(r,0) \sim r^{-\beta_s}\cos(2k_{\rm F}r) \qquad \beta_s = 4/\alpha_c + \alpha_c/4\,. \tag{3.22}$$

The triplet pair has the leading uniform term

$$P_t(r,0) \sim r^{-\beta_t} \qquad \beta_t = 1 + 4/\alpha_c\,. \tag{3.23}$$

Notice that the singlet pair correlation also has the uniform piece with the same exponent β_t. The exponents β_s and β_t are plotted in figure 4, from which we observe that the superconducting correlations get more enhanced as holes are doped into the half-filled band [9,31]. It is interesting to notice that even in the t–J model the superconducting correlations never overwhelm the spin correlation since β_t and β_s are always larger than α_s for arbitrary electron filling.

3.2. Magnetic field dependence

Let us investigate how the correlation exponents behave when we turn on the external magnetic field. For simplicity we consider two typical cases: just at half-filling and near half-filling, on the basis of which we will be able to clarify the essential properties of the field dependence. At half-filling there is no massless excitation associated with the charge fluctuation since the strong correlation effect opens the very large Hubbard

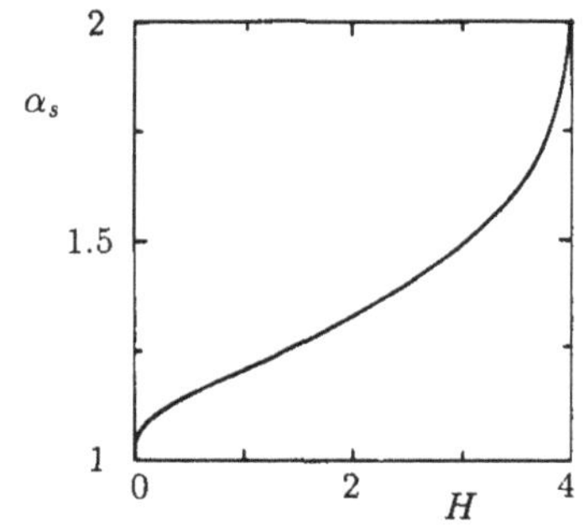

Figure 5. The $2k_F$ exponent α_s in the spin correlator as a function of H at half-filling.

gap. The spin excitation remains massless, which can be described by the $c = 1$ SU(2) Kac–Moody theory. The long-distance behaviour of the spin correlator is thus equivalent to that for the antiferromagnetic Heisenberg model in [32]

$$\chi(r,0) \sim \mathcal{M}^2 + B_0 r^{-2} + B_2 r^{-\alpha_s} \cos(2k_{F\downarrow} r) \tag{3.24}$$

where $\mathcal{M}$ is the magnetization. We plot in figure 5 the magnetic field dependence of the $2k_F$ spin exponent $\alpha_s = 2\xi_s(B)^2$, where $\xi_s(B)$ is given in (3.26) below.

In the metallic phase away from half-filling, the holon becomes massless as in zero field. An essential difference from zero-field case is that the holon is no longer treated as a spinless hole because it acquires the effective spin induced by the magnetic field. Similarly the spinon may get electrically charged.

These effective spin and charge are computed by creating the holon and spinon excitations in magnetic fields [33]. We then observe that they are nothing but the elements of the dressed charge matrix introduced in section 2. The physical meaning of each element is that Z_{cc} and $(\frac{1}{2} - Z_{sc})$ are the effective charge and spin of the holon, and Z_{ss} and Z_{cs} are the effective spin and charge of the spinon. Approaching half-filling ($Q \to 0$), the effective charge of the holon is independent of field and becomes unity. The spinon is not charged even in the metallic phase, i.e. $Z_{cs} = 0$. Furthermore the field dependence of the effective spin of the spinon is given by that for the Heisenberg model (corresponding to half-filling). The dressed charge matrix thus turns out to be

$$\begin{pmatrix} Z_{cc} & Z_{cs} \\ Z_{sc} & Z_{ss} \end{pmatrix} = \begin{pmatrix} 1 & 0 \\ \frac{1}{2} - s_h(B) & \xi_s(B) \end{pmatrix} \tag{3.25}$$

where $\xi_s(B)$ is equal to the dressed charge (or effective spin) for the spin-$\frac{1}{2}$ Heisenberg chain obeying

$$\xi_s(k) = \frac{1}{2} + \int_{|k'|\geqslant B} \mathrm{d}k'\, R(k-k')\xi_s(k') \tag{3.26}$$

with the kernel $R(x)$ given in (3.15) and the effective spin of the holon is

$$s_h(B) = \frac{1}{2}\int_{|k|\geqslant B} \mathrm{d}k\, \mathrm{sech}\,(\pi k)\xi_s(k)\,. \tag{3.27}$$

For comparison let us quote the zero-field dressed charge matrix near half-filing

$$\begin{pmatrix} 1 & 0 \\ \frac{1}{2} & 1/\sqrt{2} \end{pmatrix} . \tag{3.28}$$

It is not difficult to verify (3.25) from (2.38) if one applies Fourier transform. The field dependence of $\xi_s(B)$ and $s_h(B)$ are depicted in figures 6 and 7, respectively. We note in figures 5 and 6 that the weak-field behaviour of α_s and ξ_s exhibit the logarithmic singularity whose origin is the same as for the spin susceptibility of the Heisenberg chain [12,34].

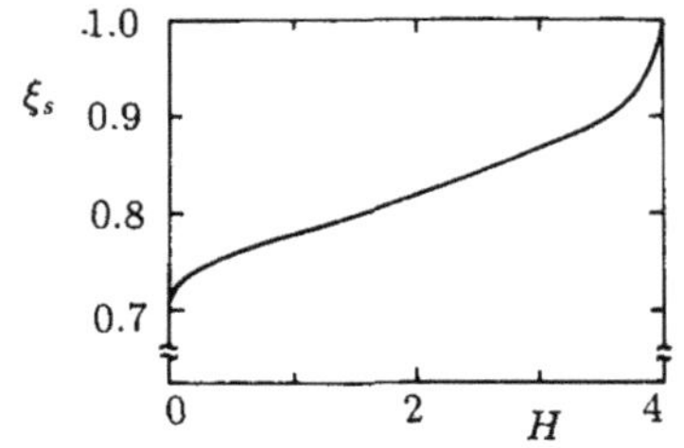

Figure 6. The dressed charge $\xi_s(B)$ of the spinon as a function of the external magnetic field H at and near half-filling.

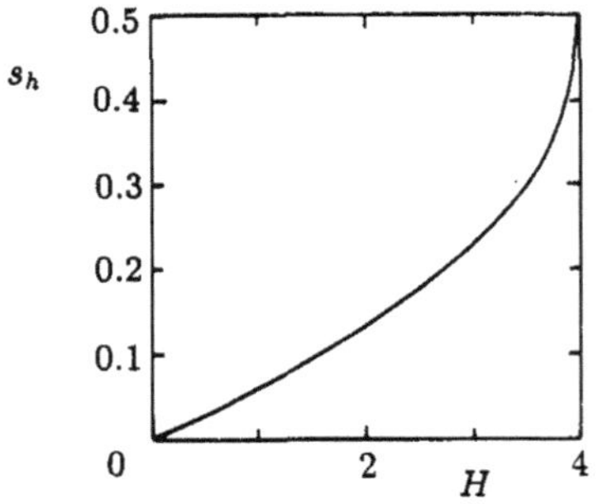

Figure 7. The effective spin $s_h(B)$ of the holon as a function of H near half-filling.

We next discuss the field dependence of critical exponents close to half-filling. The exponent of the $4k_F(=2k_{F\uparrow}+2k_{F\downarrow})$ oscillation piece in the charge correlator takes the value $\alpha_c = 2$ irrespective of magnetic fields owing to the fact that it is controlled by charge excitation alone. Since the charge density operator $n(r) = n_\uparrow(r) + n_\downarrow(r)$ the $2k_F$ part splits into two pieces with the momentum $2k_{F\downarrow}$ and $2k_{F\uparrow}$, the exponents of which are given by $\alpha_{s\downarrow} = 2Z_{ss}^2+2(1-Z_{sc})^2$ and $\alpha_{s\uparrow} = 2Z_{ss}^2+2Z_{sc}^2$, respectively. Note that these exponents have the magnetic-field dependence only through the effective spins of the spinon and holon. The values of $(\alpha_{s\downarrow}, \alpha_{s\uparrow})$ are increased from $(\frac{3}{2}, \frac{3}{2})$ to $(4, 2)$ as the field increases.

The singlet pairing exponent has the form

$$\beta_s = \frac{5}{2} + \frac{1}{2}\left(\frac{1-2Z_{sc}}{Z_{ss}}\right)^2 \tag{3.29}$$

while the triplet one reads

$$\beta_t = 2 + 2\left(\frac{1-Z_{sc}}{Z_{ss}}\right)^2 . \tag{3.30}$$

With the increase of the magnetic field, the values of (β_s, β_t) are monotonically increased from $(\frac{5}{2}, 3)$ to $(3, 4)$, respectively. Therefore the superconducting correlation is suppressed in the presence of the magnetic field, as might be expected.

The Luttinger anomaly exponent for the momentum distribution is given by

$$\begin{aligned} \theta_\uparrow &= \frac{1}{2}\left[-1 + Z_{sc}^2 + Z_{ss}^2 + \left(\frac{1 - Z_{sc}}{Z_{ss}}\right)^2\right] && \text{for } k_{F\uparrow} \\ \theta_\downarrow &= \frac{1}{2}\left[-1 + (1 - Z_{sc})^2 + Z_{ss}^2 + \left(\frac{Z_{sc}}{Z_{ss}}\right)^2\right] && \text{for } k_{F\downarrow}\,. \end{aligned} \tag{3.31}$$

Both of the exponents $\theta_{\uparrow,\downarrow}$ increase up to $\frac{1}{2}$ with the increase of magnetic fields. The momentum distribution around the Fermi momentum is therefore smoothed by the magnetic field.

We mention that Ogata *et al* analysed numerical data on the magnetic field dependence in the $U \to \infty$ limit of the Hubbard model [35] comparing with the analytic result obtained by Frahm and Korepin [12]. Our present results for the magnetic field effect in the t–J model are essentially the same as theirs in the metallic system very close to half-filling. It is worth noting that in this regime the effective spin of the holon defined here is nothing but the magnetization $\mathcal{M}$ (2.22) of the system. The magnetization in this limit is of course equal to that in the Heisenberg model calculated by Griffiths [34]. We point out, however, that this relation holds only for highly correlated systems. In generic cases they are not equivalent. This will be seen explicitly in subsection 4.1, where the effect of the finite Coulomb interaction is discussed using the Hubbard model.

4. Luttinger liquid properties

According to Haldane, the idea of Luttinger liquids applies to the low-energy excitations in a variety of 1D metallic systems [1,2]. His demonstration is mainly based on the systems containing the one-component massless excitation, such as the Heisenberg model, the bose gas model *etc.* The low-energy spectrum of Luttinger liquids contains the three spectral parameters, v_F, v_J and v_N. These are all velocities associated with the excitations of particle–hole pairs (v_F), of the $2k_F$ momentum transfer (v_J), and of the particle number change (v_N). Here v_F is the usual sound velocity. The crucial point is that these velocities are not mutually independent but connected through the universal relation $v_F = (v_J v_N)^{1/2}$. Hence one can write

$$v_J = \exp(2\psi) v_F \qquad v_N = \exp(-2\psi) v_F \tag{4.1}$$

where the parameter $\exp(2\psi)$ is non-universal and depends on the details of the interactions of underlying microscopic models. All the correlation exponents are essentially determined by this parameter. In short, what Haldane claims is that the low-energy massless excitations in 1D metallic systems are all solved by the procedure of bosonization.

In the t–J model, as we have seen, there exist two massless degrees of freedom, the holon and spinon. These excitations are decoupled and described by two independent $c = 1$ conformal theories, i.e. Gaussian theories. Thus the low-energy action reads

$$S = \frac{1}{2} \sum_{\alpha=c,s} v_\alpha \int \mathrm{d}t \int_0^{2\pi} \mathrm{d}x\, \partial_\mu \phi_\alpha \partial_\mu \phi_\alpha \tag{4.2}$$

where free boson fields are periodic $\phi_\alpha(t,x) = \phi_\alpha(t,x+2\pi)+2\pi N_\alpha R_\alpha$ with $N_\alpha \in \mathbf{Z}/2$. The conformal weights (3.16) are characteristic of the Gaussian theory [36]. For the charge sector, therefore, the field periodicity R_c is parametrized as $\sqrt{\pi}R_c = \xi_c(Q)^{-1}$, i.e. it depends continuously on the electron concentration. The spin sector has the periodicity $\sqrt{\pi}R_s = Z_{ss} = 1/\sqrt{2}$ for any electron density. This implies that the spin sector is described by the level-1 SU(2) Kac–Moody theory just like the spin-$\frac{1}{2}$ antiferromagnetic Heisenberg chain [13]. It is also instructive to compare the formula (3.1) for the energy gaps $2\pi v_\alpha(\Delta_\alpha^+ + \Delta_\alpha^-)/N$ with Haldane's result (see equation (6) of [2]). They are in fact equivalent under the identification $e^\psi = \xi_c(Q)$ (or $e^\psi = Z_{ss} = 1/\sqrt{2}$) for the charge (or spin) sector. In the presence of the magnetic field the quantity e^ψ is generalized to the dressed charge matrix. Hence the critical properties of the t–J model nicely fit in with the Luttinger liquid picture.

In comparison with the Fermi liquid theory the most striking feature of the Luttinger liquid is the power-law singularity of the momentum distribution function (3.20) near $k = k_\mathrm{F}$. This reflects the fact that the low-energy excitation is not of the quasiparticle type, but of the collective type. The power-law anomaly (in view of the Fermi liquid theory) was first discovered in the Tomonaga–Luttinger model which essentially describes a weakly correlated electron system [37–39]. As for highly correlated systems this behaviour has been established only recently in the repulsive Hubbard model [6, 8, 10–12, 40]. We now have shown that the same conclusion holds for the t–J model.

In order for these systems to be classified as Luttinger liquids it has been crucial that the charge and spin degrees of freedom are separated and described by two independent $c = 1$ conformal field theories. Universal scaling relations (3.18), (3.21) and (3.23) are then valid for these metallic models. Each exponent, however, depends on the non-universal microscopic property of the theory due to the existence of the marginal operator. To clarify this point we would like to compare the t–J model with the repulsive Hubbard model in the next subsection.

4.1. Comparison with the Hubbard model

The 1D Hubbard chain describes a system of itinerant electrons feeling the on-site Coulomb repulsion U. The Hamiltonian takes the form

$$\mathcal{H} = -t\sum_{i,\sigma}(c_{i\sigma}^\dagger c_{i+1\sigma} + c_{i+1\sigma}^\dagger c_{i\sigma}) + U\sum_i n_{i\uparrow}n_{i\downarrow} \qquad U > 0\,. \tag{4.3}$$

As mentioned before, in the strong correlation limit ($U \gg t$) the model reduces to the t–J model in the region $J \sim 2t^2/U \ll t$.

The finite-size corrections in the Hubbard model have been analysed by Woynarovich [28]. For vanishing magnetic field the critical exponents α_s, θ, β_s and β_t are all expressed in terms of α_c just through the same scaling relations (3.18) and (3.21)–(3.23) as in the t–J model [10–12]. The $4k_\mathrm{F}$ oscillation exponent α_c is determined through $\alpha_c = 2\eta_c(Q)^2$, where the dressed charge function $\eta_c(k)$ of the holon is the solution to the integral equation [11, 12]

$$\eta_c(k) = 1 + \int_{-Q}^{Q} \mathrm{d}k' \, \cos(k')G(\sin k - \sin k')\eta_c(k') \tag{4.4}$$

with the kernel being

$$G(x) = \int_{-\infty}^{\infty} \frac{\mathrm{d}\omega}{2\pi} \frac{\exp(-i\omega x)}{1 + \exp(U|\omega|/(2t))}\,. \tag{4.5}$$

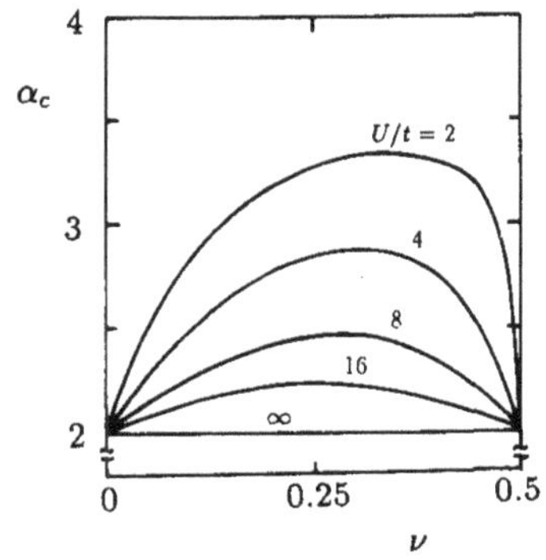

Figure 8. The charge density $4k_F$ exponent α_c as a function of ν in the Hubbard model.

Here the Fermi level Q is fixed by the electron concentration.

In figure 8 we show the exponent α_c. Strong dependence of α_c on the Coulomb interaction as well as the electron filling is clearly observed. As $U \to \infty$, α_c approaches 2 in agreement with the result for the spinless fermion. In the opposite limit $U \to 0$, α_c converges to 4 for the electron concentration $0 < \nu < \frac{1}{2}$, which is consistent with the result of the Tomonaga–Luttinger model. It should be noticed that α_c takes the value close to 2 near half-filling as long as the Coulomb interaction exists. Recall that at half-filling the Hubbard model is an insulator for all $U \neq 0$, since the Umklapp interaction becomes relevant, thereby the charge excitation possesses the gap. The gap formation strongly affects the properties of the charge excitation so that the holon behaves like the spinless fermion, resulting in the $\alpha_c = 2$ near half-filling. From α_c one can evaluate the Luttinger anomaly exponent θ for the momentum distribution and the superconducting correlation exponents through (3.21)–(3.23) [10–12]. The results are plotted in figures 9 and 10.

Let us compare the present result for the t–J model with the large-U behaviour of the Hubbard model. In the vicinity of the half-filled band the exponents of the t–J model take the values expected in the strong correlation limit of the Hubbard model, for instance $\alpha_c = 2$. This is because the exclusion of the double occupation gives the most dominant effect near half-filling, which makes the motion of doped holes behave like spinless fermions as in the Hubbard model. In the $U \to \infty$ Hubbard model, as ν decreases from half-filling α_c $(= 2)$ stays constant, and hence $\theta = \frac{1}{8}$ for any filling [10–12, 41]. On the other hand, in the t–J model the critical exponents take the values for the non-interacting system such as α_c $(= 4)$ in the low-density limit $\nu \to 0$.

This non-interacting behaviour of the t–J model for $\nu \to 0$ seems to be a bit peculiar since the model is originally supposed to describe a highly correlated system. Our result implies that the hole motion in the t–J model is not like spinless fermions for large hole-doping, but is considerably influenced by the spin fluctuation through the strong antiferromagnetic coupling J. We think that the large antiferromagnetic coupling favours the antiparallel-spin electron pairs to sit on the nearest-neighbour sites, which renders the hole motion quite different from spinless particles. In the low-density limit this configuration will be so dominant that the exclusion of the double occupancy becomes less important.

Turning to the superconducting correlations we see that the large spin coupling

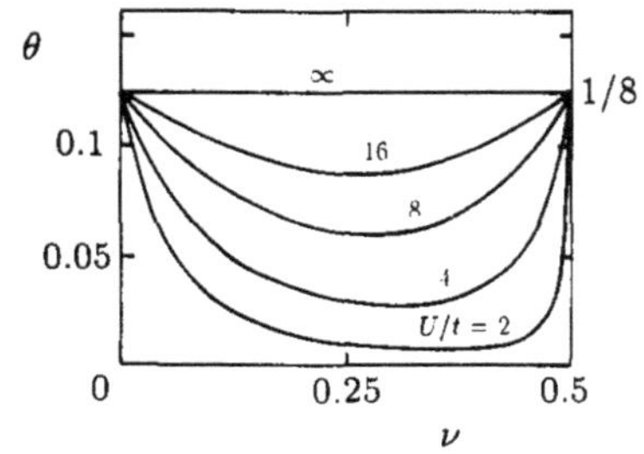

Figure 9. The exponent θ for the momentum distribution as a function of ν in the Hubbard model.

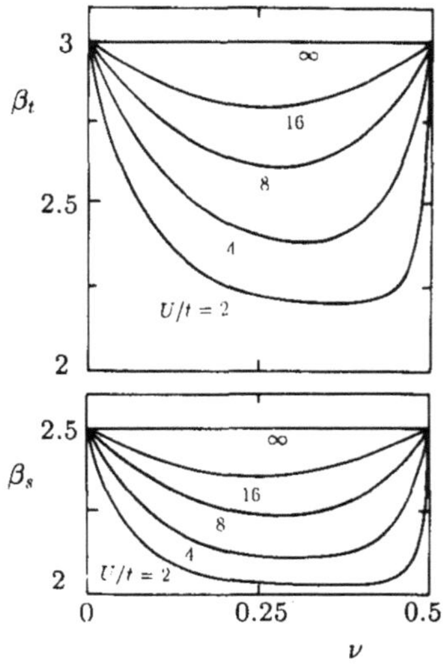

Figure 10. The superconducting correlation exponents as a function of ν in the Hubbard model. β_s and β_t are for the singlet and triplet pair, respectively.

as well as the hole doping in the t–J model play a conspicuous role to enhance the superconducting correlation. This is not the case for the strong correlation limit of the Hubbard model. Thus the t–J model tends to stabilize the superconducting state. In spite of this fact, however, the spin correlation always dominates the superconducting correlations for arbitrary electron filling, as pointed out in section 3.

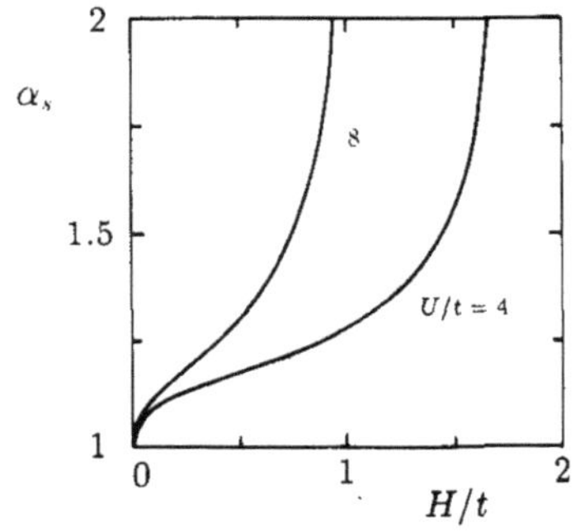

Figure 11. The $2k_{\rm F}$ exponent α_s in the spin correlator as a function of H in the Hubbard model at half-filling.

Finally we discuss the magnetic field dependence. In [11, 12] the exponent α_s for the $2k_{\rm F}$ oscillation piece in the spin correlator just at half-filling has been expressed as $\alpha_s = 2\eta_s(B)^2$, where $\eta_s(B)$ is the dressed charge explained below. We present the field dependence of α_s in figure 11. In the vicinity of the half-filled band, the dressed charge matrices for zero and for non-zero field take the same form as (3.25) and (3.28). The dressed charge function (or effective spin) η_s of the spinon satisfies the integral equation (3.26) but with the kernel replaced by $G(x)$ defined in (4.5). The effective spin $\eta_s(B)$ of the spinon is plotted in figure 12 for several values of U/t. We also depict the effective spin s_h of the holon in figure 13.

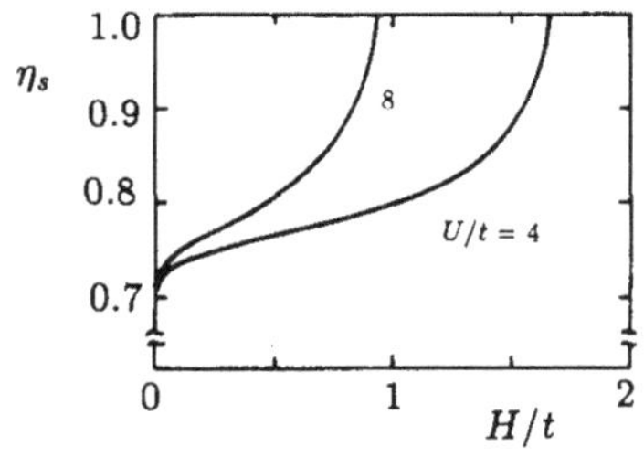

Figure 12. The dressed charge (or effective spin) $\eta_s(B)$ of the spinon as a function of H in the Hubbard model at and near half-filling.

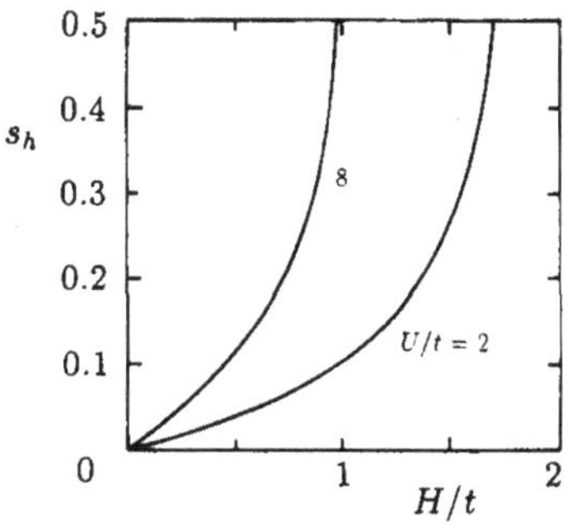

Figure 13. The effective spin $s_h(B)$ of the holon as a function of H in the Hubbard model near half-filling (In this figure $U/t = 2$ should read $U/t = 4$).

All the critical exponents are obtained in terms of the effective spins of the spinon and holon. We shall refrain from giving explicit formulae since one can readily check the field dependence of exponents using the formulae given in subsection 3.2.

Let us conclude this section by making a brief comment on the effective spin s_h of the holon. For $U/t \gg 1$ the s_h curve closely follows the magnetization curve of the Heisenberg chain, as observed in the t–J model. This observation is understood in the following way: In the strongly correlated regime near half-filling the spin state is almost degenerate, and hence all the band electrons contribute equally to the magnetization under non-zero field. Therefore making a hole in the ground-state Λ-distribution amounts to losing magnetization per lattice site. This in turn gives rise to the effective spin of the holon. Notice, however, that such a simple situation no longer holds as U/t becomes small. Therefore it should be realized that the effective spin of the holon has a different field dependence from the magnetization generically.

4.2. Relations to bulk quantities

Another interesting aspect of the Luttinger liquids is that the critical exponents can be expressed in terms of the bulk quantities. This kind of relation between the bulk quantity and the dressed charge was first noticed in [1,32]. We consider the three typical bulk quantities, the spin susceptibility χ_s, the compressibility χ_c and the specific heat coefficient γ, in the t–J model.

As shown in appendix B, the compressibility and the spin susceptibility are obtained as

$$\begin{aligned} \chi_c &= \xi_c(Q)^2/(\pi v_c) \\ \chi_s &= (g\mu_B)^2\xi_s^2/(\pi v_s) \qquad \xi_s = 1/\sqrt{2}\,. \end{aligned} \tag{4.6}$$

The low-temperature expansion of the free energy gives

$$\gamma = \frac{\pi}{3}\left(\frac{1}{v_c} + \frac{1}{v_s}\right) \tag{4.7}$$

which corresponds to two $c = 1$ conformal theories [30]. We thus find

$$\alpha_c = 4\tilde{\chi}_c/(2\tilde{\gamma} - \tilde{\chi}_s) \tag{4.8}$$

where we have renormalized the bulk quantities so that $\tilde{\gamma} = \tilde{\chi}_s = \tilde{\chi}_c = 1$ in the non-interacting limit. Note that this formula is also valid for the Hubbard model [10–12]. In the band bottom all these bulk quantities exhibit the divergent behaviour due to the dispersion relation in 1D electron systems. Approaching half-filling χ_s remains finite (a constant value of the Heisenberg model), while χ_c diverges as

$$\chi_c \simeq \frac{8(\ln 2)^2}{3\pi^2\zeta(3)}(\tfrac{1}{2} - \nu)^{-1} \tag{4.9}$$

due to the diverging density of states (see appendix B) [42], where ζ is the Riemann zeta function. Since γ is also divergent like $(\frac{1}{2} - \nu)^{-1}$ we have $\alpha_c \to 2\tilde{\chi}_c/\tilde{\gamma}$ for $\nu \to \frac{1}{2}$.

Let us next discuss an important role played by boundary conditions. Imposing twisted boundary conditions on the Bethe wavefunction does not ruin the exact integrability by virtue of the $U(1)$ symmetry of the system. Shastry and Sutherland then noticed that this was an efficient way to evaluate the effective current-carrying mass (transport mass) [43]. Under twisted boundary conditions with the twisting phase ϕ the shift of the ground-state energy from the periodic case (ϕ=0) is

$$E_0(\phi) - E_0(0) = \mathcal{D}_c\phi^2/N + \mathrm{O}(\phi^4)\,. \tag{4.10}$$

The interesting point is that the charge stiffness $\mathcal{D}_c$ is directly related to the DC part of the conductivity $\sigma(\omega)$

$$\mathrm{Re}\,\sigma(\omega) = \frac{2\pi e^2}{\hbar}\mathcal{D}_c\delta(\hbar\omega)\,. \tag{4.11}$$

For free electrons the coefficient of $\delta(\hbar\omega)$ is proportional to m^{-1} with m being the electron mass. Therefore it is legitimate to define the effective mass m^* through $m^*/m \propto \mathcal{D}_c^{-1}$ [43]. In view of conformal theories the energy shift due to twisted boundary conditions by ϕ is attributed to the excitation $I_c = I_s = D_s = 0$ and $D_c = \phi/2\pi$ [12]. From (3.1), (3.16) and (4.6)–(4.8) one can easily express the enhancement factor of the current-carrying mass in terms of the bulk quantities [44]

$$m^*/m = (2\tilde{\gamma} - \tilde{\chi}_s)^2/\tilde{\chi}_c\,. \tag{4.12}$$

Then, for instance, in the t–J model near half-filling the effective mass is extremely enhanced as

$$m^*/m \sim \frac{8(\ln 2)^2}{3\zeta(3)}(\tfrac{1}{2} - \nu)^{-1} \tag{4.13}$$

which corresponds to the fact that the system approaches the insulating phase.

To conclude this section we emphasize that formulae (4.8) and (4.12) are valid for any 1D correlated electron system, and hence characterize the universal properties of Luttinger liquids.

5. Conclusions

In this paper the long-distance properties of the t–J model at $t = J$ for arbitrary electron filling have been studied using the Bethe *ansatz* solution and the finite-size scaling method in conformal theory. The results are compared with the repulsive Hubbard model in detail. Starting with microscopic models we have shown explicitly that the electron behaviour in these highly correlated systems is characterized as the Luttinger liquid. The separation of the charge and spin degrees of freedom is quite essential. Consequently the charge sector is described by the Gaussian theory and the spin sector by the $c = 1$ SU(2) current algebra. The dressed charge matrix introduced in the Bethe *ansatz* calculation provides us with the precise link between the characteristic parameter of Luttinger liquids (i.e. Gaussian field periodicity) and the microscopic parameters in the theory. Notice that this is the most difficult step in a conventional bosonization approach. In conclusion we have presented the microscopic foundation of the concept of Luttinger liquids á la Haldane on the basis of conformal field theory and Bethe *ansatz* solutions.

Acknowledgments

We thank H Frahm and V E Korepin for useful communications.

Appendix A

In this appendix we present our calculation of the finite-size corrections in section 2 in such a way that it can be applied to generic nested Bethe *ansatz* solutions. Let us start with the Bethe *ansatz* equations

$$Np^0_\alpha(\lambda_j) = 2\pi I^\alpha_j - \sum_{\beta=1}^{l}\sum_{k=1}^{N_\beta}\phi_{\alpha\beta}(\lambda^\alpha_j - \lambda^\beta_k) \qquad \alpha = 1,\dots,l \qquad j = 1,\dots,N_\alpha \tag{A1}$$

where N denotes the system size, N_α is the number of 'particles' of the type α $(= 1,\dots,l)$ and $p^0_\alpha(\lambda)$ are the bare momenta. The phase shifts $\phi_{\alpha\beta}(\lambda)$ are assumed to obey $\phi_{\alpha\beta}(\lambda) = \phi_{\beta\alpha}(\lambda) = -\phi_{\alpha\beta}(-\lambda)$. We consider the case in which the quantum numbers I^α_j belong to the interval $[I^+_\alpha, I^-_\alpha]$ so that

$$I^+_\alpha - I^-_\alpha = N_\alpha \qquad I^+_\alpha + I^-_\alpha = 2D_\alpha\,. \tag{A2}$$

Define

$$z_{\alpha,N}(\lambda) = \frac{1}{2\pi}p^0_\alpha(\lambda) + \frac{1}{2\pi N}\sum_{\beta=1}^{l}\sum_{k=1}^{N_\beta}\phi_{\alpha\beta}(\lambda - \lambda^\beta_k) \tag{A3}$$

$$\rho_{\alpha,N}(\lambda) = \frac{\partial z_{\alpha,N}(\lambda)}{\partial\lambda} \qquad z_{\alpha,N}(q^\pm_\alpha) = \frac{I^\pm_\alpha}{N}\,. \tag{A4}$$

For $N \to \infty$ with $N_\alpha/N = \nu_\alpha$ and $D_\alpha/N = \delta_\alpha$ being fixed, the rapidity distribution functions satisfy

$$\rho_\alpha(\lambda|q^\pm) = \frac{1}{2\pi}a_\alpha(\lambda) + \sum_\beta \int_{q_\beta^-}^{q_\beta^+} \frac{d\lambda'}{2\pi} K_{\alpha\beta}(\lambda - \lambda')\rho_\beta(\lambda'|q^\pm) \tag{A5}$$

where $a_\alpha(\lambda) = p_\alpha^{0\prime}(\lambda)$ and $K_{\alpha\beta}(\lambda) = \phi'_{\alpha\beta}(\lambda)$. Then, for $z_\alpha(\lambda) = \lim_{N\to\infty} z_{\alpha,N}(\lambda)$, we have

$$z_\alpha(\lambda) = \frac{1}{2\pi}p_\alpha^0(\lambda) + \sum_\beta \int_{q_\beta^-}^{q_\beta^+} \frac{d\lambda'}{2\pi}\phi_{\alpha\beta}(\lambda - \lambda')\rho_\beta(\lambda'|q^\pm). \tag{A6}$$

Let us first calculate

$$\frac{\partial \nu_\alpha}{\partial q_\beta^+} = \frac{\partial}{\partial q_\beta^+}(z_\alpha(q_\alpha^+) - z_\alpha(q_\alpha^-)) = \frac{\partial}{\partial q_\beta^+}\int_{q_\alpha^-}^{q_\alpha^+} d\lambda\, \rho_\alpha(\lambda|q^\pm)\,. \tag{A7}$$

Introducing the dressed charge functions

$$\xi_{\alpha\beta}(\lambda_\beta) = \delta_{\alpha\beta} + \sum_\gamma \int_{-q_\gamma}^{q_\gamma} \frac{d\lambda'}{2\pi}\xi_{\alpha\gamma}(\lambda')K_{\gamma\beta}(\lambda' - \lambda_\beta) \tag{A8}$$

we obtain

$$\left.\frac{\partial \nu_\alpha}{\partial q_\beta^+}\right| = \rho_\beta(q_\beta)Z_{\alpha\beta} \tag{A9}$$

where the vertical bar is meant to put $q_\alpha^\pm = \pm q_\alpha$ and the $l \times l$ dressed charge matrix $\mathbf{Z}$ is given by $Z_{\alpha\beta} = \xi_{\alpha\beta}(q_\beta)$.

We next calculate

$$2\frac{\partial \delta_\alpha}{\partial q_\beta^+} = \frac{\partial z_\alpha(q_\alpha^+)}{\partial q_\beta^+} + \frac{\partial z_\alpha(q_\alpha^-)}{\partial q_\beta^-}\,. \tag{A10}$$

After some manipulations we get

$$\left.\frac{\partial \delta_\alpha}{\partial q_\beta^+}\right| = \rho_\beta(q_\beta)\tfrac{1}{2}(\mathbf{Z}^t)^{-1}_{\alpha\beta}\,. \tag{A11}$$

In a similar way one can show that

$$\left.\frac{\partial \nu_\alpha}{\partial q_\beta^-}\right| = -\left.\frac{\partial \nu_\alpha}{\partial q_\beta^+}\right| \qquad \left.\frac{\partial \delta_\alpha}{\partial q_\beta^-}\right| = \left.\frac{\partial \delta_\alpha}{\partial q_\beta^+}\right|. \tag{A12}$$

We thus find in matrix notation that

$$\begin{pmatrix} \rho d\mathbf{q}^+ \\ \rho d\mathbf{q}^- \end{pmatrix} = \begin{pmatrix} \frac{1}{2}\mathbf{Z}^{-1} & \mathbf{Z}^t \\ -\frac{1}{2}\mathbf{Z}^{-1} & \mathbf{Z}^t \end{pmatrix} \begin{pmatrix} d\nu \\ d\delta \end{pmatrix}. \tag{A13}$$

where $(\rho \mathrm{d}q^{\pm})_\alpha = \rho_\alpha(q_\alpha)\mathrm{d}q^{\pm}_\alpha(\mathrm{d}\nu)_\alpha = \mathrm{d}\nu_\alpha$ and $(\mathrm{d}\delta)_\alpha = \mathrm{d}\delta_\alpha$.

We are now ready to express the finite-size corrections in the energy

$$\varepsilon = \sum_\alpha \int_{q_\alpha^-}^{q_\alpha^+} \mathrm{d}\lambda\, \epsilon_\alpha^0(\lambda)\rho_\alpha(\lambda|q^\pm) \tag{A14}$$

in terms of the matrix $\mathbf{Z}$. Let us define the dressed energy functions

$$\epsilon_\alpha(\lambda|q^\pm) = \epsilon_\alpha^0(\lambda) + \sum_\beta \int_{q_\beta^-}^{q_\beta^+} \frac{\mathrm{d}\lambda'}{2\pi}\epsilon_\beta(\lambda'|q^\pm)K_{\beta\alpha}(\lambda'-\lambda) \tag{A15}$$

with the condition

$$\epsilon_\alpha(q_\alpha^\pm|q^\pm) = 0\,. \tag{A16}$$

This condition ensures the stationary condition

$$0 = \frac{\partial\varepsilon}{\partial q_\alpha^\pm} = \pm\epsilon_\alpha(q_\alpha^\pm|q^\pm)\rho_\alpha(q_\alpha^\pm|q^\pm)\,. \tag{A17}$$

Another basic relations are

$$\frac{1}{\rho_\alpha(q_\alpha)^2}\left(\frac{\partial}{\partial q_\alpha^\pm}\right)^2 \varepsilon\,\bigg| = 2\pi v_\alpha \tag{A18}$$

where v_α are the Fermi velocities.

Expanding $\varepsilon(q^\pm)$ to second order in $\mathrm{d}q_\alpha^\pm = q_\alpha^\pm \mp q_\alpha$ and substituting (A13) and (A18) we finally obtain

$$\varepsilon(q^\pm) = \varepsilon(\pm q) + 2\pi\left(\frac{1}{4}(\mathrm{d}\nu)^t(\mathbf{Z}^{-1})^t\mathbf{V}\mathbf{Z}^{-1}\mathrm{d}\nu + (\mathrm{d}\delta)^t\mathbf{Z}\mathbf{V}\mathbf{Z}^t\mathrm{d}\delta\right) \tag{A19}$$

where $V_{\alpha\beta} = v_\alpha\delta_{\alpha\beta}$. Note that $\mathrm{d}\delta_\alpha = D_\alpha/N$ and $\mathrm{d}\nu_\alpha = N_\alpha/N - \nu_\alpha^0$ where ν_α^0 is the value for the ground state. Hence $\epsilon(q^\pm) - \epsilon(\pm q)$ is of order N^{-2}.

Appendix B

We express the compressibility $\chi_c = \partial n_c^0/\partial\mu$ in terms of the dressed charge $\xi_c(Q)$ in (3.14). First notice the chain rule

$$\frac{\partial n_c^0}{\partial\mu} = \frac{\partial n_c^0}{\partial Q}\frac{\partial Q}{\partial\mu}\,. \tag{B1}$$

For zero magnetic field $(B \to +\infty)$ the rapidity distribution in the ground state satisfies

$$\rho_c(\Lambda) = R(\Lambda) + \int_{-Q}^{Q} \mathrm{d}\Lambda' R(\Lambda - \Lambda')\rho_c(\Lambda')\,. \tag{B2}$$

With the aid of an auxiliary function $F(\Lambda|\Lambda')$ obeying

$$F(\Lambda|\Lambda') = R(\Lambda - \Lambda') + \int_{-Q}^{Q} \mathrm{d}\nu R(\Lambda - \nu) F(\nu|\Lambda') \tag{B3}$$

we find from (3.14) and (B2) that

$$\begin{aligned} \xi_c(\Lambda) &= 1 + \int_{-Q}^{Q} F(\Lambda|\Lambda')\,\mathrm{d}\Lambda' \\ \frac{\partial \rho_c(\Lambda)}{\partial Q} &= \rho_c(Q)(F(\Lambda|Q) + F(\Lambda|-Q)). \end{aligned} \tag{B4}$$

It is now straightforward to show from (2.35) that

$$\frac{\partial n_c^0}{\partial Q} = -2\rho_c(Q)\xi_c(Q). \tag{B5}$$

The dressed energy function (2.33) for zero field obeys

$$\epsilon_c(\Lambda) = 2 + \mu - 2\pi R(\Lambda) + \int_{-Q}^{Q} \mathrm{d}\Lambda' R(\Lambda - \Lambda')\epsilon_c(\Lambda'). \tag{B6}$$

This function is subject to the condition $\epsilon_c(\pm Q) = 0$, according to which we obtain $\xi_c(Q) = -(\partial Q/\partial\mu)\epsilon_c'(Q)$. Using (B5) and (2.29) we thus verify the relation for χ_c in (4.6). The expression for the spin susceptibility $\chi_s = \partial\mathcal{M}/\partial H$ in (4.6) can be derived in a similar way by examining the asymptotic behaviour for $B \gg 1$.

Let us now check (4.9). Approaching half-filling we have $Q \to 0$, and hence from (2.35)

$$n_c^0 \simeq 1 - 2Q\rho_c(0). \tag{B7}$$

Equation (B2) yields $\rho_c(0) \simeq R(0)$. Thus $Q \simeq (1 - n_c^0)/(2R(0))$. The dressed charge behaves as $\xi_c(Q) \simeq 1 + (1 - n_c^0)$. Similarly it is seen from (B6) that $\epsilon_c'(Q) \simeq -\pi(1 - n_c^0)R''(0)/R(0)$. After all this we get

$$\chi_c \simeq -\frac{2}{\pi}\frac{R(0)^2}{R''(0)}(1 - n_c^0)^{-1}. \tag{B8}$$

Inserting $R(0) = (1/\pi)\ln 2$, $R''(0) = -(3/2\pi)\zeta(3)$ with ζ being the Riemann zeta function, and $\nu = n_c^0/2$ we obtain (4.9) in the text.

References

[1] Haldane F D M 1980 *Phys. Rev. Lett.* **45** 1358; 1981 *J. Phys. C: Solid State Phys.* **14** 2585; 1981 *Phys. Lett.* **81A** 153

[2] Haldane F D M 1981 *Phys. Rev. Lett.* **47** 1840

[3] See, for reviews:
Solyom J 1979 *Adv. Phys.* **28** 201

Emery V J 1979 *Highly Conducting One-Dimensional Solids* ed J T Devreese *et al* (New York: Plenum) p 247
Fukuyama H and Takayama H 1985 *Electronic Properties of Inorganic Quasi-One-Dimensional Compounds* ed P Monceau (Dordrecht: Reidel) p 41
[4] Anderson P W 1987 *Science* **235** 1196
[5] Sorella S, Tosatti E, Baroni S, Car R and Parrinello M 1988 *Int. J. Mod. Phys.* B **2** 993
[6] Sorella S, Parola A, Parrinello M and Tosatti E 1990 *Europhys. Lett.* **12** 721
[7] Imada M and Hatsugai Y 1989 *J. Phys. Soc. Japan* **58** 3752
[8] Ogata M and Shiba H 1990 *Phys. Rev.* B **41** 2326
[9] Kawakami N and Yang S-K 1990 *Phys. Rev. Lett.* **65** 2309
[10] Schulz H J 1990 *Phys. Rev. Lett.* **64** 2831
[11] Kawakami N and Yang S-K 1990 *Phys. Lett.* **148A** 359
[12] Frahm H and Korepin V E 1990 *Phys. Rev.* B **42** 10553
[13] Affleck I 1989 *Fields, Strings and Critical Phenomena* ed E Brezin and J Zinn-Justin (Amsterdam: Elsevier) p 563
[14] Affleck I 1989 *Banff Lectures* (unpublished)
[15] Schlottmann P 1987 *Phys. Rev.* B **36** 5177
[16] Lieb E H and Wu F Y 1968 *Phys. Rev. Lett.* **20** 1445
[17] Belavin A A, Polyakov A A and Zamolodchikov A B 1984 *Nucl. Phys.* B **241** 333
See for a collection of reprints: Itzykson C, Saleur H and Zuber J-B (ed.) 1988 *Conformal Invariance and Applications in Statistical Mechanics* (Singapore: World Scientific)
[18] Cardy J L 1986 *Nucl. Phys.* B **270** [FS16] 186
[19] de Vega H J and Woynarovich F 1985 *Nucl. Phys.* B **251** [FS13] 439
Woynarovich F and Eckle H-P 1987 *J. Phys. A: Math. Gen.* **20** L97
[20] Pokrovskii S V and Tsvelik A M 1987 *Sov. Phys.–JETP* **66** 1275
[21] Suzuki J 1988 *J. Phys. A: Math. Gen.* **21** L1175
[22] Izergin A G, Korepin V E and Reshetikhin N Yu 1989 *J. Phys. A: Math. Gen.* **22** 2615
[23] Frahm H and Yu N C 1990 *J. Phys. A: Math. Gen.* **23** 2115
[24] Zhang F C and Rice T M 1988 *Phys. Rev.* B **37** 3759
[25] Sutherland B 1975 *Phys. Rev.* B **12** 3795
[26] Bares P A and Blatter G 1990 *Phys. Rev. Lett.* **64** 2567
[27] Kawakami N and Okiji A 1981 *Phys. Lett.* **86A** 483
[28] Woynarovich F 1989 *J. Phys. A: Math. Gen.* **22** 4243
[29] Woynarovich F, Eckle H-P and Truong T T 1989 *J. Phys. A: Math. Gen.* **22** 4027
[30] Blöte H W J, Cardy J L and Nightingale M P 1986 *Phys. Rev. Lett.* **56** 742
Affleck I 1986 *Phys. Rev. Lett.* **56** 746
[31] Imada M 1990 *J. Phys. Soc. Japan* **59** 4121
[32] Bogoliubov N M, Izergin A G and Korepin V E 1986 *Nucl. Phys.* B **275** 687
[33] Kawakami N and Okiji A 1990 *Prog. Theor. Phys.* Suppl. **101** 429
[34] Griffiths R B 1964 *Phys. Rev.* **133** A768
[35] Ogata M, Sugiyama T and Shiba H 1991 *Phys. Rev.* B **43** 8401
[36] Kadanoff L P and Brown A C 1979 *Ann. Phys., NY* **121** 318
Yang S-K 1987 *Nucl. Phys.* B **285** [FS19] 183
[37] Luttinger J M 1963 *J. Math. Phys.* **4** 1154
[38] Gutfreund H and Schick M 1968 *Phys. Rev.* **168** 418
[39] Luther A and Peschel I 1974 *Phys. Rev.* B **9** 2911
[40] Parola A and Sorella S 1990 *Phys. Rev. Lett.* **64** 1831
[41] Anderson P W 1990 *Phys. Rev. Lett.* **64** 1839
[42] Usuki T, Kawakami N and Okiji A 1989 *Phys. Lett.* **135A** 476
Kawakami N, Usuki T and Okiji A 1989 *Phys. Lett.* **137A** 287
[43] Shastry B S and Sutherland B 1990 *Phys. Rev. Lett.* **65** 243
[44] Kawakami N and Yang S-K 1990 *Phys. Rev. Lett.* **65** 3063

Nuclear Physics B396 (1993) 611–638
North-Holland

NUCLEAR PHYSICS B

Algebraic properties of the Bethe ansatz for an spl(2,1)-supersymmetric t–J model

Angela Foerster [1] and Michael Karowski [2]

Institut für Theoretische Physik, Freie Universität Berlin, Berlin, Germany

Received 24 August 1992
Accepted for publication 6 November 1992

We investigate the algebraic structure of the supersymmetric $t-J$ model in one dimension. We prove that the Bethe ansatz states are highest-weight vectors of an spl(2,1) superalgebra. By acting with shift operators we construct a complete set of states for this model. In addition we analyse the multiplet structure of the anti-ferromagnetic ground state and some low-lying excitations. It turns out that the ground state is a member of a quartet.

1. Introduction

Since the pioneering work of Bethe [1] and a subsequent work of Faddeev and Takhtajan [2] on the isotropic Heisenberg model, it is known that the Bethe ansatz alone does not provide a complete set of states instead it only determines the highest-weight vectors of multiplets of the underlying SU(2) symmetry group. Recently, Essler et al. [3] proved that for the one-dimensional Hubbard model the Bethe ansatz states are lowest-weight vectors with respect to the SO(4) symmetry. In this paper we show that this feature, which is essential to construct a complete set of states, also appears in the context of a supersymmetric integrable model. However, the algebraic structure is more complicated and exhibits new interesting properties, e.g. the anti-ferromagnetic ground state is not a singlet but a member of a higher multiplet.

We investigate a model of classical statistical physics in two dimensions, an spl(2,1)-supersymmetric 15-vertex model, which is a generalization of the 6-vertex model. Each link in the lattice can assume one of three states where two are

[1] Supported by Conselho Nacional de Desenvolvimento Científico e Tecnológico (CNPq, Brasil) and Deutscher Akademischer Austauschdienst (DAAD).
[2] Supported by Deutsche Forschungsgemeinschaft: Sonderforschungsbereich 288 "Differentialgeometrie und Quantenphysik".

bosonic and one is fermionic. The results for the spl(2,1)-supersymmetric 15-vertex model are easily translated to the one-dimensional t–J model (for special values of the couplings t and J). Recently this model has attracted much interest in connection with high-T_c superconductivity. It describes a quantum system of electrons on a one-dimensional chain, where at a lattice point there may be an electron with spin up or spin down or a hole. The hamiltonian for a lattice of L sites is given by [4]

$$\mathscr{H} = P\left\{ -t \sum_{j,\sigma} \left(c^{\dagger}_{j,\sigma} c_{j+1,\sigma} + c^{\dagger}_{j+1,\sigma} c_{j,\sigma} \right) \right\} P + J \sum_{j} \left(S_j \cdot S_{j+1} - \frac{n_j n_{j+1}}{4} \right), \quad (1.1)$$

where the projector $P = \prod_{j=1}^{L} (1 - n_{j\uparrow} n_{j\downarrow})$ restricts the Hilbert space by the constraint of no double occupancy at one lattice point.

We present an explicit construction of the eigenvalues and eigenvectors of the transfer matrix of the spl(2,1)-supersymmetric 15-vertex model using the algebraic nested Bethe ansatz method [5,6]. By this procedure the problem of finding the spectrum is reduced to the problem of solving a system of coupled transcendental equations, the Bethe ansatz equations (BAE). We find three different kinds of BAE, which correspond to three different possible choices of pseudovacua. Two of these forms of BAE were already obtained by Lai [7], Schlottmann [8], Sutherland [9] and Sarkar [10] using similar methods. Moreover, we analyse in detail the algebraic structure of the eigenvectors obtained by this nested construction. From the invariance of the transfer matrix (and consequently of the one-dimensional t–J hamiltonian) with respect to the spl(2,1) superalgebra it follows that the eigenstates are classified in terms of supermultiplets corresponding to irreducible representations of this superalgebra. We analyse the structure of these representations. In addition, we prove that the Bethe ansatz states are highest-weight vectors of the spl(2,1) superalgebra, which was investigated by Scheunert et al. [11]. Therefore, by acting with the spl(2,1) lowering operators on the Bethe states we obtain additional eigenvectors. Finally, the total number of orthogonal eigenvectors generated by this procedure leads to a complete set of states. This result has been already announced in ref. [12].

The paper is organized as follows. In sect. 2 the spl(2,1) vertex model, as well as its transfer matrix, is defined on a two-dimensional lattice. We also give the relation between the transfer matrix and the one-dimensional supersymmetric t–J model. In sect. 3 we diagonalize the transfer matrix using the quantum inverse-scattering method. In sect. 4 the algebraic structure of the Bethe vectors is investigated. Our results for lattices with small and large number of sites are illustrated in sect. 5, where the structure of the ground state is also discussed. In sect. 6 we give details of the proof of the completeness problem of the Bethe states of this model and sect. 7 contains a summary of the main results.

2. The spl(2,1) vertex model and Yang–Baxter algebra

The graded 15-vertex model is a lattice model of classical statistical physics in two dimensions. Its partition function on a $L \times L'$ (L columns and L' rows) periodic square lattice is given as

$$\mathcal{Z} = \sum_{\text{conf.}} \prod_{x \in L \times L'} S(x), \tag{2.1}$$

where the sum extends over all allowed "bond configurations". Each bond can accept one of three states characterised by $\alpha = 1, 2, 3$, which can be bosonic (B) or fermionic (F). In what follows we will adopt the convention $1 = \text{B}$, $2 = \text{B}$, $3 = \text{F}$. We follow the general strategy of the algebraic Bethe ansatz of Faddeev et al. [5]. The vertex weights $S(x)$ are determined by 15 bond configurations at the lattice site x, and take the following values:

$$S(v)^{\gamma\delta}_{\alpha\beta} = \gamma \underset{\beta}{\overset{\delta}{-\!\!\!+\!\!\!-}} \alpha = \sigma_{\gamma\delta} \delta^{\gamma}_{\alpha} \delta^{\delta}_{\beta} - \frac{2}{v} \delta^{\gamma}_{\beta} \delta^{\delta}_{\alpha}. \tag{2.2}$$

The parametrization in terms of the spectral parameter "v" has been introduced for later convenience (see eq. (2.11)). The sign factor σ takes care of the statistics,

$$\sigma_{\gamma\delta} = \begin{cases} -1, & \text{if } \gamma = \delta = 3 \text{ (fermionic)} \\ 1, & \text{otherwise.} \end{cases} \tag{2.3}$$

S can be considered as a matrix acting in the tensor product of two three-dimensional auxiliary spaces $\mathbb{C}^3 \times \mathbb{C}^3$ and can be arranged as a 9×9 matrix,

$$S^{\gamma\delta}_{\alpha\beta}(v) = \left(\begin{array}{ccc|ccc|ccc} a & 0 & 0 & 0 & 0 & 0 & 0 & 0 & 0 \\ 0 & b & 0 & c & 0 & 0 & 0 & 0 & 0 \\ 0 & 0 & b & 0 & 0 & 0 & c & 0 & 0 \\ \hline 0 & c & 0 & b & 0 & 0 & 0 & 0 & 0 \\ 0 & 0 & 0 & 0 & a & 0 & 0 & 0 & 0 \\ 0 & 0 & 0 & 0 & 0 & b & 0 & c & 0 \\ \hline 0 & 0 & c & 0 & 0 & 0 & b & 0 & 0 \\ 0 & 0 & 0 & 0 & 0 & c & 0 & b & 0 \\ 0 & 0 & 0 & 0 & 0 & 0 & 0 & 0 & w \end{array} \right), \tag{2.4}$$

where

$$a = 1 - \frac{2}{v}, \quad b = 1, \quad c = -\frac{2}{v}, \quad w = -1 - \frac{2}{v}. \tag{2.5}$$

We define the monodromy matrix as the matrix product over the S's in the following way:

$$T_{\alpha\{\beta\}}^{\gamma\{\delta\}}(v) = S_{\alpha_2\beta_1}^{\gamma\,\delta_1}(v)\,S_{\alpha_3\beta_2}^{\alpha_2\delta_2}(v)\ldots S_{\alpha\,\beta_L}^{\alpha_L\delta_L}(v),$$

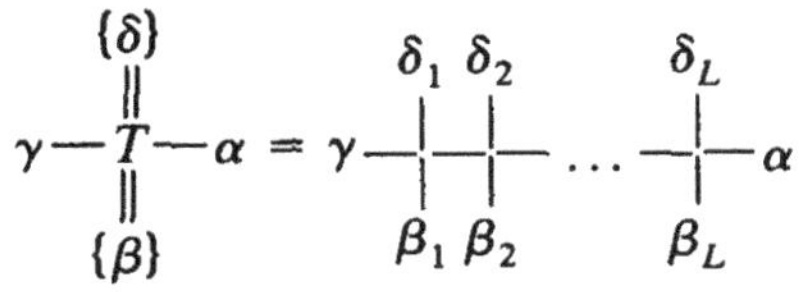

(2.6)

This monodromy matrix acts in the tensor product of an auxiliary space and a "quantum space" $\mathbb{C}^3 \times \mathbb{C}^{3L}$ and can be regarded as a 3×3 matrix of matrices acting in the "quantum space",

$$T_{\alpha}^{\gamma}(v) = \begin{pmatrix} A & B_2 & B_3 \\ C_2 & D_1 & D_2 \\ C_3 & D_3 & D_4 \end{pmatrix}. \tag{2.7}$$

The transfer matrix is defined as a trace of the monodromy matrix in the auxiliary space,

$$\tau_{\{\beta\}}^{\{\delta\}}(v) = \sum_{\alpha} \tilde{T}_{\alpha\{\beta\}}^{\alpha\{\delta\}}(v) = \sum_{\alpha} \sigma_{\alpha\alpha}\sigma_{\alpha\{\delta\}} T_{\alpha\{\beta\}}^{\alpha\{\delta\}}(v), \tag{2.8}$$

where

$$\sigma_{\alpha\{\delta\}} = \prod_i \sigma_{\alpha\delta_i}. \tag{2.9}$$

Here the σ-factors take into account the fact that we are dealing with bosons and fermions.

The thermodynamic properties of the vertex model can be obtained from the solutions of the eigenvalue problem of the transfer matrix,

$$\tau\Psi = \lambda\Psi. \tag{2.10}$$

This eigenvalue problem will be solved in sect. 3 by means of the nested Bethe ansatz.

It can easily be shown that the matrix S given by eq. (2.2) fulfills the Yang–Baxter equation

$$S_{\alpha'\,\beta'}^{\alpha''\beta''}(v-v')\,S_{\alpha\,\gamma'}^{\alpha'\gamma''}(v)\,S_{\beta\,\gamma}^{\beta'\gamma'}(v') = S_{\beta'\,\gamma'}^{\beta''\gamma''}(v')\,S_{\alpha'\,\gamma}^{\alpha''\gamma'}(v)\,S_{\alpha\,\beta}^{\alpha'\beta'}(v-v'). \tag{2.11}$$

By means of iterations we can also prove the Yang–Baxter relation for the monodromy matrix T,

$$S^{\alpha''\beta''}_{\alpha'\beta'}(v-v')T^{\alpha'\{\gamma''\}}_{\alpha\{\gamma'\}}(v)T^{\beta'\{\gamma'\}}_{\beta\{\gamma\}}(v') = T^{\beta''\{\gamma''\}}_{\beta'\{\gamma'\}}(v')T^{\alpha''\{\gamma'\}}_{\alpha'\{\gamma\}}(v)S^{\alpha'\beta'}_{\alpha\beta}(v-v'). \tag{2.12}$$

In addition conservation of fermions imply the following property of the T-matrix:

$$\sigma_{\alpha\beta''}\sigma_{\alpha\{\gamma''\}}T^{\beta''\{\gamma''\}}_{\beta\{\gamma\}}(v) = \alpha_{\alpha\beta}\sigma_{\alpha\{\gamma\}}T^{\beta''\{\gamma''\}}_{\beta\{\gamma\}}(v), \tag{2.13}$$

for all $\alpha = 1$, 2 or 3.

The Yang–Baxter equation for the monodromy matrix (2.12) together with property (2.13) imply the commutativity of the transfer matrix for different spectral parameters,

$$[\tau(v), \tau(v')] = 0. \tag{2.14}$$

This reflects the integrability of the model. In fact, the eigenvalue problem (2.10) can be solved exactly by the Bethe ansatz method.

At the end of this section we will show that the above defined transfer matrix is related to the one-dimensional supersymmetric t–J model, such that if we solve the eigenvalue problem of the transfer matrix τ we will automatically diagonalize the hamiltonian of the one-dimensional supersymmetric t–J model.

The hamiltonian of the t–J model for a one-dimensional lattice of L sites is given as [4]

$$\mathscr{H} = P\left\{-t\sum_{j,\sigma}\left(c^{\dagger}_{j,\sigma}c_{j+1,\sigma} + c^{\dagger}_{j+1,\sigma}c_{j,\sigma}\right)\right\}P + J\sum_{j}\left(\boldsymbol{S}_j\cdot\boldsymbol{S}_{j+1} - \frac{n_j n_{j+1}}{4}\right), \tag{2.15}$$

where the $c^{(\dagger)}_{j\pm}$ are spin up or down annihilation (creation) operators, the $\boldsymbol{S}_j$ spin matrices and the n_j occupation numbers of electrons at lattice site j. The projector $P = \prod_{j=1}^{L}(1 - n_{j\uparrow}n_{j\downarrow})$ restricts the Hilbert space by the constraint of no double occupancy at one lattice point. Therefore, at each lattice site we have three possibilities $(1, 2, 3) \equiv (\uparrow, \downarrow, 0)$, i.e. an electron with spin up or down or no electron (hole). This hamiltonian can be rewritten in terms of Hubbard's projection operators [13],

$$X_j^{\alpha\beta} = |\alpha_j\rangle\langle\beta_j| \quad (\alpha, \beta = 1, 2, 3), \tag{2.16}$$

where $|1_j(2_j)\rangle$ denotes an electron with spin up (down) and $|3_j\rangle$ a hole at site j. Using (2.16), up to a chemical potential the hamiltonian reads

$$\mathscr{H} = -t\sum_{\alpha=1}^{2}\sum_{j=1}^{L}\left(X_j^{\alpha 3}X_{j+1}^{3\alpha} + X_{j+1}^{\alpha 3}X_j^{3\alpha}\right) + \tfrac{1}{2}J\sum_{j=1}^{L}\left(\sum_{\alpha,\beta=1}^{2} X_j^{\alpha\beta}X_{j+1}^{\beta\alpha} - X_j^{33}X_{j+1}^{33}\right). \tag{2.17}$$

For convenience we will consider the hole operators as fermions and the spin operators as bosons. In fact, this choice is possible since in one-dimension there exists a transformation exchanging bosons and fermions. Therefore, the spectrum of the t–J model with two fermions and one boson is equivalent to the spectrum of the t–J model with two bosons and one fermion (for even L) [10].

For $J = 2t$ the t–J model is "supersymmetric" and connected to the previously defined vertex model through the relation

$$\mathscr{H} = -2\frac{\partial}{\partial v}\ln(v^L\tau(v))\bigg|_{v=0}. \tag{2.18}$$

The proof of this identity is analogous to the one for the isotropic Heisenberg model [14].

3. Construction of Bethe eigenvectors

The main subject of this section will be solving the eigenvalue problem of the transfer matrix

$$\tau\Psi = \lambda\Psi \tag{3.1}$$

through an algebraic construction [5] based on the Yang–Baxter algebra of the monodromy matrices

$$S_{\alpha'\beta'}^{\alpha''\beta''}(v-v')T_{\alpha\ \{\gamma'\}}^{\alpha'\{\gamma''\}}(v)T_{\beta\ \{\gamma\}}^{\beta'\{\gamma'\}}(v') = T_{\beta'\{\gamma'\}}^{\beta''\{\gamma''\}}(v')T_{\alpha'\{\gamma\}}^{\alpha''\{\gamma'\}}(v)S_{\alpha\beta}^{\alpha'\beta'}(v-v'). \tag{3.2}$$

The monodromy matrix T can be written as a 3×3 matrix,

$$\begin{pmatrix} A & B_2 & B_3 \\ C_2 & D_1 & D_2 \\ C_3 & D_3 & D_4 \end{pmatrix}. \tag{3.3}$$

This suggests solving the problem by means of the nested Bethe ansatz with two levels [6]. The transfer matrix is given by a trace of the monodromy matrix T (see eq. (2.8)). For the first-level Bethe ansatz the operators B_α (C_α) $(\alpha = 2, 3)$ play the role of creation (annihilation) operators of "pseudoparticles". The first-level "pseudovacuum" Φ is defined by the equation

$$C_{\gamma\{\beta'\}}^{\ \{\beta\}}\,\Phi^{\{\beta'\}} = 0,$$

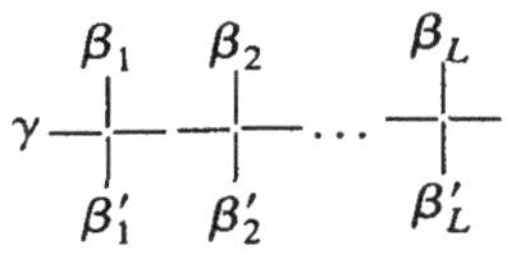

$\Phi^{\{\beta'\}} = 0$ for $\gamma = 2, 3$. (3.4)

Since at a vertex a generalized "ice rule" holds (see eq. (2.2)) the solution of this equation is

$$\Phi^{\{\beta\}} = \prod_{i=1}^{L} \delta_{\beta_i 1} = \begin{array}{cccc} \beta_1 & \beta_2 & & \beta_L \\ | & | & \cdots & | \\ 1 & 1 & & 1 \end{array} . \tag{3.5}$$

This pseudovacuum is an eigenstate of A,

$$A^{\{\beta\}}_{\{\beta'\}}(v)\,\Phi^{\{\beta'\}} = a^L(v)\Phi^{\{\beta\}},$$

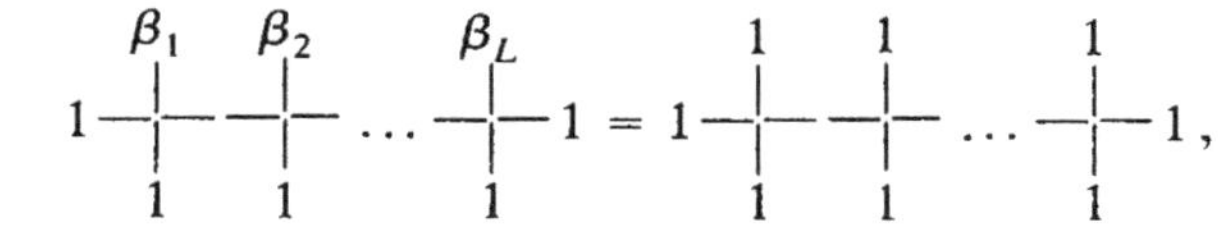

(3.6)

and also of D_1 and D_4,

$$D_{1(4)\{\beta'\}}^{\{\beta\}}(v)\,\Phi^{\{\beta'\}} = b^L(v)\Phi^{\{\beta\}},$$

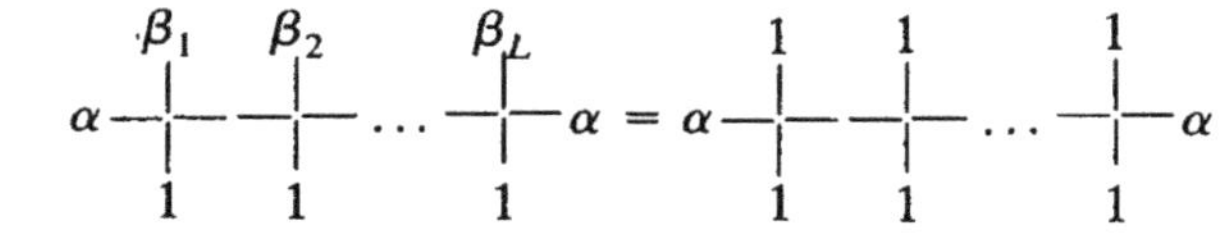

(3.7)

($\alpha = 2$ and 3, respectively). Because of the special form of the matrix S of eq. (2.2) the summations over the internal lines in eqs. (3.6) and (3.7) are trivial. In eq. (3.6) they can assume only the value 1, and in eq. (3.7) only the fixed value $\alpha = 2$ or 3, respectively. The action of B_α ($\alpha = 2$ or 3) on the "pseudovacuum" yields new states. So, the $\{B_\alpha\}$ can be considered as "creation operators" and the eigenvector of the transfer matrix can be obtained by successive application of the B's according to the first-level Bethe ansatz

$$\Psi^{\{\beta\}} = B_{\alpha_1\{\eta_1\}}^{\{\beta\}}(v_1)\,B_{\alpha_2\{\eta_2\}}^{\{\eta_1\}}(v_2)\dots B_{\alpha_N\{\eta_N\}}^{\{\eta_{N-1}\}}(v_N)\,\Phi^{\{\eta_N\}}\Psi_{(1)}^{\{\alpha\}}, \tag{3.8}$$

where the summations over the α_i ($i = 1,\dots, N$) are restricted to $\alpha_i = 2$, 3. The coefficients $\Psi_{(1)}^{\{\alpha\}}$ are to be to be determined by the second-level Bethe ansatz. This means the eigenvalue problem of the transfer matrix (3.1) will be solved in a recurrent way (nested Bethe ansatz method). The requirement that Ψ is an eigenvector of τ leads to another eigenvalue problem for a new transfer matrix $\tau_{(1)}$, as will be shown later. Now we start to solve eq. (3.1). Following the general strategy of the algebraic Bethe ansatz [5] we apply the transfer matrix $\tau(v)$ (2.8) to the state Ψ given by eq. (3.8),

$$\tau^{\{\beta''\}}_{\{\beta'\}}(v)\Psi^{\{\beta'\}} = \left(A^{\{\beta''\}}_{\{\beta'\}}(v) + \tau_{D\{\beta'\}}^{\{\beta''\}}(v)\right)\Psi^{\{\beta'\}}, \tag{3.9}$$

where

$$\tau_{D\{\beta'\}}^{\{\beta''\}}(v) = \sum_{\alpha=2}^{3} \tilde{T}_{\alpha\{\beta'\}}^{\alpha\{\beta''\}}(v) = \sum_{\alpha=2}^{3} \sigma_{\alpha\alpha}\sigma_{\alpha\{\beta''\}}T_{\alpha\{\beta'\}}^{\alpha\{\beta''\}}(v). \tag{3.10}$$

In order to commute $A(v)$, $D_1(v)$ and $D_4(v)$ through all $B(v_i)$ towards Φ and then apply (3.6) and (3.7) we use the property (2.13) and the following commutation rules, derived from the Yang–Baxter relation (3.2):

$$A(v)B_\alpha(v') = \frac{a(v'-v)}{b(v'-v)}B_\alpha(v')A(v) - \frac{c(v'-v)}{b(v'-v)}B_\alpha(v)A(v'), \tag{3.11}$$

and

$$T_{\gamma'}^{\gamma}(v)B_\alpha(v') = \frac{1}{b(v-v')}\big(B_{\alpha'}(v')T_{\gamma''}^{\gamma}(v)S_{\gamma'\ \alpha}^{\gamma''\alpha'}(v-v')$$

$$-c(v-v')B_{\gamma'}(v)T_\alpha^\gamma(v')\big) \tag{3.12}$$

$$B_\alpha(v)\ B_\beta(v') = \frac{1}{a(v-v')}B_{\beta'}(v')\ B_{\alpha'}(v)S_{\alpha\ \beta}^{\alpha'\beta'}(v-v'). \tag{3.13}$$

All indices of the auxiliary space in eqs. (3.11), (3.12) and (3.13) assume only the values 2 and 3. Using eq. (3.11) two types of terms arise when A is commuted through B_α. In the first type A and B_α preserve their arguments and in the second type their arguments are exchanged. The first kind of terms are called "wanted terms", since they will give a vector proportional to Ψ and the second type are the "unwanted terms (u.t.)". Then, using eqs. (3.9), (3.8), (2.13), (3.11) and (3.6), we get

$$A_{\{\beta'\}}^{\{\beta''\}}(v)\Psi^{\{\beta'\}} = \lambda_A(v)\Psi^{\{\beta''\}} + \text{u.t.}(A), \tag{3.14}$$

where the coefficient λ_A is given by

$$\lambda_A(v) = a^L(v)\prod_{i=1}^{N}\frac{a(v_i-v)}{b(v_i-v)}. \tag{3.15}$$

Correspondingly we obtain from eq. (3.12) the wanted and unwanted terms in the form

$$\tau_{D\{\beta'\}}^{\{\beta''\}}(v)\Psi^{\{\beta'\}} = b^L(v)\prod_{i=1}^{N}\frac{1}{b(v-v_i)}B_{\alpha_1\{\eta_1\}}^{\{\beta''\}}(v_1)B_{\alpha_2\{\eta_2\}}^{\{\eta_1\}}(v_2)\cdots$$

$$\times B_{\alpha_N\{\eta_N\}}^{\{\eta_{N-1}\}}(v_N)\Phi^{\{\eta_N\}}\tau_{(1)\{\alpha\}}^{\{\alpha'\}}(v,\{v_i\})\Psi_{(1)}^{\{\alpha\}} + \text{u.t.}(D), \tag{3.16}$$

where we have introduced a new (the second-level) transfer matrix

$$\tau_{(1)\{\alpha\}}^{\{\alpha'\}}(v, \{v_i\}) = \sum_{\beta=2}^{3} \sigma_{\beta\beta}\sigma_{\beta\{\alpha'\}} T_{(1)\beta\{\alpha\}}^{\beta\{\alpha'\}}(v, \{v_i\}) \tag{3.17}$$

as a trace (over only $\beta = 2$ and 3) of the second-level monodromy matrix. This is given by $T_{(1)} = S(v - v_N) \dots S(v - v_1)$ in analogy to eq. (2.6). Now, however, all indices (the external and the internal ones) assume only the values 2 and 3, as in the internal block of the matrix T denoted in eq. (3.3). In order to obtain in eq. (3.16) a "wanted term" proportional to Ψ, the vector $\Psi_{(1)}$ has to fulfill the eigenvalue equation

$$\tau_{(1)\{\alpha\}}^{\{\alpha'\}}(v, \{v_i\}) \Psi_{(1)}^{\{\alpha\}} = \lambda_{(1)}(v, \{v_i\}) \Psi_{(1)}^{\{\alpha'\}}, \tag{3.18}$$

which is solved by the second-level Bethe ansatz. The monodromy matrix $T_{(1)}$ belongs to an SL(1,1) 6-vertex model slightly modified compared to the SU(2) one due to the presence of fermions. If we identify $T_{(1)2}^{\ 2} \equiv A_{(1)}$, $T_{(1)3}^{\ 2} \equiv B_{(1)}$, $T_{(1)2}^{\ 3} \equiv C_{(1)}$ and $T_{(1)3}^{\ 3} \equiv D_{(1)}$ again $B_{(1)}$ $(C_{(1)})$ can be interpreted as a creation (annihilation) operator with respect to the "pseudovacuum" $\Phi_{(1)}$, which is now of the form

$$\Phi_{(1)}^{\{\alpha\}} = \prod_{i=1}^{N} \delta_{\alpha_i,2} = \begin{matrix} \alpha_N \\ | \\ 2 \end{matrix} \ \cdots \ \begin{matrix} \alpha_2 \\ | \\ 2 \end{matrix} \ \begin{matrix} \alpha_1 \\ | \\ 2 \end{matrix} \ . \tag{3.19}$$

It is an eigenstate of $A_{(1)}$ and $D_{(1)}$, satisfying

$$A_{(1)\{\alpha'\}}^{\{\alpha\}}(v, \{v_i\}) \Phi_{(1)}^{\{\alpha'\}} = \prod_{i=1}^{N} a(v - v_i) \Phi_{(1)}^{\{\alpha\}},$$

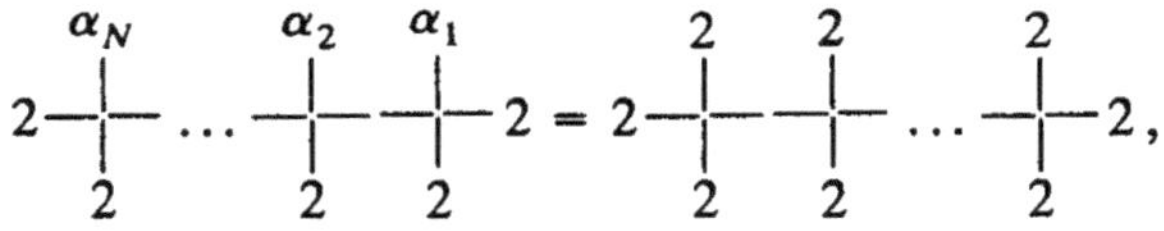

(3.20)

$$D_{(1)\{\alpha'\}}^{\{\alpha\}}(v, \{v_i\}) \Phi_{(1)}^{\{\alpha'\}} = \prod_{i=1}^{N} b(v - v_i) \Phi_{(1)}^{\{\alpha\}},$$

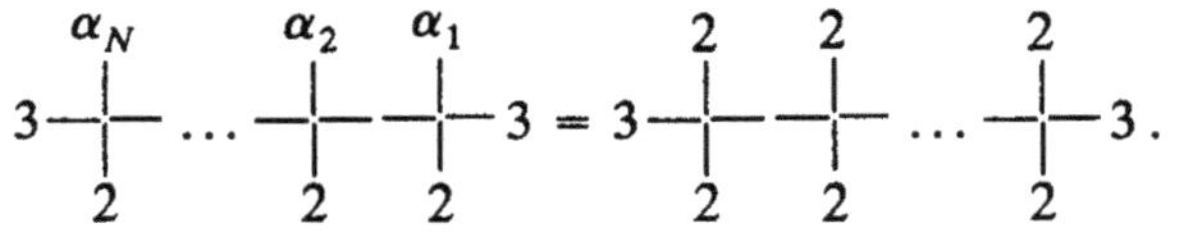

(3.21)

The summations over the internal lines in eqs. (3.20) and (3.21) are only over the values 2 and 3, respectively. The eigenvector $\Psi_{(1)}$ of $\tau_{(1)}$ is given by the second-level

Bethe ansatz

$$\Psi_{(1)}^{\{\alpha\}} = B_{(1)\{\eta_1\}}^{\{\alpha\}}(\gamma_1, \{v_i\}) B_{(1)\{\eta_2\}}^{\{\eta_1\}}(\gamma_2, \{v_i\}) \dots B_{(1)\{\eta_M\}}^{\{\eta_{M-1}\}}(\gamma_M, \{v_i\}) \Phi_{(1)}^{\{\eta_M\}}. \quad (3.22)$$

Following a strategy analogous to the one above, we apply $\tau_{(1)}$ to the state $\Psi_{(1)}$ and commute $A_{(1)}(v, \{v_i\})$ and $D_{(1)}(v, \{v_i\})$ through the $B_{(1)}(\gamma_\alpha, \{v_i\})$ towards $\Phi_{(1)}$ and then use eqs. (3.20) and (3.21). Since the Yang–Baxter algebra for the monodromy matrices (3.2) is also valid in the inhomogeneous cases when $T(v)$ is replaced by $T(v, \{v_i\})$ [15], we derive the following commutation relations:

$$A_{(1)}(v, \{v_i\}) B_{(1)}(v', \{v_i\}) = \frac{a(v'-v)}{b(v'-v)} B_{(1)}(v', \{v_i\}) A_{(1)}(v, \{v_i\})$$

$$- \frac{c(v'-v)}{b(v'-v)} B_{(1)}(v, \{v_i\}) A_{(1)}(v', \{v_i\}), \quad (3.23)$$

$$D_{(1)}(v, \{v_i\}) B_{(1)}(v', \{v_i\}) = \frac{w(v-v')}{b(v-v')} B_{(1)}(v', \{v_i\}) D_{(1)}(v, \{v_i\})$$

$$- \frac{c(v-v')}{b(v-v')} B_{(1)}(v, \{v_i\}) D_{(1)}(v', \{v_i\}). \quad (3.24)$$

$$B_{(1)}(v, \{v\}) B_{(1)}(v', \{v\}) = \frac{w(v-v')}{a(v-v')} B_{(1)}(v', \{v\}) B_{(1)}(v, \{v\}). \quad (3.25)$$

Using eqs. (3.17), (3.22), (2.13), (3.23), (3.24), (3.20) and (3.21) as above we obtain again wanted and unwanted terms,

$$\tau_{(1)\{\alpha\}}^{\{\alpha'\}}(v, \{v_i\}) \Psi_{(1)}^{\{\alpha\}} = \left(\lambda_{A_{(1)}}(v, \{v_i\}) + \lambda_{D_{(1)}}(v, \{v_i\})\right) \Psi_{(1)}^{\{\alpha'\}} + \text{u.t.}(A_{(1)}) + \text{u.t.}(D_{(1)}), \quad (3.26)$$

where

$$\lambda_{A_{(1)}} = \prod_{i=1}^{N} a(v-v_i) \prod_{\beta=1}^{M} \frac{a(\gamma_\beta - v)}{b(\gamma_\beta - v)}, \quad (3.27)$$

$$\lambda_{D_{(1)}} = -(-1)^M \prod_{i=1}^{N} b(v-v_i) \prod_{\beta=1}^{M} \frac{w(v-\gamma_\beta)}{b(v-\gamma_\beta)}. \quad (3.28)$$

Substituting these equations in (3.16) and taking (3.8) into account we get, in case the unwanted terms u.t.$(A_{(1)})$ and u.t.$(D_{(1)})$ cancel,

$$\tau_{D\{\beta'\}}^{\{\beta''\}}(v) \Psi^{\{\beta'\}} = \left(\lambda_{D_{\rm I}}(v) + \lambda_{D_{\rm II}}(v)\right) \Psi^{\{\beta''\}} + \text{u.t.}(D), \quad (3.29)$$

where λ_{D_I} and $\lambda_{D_{II}}$ are given by

$$\lambda_{D_I}=b^L(v)\prod_{i=1}^{N}\frac{a(v-v_i)}{b(v-v_i)}\prod_{\beta=1}^{M}\frac{a(\gamma_\beta-v)}{b(\gamma_\beta-v)}, \tag{3.30}$$

$$\lambda_{D_{II}}=-(-1)^M b^L(v)\prod_{\beta=1}^{M}\frac{w(v-\gamma_\beta)}{b(v-\gamma_\beta)}. \tag{3.31}$$

Finally, combining eqs. (3.14) and (3.29) we have, again if the unwanted terms u.t.(A) and u.t.(D) cancel,

$$\tau_{\{\beta'\}}^{\{\beta''\}}(v)\Psi^{\{\beta'\}}=\lambda(v)\Psi^{\{\beta''\}}, \tag{3.32}$$

where

$$\lambda(v)=\lambda_A(v)+\lambda_{D_I}(v)+\lambda_{D_{II}}(v). \tag{3.33}$$

The cancellation of all unwanted terms ensure that Ψ, as given by eq. (3.8), is an eigenstate of the transfer matrix τ (2.8) with eigenvalue $\lambda(v)$ of eq. (3.33).

In appendix A we show that the unwanted terms indeed vanish if the Bethe ansatz equations hold,

$$\left(\frac{a(v_j)}{b(v_j)}\right)^L\prod_{i=1}^{N}\frac{a(v_i-v_j)}{b(v_i-v_j)}\frac{b(v_j-v_i)}{a(v_j-v_i)}\prod_{\beta=1}^{M}\frac{b(\gamma_\beta-v_j)}{a(\gamma_\beta-v_j)}=-1,\quad j=1,\dots,N, \tag{3.34}$$

$$(-1)^M\prod_{i=1}^{N}\frac{a(\gamma_\alpha-v_i)}{b(\gamma_\alpha-v_i)}\prod_{\beta=1}^{M}\frac{a(\gamma_\beta-\gamma_\alpha)}{b(\gamma_\beta-\gamma_\alpha)}\frac{b(\gamma_\alpha-\gamma_\beta)}{w(\gamma_\alpha-\gamma_\beta)}=1,\quad \alpha=1,\dots,M, \tag{3.35}$$

where N is the number of holes plus down spins and M is the number of holes. Another way to obtain these equations is to require that the eigenvalue $\lambda(v)$ (3.33) has no poles at $v=v_i$ and $v=\gamma_\beta$. Using (2.5) and making the change of variables $v\to iv+1$, $\gamma\to i\gamma+2$ we obtain

$$\left(\frac{v_j+i}{v_j-i}\right)^L=-\prod_{k=1}^{N}\frac{v_j-v_k+2i}{v_j-v_k-2i}\prod_{\beta=1}^{M}\frac{v_j-\gamma_\beta-i}{v_j-\gamma_\beta+i},\quad j=1,\dots,N, \tag{3.36}$$

$$\prod_{j=1}^{N}\frac{\gamma_\alpha-v_j+i}{\gamma_\alpha-v_j-i}=1,\quad \alpha=1,\dots,M. \tag{3.37}$$

This form of the Bethe ansatz equations (BAE) was previously derived by Sutherland [9] and later by Sarkar using a generalized permutation operator [10]. We stress that this procedure could be repeated with two other choices of the pseudovacuum leading to two other forms of the BAE. The pseudovacua of both levels of the Bethe ansatz Φ and $\Phi_{(1)}$ (see eqs. (3.5) and (3.19)), which we used above, consist of states of kind 1 = B and 2 = B, respectively. Basically, the change of pseudovacuum is determined by altering the initial convention (1 = B, 2 = B, 3 = F). Using (1 = F, 2 = B, 3 = B) we get

$$\left(\frac{v_j+i}{v_j-i}\right)^L = \prod_{\beta=1}^{M} \frac{v_j-\gamma_\beta+i}{v_j-\gamma_\beta-i}, \quad j=1,\dots,N, \tag{3.38}$$

$$\prod_{j=1}^{N} \frac{\gamma_\alpha-v_j-i}{\gamma_\alpha-v_j+i} = \prod_{\beta=1}^{M} \frac{\gamma_\alpha-\gamma_\beta-2i}{\gamma_\alpha-\gamma_\beta+2i}, \quad \alpha=1,\dots,M, \tag{3.39}$$

where N is the total number of spins and M is the number of spins down. These equations were already obtained by Lai [7] and Schlottmann [8] using the coordinate Bethe ansatz method.

Finally, the choice (1 = B, 2 = F, 3 = B) leads to a new form of the BAE *,

$$\left(\frac{v_j-i}{v_j+i}\right)^L = \prod_{\beta=1}^{M} \frac{v_j-\gamma_\beta-i}{v_j-\gamma_\beta+i}, \quad j=1,\dots,N, \tag{3.40}$$

$$\prod_{j=1}^{N} \frac{\gamma_\alpha-v_j+i}{\gamma_\alpha-v_j-i} = 1, \quad \alpha=1,\dots,M, \tag{3.41}$$

where N is the number of holes plus spin downs and M is the number of spins down. In the following we will work with the BAE's (3.36) and (3.37), since this is the most convenient form for the present investigation.

We have reduced the eigenvalue problem of the transfer matrix (3.1) to a system of coupled algebraic equations for the parameters $\{v_j\}$ $(j=1,\dots,N)$ and $\{\gamma_\alpha\}$ $(\alpha=1,\dots,M)$. The basic procedure to solve eqs. (3.36) and (3.37) is to adopt the string-conjecture, which means that the v's appear as strings and all roots γ's are real,

$$\begin{aligned} v^n_{\alpha j} &= v^n_\alpha + i(n+1-2j), && j=1,\dots,n, \quad \alpha=1,\dots,N_n, \quad n=1,2,\dots, \\ \gamma_\beta &= \text{real}, && \beta=1,\dots,M, \end{aligned} \tag{3.42}$$

where v^n_α is the position of the center of the string on the real v-axis. The number of n-strings N_n satisfy the relation

$$N = \sum_n nN_n. \tag{3.43}$$

* When this paper was in preparation the authors were informed about a preprint of Essler, Korepin and Schoutens where this new form of the BAE also was obtained.

This hypothesis for the v's can be easily understood by heuristic arguments, analogously to the isotropic Heisenberg model [2,16]. To understand absence of complex roots for the γ's we apply the following argument, which is similar to that one developed by Takahashi for the one-dimensional electron gas with a repulsive delta function [17]. If all v_i are real or appear as complex conjugate pairs, $\operatorname{Im}\gamma_\alpha > 0$ implies that the absolute value of the left-hand side of eq. (3.37) is larger than unity. Therefore, $\operatorname{Im}\gamma_\alpha > 0$ is not possible. In the same way we can prove that $\operatorname{Im}\gamma_\alpha < 0$ is not possible. We can see here the great advantage of using this form of BAE. In the other two forms not only the parameters v but also the roots γ appear as strings. This means that counting the states is much more complicated. Although we are not able to prove the string-conjecture rigorously, we will assume it to be valid. Since Bethe [1], assumptions of this kind have been widely used by many authors (ref. [16] and references therein). Applying this conjecture in (3.36) and (3.37) and taking its logarithm we obtain the coupled equations for the v_α^n and γ_β,

$$L\theta\left(\frac{v_\alpha^n}{n}\right) - \underbrace{\sum_m \sum_{\beta=1}^{N_m}}_{(m,\beta)\neq(n,\alpha)} \Theta_{nm}(v_\alpha^n - v_\beta^m) + \sum_{\beta=1}^{M} \theta\left(\frac{v_\alpha^n - \gamma_\beta}{n}\right) = 2\pi I_\alpha^n, \quad (3.44)$$

$$\sum_n \sum_{\alpha=1}^{N_n} \theta\left(\frac{v_\alpha^n - \gamma_\beta}{n}\right) = 2\pi J_\beta, \quad (3.45)$$

where $\theta(x) = 2\arctan x$ and

$$\Theta_{nm}(x) = \begin{cases} \theta\left(\frac{x}{|n-m|}\right) + 2\theta\left(\frac{x}{|n-m|+2}\right) + \ldots + 2\theta\left(\frac{x}{n+m-2}\right) \\ \quad +\theta\left(\frac{x}{n+m}\right) \quad \text{for } n \neq m \\ 2\theta\left(\frac{x}{2}\right) + 2\theta\left(\frac{x}{4}\right) + \ldots + 2\theta\left(\frac{x}{2n-2}\right) + \theta\left(\frac{x}{2n}\right) \quad \text{for } n = m. \end{cases} \quad (3.46)$$

Hence the solutions of eqs. (3.36) and (3.37) are parametrized in terms of the numbers I_α^n and J_β. Here, the I_α^n are integers (half-integers) if $L + M - N_n$ is odd (even) and the J_β are integers (half-integers) if $\sum_n N_n$ is even (odd). In addition they are limited to the intervals

$$|I_\alpha^n| \leq I_{\max}^n = \tfrac{1}{2}\left(L + M - \sum_m t_{nm} N_m - 1\right), \quad (3.47)$$

$$|J_\beta| \leq J_{\max} = \tfrac{1}{2}\left(\sum_n N_n - 2\right), \quad (3.48)$$

where $t_{nm} = 2\min(n, m) - \delta_{nm}$. In fact, all sets $\{I_\alpha^n, J_\beta\}$ where the I's and J's are pairwise different specify all the Bethe vectors ($|\psi_{\text{Bethe}}\rangle_{N,M}$). They are highest-weight vectors of an spl(2,1) superalgebra, as we will show in sect. 4.

In order to avoid misunderstandings we should add some general remarks on the string-conjecture (3.42) and the bounds $I_{\max}^n$ and $J_{\max}$ given by eqs. (3.47) and (3.48). Both statements are to be considered as assumptions, they cannot be proven rigorously. In fact they are not exact. There are finite-size corrections of the string configurations of order $O(e^{-L})$ for fixed string centers v_α^n and of order O(1) near to the boundary $v_{\max}^n$ (given by $I_{\max}^n$), producing "exotic solutions". On the other hand a naive estimate of $I_{\max}^n$ from eq. (3.44) would suggest additional solutions (for $n \geqslant 2$) which are cancelled by assumption (3.47). However, both assumptions together lead to the correct number of states, as is well known for the SU(2) case [2] and will be proven below for the spl(2,1) case. Obviously, the effects of the two phenomena mentioned above compensate for this computation. In addition to the "exotic solutions" mentioned above, there exist also "wide pairs" and "quartets" if the density of real roots is large enough. It is believed that these problems may be avoided and exotic effects may be neglected, if one considers the following thermodynamic limit. Introduce a symmetry breaking magnetic field B and take first the limit $L \to \infty$ and then $B \to 0$. It should be stressed that many features of the Bethe ansatz are not well understood.

In the thermodynamic limit the BAE's are written in terms of densities of roots ($\rho_n(v)$, $\sigma(\lambda)$) and BA-holes * ($\rho_n^{\mathrm{h}}(v)$, $\sigma^{\mathrm{h}}(\lambda)$), such that eqs. (3.44) and (3.45) can be replaced by integral equations for the densities.

At the end of this section we apply the results obtained for the spl(2,1) vertex model to the supersymmetric t–J model. Using the identity (2.18) it is possible to obtain the energy eigenvalues of the t–J model from the eigenvalues of the transfer matrix (3.33). The terms $\lambda_{D_{\mathrm{I}}, D_{\mathrm{II}}}$ given by eqs. (3.30) and (3.31) do not contribute and from eq. (3.15) we find

$$E = L - \sum_{j=1}^{N} \frac{4}{1 + v_j^2}. \tag{3.49}$$

Thermodynamic properties of the model were investigated in ref. [8] using the second form of the BAE (3.38), (3.39). The ground state and the excitation spectrum were discussed in ref. [18] using the first and second form of the BAE.

* Unfortunately, in this paper the meaning of the term hole is ambiguous: A "hole", as denoted above is a physical hole, i.e. a lattice site with no electron. A "BA-hole" corresponds to a non-occupied place in the set of numbers $\{I^n\}$ or $\{J\}$ for a solution of the BAE (see sect. 5 for examples).

4. Algebraic properties of the Bethe states

In this section we analyse the algebraic properties of the Bethe states. By asymptotic expansion ($v \to \infty$) we obtain the generators of spl(2,1) as matrix elements of a matrix $\tilde{M}$ of operators in the "quantum space" defined as follows:

$$T_{\alpha}^{\alpha''\{\gamma''\}}{}_{\{\gamma\}}(v) = \sigma_{\alpha''\{\gamma''\}}\delta_{\alpha}^{\alpha''}\delta_{\{\gamma\}}^{\{\gamma''\}} - \frac{2}{v}\sigma_{\alpha''\alpha}\sigma_{\alpha''\{\gamma''\}}\tilde{M}_{\alpha}^{\alpha''\{\gamma''\}}{}_{\{\gamma\}} + O(v^{-2}). \tag{4.1}$$

We prove the commutation relations of the entries of $\tilde{M}$ using the Yang–Baxter relation (3.2) for the monodromy matrix and the property (2.13). For $v \to \infty$ we have (in what follows we will omit the quantum space indices and write them only whenever necessary)

$$\begin{aligned}\tilde{M}_{\alpha}^{\alpha''}\tilde{T}_{\beta}^{\beta''}(v') - \Sigma(\alpha'', \beta'', \alpha, \beta)\tilde{T}_{\beta}^{\beta''}(v')\tilde{M}_{\alpha}^{\alpha''}\\ = \tilde{T}_{\alpha}^{\beta''}(v')\delta_{\beta}^{\alpha''} - \Sigma(\alpha'', \beta'', \alpha, \beta)\delta_{\alpha}^{\beta''}\tilde{T}_{\beta}^{\alpha''}(v').\end{aligned} \tag{4.2}$$

Here the sign function Σ is given by

$$\Sigma(\alpha'', \beta'', \alpha, \beta) = \sigma_{\alpha''\beta''}\sigma_{\alpha''\beta}\sigma_{\alpha\beta''}\sigma_{\alpha\beta}. \tag{4.3}$$

Furthermore, taking $v' \to \infty$ we get

$$\begin{aligned}\tilde{M}_{\alpha}^{\alpha''}\tilde{M}_{\beta}^{\beta''} - \Sigma(\alpha'', \beta'', \alpha, \beta)\tilde{M}_{\beta}^{\beta''}\tilde{M}_{\alpha}^{\alpha''}\\ = \tilde{M}_{\alpha}^{\beta''}\delta_{\beta}^{\alpha''} - \Sigma(\alpha'', \beta'', \alpha, \beta)\delta_{\alpha}^{\beta''}\tilde{M}_{\beta}^{\alpha''}.\end{aligned} \tag{4.4}$$

This relation represents the commutation and anti-commutation rules of the spl(2,1) superalgebra [11]. The generators $\tilde{M}_3^{\alpha}$, $\tilde{M}_{\alpha}^3$ ($\alpha \neq 3$) are fermionic, whereas the $\tilde{M}_3^3$ and $\tilde{M}_{\beta}^{\alpha}$ ($\alpha, \beta \neq 3$) are bosonic. The sign factors Σ take into account the statistics, i.e. $\Sigma = -1$ (1) if we are dealing with odd (even) generators. Eq. (4.4) can be written in the compact form

$$\left[\tilde{M}_{\alpha}^{\alpha''}, \tilde{M}_{\beta}^{\beta''}\right]_{\pm} = \tilde{M}_{\alpha}^{\beta''}\delta_{\beta}^{\alpha''} \pm \delta_{\alpha}^{\beta''}\tilde{M}_{\beta}^{\alpha''}. \tag{4.5}$$

In addition, from eq. (4.2) it is easy to see that the transfer matrix τ (2.8) is invariant with respect to the spl(2,1) superalgebra, i.e.

$$\left[\tilde{M}_{\alpha}^{\alpha''}, \tau(v')\right] = 0. \tag{4.6}$$

Notice that the results (4.2), (4.4), (4.5) and (4.6) are also valid if we change the convention (1 = B, 2 = B, 3 = F). The position of the fermion simply determine which are the odd generators.

Let us now consider the matrix $\tilde{M}$,

$$\tilde{M} = \begin{pmatrix} W_1 & \tilde{M}_2^1 & \tilde{M}_3^1 \\ \tilde{M}_1^2 & W_2 & \tilde{M}_3^2 \\ \tilde{M}_1^3 & \tilde{M}_2^3 & W_3 \end{pmatrix}. \tag{4.7}$$

The diagonal elements W_α ($\alpha = 1, 2, 3$) generate the Cartan subalgebra with weights w_α ($\alpha = 1, 2, 3$),

$$W_\alpha \Psi = w_\alpha \Psi. \tag{4.8}$$

In terms of the t–J model the weights are related to the z-component of the SU(2)-spin $S_z = \frac{1}{2}(w_1 - w_2)$ and the number of electrons $Q = w_1 + w_2$. In order to calculate these weights for Bethe ansatz states we substitute (2.5) in eqs. (3.14), (3.15), (3.29), (3.30) and (3.31) and obtain with eq. (4.1) and (4.7) for $v \to \infty$

$$\begin{aligned} \left(1 - \frac{2}{v}W_1\right)\Psi + O(v^{-2}) &= \left(1 - \frac{2}{v}(L-N)\right)\Psi + O(v^{-2}), \\ \left(1 - \frac{2}{v}W_2\right)\Psi + O(v^{-2}) &= \left(1 - \frac{2}{v}(N-M)\right)\Psi + O(v^{-2}), \\ \left(-1 - \frac{2}{v}W_3\right)\Psi + O(v^{-2}) &= \left(-1 - \frac{2}{v}M\right)\Psi + O(v^{-2}). \end{aligned} \tag{4.9}$$

Therefore, the weights can be expressed in terms of the quantities L (= number of sites), N (= number of first-level roots) and M (= number of second-level roots),

$$w_1 = n_\uparrow = L - N, \qquad w_2 = n_\downarrow = N - M, \qquad w_3 = n_h = M, \tag{4.10}$$

where $n_\uparrow$, $n_\downarrow$, n_h are the numbers of up-spins, down-spins and holes, respectively. At the end of this section we will derive inequalities between these weights and give a physical interpretation.

Next we show that the Bethe vectors are highest-weight vectors with respect to the spl(2,1) superalgebra, i.e.

$$\tilde{M}_\alpha^\beta \Psi = 0, \quad \beta > \alpha. \tag{4.11}$$

For $\alpha = 1$, $\beta = 2$ or 3 we have, after using eqs. (3.8) and (4.2),

$$\begin{aligned} \tilde{M}_1^\beta \Psi = \sum_{j=1}^{N} & \sigma_{\beta(\alpha_1,\dots,\alpha_{j-1})} B_{\alpha_1}(v_1) B_{\alpha_2}(v_2) \dots B_{\alpha_{j-1}}(v_{j-1}) \\ & \times \left[\tilde{M}_1^\beta, B_{\alpha_j}(v_j)\right]_\pm B_{\alpha_{j+1}}(v_{j+1}) \dots B_{\alpha_N}(v_N) \Phi \Psi_{(1)}^{\{\alpha\}}, \end{aligned} \tag{4.12}$$

where

$$\left[\tilde{M}_1^\beta, B_\alpha(v)\right]_\pm = \tilde{M}_1^\beta B_\alpha(v) - \sigma_{\beta\alpha} B_\alpha(v)\tilde{M}_1^\beta = \delta_\alpha^\beta A(v) - \sigma_{\beta\alpha}\tilde{T}_\alpha^\beta(v). \quad (4.13)$$

In order to commute $A(v_j)$ and $\tilde{T}_\alpha^\beta(v_j)$ through the B_α's toward Φ we use the commutation rules (3.11), (3.12) and the property (2.13). Although many terms appear, it is possible to arrange them as follows:

$$\tilde{M}_1^\beta \Psi = \sum_{j=1}^{N} Y_{j\alpha_j}^\beta(v_j, \{v_i\}) B_{\alpha_1}(v_1) B_{\alpha_2}(v_2) \dots B_{\alpha_{j-1}}(v_{j-1})$$

$$\times B_{\alpha_{j+1}}(v_{j+1}) \dots B_{\alpha_N}(v_N) \Phi \Psi_{(1)}^{\{\alpha\}}, \quad (4.14)$$

with yet unknown coefficients $Y_{j\alpha_j}^\beta$. The first coefficient, $Y_{1\alpha_1}^\beta$, can be obtained by using the first term in (3.11) and (3.12) when commuting $A(v_1)$ and $\tilde{T}_\alpha^\beta(v_1)$ with $B_{\alpha_2}(v_2)B_{\alpha_3}(v_3)\dots B_{\alpha_N}(v_N)$, since otherwise the argument v_1 re-appears in the B_α. The contribution of the $A(v_1)$ term to $Y_{1\alpha_1}^\beta$ is straightforward, whereas for the $\tilde{T}_\alpha^\beta(v_1)$ term we shall use the relation

$$-\tfrac{1}{2} \operatorname*{Res}_{v'=v} S_{\alpha\beta}^{\gamma\delta}(v - v') = \delta_\beta^\gamma \delta_\alpha^\delta \quad (4.15)$$

to get the eigenvalue problem for the transfer matrix $\tau_{(1)}(v_1, \{v_i\})$ (3.17). Once again, we just take the first term in eqs. (3.23) and (3.24) when commuting $A_{(1)}$ and $D_{(1)}$ with the $B_{(1)}$'s. Then, after some manipulations we have

$$Y_{1\alpha_1}^\beta = \delta_{\alpha_1}^\beta \left(a^L(v_1) \prod_{i=2}^{N} \frac{a(v_i - v_1)}{b(v_i - v_1)} - b^L(v_1) \prod_{i=2}^{N} \frac{a(v_1 - v_i)}{b(v_1 - v_i)} \prod_{\beta=1}^{M} \frac{a(\gamma_\beta - v_1)}{b(\gamma_\beta - v_1)} \right). \quad (4.16)$$

Analogous expressions follow for the other coefficients $Y_{j\alpha_j}^\beta$ $(j \geq 2)$,

$$Y_{j\alpha_j}^\beta \propto \delta_{\alpha_j}^\beta \left(a^L(v_j) \prod_{i \neq j}^{N} \frac{a(v_i - v_j)}{b(v_i - v_j)} - b^L(v_j) \prod_{i \neq j}^{N} \frac{a(v_j - v_i)}{b(v_j - v_i)} \prod_{\beta=1}^{M} \frac{a(\gamma_\beta - v_j)}{b(\gamma_\beta - v_j)} \right),$$

$$j = 1, \dots, N. \quad (4.17)$$

We observe that the requirement $Y_{j\alpha_j}^\beta = 0$ $(j = 1 \dots N,\ \beta = 2, 3)$ is equivalent to the Bethe ansatz equations (3.34), therefore Bethe states fulfill the highest-weight condition $\tilde{M}_1^\beta \Psi = 0$ $(\beta = 2$ or $3)$.

To calculate $\tilde{M}_2^3\Psi$ we use the relation

$$\tilde{M}_2^3\Psi = \sigma_{3\{\alpha'\}}B_{\alpha'_1}(v_1)B_{\alpha'_2}(v_2)\dots B_{\alpha'_N}(v_N)\Phi\tilde{M}_{(1)2\{\alpha\}}^{\;3\{\alpha'\}}\Psi_{(1)}^{\{\alpha\}}, \tag{4.18}$$

which follows from (4.2). $\tilde{M}_{(1)}$ is defined by asymptotic expansion of the monodromy $T_{(1)}$, in analogy with $\tilde{M}$ given by eq. (4.1). From (3.22) and commutation relations for $\tilde{M}_{(1)}$ and $\tilde{T}_{(1)}$ analogous to eq. (4.2) we get

$$\tilde{M}_{(1)2}^{\;3}\Psi_{(1)} = \sum_{\beta=1}^{M}(-1)^{\beta-1}B_{(1)}(\gamma_1,\{v_i\})\dots B_{(1)}(\gamma_{\beta-1},\{v_i\})$$
$$\times\left[\tilde{M}_{(1)2}^{\;3},\,B_{(1)}(\gamma_\beta,\{v_i\})\right]_+B_{(1)}(\gamma_{\beta+1},\{v_i\})\dots B_{(1)}(\gamma_M,\{v_i\})\Phi_{(1)}, \tag{4.19}$$

where

$$\left[\tilde{M}_{(1)2}^{\;3},\,B_{(1)}(\gamma)\right]_+ = A_{(1)}(\gamma)+\tilde{T}_{(1)3}^{\;3}(\gamma). \tag{4.20}$$

Analogously, by commuting $A_{(1)}+\tilde{T}_{(1)3}^{\;3}$ through the $B_{(1)}$'s we have

$$\tilde{M}_{(1)2}^{\;3}\Psi_{(1)} = \sum_{\beta=1}^{M}Y_{(1),\beta}(\gamma_\beta,\{\gamma_\alpha\},\{v_i\})B_{(1)}(\gamma_1,\{v_i\})\dots$$
$$\times B_{(1)}(\gamma_{\beta-1},\{v_i\})B_{(1)}(\gamma_{\beta+1},\{v_i\})\dots B_{(1)}(\gamma_M,\{v_i\})\Phi_{(1)}. \tag{4.21}$$

The coefficients $Y_{(1),\beta}$ can be derived in a straightforward way by taking the first terms of the commutation relations (3.23) and (3.24). We get

$$Y_{(1),\beta} = \prod_{i=1}^{N}a(\gamma_\beta-v_i)\prod_{\alpha\neq\beta}^{M}\frac{a(\gamma_\alpha-\gamma_\beta)}{b(\gamma_\alpha-\gamma_\beta)}$$
$$+(-1)^M\prod_{i=1}^{N}b(\gamma_\beta-v_i)\prod_{\alpha\neq\beta}^{M}\frac{w(\gamma_\beta-\gamma_\alpha)}{b(\gamma_\beta-\gamma_\alpha)},$$
$$\beta=1,\dots,M. \tag{4.22}$$

The requirement $Y_{(1),\beta}=0$ $(\beta=1,\dots,M)$ is equivalent to the Bethe ansatz equations (3.35), which implies $\tilde{M}_2^3\Psi=0$. We stress that the property (4.11) can also be proved for the other two choices of pseudovacuum in a similar way.

At the end of this section we derive some inequalities between the weights w_α $(\alpha=1,2,3)$. From eq. (4.5) we have

$$\left[\tilde{M}_\alpha^\beta,\,\tilde{M}_\beta^\alpha\right]_\pm = W_\alpha^\alpha \pm W_\beta^\beta,\quad \beta>\alpha. \tag{4.23}$$

Using $(\tilde{M}_\alpha^\beta)^\dagger = \tilde{M}_\beta^\alpha$ and the highest-weight property of the Bethe vectors (4.11) we obtain

$$w_1 \geqslant w_2 \geqslant -w_3. \tag{4.24}$$

Combining (4.10) with $w_i \geqslant 0$ $(i = 1, 2, 3)$ and (4.24) we find conditions for the numbers N and M of roots in the first- and second-level Bethe ansatz, respectively,

$$M \leqslant N \leqslant \frac{L+M}{2}, \qquad 0 \leqslant M \leqslant L. \tag{4.25}$$

This means in terms of physical quantities that the magnetization $S_z = \frac{1}{2}(n_\uparrow - n_\downarrow) = \frac{1}{2}(L - 2N + M)$ and the number of electrons $Q = n_\uparrow + n_\downarrow = L - M$ are restricted to $0 \leqslant S_z \leqslant Q/2 \leqslant L/2$.

5. Results for small and large lattices

In this section we illustrate the algebraic properties of the Bethe states. We begin with a lattice of two sites and then discuss the case of lattices with a large number of sites.

The simple case of one lattice point corresponds to the fundamental representation of spl(2,1) which is given by the following weight diagram in the (S_z, Q) plane, where Q is the number of electrons and S_z the total magnetization of the system:

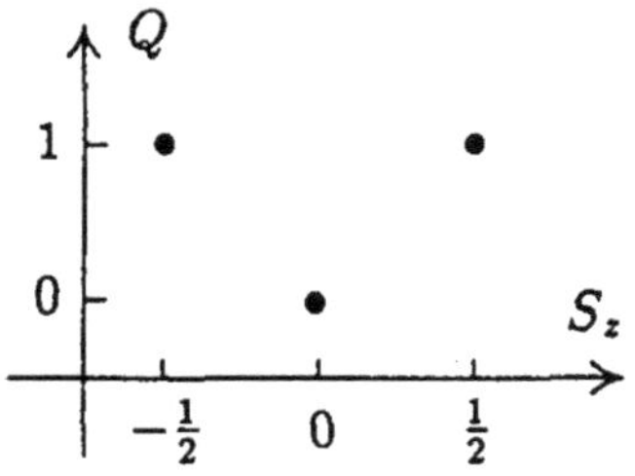

By diagonalization of the t–J hamiltonian (2.17) (or of the transfer matrix τ) on a lattice with two sites we obtain

$$\Psi_1 = |\uparrow\uparrow\rangle, \qquad E = 2,$$

$$\Psi_2 = \frac{1}{\sqrt{2}}(|\uparrow\downarrow\rangle + |\downarrow\uparrow\rangle), \quad E = 2,$$

$$\Psi_3 = |\downarrow\downarrow\rangle, \qquad E = 2,$$

$$\Psi_4 = \frac{1}{\sqrt{2}}(\,|0\uparrow\rangle + |\uparrow 0\rangle), \qquad E = 2,$$

$$\Psi_5 = \frac{1}{\sqrt{2}}(\,|0\downarrow\rangle + |\downarrow 0\rangle), \qquad E = 2, \tag{5.1}$$

$$\Psi_6 = \frac{1}{\sqrt{2}}(\,|\uparrow\downarrow\rangle - |\downarrow\uparrow\rangle), \qquad E = -2,$$

$$\Psi_7 = \frac{1}{\sqrt{2}}(\,|0\uparrow\rangle - |\uparrow 0\rangle), \qquad E = -2,$$

$$\Psi_8 = \frac{1}{\sqrt{2}}(\,|0\downarrow\rangle - |\downarrow 0\rangle), \qquad E = -2,$$

$$\Psi_9 = |00\rangle, \qquad E = -2, \tag{5.1'}$$

where 0 denotes a hole. This result can be visualized in terms of the following spl(2,1) weight diagrams in the Clebsch–Gordan series $3 \otimes 3 = 5 \oplus 4$:

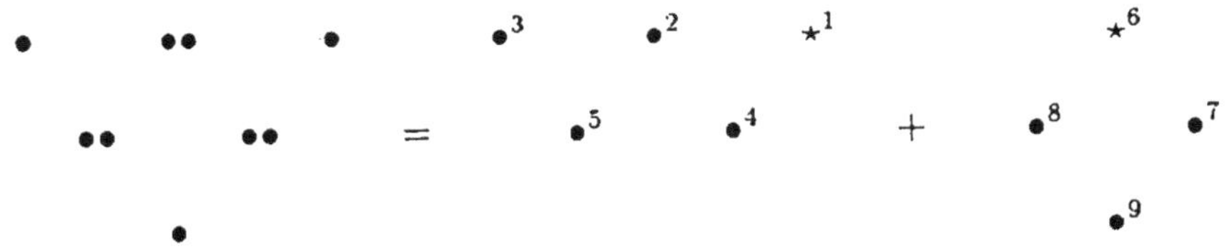

The numbers in the weight diagrams specify the eigenvectors according to eqs. (5.1) and (5.1′). The symbol ⋆ denotes the highest-weight vectors according to eq. (5.2) below. Notice that the ground state is degenerate and given by a quartet. All states of an irreducible representation can be generated by repeated application of the shift operators $\tilde{M}_\alpha^\beta$ $(\beta \neq \alpha)$ to any one of the states. Graphically, the effect of the shift operators on a general state of a representation of spl(2,1) is given by

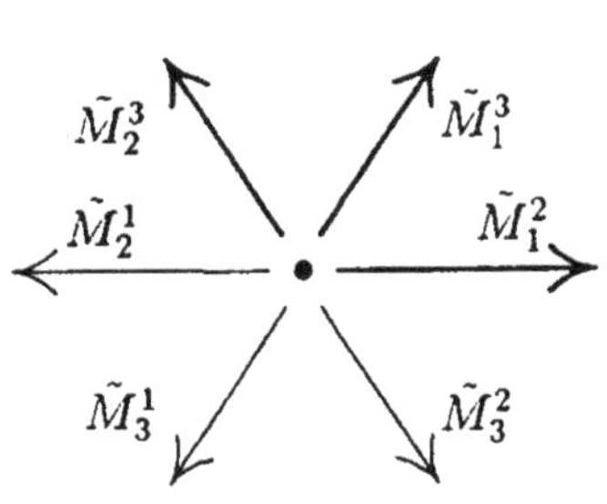

On the other hand, if we solve the Bethe ansatz equations (3.36) and (3.37) for two sites we obtain only two eigenvectors, $\Psi_1 = \Phi$ and $\Psi_6 = B_{\alpha_1}(v_1 = 0)\Phi\Phi_{(1)}^{\alpha_1}$, with energy eigenvalues 2 and -2, respectively (see eq. (3.49)). In the language of the nested Bethe ansatz Φ and $\Phi_{(1)}$ are the first- and second-level pseudoground states, respectively. We can easily check that these eigenvectors are highest-weight vectors of the spl(2,1) superalgebra, in agreement with our general proof in sect. 4.

$$\tilde{M}_\alpha^\beta \Psi_1 = \tilde{M}_\alpha^\beta \Psi_6 = 0, \quad \beta > \alpha. \tag{5.2}$$

Furthermore, the seven missing eigenvectors can be obtained by successive applications of the shift operators,

$$\begin{aligned}
\Psi_2 &= \tilde{M}_2^1 \Psi_1, \\
\Psi_3 &= \left(\tilde{M}_2^1\right)^2 \Psi_1, \\
\Psi_4 &= \tilde{M}_3^1 \Psi_1, \\
\Psi_5 &= \tilde{M}_2^1 \tilde{M}_3^1 \Psi_1, \\
\Psi_7 &= \tilde{M}_3^2 \Psi_6, \\
\Psi_8 &= \tilde{M}_3^1 \Psi_6, \\
\Psi_9 &= \tilde{M}_3^1 \tilde{M}_3^2 \Psi_6.
\end{aligned} \tag{5.3}$$

Therefore, the Bethe ansatz together with the supersymmetry of the model provide all 9 eigenvectors for the two-sites model.

We remark that by solving all three different forms of the BAE we get all highest-weight vectors of the SU(2) algebra. Solving eqs. (3.38) and (3.39) we get the eigenvectors Ψ_4 and Ψ_9 and from eqs. (3.40) and (3.41) we obtain the eigenvectors Ψ_1 and Ψ_7.

In the case of lattices with a large number of sites the Bethe ansatz method turns out to be crucial, since the effort of an exact diagonalization growths exponentially with the number of sites L. As already pointed out in sects. 3 and 4, by this method, the problem of finding the spectrum of the t–J hamiltonian reduces to the solution of the BAE's (3.36) and (3.37) for the parameters v's and γ's. Adopting the string-conjecture, which has an accuracy of $O(e^{-L})$, the solutions of the BAE's are parametrized in terms of the numbers I_α^n and J_β. Moreover, each set $\{I_\alpha^n, J_\beta\}$ where the I's and J's are pairwise different specify a Bethe vector, which is the leading vector of an spl(2,1) multiplet.

Now we illustrate our results for the ground state and some elementary excitations at "half-filling" $F = Q/L = 1$. The following holds true for any lattice size, especially in the thermodynamic limit $L \to \infty$. The ground state involves only real roots. This can be proved as usual by minimizing the free energy for finite temperature T and taking $T \to 0$ [18]. For example, for a lattice of size $L = 40$ we find $N = 20$ first-level real roots and no BA-holes. There are no second-level real roots ($M = 0$), but 19 BA-holes. Therefore, we have the following distribution of I's and J's:

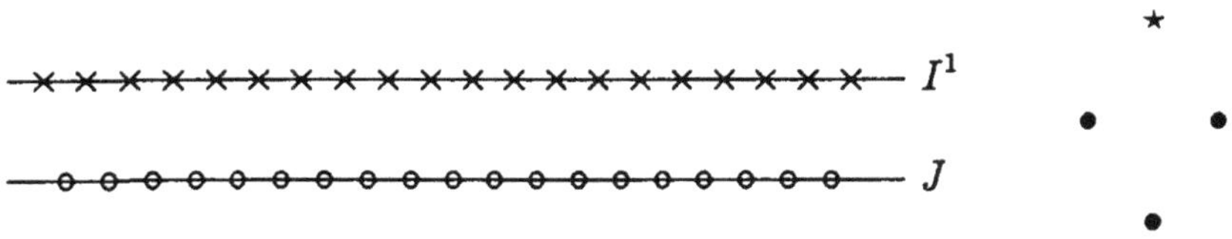

where the numbers corresponding to roots are denoted by × and those corresponding to BA-holes by ○. On the r.h.s. the associated spl(2,1) representation is shown. The quantum numbers of the ground state are $S_z = 0$ and $Q = L = 40$, which means vanishing magnetization and half-filling $F = 1$. The Bethe vector specified by this set of numbers $\{I_\alpha^1\}$ is the highest-weight vector of the irreducible representation of dimension 4, depicted by ⋆. Notice that the ground state is not a singlet but a member of an spl(2,1) quartet. Of course, the state is a singlet with respect to the SU(2) subgroup.

One kind of elementary excitation over the ground state is the "spinon". It is obtained by removing a root from the I^1-axis or introducing a first-level BA-hole, which corresponds to a spin flip. For a lattice of size $L = 41$ * we have $N = 20$ first-level roots and one BA-hole, $M = 0$ second-level roots and 19 BA-holes. The distribution of I's and J's and the corresponding irreducible representation generated by the Bethe vector (⋆) determined by this set of I's and J's are for example illustrated by

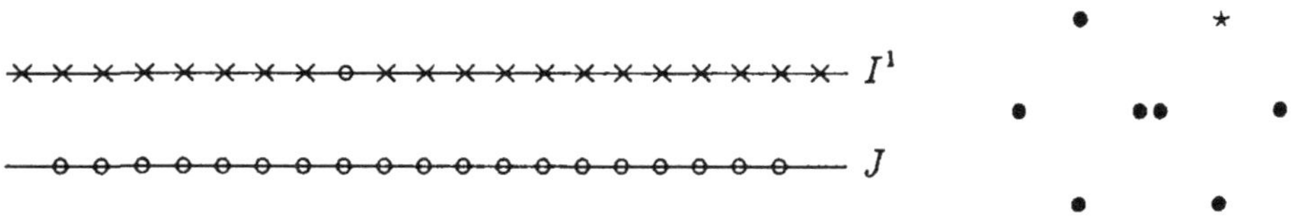

The quantum numbers of this state are $S_z = 1/2$ and $Q = L = 41$, which means a spinon is a particle-like excitation with spin 1/2 and charge 0.

Another excitation is given by the presence of a two-string in the complex v-plane. For $L = 40$ we have $N^1 = 18$, $N^2 = 1$ and $M = 0$,

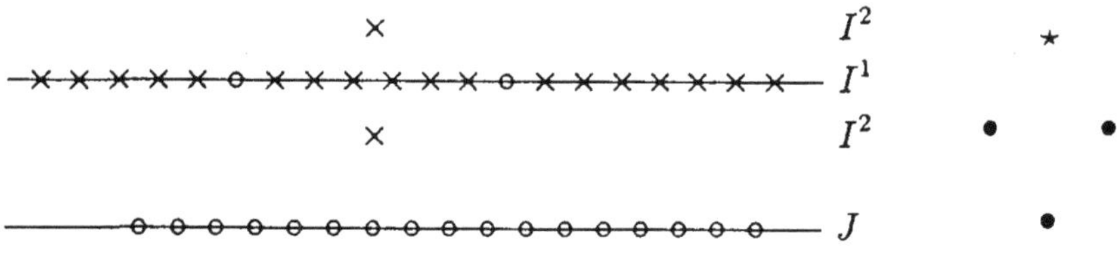

* Note that a one-spinon state exists only on lattices with an odd number of lattice sites, otherwise spinons appear pairwise.

(To support the visual perception we have drawn the I^2-axis twice in order to obtain a picture similar to the corresponding one in the complex v-plane). The quantum numbers are $S_z = 0$ and $Q = L = 40$, which means vanishing magnetization and half-filling $F = 1$. The interpretation of this state is that we have the spin-0 contribution of a two-spinon state.

By filling a vacancy in the J-axis we get another excitation called "holon", i.e. we are removing an electron from the system or introducing a physical hole. For $L = 41$ we have $N^1 = 21$ first-level roots, $M = 1$ second-level root and 19 holes,

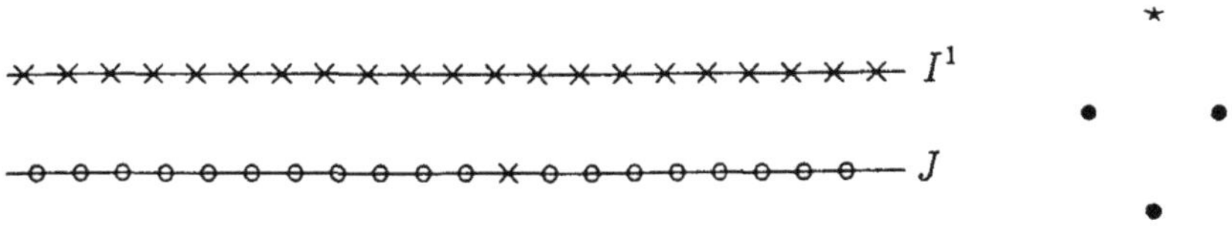

The quantum numbers of this state are $S_z = 0$ and $Q = 40 = L - 1$, which means a holon is a particle-like excitation with spin 0 and charge -1.

At arbitrary filling $F < 1$, for the ground state the distribution of the roots in the v-plane also involves only real roots. In contrast to the half-filling case there is now in addition a "sea" of real roots in the γ-plane, such that there appears a nontrivial Fermi level. For example, for a lattice of $L = 40$ sites we find a Bethe ansatz state with $N = 25$ first-level real roots and $M = 10$ second-level real roots and 14 BA-holes,

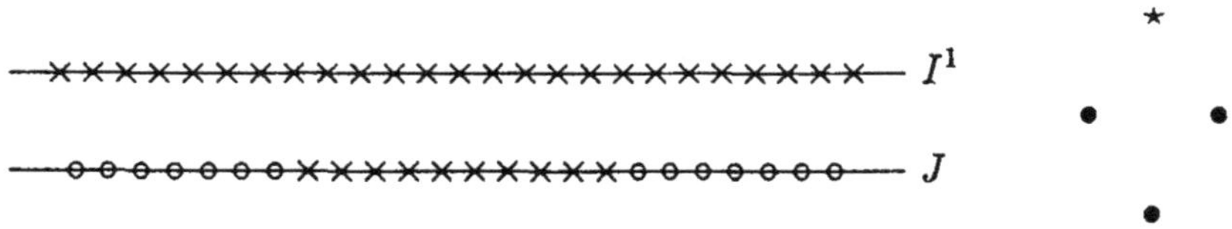

Also here the ground state is member of a quartet. The quantum numbers are $S_z = 0$ and $Q = 30$, which means spin 0 and filling $F = 1 - 10/40 = 0.75$. Due to the nontrivial Fermi level there exist "holon–antiholon" excitations in this case.

6. Completeness of the Bethe vectors

In this section we show how to construct a complete set of eigenvectors of the t–J hamiltonian for arbitrary chain of length L. This is obtained by combining the Bethe ansatz with the supersymmetry of the model.

From the sect. 3 we know that all collections $\{I_\alpha^n, J_\beta\}$ where the I's and J's are pairwise different specify all the Bethe vectors $(|\psi_{\text{Bethe}}\rangle_{N,M})$. The number of admissible values for the I_α^n and the J_β (for fixed $\{N_n\}$ and M) is $(2I_{\max}^n + 1)$ and $(2J_{\max} + 1)$, respectively. $I_{\max}^n$ and $J_{\max}$ are given by eqs. (3.47) and (3.48). Taking into account that many different string configurations N_n give the same number of roots N (see eq. (3.43)), the number of possible Bethe vectors for fixed N, M is given by

$$Z(N,M)=\sum_{\{N_n\}}\binom{2J_{\max}+1}{M}\prod_n\binom{2I^n_{\max}+1}{N_n}, \tag{6.1}$$

where the sum over $\{N_n\}$ is constrained to $\Sigma_n\, nN_n=N$. It is convenient to introduce the quantity $q=\Sigma_n N_n$. Using eqs. (3.47) and (3.48) we write this sum as

$$Z(N,M)+\sum_{q=0}^{N}\binom{q-1}{M}\sum_{\{N_n\}}\prod_n\binom{L-\Sigma_m t_{nm}N_m+M}{N_n}, \tag{6.2}$$

where the inner sum is constrained to fixed values of N and q. This expression resembles the one calculated by Bethe in the isotropic Heisenberg model [1,2] and can be simplified to

$$Z(N,M)=\sum_{q=0}^{N}\frac{L+M-2N+1}{L+M-N+1}\binom{q-1}{M}\binom{L+M-N+1}{q}\binom{N-1}{q-1}. \tag{6.3}$$

The total number of Bethe vectors is obtained by summing $Z(N,M)$ over all N, M restricted to (4.25). However, this number is less than 3^L, so that the Bethe ansatz does not yield all the states of the model. In order to construct a complete set we shall invoke the supersymmetry of the transfer matrix. First, from eq. (4.6) it follows that the Bethe vectors are classified by multiplets corresponding to irreducible representations of the superalgebra spl(2,1). Furthermore, from eq. (4.11) follows that the Bethe vectors are highest-weight vectors. Then by acting with the spl(2,1) lowering operators $\tilde{M}^{\beta}_{\alpha}(\beta<\alpha)$ on the Bethe states we obtain additional states. Each Bethe state (with fixed N, M in the interval (4.25)) is the highest-weight vector in a multiplet of dimension [11]

$$d(N,M)=\begin{cases}4S_z+1=2L+1 & \text{if } N=M=0\\ 8(S_z+1/2)=4(L-2N+M+1) & \text{otherwise.}\end{cases} \tag{6.4}$$

With these considerations, the total number of eigenvectors is

$$\begin{aligned}Z&=\sum_{M=0}^{L}\sum_{N=M}^{(L+M)/2}d(N,M)Z(N,M)=2L+1+Z_1-1+Z_2\\ &=2L+1+4\sum_{N=1}^{L/2}(L-2N+1)\frac{L-2N+1}{L-N+1}\sum_{q=1}^{N}\binom{L-N+1}{q}\binom{N-1}{q-1}\end{aligned}$$

$$+4\sum_{M=1}^{L}\sum_{N=M}^{(L+M)/2}(L-2N+M+1)\frac{L+M-2N+1}{L+M-N+1}$$

$$\times\sum_{q=1}^{N}\binom{q-1}{M}\binom{L+M-N+1}{q}\binom{N-1}{q-1}. \tag{6.5}$$

The first sum in eq. (6.5) can be performed (see ref. [2]) to give

$$Z_1 = 4\cdot 2^L - 4(L+1). \tag{6.6}$$

The second sum Z_2 deserves special attention. We present the main necessary steps for its evaluation. First, performing the sum over q we get

$$Z_2 = 4\sum_{M=1}^{L}\sum_{N=M}^{(L+M)/2}(L-2N+M+1)$$

$$\times\frac{L+M-2N+1}{L+M-N+1}\binom{N-1}{M}\binom{L}{N}. \tag{6.7}$$

Employing some combinatorics and making the substitution $N\to x=N-M$ we obtain

$$Z_2 = 4\sum_{M=1}^{L}\sum_{x=0}^{(L-M)/2}(L-2x-M+1)$$

$$\times\left[\binom{L}{x+M}\binom{x+M-1}{M}-\binom{L}{x-1}\binom{L-x}{M}\right]. \tag{6.8}$$

After some re-arrangements this expression can be rewritten as

$$Z_2 = 4\sum_{M=1}^{L}\sum_{x=0}^{(L-M)/2}L\left[\left[\binom{L-1}{x+M}\binom{x+M-1}{M}+\binom{L-1}{x-2}\binom{L-x}{M}\right]\right.$$

$$\left.-(M+1)\left[\binom{L}{x+M}\binom{x+M-1}{x-2}+\binom{L}{x-1}\binom{L-x}{M+1}\right]\right]. \tag{6.9}$$

Substituting $x\to L-x-M+1$ in the second and fourth terms of eq. (6.9) we get

$$Z_2 = 4\sum_{M=1}^{L}\sum_{x=0}^{L-M+1}\left[L\binom{L-1}{x+M}\binom{x+M-1}{M}-(M+1)\left[\binom{L}{x+M}\binom{x+M-1}{M+1}\right]\right]. \tag{6.10}$$

Using the binomial formula we obtain after some re-arrangements

$$Z_2 = 4L! \sum_{M=1}^{L} \frac{1}{M!(L-2-M)!} \int_0^1 p^M \left[(p+1)^{L-M-2} (1-p) \right] \mathrm{d}p. \quad (6.11)$$

Interchanging the sum and the integral and performing the sum gives

$$Z_2 = \frac{4L!}{(L-2)!} \int_0^1 (1-p) \left[(1+2p)^{L-2} - (1+p)^{L-2} \right] \mathrm{d}p. \quad (6.12)$$

This integral can be easily performed, resulting in

$$Z_2 = 3^L - 4 \cdot 2^L + 2L + 3. \quad (6.13)$$

Substituting eqs. (6.6) and (6.13) into (6.5) we get

$$Z = 3^L. \quad (6.14)$$

Thus we have shown that the number of eigenvectors of the t–J hamiltonian is 3^L, which is precisely the number of states in the Hilbert space of a chain of length L, where at each site there may be either a spin-up or a spin-down electron or a hole.

7. Conclusions

In this paper we have shown that the Bethe ansatz states for the one-dimensional supersymmetric t–J model are highest-weight vectors of an spl(2,1) superalgebra. Then by acting with the spl(2,1) lowering operators on the Bethe states we have obtained a complete set of eigenvectors of the t–J hamiltonian.

An interesting extension of this work is an analysis of the $\mathrm{spl}_q(2,1)$ "quantum-group" structure of a "q-deformed" version of this model (see also ref. [19]). This is presently under investigation.

Appendix A

In this appendix we show that the cancellation conditions of the "unwanted terms" u.t. = u.t.(A) + u.t.(D) and $\text{u.t.}_{(1)} = \text{u.t.}(A_{(1)}) + \text{u.t.}(D_{(1)})$ are equivalent to the Bethe ansatz equations (3.34) and (3.35). As already pointed out in sect. 3 all terms whose arguments are exchanged when $A(v)$ and $\tau_D(v)$ is commuted through

$\prod_{j=1}^{N} B_{\alpha_j}(v_j)$ using eqs. (3.11) and (3.12) are called u.t.(A) and u.t.(D), respectively. They can be arranged as follows [15]:

$$\text{u.t.}(A) = \sum_{j=1}^{N} K_j^{(A)}(v_j, \{v_i\}) B_{\alpha_1}(v_1) B_{\alpha_2}(v_2) \dots B_{\alpha_{j-1}}(v_{j-1}) B_{\alpha_j}(v)$$

$$\times B_{\alpha_{j+1}}(v_{j+1}) \dots B_{\alpha_N}(v_N) \Phi \Psi_{(1)}^{\{\alpha\}}, \tag{A.1}$$

$$\text{u.t.}(D) = \sum_{j=1}^{N} K_j^{(D)}(v_j, \{v_i\}) B_{\alpha_1}(v_1) B_{\alpha_2}(v_2) \dots B_{\alpha_{j-1}}(v_{j-1}) B_{\alpha_j}(v)$$

$$\times B_{\alpha_{j+1}}(v_{j+1}) \dots B_{\alpha_N}(v_N) \Phi \Psi_{(1)}^{\{\alpha\}}. \tag{A.2}$$

Here $K_j^{(A)}$ and $K_j^{(D)}$ ($j = 1, \dots, N$) are coefficients to be determined. The first coefficient of eq. (A.1) can be computed using the second term in (3.11) when commuting $A(v)$ with $B_{\alpha_1}(v_1)$ and then using the first term in eq. (3.11) when commuting $A(v_1)$ with the remaining B_α's, since otherwise the argument v_1 reappears in the B_α's. We get

$$K_1^{(A)} = -a^L(v_1) \frac{c(v_1 - v)}{b(v_1 - v)} \prod_{i \neq 1}^{N} \frac{a(v_i - v_1)}{b(v_i - v_1)}. \tag{A.3}$$

In order to calculate $K_j^{(D)}$ we rewrite the second term of eq. (3.12) as

$$-\frac{1}{v - v'} \operatorname*{Res}_{v''=v'} \left(S_{\gamma' \alpha}^{\gamma'' \alpha'}(v'' - v') B_{\alpha'}(v) T_{\gamma''}^{\gamma'}(v') \right), \tag{A.4}$$

by means of eqs. (2.5) and (4.15). Then, proceeding along the same lines as in the calculation of $K_1^{(A)}$ we get the eigenvalue problem for the transfer matrix $\tau_{(1)}$ (3.17). In addition, just taking the first term in eqs. (3.23) and (3.24) when passing $A_{(1)}$ and $\tilde{T}_3^3$ through the $B_{(1)}$'s we obtain, after some re-arrangements,

$$K_1^{(D)} = -b^L(v_1) \frac{c(v - v_1)}{b(v - v_1)} \prod_{i \neq 1}^{N} \frac{a(v_1 - v_i)}{b(v_1 - v_i)} \prod_{\beta=1}^{M} \frac{a(\gamma_\beta - v_1)}{b(\gamma_\beta - v_1)} \tag{A.5}$$

To get the other coefficients $K_j^{(A)}$ and $K_j^{(D)}$ ($j = 2, \dots, N$) we use the commutation rule for the B_α's (3.13) and put $B_{\alpha_j}(v)$ in the first place. Then, repeating the same procedure we obtain analogous expressions with j in the place of 1. Furthermore, the requirement $K_j^{(A)} + K_j^{(D)} = 0$ ($j = 1, \dots, N$) together with the fact that $c(v)/b(v)$ is an odd function (see eq. (2.5)) leads to the Bethe ansatz equation (3.34).

The "unwanted terms" that appear in the second level of the Bethe ansatz method can be arranged as follows:

$$\text{u.t.}_{(1)} = \sum_{\beta=1}^{M} \left(K_\beta^{(A_{(1)})} + K_\beta^{(D_{(1)})}\right) B_{(1)}(\gamma_1, \{v_i\}) B_{(1)}(\gamma_2, \{v_i\}) \ldots$$

$$\times B_{(1)}(\gamma_{\beta-1}, \{v_i\}) B_{(1)}(v, \{v_i\}) B_{(1)}(\gamma_{\beta+1}, \{v_i\}) \ldots B_{(1)}(\gamma_M, \{v_i\}) \Phi_{(1)}. \quad \text{(A.6)}$$

By similar arguments as above, the coefficients $K_1^{(A_{(1)})} + K_1^{(D_{(1)})}$ can be computed using the second term in eqs. (3.23) and (3.24) when commuting $A_{(1)}(v, \{v_i\})$ and $\tilde{T}_{(1)3}^{3}(v, \{v_8 \text{ii}\})$ through $B_{(1)}(\gamma_1, \{v_i\})$ and then using the first term in (3.23) and (3.24) when commuting $A_{(1)}(\gamma_1, \{v_i\})$ and $\tilde{T}_{(1)3}^{3}(\gamma_1, \{v_i\})$ with the remaining $B_{(1)}$'s.

$$K_\beta^{(A_{(1)})} + K_\beta^{(D_{(1)})} = -\prod_{i=1}^{N} a(\gamma_\beta - v_i) \frac{c(\gamma_\beta - v)}{b(\gamma_\beta - v)} \prod_{\alpha \neq \beta}^{M} \frac{a(\gamma_\alpha - \gamma_\beta)}{b(\gamma_\alpha - \gamma_\beta)}$$

$$-(-1)^M \frac{c(v - \gamma_\beta)}{b(v - \gamma_\beta)} \prod_{i=1}^{N} b(\gamma_\beta - v_i) \prod_{\alpha \neq \beta}^{M} \frac{w(\gamma_\beta - \gamma_\alpha)}{b(\gamma_\beta - \gamma_\alpha)}. \quad \text{(A.7)}$$

Once again, the other coefficients can be obtained using the commutation rules (3.25). The requirement $K_\beta^{(A_{(1)})} + K_\beta^{(D_{(1)})} = 0$ $(\beta = 1, \ldots, M)$ together with the fact that $c(v)/b(v)$ is an odd function yields the Bethe ansatz equation (3.35).

References

[1] H. Bethe, Z. Phys. 71 (1931) 205
[2] L.D. Faddeev and L. Takhtajan, Zap. Nauch. Semin. LOMI 109 (1981) 134
[3] F. Essler, V. Korepin and K. Schoutens, Phys. Rev. Lett. 67 (1991) 3848; Nucl. Phys. B372 (1992) 559
[4] P.W. Anderson, Science 235 (1987) 1196;
F.C. Zhang and T.M. Rice, Phys. Rev. B37 (1988) 3759
[5] L.A. Takhtajan and L.D. Faddeev, Russian Math. Surveys 34 (1979) 186
[6] O. Babelon, H.J. de Vega and C.M. Viallet, Nucl. Phys. B200 (1982) 266;
H.J. de Vega and M. Karowski, Nucl. Phys. B280 (1987) 225
[7] C.K. Lai, J. Math. Phys. 15 (1974) 1675
[8] P. Schlottmann, Phys. Rev. B36 (1987) 5177
[9] B. Sutherland, Phys. Rev. B12 (1975) 3795
[10] S. Sarkar, J. Phys. A23 (1990) L409; A24 (1991) 1137
[11] M. Scheunert, W. Nahm and V. Rittenberg, J. Math. Phys. 18 (1977) 146
[12] A. Foerster, M. Karowski, Phys. Rev. B46 (1992) 9234
[13] D. Förster, Phys. Rev. Lett. 63 (1989) 2140
[14] R.J. Baxter, Exactly solved models in statistical mechanics (Academic Press, New York, 1989)
[15] L.A. Takhtajan, Exactly solved problems in condensed matter and relativistic field theory (Springer, Berlin, 1985) p. 175
[16] M. Takahashi, Prog. Theor. Phys. 46 (1971) 401
[17] M. Takahashi, Prog. Theor. Phys. 46 (1971) 1388
[18] P. Bares and G. Blatter, Phys. Rev. Lett. 64 (1990) 2567; Phys. Rev. B44 (1991) 130
[19] A. Klümper, A. Schadschneider and J. Zittartz, J. Phys. A24 (1991) L955

III

Other models of strongly-correlated electrons

III. OTHER MODELS OF STRONGLY CORRELATED ELECTRONS

A. MODELS WITH LOCAL INTERACTION

The first electronic model that was exactly solved by means of the nested Bethe *Ansatz* is the non-relativistic continuum model of electrons with local interaction. This model is equivalent to the many-body problem of spin-$\frac{1}{2}$ fermions with δ-function interaction (the Bose gas with δ-function interaction was solved by E. H. Lieb and W. Liniger in Refs. 163, 164. Preceding the exact Bethe *Ansatz* solution were studies by J. M. McGuire[165] and by M. Flicker and E. H. Lieb[166], who investigated the spectrum and the ground state in the sectors with one and two down spins respectively. Especially important was Flicker's and Lieb's paper with the solution of the two down spin problem. In 1967 C. N. Yang[3] derived expressions for the eigenfunctions of the hamiltonian of the system for an arbitrary number of down spins (see also M. Gaudin Ref. 4). Yang solved the general case of a lattice gas of identical particles and identical vacancies and gave a detailed account of the nested Bethe *Ansatz*. His paper is reprinted as [repr.III.A.1]. B. Sutherland extended Yang's results to the case of distinguishable particles and identical vacancies[139] (the results of this paper were used by Sutherland in his solution of the t-J model in [repr.II.1]).

The N body S-matrix of the model of non-relativistic electrons with local interaction was investigated by various authors[167–170]. The thermodynamics of the model was investigated by M. Takahashi[171] and by C. K. Lai[172] in 1971. Both authors derived an infinite set of coupled integral equations, the solutions of which determine all thermodynamic quantities. They also studied the special limits of zero temperature, and of zero and infinite coupling constant. Takahashi's more detailed paper is reprinted as [repr.III.A.2].

The low-energy spectrum of Bethe *Ansatz* solvable models (which is essential for the asymptotic behaviour of correlations) quite generally can be described in terms of the so-called Luttinger-liquid theory, which was invented by F. D. M. Haldane in Refs. 90–94. Nowadays Luttinger-liquid theory is one of the main approaches to strongly correlated systems of condensed matter physics. The main idea of Luttinger-liquid theory is to base the low-energy effective hamiltonians of solvable models on the spectrum of the Luttinger model[173] (for a review of the Luttinger model see Ref. 174) plus residual anharmonic couplings, which vanish at low energies. The Luttinger model itself is an exactly solvable fermionic model. The exact spectrum, free energy and dielectric constant were determined by D. C. Mattis and E. H. Lieb in [repr.III.A.3][175]. The most striking feature of the Luttinger-liquid approach is that it allows the evaluation of the power-law singularities in the correlation functions (critical exponents) of many solvable models†. In his series of papers F. D. M. Haldane applied his ideas to a variety of models, e.g. one-dimensional interacting spinless fermions[91], and models solvable by a Bethe *Ansatz* without a nesting[92] [repr.III.A.4, III.A.5].

Quite recently two new models of strongly correlated electrons have been proposed in connection with high-T_c superconductivity. The first model arose in R. Z. Bariev's Bethe *Ansatz* solution of two coupled xy-spin chains upon fermionization by means of a Jordan-Wigner transformation[176]. In [repr.III.A.6] Bariev introduced the model and constructed the Bethe *Ansatz*. He found that the ground state exhibits a finite magnetization along the quantization axis. In a second paper R. Z. Bariev, A. Llümper, A. Schadschneider and J. Zittartz related the model to J. E. Hirsch's model of hole superconductivity and computed the critical exponents, conductivity and effective transport mass[177] [repr.III.A.7].

Another model of strongly correlated electrons was proposed by F. H. L. Eßler, V. E. Korepin and K. Schoutens[178] [repr.III.A.8]. Their $SU(2|2)$-supersymmetric model contains the supersymmetric t-J model as a submodel and can be interpreted as the Hubbard model plus moderate nearest-neighbour interactions. These interactions include bond-charge repulsion[179] and pair-hopping[180–182] terms. In one dimension the model is integrable. The hamiltonian is of the permutation type[138] already encountered for the supersymmetric t-J model. In Ref. 2 they determined the ground state wavefunction in any dimension and proved it to exhibit off-diagonal long-range order (ODLRO), thus establishing the superconducting nature of the model. The Bethe *Ansatz* solutions in one dimension, a highest weight theorem for the Bethe *Ansatz* states with respect to the $SU(2|2)$-supersymmetry and higher conservation laws, were derived in Ref. 183.

The relation of $U(M|N)$-supersymmetries and hamiltonians of systems of interacting fermions was addressed by K. Okumura[184,185]. A generalization of the t-J model was considered by P. Schlottmann in Ref. 186. He constructed the Bethe Ansatz for a two-band (supersymmetric) t-J model with an interband splitting and interaction within and between the two bands. He also derived the excitation spectrum.

†For an application of Luttinger-liquid theory to the Hubbard model see [repr.I.18]

VOLUME 19, NUMBER 23 PHYSICAL REVIEW LETTERS 4 DECEMBER 1967

SOME EXACT RESULTS FOR THE MANY-BODY PROBLEM IN ONE DIMENSION WITH REPULSIVE DELTA-FUNCTION INTERACTION*

C. N. Yang
Institute for Theoretical Physics, State University of New York, Stony Brook, New York
(Received 2 November 1967)

The repulsive δ interaction problem in one dimension for N particles is reduced, through the use of Bethe's hypothesis, to an eigenvalue problem of matrices of the same sizes as the irreducible representations R of the permutation group S_N. For some R's this eigenvalue problem itself is solved by a second use of Bethe's hypothesis, in a generalized form. In particular, the ground-state problem of spin-$\frac{1}{2}$ fermions is reduced to a generalized Fredholm equation.

(1) Consider the one-dimensional N-body problem

$$H=-\sum_1^N \partial^2/\partial x_i^2+2c\sum_{i<j}\delta(x_i-x_j),\quad c>0, \tag{1}$$

with no limitation on the symmetry of the wave function ψ. For a given irreducible representation R_ψ of the permutation group S_N of the N coordinates x_i, we want to determine the wave function ψ. Assume Bethe's hypothesis[1] to be valid: Let $p_1,\cdots,p_N$ = a set of unequal numbers. For $0<x_{Q1}<x_{Q2}<\cdots<x_{QN}<L$,

$$\psi=\sum_P[Q,P]\exp i[p_{P1}x_{Q1}+\cdots+p_{PN}x_{QN}], \tag{2}$$

where $P=[P1,P2,\cdots,PN]$ and $Q=[Q1,Q2,\cdots,QN]$ are two permutations of the integers $1,2,\cdots,N$. $[Q,P]$ can be arranged as a $N!\times N!$ matrix. Denote the columns of this matrix by ξ_P. To satisfy the continuity of ψ and the proper discontinuity of its derivative as required by (1) at $x_{Q3}=x_{Q4}$, it is sufficient to have

$$\xi_{\dots ij\dots}=Y_{ji}^{34}\xi_{\dots ji\dots}, \tag{3}$$

where the subscripts for ξ on the two sides represent any two permutation P and P' so that $P1=P'1$, $P2=P'2$, $P3=i=P'4$, $P4=j=P'3$, etc. The operator Y is defined by

$$Y_{ij}^{34}=(y_{ij}^{-1}-1)+y_{ij}^{-1}P_{34}=Y_{ij}^{43}, \tag{4}$$

where

$$y_{ij}=1+x_{ij}, \tag{5}$$

$$x_{jk}=ic(p_j-p_k)^{-1}=-x_{kj}, \tag{6}$$

and P_{34} = the permutation operator on ξ so that it interchanges $Q3$ and $Q4$. Altogether there are $N!(N-1)$ equations of the form (3). Are they mutually consistent? The answer is yes for any set of unequal p's. This can be seen with the aid of the following identities:

$$Y_{ij}^{ab}Y_{ji}^{ab}=1, \tag{7}$$

and

$$Y_{jk}^{ab}Y_{ik}^{bc}Y_{ij}^{ab}=Y_{ij}^{bc}Y_{ik}^{ab}Y_{jk}^{bc}, \tag{8}$$

which are easily verified. Thus given a set of unequal p's, and $\xi_0=\xi_P$ for P = identity, all ξ_P's are determined.

(2) The imposition of the periodic boundary conditions leads to equations which, upon expressing ξ_P in terms of ξ_0, become

$$\lambda_j\xi_0=X_{(j+1)j}\times X_{(j+2)j}\cdots X_{Nj}X_{1j}X_{2j}\cdots X_{(j-1)j}\xi_0,\quad j=1,\cdots,N, \tag{9}$$

where

$$\lambda_j=\exp(ip_jL), \tag{10}$$

and

$$X_{ij}=P_{ij}Y_{ij}^{ij}=(1-P_{ij}x_{ij})(1+x_{ij})^{-1}. \tag{11}$$

The N Eqs. (9) say that ξ_0 is simultaneously an eigenvector of N operators. These N operators can be shown to commute with each other, using

$$X_{ij}X_{ji}=1,\quad X_{jk}X_{ik}X_{ij}X_{kj}X_{ki}X_{ji}=1,$$
$$X_{ij}X_{kl}=X_{kl}X_{ij};\quad i,j,k,\text{ and } l \text{ all unequal}. \tag{12}$$

(3) The operators P_{ij} on ξ form a $N!\times N!$ representation of S_N. To find the eigenfunctions ξ_0 in (9) we can first reduce this representation to irreducible ones. Choosing one specific irreducible representation R reduces the

eigenvalue problem (9) to one of smaller dimensions. It can be shown that the resultant wave function (2) would have a permutation symmetry R_ψ which is the same as R. For example, if R = identity representation = $[N]$, then $P_{ij}=1$, and (9) becomes 1×1 matrix equations and the result is precisely the well-known boson result.[2] If R = antisymmetric representation = $[1^N]$, then $P_{ij}=-1$, and $X_{ij}=1$, so that (9) and (10) reduce to $\exp(ip_jL)=1$, showing there is no interaction, a result to be expected for the antisymmetrical wave function.

(4) The λ_j's are functions of the p's, c, and R. It is easily seen (that R and $\tilde{R}$ being conjugate representations)

$$\lambda_j(p;c;R)=\prod_{i\neq j}\left(\frac{1-x_{ij}}{1+x_{ij}}\right)\lambda_j(p;-c;\tilde{R}). \tag{13}$$

(5) Define $\mu_j(p;c;R)$ by

$$\mu_j\Phi=X_{(j+1)j}'X_{(j+2)j}'\cdots$$
$$\times X_{Nj}'X_{1j}'X_{2j}'\cdots X_{(j-1)j}'\Phi, \tag{14}$$

where

$$X_{ij}'=(1+P_{ij}x_{ij})(1+x_{ij})^{-1}. \tag{15}$$

Clearly

$$\mu_j(p;c;\tilde{R})=\lambda_j(p;c;R). \tag{16}$$

(6) We now evaluate λ_j for $R_\psi=R=[2^M1^{N-2M}]$. By (16) we need to find $\mu_j(p;c;[N-M,M])$. To do this we first define a convenient representation for P_{ij} of (15):

Consider N spin-$\frac{1}{2}$ particles, and consider the spin wave functions Φ for total z spin $=\frac{1}{2}(N-2M)$. These spin wave functions transform under S_N according to a sum of irreducible representations,

$$[N]+[N-1,1]+[N-2,2]+\cdots+[N-M,M]. \tag{17}$$

We consider the P_{ij}'s of (15) as operating on these spin wave functions Φ. The eigenvalue equations (14) for μ_j are then to be solved for a Φ that belongs to the symmetry $[N-M,M]$.

(7) Consider the N spins as forming a cyclic chain. The wave function Φ has $C_M{}^N$ components [$N-M$ spins up, M spins down]. The eigenvalue problem (14) can be solved with a generalized Bethe's hypothesis:

$$\Phi=\sum_P A_P F(\Lambda_{P1},y_1)$$
$$\times F(\Lambda_{P2},y_2)\cdots F(\Lambda_{PM},y_M), \tag{18}$$

where $y_1<y_2<\cdots<y_M$ are the "coordinates," along the chain, of the M down spins, and Λ_1, $\Lambda_2,\cdots,\Lambda_M$ are a set of unequal numbers. With this hypothesis, one finds

$$F(\Lambda,y)=\prod_{j=1}^{y-1}\frac{ip_j-i\Lambda-c'}{ip_{j+1}-i\Lambda+c'}\quad(c'=\tfrac{1}{2}c); \tag{19}$$

$$-\prod_j\frac{ip_j-i\Lambda_\alpha-c'}{ip_j-i\Lambda_\alpha+c'}=\prod_\beta\frac{-i\Lambda_\beta+i\Lambda_\alpha+c}{-i\Lambda_\beta+i\Lambda_\alpha-c}; \tag{20}$$

and

$$\mu_j(p;c;[N-M,M])=\prod_\beta\frac{ip_j-i\Lambda_\beta-c'}{ip_j-i\Lambda_\beta+c'}. \tag{21}$$

(8) Thus for the $R_\psi=[2^M1^{N-2M}]$ symmetry, we need to solve

$$\exp(ip_jL)=\text{right-hand side of (21)}, \tag{22}$$

together with (20). In taking the logarithm of (20) and (22) care must be taken to add terms $2\pi i$(integer). The value of the integer can be determined by going to the limit $c\to+\infty$. One obtains, for the ground state with the symmetry $R_\psi=[2^M1^{N-2M}]$, for the case N = even, M = odd,

$$-\sum_p\theta(2\Lambda-2p)=2\pi J_\Lambda-\sum_{\Lambda'}\theta(\Lambda-\Lambda'), \tag{23a}$$

$$Lp=2\pi I_p+\sum_\Lambda\theta(2p-2\Lambda), \tag{23b}$$

where the p's are a set of N ascending real numbers, the Λ's a set of M ascending real numbers,

$$\theta(p)=-2\tan^{-1}(p/c)\quad(-\pi\leq\theta<\pi), \tag{24}$$

and

$$J_\Lambda=\text{successive integers from}\quad -\tfrac{1}{2}(M-1)\text{ to }+\tfrac{1}{2}(M-1), \tag{24a}$$

$$\tfrac{1}{2}+I_p=\text{successive integers from}\quad 1-\tfrac{1}{2}N\text{ to }\tfrac{1}{2}N. \tag{24b}$$

Equation (23a) differs from that given in a re-

VOLUME 19, NUMBER 23 PHYSICAL REVIEW LETTERS 4 DECEMBER 1967

cent paper,[3] in the definition of θ and our introduction of J_Λ. The present equation allows for a natural discussion of the limit $c \to +\infty$ (not $c \to 0$!) and hence the values of J_Λ.

(9) We can now approach the limit $N \to \infty$, $M = \infty$, $L \to \infty$ proportionally, obtaining

$$-\int_{-Q}^{Q} \theta(2\Lambda - 2p)\rho(p)dp = 2\pi g - \int_{-B}^{B} \theta(\Lambda - \Lambda')\sigma(\Lambda')d\Lambda', \quad (25a)$$

$$p = 2\pi f + \int_{-B}^{B} \theta(2p - 2\Lambda)\sigma(\Lambda)d\Lambda, \quad (25b)$$

$$dg/d\Lambda = \sigma, \quad df/dp = \rho. \quad (25c)$$

Or, after differentiation,

$$2\pi\sigma = -\int_{-B}^{B} \frac{2c\sigma(\Lambda')d\Lambda'}{c^2 + (\Lambda - \Lambda')^2} + \int_{-Q}^{Q} \frac{4c\rho dp}{c^2 + 4(p - \Lambda)^2}, \quad (26a)$$

$$2\pi\rho = 1 + \int_{-B}^{B} \frac{4c\sigma d\Lambda}{c^2 + 4(p - \Lambda)^2}, \quad (26b)$$

$$N/L = \int_{-Q}^{Q} \rho dp, \quad M/L = \int_{-B}^{B} \sigma d\Lambda, \quad (27a)$$

and

$$E/L = \int_{-Q}^{Q} p^2 \rho(p) dp. \quad (27b)$$

(10) Equations (26) are generalized Fredholm equations with a symmetrical kernel. It is easy to show that the equations are nonsingular by first studying the eigenvalues of the kernel in the limit $B = Q = \infty$.

(11) Equations (26) and (27) yield the ground-state energy per particle for spatial wave functions with the symmetry $[2^M 1^{N-2M}]$, at a given density N/L. For N fermions with spin $\frac{1}{2}$ interacting through the Hamiltonian (1), this spatial wave function is coupled to a spin wave function of conjugate symmetry $[N-M, M]$, i.e., the total spin of the system is $\frac{1}{2}N - M$.

(12) For $B = \infty$, integration of (26a) over all Λ yields $N = 2M$. Thus for the fermion problem with spin $\frac{1}{2}$, $B = \infty$ gives the ground state for states with total spin = 0. This state is also the absolute ground state for the problem, by a theorem due to Lieb and Mattis.[4]

(13) For the case $B \cong 0$, M/L is proportional to B. One can readily expand all quantities in powers of B, obtaining, for fixed $r = N/L$,

$$\frac{E}{L} = \text{const.} + \frac{M}{L}\left[cr - \left(\frac{c^2}{2\pi} + 2\pi r^2\right)\tan^{-1}\frac{2\pi r}{c}\right] + \cdots. \quad (28)$$

This result is in agreement with results already obtained by McGuire[5] for the case $M = 1$ and by Flicker and Lieb[6] for the case $M = 2$.

(14) For each symmetry R_ψ of spatial wave function ψ, the excited states near the ground state can be obtained in a similar way as in the boson case.[7] More quantum numbers are, however, necessary to designate the excitations than in the boson case, because of the existence of the integers J_Λ (which are in fact quantum numbers). Details will be published elsewhere.

(15) For the boson problem the thermodynamics and excitations for finite T were treated by Yang and Yang.[8] Extension to the present problem presents no difficulty. Details will be published elsewhere.

(16) Using (13) one could generalize all the considerations above to the case of $R_\psi = [N-M, M]$. Details will be published elsewhere. The main change is that while all Eqs. (26) and (27) remain the same, (26b) is replaced by

$$2\pi\rho = 1 - \int_{-B}^{B} \frac{4c\sigma d\Lambda}{c^2 + 4(p - \Lambda)^2} + \int_{-Q}^{Q} \frac{2c\rho(p')dp'}{c^2 + (p - p')^2}. \quad (26b')$$

It is a pleasure to acknowledge useful discussions with J. B. McGuire in 1963 at University of California, Los Angeles, with T. T. Wu in 1964 at Brookhaven National Laboratory, with C. P. Yang in 1966 in Princeton, and with B. Sutherland in 1967 in Stony Brook.

---

*Research partly supported by U. S. Atomic Energy Commission under Contract No. AT(30-1)-3668B.

[1]H. A. Bethe, Z. Physik 71, 205 (1931), first used the hypothesis for the spin-wave problem. E. Lieb and W. Linger, Phys. Rev. 130, 1605 (1963), and J. B. McGuire, J. Math. Phys. 5, 622 (1964), first used the same hypothesis for the δ-function interaction problem. The present author believes that for all states (excited as well as the ground state) with periodic boundary condition the hypothesis is valid. Justification of this belief for some special cases is found in C. N. Yang and C. P. Yang, Phys. Rev. 150, 321 (1966), and to be published.

[2]Lieb and Linger, Ref. 1.

[3]M. Gaudin, Phys. Letters 24A, 55 (1967).

[4]E. Lieb and D. Mattis, Phys. Rev. 125, 164 (1962).

[5]J. B. McGuire, J. Math. Phys. 6, 432 (1965), and 7,

123 (1966).
[6]M. Flicker and E. H. Lieb, Phys. Rev. 161, 179 (1967).
[7]E. Lieb, Phys. Rev. 130, 1616 (1963).
[8]C. N. Yang and C. P. Yang, to be published. See also C. N. Yang, in Proceedings of Eastern Theoretical Conference, November, 1966, Brown University (W. A. Benjamin, Inc., New York, 1967), p. 215.

1388

Progress of Theoretical Physics, Vol. 46, No. 5, November 1971

One-Dimensional Electron Gas with Delta-Function Interaction at Finite Temperature

Minoru TAKAHASHI

Department of Physics, University of Tokyo, Tokyo

(Received May 28, 1971)

Integral equations which describe the thermodynamic properties of a one-dimensional electron gas with repulsive and attractive delta-function interactions are obtained. From these equations one can calculate the energy, entropy, magnetization, particle density and pressure at given temperature, magnetic field and chemical potential.

§ 1. Introduction

In recent papers Gaudin[1)] and Yang[2)] gave the ground state energy of a one-dimensional electron gas with a delta-function interaction[3)~6)] as a solution of a set of coupled integral equations. We try to treat the thermodynamic properties of this system as a one-dimensional Bose gas and a one-dimensional Heisenberg model.[8)] For this purpose it is necessary to obtain all of the energy eigenvalues of the Hamiltonian. In § 2 we review the work of Gaudin and Yang on the wave function. There appear two kinds of parameters k and Λ. In § 3 we make conjectures on the distributions of k's and Λ's in the complex plane. In § 4 the energy spectrum of the Hamiltonian for repulsive interaction is obtained and the integral equations which describe the thermodynamic properties are derived. In § 5 these integral equations are solved for some special cases. In §§ 6 and 7 we treat the electron gas with an attractive delta-function interaction.

§ 2. Wave function

We consider the eigenvalue problem of the Hamiltonian

$$\mathcal{H}=-\sum_{i=1}^{N}\frac{\partial^2}{\partial x_i^2}+4c\sum_{i<j}\delta(x_i-x_j)+\mu_0 H(2M-N), \qquad (2\cdot1)$$

where N is the number of electrons and M is the number of down-spin electrons. The wave function has the following form:

$$\Psi(x_1s_1, x_2s_2, \cdots, x_Ns_N)=\sum_j {}_M\Phi_j(x_1, x_2, \cdots, x_N)G_j^M. \qquad (2\cdot2)$$

Here x_i and s_i are the coordinate and spin-coordinate of the i-th electron, respectively. For a spin-$\frac{1}{2}$ electron, s_i is $+$ or $-$. G_j^M is a spin function of which a typical one is

$$G_1^M = (\underbrace{+ + \cdots + +}_{N-M} \underbrace{- - \cdots -}_{M}).$$

${}_M\Phi_1$ is an eigenfunction of (2·1) which is antisymmetric to the permutation of $x_1, x_2, \cdots, x_{N-M}$ and to the permutation of $x_{N-M+1}, \cdots, x_N$, satisfying the condition

$$(1 - \sum_{j=N-M+1}^{N} P_{N-M,j})\, {}_M\Phi_1 = 0\,. \tag{2·3}$$

Here $P_{i,j}$ is an operator which changes x_i and x_j. We can construct a full wave function Ψ using the fact that Ψ is totally antisymmetric. Gaudin and Yang gave the solution for this problem as follows:

$${}_M\Phi_1 = \sum_P [Q, P] \exp(i \sum_{j=1}^{M} k_{Pj} x_{Qj}), \tag{2·4}$$

in the region $x_{Q1} < x_{Q2} < \cdots < x_{QN}$. Here Q and P are permutations of 1, 2, $\cdots$, N and $[Q, P]$ are $N! \times N!$ coefficients which are given by

$$[Q, P] = \varepsilon(Q_1)\varepsilon(Q_2) \sum_R A_R \prod_{j=1}^{M} F_P(\Lambda_{Rj}, y_j), \tag{2·5}$$

$$F_P(\Lambda, y) = \prod_{j=1}^{y-1} \frac{k_{Pj} - \Lambda + ic}{k_{P(j+1)} - \Lambda - ic}, \tag{2·6}$$

$$A_R = \prod_{\substack{i<j \\ Ri>Rj}} e\left(\frac{\Lambda_{Rj} - \Lambda_{Ri}}{2c}\right), \tag{2·7}$$

$$e(x) \equiv (x+i)/(x-i),$$

where $y_1 < y_2, \cdots < y_M$ are coordinates of $x_{N-M+1}, x_{N-M+2}, \cdots x_N$ along the chain, Q_1 and Q_2 signify the orders of 1, 2, $\cdots$, $N-M$ and $N-M+1$, $\cdots$, N in the permutation Q. The parameters $\Lambda_1, \Lambda_2, \cdots, \Lambda_M$ are newly introduced. The periodic boundary condition

$${}_M\Phi_1(x_1, x_2, \cdots, x_i, \cdots, x_N) = {}_M\Phi_1(x_1, \cdots, x_i + L, \cdots, x_N), \qquad i = 1, 2, \cdots, N,$$

gives an equation for k's and Λ's as follows:

$$e^{ik_jL} = \prod_{\alpha=1}^{M} e\left(\frac{k_j - \Lambda_\alpha}{c}\right), \tag{2·8}$$

$$\prod_{j=1}^{N} e\left(\frac{\Lambda_\alpha - k_j}{c}\right) = \prod_{\beta \neq \alpha} e\left(\frac{\Lambda_\alpha - \Lambda_\beta}{2c}\right). \tag{2·9}$$

§ 3. Conjectures on the distribution of k's and Λ's in the complex plane

In this section we make three conjectures which are essential in the later sections of this paper.

Conjecture 1. If a set of solutions $(k_1, k_2, \cdots, k_N; \Lambda_1, \Lambda_2, \cdots, \Lambda_M)$ of (2·8) and (2·9) contain a complex k (or Λ), $\bar{k}$ (or $\bar{\Lambda}$) is also contained in the set of k's (or Λ's).

Corollary 1. At $c>0$, k's are real.
Proof: The conjecture 1 demands that the distributions of k's and Λ's are symmetric with respect to real axis. So we see that if $\mathrm{Im}\, k_j>0$, the absolute value of the right-hand side of (2·8) is larger than unity. On the other hand the left-hand side is smaller than unity because $\mathrm{Im}\, k_j>0$. So $\mathrm{Im}\, k_j>0$ is impossible. In the same way we can prove that $\mathrm{Im}\, k_j<0$ is also impossible. [Q.E.D.]
Conjecture 2. Complex Λ always forms a bound state with several other Λ's. In this set of Λ's the real parts of these Λ's are the same and the imaginary parts are $(n-1)ci$, $(n-3)ci$, $\cdots$, $-(n-1)ci$ for the bound state of $n-\Lambda$'s within the accuracy of $O(\exp(-\delta N))$, where δ is a positive number.
Conjecture 3. In the case $c<0$, complex k_α makes a pair with its complex conjugate $\bar{k}_\alpha$ and a real Λ, which we write as Λ_α'. The real parts of k_α, $\bar{k}_\alpha$ and Λ_α' are the same and the imaginary parts of k_α and $\bar{k}_\alpha$ are c and $-c$ within the accuracy of $O(\exp(-\delta L))$.

§ 4. Derivation of integral equations for the case of a repulsive interaction

In this case all k's are real by the corollary 1 in § 3. But Λ's are not necessarily real. We write Λ's as $\Lambda_\alpha^{n,j}$. Here n means that this belongs to a bound state of $n-\Lambda$'s, j specifies the imaginary part and α is the number of this bound state in the bound states of $n-\Lambda$'s. We write the real part of $\Lambda_\alpha^{n,j}$ by Λ_α^n. By the conjecture 2 we have

$$\Lambda_\alpha^{n,j}=\Lambda_\alpha^n+(n+1-2j)ci+O(\exp(-\delta N)),\ j=1,2,\cdots,n\,. \tag{4·1}$$

In the case of M_n bound states of n spins and N electrons we derive equations for Λ_α^n's and k_j's from Eqs. (2·8) and (2·9). Equations (2·8) can be rewritten as

$$e^{ik_jL}=\prod_{n=1}^{\infty}\prod_{\alpha=1}^{M_n}e\left(\frac{k_j-\Lambda_\alpha^n}{nc}\right),\qquad j=1,2,\cdots,N\,. \tag{4·2a}$$

Let us consider a product

$$\prod_{j=1}^{N}e\left(\frac{\Lambda_\alpha^n-k_j}{nc}\right).$$

By (4·1) this is transformed as

$$\prod_{j=1}^{N}\prod_{l=1}^{n}e\left(\frac{\Lambda_\alpha^{n,l}-k_j}{c}\right),$$

and by (2·9)

$$=\prod_{l=1}^{n}\left\{-\prod_{(m,\beta)}\prod_{h=1}^{m}e\left(\frac{\Lambda_\alpha^{n,l}-\Lambda_\beta^{m,h}}{2c}\right)\right\}=\prod_{l=1}^{n}\left\{\prod_{(m,\beta)\neq(n,\alpha)}\prod_{h=1}^{m}e\left(\frac{\Lambda_\alpha^{n,l}-\Lambda_\beta^{m,h}}{2c}\right)\right\}.$$

Substituting (4·1) we have finally

$$\prod_{j=1}^{N} e\left(\frac{\Lambda_\alpha{}^n - k_j}{nc}\right) = -\prod_{m=1}^{\infty}\prod_{\beta=1}^{M_m} E_{nm}\left(\frac{\Lambda_\alpha{}^n - \Lambda_\beta{}^m}{c}\right), \qquad \begin{array}{l} n=1, 2, \cdots, \\ \alpha=1, 2, \cdots, M_n, \end{array} \tag{4·2b}$$

where

$$E_{nm}(x) \equiv \begin{cases} e\left(\frac{x}{|n-m|}\right) e^2\left(\frac{x}{|n-m|+2}\right) e^2\left(\frac{x}{|n-m|+4}\right) \cdots e^2\left(\frac{x}{n+m-2}\right) e\left(\frac{x}{n+m}\right) \\ \qquad\qquad\qquad\qquad \text{for } n \neq m, \\ e^2\left(\frac{x}{2}\right) e^2\left(\frac{x}{4}\right) \cdots e^2\left(\frac{x}{2n-2}\right) e\left(\frac{x}{2n}\right) \qquad \text{for } n=m. \end{cases}$$

The logarithms of these equations give

$$k_j L = 2\pi I_j - \sum_{n=1}^{\infty}\sum_{\alpha=1}^{M_n} \theta\left(\frac{k_j - \Lambda_\alpha{}^n}{nc}\right), \qquad j=1, 2, \cdots, N, \tag{4·3a}$$

$$\sum_{j=1}^{N} \theta\left(\frac{\Lambda_\alpha{}^n - k_j}{nc}\right) = 2\pi J_\alpha{}^n + \sum_{m=1}^{\infty}\sum_{\beta=1}^{M_m} \Theta_{nm}\left(\frac{\Lambda_\alpha{}^n - \Lambda_\beta{}^m}{c}\right), \qquad \begin{array}{l} n=1, \cdots, \\ \alpha=1, 2, \cdots, M_n \end{array} \tag{4·3b}$$

where $\theta(x) \equiv 2\tan^{-1}x$, $-\pi < \theta \leq \pi$ and

$$\Theta_{nm}(x) \equiv \begin{cases} \theta\left(\frac{x}{|n-m|}\right) + 2\theta\left(\frac{x}{|n-m|+2}\right) + 2\theta\left(\frac{x}{|n-m|+4}\right) \\ \qquad + \cdots + 2\theta\left(\frac{x}{n+m-2}\right) + \theta\left(\frac{x}{n+m}\right) \qquad \text{for } n \neq m, \\ 2\theta\left(\frac{x}{2}\right) + 2\theta\left(\frac{x}{4}\right) + \cdots + 2\theta\left(\frac{x}{2n-2}\right) + \theta\left(\frac{x}{2n}\right) \qquad \text{for } n=m. \end{cases}$$

I_j's are different integers (half-odd integer) for even (odd) $M_1 + M_2 + \cdots$. This can be written as

$$I_j \equiv M_1 + M_2 + \cdots \qquad (\text{mod } 1). \tag{4·4a}$$

$J_\alpha{}^n$'s are different integers and satisfy the conditions

$$J_\alpha{}^n \equiv N - M_n + \tfrac{1}{2} \qquad (\text{mod } 1), \tag{4·4b}$$

$$|J_\alpha{}^n| \leq \tfrac{1}{2}(N - 1 - \sum t_{nm} M_m), \tag{4·4c}$$

where

$$t_{nm} = 2\,\text{Min}(n, m) - \delta_{nm}.$$

Giving a set of integers $\{I_j, J_\alpha{}^n\}$ which satisfies Eqs. (4·4), we can determine a set of k_j and $\Lambda_\alpha{}^n$ through Eqs. (4·3). For a set of integers $\{I_j, J_\alpha{}^n\}$ there is a set of omitted integers which satisfy Eqs. (4·4) and are not contained in $\{I_j, J_\alpha{}^n\}$. We define holes of k and holes of Λ^n as solutions of

$$Lh(k) = 2\pi \times (\text{omitted } I),$$

$$Lj_n(\Lambda) = 2\pi \times (\text{omitted } J_\alpha{}^n),$$

where

$$h(k)\equiv k+\frac{1}{L}\sum_{n=1}^{\infty}\sum_{\alpha=1}^{M_n}\theta\left(\frac{k-\Lambda_\alpha^n}{nc}\right), \tag{4·5a}$$

$$j_n(\Lambda)\equiv\frac{1}{L}\sum_{j=1}^{N}\theta\left(\frac{\Lambda-k_j}{nc}\right)-\frac{1}{L}\sum_{m=1}^{\infty}\sum_{\alpha=1}^{M_m}\Theta_{nm}\left(\frac{\Lambda-\Lambda_\alpha^m}{c}\right). \tag{4·5b}$$

Let us consider the case of a very large system. We put the distribution functions of k's and Λ^n's as $\rho(k)$ and $\sigma_n(k)$, and those of holes as $\rho^h(k)$ and $\sigma_n^h(k)$. By the definition of holes it is clear that

$$\frac{d}{dk}h(k)=2\pi(\rho(k)+\rho^h(k)), \tag{4·6a}$$

$$\frac{d}{dk}j_n(k)=2\pi(\sigma_n(k)+\sigma_n^h(k)). \tag{4·6b}$$

Equations (4·5a) and (4·5b) are rewritten as

$$h(k)=k+\sum_{n=1}^{\infty}\int\theta\left(\frac{k-k'}{nc}\right)\sigma_n(k')\,dk',$$

$$j_n(k)=\int\theta\left(\frac{k-k'}{nc}\right)\rho(k')\,dk'-\int\Theta_{nm}\left(\frac{k-k'}{c}\right)\sigma_m(k')\,dk'.$$

Hereafter we put that $\int dk$ means $\int_{-\infty}^{\infty}dk$. Substituting these into Eqs. (4·6) we have

$$\frac{1}{2\pi}=\rho(k)+\rho^h(k)-\sum_{n=1}^{\infty}[n]\sigma_n(k), \tag{4·7a}$$

$$[n]\rho(k)=\sigma_n^h(k)+\sum_{m=1}^{\infty}A_{nm}\sigma_m(k), \tag{4·7b}$$

where $[n]$ is an operator defined by

$$[n]f(k)\equiv\frac{1}{\pi}\int_{-\infty}^{\infty}\frac{n|c|}{(nc)^2+(k-k')^2}f(k')\,dk',$$

$$[0]f(k)\equiv f(k),$$

and

$$A_{nm}\equiv[|n-m|]+2[|n-m|+2]+2[|n-m|+4]+\cdots+2[n+m-2]+[n+m].$$

The energy per unit length is

$$E/L=\int(k^2-\mu_0H)\rho(k)\,dk+\sum_{n=1}^{\infty}2n\mu_0H\int\sigma_n(k)\,dk. \tag{4·8a}$$

The entropy per unit length is

$$S/L=\int\{(\rho+\rho^h)\ln(\rho+\rho^h)-\rho\ln\rho-\rho^h\ln\rho^h\}\,dk$$

$$+\sum_{n=1}^{\infty}\int\{(\sigma_n+\sigma_n^h)\ln(\sigma_n+\sigma_n^h)-\sigma_n\ln\sigma_n-\sigma_n^h\ln\sigma_n^h\}dk\,. \tag{4·8b}$$

The particle density is

$$N/L=\int\rho dk\,. \tag{4·8c}$$

The magnetization to the z-direction is

$$S_z/L=\tfrac{1}{2}\int\rho dk-\sum_n n\int\sigma_n dk\,. \tag{4·8d}$$

At the equilibrium state the thermodynamic potential $\Omega\equiv E-AN-TS$ should be minimized. So the variation of Ω is zero:

$$\begin{aligned}0=\delta\Omega/L=&\int(k^2-A-\mu_0H)\delta\rho(k)dk+\sum_{n=1}^{\infty}2n\mu_0H\int\delta\sigma(k)dk\\&-T\int\left\{\delta\rho\ln\left(\frac{\rho+\rho^h}{\rho}\right)+\delta\rho^h\ln\left(\frac{\rho+\rho^h}{\rho^h}\right)\right\}dk\\&-T\int\left\{\delta\sigma_n\ln\left(\frac{\sigma_n+\sigma_n^h}{\sigma_n}\right)+\delta\sigma_n^h\ln\left(\frac{\sigma_n+\sigma_n^h}{\sigma_n^h}\right)\right\}dk\,.\end{aligned} \tag{4·9}$$

From Eq. (4·7) we have

$$\delta\rho^h=-\delta\rho+\sum_{n=1}^{\infty}[n]\delta\sigma_n\,,$$

$$\delta\sigma_n^h=[n]\delta\rho-\sum_{m=1}^{\infty}A_{nm}\delta\sigma_m\,.$$

Substituting these into Eq. (4·9) we have

$$\begin{aligned}\frac{\delta\Omega}{TL}=&\int\left\{\frac{k^2-A-\mu_0H}{T}-\ln\left(\frac{\rho^h}{\rho}\right)-\sum_{n=1}^{\infty}[n]\ln\left(1+\frac{\sigma_n}{\sigma_n^h}\right)\right\}\delta\rho dk\\&+\sum_{n=1}^{\infty}\int\left\{\frac{2n\mu_0H}{T}-\ln\left(1+\frac{\sigma_n^h}{\sigma_n}\right)-[n]\ln\left(1+\frac{\rho}{\rho^h}\right)\right.\\&\left.+\sum_{m=1}^{\infty}A_{nm}\ln\left(1+\frac{\sigma_m^h}{\sigma_m}\right)\right\}\delta\sigma_n dk\,.\end{aligned}$$

Then we have a set of coupled nonlinear integral equations for $\zeta(k)\equiv\rho^h(k)/\rho(k)$ and $\eta_n(k)\equiv\sigma_n^h(k)/\sigma_n(k)$ as follows:

$$\ln\zeta(k)=\frac{k^2-A-\mu_0H}{T}-\sum_{n=1}^{\infty}[n]\ln(1+\eta_n^{-1}(k))\,, \tag{4·10a}$$

$$\ln(1+\eta_n(k))=\frac{2n\mu_0H}{T}-[n]\ln(1+\zeta^{-1}(k))+\sum_{m=1}^{\infty}A_{nm}\ln(1+\eta_m^{-1}(k))\,. \tag{4·10b}$$

Equations (4·7) are rewritten as

$$(1+\zeta(k))\rho(k)=\frac{1}{2\pi}+\sum_{n=1}^{\infty}[n]\sigma_n(k), \tag{4·11a}$$

$$[n]\rho(k)=\eta_n(k)\sigma_n(k)+\sum_{m=1}^{\infty}A_{nm}\sigma_m(k). \tag{4·11b}$$

From thermodynamics the pressure is given by

$$P=-\Omega/L.$$

Using (4·10) and (4·11) one obtains

$$P=T\int \ln(1+\zeta^{-1}(k))\frac{dk}{2\pi}. \tag{4·12}$$

This expression for the pressure is the same as that for bosons obtained by Yang and Yang.[7)]

If we can solve Eqs. (4·10a), (4·10b), (4·11a) and (4·11b), we can determine the energy, entropy, particle density, magnetization and pressure for given temperature, chemical potential and magnetic field using (4·8a), (4·8b), (4·8c), (4·8d) and (4·12).

Equations (4·10a) and (4·10b) are equivalent to

$$[1]\{\ln(1+\eta_2)-\ln(1+\zeta^{-1})\}=([0]+[2])\ln\eta_1, \tag{4·13a}$$

$$[1]\{\ln(1+\eta_{n-1})+\ln(1+\eta_{n+1})\}=([0]+[2])\ln\eta_n,\quad n=2,3,\cdots, \tag{4·13b}$$

$$\ln(1+\eta_1)=\frac{2\mu_0 H}{T}-[1]\ln(1+\zeta^{-1})+\sum_{m=1}^{\infty}A_{1m}\ln(1+\eta_m{}^{-1}), \tag{4·13c}$$

$$\ln\zeta=\frac{k^2-A-\mu_0 H}{T}-\sum_{m=1}^{\infty}[m]\ln(1+\eta_m{}^{-1}). \tag{4·13d}$$

(4·13a) is obtained by $[1]\times$(first formula of (4·10b)) $-([0]+[2])\times$(4·10a). (4·13b) is obtained by $[1]\times\{(n-1$-th formula of (4·10b)) $+(n+1$-th formula of (4·10b))$\}-([0]+[2])\times(n$-th formula of (4·10b)). In the same way we can prove easily that Eqs. (4·11a) and (4·11b) are equivalent to

$$[1](\rho+\eta_2\sigma_2)=([0]+[2])(\eta_1+1)\sigma_1, \tag{4·14a}$$

$$[1](\eta_{n-1}\sigma_{n-1}+\eta_{n+1}\sigma_{n+1})=([0]+[2])(\eta_n+1)\sigma_n,\quad n=2,3,\cdots, \tag{4·14b}$$

$$[1]\rho=\eta_1\sigma_1+\sum_{m=1}^{\infty}A_{1m}\sigma_m, \tag{4·14c}$$

$$(1+\zeta)\rho=\frac{1}{2\pi}+\sum_{m=1}^{\infty}[m]\sigma_m. \tag{4·14d}$$

§ 5. Special cases for c>0

1) *The limit c→0*

In this limit we can put

$$[n]f(k)=f(k)$$

for an arbitrary function $f(k)$. Then Eqs. (4·12) are written as

$$(1+\eta_2)/(1+\zeta^{-1})=\eta_1^2, \tag{5·1a}$$

$$(1+\eta_{n-1})(1+\eta_{n+1})=\eta_n^2, \tag{5·1b}$$

$$1+\eta_1=z^{-2}(1+\zeta^{-1})^{-1}\prod_{n=2}^{\infty}(1+\eta_n^{-1})^2, \tag{5·1c}$$

$$\zeta=e^{(k^2-A)/T}z\prod_{n=1}^{\infty}(1+\eta_n^{-1})^{-1}, \tag{5·1d}$$

where

$$z=\exp(-\mu_0H/T).$$

The general solution of (5·1a) and (5·1b) is

$$\eta_n=f^2(n)-1\,, \qquad \zeta=\frac{f^2(0)}{1-f^2(0)}\,,$$

where

$$f(n)=(ba^n-b^{-1}a^{-n})/(a-a^{-1}).$$

The parameters a and b are functions of k and determined by (5·1c) and (5·1d). The results are

$$a=z \qquad \text{and} \qquad b=\sqrt{\left(1+z\exp\frac{k^2-A}{T}\right)\Big/\left(1+z^{-1}\exp\frac{k^2-A}{T}\right)}. \tag{5·2}$$

Equations (4·13) are transformed as

$$\rho+\eta_2\sigma_2=2(\eta_1+1)\sigma_1\,, \tag{5·3a}$$

$$\eta_{n-1}\sigma_{n-1}+\eta_{n+1}\sigma_{n+1}=2(\eta_n+1)\sigma_n\,, \tag{5·3b}$$

$$\rho=\eta_1\sigma_1+\sum_{m=1}^{\infty}A_{1m}\sigma_m\,, \tag{5·3c}$$

$$(1+\zeta)\rho=\frac{1}{2\pi}+\sum_{n=1}^{\infty}\sigma_n\,. \tag{5·3d}$$

The solution is

$$\rho=\frac{1}{2\pi}(b+b^{-1})\{-f(-1)\}, \tag{5·4a}$$

$$\sigma_n=\frac{1}{2\pi}f(0)\{-f(-1)\}\left(\frac{1}{f(n-1)f(n)}-\frac{1}{f(n)f(n+1)}\right). \tag{5·4b}$$

Substituting (5·2) into (5·4a) we have

$$\rho(k)=\frac{1}{2\pi}\left(\left(1+\exp\frac{k^2-A-\mu_0H}{T}\right)^{-1}+\left(1+\exp\frac{k^2-A+\mu_0H}{T}\right)^{-1}\right). \tag{5·5}$$

In the limit $c\to 0$ the quasi-momenta are real momenta. So (5·5) coincides with the well-known result.

2) *The limit* $c\to\infty$

In this limit Eqs. (4·10a), (4·10b), (4·11a) and (4·11b) become

$$\ln\zeta(k)=\frac{k^2-A-\mu_0H}{T}-\sum_{n=1}^{\infty}[n]\ln(1+\eta_n^{-1}(k)),\tag{5·6a}$$

$$\ln(1+\eta_n(k))=\frac{2n\mu_0H}{T}+\sum_{m=1}^{\infty}A_{nm}\ln(1+\eta_m^{-1}(k))+O\left(\frac{1}{c}\right),\tag{5·6b}$$

$$(1+\zeta(k))\rho(k)=\frac{1}{2\pi}+O\left(\frac{1}{c}\right),\tag{5·6c}$$

$$[n]\rho(k)=\eta_n(k)\sigma_n(k)+\sum_{m=1}^{\infty}A_{nm}\sigma_m(k).\tag{5·6d}$$

Equation (5·6b) are easily solved because $\eta_n(k)$ are all constants. The solution is

$$\eta_n=f^2(n)-1\,,\tag{5·7a}$$

where

$$f(n)=(z^{n+1}-z^{-n-1})/(z-z^{-1}),\qquad z=\exp(-\mu_0H/T).$$

From (5·6a) we have

$$\zeta(k)=e^{(k^2-A)/T}/(z+z^{-1}),\tag{5·7b}$$

and from (5·6c) we have

$$\rho(k)=\frac{1}{2\pi}\,\frac{z+z^{-1}}{z+z^{-1}+e^{(k^2-A)/T}}\,.\tag{5·7c}$$

Using (5·7a) one obtains

$$\sigma_n(k)=\frac{1}{z+z^{-1}}\left\{\frac{1}{f(n-1)f(n)}[n]-\frac{1}{f(n)f(n+1)}[n+2]\right\}\rho(k).\tag{5·7d}$$

So we have

$$\begin{aligned}S_z/L&=N/2L-\sum_{n=1}^{\infty}n\int\sigma_n dk\\&=\frac{N}{L}\left[\frac{1}{2}-\sum_{n=1}^{\infty}\frac{n}{z+z^{-1}}\left\{\frac{1}{f(n-1)f(n)}-\frac{1}{f(n)f(n+1)}\right\}\right]=\frac{1}{2}\left(\frac{N}{L}\right)\tanh\frac{\mu_0H}{T}\,.\end{aligned}\tag{5·7e}$$

This shows that the magnetization of the one-dimensional electron gas behaves as that of ½-spins which are free each other when c is infinity.

3) *The limit* $T\to 0$

We put $\varepsilon_n(k)\equiv T\ln\eta_n(k)$ and $\kappa(k)\equiv T\ln\zeta(k)$. One can derive that

$$\varepsilon_n(k) = 2\mu_0 H + [2] T \ln\left(1+\exp -\frac{\varepsilon_n(k)}{T}\right) + [1] T \ln\left(1+\exp \frac{\varepsilon_{n-1}(k)}{T}\right)$$

$$+ ([0]+[2]) \sum_{j=1}^{\infty} [j] T \ln\left(1+\exp -\frac{\varepsilon_{j+n}(k)}{T}\right), \quad n=2,3,\cdots,$$

from Eqs. (4·10b). Therefore ε_2, ε_3, $\cdots$ are always positive. So in the limit $T \to 0$ we have a set of equations

$$\kappa(k) = k^2 - A - \mu_0 H + [1]\varepsilon_1^-(k), \tag{5·8a}$$

$$\varepsilon_1(k) = 2\mu_0 H + [1]\kappa^-(k) - [2]\varepsilon_1^-(k), \tag{5·8b}$$

where the suffices (+) and (−) mean

$$f^+(k) \equiv \begin{cases} f(k) & \text{at } f(k)>0, \\ 0 & \text{at } f(k)\leqq 0, \end{cases} \qquad f^-(k) \equiv \begin{cases} 0 & \text{at } f(k)\geqq 0, \\ f(k) & \text{at } f(k)<0. \end{cases}$$

In Appendix A we prove that ε_1 and κ are increasing functions of k^2. So ε_1 and κ are negative in the regions $[B, -B]$ and $[Q, -Q]$, respectively. Then Eqs. (4·11) give

$$\rho(k) = \frac{1}{2\pi} + \frac{1}{\pi}\int_{-B}^{B} \frac{c}{c^2+(k-k')^2}\sigma_1(k')\,dk', \tag{5·9a}$$

$$\frac{1}{\pi}\int_{-Q}^{Q} \frac{c\rho(k')\,dk'}{c^2+(k-k')^2} = \sigma_1(k) + \frac{1}{\pi}\int_{-B}^{B} \frac{2c\sigma_1(k')\,dk'}{4c^2+(k-k')^2}. \tag{5·9b}$$

The energy, particle number and magnetization per unit length are given by

$$E/L = \int_{-Q}^{Q} k^2 \rho(k)\,dk, \tag{5·9c}$$

$$N/L = \int_{-Q}^{Q} \rho(k)\,dk, \tag{5·9d}$$

$$S_z/L = \tfrac{1}{2}\int_{-Q}^{Q} \rho(k)\,dk - \int_{-B}^{B} \sigma_1(k)\,dk. \tag{5·9e}$$

These integral equations coincide with those which were obtained by Gaudin[1)] and Yang.[2)]

§ 6. Derivation of integral equations for the case of an attractive interaction

If there are pairs of two complex k's each of which has a parameter Λ on real axis. We designate these Λ's as Λ_α' and corresponding k's as k_α^1 and k_α^2. By the conjecture 3 we have

$$k_\alpha^1 = \Lambda_\alpha' + i|c| + O(\exp(-\delta L)),$$

$$k_\alpha^2 = \Lambda_\alpha' - i|c| + O(\exp(-\delta L)).$$

From Eq. (2·8) we have

$$\exp i(k_\alpha^1+k_\alpha^2)L=\prod_{\beta\neq\alpha} e\left(\frac{\Lambda_\alpha'-\Lambda_\beta'}{-2|c|}\right)\left\{\prod_{m,\beta,j} e\left(\frac{\Lambda_\alpha'-\Lambda_\beta^{n,j}}{-2|c|}\right)\right\} e\left(\frac{k_\alpha^1-\Lambda_\alpha'}{-|c|}\right) e\left(\frac{k_\alpha^2-\Lambda_\alpha'}{-|c|}\right). \tag{6·1}$$

From Eq. (2·9) we have

$$e\left(\frac{k_\alpha^1-\Lambda_\alpha'}{-|c|}\right) e\left(\frac{k_\alpha^2-\Lambda_\alpha'}{-|c|}\right)=\prod_{j=1}^{N-2M'} e\left(\frac{\Lambda_\alpha'-k_j}{-|c|}\right)\left\{\prod_{n,\beta,j} e\left(\frac{\Lambda_\alpha'-\Lambda_\beta^{n,j}}{2|c|}\right)\right\}.$$

Substituting this into Eq. (6·1) we have

$$e^{2i\Lambda_\alpha' L}=-\prod_{j=1}^{N-2M'} e\left(\frac{\Lambda_\alpha'-k_j}{-|c|}\right)\prod_{\beta=1}^{M'} e\left(\frac{\Lambda_\alpha'-\Lambda_\beta'}{-2|c|}\right). \tag{6·2}$$

Here we have represented unpaired k as k_j. From Eq. (2·8) one obtains

$$e^{ik_jL}=\prod_{\alpha=1}^{M'} e\left(\frac{k_j-\Lambda_\alpha'}{-|c|}\right)\prod_{n=1}^{\infty}\prod_{\alpha=1}^{M_n} e\left(\frac{k_j-\Lambda_\alpha^n}{-n|c|}\right). \tag{6·3}$$

And from Eq. (2·9) we have

$$\prod_{j=1}^{N-2M'} e\left(\frac{\Lambda_\alpha^n-k_j}{n|c|}\right)=-\prod_{m=1}^{\infty}\prod_{\beta=1}^{M_m} E_{nm}\left(\frac{\Lambda_\alpha^n-\Lambda_\beta^m}{|c|}\right). \tag{6·4}$$

The logarithms of Eqs. (6·2), (6·3) and (6·4) are

$$2\Lambda_\alpha'L=2\pi J_\alpha'+\sum_{j=1}^{N-2M'}\theta\left(\frac{\Lambda_\alpha'-k_j}{|c|}\right)+\sum_{\beta=1}^{M'}\theta\left(\frac{\Lambda_\alpha'-\Lambda_\beta'}{2|c|}\right),\quad \alpha=1,2,\cdots,M', \tag{6·5a}$$

$$k_jL=2\pi I_j+\sum_{\alpha=1}^{M'}\theta\left(\frac{k_j-\Lambda_\alpha'}{|c|}\right)+\sum_{n=1}^{\infty}\sum_{\alpha=1}^{M_n}\theta\left(\frac{k_j-\Lambda_\alpha^n}{n|c|}\right),\quad j=1,\cdots,N-2M', \tag{6·5b}$$

$$\sum_{j=1}^{N-2M'}\theta\left(\frac{\Lambda_\alpha^n-k_j}{n|c|}\right)=2\pi J_\alpha^n+\sum_{m=1}^{\infty}\sum_{\beta=1}^{M_m}\Theta_{nm}\left(\frac{\Lambda_\alpha^n-\Lambda_\beta^m}{|c|}\right),\quad \begin{array}{l}\alpha=1,2,\cdots,M_n,\\ n=1,2,\cdots.\end{array} \tag{6·5c}$$

Here J_α' is integer (half-odd integer) for $N-M'$ odd (even), I_j is integer (half-odd integer) for $M'+M_1+M_2+\cdots$ even (odd) and J_α^n is integer (half-odd integer) for $N-M_n$ odd (even). J_α^n should satisfy the condition

$$|J_\alpha^n|\leq\tfrac{1}{2}\Big(N-2M'-\sum_{m=1}^{\infty}t_{nm}M_m\Big).$$

Following § 4 we define $j'(\Lambda')$, $h(k)$ and $j_n(\Lambda^n)$:

$$j'(\Lambda')\equiv2\Lambda'-\frac{1}{L}\left\{\sum_{j=1}^{N-2M'}\theta\left(\frac{\Lambda'-k_j}{|c|}\right)+\sum_{\beta=1}^{M'}\theta\left(\frac{\Lambda'-\Lambda_\beta'}{2|c|}\right)\right\}, \tag{6·6a}$$

$$h(k)\equiv k-\frac{1}{L}\left\{\sum_{\alpha=1}^{M'}\theta\left(\frac{k-\Lambda_\alpha'}{|c|}\right)+\sum_{n=1}^{\infty}\sum_{\alpha=1}^{M_n}\theta\left(\frac{k-\Lambda_\alpha^n}{n|c|}\right)\right\}, \tag{6·6b}$$

$$j_n(\Lambda^n)\equiv\frac{1}{L}\left\{\sum_{j=1}^{N-2M'}\theta\left(\frac{\Lambda^n-k_j}{n|c|}\right)-\sum_{m=1}^{\infty}\sum_{\beta=1}^{M_m}\Theta_{nm}\left(\frac{\Lambda^n-\Lambda_\beta^m}{|c|}\right)\right\}. \tag{6·6c}$$

Holes of Λ', k and Λ^n are defined as solutions of

$$j'(\Lambda')=2\pi\times(\text{omitted } J'),$$

$$h(k)=2\pi\times(\text{omitted } I),$$

$$j_n(\Lambda^n)=2\pi\times(\text{omitted } J^n).$$

In the limit of a very large system we define the distribution functions of Λ', k_j, $\Lambda_\alpha{}^n$ as $\sigma'(k)$, $\rho(k)$, $\sigma_n(k)$ and those of holes as $\sigma'^h(k)$, $\rho^h(k)$, $\sigma_n{}^h(k)$. Using the relations

$$\frac{dj'(k)}{dk}=2\pi(\sigma'(k)+\sigma'^h(k)),$$

$$\frac{dh(k)}{dk}=2\pi(\rho(k)+\rho^h(k)),$$

$$\frac{dj_n(k)}{dk}=2\pi(\sigma_n(k)+\sigma_n{}^h(k)),$$

we have equations for σ', ρ, σ_n, σ'^h, ρ^h and $\sigma_n{}'$:

$$\frac{1}{\pi}=\sigma'+\sigma'^h+[2]\sigma'+[1]\rho\,, \qquad (6\cdot 7\text{a})$$

$$\frac{1}{2\pi}=\rho+\rho^h+[1]\sigma'+\sum_n[n]\sigma_n\,, \qquad (6\cdot 7\text{b})$$

$$[n]\rho=\sigma_n{}^h+\sum_m A_{nm}\sigma_m\,. \qquad (6\cdot 7\text{c})$$

The definitions of $[n]$ and A_{nm} were given in § 4.

The energy per unit length is

$$E/L=\int(k^2-\mu_0H)\rho dk+\int 2(k^2-c^2)\sigma'dk'+\sum_{n=1}^{\infty}2n\mu_0H\int\sigma_n dk\,. \qquad (6\cdot 8\text{a})$$

The entropy per unit length is

$$S/L=\int\{(\rho+\rho^h)\ln(\rho+\rho^h)-\rho\ln\rho-\rho^h\ln\rho^h\}dk$$

$$+\int\{(\sigma'+\sigma'^h)\ln(\sigma'+\sigma'^h)-\sigma'\ln\sigma'-\sigma'^h\ln\sigma'^h\}dk$$

$$+\sum_n\int\{(\sigma_n+\sigma_n{}^h)\ln(\sigma_n+\sigma_n{}^h)-\sigma_n\ln\sigma_n-\sigma_n{}^h\ln\sigma_n{}^h\}dk\,. \qquad (6\cdot 8\text{b})$$

The magnetization per unit length is

$$S_z/L=\tfrac{1}{2}\int\rho dk-\sum_{n=1}^{\infty}n\int\sigma_n dk\,. \qquad (6\cdot 8\text{c})$$

The particle density is

$$N/L=\int \rho dk+2\int \sigma' dk\,. \tag{6·8d}$$

The thermodynamic potential $\Omega \equiv E-TS-AN$ should be minimized. So we have

$$\begin{aligned}0=\delta\Omega/L=&\int 2(k^2-c^2-A)\delta\sigma' dk+\int (k^2-A)\delta\rho dk\\&+\sum_{n=1}^{\infty}2n\mu_0 H\int \delta\sigma_n dk-T\int\left\{\delta\rho\ln\left(1+\frac{\rho^h}{\rho}\right)+\delta\rho^h\ln\left(1+\frac{\rho}{\rho^h}\right)\right\}dk\\&-T\int\left\{\delta\sigma'\ln\left(1+\frac{\sigma'^h}{\sigma'}\right)+\delta\sigma'^h\ln\left(1+\frac{\sigma'}{\sigma'^h}\right)\right\}dk\\&-T\sum_{n=1}^{\infty}\int\left\{\delta\sigma_n\ln\left(1+\frac{\sigma_n^h}{\sigma_n}\right)+\delta\sigma_n^h\ln\left(1+\frac{\sigma_n}{\sigma_n^h}\right)\right\}dk\,.\end{aligned} \tag{6·9}$$

From Eqs. (6·7a), (6·7b) and (6·7c) we have

$$\delta\sigma'^h=-\delta\sigma'-[2]\delta\sigma'-[1]\delta\rho\,,$$

$$\delta\rho^h=-\delta\rho-[1]\delta\sigma'-\sum_{n=1}^{\infty}[n]\delta\sigma_n\,,$$

$$\delta\sigma_n^h=[n]\delta\rho-\sum_{m=1}^{\infty}A_{nm}\delta\sigma_m\,.$$

Substituting these into Eq. (6·9) we have a set of coupled-nonlinear integral equations for $\zeta=\rho^h/\rho$, $\eta'=\sigma'^h/\sigma'$ and $\eta_n=\sigma_n^h/\sigma_n$ as follows:

$$\ln\eta'=\frac{2(k^2-A-c^2)}{T}+[2]\ln(1+\eta'^{-1})+[1]\ln(1+\zeta^{-1}), \tag{6·10a}$$

$$\ln\zeta=\frac{k^2-A-\mu_0 H}{T}+[1]\ln(1+\eta'^{-1})-\sum_{n=1}^{\infty}[n]\ln(1+\eta_n^{-1}), \tag{6·10b}$$

$$\ln(1+\eta_n)=\frac{2n\mu_0 H}{T}+[n]\ln(1+\zeta^{-1})+\sum_{m=1}^{\infty}A_{nm}\ln(1+\eta_m^{-1}). \tag{6·10c}$$

Equations (6·7) are rewritten as

$$\frac{1}{\pi}=(1+\eta')\sigma'+[2]\sigma'+[1]\rho\,, \tag{6·11a}$$

$$\frac{1}{2\pi}=(1+\zeta)\rho+[1]\sigma'+\sum_{n=1}^{\infty}[n]\sigma_n\,, \tag{6·11b}$$

$$[n]\rho=\eta_n\sigma_n+\sum_m A_{nm}\sigma_m\,. \tag{6·11c}$$

The pressure P and thermodynamic potential Ω are given by

$$P=-\Omega/L=T\int\ln(1+\eta'^{-1})\frac{dk}{\pi}+T\int\ln(1+\zeta^{-1})\frac{dk}{2\pi}\,. \tag{6·12}$$

Here we have used Eqs. (6·10) and (6·11).

The integral equations (6·10) are transformed as

$$[1]\{\ln(1+\eta_1)-\ln(1+\eta')\}=([0]+[2])\ln\zeta^{-1}, \tag{6·13a}$$

$$[1]\{\ln(1+\zeta^{-1})+\ln(1+\eta_2)\}=([0]+[2])\ln\eta_1, \tag{6·13b}$$

$$[1]\{\ln(1+\eta_{n-1})+\ln(1+\eta_{n+1})\}=([0]+[2])\ln\eta_n, \quad n=2,3,\cdots, \tag{6·13c}$$

$$\ln\eta'=\frac{2(k^2-A-c^2)}{T}+[2]\ln(1+\eta'^{-1})+[1]\ln(1+\zeta^{-1}), \tag{6·13d}$$

$$\ln\zeta=\frac{k^2-A-\mu_0H}{T}+[1]\ln(1+\eta'^{-1})-\sum_{n=1}^{\infty}[n]\ln(1+\eta_n^{-1}). \tag{6·13e}$$

Equations (6·11) are transfomed as follows:

$$[1](\eta'\sigma'+\eta_1\sigma_1)=([0]+[2])(1+\zeta)\rho, \tag{6·14a}$$

$$[1](\rho+\eta_2\sigma_2)=([0]+[2])(1+\eta_1)\sigma_1, \tag{6·14b}$$

$$[1](\eta_{n-1}\sigma_{n-1}+\eta_{n+1}\sigma_{n+1})=([0]+[2])(1+\eta_n)\sigma_n, \quad n=2,3,\cdots, \tag{6·14c}$$

$$\frac{1}{\pi}=(1+\eta'+[2])\sigma'+[1]\rho, \tag{6·14d}$$

$$\frac{1}{2\pi}=(1+\zeta)\rho+[1]\sigma'+\sum_{n=1}^{\infty}[n]\sigma_n. \tag{6·14e}$$

§ 7. Special cases for $c<0$

1) *The limit $c\to 0$*

In this limit we can put $[n]=[0]$. Therefore Eqs. (6·13) become

$$(1+\eta_1)/(1+\eta')=\zeta^{-2},$$

$$(1+\zeta^{-1})(1+\eta_2)=\eta_1^2,$$

$$(1+\eta_{n-1})(1+\eta_{n+1})=\eta_n^2, \quad n=2,3,\cdots,$$

$$\eta_1^2/(1+\eta_1)=e^{2(k^2-A)/T}(1+\zeta^{-1}),$$

$$\xi=ze^{(k^2-A)/T}(1+\eta'^{-1})/\prod_{n=1}^{\infty}(1+\eta_n^{-1}).$$

And the solutions are

$$\eta_n=f^2(n)-1, \tag{7·1a}$$

$$\zeta=(f^2(0)-1)^{-1}, \tag{7·1b}$$

$$\eta'=f^2(-1)-1, \tag{7·1c}$$

where

$$f(n)=\frac{bz^n-b^{-1}z^{-n}}{z-z^{-1}}, \qquad b=z^2\sqrt{\frac{1+z^{-1}e^{(k^2-A)/T}}{1+ze^{(k^2-A)/T}}}.$$

Equations (6·13) become

$$\eta'\sigma'+\eta_1\sigma_1=2(1+\zeta)\rho,$$
$$\rho+\eta_2\sigma_2=2(1+\eta_1)\sigma_1,$$
$$\eta_{n-1}\sigma_{n-1}+\eta_{n+1}\sigma_{n+1}=2(1+\eta_n)\sigma_n, \qquad n=2,3,\cdots,$$
$$\frac{1}{\pi}=(2+\eta')\sigma'+\rho,$$
$$\frac{1}{2\pi}=(1+\zeta)\rho+\sigma'+\sum_{n=1}^{\infty}\sigma_n.$$

Using (7·1a), (7·1b) and (7·1c) we have a solution for these linear equations:

$$\rho=\frac{1}{2\pi}\frac{(b+b^{-1})f(-1)(-f(-2))}{f(0)}, \tag{7·2a}$$

$$\sigma'=\frac{1}{2\pi}\frac{f^2(-1)}{f(0)}(bz^{-1}+b^{-1}z), \tag{7·2b}$$

$$\sigma_n=\frac{1}{2\pi}f(-1)(-f(-2))\left\{\frac{1}{f(n-1)f(n)}-\frac{1}{f(n)f(n+1)}\right\}. \tag{7·2c}$$

One can calculate $\rho+2\sigma'$ which is the distribuion of real momenta in the limit $c\to 0$:

$$\rho+2\sigma'=\frac{1}{2\pi}((e^{(k^2-A-\mu_0H)/T}+1)^{-1}+(e^{(k^2-A-\mu_0H)/T}+1)^{-1}). \tag{7·3}$$

This result coincides with the well-known facts and suggests that our theory is correct.

2) *The limit $T\to 0$*

We prove that $\varepsilon_n(k)>0$ from Eq. (6·10c). Therefore in the limit $T\to 0$ we see $\eta_n=\infty$ and $\sigma_n=0$ for $n=1, 2, \cdots$. ε' and κ are determined by

$$\varepsilon'(k)=2(k^2-A-c^2)-[2]\varepsilon'^-(k)-[1]\kappa^-(k), \tag{7·4a}$$

$$\kappa(k)=k^2-A-\mu_0H-[1]\varepsilon'^-(k), \tag{7·4b}$$

and are monotonically increasing functions of k^2 as will be shown in Appendix B. We define the parameters B and Q by $\varepsilon'(B)=0$ and $\kappa(Q)=0$. η' and ζ are zero in the region $[B, -B]$ and $[Q, -Q]$, respectively, and infinity outside these regions. So one obtains a set of coupled linear equations in the limt $T\to 0$.

$$\frac{1}{\pi}=\sigma'(k)+\frac{1}{\pi}\int_{-B}^{B}\frac{2|c|\sigma'(k')dk'}{4|c|^2+(k-k')^2}+\frac{1}{\pi}\int_{-Q}^{Q}\frac{|c|\rho(k')dk'}{|c|^2+(k-k')^2},$$

$$\frac{1}{2\pi}=\rho(k)+\frac{1}{\pi}\int_{-B}^{B}\frac{|c|\sigma'(k')dk'}{|c|^2+(k-k')^2},$$

$$E/L=\int_{-B}^{B}2(k^2-c^2)\sigma'(k)\,dk+\int_{-Q}^{Q}k^2\rho(k)\,dk\,,$$

$$N/L=2\int_{-B}^{B}\sigma'(k)\,dk+\int_{-Q}^{Q}\rho(k)\,dk\,,$$

$$S_z/L=\int_{-Q}^{Q}\rho(k)\,dk\,.$$

These equations are equivalent to those which were obtained by Gaudin.[1]

§ 8. Discussion

Our equations are non-linear and have infinite unknown functions. But the author believes that the numerical calculation of physical quantities can be done if we use a high-speed computer.

It is possible to calculate the excitation spectra from the thermodynamically equilibrium state as Yang and Yang[7] did for one-dimensional bosons.

In Ref. 6) the author discussed the analytic properties of the energy at zero temperature. But if one uses our integral equations it is possible to investigate the analytic properties of thermodynamic quantities at finite temperature.

Our theory is based on the three conjectures of § 3. So it is necessary to prove these conjectures strictly.

We have obtained the integral equations for two-component bosons, namely, the case of a wave function which transforms as an irreducible representation of S_N with two rows. The integral equations for the ground state energy was derived by Yang.[3] We put a chemical potential for first-kind of bosons as $A+\mu_0H$ and one for second-kind of bosons as $A-\mu_0H$. In the case of repulsive interactions (4·10a), (4·10b) and (4·11a) are replaced by

$$\ln\zeta+[2]\ln(1+\zeta^{-1})=\frac{k^2-A-\mu_0H}{T}-\sum_{n=1}^{\infty}[n]\ln(1+\eta_n^{-1}),$$

$$\ln(1+\eta_n)=\frac{2n\mu_0H}{T}+[n]\ln(1+\zeta^{-1})+\sum_{m=1}^{\infty}A_{nm}\ln(1+\eta_m^{-1}),$$

$$(1+\zeta)\rho=\frac{1}{2\pi}+[2]\rho-\sum_{n=1}^{\infty}[n]\sigma_n\,,$$

respectively.

Appendix A

Equation (5·8) is transformed as

$$\varepsilon_1(k)=\mu_0H+\int\frac{1}{4c}\operatorname{sech}\frac{\pi(k-k')}{2c}\,\kappa^-(k')\,dk'+\int\frac{1}{c}R\Big(\frac{k-k'}{c}\Big)\varepsilon_1^+(k')\,dk'. \tag{A1}$$

So we consider a series of functions $\{\varepsilon_1^{(n)}\}$ and $\{\kappa^{(n)}\}$:

$$\varepsilon_1^{(n+1)}(k)=\mu_0 H+\int\frac{1}{4c}\operatorname{sech}\frac{\pi(k-k')}{2c}\kappa^{(n)-}(k')dk'+\int\frac{1}{c}R\Big(\frac{k-k'}{c}\Big)\varepsilon_1^{(n)+}(k')dk', \tag{A2}$$

$$\kappa^{(n+1)}(k)=k^2-A-\mu_0H+[1]\varepsilon_1^{(n)-}(k), \tag{A3}$$

$$\kappa^{(1)}(k)=k^2-A-\mu_0H, \tag{A4}$$

$$\varepsilon_1^{(1)}(k)=2\mu_0H. \tag{A5}$$

We prove the following lemma by mathematical induction.
Lemma 1.

a) $\varepsilon_1^{(n)}\geqq -A-\mu_0H,\ \kappa_1^{(n)}\geqq -2A-2\mu_0H.$

b) $\varepsilon_1^{(n)}\geqq\varepsilon_1^{(n+1)},\ \kappa^{(n)}\geqq\kappa^{(n+1)}.$

c) $\varepsilon_1^{(n)}$ and $\kappa^{(n)}$ are monotonically increasing functions (MIF) of k^2.

[Proof] It is clear from (A4) and (A5) that a) and c) are valid for $n=1$. From

$$\varepsilon_1^{(2)}=2\mu_0H+\int\frac{1}{4c}\operatorname{sech}\frac{\pi(k-k')}{2c}\kappa^{(n)-}(k')dk'\leqq\varepsilon_1^{(1)}$$

and

$$\kappa^{(1)}=\kappa^{(2)}.$$

We see that b) is valid for $n=1$. It is clear from (A2) and (A3) that if a), b) and c) is valid for $n=k$, a), b) and c) are valid for $n=k+1$. [Q.E.D.]

From a) and b) we see that the limit $\varepsilon_1=\lim_{n\to\infty}\varepsilon_1^{(n)}$ and $\kappa=\lim_{n\to\infty}\kappa^{(n)}$ exist. These two functions ε_1 and κ are solutions of (5·8a) and (5·8b) and MIF's of k^2.

Appendix B

The Equations (7·4a) and (7·4b) are transformed as

$$\kappa(k)=-\mu_0H+\int\frac{dk'}{4|c|}\operatorname{sech}\frac{\pi(k-k')}{2|c|}\varepsilon'^{+}(k')+\int\frac{dk'}{|c|}R\Big(\frac{k-k'}{|c|}\Big)\kappa^{-}(k'), \tag{B1}$$

$$\varepsilon'(k)=k^2-A-2c^2+\mu_0H+\{k^2+[1](\kappa^{+}(k)-k^2)\}. \tag{B2}$$

So we consider the series of functions defined by

$$\varepsilon'^{(1)}(k)=2(k^2-A-c^2), \tag{B3}$$

$$\kappa^{(1)}(k)=k^2-A-\mu_0H, \tag{B4}$$

$$\varepsilon'^{(n+1)}(k)=k^2-A-2c^2+\mu_0H+\{k^2+[1](\kappa^{(n)+}(k)-k^2)\}, \tag{B5}$$

$$\kappa^{(n+1)}(k)=-\mu_0H+\int\frac{1}{|c|}R\Big(\frac{k-k'}{|c|}\Big)\kappa^{(n)-}(k')dk' + \int\frac{1}{4|c|}\operatorname{sech}\frac{\pi(k-k')}{2|c|}\varepsilon'^{(n+1)+}(k')dk'. \tag{B6}$$

Lemma 2.

a) $\kappa^{(n)} \leqq \kappa^{(n+1)}$, $\varepsilon'^{(n)} \leqq \varepsilon'^{(n+1)}$.

b) $\kappa^{(n)}(k) \leqq k^2 + c^2$, $\varepsilon'^{(n)}(k) \leqq 2k^2 + 2\mu_0 H$.

c) $\kappa^{(n)}$ and $\varepsilon'^{(n)}$ are MIF's of k^2.

[Proof] For $n=1$ b) and c) are easily proved by (B3) and (B4). From (B5) we have

$$\varepsilon'^{(2)}(k) \geqq k^2 - A - 2c^2 + \mu_0 H + \{k^2 + [1](\kappa^{(1)}(k) - k^2)\} = \varepsilon'^{(1)}(k). \qquad \text{(B7)}$$

Substituting (B5) into (B6) we have

$$\kappa^{(2)}(k) = k^2 - \frac{A+\mu_0 H}{2} + \int \frac{1}{|c|} R\left(\frac{k-k'}{|c|}\right)(\kappa^{(1)}(k') - k'^2)\,dk'$$
$$- \int \frac{1}{4|c|} \operatorname{sech} \frac{\pi(k-k')}{2|c|} \varepsilon'^{(2)-}(k')\,dk' \geqq k^2 - \frac{A+\mu_0 H}{2}$$
$$+ \int \frac{1}{|c|} R\left(\frac{k-k'}{|c|}\right)(\kappa^{(1)}(k') - k'^2)\,dk' = \kappa^{(1)}(k).$$

From Eqs. (B5) and (B6) it can be easily proved that a) and b) are valid for $n=k+1$ if they are for $n=k$. One can prove that

$$\int \frac{1}{|c|} R\left(\frac{k-k'}{|c|}\right) f(k')\,dk',$$
$$\int \frac{1}{4|c|} \operatorname{sech} \frac{\pi(k-k')}{2|c|} f(k')\,dk'$$

and

$$k^2 + [1](f(k) - k^2)$$

are MIF's of k^2 if $f(k)$ is an MIF of k^2. From this fact we have that c) is valid for $n=k+1$ if it is for $n=k$. [Q.E.D.]

From a) and b) we see that there exist the limits

$$\varepsilon'(k) = \lim_{n\to\infty} \varepsilon'^{(n)}(k) \qquad \text{and} \qquad \kappa(k) = \lim_{n\to\infty} \kappa^{(n)}(k).$$

It is clear that these two functions satisfy (7·4a) and (7·4b) are MIF's of k^2.

References

1) M. Gaudin, Phys. Letters **24A** (1967), 55; thesis (University of Paris, 1967), unpublished.
2) C. N. Yang, Phys. Rev. Letters **19** (1967), 1312.
3) J. B. McGuire, J. Math. Phys. **5** (1964), 622; **6** (1965), 432; **7** (1966), 123.
4) M. Flicker and E. H. Lieb, Phys. Rev. **161** (1967), 179.
5) M. Suzuki, Prog. Theor. Phys. **42** (1969), 1076.

6) M. Takahashi, Prog. Theor. Phys. **44** (1970), 348.
7) C. N. Yang and C. P. Yang, J. Math. Phys. **10** (1969), 1115.
8) M. Takahashi, Prog. Theor. Phys. **46** (1971), 401.
M. Gaudin, Phys. Rev. Letters **26** (1971), 1301.

Note added in proof: C. K. Lai gave equations which are equivalent to those of ours for fermions with repulsive and attractive interactions. See C. K. Lai, Phys. Rev. Letters **26** (1971), 1472.

Reprinted from *Printed in U.S.A.*
JOURNAL OF MATHEMATICAL PHYSICS VOLUME 6, NUMBER 2 FEBRUARY 1965

Exact Solution of a Many-Fermion System and Its Associated Boson Field

DANIEL C. MATTIS
International Business Machines Corp., Thomas J. Watson Research Center, Yorktown Heights, New York
AND
ELLIOTT H. LIEB*
Belfer Graduate School of Science, Yeshiva University, New York, New York
(Received 22 September 1964)

Luttinger's exactly soluble model of a one-dimensional many-fermion system is discussed. We show that he did not solve his model properly because of the paradoxical fact that the density operator commutators $[\rho(p), \rho(-p')]$, which always vanish for any finite number of particles, no longer vanish in the field-theoretic limit of a filled Dirac sea. In fact the operators $\rho(p)$ *define* a boson field which is *ipso facto* associated with the Fermi-Dirac field. We then use this observation to solve the model, and obtain the exact (and now nontrivial) spectrum, free energy, and dielectric constant. This we also extend to more realistic interactions in an Appendix. We calculate the Fermi surface parameter $\bar{n}_k$, and find: $\partial\bar{n}_k/\partial k|_{k_F} = \infty$ (i.e., there exists a sharp Fermi surface) only in the case of a sufficiently weak interaction.

I. INTRODUCTION

THE search for a soluble but realistic model in the many-electron problem has been just about as unfruitful as the historic quest for the philosopher's stone, but has equally resulted in valuable byproducts. For example, 15 years ago Tomonaga[1] published a theory of interacting fermions which was soluble only in one dimension with the provision that certain truncations and approximations were introduced into his operators. Nevertheless he had success in showing approximate boson-like behavior of certain collective excitations, which he identified as "phonons." (Today we would denote these as "plasmons," following the work of Bohm and Pines.[2]) Lately, Luttinger[3] has revived interest in the subject by publishing a variant model of spinless and massless one-dimensional interacting fermions, which demonstrated a singularity at the Fermi surface, compatible with the results of the modern many-body perturbation theory.[4]

Unfortunately, in calculating the energies and wavefunctions of his model Hamiltonian, Luttinger fell prey to a subtle paradox inherent in quantum field theory[5] and *therefore did not achieve a correct solution of the problem he himself had posed.* In the present paper we shall give the solution to his interesting problem and calculate the free energy. We shall show the existence of collective plasmon modes, and shall calculate the singularity at the Fermi surface (which may in fact disappear if the interaction is strong enough), the energy of the plasmons, and the (nontrivial) dielectric constant of the system. In an Appendix we shall show how the model may be generalized in such a manner as to remove certain restrictions on the interactions which Luttinger had found necessary to impose.

It is fortunate that solid-state and many-body theorists have so far been spared the plagues of quantum field theory. Second quantization has been often just a convenient bookkeeping arrangement to save us from writing out large determinantal wavefunctions. However there is a difference between very large determinants and *infinitely* large ones; we shall show that one of the important differences *is the failure of certain commutators to vanish* in the field-theoretic limit when common sense and experience based on finite N tells us they *should* vanish! (Here N refers to the number of particles in the field.)

* Research supported by the U. S. Air Force Office of Scientific Research.

1 S. Tomonaga, Progr. Theoret. Phys. (Kyoto) **5**, 544 (1950).

2 D. Bohm and D. Pines, Phys. Rev. **92**, 609 (1953).

3 J. M. Luttinger, J. Math. Phys. **4**, 1154 (1963). Note that we set his $v_0 = 1$, thereby fixing the unit of energy. References to this paper will be frequent, and will be denoted by L (72), for example, signifying his Eq. (72).

4 J. M. Luttinger and J. C. Ward, Phys. Rev. **118**, 1417 (1960).

5 Luttinger made a transformation, L (8), which was canonical in appearance only. But in the language of G. Barton [*Introduction to Advanced Field Theory,* (Interscience Publishers, Inc., New York, 1963), pp. 126 *et seq.*] this transformation connected two "unitarily inequivalent" Hilbert spaces, which has as a consequence that commutators, among other operators, must be reworked so as to be well-ordered in fermion field operators. It was first observed by Julian Schwinger [Phys. Rev. Letters **3**, 296 (1959)] that the very fact that one postulates the existence of a ground state (i.e., the filled Fermi sea) *forces* certain commutators to be nonvanishing even though in first quantization they automatically vanish. The "paradoxical contradictions" of which Schwinger speaks seem to anticipate the difficulties in the Luttinger model.

We shall show that these nonvanishing commutators *define* boson fields which must *ipso facto* always be associated with a Fermi–Dirac field, and we shall use the ensuing commutation relations to solve Luttinger's model exactly. Because this model is soluble both in the Hilbert space of finite N and also in the Hilbert space $N = \infty$, with different physical behavior in each, we believe it has applications to the *theory of fields* which go beyond the study of the many-electron problem. The model can be extended to the case of electrons with spin. This has interesting consequences in the band *theory of ferromagnetism*, as will be discussed in some detail in an article under preparation.[5a]

II. MODEL HAMILTONIAN

We recall Luttinger's Hamiltonian[3] and recapitulate some of his results:

$$H = H_0 + H', \tag{2.1}$$

where the "unperturbed" part is

$$H_0 = \int_0^L dx\ \psi^+(x)\sigma_3 p\psi(x) \tag{2.2a}$$

$$= \sum_k (a_{1k}^* a_{1k} - a_{2k}^* a_{2k})k, \tag{2.2b}$$

and the interaction is

$$H' = 2\lambda \iint_0^L dx\, dy\ \psi_1^+(x)\psi_1(x) \times V(x-y)\psi_2^+(y)\psi_2(y) \tag{2.3a}$$

$$= \frac{2\lambda}{L}\sum \delta_{k_1+k_3,k_2+k_4} v(k_3-k_4) \times a_{1k_1}^* a_{1k_2} a_{2k_3}^* a_{2k_4}. \tag{2.3b}$$

Here ψ is a two-component field and the form (b) of the operator is obtained from (a) by setting

$$\psi = \frac{1}{\sqrt{L}}\sum_k e^{ikx}\begin{pmatrix} a_{1k} \\ a_{2k}\end{pmatrix}$$

and

$$\psi^+ = \frac{1}{\sqrt{L}}\sum_k e^{-ikx}(a_{1k}^*,\ a_{2k}^*), \tag{2.4}$$

with a_{jk}'s defined to be anticommuting fermion operators which obey the usual relations

$$\begin{aligned} a_{jk}a_{j'k'} + a_{j'k'}a_{jk} &\equiv \{a_{jk}, a_{j'k'}\} = 0 \\ \{a_{j,k}^*, a_{j'k'}^*\} = 0, &\text{ and } \{a_{jk}, a_{j'k'}^*\} = \delta_{jj'}\delta_{kk'}. \end{aligned} \tag{2.5}$$

Luttinger noted that for an appropriate operator S_0, the canonical transformation

$$\tilde{H} = e^{i\lambda S_0}He^{-i\lambda S_0} \tag{2.6}$$

gave the result that

$$\tilde{H} = H_0, \tag{2.7}$$

and consequently that *the spectrum of* $H = H_0 + H'$ *was the same as that of* H_0, *independent of the interaction* $V(x-y)$. This can be explicitly verified for his choice of

$$S_0 = \iint_0^L dx\, dy\ \psi_1^+(x)\psi_1(x)E(x-y)\psi_2^+(y)\psi_2(y), \tag{2.8}$$

where $E(x)$, not to be confused with the energy E, is defined by:

$$\partial E(x-y)/\partial x \equiv V(x-y), \tag{2.9}$$

assuming that

$$\bar{V} \equiv \frac{1}{L}\int_0^L V(x)\,dx = 0. \tag{2.10}$$

In the Appendix we shall show among other things how to generalize to $\bar{V} \neq 0$. It is also simple and instructive to verify Eqs. (2.6) and (2.7) somewhat differently by using the *first* quantization,

$$H_0 = -i\sum_{n=1}^N \frac{\partial}{\partial x_n} + i\sum_{m=1}^M \frac{\partial}{\partial y_m} \tag{2.11}$$

and

$$H' = 2\lambda \sum_{n=1}^N \sum_{m=1}^M V(x_n - y_m), \tag{2.12}$$

where N and M are, respectively, the total number of "1" particles and "2" particles, with coordinates x_n and y_m, respectively. The properly antisymmetrized wavefunctions are given by

$$\Psi = \det|e^{ik_ix_j}|\det|e^{iq_iy_j}| \times \exp\left\{\sum_{n=1}^N\sum_{m=1}^M iE(x_n-y_m)\right\}. \tag{2.13}$$

Using Eqs. (2.9) and (2.10), Ψ is readily seen to obey Schrödinger's equation

$$H\Psi = E\Psi \tag{2.14}$$

with just the unperturbed eigenvalue

$$E = \sum_{n=1}^N k_n - \sum_{m=1}^M q_m. \tag{2.15}$$

The wavenumbers are of the form

$$k_i \text{ or } q_i = 2\pi \text{ integer}/L, \tag{2.16}$$

as required for periodic boundary conditions. This is in exact agreement with the results of Ref. 3, and can also be checked in perturbation theory; first-

[5a] D. Mattis, Physics 1, 184 (1964).

order perturbation theory also gives vanishing results, and indeed, it is easy to verify that to every order in λ the cancellation is complete, in accordance with the exact result given above.

Up to this point, Luttinger's analysis (which we have briefly summarized) is perfectly correct. It is the next step that leads to difficulty. The Hamiltonian discussed so far has no ground-state energy; in order to remove this obstacle, and thereby establish contact with a real electron gas, Luttinger proposed modifying the model by "filling the infinite sea" of negative energy levels (i.e., all states with $k_1 <$ and $q_2 > 0$). Following L(8) we define b's and c's obeying the usual anticommutators, such that

$$a_{1k} = \begin{cases} b_k & k \geq 0 \\ c^*_{k} & k < 0, \end{cases}$$

and

$$a_{2k} = \begin{cases} b_k & k < 0 \\ c^*_k & k \geq 0. \end{cases} \tag{2.17}$$

Using this notation the total particle-number operator becomes

$$\mathfrak{N} = \sum_{\text{all } k} b_k^+ b_k - c_k^+ c_k \tag{2.17a}$$

(i.e., the number of particles minus the number of holes).

Since the Hamiltonian commutes with $\mathfrak{N}$ we can demand that $\mathfrak{N}$ have eigenvalue N_0. In the noninteracting ground state there are no holes and the b particles are filled from $-k_F$ to k_F where $k_F = \pi(N_0/L) = \pi\rho$. The noninteracting ground-state energy is $N_0\pi\rho$ + energy of the filled sea (W).

The kinetic energy assumes the form

$$H_0 = \sum_{\text{all } k} (b_k^* b_k + c_k^* c_k)\,|k| + W, \tag{2.18}$$

where

$$W = (\sum_{k<0} k - \sum_{k>0} k) \tag{2.18a}$$

is the infinite energy of the filled sea, an uninteresting c number which we drop henceforth in accordance with Luttinger's prescription. The interaction [H', Eq. (2.3) and the operator S_0, Eq. (2.8)] can also be expressed in the new language by means of the substitution (2.17). The reader will no doubt be surprised, as indeed we were, to find that now with the new operators, *Eq.* (2.7), *with* $\bar{H}$ *defined in* (2.6), *is no longer obeyed.*

Upon further reflection one sees that this must be so, on the basis of very general arguments. In the new Hilbert space defined by the transformation to the particle–hole language (2.17), H is no longer unbounded from below and now has a ground state. A general and inescapable *concavity theorem* states that if $E_0(\lambda)$ is the ground-state energy in the presence of interactions, (2.3), then

$$\partial^2 E_0(\lambda)/\partial\lambda^2 < 0. \tag{2.19}$$

This inequality is incompatible with the previous result, viz. all E = independent of λ, which was possible only in the strange case of a system without a ground state.

The same thing can be seen more trivially using second-order perturbation theory (first-order perturbation theory vanishes). It is easily seen that

$$E_0^{(2)} = -\left(\frac{2\lambda}{L}\right)^2 \sum_k \frac{|v(k)|^2}{2k} n_1(k) n_2(-k), \tag{2.20}$$

where $n_1(k)$ and $n_2(k)$ are the number of ways of shifting a particle of type "1" and type "2" respectively by an amount k to an unoccupied state. A simple geometric exercise will convince the reader of the following facts: (1) if we start with a state having a finite number of particles, then n_1 and n_2 are *always* even functions of k (i.e., there are just as many ways to increase the momentum by k as to decrease it by the same amount.) (2) If we start with a filled infinite sea then there is no way to decrease the momentum of the "1" particles nor to increase the momentum of "2" particles. Hence for this second case $n_1(k)n_2(-k)$ is nonzero only for $k > 0$. Thus $E_0^{(2)}$ vanishes for a state with a finite number of particles, but it is negative for a filled sea

If the reader is unconvinced by perturbation theory, then he can easily prove that E_0 is lowered by doing a variational calculation.

What has gone wrong? We turn to some algebra to resolve this paradox, and following this, present a solution of the field-theoretic problem defined by $H_0 + H'$ in the representation of b's and c's.

III. CASE OF THE FILLED DIRAC SEA

The various relevant operators are given below; the form (a) of each equation will *not* be used in the bulk of the paper, and is just given here for completeness. In the following equations, $p > 0$.

$$\rho_1(+p) \equiv \sum_k a_{1\,k+p}^* a_{1\,k} \tag{3.1a}$$

$$= \sum_{k<-p} c_{k+p} c_k^* + \sum_{-p\leq k<0} b_{k+p}^* c_k^* + \sum_{k\geq 0} b_{k+p}^* b_k, \tag{3.1b}$$

$$\rho_1(-p) \equiv \sum_k a_{1\,k}^* a_{1\,k+p} \tag{3.2a}$$

$$= \sum_{k<-p} c_k c_{k+p}^* + \sum_{-p\leq k<0} c_k b_{k+p} + \sum_{k\geq 0} b_k^* b_{k+p}, \tag{3.2b}$$

$$\rho_2(+p) \equiv \sum_k a^*_{2\,k+p}a_{2\,k} \tag{3.3a}$$

$$= \sum_{k<-p} b^*_{k+p}b_k + \sum_{-p\le k<0} c_{k+p}b_k + \sum_{k\ge 0} c_{k+p}c^*_k, \tag{3.3b}$$

$$\rho_2(-p) \equiv \sum_k a^*_{2\,k}a_{2\,k+p} \tag{3.4a}$$

$$= \sum_{k<-p} b^*_k b_{k+p} + \sum_{-p\le k<0} b^*_k c^*_{k+p} + \sum_{k>0} c_k c^*_{k+p}. \tag{3.4b}$$

Equations (3.1a)–(3.4a) give the density operators in the original representation, so let us calculate in this language a commutator such as (assume $p \ge p' \ge 0$ for definiteness)

$$[\rho_1(-p), \rho_1(p')] = \sum_{k,k'} [a^*_{1\,k}a_{1\,k+p}, a^*_{1\,k'+p'}a_{1\,k'}]$$

$$= \sum_{k=-\infty}^{+\infty} a^*_{1\,k}a_{1\,k+p-p'} - \sum_{k=-\infty}^{+\infty} a^*_{1\,k+p'}a_{1\,k+p} = 0. \tag{3.5}$$

The zero result could have been expected by writing the operators in first quantization:

$$\rho_1(-p) = \sum_n e^{-ipx_n} \quad \text{and} \quad \rho_2(p) = \sum_m e^{ipy_m}, \tag{3.6}$$

whence they evidently commute. Nevertheless, the zero result is achieved in (3.5) only through the almost "accidental" cancellation of two operators, each of which may diverge in the field-theory limit when $N = \infty$. We now show that in that limit the operators in fact no longer cancel, by evaluating the commutator using form (b) for the density operators. It is a matter of only some minor manipulation to obtain the important new result:

$$[\rho_1(-p), \rho_1(p')] = [\rho_2(p), \rho_2(-p')]$$

$$= \delta_{p,p'} \sum_{-p<k<0} 1 = \frac{pL}{2\pi}\delta_{p,p'}, \quad (p' > 0). \tag{3.7a}$$

In addition,

$$[\rho_1(p), \rho_2(p')] = 0. \tag{3.7b}$$

A quick check is provided by evaluating the vacuum expectation value

$$\langle 0|\, [\rho_1(-p), \rho_1(p)]\, |0\rangle$$

$$= \sum_{-p<k,k'<0} \langle 0|\, c_k b_{k+p} b^*_{k'+p} c^*_{k'}\, |0\rangle = pL/2\pi, \tag{3.8}$$

which is exactly what is expected on the basis of the previous equation. Evidently the form (b) of the operators $(2\pi/pL)^{+\frac{1}{2}}\rho_1(+p)$ and $(2\pi/pL)^{+\frac{1}{2}}\rho_2(-p)$ have properties of boson raising operators [call them $A^*(p)$ and $B^*(-p)$] and $(2\pi/pL)^{+\frac{1}{2}}\rho_1(-p)$ and $(2\pi/pL)^{+\frac{1}{2}}\rho_2(+p)$ have properties of boson lowering operators [$A(p)$ and $B(-p)$], i.e.,

$$[A, B] = [A^*, B] = 0,$$
$$[A(p), A^*(p')] = [B(-p), B^*(-p)] = \delta_{p,p'}. \tag{3.9}$$

The B field is the continuation of the A field to negative p; therefore together they form a *single* boson field defined for all p.

The relationship of the $\rho(p)$'s to Luttinger's $N(x)$'s, L(25), is obtained by using (2.4):

$$N_1(x) = \psi^*_1(x)\psi_1(x) = \frac{1}{L}\sum_p \rho_1(p)e^{-ipx},$$
$$N_2(x) = \psi^*_2(x)\psi_2(x) = \frac{1}{L}\sum_p \rho_2(p)e^{-ipx}. \tag{3.10}$$

IV. SOLUTIONS OF THE MODEL HAMILTONIAN

Before making use of the results of the previous section, we remark that $\rho_1(+p)$ and $\rho_2(-p)$ are exact raising operators of H_0, and $\rho_1(-p)$ and $\rho_2(p)$ are exact lowering operators of H_0 corresponding to excitation energies p. That is,

$$[H_0, \rho_1(\pm p)] = \pm p\rho_1(\pm p),$$
$$[H_0, \rho_2(\pm p)] = \mp p\rho_2(\pm p). \tag{4.1}$$

The identification of the ρ's with boson operators made in the previous section suggested to us the possibility of constructing a new operator T which obeys the same equations (4.1), as H_0. This is indeed possible, if we define T as follows:

$$T \equiv \frac{2\pi}{L}\sum_{p>0} \{\rho_1(p)\rho_1(-p) + \rho_2(-p)\rho_2(p)\} \tag{4.2}$$

[the ρ's being defined here and in the remainder of the paper by Eqs. (3.1b)—(3.4b), i.e., always in the hole-particle representation]. It follows that

$$[T, \rho_1(\pm p)] = \pm p\rho_1(\pm p) \tag{4.3}$$

as required, and similarly for $\rho_2(\mp p)$. Therefore, let us decompose H into two parts

$$H = H_1 + H_2 \tag{4.4}$$

with

$$H_1 = H_0 - T = \Big\{\sum_k |k|\,(b^*_k b_k + c^*_k c_k) - \frac{2\pi}{L}\sum_{p>0}\{\rho_1(p)\rho_1(-p) + \rho_2(-p)\rho_2(p)\}\Big\}, \tag{4.5}$$

and

$$H_2 = H' + T$$
$$= \frac{1}{L}\Big[2\lambda\sum_{p>0}\{v(p)\rho_1(-p)\rho_2(p) + v(-p)\rho_1(p)\rho_2(-p)\} + 2\pi\sum_{p>0}\{\rho_1(p)\rho_1(-p) + \rho_2(-p)\rho_2(p)\}\Big] \tag{4.6}$$

with $v(p)$ = real, even function of p. By actual construction, all the ρ operators which appear in H_2

commute with H_1. This will be an important feature in constructing an exact solution of the model. We define an Hermitian operator S,

$$S = \frac{2\pi i}{L} \sum_{\text{all } p} \frac{\varphi(p)}{p} \rho_1(p)\rho_2(-p), \quad (4.7)$$

where $\varphi(p)$ is also a real, even, function of p to be determined subsequently by imposing a condition that the unitary transformation e^{iS} diagonalize H_2. First we evaluate the effect of such a transformation on various operators. It commutes with H_1,

$$e^{iS}H_1e^{-iS} = H_1 = H_0 - T, \quad (4.8)$$

because both ρ_1 and ρ_2 appearing in S commute with H_1, as noted above. In the following, p can have either sign:

$$e^{iS}\rho_1(p)e^{-iS} = \rho_1(p)\cosh\varphi(p) + \rho_2(p)\sinh\varphi(p), \quad (4.9)$$

$$e^{iS}\rho_2(p)e^{-iS} = \rho_2(p)\cosh\varphi(p) + \rho_1(p)\sinh\varphi(p). \quad (4.10)$$

We have verified that this transformation is a proper unitary transformation and preserves commutation relations (3.7) as well as anticommutation relations (2.5), and the reader may easily check this point. H_2 is brought into canonical form by requiring that in (exp iS) H_2 (exp $-iS$) there be no cross terms such as $\rho_1(p)\rho_2(-p)$. This leads to the equation

$$\tanh 2\varphi = -\lambda v(p)/\pi, \quad (4.11)$$

which cannot be obeyed unless

$$|\lambda v(p)| < \pi \quad \text{for all} \quad p. \quad (4.12)$$

Equation (4.12) serves to limit the magnitude of potentials capable of having well-behaved solutions (e.g., a real ground-state energy). For the more realistic potentials discussed in the Appendix, there is also a more realistic bound on $v(p)$: there, $v(p)$ may not be *too* attractive, but it can have any magnitude when it is repulsive, i.e., positive.

With the choice of φ in (4.11), the evaluation of H_2 becomes

$$e^{iS}H_2e^{-iS} = \frac{2\pi}{L}\sum_{p>0} \operatorname{sech} 2\varphi(p)\{\rho_1(p)\rho_1(-p) + \rho_2(-p)\rho_2(p)\} - \sum_{p>0} p(1 - \operatorname{sech} 2\varphi). \quad (4.13a)$$

The second term is the vacuum renormalization energy

$$W_1 = -\sum_{p>0} p(1 - \operatorname{sech} 2\varphi) = \frac{L}{2\pi}\int_0^\infty dp\, p\left\{\left(1 - \frac{\lambda^2 v^2(p)}{\pi^2}\right)^{\frac{1}{2}} - 1\right\}. \quad (4.13b)$$

It may be expanded in powers of λ to effect a comparison with Goldstone's many-body perturbation theory[4]; we have checked that they agree to third order.

The problem is now formally solved, for we can find all the eigenfunctions and eigenvalues by studying Eqs. (4.4), (4.8), and (4.13). First notice that the operator T does not depend upon the interaction and that if there is *no interaction* we could write the Hamiltonian either as

$$H = H_0, \quad (4.14a)$$

or as

$$H = (H_0 - T) + T = H_1 + H_2. \quad (4.14b)$$

Since H_1 and H_2 commute, every eigenstate, Ψ, of H may be assumed to be an eigenfunction of H_1 and H_2 separately. Moreover, Ψ may also be assumed to be an eigenfunction of each $\alpha_p = A_p^+A_p$ and $\beta_p = B_{-p}^+B_{-p}$ for all $p > 0$, since these operators commute with H and $\mathfrak{N}$.

Evidently (4.14a) and (4.14b) provide two different ways of viewing the noninteracting spectrum. H_0 is quite degenerate: the raising operators of H_0 are the b^+'s and c^+'s. By requiring that Ψ also be an eigenstate of α_p, β_p and H, we are merely attaching quantum numbers to the degenerate levels of H_0. If $\alpha_p\Psi = n_p\Psi$ and $\beta_p\Psi = m_p\Psi$ (where n_p and m_p are of course integers), we say that we have n_p plasmons of momentum p and m_p plasmons of momentum $-p$. With no interaction the energy of a plasmon is

$$\epsilon(p) = |p|. \quad (4.15)$$

We may speak of H_1 as the quasiparticle part of the Hamiltonian; in H_1 the operator T plays the role of subtracting the plasmon part of the energy from H_0.

When we turn on the interaction, the above description of the energy levels is still valid, except that now we are *forced* to use the form (4.14b) because H_2 is no longer T. The degeneracy of H is partially removed by the interaction, because now the energy of a plasmon is

$$\epsilon'(p) = |p| \operatorname{sech} 2\varphi(p). \quad (4.16)$$

Notice that the plasmon energy is always *lowered* [and therefore the plasmons cannot propagate faster than the speed of light $c = 1$, i.e., $d\epsilon'/dp \leq 1$. In the more realistic case discussed in the Appendix, the plasmon energy *can* be increased by the interaction although $d\epsilon'/dp \leq 1$ is always obeyed.] by the interaction; if (4.12) is violated the plasmon energy is no longer real and the system becomes unstable. Note, there are no plasmons in the ground state, so that W_1 (4.13), is the shift in the ground-state energy of the system.

There is one important point, however, that requires some elucidation. We would like to be able to say that in view of the fact that H_1, $\alpha(p)$, and $\beta(p)$

conserve particle number, the most general energy level of H (fixed N_0) is the sum of *any* energy of H_1 (same N_0, and no plasmons) plus *any* (plasmon) energy of H_2 (note: the plasmon spectrum is independent of N_0). Were we dealing with a finite-dimensional vector space, such a statement would not be true, for even though H_1 and H_2 commute they could not possibly be independent. Thus, if H_2 had n eigenvalues $e_1, \cdots, e_n$, and if H_1 had an equal number $E_1, \cdots, E_n$ the general total eigenvalue would not be *any* combination of $e_i + E_i$ for this would give too many values (viz. n^2 instead of n.) But we are dealing with an infinite-dimensional Hilbert space and the additivity hypothesis is in fact true for the present model.

To prove this assertion we consider any eigenstate Ψ which is necessarily parameterized by the integers n_p and m_p. Consider the state $\Phi = \{\prod_p (A_p)^{n_p}(B_p)^{m_p}\}\Psi$. The state Φ is nonvanishing and has quantum numbers $n_p = 0 = m_p$. It is also an eigenstate of H_1 with energy $E_1(\Psi)$. In addition (and this is the important point) the state Ψ may be recovered from Φ by the equation

$$\Psi = \text{const} \times \{\prod_p (A_p^+)^{n_p}(B_p^+)^{m_p}\}\Phi.$$

To every state Ψ, therefore, there corresponds a *unique* state Φ from which it may be obtained using raising operators. Conversely, to any eigenstate of H_1 (for fixed N_0) we may apply raising operators as often as we please and obtain a new (nonvanishing) eigenstate. Thus the general energy is an arbitrary sum of quasiparticle and plasmon energies.

It may be wondered where we used the fact that the Hilbert space is infinite-dimensional in the above proof. The answer lies in the boson commutation relations of the A's and B's. It is impossible to have such relations in a finite-dimensional vector space.

The eigenvalues corresponding to these states Φ will be labeled in some order, E_i $(i = 1, 2, \cdots)$, so that the total canonical partition function $Z(\lambda)$ and the free energy $F(\lambda)$ are given by

$$\begin{aligned} Z(\lambda) &= e^{-F(\lambda)/kT} \\ &= \left(\sum_i e^{-E_i/kT}\right)(e^{-W_1/kT}) \prod_{\substack{\text{all } p \\ \neq 0}} \left(\sum_{n=0}^{\infty} e^{-n\epsilon'(p)/kT}\right). \end{aligned} \tag{4.17}$$

The first factor is difficult to evaluate directly. However it can be obtained circuitously by noting that the energies E_i are independent of λ and therefore

$$\begin{aligned} Z(0) &= e^{-F(0)/kT} \\ &= \left(\sum_i e^{-E_i/kT}\right) \prod_{\substack{\text{all } p \\ \neq 0}} \left(\sum_{n=0}^{\infty} e^{-n\epsilon(p)/kT}\right). \end{aligned} \tag{4.18}$$

But the second factor can be trivially evaluated, as can $F(0)$ = free energy of noninteracting fermions. Therefore we use (4.18) to eliminate the trace involving the E_i's in (4.17), with the final result:

$$\begin{aligned} F(\lambda) &= F(0) + W_1 \\ &+ 2kT \sum_{p>0} \ln \{(1 - e^{-\epsilon'(p)/kT})/(1 - e^{-\epsilon(p)/kT})\}, \end{aligned} \tag{4.19}$$

where ϵ and ϵ' are given in (4.15) and (4.16). It is noteworthy that the ground state and free energy both diverge in the case of a δ-function potential.

V. EVALUATION OF THE MOMENTUM DISTRIBUTION

In this section we calculate the mean number of particles with momentum k. This quantity is $\bar{n}_k$ and is the expectation value of

$$n_k = b_k^+ b_k \tag{5.1}$$

in the ground state. Since $\bar{n}_k$ is an even function of k we need only consider $k > 0$, and it is further convenient to introduce a Fourier transform so that [using (2.4)]

$$\bar{n}_k = \frac{1}{L} \iint_0^L ds\, dt\, e^{ik(s-t)} I(s, t). \tag{5.2}$$

Here

$$\begin{aligned} I(s, t) &= \langle\Psi|\ \psi_1^+(s)\psi_1(t)\ |\Psi\rangle \\ &= \langle\Psi_0|\ e^{iS}\psi_1^+(s)e^{-iS}e^{iS}\psi_1(t)e^{-iS}\ |\Psi_0\rangle, \end{aligned} \tag{5.3}$$

where S is given by (4.7), Ψ is the new ground state, and Ψ_0 is the noninteracting ground state which is filled with b particles between $-k_F$ and k_F and has no holes (or c particles). This assignment depends on there having been no level crossing, which can be readily verified using (4.7)–(4.13).

In order to calculate the quantity $e^{iS}\psi_1(t)e^{-iS}$ we introduce the auxiliary operator

$$f_\sigma(t) = e^{i\sigma S}\psi_1(t)e^{-i\sigma S}, \tag{5.4}$$

where σ is a c number. We observe that $f_1(t)$ is the desired quantity while

$$f_0(t) = \psi_1(t). \tag{5.5}$$

In addition,

$$\begin{aligned} \partial f/\partial\sigma &= e^{i\sigma S} i[S, \psi_1(t)]e^{-i\sigma S} \\ &= e^{i\sigma S}[2\pi/L \sum_p \rho_2(-p)\varphi(p)p^{-1}e^{ipt}]e^{-i\sigma S} f_\sigma(t), \end{aligned} \tag{5.6}$$

where we have used the commutation relations (3.7) as well as the fact that ψ_1 commutes with ρ_2. Equation (5.6) is a differential equation for $f_\sigma(t)$ and (5.5) is the boundary condition. The solution is

$$f_\sigma(t) = W_\sigma(t)R_\sigma(t)\psi_1(t), \tag{5.7}$$

where

$$W_\sigma(t) = \exp\{2\pi/L \sum_{p>0} [\rho_1(-p)e^{ipt} - \rho_1(p)e^{-ipt}]p^{-1}[\cosh \sigma\varphi(p) - 1]\} \tag{5.8}$$

and

$$R_\sigma(t) = \exp\{2\pi/L \sum_{p>0} [\rho_2(-p)e^{ipt} - \rho_2(p)e^{-ipt}]p^{-1} \sinh \sigma\varphi(p)\} \tag{5.9}$$

The reader may verify that (5.7) satisfies (5.5) and (5.6) by using the commutation relations (3.7). We recall the well-known rule that

$$\exp(A + B) = \exp(A)\exp(B)\exp(-1/2[A, B]) \tag{5.10}$$

when $[A, B]$ commutes with A and B. From here on we shall set $\sigma = 1$ and drop it as a subscript. We note that since $\rho_1(p)^+ = \rho_1(-p)$ and $\rho_2(p)^+ = \rho_2(-p)$,

$$R^+(t) = R^{-1}(t) \quad \text{and} \quad W^+(t) = W^{-1}(t). \tag{5.11}$$

We also note that R and W commute with each other. Thus, (5.3) becomes

$$\begin{aligned} I(s, t) &= \langle\Psi_0|\ \psi_1^+(s)R^{-1}(s)W^{-1}(s)W(t)R(t)\psi_1(t)\ |\Psi_0\rangle \\ &= I_1(s, t)I_2(s, t), \end{aligned} \tag{5.12}$$

where

$$\begin{aligned} I_1(s, t) &= \langle\Psi_1|\ \psi_1^+(s)W^{-1}(s)W(t)\psi_1(t)\ |\Psi_1\rangle, \\ I_2(s, t) &= \langle\Psi_2|\ R^{-1}(s)R(t)\ |\Psi_2\rangle. \end{aligned} \tag{5.13}$$

We have used the fact that the ground state is a product state: $\Psi_0 = \Psi_1 * \Psi_2$ where Ψ_1 is a state of the "1" field and Ψ_2 is a state of the "2" field. Ψ_1 is filled with b particles up to $+k_F$ and has no c particles; Ψ_2 is filled with b particles down to $-k_F$ and has no c particles.

Now, using the definition (5.8) and the rule (5.10) we easily find that

$$W^{-1}(s)W(t) = W_-(s, t)W_+(s, t)Z_1(s, t), \tag{5.14}$$

with

$$W_+(s, t) = \exp\{2\pi/L \sum_{p>0} \rho_1(-p)[\cosh \varphi(p) - 1] \times p^{-1}(e^{ipt} - e^{ips})\},$$

$$W_-(s, t) = \exp\{2\pi/L \sum_{p>0} \rho_1(p)[\cosh \varphi(p) - 1] \times p^{-1}(e^{-ips} - e^{-ipt})\},$$

$$Z_1(s, t) = \exp\{2\pi/L \sum_{p>0} [\cosh \varphi(p) - 1]^2 \times p^{-1}(e^{ip(s-t)} - 1)\}. \tag{5.15}$$

Likewise,

$$R^{-1}(s)R(t) = R_-(s, t)R_+(s, t)Z_2(s, t), \tag{5.16}$$

with

$$R_+(s, t) = \exp\{2\pi/L \sum_{p>0} \rho_2(p)[\sinh \varphi(p)] \times p^{-1}(e^{-ips} - e^{-ipt})\},$$

$$R_-(s, t) = \exp\{2\pi/L \sum_{p>0} \rho_2(-p)[\sinh \varphi(p)] \times p^{-1}(e^{ipt} - e^{ips})\},$$

$$Z_2(s, t) = \exp\{2\pi/L \sum_{p>0} [\sinh \varphi(p)]^2 \times p^{-1}(e^{ip(t-s)} - 1)\}. \tag{5.17}$$

We see at once from the definition (3.1b), (3.2b), of $\rho_1(p)$ that, for $p > 0$, $\rho_1(-p)\ |\Psi_1\rangle = 0$. Similarly $\langle\Psi_1|\ \rho(p) = 0$, $\rho_2(p)\ |\Psi_2\rangle = 0$, and $\langle\Psi_2|\ \rho_2(-p) = 0$. Hence,

$$I_2(s, t) = Z_2(s, t)$$

and

$$I_1(s, t) = Z_1(s, t)\langle\Psi_1|\ W_-^{-1}\psi_1^+(s)W_-W_+\psi_1(t)W_+^{-1}\ |\Psi_1\rangle. \tag{5.18}$$

If we now define

$$h_+(y) = 2\pi/L \sum_{p>0} [\cosh \varphi(p) - 1] \times p^{-1}(e^{ipt} - e^{ips})e^{-ipy},$$

$$h_-(y) = 2\pi/L \sum_{p>0} [\cosh \varphi(p) - 1] \times p^{-1}(e^{-ipt} - e^{-ips})e^{ipy}, \tag{5.19}$$

combining (3.10) and (5.15) we have that

$$\begin{aligned} W_+(s, t) &= \exp \int_0^L N_1(y)h_+(y)\, dy, \\ W_-(s, t) &= \exp -\int_0^L N_1(y)h_-(y)\, dy. \end{aligned} \tag{5.20}$$

Since

$$\begin{aligned} [\psi_1(x), N_1(y)] &= \delta(x - y)\psi_1(x), \\ [\psi_1^+(x), N_1(y)] &= -\delta(x - y)\psi_1^+(x), \end{aligned} \tag{5.21}$$

it follows that

$$\begin{aligned} W_+(s, t)\psi_1(t)W_+^{-1}(s, t) &= \psi_1(t)\exp[-h_+(t)] \\ W_-^{-1}(s, t)\psi_1^+(s)W_-(s, t) &= \psi_1^+(s)\exp[+h_-(s)]. \end{aligned} \tag{5.22}$$

Finally,

$$\langle\Psi_1|\ \psi_1^+(s)\psi_1(t)\ |\Psi_1\rangle = 1/L \sum_{p\le k_F} e^{ip(t-s)} \equiv Z_3(s,t). \quad (5.23)$$

Combining all these results, we conclude that

$$I(s,t) = Z_0(s,t)Z_1(s,t)Z_2(s,t)Z_3(s,t), \quad (5.24)$$

where

$$Z_0(s,t) = \exp\,(h_-(s) - h_+(t)) = \exp\,\{-4\pi/L \sum_{p>0} [\cosh\varphi(p) - 1] \times (1 - e^{ip(s-t)})\}. \quad (5.25)$$

In order to make a comparison with Luttinger's calculation of $\bar{n}_k$, we first observe that the functions $Z_i(s,t)$ are really functions of $r = s - t$ and that they are periodic in s and t in $(0, L)$. We then define the functions $G(r)$ and $Q(r)$ as follows:

$$\exp\,[-Q(r)] \equiv G(r) \equiv Z_0(r)Z_1(r)Z_2(r). \quad (5.26)$$

Substituting (5.26), (5.24), and (5.23) into (5.2) we obtain

$$\bar{n}_k = 2\pi/L \sum_{p\le k_F} F(k-p), \quad (5.27)$$

where

$$F(k) = 1/2\pi \int_{-\frac{1}{2}L}^{\frac{1}{2}L} dr\, e^{ikr} e^{-Q(r)} \quad (5.28)$$

$$\cong 1/2\pi \int_{-\infty}^{\infty} dr\, e^{ikr} e^{-Q(r)}. \quad (5.29)$$

In (5.29) we have passed to the bulk limit N, $L \to \infty$, not an approximation.

At this point our expression for $\bar{n}_k$ is formally the same as Luttinger's [cf. L (52), L (69)]. The difference is that our Q is different from his. He obtains Q by evaluating an infinite Toeplitz determinant with the result that [L (70)]

$$Q(r) = \lambda^2/2\pi^2 \int_0^\infty dp\, \frac{1 - \cos pr}{p}\, |v(p)|^2. \quad \text{(Luttinger)} \quad (5.30)$$

Our Q, which is the correct one to use, is obtained by combining (5.15), (5.17), and (5.25), replacing sums by integrals in the usual way, and using the definition (4.11) of $\varphi(p)$. The result is

$$Q(r) = \lambda^2/2\pi^2 \int_0^\infty dp\, \frac{1 - \cos pr}{p}\, |u(p)|^2, \quad (5.31)$$

where

$$|u(p)|^2 = (2\pi^2/\lambda^2)\{(1 - (\lambda v(p)/\pi)^2)^{-\frac{1}{2}} - 1\}. \quad (5.32)$$

It is worth noting that (5.30) agrees with (5.31) to leading order in λ^2.

Since we have not yet specified $v(p)$, we may now follow Luttinger's discussion from this point on with the proviso that we use the correct (λ dependent) $u(p)$ instead of $v(p)$. The reader is referred to pages 1159 and 1160 of Luttinger's paper.

There are two main conclusions one can draw. The first is that if we start with a δ-function interaction [so that $v(p)$ and hence $u(p)$] are constants, it can be shown that $\bar{n}_k = \frac{1}{2}$ for all k. Such a result is quite unphysical, but it is not unreasonable because the ground-state energy W (4.13a) diverges when $v(p) =$ constant at large p. Also, the result would be the same if we started with the more physical interaction

$$H' = 1/L \sum_p \{\rho_1(p) + \rho_2(p)\}\{\rho_1(-p) + \rho_2(-p)\}v(p)$$

discussed in the Appendix. This is indeed unfortunate, because relativistic field theories usually begin with local (δ-function) interactions.

The second conclusion is that if one makes a reasonable assumption about $v(p)$, and hence about $u(p)$ and $Q(r)$, one finds that for k in the vicinity of k_F, $\bar{n}_k$ behaves like

$$\bar{n}_k \sim d - e\,|k - k_F|^{2\alpha}\,\sigma(k - k_F), \quad (5.33)$$

where

$$\sigma(k) = 1, \quad k > 0; \qquad = -1, \quad k < 0 \quad (5.34)$$

and d, e, and α are certain positive constants. Now in Luttinger's calculation

$$\alpha = \lambda^2/4\pi^2 v(0)^2, \quad \text{(Luttinger)} \quad (5.35)$$

[cf. L(75)], where $v(0) \equiv \lim_{p\to 0} v(p)$.

If $2\alpha < 1$, then the conclusion to be drawn is that although the interaction removes the discontinuity in $\bar{n}_k$ at the Fermi surface, we are left with a function that has an infinite slope there. There is, so to speak, a residual Fermi surface. In Sec. IV of his paper, Luttinger shows that at least for one example of $v(p)$ perturbation theory gives the same qualitative result as (5.33) with the same value of α, (5.35).

If, on the other hand, $2\alpha > 1$ then there is no infinite derivative at the Fermi surface. $\bar{n}_k$ is perfectly smooth there (although, technically speaking, it is nonanalytic unless $2\alpha =$ odd integer.) In this case virtually all trace of the Fermi surface has been eliminated. But notice that the correct α to use is obtained by replacing $v(0)$ by $u(0) \equiv \lim_{p\to 0} u(p)$ in (5.35), i.e.,

$$2\alpha = \{1 - [\lambda v(0)/\pi]^2\}^{-\frac{1}{2}} - 1. \qquad (5.36)$$

Thus, even subject to the requirement that $|\lambda v(0)|$ be less than π, 2α can become as large as one pleases. Yet perturbation theory predicts (5.35) which yields 2α always less than $\frac{1}{2}$.

We may conclude that a strong enough interaction can eliminate the Fermi surface, while perturbation theory predicts that is always there.

VI. DIELECTRIC CONSTANT

Because the response to external fields of wave vector q only depends on an interaction expression linear in the density operators, we can immediately obtain for the generalized static susceptibility function or *dielectric constant* (response ÷ driving force), for any temperature, T

$$\chi_\lambda(q, T) = \chi_0(q, T)\{\sinh\varphi(q) + \cosh\varphi(q)\}^2 \cosh 2\varphi_q$$
$$= \chi_0(q, T)\frac{1}{1 + \lambda v(q)/\pi} \qquad (6.1)$$

in terms of the "unperturbed" susceptibility $\chi_0(q, T)$. It is also a simple exercise to calculate exactly the time dependent susceptibility in terms of the "unperturbed" quantity.

It is interesting to note that the susceptibility can diverge (which is symptomatic of a phase transformation) only for

$$\lambda v(q) \to -\pi, \qquad (6.2)$$

i.e. only for sufficiently *attractive* interactions and not for repulsive $[v(q) > 0]$ interactions.

Recently Ferrell[6] advanced plausible arguments why a one-dimensional metal cannot become superconducting. We can prove this rigorously in the present model. The electron–phonon interaction is

$$H_{\mathrm{el-ph}} = \sum_p g(p)[\rho_1(p) + \rho_2(p)]\cdot[\xi_p + \xi^+_{-p}], \qquad (6.3)$$

where ξ and ξ^+ are the phonon field operators. In the "filled-sea" limit this coupling is bilinear in harmonic-oscillator operators, and therefore the Hamiltonian continues to be exactly diagonalizable. The new normal modes can be calculated and there is found to be no phase transition at any finite temperature.

APPENDIX

We shall be interested in extending Luttinger's model in two ways. Firstly, we note that the restriction $\bar{V} = 0$ is really not necessary. Turning back to Eqs. (2.13) *et seq.* we impose periodic boundary conditions $\Psi(\cdots, x_i + L, \cdots) = \Psi(\cdots, x_i, \cdots)$, and find that

$$(q + N\lambda\bar{V}) \text{ and } (k + M\lambda\bar{V}) = 2\pi/L \times \text{integer} \quad (A1)$$

replace the usual condition (2.16), where N = number of "1" particles and M = number of "2" particles. However, when $N, M \to \infty$ in the field-theoretic limit the problem evidently becomes ill-defined unless $\bar{V} \equiv 0$.

A less trivial observation concerns the form of the interaction potential. There is no reason to restrict it to the form $\propto \rho_1\rho_2$, and in fact the more realistic two-body interaction

$$H' = \frac{\lambda}{L}\sum_p v(p)\{\rho_1(-p) + \rho_2(-p)\}\{\rho_1(p) + \rho_2(p)\} \qquad (A2)$$

is fully as soluble as the one assumed in the text, for any strength positive $v(p)$, and provided only

$$\lambda v(p) > -\tfrac{1}{2}\pi, \qquad (A3)$$

i.e. provided no Fourier component is *too* attractive. The shift in the ground-state energy is now given by

$$W_2 = \sum_{p>0} p\left\{\left(1 + \frac{2\lambda v(p)}{\pi}\right)^{\frac{1}{2}} - 1\right\}. \qquad (A4)$$

The plasmon energy is now

$$\epsilon''(p) \equiv |p|\,(1 + 2\lambda v(p)/\pi)^{\frac{1}{2}} \qquad (A5)$$

and *for the important case of the Coulomb repulsion*, $v(p) = p^{-2}$, the plasmons describe a relativistic boson field with mass

$$m^* \equiv (2\lambda/\pi)^{\frac{1}{2}} \qquad (A6)$$

and dispersion

$$\epsilon''(p) = (p^2 + m^{*2})^{\frac{1}{2}}. \qquad (A7)$$

Here, too, $d\epsilon''/dp < 1$.

ACKNOWLEDGMENTS

We thank D. Jepsen, L. Landovitz, and T. Schultz for interesting discussions during various stages of this work. We especially thank J. Luttinger for performing a detailed review of his own work and ours, and thus corroborating the results of the present article. While reading our manuscript, Professor Luttinger called our attention to Schwinger's relevant, indeed prophetic remarks (cf. Ref. 5), and we thank him for this reference. We wish to thank Dr. A. S. Wightman, for pointing out to us that P. Jordan first discovered the bosons associated with fermion fields (in his attempts to construct a neutrino theory of light), and for a list of references.[7]

[6] R. A. Ferrell, Phys. Rev. Letters **13**, 330 (1964).

[7] P. Jordan, Z. Physik **93**, 464 (1935); **98**, 759 (1936); **99**, 109 (1936); **102**, 243 (1936); **105**, 114 (1937); **105**, 229 (1937). M. Born and N. Nagendra-Nath, Proc. Ind. Acad. Sci. **3**, 318 (1936). A. Sokolow, Phys. Z. der Sowj. **12**, 148 (1937).

J. Phys. C: Solid State Phys., **14** (1981) 2585–2609. Printed in Great Britain

'Luttinger liquid theory' of one-dimensional quantum fluids: I. Properties of the Luttinger model and their extension to the general 1D interacting spinless Fermi gas

F D M Haldane†

Institut Laue–Langevin, 156X, 38042 Grenoble Cedex, France

Received 6 October 1980, in final form 26 January 1981

Abstract. The explicitly soluble Luttinger model is used as a basis for the description of the general interacting Fermi gas in one dimension, which will be called 'Luttinger liquid theory', by analogy with Fermi liquid theory. The excitation spectrum of the Luttinger model is described by density-wave, charge and current excitations; its spectral properties determine a characteristic parameter that controls the correlation function exponents. These relations are shown to survive in non-soluble generalisations of the model with a non-linear fermion dispersion. It is proposed that this low-energy structure is universal to a wide class of 1D systems with conducting or fluid properties, including spin chains.

1. Introduction

This paper is the first in a series that will present a general description of the low-energy properties of a wide class of one-dimensional quantum many-body systems, which I will call 'Luttinger liquids'. The work to be described was originally motivated by the search for a replacement for Fermi liquid theory in one dimension, where it fails because of the infrared divergence of certain vertices it assumes to remain finite; these divergences make an approach based on conventional fermion many-body perturbation theory useless. However, there is a certain model of an interacting one-dimensional spinless fermion system, the Luttinger model (Luttinger 1963), which has been explicitly solved (Mattis and Lieb 1965). This solution, by a Bogoliubov transformation, in effect resums all the divergences encountered in perturbation theory. The excitation spectrum of the diagonalised model is described in terms of *non-interacting* boson collective modes.

The feature of the Luttinger model that allows its solution is its exactly linear fermion dispersion. What will be demonstrated in this paper is that correction terms representing non-linearity of the fermion dispersion can be added to the model, and give rise to non-linear boson couplings between the collective modes. A *boson* many-body perturbation expansion in these terms is shown to be completely regular, so the Bogoliubov transformation technique that solves the Luttinger model is shown to provide a general method for resumming *all* the infrared divergences present, at least in the spinless

† Address from September 1981: Department of Physics, University of Southern California, Los Angeles, CA90007, U.S.A.

Fermi gas. The name 'Luttinger liquid' has been chosen to reflect the idea that such systems have a low-energy excitation spectrum similar to the Luttinger model spectrum, but with *interactions* between the elementary excitations. This resembles the relation between the Fermi liquid theory and the soluble model on which it is based, the free Fermi gas.

This paper is perhaps the most technical of the planned series. It sets up the essential machinery for working with the Luttinger model and its generalisations, and uses it to discuss the effects of a non-linear fermion dispersion. The previous treatments of the model in the literature are often ambiguous on certain points, and there has been a certain amount of confusion, particularly associated with the role of cut-offs. I have therefore aimed to present a completely self-contained and precise treatment of the original Luttinger model, in particular emphasing the key role played by charge and current excitations (as opposed to collective density wave modes) which have in general been neglected in previous treatments. It was attention to these details that allowed the identification of a key part of the underlying structure of the solution that proved to remain valid in the 'Luttinger liquid' generalisation, with applications to be described in future papers.

The characteristic properties of a 'Luttinger liquid' that have emerged are: (i) a conserved charge; (ii) a characteristic 'Kohn anomaly' wavevector '$2k_F$', varying linearly with charge density; (iii) persistent currents at low temperatures, quantised in units that carry momentum $2k_F$; (iv) a spectrum of collective density wave elementary excitations, with a dispersion linear in $|q|$ at long wavelengths that defines a sound velocity v_S; (v) two additional velocities, v_N and v_J, associated with charge and current excitations, obeying $v_S = (v_N v_J)^{1/2}$; (vi) power-law decay of correlation functions at $T = 0$, with coupling-strength-dependent exponents that depend only on $\exp(-2\varphi)$, where $v_N = v_S \exp(-2\varphi)$ and $v_J = v_S \exp(2\varphi)$. It should be emphasised that this means that $\exp(-2\varphi)$ is a measure of the essential renormalised coupling constant, and can thus be obtained from knowledge of v_S and the change of ground state energy with charge, which gives v_N.

1D systems with this Luttinger liquid structure so far identified include: (*a*) interacting spinless fermions; (*b*) interacting spin-½ fermions (and those with higher internal symmetries; (*c*) the Bose fluid (including systems with internal symmetries); (*d*) the finite-density gas of solitons of the Sine–Gordon theory; (*e*) uniaxially anisotropic spin systems (the 'charge' here is azimuthal spin)—antiferromagnets (only in the case of finite azimuthal magnetisation in the easy-axis case) and ferromagnets (easy-plane only). For many of these classes there exist models exactly soluble by the Bethe *ansatz* (Bethe 1930), and the Luttinger liquid structure can then be explicitly tested and verified (Haldane 1981). Subsequent papers will present such 'case studies' (see also Haldane 1980).

The Luttinger liquid has a characteristic instability if a multiple of its fundamental momentum $2k_F$ is equal to a reciprocal lattice vector reflecting an underlying periodicity. For large enough values of the parameter $\exp(-2\varphi)$, a gap opens in the spectrum, and the system becomes insulating. This instability can be studied in detail using the precise operator machinery set up in this paper, and will be the subject of paper II in this series. A universal description of the behaviour of the strongly renormalised Luttinger liquid near this instability emerges. The precise agreement between the predictions of this description and the features found in many of the 'test case' models solvable by the Bethe *ansatz* will provide strong evidence for the universality of the 'Luttinger liquid' description.

The organisation of this paper is as follows. To avoid confusion, it deals only with the *spinless* form of the model. The necessary generalisation to spin-½ fermions, and from these to the Bose fluid and spin systems, will be dealt with in subsequent articles. In § 2 there is a brief introduction to specifically one-dimensional features of the Fermi gas. Section 3 contains the bulk of the technical development, and describes the structure of the *non-interacting* Luttinger model. The machinery set up in § 3 leads quickly to the solution of the *interacting*-fermion Luttinger model in § 4. Section 5 uses the machinery to discuss the effects of a non-linear fermion dispersion. Finally, § 6 summarises the results, and formulates the hypothesis that the 'Luttinger liquid' structure is universal to conducting spinless fermion systems in 1D.

2. The one-dimensional Fermi gas

Fermion systems in one dimension have features quite distinct from those in higher dimensions. This is because the one-dimensional Fermi surface consists of two discrete points, while in higher dimensions it is continuous. The special spectral structure resulting from this can be seen by examining the full spectrum of excited states above a ground state with Fermi wavevector k_F. Figures 1(*a*) and (*b*) show the single-particle dispersion (and ground state occupancy) and the particle–hole pair spectrum of the (spinless) 1D Fermi gas with periodic boundary conditions on a length L. The distinctive one-dimensional feature of the pair spectrum is the non-existence of low-energy pairs for $0 < |k| < 2k_F$; in higher dimensions this region of 'missing' states is filled in. The full spectrum of excited states with zero excited charge (with respect to a ground state with odd charge $N_0 = k_F L/\pi$) is obtained by using figure 1(*b*) to determine the allowed energies of multiple-pair states, and is shown in figure 1(*c*).

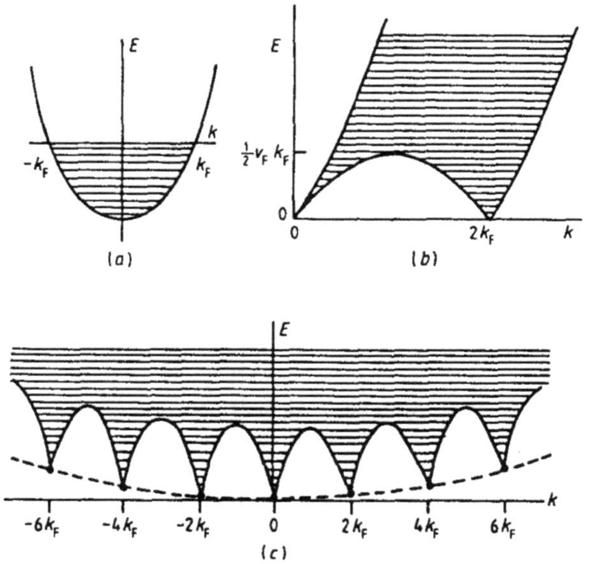

Figure 1. (*a*) Single-particle spectrum of the free Fermi gas in 1D; (*b*) Particle–hole pair spectrum; (c) full zero-charge (multiple particle–hole) excitation spectrum (energy differences $E(n) = 2\pi v_F n^2/L$ of extremal states at $k = 2nk_F$ greatly exaggerated).

At low energies $E \ll v_F k_F$, where v_F is the Fermi velocity $d\varepsilon(k_F)/dk$, the spectrum splits up into separate sectors that can be labelled by an even integer J, and can be described as excitations about a set of extremal states with momentum $k_F J$. These states have an energy $\frac{1}{2}(\pi/L)v_F J^2$, and this quadratic energy dependence (valid for energies $\ll v_F k_F$) is shown on a greatly exaggerated scale in figure 1(*c*). The spectrum of excitations with non-zero but even values of extra charge $(N - N_0)$ is similar, but with k_F replaced by $(k_F + \pi(N - N_0)/L)$, and the addition of a term $\frac{1}{2}(\pi/L)v_F(N - N_0)^2$ to the energy. The spectrum of excitations with odd values of $(N - N_0)$ differs only in that extremal states correspond to odd values of J. The form of the excitation spectrum suggests that at low energies it can be described by linear boson ('sound wave') excitations about the extremal states labelled by integers N and J. Such a classification breaks down at higher energies not only because of the non-linearity of the electron dispersion, but also because there is no longer any unambiguous operational way of assigning the quantum number J.

These observations can be summarised by the hypothesis that the *low-energy* spectrum can be represented by the form (where $b_q^\dagger$ are boson creation operators)

$$H = v_S \sum_q |q| b_q^\dagger b_q + \tfrac{1}{2}(\pi/L)[v_N(N - N_0)^2 + v_J J^2] \tag{2.1}$$

$$P = [k_F + (\pi/L)(N - N_0)]J + \sum_q q b_q^\dagger b_q; \qquad k_F = \pi N_0/L, \tag{2.2}$$

where $qL/2\pi = \pm1, \pm2, \ldots$; N and J are integers, subject to the selection rule (for periodic boundary conditions)

$$(-1)^J = -(-1)^N. \tag{2.3}$$

The parameters are identified as

$$v_S = v_N = v_J = v_F, \tag{2.4}$$

where v_F is the Fermi velocity. Such a spectral form is obviously compatible with the above discussion, but only has the status of a plausible hypothesis until it has been verified that it gives the correct multiplicity of states. This can in fact be verified, as is shown in the next section, by examination of the *non-interacting Luttinger model*, for which the spectral form (2.1)–(2.3) *holds exactly at all energies*.

Though the three parameters v_S, v_N and v_J are all equal to the Fermi velocity in the case of the non-interacting Fermi gas, they describe quite distinct properties of the spectrum. It is thus natural to wonder whether in fact *interacting* gapless fermion systems also have a low-energy spectrum described by (2.1)–(2.3), but with renormalised and unequal values of the three velocity parameters v_S, v_N and v_J. This can in fact be confirmed by the study of the *interacting Luttinger model* (described in § 4, which is explicitly soluble. Though it has the feature that v_S, v_N and v_J are no longer equal, they are not independent, and their ratios are determined by a parameter that characterises the essential interaction strength and low-energy physical properties such as the asymptotic forms of the various correlation functions. I will argue that these relations, together with the spectral form (2.1)–(2.3) are universally valid for the description of the low-energy properties of gapless interacting one-dimensional spinless fermion systems. The assignment of different values to the parameters v_S, v_N and v_J in this 'Luttinger liquid theory' will be analogous to the assignment of different effective masses to the quasiparticles for the characterisation of different physical properties in Fermi liquid theory.

3. The Luttinger model and its solution: I. The non-interacting limit

3.1. Historical development

The Luttinger (1963) model is an exactly soluble model of interacting fermions in one dimension with the following key features:

(i) its elementary excitations are non-interacting bosons;
(ii) the mean fermion current j is a good quantum number;
(iii) all its correlation functions can be explicitly evaluated.

The complete solubility of this model only emerged over a period of a decade; because the resolution of certain ambiguities in versions of the solution developed in the literature over this period turned out to be a key step in the work reported in this series of papers, I will merely cite some of the key papers in the literature, and then present a detailed version of the solution without further reference to its historical development.

The model was proposed by Luttinger (1963), but this first step in its correct solution was taken by Mattis and Lieb (1965), who discovered the free boson elementary excitations. Soon after, Overhauser (1965) pointed out that these bosons could be used to construct a complete set of eigenstates. Theumann (1967) and Dover (1968) gave early calculations of the single-particle correlation function, but the systematic calculation of correlation functions became trivial after the simultaneous discovery of the existence of a simple representation of the fermion operators in terms of the boson fields by Mattis (1974) and Luther and Peschel (1974). In fact, these fields are not on their own sufficient for the full construction of fermion operators in the diagonal basis, and both these early forms have problems associated with the characterisation of $q = 0$ modes. In particular, Luther and Peschel (1974) introduced a certain cut-off parameter α in their version, with the stipulation that it only became an exact operator identity in the limit $\alpha \to 0$. The necessity for any such limiting procedure has been entirely eliminated in the exact formulation reviewed below. The first completely precise formulation in the solid-state literature (though from a field-theory viewpoint) was given by Heidenreich *et al* (1975), though there has been an entirely parallel development in the field-theoretical literature on the related 'massless Thirring model' which I will not review here. The first construction of the important unitary charge-raising operators in terms of the bare fermions was apparently given by Haldane (1979). An important paper essentially parallel to, but not part of, the above developments is that of Dzyaloshinskii and Larkin (1973) who studied the spin-$\frac{1}{2}$ version of the (originally spinless-fermion) model, and provided an interpretation of the Mattis–Lieb solution from the point of view of conventional many-body diagrammatic perturbation theory. Similarly, Everts and Schulz (1974) have shown how the power-law character of the correlation functions can be simply recovered by the standard equation-of-motion techniques. Below, I give a description of the spinless fermion form of the model; the simple extension to the spin-$\frac{1}{2}$ case will be discussed elsewhere.

3.2. The fermion description

It is useful to begin a discussion of the Luttinger model by characterising its Hilbert space: this is *not* the usual electron Hilbert space, but has been expanded to include a branch of *'positron'* states as well. This second, unphysical set of fermions will require high energies for their excitation, so will not qualitatively affect low-energy properties, but are absolutely necessary for the construction of the new basis of eigenstates given here. Note that definition of the Hilbert space does *not* require any precise specification of the electron and positron dispersions, only that these energies are bounded below, and increase without limit as the momentum $|k| \rightarrow \infty$. The model is defined on a finite ring of length L; only periodic fermion boundary conditions will be considered. It is then useful in developing the formalism to take the ground-state charge $N_0 = k_F L/\pi$ to be *odd*, so the ground state is non-degenerate (this restriction is eventually dropped). The Hilbert space worked in is spanned by the set of finite-energy eigenstates of the free Luttinger model, measured from a ground state with electron states from $-k_F$ to k_F filled, and all positron states empty.

A correct definition of the Hilbert space is required before any operator acting in it, such as the Hamiltonian, is defined. It gives meaning to 'operator identities' such as $\hat{A} = \hat{B}$, shorthand for $\langle\alpha|\hat{A}|\beta\rangle = \langle\alpha|\hat{B}|\beta\rangle$ for all $|\alpha\rangle$, $|\beta\rangle$ forming a set that spans the Hilbert space. Operators are only well defined if $\langle\alpha|\hat{A}|\beta\rangle$ is finite for all α, β; the problem of ill-defined operators does not arise in finite-dimensional Hilbert spaces such as in lattice systems, but problems can arise with infinite-dimensional spaces arising from continuum problems. This type of problem flawed Luttinger's original solution of the model. One standard way to ensure all operators worked with are finite is to consider only quantities that are normal-ordered in a set of creation operators that create excited states out of the ground state.

Instead of working directly with charge +1 electron states, and charge −1 positron states, it is useful to describe the Luttinger model in terms of charge +1 'right-' and 'left-moving' fermions labelled by $p = \pm 1$ (note that this label should *not* be confused with a momentum label, for which k is used here). The kinetic part of the Luttinger Hamiltonian is then given (using units where $\hbar = 1$) by

$$H^0 = v_F \sum_{kp} (pk - k_F)(n_{kp} - \langle n_{kp}\rangle_0) \qquad \langle n_{kp}\rangle_0 = \theta(k_F - pk) \tag{3.1}$$

$$= v_F \int_0^L dx \sum_p : \psi_p^\dagger(x)(ip\nabla - k_F)\psi_p(x): \tag{3.2}$$

where :(...): means fermion normal-ordering with respect to the ground state of (3.1). The term k_F is essentially a chemical potential to fix the ground state charge. The spectral diagrams corresponding to figure 1 for the non-interacting Luttinger model (3.1) are shown in figure 2. The fermion field $\psi_p^\dagger(x)$ is given by

$$\psi_p^\dagger(x) = \lim_{\varepsilon \rightarrow 0^+} \left[L^{-1/2} \sum_k e^{ikx} \exp(-\varepsilon|kL/2\pi|)\, c_{kp}^\dagger \right]. \tag{3.3}$$

The limiting procedure $\varepsilon \rightarrow 0^+$ is usually left implicit, but has been explicitly included here to emphasise the similarity with a analogous construction that will appear later. It is necessary for the definition of the periodic delta function generated by the anticommutation relations:

$$\{\psi_p^\dagger(x), \psi_{p'}(x')\} = \delta_{pp'} \left[\lim_{\varepsilon \rightarrow 0^+} \left(L^{-1} \sum_k \exp[-k(x - x')] \exp(-\varepsilon|kL/\pi|) \right) \right]$$
$$= L^{-1}\delta_{pp'} \lim_{\varepsilon \rightarrow 0^+} [(1 - ze^{-2\varepsilon})^{-1} + (1 - z^* e^{-2\varepsilon})^{-1} - 1]$$
$$= \delta_{pp'} \sum_{n=-\infty}^{\infty} \delta(x - x' - nL). \tag{3.4}$$

where $z = \exp(2\pi ix/L)$, and allowed k-values in the sum satisfy $\exp(ikL) = 1$.

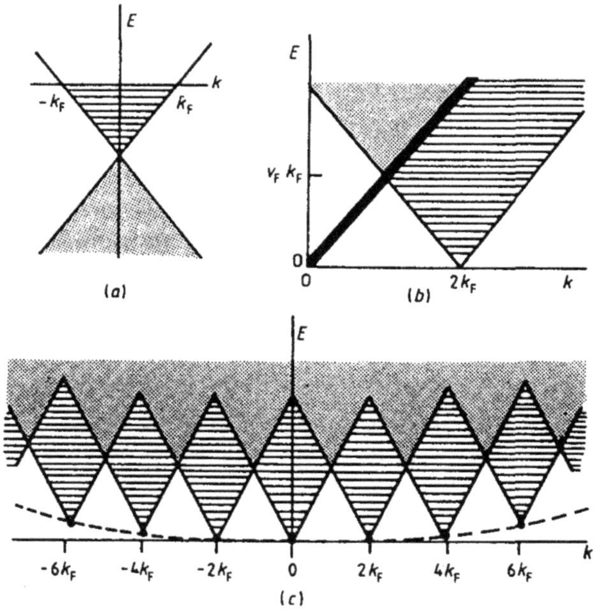

Figure 2. Diagrams corresponding to figure 1, this time for the spectrum of the non-interacting Luttinger model (3.1). Dotted areas indicate the presence of 'unphysical' states involving excited 'positrons'.

Note that the quantity ε appears as a dimensionless infinitesimal quantity necessary for controlling the sums over the infinite range of values of k, and in no way plays the role of a 'cut-off length'.

The fundamental electron and positron fields are related to the fields $\psi_p^\dagger(x)$ in a *non-local* way: in terms of the c_{kp}, they are given by

$$\psi^\dagger(x) = L^{-1/2} \sum_{kp} \theta(kp)\, \mathrm{e}^{\mathrm{i}kx}\, c_{kp}^\dagger$$

$$\overline{\psi}^\dagger(x) = L^{-1/2} \sum_{kp} \theta(-kp)\, \mathrm{e}^{-\mathrm{i}kx}\, c_{kp}. \tag{3.5}$$

When expressed in terms of $\psi_p^\dagger(x)$, the electron field $\psi^\dagger(x)$ is given by

$$\psi^\dagger(x) = \frac{1}{2\pi\mathrm{i}} \sum_p p \int_{-\frac{1}{2}L}^{\frac{1}{2}L} \mathrm{d}y\, K(y)\, \psi_p^\dagger(x+y), \tag{3.6}$$

where $K(y) = (\pi/L)[\tan(\pi y/L)]^{-1}$.

3.3. The boson description: construction from fermion operators

A central role in the theory of the Luttinger model is played by the density operator for type-p fermions:

$$\rho_{qp} = \sum_k c_{k+qp}^\dagger c_{kp} \qquad (q \neq 0)$$

$$\equiv N_p = \sum_k n_{kp} - \langle n_{kp} \rangle_0 \qquad (q = 0). \tag{3.7}$$

Note that the subtraction of the (infinite) ground state density of type-p fermions means that the $q = 0$ component of ρ_{qp} is a well defined operator in the sense discussed above: this procedure is equivalent to normal-ordering in the fermion variables. The commutation relations of the ρ_{qp} are

$$[\rho_{qp}, \rho_{-q'p'}] = \delta_{pp'}\delta_{qq'}(Lpq/2\pi). \tag{3.8}$$

This is easily established by direct evaluation of the commutator, then writing pair operators $c^\dagger_{kp}c_{k'p}$ as $(c^\dagger_{kp}c_{k'p} - \langle c^\dagger_{kp}c_{k'p}\rangle_0) + \langle c^\dagger_{kp}c_{k'p}\rangle_0$, where $\langle c^\dagger_{kp}c_{k'p}\rangle_0 \equiv \delta_{kk'}\langle n_{kp}\rangle_0$: this guarantees that operator quantities are effectively normal-ordered, and hence well defined, so they can be manipulated safely. The commutator (3.8) trivially vanishes when $p \neq p'$; for equal indices p, it is given by

$$\hat{O}_{pqq'} - \hat{O}_{pqq'} + \delta_{qq'}\sum_k (\langle n_{k+qp}\rangle_0 - \langle n_{kp}\rangle_0);$$

$$\hat{O}_{pqq'} \equiv \sum_k (c^\dagger_{k+q-q'p}c_{kp} - \langle c^\dagger_{k+q-q'p}c_{kp}\rangle_0).$$

Being well defined, the operators $\hat{O}_{pqq'}$ can safely be cancelled, and the remaining c-number term gives the RHS of (3.8). A second important commutation relation is that with the fermion fields:

$$[\rho_{qp}, \psi^\dagger_{p'}(x)] = \delta_{pp'}\,e^{-iqx}\psi^\dagger_p(x). \tag{3.9}$$

It will prove useful to define the following partial Fourier transforms:

$$\rho^{(\pm)}_p(x) = L^{-1}\sum_q \theta(\pm pq)\,e^{iqx}\rho_{qp}. \tag{3.10}$$

These operators have the property that $(\rho^{(-)}_p(x))^\dagger = \rho^{(+)}_p(x)$, and that $\rho^{(-)}_p(x)$ annihilates the vacuum state of (3.1). They satisfy periodic boundary conditions, $\rho^{(\pm)}_p(x+L) = \rho^{(\pm)}_p(x)$. The local density of p-type fermions, with respect to the ground state density, is

$$\lim_{a\to 0}[\psi^\dagger_p(x+a)\psi_p(x) - \langle\psi^\dagger_p(x+a)\psi_p(x)\rangle_0] \equiv \rho_p(x) = [\rho^{(+)}_p(x) + \rho^{(-)}_p(x)]. \tag{3.11}$$

The limiting procedure $a \to 0$ is to avoid direct reference to the infinite quantity $\langle\psi^\dagger_p(x)\psi_p(x)\rangle_0$.

The commutation algebra (3.8) immediately suggests the construction of boson operators. For $q \neq 0$,

$$a^\dagger_q = (2\pi/L|q|)^{1/2}\sum_p \theta(pq)\rho_{qp} \qquad (q \neq 0; q = 2\pi n/L, n = \pm 1, \pm 2, \ldots). \tag{3.12}$$

These obey exact boson commutation relations, and have the property that a_q annihilates the ground state of (3.1). There is no $q = 0$ boson mode (indeed, the form (3.12) is undefined at $q = 0$); the $q = 0$ mode is represented by the number operator N_p, which commutes with the bosons $a^\dagger_q$. The density operators are then expressed by

$$\rho_{qp} = N_p\delta_{p0} + (L|q|/2\pi)^{1/2}\{\theta(pq)a^\dagger_q + \theta(-pq)a_{-q}\}. \tag{3.13}$$

The algebra of the operators $(a^\dagger_q, a_q, N_p)$ that have been constructed so far is incomplete: it lacks a ladder operator U_p that raises the fermion charge N_p in unit steps, while commuting with the bosons a_q. The ladder of allowed values of N_p has no upper or lower limit, so (in contrast to the case of boson or finite spin ladder algebras) the number operator cannot be expressed in terms of the raising operator and its conjugate lowering operator. The raising operator can be chosen to be *unitary*, $U_p^{-1} = (U_p)^\dagger$. Finally, the fermion nature of the ladder operators means that U_p will *anti*commute with U_{-p} and U_{-p}^{-1}.

It is useful to study the construction of the ladder operators U_p in detail. It is important that these operators be given in a well defined form. A heuristic understanding of their form can be gained from the following argument. A special subset of eigenstates of (3.1) are those with occupations $n_{kp} = \theta(k_F + (2\pi N_p/L) - pk)$; these states $|\{N_p\}\rangle$ include the vacuum, and share with it the property that they are annihilated by a_q. The ladder operator U_p must have the property that

$$U_p\,|N_p, N_{-p}\rangle = \eta(p, N_p, N_{-p})\,|N_p + 1, N_{-p}\rangle, \quad \eta = \pm 1.\dagger$$

A construction with this property is

$$\sum_k c^{\dagger}_{kp}\delta(pk - [k_F + (2N_p + 1)\pi/L]).$$

Writing this in a more symmetrical form, using an integral representation of the Kronecker delta function, this becomes:

$$U_p = L^{-1/2}\int_0^L \mathrm{d}x\, \exp(-\mathrm{i}pk_F x)\exp[-\mathrm{i}\phi_p^{\dagger}(x)]\psi_p^{\dagger}(x)\exp[-\mathrm{i}\phi_p(x)];$$

$$\phi_p(x) = p(\pi x/L)N_p. \tag{3.14}$$

The ladder operator U_p must also have the property that it commutes with the boson operators a_q, or equivalently with the density operators ρ_{qp}, when $q \neq 0$. The commutation relation (3.9) means that the trial form above does *not* have this property. However, it would if the operators $\phi_p(x)$ were modified so that

$$[\rho_{qp}, \phi_p(x)] = -\mathrm{i}\delta_{pp'}\theta(pq)\,\mathrm{e}^{-\mathrm{i}qx}(1 - \delta_{q0}). \tag{3.15}$$

Because this commutator is a *c*-number,

$$[\rho_{qp}, \exp(-\mathrm{i}\phi_{p'}(x))] = -\theta(pq)(\delta_{pp'}\,\mathrm{e}^{-\mathrm{i}qx})\exp\{-\mathrm{i}\phi_p(x)\}(1 - \delta_{q0}),$$
$$[\rho_{pq}, \exp\{-\mathrm{i}\phi^{\dagger}_{p'}(x)\}] = -\theta(-pq)(\delta_{pp'}\,\mathrm{e}^{-\mathrm{i}qx})\exp\{-\mathrm{i}\phi_p^{\dagger}(x)\}(1 - \delta_{q0}). \tag{3.16}$$

When $q \neq 0$, these terms exactly counterbalance the commutator (3.9), so when U_p is given by the form (3.14) with a $\phi_p(x)$ that satisfies (3.15),

$$[\rho_{qp}, U_{p'}] = \delta_{pp'}\delta_{q0}U_p. \tag{3.17}$$

The explicit construction of the quantities $\phi_p(x)$ is now easily given; using the property (3.8):

$$\phi_p(x) = (2\pi p/L)\left(\tfrac{1}{2}xN_p + \mathrm{i}\sum_{q=0}\theta(-pq)(\mathrm{e}^{-\mathrm{i}qx}/q)\rho_{qp}\right). \tag{3.18}$$

The first term in this is just the $q = 0$ component of the sum, with the limit $q \to 0$ properly taken, so this can be re-expressed as

$$\phi_p(x) = \lim_{\varepsilon\to 0^+}\left((2\pi \mathrm{i}p/L)\sum_q \theta(-pq)\,\mathrm{e}^{-\mathrm{i}qx}/q)\exp[-\varepsilon(|q|L/2\pi)]\,\rho_{qp}\right). \tag{3.19}$$

The limiting procedure $\varepsilon \to 0^+$ has been included in order to properly define the sum in much the same spirit as in equation (3.3), and it will be needed to properly define the periodic delta function. ε is a positive dimensionless infinitesimal, and in no way should

† η depends on the ordering convention used in constructing $|\{N_p\}\rangle$.

be interpreted as a cut-off length, despite the formal similarity to the length parameter α introduced by Luther and Peschel (1974). The field $\phi_p(x)$ is easily found to have the following properties:

$$\nabla\phi_p(x) = 2\pi p\rho_p^{(-)}(x) \tag{3.20}$$

$$[\phi_p(x), \phi_{p'}(x')] = [\phi_p^\dagger(x), \phi_{p'}^\dagger(x')] = 0;$$

$$[\phi_p(x), \phi_{p'}^\dagger(x')] = \lim_{\varepsilon\to 0^+} \{\delta_{pp'}(-\ln(1 - e^{-2\varepsilon} z^{-2p}))\} \tag{3.21}$$

where $z \equiv \exp(i\pi(x - x')/L)$. $\phi_p(x)$ has the property that it annihilates the vacuum state of (3.1); this is easily seen when $\phi_p(x)$ is written in terms of N_p and a_q:

$$\phi_p(x) = p(\pi x/L)N_p + i\sum_{q\neq 0} \theta(pq)(2\pi/L|q|)^{1/2} e^{-iqx} a_q. \tag{3.22}$$

This means that the operator U_p defined by (3.14) plus (3.19) is normal-ordered in the boson operators that annihilate the vacuum state of (3.1). As will shortly be seen, this guarantees that this construction of U_p defines a well defined operator.

It is useful to give a representation of the unitary operators U_p in terms of Hermitian phase variables $\bar\theta_p = \bar\theta_p^\dagger$, conjugate to N_p:

$$U_p = (-1)^{(\frac{1}{2}pN_{-p})}\bar U_p, \qquad \bar U_p = \exp(i\bar\theta_p); \tag{3.23}$$

$$[N_p, \bar\theta_{p'}] = i\delta_{pp'}, \qquad [\bar\theta_p, \bar\theta_{p'}] = [N_p, N_{p'}] = 0. \tag{3.24}$$

The prefactor $(-1)^{(\frac{1}{2}pN_{-p})}$ ensures that U_p and U_{-p} anticommute, so that the unitary operators $\bar U_p$ and $\bar U_{-p}$ can commute. This choice of anticommutation factor corresponds to a particular ordering convention in constructing states $|\{N_p\}\rangle$ in terms of fermion operators; other choices are possible.

For some purposes, it is useful to introduce a *local* phase field $\theta_p(x)$:

$$\theta_p(x) = \bar\theta_p + \phi_p(x) + \phi_p^\dagger(x). \tag{3.25}$$

This has the property

$$\nabla\theta_p(x) = 2\pi p\rho_p(x). \tag{3.26}$$

It can easily be seen that $\rho_p(x)$ and $\theta_p(x)$ are canonical conjugate fields:

$$[\rho_p(x), \theta_{p'}(x')] = i\delta_{pp'}\sum_n \delta(x - x' + nL). \tag{3.27}$$

3.4. Boson form of the Hamiltonian

So far, no use whatsoever has been made of the fact that the Hamiltonian (3.1) has a linear fermion dispersion, and the above discussion has only depended on identification of its vacuum state and the structure of the associated Hilbert space. The linear spectral property will now be used to construct a new basis of eigenstates that will be shown to be complete and hence to span the Hilbert space defined by all finite-energy eigenstates of (3.1). The operators N_p commute with (3.1), but $a_q^\dagger$ and U_p do not: first note the commutator of the density operator with the Hamiltonian:

$$[H^0, \rho_{qp}] = v_F pq\rho_{qp}. \tag{3.28}$$

With this, the definition (3.12) of the boson operators a_q leads to

$$[H^0, a_q^\dagger] = v_F |q| a_q^\dagger. \tag{3.29}$$

Instead of attempting to directly evaluate the commutator $[H^0, U_p]$, I use the following argument: the special set of states $|\{N_p\}\rangle$ can be constructed from the vacuum by acting on it with U_p:

$$|\{N_p\}\rangle = \pm \prod_p (U_p)^{N_p}|0\rangle; \tag{3.30}$$

this is verified by explicitly showing that U_p as constructed indeed has the property $U_p|N_p, N_{-p}\rangle = \eta|N_p + 1, N_p\rangle$. Because a_q annihilates the states $|\{N_p\}\rangle$, (3.22) implies that

$$U_p|\{N_p\}\rangle = L^{-1}\sum_k \int_0^L \mathrm{d}x \exp[\mathrm{i}(k - pk_F)x] \exp[-\mathrm{i}\phi_p^\dagger(x)]c_{kp}^\dagger \exp[-\mathrm{i}p(\pi x/L)N_p]|\{N_p\}\rangle$$

$$= \prod_{q\neq 0}\left(1 + \theta(pq)\sum_{n_q=1}^{\infty}(2\pi/L|q|)^{\frac{1}{2}n_q}(n_q!)^{-1}(a_q^\dagger)^{n_q}\right)$$

$$\times \sum_k c_{kp}^\dagger \delta\left(pk - [k_F + (2N_p + 1)\pi/L] + \sum_q |q|n_q\right)|\{N_p\}\rangle.$$

The Kronecker delta ensures that no states containing boson excitations survive in this sum, and (3.30) is verified.

The energies of the eigenstates $|\{N_p\}\rangle$ are easily obtained by examining their construction: $E(\{N_p\}) = v_F(\pi/L)\,\Sigma_p(N_p)^2$. A larger set of eigenstates is obtained by acting these with the boson operators:

$$|\{N_p\}, \{n_q\}\rangle = \prod_{q\neq 0}\left(\frac{(a_q^\dagger)^{n_q}}{(n_q!)^{1/2}}\right)\prod_p (U_p)^{N_p}|0\rangle. \tag{3.31}$$

In its action on *these* eigenstates, the Hamiltonian is given by

$$H^0 = v_F\left[\sum_q |q| a_q^\dagger a_q + (\pi/L)\sum_p (N_p)^2\right]. \tag{3.32}$$

This can be written in terms of the phase fields $\theta_p(x)$ as

$$H^0 = v_F\frac{1}{\pi}\int_0^L \mathrm{d}x \sum_p :(\nabla\theta_p(x))^2:, \tag{3.33}$$

where *boson* normal-ordering is implied. The momentum operator is similarly given by

$$P = \sum_p p[k_F + (\pi/L)N_p]N_p + \sum_q q a_q^\dagger a_q \tag{3.34}$$

$$= \frac{1}{\pi}\int_0^L \mathrm{d}x \sum_p p[k_F\nabla\theta_p(x) + :(\nabla\theta_p(x))^2:]. \tag{3.35}$$

The question arises: *are the eigenstates* (3.31) *a complete set?* If so, U_p is proved to be a well defined operator, and (3.32) and (3.34) have the status of identities in the full Hilbert space based on the vacuum of (3.1). The possibility of two such different sets of eigenstates of the free Luttinger Hamiltonian (3.1) arises because of the high degree of degeneracy of the spectrum due to the linear dispersion: all eigenstates with even (odd) fermion charge $N - N_0$ have energies that are even (odd) multiples of $\pi v_F/L$ with respect to the ground state. One way to check the completeness of the set (3.31) is to directly investigate the degeneracy of states at a given energy. An equivalent, more elegant, way is to compute the grand partition sum of the Hamiltonian at arbitrary inverse temperature β, first using the 'obvious' set of fermion excitation states, then the set (3.31). This is a sum over positive definite quantities, so if any states were missing from (3.31), the result of the second calculation would be less than the first.

Defining $w = \exp(-\beta\pi v_F/L)$, the direct evaluation of the partition function using the free fermion basis gives

$$Z(w) = \left(\prod_{n=1}^{\infty} (1 + w^{2n-1})^2\right)^2. \tag{3.36a}$$

Using the set (3.31), one obtains

$$Z(w) = \left(\prod_{n=1}^{\infty} (1 - w^{2n})^{-1}\right)^2 \left(\sum_{m=-\infty}^{\infty} w^{(n^2)}\right)^2. \tag{3.36b}$$

These apparently different expressions are in fact both equal, since the elliptic theta function $\vartheta_3(0; w)$ (Gradsteyn and Ryzhik 1965, p 921) has both a series and a product representation:

$$\vartheta_3(0; w) = \sum_{n=-\infty}^{\infty} w^{(n^2)} = \prod_{n=1}^{\infty} (1 + w^{2n-1})^2(1 - w^{2n}).$$

The set (3.31) is thus complete, and spans the full Hilbert space.

3.5. *Boson form of fermion operators*

With the completeness of the set of eigenstates (3.31) established, the remaining task is to construct the representation of the fermion operators $\psi_p^\dagger(x)$ in this basis. The ground work has been laid: $\psi_p^\dagger(x)$ is trivially obtained by inverting the expression (3.14) for U_p:

$$\begin{aligned}\psi_p^\dagger(x) &= L^{-1/2} \exp(ipk_F x)\{\exp[i\phi_p^\dagger(x)]U_p \exp[i\phi_p(x)]\}\\ &= (-1)^{(\frac{1}{2}pN-p)}L^{-1/2} \exp(ipk_F x)\{\exp[i\phi_p^\dagger(x)] \exp(i\bar{\theta}_p) \exp[i\phi_p(x)]\},\end{aligned} \tag{3.37}$$

where $\phi_p(x)$ is now defined directly by (3.22). This is an explicitly well defined operator, since it is normal-ordered in terms of the bosons a_q. The anticommutation relations can be explicitly verified; here the limiting procedure defined in (3.19) and (3.21) is required: the procedure is to construct the anticommutators, and then re-normal-order the resulting products in terms of the bosons, so they become explicitly well defined operators that can be manipulated and cancelled. The anticommutation of fields with different labels p is trivially assured by the anticommuting properties of U_p; for equal p, the anticommutator $\{\psi_p^\dagger(x), \psi_p^\dagger(x')\}$ is given by

$$\begin{aligned}&L^{-1} \exp[ipk_F(x + x')] \exp[i\hat{O}_1^\dagger(x, x')]U_p^2 \exp[i\hat{O}_1(x, x')]F_1(x, x')\\ &\hat{O}_1(x, x') = \phi_p(x) + \phi_p(x')\\ &F_1(x, x') = G_1(x - x') + G_1(x' - x)\\ &G_1(x - x') = \exp[i\pi p(x - x')/L] \exp\{-[\phi_p(x), \phi_p^\dagger(x')]\}.\end{aligned} \tag{3.38}$$

The c-number function $F_1(x, x')$ is multiplying a well defined (i.e. normal-ordered) operator expression. $F_1(x, x')$ can be evaluated using equation (3.21): setting $z = \exp[i\pi(x - x')/L]$,

$$F_1 = \lim_{\varepsilon\to 0^+} \{z^p(1 - e^{-2\varepsilon}z^{-2p}) + z^{-p}(1 - e^{-2\varepsilon}z^{2p})\} = 0. \tag{3.39}$$

The anticommutator $\{\psi_p^\dagger(x), \psi_p^\dagger(x')\}$ thus vanishes correctly. The anticommutator $\{\psi_p(x), \psi_p^\dagger(x')\}$ is given by

$$\begin{aligned} &L^{-1}\exp[ipk_F(x' - x)]\exp[i\hat{O}_2^\dagger(x, x')]\exp[i\hat{O}_2(x, x')]F_2(x, x') \\ &\hat{O}_2^\dagger(x, x') = \phi_p(x) - \phi_p(x') \\ &F_2(x, x') = G_2(x - x') + G_2(x' - x) \\ &G_2(x - x') = \exp[i\pi p(x - x')/L]\exp\{+[\phi_p(x), \phi_p^\dagger(x')]\}. \end{aligned} \tag{3.40}$$

Again this is a normal-ordered operator expression, times a c-number function $F_2(x, x')$. Again using (3.21),

$$\begin{aligned} F_2 &= \lim_{\varepsilon\to 0^+} \{z^p(1 - e^{-2\varepsilon}z^{-2p})^{-1} + z^{-p}(1 - e^{-2\varepsilon}z^{2p})^{-1}\} \\ &= L \sum_{n=-\infty}^{\infty} (-1)^n \delta(x - x' + nL). \end{aligned} \tag{3.41}$$

when $x - x' = nL$, the operator-valued expression that multiplies F_2 in (3.41) takes the simple c-number values $L^{-1}\exp(-inpk_F L) = L^{-1}(-1)^n$. The anticommutator is thus correctly given by the periodic delta function as in (3.3).

This completes the derivation of the operator algebra needed to describe the model using the alternative basis set of eigenstates (3.31). This algebra is a precise tool, and I now use it to recover the expressions (3.32) and (3.34) for H^0 and P directly from the fermion representation (3.37). Consider the quantity

$$\int_0^L dx \exp(-ipk_F a)\, \psi_p^\dagger(x + \tfrac{1}{2}a)\psi_p(x - \tfrac{1}{2}a). \tag{3.42}$$

Using the expression (3.3) for $\psi_p(x)$, this is easily found to be

$$L(2\pi i p a')^{-1} + \sum_k \exp[i(k - pk_F)a](n_{kp} - \langle n_{kp}\rangle_0) \tag{3.43}$$

where $a' = (L/\pi)\sin(\pi a/L)$. Using the alternative expression (3.37), and then normal-ordering, it is found to be

$$\begin{aligned} &(2\pi i p a')^{-1}\left(L + \int_0^L dx\{\exp[2\pi i p(a/L)N_p]\exp[i\Phi_p^\dagger(x)]\exp[i\Phi_p(x)] - 1\}\right) \\ &\Phi_p(x) = i\sum_q (2\pi/L\,|q|)^{1/2} e^{-iqx}\theta(qp)\, 2\sin(\tfrac{1}{2}qa)a_q. \end{aligned} \tag{3.44}$$

Cancelling the divergent term $L(2\pi i p a')^{-1}$ and comparing the term $O(a)$ in the expansions of the two expressions, one directly obtains

$$\sum_k (pk - k_F)(n_{kp} - \langle n_{kp}\rangle_0) = (\pi/L)N_p^2 + \sum_q pq\,\theta(qp)a_q^\dagger a_q. \tag{3.45}$$

The expressions for H^0 and P are now trivial to obtain.

3.6. The charge and current formalism

So far, the formalism has been developed in terms of operators labelled by $p = \pm 1$, corresponding to the right- and left-going fermions. It is convenient for some purposes to introduce the symmetric and antisymmetric combinations, labelled by N and J respectively, which will be related to charge and current variables. It is also useful to include the ground state electronic charge (number of electrons minus number of positrons) $N_0 = k_F L/\pi$ in the charge variables. The following combinations are defined:

$$N = N_0 + \sum_p N_p \qquad J = \sum_p pN_p \tag{3.46}$$

$$\bar{\theta}_N = \sum_p \bar{\theta}_p \qquad \bar{\theta}_J = \sum_p p\bar{\theta}_p \tag{3.47}$$

$$\rho_N(x) = (N_0/L) + \sum_p \rho_p(x) \qquad \rho_J(x) = \sum_p p\rho_p(x) \tag{3.48}$$

$$\phi_N(x) = \sum_p \phi_p(x) \qquad \phi_J(x) = \pi(N_0/L)x + \sum_p p\phi_p(x). \tag{3.49}$$

Phase fields $\theta_N(x)$ and $\theta_J(x)$ are then defined by, e.g.,

$$\theta_N(x) = \bar{\theta}_N + \phi_N(x) + \phi_N^\dagger(x). \tag{3.50}$$

The following relations are found:

$$\nabla\theta_N(x) = 2\pi\rho_J(x) \qquad \nabla\theta_J(x) = 2\pi\rho_N(x) \tag{3.51}$$

$$[\rho_N(x), \theta_N(x')] = [\rho_J(x), \theta_J(x')] = \mathrm{i}\sum_n \delta(x - x' + nL); \tag{3.52}$$

$$[\rho_N(x), \theta_J(x')] = [\rho_J(x), \theta_N(x')] = 0. \tag{3.53}$$

The fields $(\rho_N(x), \theta_N(x))$ and $(\rho_J(x), \theta_J(x))$ are canonically conjugate pairs. Note however that $[\rho_N(x), \rho_J(x')]$ and $[\theta_J(x), \theta_N(x')]$ do *not* vanish, except at equal positions, $x = x'$. On the other hand, $[\theta_N(x), \theta_N(x')]$, $[\rho_N(x), \rho_N(x')]$, etc, *do* vanish.

When the quantities $\rho_N(x)$, $\phi_N(x)$, etc are expressed in terms of boson variables, they are explicitly given by

$$\rho_N(x) = (N/L) + \sum_q (|q|/2\pi L)^{1/2}\, \mathrm{e}^{\mathrm{i}qx}\, (a_q^\dagger + a_{-q})$$

$$\phi_N(x) = \pi(J/L)x + \mathrm{i}\sum_q (2\pi/L\,|q|)^{1/2}\, \mathrm{e}^{-\mathrm{i}qx} a_q$$

$$\rho_J(x) = (J/L) + \sum_q (|q|/2\pi L)^{1/2}\, \mathrm{sgn}(q)\, \mathrm{e}^{\mathrm{i}qx}(a_q^\dagger - a_{-q})$$

$$\phi_J(x) = \pi(N/L)x + \mathrm{i}\sum_q (2\pi/L\,|q|)^{1/2}\, \mathrm{sgn}(q)\, \mathrm{e}^{-\mathrm{i}qx} a_q. \tag{3.54}$$

Note how $\mathrm{sgn}(q)$ characteristically appears in the boson part of J-labelled quantities.

The commuting unitary operators $\bar{U}_N = \exp(\mathrm{i}\bar{\theta}_N)$ and $\bar{U}_J = \exp(\mathrm{i}\bar{\theta}_J)$ respectively raise N and J by one. In this basis, the fermion field operator $\psi_p^\dagger(x)$ becomes

$$\psi_p^\dagger(x) = L^{-\frac{1}{2}}(-1)^{\frac{1}{2}(pJ-N)}\{\exp[\tfrac{1}{2}\mathrm{i}(p\phi_J^\dagger + \phi_N^\dagger)]\exp[\tfrac{1}{2}\mathrm{i}(p\bar{\theta}_J + \bar{\theta}_N)]$$
$$\times \exp[\tfrac{1}{2}\mathrm{i}(p\phi_J + \phi_N)]\}. \tag{3.55}$$

The dependence on k_F has been absorbed into the definition of $\phi_J(x)$.

The Hamiltonian takes the form

$$H^0 = v_F\left(\sum_q |q|\, a_q^\dagger a_q + \tfrac{1}{2}(\pi/L)((N-N_0)^2 + J^2)\right) \qquad (-1)^J = -(-1)^N; \tag{3.56}$$

$$P = [k_F + \pi(N-N_0)/L]\,J + \sum_q q a_q^\dagger a_q \qquad k_F = \pi N_0/L. \tag{3.57}$$

This is just the form postulated in § 2. The selection rule linking allowed values of J and N arises because N_p and N_{-p} are both integral.

In the phase field variables, the Hamiltonian can be written

$$H = v_F \frac{1}{\pi}\int_0^L \mathrm{d}x :(\nabla\theta_N(x))^2 + (\nabla\theta_J(x))^2:$$

$$P = \frac{1}{\pi}\int_0^L \mathrm{d}x :\nabla\theta_N(x)\nabla\theta_J(x): + \text{HC}. \tag{3.58}$$

Since $\rho_N(x) = (2\pi)^{-1}\nabla\theta_J(x)$ is the canonical conjugate to $\theta_N(x)$, and $\rho_J(x)$ to $\theta_J(x)$, the Hamiltonian (3.58) can be written as a Klein–Gordon field Hamiltonian in either the N or the J variables. The periodic fermion boundary conditions that must be satisfied by (3.55) imply that $\exp[\mathrm{i}\theta_N(x+L)] = \exp[\mathrm{i}\theta_N(x)]$, etc, so $\theta_N(x+L) = 2\pi J + \theta_N(x+L)$, and $\theta_J(x+L) = 2\pi N + \theta_J(x)$. The quantum numbers N and J thus can be related to *topological* excitations of the phase fields $\theta_N(x)$ and $\theta_J(x)$, while the bosons relate to their small fluctuations.

The physical interpretation of the quantum number N is simple: it is just the total electronic charge (electrons minus positrons). Similarly, it will now be shown that J is proportional to the mean *current*. It would be tempting to identify $\rho_N(x)$ with the *local* charge density operator $\rho(x)$; unfortunately, this is not correct, due to the non-local relation between the electron field and $\psi_p^\dagger(x)$. The fundamental definition of the local electronic density in terms of the electrons and positrons leads to

$$\rho(x) = \rho_N(x) + \tau(x),$$

$$\tau(x) = \sum_{kk'pp'} \exp[\mathrm{i}(k-k')x](-pp')\theta(-kk')c_{kp}^\dagger c_{k'p'}. \tag{3.59}$$

At low energies, the extra term $\tau(x)$ involves only fluctuations with $q \sim 2k_F$. The fundamental definition of the current $j(x)$ is through the continuity equation for local charge:

$$\frac{\mathrm{d}}{\mathrm{d}t}\rho(x) \equiv \mathrm{i}[H, \rho(x)] = \nabla j(x). \tag{3.60}$$

The mean current j is then given by

$$j \equiv L^{-1}\int_0^L \mathrm{d}x\, j(x) = \lim_{q\to 0}\{(qL)^{-1}[H, \rho_q]\}, \tag{3.61}$$

ρ_q being the Fourier transform of $\rho(x)$. In a low-energy subspace, and provided k_F is finite, the contribution from $\tau(x)$ can be neglected, and ρ_{Nq} substituted for ρ_q in (3.61). Then it is easily found that

$$j = v_F(J/L). \tag{3.62}$$

Actually, this is exact in the case of the free Luttinger model, but in a more general model extra terms will be present, and a linear relation like (3.62) will only be valid in a low-energy subspace, where the presence of long-wavelength boson excitations does not affect the current.

Finally, I note a useful low-energy, finite k_F *approximation* for $\tau(x)$:

$$\tau(x) \sim \sum_p \psi_p^\dagger(x)\psi_{-p}(x). \tag{3.63}$$

4. The Luttinger model and its solution: II. The interacting model

The full Luttinger model is obtained by taking the kinetic term (3.1), (3.32) and adding the fermion two-particle interaction:

$$H^1 = (\pi/L) \sum_{pp'q} (V_{1q}\delta_{pp'} + V_{2q}\delta_{p,-p'})\rho_{qp}\rho_{-qp'}. \tag{4.1}$$

The density operators ρ_{qp} are defined by (3.7) and (3.13). The coupling constants $V_{1q} \equiv V_1(|q|R)$ and $V_{2q} \equiv V_2(|q|R)$ have dimensions of velocity. They will be required to satisfy the following conditions:

(i) $V_1(0)$, $V_2(0)$ are finite;
(ii) $V_{2q}/(v_F + V_{1q}) \to 0$ as $|q| \to \infty$, faster than $|q|^{-\frac{1}{2}}$;
(iii) $|V_{2q}| < (v_F + V_{1q})$ for all q.

Conditions (i) could be relaxed somewhat, but this would alter the physics of the model. Conditions (ii) and (iii) are necessary to ensure that the Hilbert space of the model $H^0 + H^1$ remains the same as that of H^0. The conditions (i) and (ii) imply the existence of some length scale R that controls the crossover from the small-q to large-q regimes. The inclusion of this length scale in (4.1) means that V_1 and V_2 can be written as functions with a dimensionless argument. R is an effective range of the interaction in real space.

Using the phase-field formalism of § 3.6, the *low-energy* ($E \ll v_F/R$) form of the Hamiltonian can be written

$$H \simeq \frac{1}{\pi}\int_0^L \mathrm{d}x :v_N(\nabla\theta_N(x))^2 + v_J(\nabla\theta_J(x))^2: \tag{4.2}$$

where

$$v_N = v_F + V_1(0) + V_2(0), \qquad v_J = v_F + V_1(0) - V_2(0). \tag{4.3}$$

It is also useful to define the quantities

$$\omega_q = |(v_F + V_{1q})^2 - (V_{2q})^2|^{1/2}|q| \tag{4.4}$$

$$\tanh(2\varphi_q) = -V_{2q}/(v_F + V_{1q}). \tag{4.5}$$

Then the quantities v_S and φ are defined by

$$v_S = \lim_{q\to 0} (\omega_q/|q|); \qquad \varphi = \lim_{q\to 0} (\varphi_q). \tag{4.6}$$

The definitions (4.2)–(4.5) imply the relations

$$v_N = v_S \exp(-2\varphi); \qquad v_J = v_S \exp(2\varphi). \tag{4.7}$$

It will be convenient to represent φ_g as $\varphi g(|q|R)$, where the function $g(y)$ has the properties

$$g(0) = 1; \qquad y^{1/2}g(y) \to 0 \quad \text{as} \quad y \to \infty. \tag{4.8}$$

The conditions (i)–(iii) assure this, and also that φ_q is finite, ω_q is positive definite (except at $q = 0$) and v_N, v_J are positive definite. The model is fully parametrised by L, k_F, ω_q, and φ_q (or φ, R and $g(y)$). R has not been defined up to a multiplicative factor: it should be chosen so the crossover in $g(y)$ is around $y \sim 1$ (a unique definition might be provided by demanding that $(g(y))^2$ is normalised, for example).

When the full Hamiltonian $H^0 + H^1$ is written out in terms of a_q and the number operators N, J, it takes the simple bilinear form

$$H = -\tfrac{1}{2}\Big(\sum_q v_F|q|\Big) + \tfrac{1}{2}(\pi/L)(v_N N^2 + v_J J^2) + \tfrac{1}{2}\sum_q |q|\,[(v_F + V_{1q})(a_q^\dagger a_q + a_q a_q^\dagger) + V_{2q}(a_q^\dagger a_{-q}^\dagger + a_q a_{-q})]. \tag{4.9}$$

This is trivially diagonalised by a Bogoliubov transformation. The new ground state is given by

$$|\text{GS}\rangle = \exp[-(A^2L/R)]\exp\Big(\sum_{q>0}\tanh(\varphi_q)a_q^\dagger a_{-q}^\dagger\Big)\,|0, 0\rangle$$

$$A^2(\varphi g(y)) = \frac{1}{2\pi}\int_0^\infty dy\,\ln[\cosh(\varphi g(y))]. \tag{4.10}$$

For the ground state to belong to the Hilbert space of H^0, the normalisation constant must be finite; this means that the limit $R \to 0$ cannot be taken. The condition (4.8) assures that the constant A^2 is finite.

The Hamiltonian is diagonal in terms of the new boson operators

$$b_q^\dagger = \cosh(\varphi_q)a_q^\dagger - \sinh(\varphi_q)a_{-q} = \sum_p \alpha(pq, -\varphi_q)\,\rho_{qp} \qquad (q \neq 0); \tag{4.11}$$

$$\alpha(q, \varphi_q) = (2\pi/L\,|q|)^{1/2}[\theta(q)\cosh(\varphi_q) + \theta(-q)\sinh(\varphi_q)]. \tag{4.12}$$

The diagonalised Hamiltonian is given by substituting these into (4.9):

$$H = E_0 + \sum_q \omega_q b_q^\dagger b_q + \tfrac{1}{2}(\pi/L)(v_N N^2 + v_J J^2),$$

$$E_0 = \tfrac{1}{2}\sum_q(\omega_q - v_F|q|). \tag{4.13}$$

The ground state energy shift E_0 may well be divergent if ω_q does not tend to $v_F|q|$ fast enough as $|q| \to \infty$; however, in contrast to the case of a divergence of the ground state normalisation parameter A^2, this divergence is subtractable, and causes no problems. The form of the momentum operator remains essentially unchanged:

$$P = [k_F + \pi(N/L)]J + \sum_q q b_q^\dagger b_q. \tag{4.14}$$

The relation between the 'true' Fermi momentum $[k_F + \pi(N/L)]$ and the total charge N is unaffected by the interactions.

In addition to the total charge N with respect to the ground state remaining a good quantum number, the current quantum number J is also conserved. This reflects an invariance of the Hamiltonian, under which it is unchanged by independent global gauge transformations of the 'right-' and 'left-moving' ('clockwise' and 'anticlockwise') fermion fields $\psi_p(x)$, $p = \pm 1$. The density operators ρ_{qp} are given by

$$\rho_{qp} = \tfrac{1}{2}(N + pJ)\,\delta_{q0} + (L|q|/2\pi)[\alpha(pq, \varphi_q)b_q^\dagger + \alpha(-qp, \varphi_q)b_{-q}],$$

$$[H, \rho_{qp}] = p\,\text{sgn}(q)\omega_q[\cosh(2\varphi_q)\rho_{qp} - \sinh(2\varphi_q)\rho_{q-p})]. \tag{4.15}$$

Following the arguments of (3.60)–(3.62), the mean current j is given by

$$j \simeq \lim_{q\to 0}\left\{(qL)^{-1}\sum_p [H, \rho_{qp}]\right\} = \lim_{q\to o}\left(L^{-1}(\omega_q/|q|)\, e^{2\varphi_q}\sum_p p\rho_{qp}\right). \tag{4.16}$$

From (4.7) the mean current is found to be $j = v_J(J/L)$; v_J thus plays the role of the renormalised Fermi velocity for fermion currents, as well as controlling the energy of $2k_F$ excitations. Note that it is somewhat unphysical for V_1 and V_2 to differ: if they are set equal, as would be the case if the model was derived as an effective Hamiltonian for a model where only the total charge density was coupled, v_J remains equal to the bare value v_F due to the kinetic term, and is not renormalised.

It is now necessary to transcribe the fermion field $\psi_p(x)$ (3.37) into a form normal-ordered in the new basis. First the definition of the quantities $\phi_p(x)$ (3.22) must be generalised:

$$\phi_p(x, \varphi_q) = p(\pi x/L)N_p + \mathrm{i}\sum_{q\neq 0} \alpha(pq, -\varphi_q)\, e^{-iqx} b_q; \tag{4.17}$$

note that the phase field $\theta_p(x)$ is still given by

$$\theta_p(x) = \bar\theta_p + \phi_p(x, \varphi_q) + \phi_p^\dagger(x, \varphi_q) \tag{4.18}$$

independent of φ_q. Then the fermion field is given by

$$\psi_p(x) = e^{-\frac{1}{2}A} L^{-\nu} R^{\nu-1/2} \exp(ipk_F x) \exp[i\phi_p^\dagger(x, \varphi_p)] U_p \exp[i\phi_p(x, \varphi_q)] \tag{4.19}$$

where $\nu = \frac{1}{2}\cosh(2\varphi)$, and the cut-off-dependent constant $\bar A(\varphi, g(y))$ and a similar quantity $\bar B$ are given by

$$\bar A = \lim_{\varepsilon\to 0^+}\left[2\sinh^2(\varphi)\left(C + \ln(\varepsilon/2\pi) + \int_\varepsilon^\infty dy\, y^{-1} 2\sinh^2(\varphi g(y))\right)\right],$$

$$\bar B = \lim_{\varepsilon\to 0^+}\left[-\sinh(2\varphi)\left(C + \ln(\varepsilon/2\pi) - \int_\varepsilon^\infty dy\, y^{-1}\sinh(2\varphi g(y))\right)\right]; \tag{4.20}$$

(C here is Euler's constant).

It is also useful to define two cut-off-dependent and φ-dependent functions $A_1(u) \equiv A_1(u; \varphi g(y'))$ and $B_1(u)$:

$$A_1(u) = \int_0^\infty dy\, y^{-1}\sinh^2(\varphi g(y))\,[2\sin(\tfrac{1}{2}uy)]^2,$$

$$B_1(u) = -\tfrac{1}{2}\int_0^\infty dy\, y^{-1}\sinh(2\varphi g(y))\,[2\sin(\tfrac{1}{2}uy)]^2. \tag{4.21}$$

The even functions $A_1(u)$ and $B_1(u)$ vanish as $u\to 0$; for large $|u|$, they behave as

$$A_1(u) \sim \bar A + 2\sinh^2(\varphi)\ln(2\pi|u|) + O(|u|^{-1}),$$

$$B_1(u) \sim \bar B - \sinh(2\varphi)\ln(2\pi|u|) + O(|u|^{-1}). \tag{4.22}$$

Together with $\bar A$ and $\bar B$, they vanish in the non-interacting limit $\varphi\to 0$.

These quantities characterise the commutation algebra of the quantities $\phi_p(x, \varphi_q)$, which I henceforth write as $\phi_p(x)$, suppressing the explicit dependence on φ_q: in the limit $L \gg R$,

$$[\phi_p(x), \phi_{p'}(x')] = [\phi_p^\dagger(x), \phi_{p'}^\dagger(x')] = 0;$$
$$[\phi_p(x), \phi_p^\dagger(x')] = \lim_{\varepsilon\to 0^+} \{-\ln[1 - e^{-2\varepsilon}\exp(-2\pi i p(x-x')/L)]\}$$
$$-A_1(d(x-x')/R) + \bar{A} - 2\sinh^2(\varphi)\ln(R/L),$$
$$[\phi_p(x), \phi_{-p}^\dagger(x')] = -B_1(d(x-x')/R) + \bar{B} + \sinh(2\varphi)\ln(R/L). \tag{4.23}$$

Here $d(x) \equiv (L/\pi)|\sin(\pi x/L)|$ is the *chord* distance between points with separation x along the circumference of the ring of length L.

The necessary mechanism for calculation of correlation functions has now been established: the desired quantity must be constructed in terms of the fermion operators (4.19), and then manipulated into normal-ordered form in boson variables. The limit $L \to \infty$ can then be taken. As an example, the *electron* single-particle correlation function is easily constructed (using (3.5) to construct the electron field in terms of $\psi_p(x)$); the finite-temperature terms are easily evaluated using the familiar property that $\langle\exp(\alpha b^\dagger)\exp(\alpha' b)\rangle = \exp(\alpha\alpha'\langle b^\dagger b\rangle)$ if $H = \omega b^\dagger b$:

$$\langle\psi^\dagger(x)\psi(0)\rangle_{T=0} = (k_F/\pi)[\sin(k_F x)/(k_F x)]\exp[-A_1(|x|/R)]; \tag{4.24}$$
$$\langle\psi^\dagger(x),\psi(0)\rangle = \langle\psi^\dagger(x)\psi(0)\rangle_{T=0}\exp[-F(|x|)],$$
$$F(x) = \int_0^\infty dq\, q^{-1}[\exp(\beta\omega_q) - 1]^{-1}\cosh(2\varphi_q)[2\sin(\tfrac{1}{2}qx)]^2. \tag{4.25}$$

At $T = 0$ the familiar free-electron result is reduced at large separations by a factor $\exp(-\bar{A})|2\pi x/R|^{1-2\nu}$. At low but finite temperatures $T \ll v_S/R$, it is further reduced at separations $|x| \gg v_S/T \gg R$ by a factor $\exp(-2\nu|x|/\xi)$, where $\xi = (v_S\pi T)$. Note that when models with the same sound velocity v_S are compared, the single-particle correlation function of the interacting model is always reduced below that of the free model.

The recipe for such calculations of correlation functions was first given by Luther and Peschel (1974). The calculation is easily extended to give the dynamic correlation functions, as shown by these authors. In table 1 I summarise the low-energy properties of the spinless fermion Luttinger model, and list the static single-particle, density, and pair correlation functions. In the Luttinger model itself, the linear relation (3.6) between the electron field $\psi(x)$ and the fields $\psi_p(x)$ means that the single-particle correlation only has a k_F oscillatory component, while the density and pair correlations only have 0 and $2k_F$ components, just as in the case of the free Fermi gas. However, in a more general model where J is not strictly conserved, interaction effects will give rise to additional periodic components with extra multiples of $2k_F$ in the period. For example, in addition to the two components $\psi_p^\dagger(x)$ and $\psi_{-p}^\dagger(x)$ making up the operator representing the electron field $\psi^\dagger(x)$, there will be admixture of terms like $\psi_p^\dagger(x)\nabla\psi_p^\dagger(x)\psi_{-p}(x)$ which adds a $3k_F$ oscillatory term to the single-particle correlation function. Charge conservation allows terms with periodicity $(2m+1)k_F$ in the single-particle correlation function, and $2mk_F$ in the density and pair correlation functions, and the relevant terms are listed in table 1.

To conclude the discussion of the Luttinger model solution, I note that the low-energy properties of the diagonalised model depend on five distinct parameters: v_S, v_N and v_J parametrise the Hamiltonian, k_F the momentum operator, and φ the fermion field operator. A fundamental result is the relations $v_N = v_S\exp(-2\varphi)$, $v_J = v_S\exp(2\varphi)$, which were deduced from the structure of the solution. The question arises: are these relations fundamental, in that they can be deduced solely from the low-energy structure

Table 1. Summary of 'Luttinger liquid' properties of the spinless 1D Fermi gas. $[\psi_p(x)]^m$ means

$$\lim_{a\to 0} [a^{-\frac{1}{2}m(m-1)}\psi_p(x)\psi_p(x+a)\ldots\psi_p(x+(m-1)a)].$$

Higher harmonics of $2k_F$ allowed by charge conservation, and likely to be present in a more general model, are also included in the list of correlation functions. The phase (cos or sin) of the asymptotic oscillations is also indicated.

1. Interaction parameter (>1 for repulsive forces): $\exp(-2\varphi)$
2. Relation of Fermi vector k_F to charge density $\rho = N/L$: $k_F = \pi\rho$
3. Density fluctuation sound velocity: v_S
4. Change of chemical potential with Fermi vector: $v_N \equiv d\mu/dk_F = v_S e^{-2\varphi}$
5. Fermi velocity (for currents): $v_J = v_S e^{2\varphi}$
6. Asymptotic form of low-temperature correlation functions:

$$\langle A^\dagger(x)A(x')\rangle \sim \sum_i C_i \begin{Bmatrix} \cos(n_i k_F(|x-x'|) \\ \sin(n_i k_F|x-x'|) \end{Bmatrix} [|x-x'|^{-1}\exp(-\pi T|x-x'|/v_S)]^{\eta_i}.$$

Correlation	$A^\dagger(x)$	Luttinger model form	n	η
Single-particle	$\psi^\dagger(x)$	$\psi_p^\dagger(x)$	1	$\frac{1}{2}e^{-2\varphi} + \frac{1}{2}e^{2\varphi}$
(sin)		$[\psi_p^\dagger(x)]^{m+1}[\psi_{-p}(x)]^m$	$(2m+1)$	$\frac{1}{2}e^{-2\varphi} + 2(m+\frac{1}{2})^2 e^{2\varphi}$
Density	$[\psi^\dagger(x)\psi(x) - \rho]$	$\rho_p(x)$	0	2
(cos)		$[\psi_p^\dagger(x)]^m[\psi_{-p}(x)]^m$	$2m(\geqslant 2)$	$2m^2e^{2\varphi}$
Pair	$\psi^\dagger(x)\nabla\psi^\dagger(x)$	$\psi_p^\dagger(x)\psi_{-p}^\dagger(x)$	0	$2e^{-2\varphi}$
(cos)		$[\psi_p^\dagger(x)]^{m+1}[\psi_{-p}(x)]^{m-1}$	$2m$	$2e^{-2\varphi} + 2m^2e^{2\varphi}$

of the diagonalised form of the Hamiltonian, without reference to the 'bare' form of the model? The answer is yes: the relations (4.7) can be obtained by considering the static response functions of the density components $\Sigma_p \rho_{pq}$ and $\Sigma_p p\rho_{pq}$; when $q \neq 0$, the calculation only involves the boson variables, and v_S and $\exp(-2\varphi)$. In the limit $q \to 0$, the results must go over into the results $1/2\pi v_N$ and $1/2\pi v_J$ calculated when $q = 0$, and the relations (4.7) are recovered.

In addition to the above five characteristic parameters, various multiplicative factors appear in the asymptotic form of the various correlation functions. These depend only on the length scale R, and the two constants A and B; however, in contrast to (4.7), the relation between these various multiplicative factors is likely to be a model-dependent feature of the Luttinger model, as R, A and B depend on the high-energy structure of the model (i.e., the cut-off function $g(y)$).

5. Generalisation to non-soluble models: the 'Luttinger liquid' concept

The complete solubility of the Luttinger model makes it a fascinating example of an interacting one-dimensional system. Nevertheless, its solubility rests on quite specific properties that are lost if the model is modified. However, I will argue that its low-energy structure still provides a model of the most important features of more general, non-soluble models. As an example I consider a generalisation of the Luttinger model that incorporates a non-linear fermion dispersion relation:

$$\varepsilon(kp) = v_F(kp - k_F) + (1/2m)(kp - k_F)^2 + \lambda(1/12m^2 v_F)(kp - k_F)^3. \tag{5.1}$$

For stability reasons, it is necessary to include the cubic term: the ground state of the non-interacting model is altered unless $\lambda > \frac{3}{4}$, when $\mathrm{sgn}(\varepsilon(kp)) = \mathrm{sgn}(kp)$; $\varepsilon(kp)$ increases monotonically if $\lambda > 1$. In general, the interacting model will remain stable for λ greater than some positive limit λ_c. This modification of the model retains the feature that J is a good quantum number; though the non-linear dispersion means that the mean current operator j is no longer simply proportional to J, it remains so in a low-energy subspace.

The procedure for translating this generalised Luttinger model into normal-ordered boson form is extremely simple. An expansion technique as in equations (3.42)–(3.45) can be used to transcribe the non-linear fermion dispersion terms. The general fermion representation (4.19) should be used, with arbitrary parameter φ_q. The final result is a boson normal-ordered Hamiltonian with quadratic boson terms that depend on N and J, plus new cubic and quartic boson interaction terms. The parameter $\varphi_q(N/L, J/L)$ is then chosen to diagonalise the quadratic boson terms, giving a Luttinger model with N- and J-dependent parameters, *plus* irreducible boson interaction terms. The dependence of the Luttinger model parameters on N and J merely reflects the change in Fermi velocity for non-zero N and J, so in order to show up more clearly the other new feature (the boson–boson interaction), I give the new Hamiltonian only in the subspace $N = J = 0$; when $J \neq 0$, the structure of the boson spectrum is slightly altered in that $\varphi_q(N/L, J/L)$ and $\omega_q(N/L, J/L)$ are no longer even functions of q because the right- and left-travelling fermions then have different Fermi velocities. The boson part of the Hamiltonian has the form

$$H(N, J = 0) = \sum_q \omega_q b_q^\dagger b_q + \sum_p \frac{1}{2\pi}\int_0^L \mathrm{d}x[:(1/6m)(\Phi_p(x))^3 + (\lambda/48m^2 v_F)(\Phi_p(x))^4:]$$

$$\Phi_p(x) = \sum_p pq\alpha(pq, -\varphi_q)(\mathrm{e}^{iqx} b_q^\dagger + \mathrm{e}^{-iqx} b_q). \tag{5.2}$$

The colons :(...): mean boson normal-ordering. The parameters ω_q and φ_q are now given by modified versions of the expressions (4.4) and (4.5), where v_F has been replaced by $\tilde{v}_{Fq} = v_F + (\lambda/4m^2 v_F)(c_1 + \frac{1}{6}q^2)$; the equation for φ_q must be solved self-consistently, since the constant term c_1 itself depends on φ_q:

$$c_1 = \frac{2\pi}{L}\sum_q |q| \sinh^2(\varphi_q) \equiv A_1''(0)/R^2.$$

The constant c_1 exists provided the large-q behaviour of the fermion interaction matrix elements is sufficiently good for $yg(y)$ to vanish as $y \to \infty$. In fact, as will be seen, the requirement that the renormalisation of the ground state of the quadratic part of (5.2) by the boson interactions be finite imposes the stronger requirement $y^3 g(y) \to 0$ as $y \to \infty$. Assuming $V_1(q)$ does not diverge as $q \to \infty$, this implies the condition $qV_2(q) \to 0$ as $q \to \infty$, a slightly stronger condition than in the absence of a non-linear dispersion ($q^{1/2}V_2(q) \to 0$).

With the explicit construction (5.2) of the boson–boson interaction terms induced by a non-linear fermion dispersion, it is possible to construct an expansion in m^{-1} for the changes in the model properties due to the modification. This is particularly interesting in the case of the correlation functions: it allows the rigorous proof, at least for this type of generalised model, that the relations (4.7) between the spectral parameters v_S, v_N, and v_J and the parameter φ, and the relation between φ and the correlation exponents, remain unchanged from those found in the unmodified Luttinger model. This provides evidence in favour of the universal nature of these relations which will be proposed in this paper.

The relation between the spectral parameters is easiest to demonstrate; I give the form of the Hamiltonian in the subspace where no boson modes are excited:

$$H(n_q = 0) = \tfrac{1}{2}(\pi/L)(v_N N^2 + v_J J^2) + (1/6m)(\pi/L)^2(N^3 + 3NJ^2)$$
$$+ (\lambda/48m^2 v_F)(\pi/L)^3(N^4 + 6N^2J^2 + J^4);$$
$$v_N = \bar{v}_{F0} + V_1(0) + V_2(0); \qquad v_J = \bar{v}_{F0} + V_1(0) - V_2(0). \tag{5.3}$$

The relations (4.7) between v_N, v_J, v_S and φ are clearly unchanged. The stability condition giving the lower bound λ_c to allowed values of λ is clearly obtained by demanding that the ground state of (5.3) has $N = J = 0$. A necessary condition is that v_N and v_J are positive definite, i.e., that $|V_2(0)| < v_F + V_1(0) + (\lambda/4m^2 v_F)c_1(\lambda)$; since $c_1(\lambda)$ is positive, this is a less restrictive condition than that in the original Luttinger model with $m^{-1} = 0$. The condition $\lambda/v_F > \max(3/4v_N, 1/v_J)$ ensures (5.2) has no stationary points other than $N = J = 0$, and is sufficient to guarantee stability.

The effect of the non-linear dispersion on the correlation functions will now be discussed. I study the single-electron correlation function $\langle\psi^\dagger(x)\psi(0)\rangle_{T=0}$ (4.24) discussed earlier, as an example. Following that discussion, this is given (after a little manipulation) by

$$\langle\psi^\dagger(x)\psi(0)\rangle_{T=0} = (k_F/\pi)(k_F x)^{-1} \exp[-A_1(x/R)] \frac{1}{2i} \sum_p p \exp(ipk_F x)$$
$$\times \langle \exp[i\chi_p^\dagger(x)] \exp[i\chi_p(x)]\rangle$$
$$\chi_p(x) = \sum_q \alpha(pq, -\varphi_q)\, 2\sin(\tfrac{1}{2}qx) b_q. \tag{5.4}$$

$\langle\psi^\dagger\psi\rangle = (k_F/\pi)$, so the relation between k_F and electron density is unaffected by the non-linear dispersion. The expectation value is of course taken in the ground state of the interacting boson system (5.2), and hence differs from unity when m^{-1} is non-zero. A perturbation expansion in m^{-1} can be developed; the ground state expansion is

$$|\mathrm{GS}\rangle = \mathcal{N}\left(1 + \frac{1}{6mv_S} \sum_{q_1+q_2+q_3=0} f(q_1, q_2, q_3) b_{q_1}^\dagger b_{q_2}^\dagger b_{q_3}^\dagger + O(m^{-2})\right)|0\rangle$$

$$f(q_1, q_2, q_3) = \frac{L}{2\pi} \frac{q_1 q_2 q_3}{|q_1| + |q_2| + |q_3|} \sum_p p \prod_{i=1}^{3} \alpha(pq_i, -\varphi_{q_i}). \tag{5.5}$$

The normalisation constant $\mathcal{N}$ is given by

$$\mathcal{N} = 1 - \tfrac{1}{2}(1/mv_S R)^2 (L/2\pi R) c_2[\varphi g(y)] + O(m^{-4});$$
$$c_2 = \frac{R^3}{6} \frac{2\pi}{L} \sum_{q_1+q_2+q_3=0} f(q_1, q_2, q_3)^2 \tag{5.6}$$

$c_2[\varphi g(y)]$ is a positive dimensionless constant that is finite provided $y^3 g(y) \to 0$ as $y \to \infty$, as mentioned earlier:

$$c_2 = \int_0^\infty \mathrm{d}x \int_0^x \mathrm{d}y\, x^{-1}(x^2 - y^2)\, h(x, y)^2$$

$$h(x, y) = \tfrac{1}{2}[cg(x + y)cg(x - y)sg(2x) + sg(x + y)sg(x - y)cg(2x)] \tag{5.7}$$

where $cg(x)$ and $sg(x)$ are $\cosh(\varphi g(y))$ and $\sinh(\varphi g(y))$. Note that c_2 vanishes in the absence of fermion interactions ($\varphi = 0$), when there is no renormalisation of the ground state by the boson interactions.

I now calculate the single-electron correlation function to $O(m^{-1})$. From (5.4) and (5.5), this is given by

$$(k_F/\pi)\exp[-A_1(|x|/R)](k_F x)^{-1}\left[\sin(k_F x) - \cos(k_F x)\left(\tfrac{1}{6}\left\langle\sum_p p(\chi_p(x))^3\right\rangle + \mathrm{HC}\right) + O(m^{-2})\right];$$

$$\left\langle\sum_p p(\chi_p(x))^3\right\rangle = -(1/mv_S x)F(|x|/R) + O(m^{-2});$$

$$F(u) = 4\int_0^\infty dx \int_0^x dy\, x^{-1}\sin x\,(\cos x - \cos y)h(x/u, y/u)^2.$$

The function $F(u)$ vanishes at $u = 0$, and remains bounded as $u \to \infty$; the corrections to the correlation function thus do not affect the asymptotic behaviour of the correlation functions. Physically, this is because the factors $|q_i|^{1/2}$ in the interaction matrix elements of (5.2) kill the effects of the boson interactions at long wavelengths. The relation between the various correlation exponents and the parameter φ is thus identical to that in the original Luttinger model; the *value* of the parameter φ, on the other hand, *is* affected by the interaction terms, and varies with the ground state charge density.

6. Discussion: the Luttinger liquid concept

To summarise the results of this paper: it has been shown that the low-energy excitation of the soluble Luttinger model of interacting fermions in one dimension consists of three parts: the well known collective density fluctuation boson modes, plus charge and current excitations, which have not previously been emphasised. Associated with these three types of excitations are three velocities, v_S, v_N and v_J, which obey the relation $v_S = (v_N v_J)^{1/2}$. The current of the Luttinger model is a good quantum number, and is quantised in units $2v_J/L$, each unit carrying momentum $2k_F$. $v_N = d\mu/dk_F$ describes the rate of change of chemical potential with the Fermi vector, which is unrenormalised by interactions, and given by the charge density, $k_F = \pi(N/L)$; v_S is the density excitation sound velocity. The relation between the three velocities defines a parameter φ: $v_N = v_S\exp(-2\varphi)$, $v_J = v_S\exp(2\varphi)$. This parameter φ is the intrinsic renormalised coupling constant of the model, and determines the non-integer power laws characterising the asymptotic behaviour of the correlation functions. The elementary excitations of the Luttinger model are non-interacting, which explains why it can be explicitly solved. An important tool for working with the model and its generalisation is the representation of the fermion fields in terms of the elementary excitations: this is given here in a fully precise form.

A generalisation of the Luttinger model with a non-linear fermion dispersion, but where the current quantum number J is still conserved, was considered here. It was shown that the characteristic low-energy structure of the Luttinger model was preserved, including the relations between its velocities and correlation exponents, but that its renormalised parameters now depend on the position of the Fermi level, and non-linear couplings appear between the elementary excitations.

On the basis of this demonstration that this structure remains valid in a much wider class of models than the Luttinger model itself, I will propose that it is generally valid for conducting spinless fermion systems in one dimension. For full generality, it is necessary to consider models where the current quantum number J is no longer a good quantum number: this will be done in the next paper in this series. What emerges is that unless a multiple of the fundamental wavevector $2k_F$ is some multiple of a reciprocal lattice vector reflecting an underlying periodicity of the system, momentum conservation eventually inactivates a non-J-conserving term at low energies (though such terms will give rise to renormalisations of the low-energy spectral parameters), and the low-energy structure is again of the form described here. If $2k_F = (n/m)G$, this remains valid provided $\exp(-2\varphi)$ is less than a critical value $\frac{1}{2}m^2$, above which an instability against an insulating pinned charge-density-wave state occurs. If such Umklapp processes are present, but $\exp(-2\varphi) < \frac{1}{2}m^2$, there is a characteristic non-analytic scaling dependence of the renormalised $\exp(-2\varphi)$ on $|2k_F - (n/m)G|$, reflecting the power laws of the correlation functions.

A very important test of the universality of the Luttinger model structure is provided by the class of models exactly soluble by the Bethe *ansatz*, mentioned in the Introduction. For these models, v_S, v_N, and v_J can be explicitly calculated, though their correlation functions have not as yet been obtained. As described in Haldane (1981), the relation $v_S = (v_J v_N)^{1/2}$ can be explicitly verified, and the parameter $\exp(-2\varphi)$ obtained from these velocities shows the characteristic behaviour due to Umklapp processes when $2k_F \sim (n/m)G$ mentioned above, providing additional confirmation that the relation between $\exp(-2\varphi)$ and the correlation exponents is valid (Haldane 1980).

It is obviously possible to generalise the discussion to the case of spin-$\frac{1}{2}$ fermions; the spin-$\frac{1}{2}$ Fermi gas has a characteristic instability against a gap opening in the spin excitation spectrum in zero magnetic field, if $2k_F$ exchange (backscattering) processes are attractive (Luther and Emery 1974); the resulting state is the one-dimensional analogue of superconductivity, though no long-range order is involved, and can be related to the 1D Bose fluid. Similarly, when Umklapp processes open up a gap in the charge density excitation spectrum, leaving gapless low-energy spin-wave modes (Emery *et al* 1976), the resulting system models the antiferromagnetic chain. This in turn can be related to a ferromagnetic chain by a sublattice rotation. In this way, the apparently diverse collection of systems mentioned in the Introduction can be brought into the framework of what I propose to call 'Luttinger liquid theory', which can be tested on those models soluble by the Bethe *ansatz*. Of course, this description of these models is only valid in those regimes where they have a gapless linear density wave excitation, and are conductors of a locally conserved charge, with associated quantised persistent currents at $T = 0$. This underlying unity explains the rather bizarre fact that spin systems and Bose fluids in one dimension have the fermion-like property of a characteristic momentum $2k_F$, as seen in the equivalence of the $S = \frac{1}{2}$ XY spin chain and hard core bose lattice gas to a spinless fermion system (Lieb *et al* 1961, Matsubara and Matsuda 1956). These generalisations will be discussed in detail in subsequent papers.

The emphasis here has been on spectral properties and correlation functions. As a final comment, I note that the approach introduced here could be used as the basis of a theory of transport processes in 'Luttinger liquids'; for example, in the Luttinger model itself, transport of energy by the boson modes would be purely ballistic, since they are non-interacting. The boson interactions due to a non-linear fermion dispersion would introduce lifetime effects and dissipative behaviour.

References

Bethe H A 1930 *Z. Phys.* **71** 205
Dover C B 1968 *Ann. Phys., NY* **50** 500
Dzyaloshinskii I E and Larkin A I 1973 *Zh. Eksp. Teor. Fiz.* **65** 411 (Engl. Transl. 1974 *Sov. Phys.–JETP* **38** 202)
Emery V J, Luther A I and Peschel I 1976 *Phys. Rev.* B **13** 1272
Everts H V and Schulz H 1974 *Solid State Commun.* **15** 1413
Gradsteyn I S and Ryzhik I M 1965 *Tables of Integrals, Series, and Products* (New York: Academic Press)
Haldane F D M 1979 *J. Phys. C: Solid State Phys.* **12** 4791
—— 1980 *Phys. Rev. Lett.* **45** 1358
—— 1981 *Phys. Lett.* **81**A 153
Heidenreich R, Schroer B, Seiler R and Uhlenbrock D 1975 *Phys. Lett.* **54**A 119
Lieb E H, Schultz T and Mattis D C 1961 *Ann. Phys., NY* **16** 407
Luther A and Emery V J 1974 *Phys. Rev. Lett.* **33** 389
Luther A and Peschel I 1974 *Phys. Rev.* **B9** 2911
Luttinger J M 1963 *J. Math. Phys.* **15** 609
Matsubara T and Matsuda H 1956 *Prog. Theor. Phys.* **16** 569
Mattis D C 1974 *J. Math. Phys.* **15** 609
Mattis D C and Lieb E H 1965 *J. Math. Phys.* **6** 304
Overhauser A W 1965 *Physics* **1** 307
Theumann A 1967 *J. Math. Phys.* **8** 2460

Volume 81A, number 2,3 PHYSICS LETTERS 12 January 1981

DEMONSTRATION OF THE "LUTTINGER LIQUID" CHARACTER OF BETHE-ANSATZ-SOLUBLE MODELS OF 1-D QUANTUM FLUIDS

F.D.M. HALDANE

Institut Laue–Langevin, 38042 Grenoble Cedex, France

Received 31 October 1980

The parameters describing the excitation spectrum of Bethe-ansatz-soluble models of 1-D quantum fluids are shown to satisfy a relation $v_S = (v_N v_J)^{1/2}$ characteristic of "Luttinger liquids"; this identification permits exact calculation of a renormalised coupling determining correlation-function exponents.

Recent work on 1-D quantum systems, based on the soluble Luttinger [1,2] model of the spinless 1-D Fermi gas has led to a unified description of the low energy properties of the 1-D quantum fluid as those of a "Luttinger liquid" [3]. This development is based on the identification of a key structure of the low-energy spectrum of the Luttinger model, and the demonstration that it survives in more general, non-soluble models [3]. A test of a theory purporting to provide a general description of 1-D quantum fluids is provided by those models soluble by the Bethe ansatz (see ref. [4] for references). In this note, I show that these models quite generally have the "Luttinger liquid" structure in their low-energy spectrum, allowing identification of the exact value of a characteristic parameter determining the coupling-dependent exponents describing the power-law asymptotic decay of their correlation exponents. The similarity of 1-D Fermi and (interacting) Bose systems can be attributed to the fact that in one dimension particle exchange necessarily involves a collision, and phase changes of the wavefunction due to statistics and scattering cannot be separated.

Systems so far identified as "Luttinger liquids" include [5] (a) the 1-D Fermi fluid, (b) the 1-D Bose fluid, (c) axially symmetric spin chains (antiferromagnetic, and easy-plane ferromagnetic), and (d) the finite density gas of sine-Gordon topological solitons. In its various guises, the $S = 1/2$ Heisenberg–Ising–XY spin chain [4] provides examples of (a)–(c); a calculation of the "Luttinger liquid" parameters of this model using the Bethe ansatz formalism that solves it has already been reported [6]. Other Bethe-ansatz-soluble models for which such calculations are in progress are the Hubbard chain [7] and sine-Gordon/massive Thirring model [8].

The low-energy spectrum of a simple "Luttinger liquid" (without internal symmetries such as spin) has the form [3,6]

$$H = v_S \sum_{q\neq 0} |q| b_q^\dagger b_q + \tfrac{1}{2}(\pi/L)[v_N(N-N_0)^2 + v_J J^2] ,$$

$$P = [k_F + (\pi/L)(N-N_0)] J + \sum_{q\neq 0} q b_q^\dagger b_q ,$$

$$(-1)^J = \exp[i\theta(N)] , \quad dk_F/dN_0 = \pi/L . \qquad (1)$$

Here N is the number of particles of the fluid, which has periodic boundary conditions on a length L, and N_0/L is the ground state density, fixed by a chemical potential. The quantum number J controls the mean current $j = v_J(J/L)$, and for a given value of N, allowed values of J differ by an even integer; in fact it is only in the details of the N-dependence of the selection rule on allowed values of J that models with underlying Fermi and Bose statistics can be distinguished [5]. The boson creation operators $b_q^\dagger$ ($qL/2\pi = \pm 1, \pm 2, ...$) characterise approximately independent collective density fluctuation modes. There is a char-

Volume 81A, number 2,3 PHYSICS LETTERS 12 January 1981

acteristic momentum $2k_F$, which varies linearly with the fluid density, that is the momentum carried by the "quantum" of current excitation.

A fundamental result of "Luttinger liquid theory" [3] is a relation between the three spectral parameters v_S, v_N, and v_J (all velocities):

$$v_S = (v_N v_J)^{1/2} . \tag{2}$$

This allows the definition of a parameter φ where

$$v_N = v_S \exp(-2\varphi) , \quad v_J = v_S \exp(2\varphi) ; \tag{3}$$

it turns out that this parameter is the essential renormalised coupling parameter, and it alone controls the exponents of the correlation functions. For a given fluid, these exponents can be calculated in terms of φ using the "boson representation" of fermions [3, 9–11] developed for the Luttinger model. The Bose fluid and spin chain results can be obtained by identification with the spin–1/2 fermion system with gaps in its spin spectrum (due to backscattering processes) or in its charge spectrum (due to umklapp processes) [5]. The aim of this note is to (i) demonstrate relation (2) in Bethe-ansatz-soluble models, and (ii) give recipes for calculating the Luttinger liquid parameters v_S and φ.

The essential part of a Bethe-ansatz calculation [4] of energies of eigenstates is the calculation of a set of "pseudomomenta" k_i, $i = 1, ..., N$, where N is the number of particles in the fluid. These are determined through a set of quantum numbers I_i, with allowed values differing by an integer; the energy E and momentum P are determined by the k_i:

$$k_i L = 2\pi I_i - \sum_{j=1}^{N} \Theta(k_i, k_j) , \quad i = 1, ..., N ,$$

$$P = \sum_i k_i , \quad E = \sum_i \epsilon(k_i) . \tag{4}$$

These equations are appropriate for systems without internal symmetry, such as the Heisenberg–Ising–XY chain, the boson–Hubbard chain [12], and the sine-Gordon soliton gas, together with any continuum limits of these; the necessary generalisations for systems with internal symmetries are simple, but will not be dealt with here. The energy function $\epsilon(k)$ and the "phase shift" function $\Theta(p, q)$ (which depends on the interaction strength) have the symmetries

$$\epsilon(k) = \epsilon(-k) ; \quad \Theta(p, q) = -\Theta(q, p) = -\Theta(-p, -q) . \tag{5}$$

The notation $\Theta_p(p, q)$ and $\Theta_q(p, q)$ for the derivatives with respect to the left and right argument will be used. The symmetry (5) of $\Theta(p, q)$ in fact arises from a more specific structural property: a change of variable puts it into difference form:

$$\Theta(p, q) \equiv \Theta(\alpha(p) - \alpha(q)), \quad \Theta(-\alpha) = -\Theta(\alpha) ,$$

$$\alpha(-p) = -\alpha(p) . \tag{6}$$

Property (6) is vital for relation (2) to be obeyed.

In the ground state of the quantum fluid, the I_i are all different and consecutive, and the k_i are real, and lie in an interval $[-\Lambda, \Lambda]$. In the thermodynamic limit, their density $\rho(k_i) = 1/(k_{i+1} - k_i)L$ is given by a linear inhomogeneous Fredholm integral equation of the second kind:

$$2\pi\rho(k) = 1 + \int_{-\Lambda}^{\Lambda} dk' \, \Theta_p(k, k')\rho(k') . \tag{7}$$

The particle density and ground state energy per unit length are then given by

$$N/L = \int_{-\Lambda}^{\Lambda} dk \, \rho(k) , \quad E/L = \int_{-\Lambda}^{\Lambda} dk \, \epsilon(k)\rho(k) . \tag{8}$$

It will be useful to define two more solutions $\sigma(k)$, $\tau(k)$ of the integral equations:

$$2\pi\sigma(k) = \Theta(k, \Lambda) - \int_{-\Lambda}^{\Lambda} dk' \, \Theta_q(k, k')\sigma(k') , \tag{9}$$

$$2\pi\tau(k) = \Theta_q(k, \Lambda) - \int_{-\Lambda}^{\Lambda} dk' \, \Theta_q(k, k')\tau(k') . \tag{10}$$

The ground state density N/L increases monotonically with Λ; the change in the system when this is increased is found by calculating $\delta\rho(k)/\delta\Lambda$, after which some manipulation is found to be given by

$$\delta\rho(k)/\delta\Lambda = \rho(\Lambda)[1 - \sigma(\Lambda) + \sigma(-1)]^{-1} \times (d/dk)[\sigma(k) - \sigma(-k)] . \tag{11}$$

Volume 81A, number 2,3 PHYSICS LETTERS 12 January 1981

Then

$$\delta(N/L)/\delta\Lambda = 2\rho(\Lambda)[1-\sigma(\Lambda)+\sigma(-\Lambda)]^{-1} \quad (12)$$

and the chemical potential $\mu \equiv \delta(E/L)/\delta(N/L)$ is found to be

$$\mu = \epsilon(\Lambda) - \int_{-\Lambda}^{\Lambda} \mathrm{d}k\, \epsilon'(k)\sigma(k)\,. \quad (13)$$

The sound wave velocity is calculated by comparing the energy of a state with quantum numbers $I_i = I_i^0$ $(i \neq N)$, $I_N = I_N^0 + m$, $1 \ll m \ll N$, with that of the ground state with quantum numbers I_i^0. The momentum of the excited state is $\Delta P = 2\pi m/L$, and $v_S = \Delta E/\Delta P$. Easy manipulations lead to

$$v_S = [2\pi\rho(\Lambda)]^{-1}\left[\epsilon'(\Lambda) - \int_{-\Lambda}^{\Lambda} \mathrm{d}k\, \epsilon'(k)\tau(k)\right]. \quad (14)$$

The parameter v_N is calculated from the relation $\delta\mu/\delta(N/L) = \pi v_N$. This requires $\delta\sigma(k)/\delta\Lambda$ which is found to be given by

$$[1-\sigma(\Lambda)]\,\tau(k) - \sigma(-\Lambda)\tau(-k)\,.$$

After a little manipulation, v_N can be expressed as

$$v_N = [1-\sigma(\Lambda)+\sigma(-\Lambda)]^2 v_S\,. \quad (15)$$

To make a current carrying state with quantum number $J \neq 0$, one chooses $I_i = I_i^0 + \frac{1}{2}J$. The relevant integral equation describing the change in the distribution of the k_i is easily obtained. To lowest order in J, the range of the k_i shifts from $[-\Lambda, \Lambda]$ to $[-\Lambda+\Lambda', \Lambda+\Lambda']$, where

$$\delta(J/L)/\delta\Lambda' = 2\rho(\Lambda)[1-\sigma(\Lambda)-\sigma(-\Lambda)]^{-1}\,. \quad (16)$$

v_J is given by $E/L \sim \frac{1}{2}\pi v_J (J/L)^2$ as $J \to 0$, and can be expressed as

$$v_J = [1-\sigma(\Lambda)-\sigma(-\Lambda)]^2 v_S\,. \quad (17)$$

For the Luttinger liquid relation (2) to be obeyed, one requires

$$X[\sigma] \equiv [1-\sigma(\Lambda)-\sigma(-\Lambda)]\,[1-\sigma(\Lambda)+\sigma(-\Lambda)] = 1\,. \quad (18)$$

Note that given only the symmetry properties (5) used so far, this is *not* provable: the eigenfunctions of the kernel of (9) fall into two classes, even and odd; the first factor in (18) can be expressed in terms of the even eigenfunctions, and the second factor in terms of the odd eigenfunctions. The even and odd sectors are in general unrelated, so (18) is not in general true.

The crucial extra fact about the kernel is the property (6): there is the change of variables $k \to \alpha$ that puts it into difference form:

$$2\pi\sigma(\alpha) = \Theta(\alpha-\alpha_0) + \int_{-\alpha_0}^{\alpha_0} \mathrm{d}\alpha'\, \Theta'(\alpha-\alpha')\sigma(\alpha')\,, \quad (19)$$

$\alpha_0 \equiv \alpha(\Lambda)$. The proof of (18) follows from the demonstration that (i) $\sigma(\pm\Lambda) \to 0$ as $\Lambda \to 0$, so $X = 1$ when $\Lambda = 0$, and (ii) $\delta x/\delta\Lambda = 0$. The first requirement is evidently satisfied by (9), since $\Theta(0,0)$ vanishes. The second is demonstrated by noting that

$$(\mathrm{d}/\mathrm{d}\alpha_0)\sigma(\pm\alpha_0) = \delta\sigma(\pm\alpha_0)/\delta\alpha_0 \pm \partial\sigma(\pm\alpha_0)/\partial\alpha\,,$$

where

$$\delta\sigma(\alpha)/\delta\alpha_0 = [1-\sigma(\alpha_0)]\,\tau(\alpha) + \sigma(-\alpha_0)\tau(-\alpha)\,,$$

$$\partial\sigma(\alpha)/\partial\alpha = [1-\sigma(\alpha_0)\tau(\alpha) - \sigma(-\alpha_0)\tau(-\alpha)$$

Substitution into the expression for $\delta x/\delta\Lambda$ gives the desired result.

The expressions (12) and (16) can thus be written as

$$\delta(N/L)/\delta\Lambda = 2\rho(\Lambda)\,\mathrm{e}^{\varphi}\,, \quad \delta(J/L)/\delta\Lambda' = 2\rho(\Lambda)\mathrm{e}^{-\varphi}\,. \quad (20)$$

In the Bethe-ansatz-soluble models, the dimensionless coupling φ thus reveals itself most characteristically in the variation of the charge and current quantum number densities with small changes of the "Fermi pseudomomenta" at $k = \pm\Lambda$.

[1] J.M. Luttinger, J. Math. Phys. 4 (1963) 1154.
[2] D.C. Mattis and E.H. Lieb, J. Math. Phys. 6 (1965) 304.
[3] F.D.M. Haldane, "Luttinger liquid theory" of one-dimensional quantum fluids (I), preprint (1980), submitted to J. Phys. C.
[4] C.N. Yang and C.P. Yang, Phys. Rev. 150 (1966) 312.
[5] F.D.M. Haldane, to be published.
[6] F.D.M. Haldane, Phys. Rev. Lett. 45 (1980) 1358.
[7] E.H. Lieb and F.Y. Wu, Phys. Rev. Lett. 20 (1968) 1445.
[8] H. Bergknoff and H.B. Thacker, Phys. Rev. D19 (1979) 3666.
[9] D.C. Mattis, J. Math. Phys. 8 (1974) 609.
[10] A. Luther and I. Peschel, Phys. Rev. B9 (1974) 2911.
[11] R. Heidenreich, B. Schroer, R. Seiler and D. Uhlenbrock, Phys. Lett. 54A (1975) 119.
[12] F.D.M. Haldane, Phys. Lett. 80A (1980) 281.

J. Phys. A: Math. Gen. 24 (1991) L549–L553. Printed in the UK

LETTER TO THE EDITOR

Integrable spin chain with two- and three-particle interactions

R Z Bariev

The Kazan Physico-Technical Institute of the Academy of Sciences of the USSR, Kazan 420029, USSR

Received 21 February 1990

Abstract. The one-dimensional model which consists of two isotropic XY chains with spins $S=\frac{1}{2}$ coupled by three-spin interactions is considered with the help of the Bethe ansatz. It is shown that the diagonalization of the Hamiltonian can be reduced to solving a set of coupled nonlinear equations. The exact solution of these equations corresponding to the ground state of system is obtained. The model considered exhibits a ground state of Anderson type with the finite magnetization along the quantization axis.

The one-dimensional quantum integrable models were intensively investigated in recent years. The investigation of these models was begun sixty years ago in the well known paper by Bethe [1, 2] concerned with the calculation of the wavefunctions for the one-dimensional Heisenberg magnet with pair interactions and spin $S=\frac{1}{2}$.

The one-dimensional Hubbard model is the other well known example of the integrable lattice system. The exact solution for the wavefunctions and ground state energy of this model was obtained by Lieb and Wu [3]. These authors used the exact solution of the one-dimensional gas of particles with repulsion given by Yang in 1967 [4]. Further development in the theory of integrable models was connected with the generalization of the Bethe ideas and their application to more complicated problems concerned with the study of systems with spin $S>\frac{1}{2}$ [5, 6]. In many cases the models with pair interactions were investigated. Of certain interest is the consideration of Bethe ideas applied to the systems in which both pair and many-particle interactions are present.

In the present letter we consider a spin one-half chain with two- and three-particle interactions. In our case these interactions are competing ones. It leads to the frustrated ground state with the finite magnetization. In this sense the ground state under consideration can be interpreted as the RVB state of Anderson type [7].

The Hamiltonian of the considering model has the following form:

$$H=-\tfrac{1}{2}\sum_{j=1}^{N}\sum_{\tau=1,2}(\sigma^x_{j(\tau)}\sigma^x_{j+1(\tau)}+\sigma^y_{j(\tau)}\sigma^y_{j+1(\tau)})(1-U\sigma^z_{j+\tau-1(\tau+1)}) \tag{1}$$

where $\sigma^\alpha_{j(\tau)}(\alpha=x,y,z)$ are Pauli spin matrices of the jth lattice site on the sublattice τ $(\tau=1,2,\sigma^\alpha_{j(3)}=\sigma^\alpha_{j(1)})$. We used the periodical boundary condition $(\sigma^\alpha_{N+1(\tau)}=\sigma^\alpha_{1(\tau)})$. The Hamiltonian possesses the obvious symmetry with respect to the change $U\Rightarrow-U$. Therefore in the following we restrict ourselves to the investigation of the case $U>0$. The case $U=0$ corresponds to the model of two non-interacting isotropic XY chains [8]. In the case $U\to\infty$ we obtain the modified XY chain [9]. By using the Jordan–Wigner

0305-4470/91/100549+05$03.50

transformation [8] the Hamiltonian (1) can be presented in terms of the creation and annihilation operators up to the boundary conditions

$$H=-\sum_{j=1}^{N}\sum_{\tau=1,2}(a_{j(\tau)}^{+}a_{j+1(\tau)}+a_{j+1(\tau)}^{+}a_{j(\tau)})(1-Va_{j+\tau-1(\tau+1)}^{+}a_{j+\tau-1(t+1)}) \quad (2)$$

where $V=2U/(1+U)$.

This Hamiltonian can be interpreted as follows. There is a one-dimensional lattice which consists of two sublattices (see figure 1). The particles move along the sites of each sublattice so that only the jumps between the neighbouring sites of the same sublattice are possible. The interaction between the sublattices means that the energy of the jump between the sites j and $j+1$ depends on whether the site in the other sublattice which corresponds to this pair is occupied or vacant. The presence of such three-site interactions contrasts the given model with the one-dimensional Hubbard model [3] in the Hamiltonian of which the four-fermion interaction is the two-site one. Thus we obtain an interesting problem of statistical mechanics in the formulation of secondary quantization operators as well.

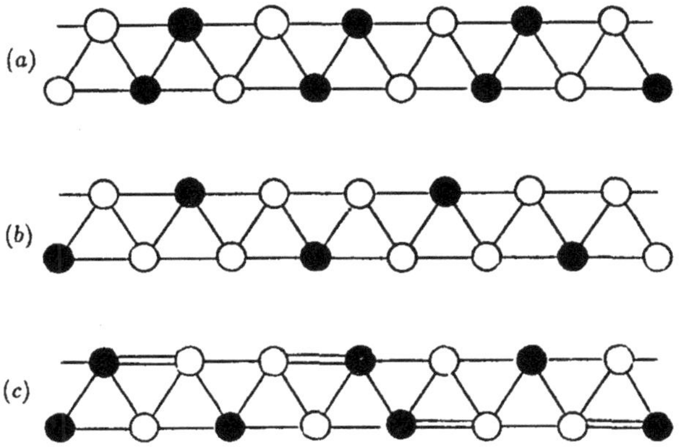

Figure 1. Symmetrical configurations of the ground state for (*a*) $U=0$; (*b*) $U=1$; (*c*) $0<U<1$.

Let us seek the amplitude of the wavefunction of the Hamiltonian (1) which corresponds to the state with m spins directed upward (↑) in the sites with coordinates $x_1<x_2<\ldots<x_m$ of the first sublattice and with $(n-m)$ spins ↑ in the sites $x_{m+1}<x_{m+2}<\ldots<x_n$ of the second sublattice in the form of the generalized Bethe ansatz [3, 4, 6, 10]

$$f(x_1,\ldots,x_n)=\sum A_{P_1\ldots P_n}^{\tau_{Q_1}\ldots\tau_{Q_n}}\prod_{j=1}^{n}\exp\{ik_{P_j}[x_{Q_j}+\tfrac{1}{2}(\tau_{Q_j}-1)]\}. \quad (3)$$

The summation in this expression is carried out over all permutations $[P_1,\ldots,P_n]$ of the numbers $1,2,\ldots,n$. The permutation $[Q_1,\ldots,Q_n]$ of the numbers $1,2,\ldots,n$ is such that $1\leqslant x_{Q_1}\leqslant x_{Q_2}\leqslant\ldots\leqslant x_{Q_n}$, so that $\tau_{Q_j}=1,2$ for $Q_j\leqslant m$ and $Q_j>m$ respectively. The eigenvalue of the Hamiltonian (1) appropriate to this wavefunction is

$$E=-2(1+U)\sum_{j=1}^{n}\cos k_j. \quad (4)$$

The wavefunction which was constructed with the help of coefficients (3) is the eigenfunction of the Hamiltonian (1) if these coefficients satisfy the following conditions

$$A^{\dots\tau_1\tau_2\dots}_{\dots P_1P_2\dots} = \sum_{\tau,\tau'=1,2} S^{\tau'\tau}_{\tau_1\tau_2}(k_{P_1}-k_{P_2})A^{\dots\tau\tau'\dots}_{\dots P_2P_1\dots} \tag{5}$$

$$A^{\tau_1\dots\tau_n}_{P_1\dots P_n} = A^{\tau_2\dots\tau_n\tau_1}_{P_2\dots P_nP_1}\exp(ik_{P_1}N) \tag{6}$$

where the non-vanishing elements of S-matrix are

$$S^{11}_{11}(k)=S^{22}_{22}(k)=-1 \qquad S^{12}_{12}(k)=S^{21}_{21}(k)=\sin\tfrac{1}{2}k/\sin(\tfrac{1}{2}k-i\alpha)$$

$$S^{12}_{21}(k)=S^{21}_{12}(k)=i sh\alpha/\sin(\tfrac{1}{2}k-i\alpha) \qquad e^{\alpha}=(1+U)/(1-U).$$

A necessary and sufficient condition for the compatibility of equation (5) is the fulfilment of the Yang-Baxter relations [4, 5]. In our case the S-matrix satisfies these relations and we may use the quantum method of the inverse problem [11] to solve equations (5), (6). As result, we have

$$\exp(ik_jN)=(-1)^{n-m+1}\prod_{\beta=1}^{m}\frac{\sin[\frac{1}{2}(k_j-\Lambda_\beta)+i\alpha']}{\sin[\frac{1}{2}(k_j-\Lambda_\beta)-i\alpha']} \qquad \alpha'=\alpha/2$$

$$(-1)^n\prod_{j=1}^{n}\frac{\sin[\frac{1}{2}(\Lambda_\beta-k_j)+i\alpha']}{\sin[\frac{1}{2}(\Lambda_\beta-k_j)-i\alpha']}=\prod_{\substack{\gamma=1\\ \gamma=\beta}}^{m}\frac{\sin[\frac{1}{2}(\Lambda_\beta-\Lambda_\gamma)+2i\alpha']}{\sin[\frac{1}{2}(\Lambda_\beta-k_j)-2i\alpha']}.$$

Taking the logarithm of both sides of these equations, we obtain

$$k_jN+\sum_{\beta=1}^{m}\theta(k_j-\Lambda_\beta,\alpha')=2\pi I_j \qquad (j=1,2,\dots,n)$$

$$\sum_{j=1}^{n}\theta(\Lambda_\beta-k_j,\alpha')-\sum_{\gamma=1}^{m}\theta(\Lambda_\beta-\Lambda_\gamma,2\alpha')=2\pi J_\beta \qquad (\beta=1,2,\dots,m)$$

$$\theta(k,\alpha')=2\arctan[\coth(\alpha')\tanh\tfrac{1}{2}k] \qquad -\pi\leqslant\theta(k,\alpha)\leqslant\pi \tag{7}$$

where I_j and J_β are half-integer (integer) numbers for even (odd) n and m, respectively.

Thus, we have obtained the energy spectrum and eigen-vectors of the Hamiltonian under consideration through the solution of the system of equations (7). In the present letter we restrict ourselves with the investigation of the ground state, which corresponds to the following values of I_j and J_β

$$I_j=j-(n+1)/2 \qquad (j=1,2,\dots,n)$$

$$J_\beta=\beta-(m+1)/2 \qquad (\beta=1,2,\dots,m)$$

A momentum of this state is

$$P=\sum_{j=i}^{n}k_j=2\pi\left(\sum_{j=i}^{n}I_j+\sum_{\beta=i}^{m}J_\beta\right)N^{-1}=0.$$

Following the pioneering works on Bethe ansatz [5, 6], we assume that in the thermodynamical limit $N\to\infty$, $n\to\infty$, $m\to\infty$ for fixed ratios n/N and m/N the values of k_j and Λ_β fill the intervals $[-Q, Q]$ and $[-B, B]$ uniformly with densities $\rho(k)$ and $\sigma(\Lambda)$, respectively. Then, instead of (7), we obtain the system of integral equations

$$2\pi\rho(k)=1+\int_{-B}^{B}\theta'(k-\Lambda;\alpha')\sigma(\Lambda)\,d\Lambda \qquad \theta'(k;\alpha')=\sinh\alpha(\cosh\alpha-\cos k)^{-1}$$

$$2\pi\sigma(\Lambda)=\int_{-Q}^{Q}\theta'(\Lambda-k;\alpha')\rho(k)\,dk-\int_{-B}^{B}\theta'(\Lambda-\Lambda';2\alpha')\sigma(\Lambda')\,d\Lambda' \tag{8}$$

$$\int_{-Q}^{Q}\rho(k)\,dk=n/N \qquad \int_{-B}^{B}\sigma(\Lambda)\,d\Lambda=m/N.$$

Similarly, going over to the continuous distribution in (4) we obtain

$$E=-2N(1+U)\int_{-Q}^{Q}\cos k\rho(k)\,dk. \tag{9}$$

From the symmetry of the system it is clear that in the ground state $m=n/2$, and it corresponds to $B=\pi$. Then, excluding the function $\sigma(\Lambda)$ equations (8), we obtain

$$2\pi\rho(k)-\int_{-Q}^{Q}\varphi(k-k')\rho(k')\,dk'=1 \qquad y=1-\int_{-Q}^{Q}\rho(k)\,dk$$
$$\varphi(k)=\tfrac{1}{2}+2\sum_{n=1}^{\infty}\cos(nk)/[1+\exp(2|\alpha|n)]. \tag{10}$$

These equations determine the ground state energy (9) as a function of the magnetization y. The analysis of (10) shows that the function $E(y)$ has a wonderful peculiarity, namely, this function has an absolute minimum for $y=0$, in this case $Q=\pi/2$

$$y_0=1-\int_{-\pi/2}^{\pi/2}\rho_0(k)\,dk \qquad \frac{1}{N}E_0=-2(1+U)\int_{-\pi/2}^{\pi/2}\rho_0(k)\cos k\,dk.$$

The function $\rho_0(k)$ satisfies equations (10) at $Q=\pi/2$. The solution of this equation can be obtained with the help of numerical integration or using the perturbation theory. For example, at large α we have

$$y_0=\frac{1}{3}-\frac{32}{9\pi^2}e^{-2|\alpha|}+O(e^{-4|\alpha|})$$

$$\frac{1}{N}E_0=-2(1+U)\left[\frac{4}{3\pi}+\frac{2}{3\pi}\left(1+\frac{8}{3\pi}\right)e^{-2|\alpha|}+O(e^{-4|\alpha|})\right].$$

The presence of this minimum shows that in the ground state in the zero field the considering systems has a finite magnetization along the z axis. The value of this magnetization is determined by the value of the interaction constant U (or V for the model (2)).

Let us interpret the obtained result using Anderson's picture of the ground state, the so-called resonating-valence-bond (RVB) state [7]. For the vanishing interaction $\alpha=0$ ($U=0$) in the ground state we have equal numbers of spins ↑ and ↓. The symmetrical configuration is presented in figure 1(a). This ground state can be envisaged as a linear combination of wavefunctions built up of singlet pairs (SP). We may think of it as a liquid-type state in which the system of pair bonds is resonating between the possible configurations. In the other limiting case $\alpha\to\infty$($U\to 1$) the number of spins ↑ is two times smaller than the number of spins ↓ (see figure 1(b)). The ground state is a superposition of states of triangles in which there are one spin ↑ and two spins ↓, i.e. these triangles (MT) have a magnetic momentum. In the general case ($0 < U < 1$) the ground state is the linear combination of SP and MT. In the symmetrical configuration MT is surrounded by an interchangeable pair (IP); in figure 1(c) IP is shown by a double line. States of spins in this pair can be interchanged without breaking short-range order. The interchanging of the spin states in IP leads to the moving of MT a distance of two lattice constants. The MT moves along the lattice with the transfer of the magnetic momentum without change of the energy of system. Thus the ground state may be presented as a magnetic liquid or as a conducting liquid if we consider the Hamiltonian (2).

Of course, this interpretation is preliminary and naive. For more detailed understanding of the model it is necessary to calculate the excitation spectrum and it will be the subject of a subsequent investigation.

Part of this work was performed during the short stay of the author at the University of Maryland, USA. He would like to thank Professor M E Fisher for the invitation and his hospitality during the visit. In addition, the author is grateful to the Scientific Council on High Temperature Superconductivity (USSR) for financial support under grant no. 344.

References

[1] Bethe H 1931 *Z. Phys.* **71** 205
[2] Yang C N and Yang C P 1966 *Phys. Rev.* **150** 321
[3] Lieb E H and Wu F Y 1968 *Phys. Rev. Lett.* **20** 1445
[4] Yang C N 1967 *Phys. Rev. Lett.* **19** 1312
[5] Baxter R J 1982 *Exactly Solved Models in Statistical Mechanics* (New York: Academic)
[6] Gaudin M 1983 *La Fonction d'Onde de Bethe* (Paris: Masson)
[7] Fazekas P and Anderson P W 1971 *Phil. Mag.* **30** 423
[8] Lieb E, Schultz T and Mattis D 1961 *Ann. Phys.* **16** 407
[9] Suzuki M 1971 *Prog. Theor. Phys.* **46** 1337
[10] Bariev R Z 1981 *Theor. Math. Phys.* **49** 1021
[11] Takhtajan L A and Faddeev L D 1979 *Russ. Math. Surveys*

J. Phys. A: Math. Gen. 26 (1993) 1249–1257. Printed in the UK

Exact solution of a one-dimensional model of hole superconductivity*

R Z Bariev†, A Klümper, A Schadschneider and J Zittartz

Institut für Theoretische Physik, Universität zu Köln, Zülpicher Strasse 77, D-5000 Köln, Federal Republic of Germany

Received 4 September 1992, in final form 7 December 1992

Abstract. A new integrable model of a strongly correlated electronic system is formulated as a model of hole superconductivity. The model is solved by using the Bethe ansatz. The critical exponents describing the decrease of correlation functions on long distances are derived. The behaviour of these correlations indicates that Cooper pairs of holes are formed in the repulsive region of the model. This conclusion is also confirmed by the calculation of the conductivity and the effective transport mass. In the attractive region, the model is a highly conducting system in which the current carriers with small effective mass are 'light fermions'.

1. Introduction

The discovery of high-temperature superconductivity [1] has greatly stimulated the interest in low-dimensional electronic systems with strong correlations. Among the relevant models the one-dimensional Hubbard [2] and the (supersymmetric) t-J models [3–7] are special since they can be treated exactly in terms of the Bethe ansatz. As exact results are highly desirable, particularly for low-dimensional systems in the strong correlation regime, these models have been extensively studied. The physical motivation for considering the Hubbard and the t-J model is the fact that the motion of electrons is strongly influenced by the on-site Coulomb repulsion and by the spin fluctuations through antiferromagnetic coupling, respectively.

Another approach to high-temperature superconductivity proposed by Hirsch [8, 9] makes it possible to formulate a new integrable model of strongly correlated systems. According to [9] the charge carriers of high-temperature superconductors are holes. The kinetic energy of hole hopping between nearest-neighbour sites depends on the occupation of these sites. In such a model the repulsive Coulomb interaction for electrons leads to an attractive interaction for holes which is strongest at low densities of holes.

In the following we shall consider a simplified version of Hirsch's model on a one-dimensional chain of even length L, closed periodically. It is given by the

* Work performed within the research program of the Sonderforschungsbereich 341, Köln–Aachen–Jülich.

† Permanent address: The Kazan Physico-Technical Institute of the Russian Academy of Sciences, Kazan 420029, Russia.

0305-4470/93/051249+09$07.50

Hamiltonian

$$H(t) = -\sum_{j=1}^{L} \sum_{\sigma=\pm1} (c_{j\sigma}^{+} c_{j+1\sigma} + c_{j+1\sigma}^{+} c_{j\sigma})(1 + t\, n_{j+(1+\sigma)/2,-\sigma}) \qquad (1)$$

with interaction parameter t, where $c_{j\sigma}$ is the annihilation operator for an electron with spin σ at site j and $n_{j\sigma} = c_{j\sigma}^{+} c_{j\sigma}$. For comparison we present the Hamiltonian $\tilde{H}$ of the original Hirsch model [9]

$$\tilde{H} = -\sum_{j=1}^{L} \sum_{\sigma=\pm1} (c_{j\sigma}^{+} c_{j+1\sigma} + c_{j+1\sigma}^{+} c_{j\sigma})[1 + \tfrac{1}{2}t(n_{j+(1+\sigma)/2,-\sigma} + n_{j+(1-\sigma)/2,-\sigma})]. \qquad (2)$$

Obviously our Hamiltonian (1) only contains half of the interaction terms of $\tilde{H}$. The advantage of model (1) is that it is solvable by the Bethe ansatz whereas model (2) is not, as a direct calculation shows that its S-matrix does not satisfy the Yang–Baxter equations [10, 11]. These are satisfied for $\tilde{H}$ only in the continuum limit, i.e. in the limit of low or high densities of particles. The last case is the most important one from the physical point of view, because here the existence of hole pairs of Cooper type can be expected. A direct comparison of the S-matrices of Hamiltonians (1) and (2) shows that the continuum versions of both models coincide. Therefore it can be expected that both models have identical critical properties. Nevertheless a proof of this claim, based for instance on the renormalization group approach, is highly desirable.

In any case, model (1) keeps the main idea of Hirsch's approach to superconductivity, namely the modulation of the hopping process by the presence of other particles as the main reason for superconductivity. Hamiltonian (1) includes the simplest form of terms describing such processes. The various arguments to support the possibility that such terms could overwhelm the direct Coulomb repulsion and may lead to superconductivity have been given in [8, 9]. This is reason enough to study the superconducting properties of model (1). In [12] another model for interacting fermions including correlated hopping terms was constructed. The ground state of this system can be given explicitly in any dimension, in one dimension the system is integrable. The model of [12] and (1) are related, the latter one however enjoys a simpler physical interpretation.

In [13] one of the authors has shown that model (1) is solvable by using the Bethe ansatz. The relevant equations have been derived and the ground-state energy has been calculated. In the present paper we shall mainly consider the correlation functions of the model in order to investigate the possibility of superconductivity.

As a model for electrons the interaction term in (1) should be negative, i.e. $t < 0$, corresponding to Coulomb repulsion. If, according to [9] this leads to an attraction of holes, this fact must be seen in the behaviour of the hole correlation functions. The corresponding Hamiltonian for holes is obtained from (1) by a particle-hole transformation U together with a sublattice rotation [14], namely $c_{j\sigma}^{+} \Rightarrow (-1)^{j} c_{j\sigma}$. Multiplying also with a suitable scale factor, $(1+t)^{-1}$, the resulting hole Hamiltonian $H(t')$

$$H(t') = (1+t)^{-1} U H(t) U^{-1} \qquad t' = -\frac{t}{1+t} \qquad (3)$$

is then of the same form as (1). Now, however, the t'-interaction is positive. As a consequence we shall study model (1) with attractive interaction, i.e. with $t > 0$, regarding its correlation functions as hole correlation functions of the repulsive model.

2. The Bethe ansatz

The Bethe ansatz for the model has been formulated in [13], from which we quote the relevant equations. The energy eigenstates are characterized by sets of wave numbers k_j for the particles and additional parameters Λ_α. Each of the latter ones is related to a particle with down spin. The Bethe ansatz wave numbers k_j and Λ_α satisfy a set of nonlinear equations derived in [13]

$$Lk_j = 2\pi I_j + \sum_{\beta=1}^{M} \Theta\left(k_j - \Lambda_\beta; \frac{\eta}{2}\right) \qquad j = 1, \ldots, N$$
$$\sum_{j=1}^{N} \Theta\left(\Lambda_\alpha - k_j; \frac{\eta}{2}\right) - \sum_{\beta=1}^{M} \Theta(\Lambda_\alpha - \Lambda_\beta; \eta) = 2\pi J_\alpha \qquad \alpha = 1, \ldots, M \tag{4}$$

with the phase shift function

$$\Theta(k;\eta) = 2\tan^{-1}\left(\coth\eta \tan \tfrac{1}{2}k\right) \qquad -\pi \leqslant \Theta < \pi \tag{5}$$

and the interaction parameter $\eta = \ln(1+t)$. Furthermore N is the total number of particles, M is the number of particles with down spin, and I_j and J_α are integers or half-odd integers depending on the parities of N and M. The energy and momentum of the corresponding state are given by

$$E = -\sum_{j=1}^{N} 2\cos k_j + \mu N$$
$$P = \sum_{j=1}^{N} k_j = \frac{2\pi}{L}\left(\sum_{j=1}^{N} I_j + \sum_{\alpha=1}^{M} J_\alpha\right) \tag{6}$$

where, from now on, the chemical potential μ has been added to control the particle number.

Equations (4) and (6) hold regardless of the sign of η, nevertheless the structure of the solutions is very different for $\eta < 0$ and $\eta > 0$. In [13] the model was considered for $\eta < 0$. Here we treat (4) for positive η in the symmetric case when there are equally many particles with spin up and spin down ($M = N/2$). The eigenstates consist of a certain number of singlet bound pairs and a certain number of free particles. The bound pairs are characterized by pairs of complex wavenumbers $k^\pm$

$$k_\alpha^\pm = \Lambda_\alpha \pm i\eta. \tag{7}$$

In the ground state we only have pairs. Using (7) the above equations are reduced to the following set of equations after some simple algebra

$$2L\Lambda_\alpha = 2\pi J_\alpha + \sum_{\beta=1}^{M} \Theta(\Lambda_\alpha - \Lambda_\beta; \eta)$$

$$E = \sum_{\alpha=1}^{M} \epsilon_0(\Lambda_\alpha) \qquad \epsilon_0(\Lambda) = 2\mu - 4\cosh\eta\cos\Lambda \tag{8}$$

$$P = 2\sum_{\alpha=1}^{M} \Lambda_\alpha .$$

The ground state is characterized by the following values of J_α

$$J_\alpha^0 = \alpha - (M+1)/2 \qquad (\alpha = 1, 2, \ldots, M). \tag{9}$$

Deviations from this distribution of J_α describe gapless excitations of particle-hole type. In the thermodynamic limit $L \to \infty$, $M \to \infty$ for fixed ratio M/L the values of Λ_α fill an interval $[-\Lambda_0, \Lambda_0]$ uniformly with density $\sigma(\Lambda)$. From (8) we then obtain the integral equation for the distribution function $\sigma(\Lambda)$

$$2\pi\sigma(\Lambda) + \int_{-\Lambda_0}^{\Lambda_0} \Theta'(\Lambda - \tilde{\Lambda}; \eta)\sigma(\tilde{\Lambda})\, \mathrm{d}\tilde{\Lambda} = 2 \qquad \Theta'(\Lambda, \eta) = \frac{\sinh 2\eta}{\cosh 2\eta - \cos\Lambda} \tag{10}$$

with the subsidiary condition

$$\int_{-\Lambda_0}^{\Lambda_0} \sigma(\Lambda)\, \mathrm{d}\Lambda = \rho \tag{11}$$

where $2\rho = 2M/L$ is the density of the hole liquid. For fixed chemical potential the parameter Λ_0 must be chosen to minimize the ground-state energy, given by

$$\begin{aligned} E_0/L &= \int_{-\Lambda_0}^{\Lambda_0} \epsilon_0(\Lambda)\sigma(\Lambda)\, \mathrm{d}\Lambda \\ &= \frac{1}{\pi}\int_{-\Lambda_0}^{\Lambda_0} \epsilon(\Lambda)\, \mathrm{d}\Lambda \end{aligned} \tag{12}$$

where in the second representation the dressed energy $\epsilon(\Lambda)$ has been used which is the solution of the integral equation

$$\epsilon(\Lambda) + \frac{1}{2\pi}\int_{-\Lambda_0}^{\Lambda_0} \Theta'(\Lambda - \tilde{\Lambda}; \eta)\epsilon(\tilde{\Lambda})\, \mathrm{d}\tilde{\Lambda} = \epsilon_0(\Lambda) \tag{13}$$

such that $\epsilon(\pm\Lambda_0) = 0$ which is the minimization condition. The solution of (13) also defines the energy of particle-hole excitations. The momentum of such an excitation is given by

$$P(\Lambda) = 2\Lambda - \int_{-\Lambda_0}^{\Lambda_0} \Theta(\Lambda - \tilde{\Lambda}; \eta)\sigma(\tilde{\Lambda})\, \mathrm{d}\tilde{\Lambda}. \tag{14}$$

The only other type of excitation consists of broken pairs. The breaking-up of one bound pair leads to the creation of two free particles. The energies and momenta of these particles are obtained from (4) and (6), for instance in the case when there are $M-1$ bound pairs and two particles with real momenta $k_{1,2}$. The energy of each free particle with momentum k is given by

$$\epsilon_{\mathrm{f}}(k) = \mu - 2\cos k - \frac{1}{2\pi}\int_{-\Lambda_0}^{\Lambda_0} \Theta'\left(\Lambda - k; \frac{\eta}{2}\right) \epsilon(\Lambda)\, \mathrm{d}\Lambda . \tag{15}$$

We content ourselves with pointing out that this excitation has a gap at $k = 0$. A detailed study of this type of excitations for a more general model will be presented in [15].

In order to study the ground-state correlations of the model, we use two different approaches. First, following [16, 17], we calculate the critical exponents and determine the long-distance behaviour of the two-point correlation functions. Secondly, we investigate the conductivity of the model as a function of the particle density. This approach is based on the calculation of the ground-state energy under twisted boundary conditions [18, 19].

3. Critical exponents of the correlation functions

To obtain the critical exponents of the correlation functions we use the predictions of conformal field theory [20, 21]. According to this theory there is a one-to-one correspondence between the conformal dimensions of the scaling operators and the finite-size corrections to the energy of the excited states of the critical Hamiltonian. Our model is critical since the gapless excitations have a linear dispersion law in the vicinity of the Fermi points. The excitations corresponding to the breaking-up of bound pairs have a gap as mentioned before. These excitations do not affect the critical properties and the finite-size corrections. The finite-size corrections to the gapless excitations (11) can be calculated in a straightforward way [22, 23]. Omitting the details of the calculation, we only present the results.

We denote the change of the number of bound pairs as ΔM and the number of pairs moved from the left to the right Fermi point as d. The $1/L$-corrections to the low-energy excitations are then

$$\Delta E = \frac{2\pi v_{\mathrm{F}}}{L}\left(\frac{(\Delta M)^2}{[2\xi(\Lambda_0)]^2} + [\xi(\Lambda_0)]^2 d^2 + N^+ + N^-\right) \tag{16}$$

where $\xi(\Lambda_0)$ is the dressed charge [24] at the Fermi surface, the dressed charge function $\xi(\Lambda)$ being defined through the modified integral equation (9)

$$\xi(\Lambda) + \frac{1}{2\pi}\int_{-\Lambda_0}^{\Lambda_0} \Theta'(\Lambda - \widetilde{\Lambda}; \eta)\xi(\widetilde{\Lambda})\, \mathrm{d}\widetilde{\Lambda} = 1 \tag{17}$$

such that $\xi \equiv \pi\sigma$ in the present case. v_{F} is the Fermi velocity

$$v_{\mathrm{F}} = \frac{\epsilon'(\Lambda_0)}{2\pi\sigma(\Lambda_0)} = \frac{\epsilon'(\Lambda_0)}{2\xi(\Lambda_0)} . \tag{18}$$

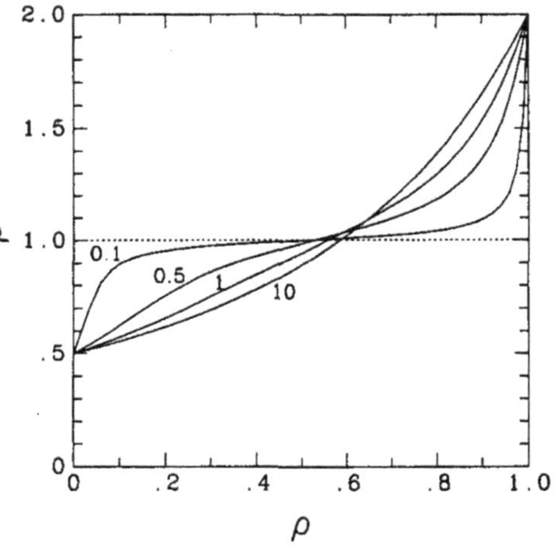

Figure 1. Depiction of the exponent β of the pair correlation function for different values of the interaction parameter $\eta = 0.1,\ 0.5,\ 1,\ 10$.

The non-negative integers $N^{\pm}$ are quantum numbers of the simple particle-hole excitations. The momentum associated with these excitations is

$$P = 2k_F d + \frac{2\pi}{L}\left(d\Delta M + N^+ - N^-\right) \qquad k_F = \pi\frac{M}{L}. \tag{19}$$

To read off the conformal dimensions $\Delta_{\pm}$ from these expressions one has to compare (16) and (19) with the predictions of conformal field theory [25, 26]. Then we have

$$\Delta_{\pm} = \frac{1}{2}\left(\frac{\Delta M}{2\xi(\Lambda_0)} \pm \xi(\Lambda_0)d\right)^2 + N^{\pm}. \tag{20}$$

We notice that in contrast to the repulsive case [27] this formula is valid for integer d without selection rule for fermions, since in the present case it corresponds to bound pairs. It is a reason why we consider the particle correlation function of the model with an attractive interaction as a hole correlation function in the repulsive case.

Now we consider the long-distance behaviour of correlation functions. The asymptotic form of the density correlation function is given by

$$\langle\rho(r)\rho(0)\rangle \simeq \rho^2 + A_1 r^{-2} + A_2 r^{-\alpha}\cos(2k_F r). \tag{21}$$

The non-oscillating part arises from the lowest particle-hole excitations. The relevant excitation for the $2k_F$ oscillation term is $(\Delta M, d, N^{\pm}) = (0,1,0)$. We thus find the critical exponent

$$\alpha = 2(\Delta_+ + \Delta_-) = 2[\xi(\Lambda_0)]^2. \tag{22}$$

The excitations relevant for the correlation function of singlet pairs are specified by $(\Delta M, d,\ N^{\pm}) = (1,0,0)$. We then obtain the asymptotic behaviour of this correlation

$$G_p(r) = \langle c^+_{r\uparrow} c^+_{r\downarrow} c_{0\downarrow} c_{0\uparrow}\rangle \simeq B r^{-\beta} \tag{23}$$

where

$$\beta = \frac{1}{2[\xi(\Lambda_0)]^2} = \alpha^{-1}. \tag{24}$$

The exponent β is plotted in figure 1 for some values of the interaction parameter η by numerically solving equations (10), (11) or (17), respectively, from which $\xi(\Lambda_0)$ is determined.

Let us discuss the results. In one-dimensional systems we have no superconductivity in the literal sense. However, the power-decay of the singlet pair correlation (23) indicates the existence of singlet pairs provided that the exponent β of this correlation is smaller than that of the density–density correlation α [16]. In this case the correlation of singlet pairs overwhelms the density correlation, and we can say that the particles are confined in pairs. From figure 1 we see that such behaviour always exists for particle concentrations $\rho \leqslant \rho_c$. The critical concentration ρ_c is defined by $\beta(\rho_c) = 1$ and varies monotonically from $1/2$ to $2-\sqrt{2} = 0.5858\ldots$ with increasing interaction η.

We remark that these results for the model with attractive interaction can be applied to the model with repulsive interaction by a particle-hole transformation. Therefore we have Cooper type singlet pairs of holes in the model of repulsive electrons.

4. Conductivity and effective transport mass

In order to substantiate the physical picture given above we now study the conductivity and the effective transport masses which can be calculated following the ideas of [18, 19]. To this end we change the periodic boundary condition leading to (8) by a twisted one with twisting angle φ. In this case instead of (8) we obtain

$$2L\Lambda_\alpha = 2\pi I_\alpha + 2\varphi + \sum_{\beta=1}^{M} \Theta(\Lambda_\alpha - \Lambda_\beta; \eta). \tag{25}$$

Physically the additional phase φ can be obtained by enclosing a magnetic flux in the ring on which the electrons can move. For small φ this leads to a change in the ground-state energy

$$\Delta E_0 = D_c \varphi^2 / L \tag{26}$$

where D_c is the charge stiffness. The conductivity of the system is directly proportional to D_c [18, 19]. In order to see the correlation effects clearly it is useful to introduce the effective transport mass m defined by the relation

$$\frac{m}{m_e} = \frac{D_c^0}{D_c} \tag{27}$$

where $D_c^0 = \frac{2}{\pi}\sin(\pi\rho)$ is the charge stiffness of the non-interacting system and m_e is the electron mass. On the other hand this change in the boundary conditions corresponds to the finite-size correction (16) for $\Delta M = 0$, $d = \varphi/\pi$ and we have for the charge stiffness of the Hamiltonian (1) with attractive interactions ($t > 0$)

$$D_c^> = \frac{2}{\pi} v_F \xi^2(\Lambda_0) = \frac{1}{\pi}\epsilon'(\Lambda_0)\xi(\Lambda_0) \tag{28}$$

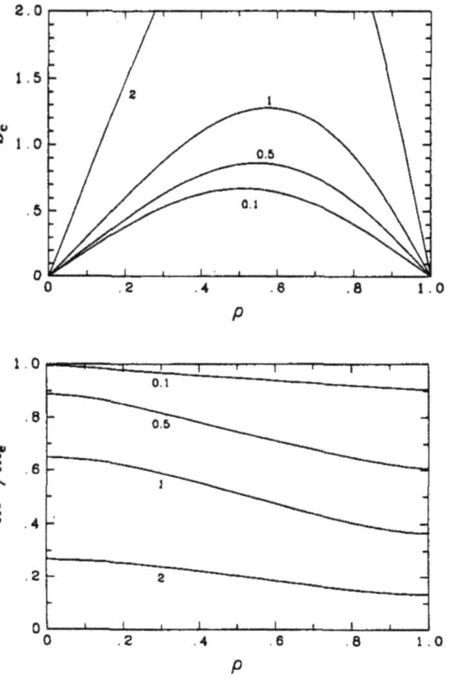

Figure 2. Dependence of charge stiffness $D_c^>$ and effective transport mass $m^>$ on ρ for different interactions η = 0.1, 0.5, 1, 2. Note that ρ is the density of particles in the attractive model.

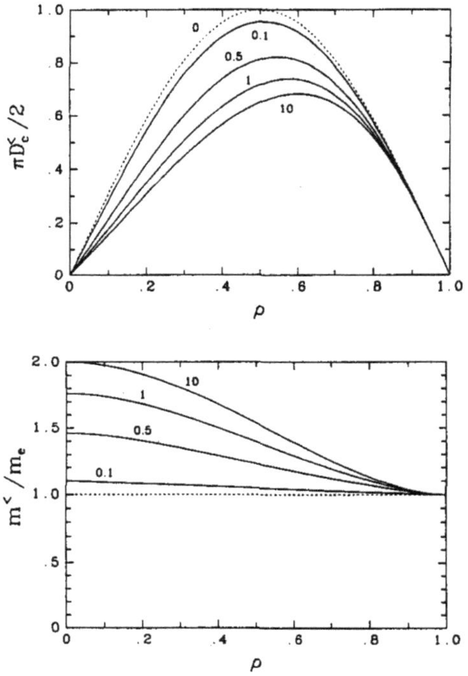

Figure 3. Depiction of charge stiffness $D_c^<$ and effective transport mass $m^<$ after rescaling using (29). Note that in this repulsive case ρ is the density of holes.

and with repulsive interactions ($t < 0$)

$$D_c^< = e^{-\eta} D_c^> = \frac{2}{\pi} v_F \xi^2(\Lambda_0) e^{-\eta}. \tag{29}$$

The result (29) can also be obtained by using results of [27] where the repulsive model has been investigated directly. In figures 2 and 3 we present the conductivity and effective transport mass as a function of the particle concentration for the Hamiltonian (1) with attractive and repulsive interactions, respectively.

We begin with the discussion of the conductivity of the repulsive model. From figure 3 it is clear that the conductivity in the high-density limit vanishes linearly as the concentration decreases ($D_c^< \sim 1 - \rho$). This is simply due to the decrease of the carrier density just as for the non-interacting case. It indicates that in this region the current carriers are the free electrons of the repulsive model. It is noteworthy that in the low-density limit the effective mass is enhanced by a factor of two. This behaviour and our previous findings for the correlation functions can be interpreted as the formation of hole pairs due to an attractive force between holes.

In the high-density limit of the attractive model we also observe a linear decrease $D_c^> \sim 1 - \rho$ (figure 2). In this case the free particles which carry the current are holes. For all densities ρ and interactions η the effective masses are reduced in comparison to the non-interacting case. Furthermore the masses decrease with increasing interaction parameter t ($m^{>}/m_e \sim t^{-1}$). As the effective mass of the current carriers become very small, one may call them 'light fermions'.

Acknowledgments

One of the authors (RZB) gratefully acknowledges the hospitality of the Institut für Theoretische Physik der Universität zu Köln as well as partial financial support by the Scientific Council on High Temperature Superconductivity (Russia) under grant No 91-151. He also thanks Professor D I Khomsky for discussions.

References

[1] Bednorz J G and Müller K A 1986 *Z. Phys.* B **64** 189
[2] Lieb E H and Wu F Y 1968 *Phys. Rev. Lett.* **20** 1445
[3] Anderson P W 1987 *Science* **235** 1196
[4] Zhang F C and Rice T M 1988 *Phys. Rev.* B **37** 3759
[5] Sutherland B 1975 *Phys. Rev.* B **12** 3795
[6] Schlottmann P 1987 *Phys. Rev.* B **36** 5177
[7] Bares P A and Blatter G 1990 *Phys. Rev. Lett.* **64** 2567
[8] Hirsch J E 1989 *Phys. Lett.* A **134** 451
[9] Hirsch J E 1989 *Physica* **158C** 326
[10] Yang C N 1967 *Phys. Rev. Lett.* **19** 1312
[11] Baxter R J 1982 *Exactly Solved Models in Statistical Mechanics* (New York: Academic)
[12] Essler F H L, Korepin V E and Schoutens K 1992 *Phys. Rev. Lett.* **68** 2960; 1993 *Phys. Rev. Lett.* **70** 73
[13] Bariev R Z 1991 *J. Phys. A: Math. Gen.* **24** L549
[14] Sutherland B *Exactly Solved Problems in Condensed Matter and Relativistic Field Theory (Lecture Notes in Physics 242)* ed B S Shastry, S S Jha and V Sigh (Berlin: Springer) p 1
[15] Bariev R Z, Klümper A, Schadschneider A and Zittartz J 1993 in preparation
[16] Bogoliubov N M and Korepin V E 1989 *Int. J. Mod. Phys.* B **3** 427
[17] Frahm H and Korepin V E 1990 *Phys. Rev.* B **42** 10553
[18] Shastry B S and Sutherland B 1990 *Phys. Rev. Lett.* **66** 243
[19] Kawakami N and Yang S-K 1991 *Phys. Rev.* B **44** 7844
[20] Belavin A A, Polyakov A M and Zamolodchikov A B 1984 *Nucl. Phys.* B **241** 333
[21] Cardy J L 1986 *Nucl. Phys.* B **270** 186
[22] Woynarovich F, Eckle H-P and Truong T T 1989 *J. Phys. A: Math. Gen.* **22** 4027
[23] Woynarovich F 1989 *J. Phys. A: Math. Gen.* **22** 4243
[24] Korepin V E 1979 *Teor. Math. Fiz.* **41** 169
[25] Bogoliubov N M, Izergin A G and Reshetikhin N Y 1986 *Z. Eksp. Teor. Fiz. Pis. Red.* **44** 405
[26] Berkovich A and Murthy G 1988 *J. Phys. A: Math. Gen.* **21** L381
[27] Bariev R Z 1992 *J. Phys. A: Math. Gen.* **25** L261

VOLUME 68, NUMBER 19 PHYSICAL REVIEW LETTERS 11 MAY 1992

New Exactly Solvable Model of Strongly Correlated Electrons Motivated by High-T_c Superconductivity

Fabian H. L. Essler, Vladimir E. Korepin, and Kareljan Schoutens
Institute for Theoretical Physics, State University of New York at Stony Brook, Stony Brook, New York 11794-3840
(Received 19 February 1992)

We present a new model describing strongly correlated electrons on a general d-dimensional lattice. It is an extended Hubbard model and it contains the t-J model as a special case. The model naturally describes local electron pairs, which can move coherently at arbitrary momentum. By using an η-pairing mechanism we can construct eigenstates of the Hamiltonian with off-diagonal long-range order. In the attractive case the exact ground state is superconducting in any number of dimensions. On a one-dimensional lattice, the model is exactly solvable by Bethe ansatz.

PACS numbers: 71.20.Ad, 75.10.Jm

The study of strongly correlated electrons on a lattice is an important tool in theoretical condensed-matter physics in general, and in the study of high-T_c superconductivity in particular. Two well-studied models are the Hubbard model and the t-J model. On a one-dimensional lattice these models are both exactly solvable by Bethe ansatz. In this Letter we propose a new model, which is again solvable in one dimension, and which combines and extends some of the interesting features of the Hubbard model and the t-J model.

Electrons on a lattice are described by operators $c_{j,\sigma}$, $j=1,\ldots,L$, $\sigma=\uparrow,\downarrow$, where L is the total number of lattice sites. These are canonical Fermi operators with anticommutation relations given by $\{c_{i,\sigma}^\dagger,c_{j,\tau}\}=\delta_{i,j}\delta_{\sigma,\tau}$. The state $|0\rangle$ (the Fock vacuum) satisfies $c_{i,\sigma}|0\rangle=0$. At a given lattice site i there are four possible electronic states:

$$|0\rangle,\quad |\uparrow\rangle_i=c_{i,\uparrow}^\dagger|0\rangle,$$
$$|\downarrow\rangle_i=c_{i,\downarrow}^\dagger|0\rangle,\quad |\uparrow\downarrow\rangle_i=c_{i,\downarrow}^\dagger c_{i,\uparrow}^\dagger|0\rangle. \tag{1}$$

By $n_{i,\sigma}=c_{i,\sigma}^\dagger c_{i,\sigma}$ we denote the number operator for electrons with spin σ on site i and we write $n_i=n_{i,\uparrow}+n_{i,\downarrow}$. The spin operators $S=\sum_{j=1}^L S_j$, $S^\dagger$, and S^z,

$$S_j=c_{j,\uparrow}^\dagger c_{j,\downarrow},\quad S_j^\dagger=c_{j,\downarrow}^\dagger c_{j,\uparrow},\quad S_j^z=\tfrac{1}{2}(n_{j,\uparrow}-n_{j,\downarrow}), \tag{2}$$

form an SU(2) algebra and they commute with the Hamiltonians that we consider below. (We shall always give local expressions $\mathcal{O}_j$ for symmetry generators, implying that the global ones are obtained as $\mathcal{O}=\sum_{j=1}^L\mathcal{O}_j$.)

The Hubbard model Hamiltonian can be written as

$$H^{\text{Hubbard}}=-\sum_{\langle jk\rangle}\sum_{\sigma=\uparrow,\downarrow}(c_{j,\sigma}^\dagger c_{k,\sigma}+c_{k,\sigma}^\dagger c_{j,\sigma})+U\sum_{j=1}^L(n_{j,\uparrow}-\tfrac{1}{2})(n_{j,\downarrow}-\tfrac{1}{2}), \tag{3}$$

where the first summation runs over all nearest-neighbor pairs $\langle jk\rangle$. It contains kinetic (hopping) terms for the electrons and an on-site interaction term for electron pairs. An interesting feature (on a bipartite periodic lattice) is the so-called η-pairing symmetry [1,2], which involves operators η_H, $\eta_H^\dagger$, and η_H^z which form another SU(2) algebra, and which commute with the Hamiltonian (3). Using this symmetry one can, starting from an eigenstate $|\psi\rangle$ of the Hamiltonian, create a new eigenstate $\eta_H^\dagger|\psi\rangle$, which contains an additional local electron pair of momentum π. The spin SU(2) algebra (2) and the η-pairing SU(2) algebras together form an SO(4) symmetry algebra. In one dimension, the Hubbard model is solvable by Bethe ansatz [3].

In the t-J model, there is a kinematical constraint which forbids the occurrence of two electrons on the same lattice site. On this restricted Hilbert space the t-J Hamiltonian (with $t=1$, $J=2$) acts as $H^{t\text{-}J}=-\sum_{\langle jk\rangle}H_{j,k}^{t\text{-}J}$, with

$$H_{j,k}^{t\text{-}J}=\sum_{\sigma=\uparrow,\downarrow}(Q_{j,\sigma}^\dagger Q_{k,\sigma}+Q_{k,\sigma}^\dagger Q_{j,\sigma})-2[S_j^zS_k^z+\tfrac{1}{2}(S_j^\dagger S_k+S_jS_k^\dagger)-\tfrac{1}{2}(1-n_j-n_k)-\tfrac{1}{4}n_jn_k], \tag{4}$$

where we defined

$$Q_{j,\uparrow}=(1-n_{j,\downarrow})c_{j,\uparrow},\quad Q_{j,\downarrow}=(1-n_{j,\uparrow})c_{j,\downarrow}, \tag{5}$$

and the operators S_j^z, S_j, and $S_j^\dagger$ are as in (2). The t-J model (4) is supersymmetric and the spin SU(2) symmetry algebra gets enlarged to the superalgebra SU(1|2) [4,5] (see [6] for the description and classification of the classical Lie superalgebras). The generators of this symmetry algebra are S, $S^\dagger$, S^z, $Q_\uparrow$, $Q_\uparrow^\dagger$, $Q_\downarrow$, $Q_\downarrow^\dagger$, and $T=2L-\sum_{j=1}^L n_j$. In one dimension the supersymmetric t-J model (4) is exactly solvable by Bethe ansatz [7–9].

Before we present the Hamiltonian of the new model, we give some motivation, which is based on what we know about the materials that exhibit high-T_c superconductivity. It has been found that the electrons in these materials form "Cooper pairs," which are spin singlets, and that these pairs are much smaller than in the traditional superconductors. As a limiting case one can consider models which have electron pairs of size zero, i.e., pairs that are localized on single lattice sites. We will call such localized electron pairs *localons*.

In the t-J model localons are ruled out by the kinematical constraint on the space of states, and in the Hubbard model only local pairs of momentum π exist. Below we shall see that in our new model localons can move coherently with arbitrary momentum. Apart from these local pairs, the new model may also have bound states that are finite-size electron pairs.

Let us now present the Hamiltonian of the new model on a general d-dimensional lattice. We write it as

$$H = H^0 + U\sum_{j=1}^{L}(n_{j,\uparrow}-\tfrac{1}{2})(n_{j,\downarrow}-\tfrac{1}{2}) + \mu\sum_{j=1}^{L} n_j + h\sum_{j=1}^{L}(n_{j,\uparrow}-n_{j,\downarrow}), \tag{6}$$

where H^0 is given by

$$H^0 = -\sum_{\langle jk\rangle} H^0_{j,k}, \tag{7}$$

$\langle j,k\rangle$ are nearest neighbors, with

$$\begin{aligned} H^0_{j,k} = {} & c^\dagger_{k,\uparrow}c_{j,\uparrow}(1-n_{j,\downarrow}-n_{k,\downarrow}) + c^\dagger_{j,\uparrow}c_{k,\uparrow}(1-n_{j,\downarrow}-n_{k,\downarrow}) + c^\dagger_{k,\downarrow}c_{j,\downarrow}(1-n_{j,\uparrow}-n_{k,\uparrow}) + c^\dagger_{j,\downarrow}c_{k,\downarrow}(1-n_{j,\uparrow}-n_{k,\uparrow}) \\ & + \tfrac{1}{2}(n_j-1)(n_k-1) + c^\dagger_{j,\uparrow}c^\dagger_{j,\downarrow}c_{k,\downarrow}c_{k,\uparrow} + c_{j,\downarrow}c_{j,\uparrow}c^\dagger_{k,\uparrow}c^\dagger_{k,\downarrow} - \tfrac{1}{2}(n_{j,\uparrow}-n_{j,\downarrow})(n_{k,\uparrow}-n_{k,\downarrow}) \\ & - c^\dagger_{j,\downarrow}c_{j,\uparrow}c^\dagger_{k,\uparrow}c_{k,\downarrow} - c^\dagger_{j,\uparrow}c_{j,\downarrow}c^\dagger_{k,\downarrow}c_{k,\uparrow} + (n_{j,\uparrow}-\tfrac{1}{2})(n_{j,\downarrow}-\tfrac{1}{2}) + (n_{k,\uparrow}-\tfrac{1}{2})(n_{k,\downarrow}-\tfrac{1}{2}). \end{aligned} \tag{8}$$

This Hamiltonian contains kinetic terms and interaction terms that combine those of the Hubbard and of the t-J model. The second term in (6) is the on-site Hubbard interaction term (notice that it also gets a contribution from H^0). The third and fourth terms in (6) introduce a chemical potential μ and a magnetic field h. Roughly speaking, the new model can be viewed as a modified Hubbard model with additional nearest-neighbor interactions similar to those in the t-J model.

The Hamiltonian H^0 is invariant under spin-reflection $c_{j,\uparrow}\leftrightarrow c_{j,\downarrow}$ and under particle-hole replacement $c^\dagger_{j,\sigma}\leftrightarrow c_{j,\sigma}$. In addition to the spin SU(2) generators (2), the following operators commute with H^0.

η-pairing SU(2).—The generators are η, $\eta^\dagger$, and η^z:

$$\eta_j = c_{j,\uparrow}c_{j,\downarrow}, \quad \eta^\dagger_j = c^\dagger_{j,\downarrow}c^\dagger_{j,\uparrow}, \quad \eta^z_j = -\tfrac{1}{2}n_j + \tfrac{1}{2}. \tag{9}$$

Together with the spin SU(2) algebra (2), this gives an SO(4) algebra which is similar to the one for the Hubbard model. This symmetry makes it possible to generalize the η-pairing mechanism, which was developed for the Hubbard model in [2], to the new model.

Supersymmetries.—There are eight supersymmetries in total: $Q_\uparrow$, $Q^\dagger_\uparrow$, $Q_\downarrow$, and $Q^\dagger_\downarrow$ given in (5) and the operators $\tilde{Q}_\sigma$ and $\tilde{Q}^\dagger_\sigma$:

$$\tilde{Q}_{j,\uparrow} = n_{j,\downarrow}c_{j,\uparrow}, \quad \tilde{Q}_{j,\downarrow} = n_{j,\uparrow}c_{j,\downarrow}. \tag{10}$$

These generators, together with the operator $\sum_{j=1}^{L} 1$ (which is constant and equal to L), form the superalgebra SU(2|2). [Like SU(4), this algebra has fifteen generators, eight of which are fermionic. In the fundamental representation, the generators can be represented as 4×4 supermatrices with vanishing supertrace [6].]

The symmetries of the Hamiltonian H^0 can be made manifest as follows. We first add one more generator to the symmetry algebra, which is

$$X = \sum_{j=1}^{L} X_j, \quad X_j = (n_{j,\uparrow}-\tfrac{1}{2})(n_{j,\downarrow}-\tfrac{1}{2}), \quad [H^0, X] = 0. \tag{11}$$

This extends the superalgebra SU(2|2) to U(2|2). We denote the generators of this algebra by J_α, where $\alpha = 1,2,\ldots,16$. We now introduce an invariant, nondegenerate two-index tensor, denoted by $K^{\alpha\beta}$, which is the inverse of $K_{\alpha\beta} = \mathrm{str}(J_\alpha J_\beta)$ (str denotes supertrace), where the J_α are 4×4 supermatrices in the fundamental representation. Using this, we can cast $H^0_{j,k}$ in a group-theoretical form, as follows:

$$H^0_{j,k} = \sum_{\alpha,\beta=1}^{16} K^{\alpha\beta} J_{j,\alpha} J_{k,\beta} \tag{12}$$

$$\begin{aligned} = {} & \sum_{\sigma=\uparrow,\downarrow} (Q^\dagger_{j,\sigma}Q_{k,\sigma} + Q^\dagger_{k,\sigma}Q_{j,\sigma} - \tilde{Q}^\dagger_{j,\sigma}\tilde{Q}_{k,\sigma} - \tilde{Q}^\dagger_{k,\sigma}\tilde{Q}_{j,\sigma}) \\ & + (2\eta^z_j\eta^z_k + \eta^\dagger_j\eta_k + \eta_j\eta^\dagger_k) \\ & - (2S^z_jS^z_k + S^\dagger_jS_k + S_jS^\dagger_k) + X_j + X_k. \end{aligned} \tag{13}$$

It is easily checked that this expression agrees with the formula for $H^0_{j,k}$ in (8). The expression (12) immediately makes it clear that H^0 commutes with all sixteen generators of U(2|2).

We would like to stress that the appearance of the algebra U(2|2) in the model is not too surprising: On each lattice site there are four electronic states (1), two of which are fermionic. The supergroup U(2|2) is the group of all unitary rotations of these four states into one another. Our Hamiltonian H^0 has been chosen such that it commutes with the entire algebra U(2|2) and is therefore very natural. The analogous construction for U(1|2) leads to the t-J Hamiltonian (4), and for U(2) it leads to the spin-$\frac{1}{2}$ XXX Heisenberg model.

The spectrum of the Hamiltonian H^0 is symmetric around zero. This follows from the discrete symmetry $c^\dagger_{j,\downarrow}\leftrightarrow c_{j,\downarrow}$, for which $H^0 \leftrightarrow -H^0$.

There is a further aspect of H^0 that deserves to be mentioned: The terms $H^0_{j,k}$ act as graded permutations of the electron states (1) at sites j and k. By "graded" we mean that there is an extra minus sign if the two states that are permuted are both (fermionic) single-electron states. For example,

$$H^0_{j,k}c^\dagger_{j,\uparrow}|0\rangle = c^\dagger_{k,\uparrow}|0\rangle,\quad H^0_{j,k}c^\dagger_{j,\uparrow}c^\dagger_{k,\downarrow}|0\rangle = -c^\dagger_{j,\downarrow}c^\dagger_{k,\uparrow}|0\rangle,\ \ldots. \tag{14}$$

In this respect, the new model generalizes the spin-$\frac{1}{2}$ XXX model and the t-J model (4). The nearest-neighbor Hamiltonians of these models have a similar interpretation as graded permutations of the basic states, which are $\{|\uparrow\rangle,|\downarrow\rangle\}$ for the spin-$\frac{1}{2}$ XXX model and $\{|0\rangle,|\uparrow\rangle,|\downarrow\rangle\}$ for the t-J model. Lattice Hamiltonians that act like (graded) permutations were first considered by Sutherland in [8].

We define the number operators $N_\uparrow, N_\downarrow$ (the number of single electrons with given spin) and N_l (the number of localons) by

$$N_\uparrow + N_l = \sum_{j=1}^{L} n_{j,\uparrow},\quad N_\downarrow + N_l = \sum_{j=1}^{L} n_{j,\downarrow},\quad N_l = \sum_{j=1}^{L} n_{j,\uparrow}n_{j,\downarrow}, \tag{15}$$

and we write $N_e = N_\uparrow + N_\downarrow$ for the total number of single electrons. The fact that H^0 is a permutation makes it clear that these number operators commute with H^0, so that H^0 can be diagonalized within a sector with given numbers $N_\uparrow$, $N_\downarrow$, and N_l. This implies that the terms proportional to U, μ, and h in (6), which break the symmetry U(2|2), will not affect the solvability of the model in one dimension.

In the sectors without localons H^0 reduces to the t-J Hamiltonian (4). (This is clear from the fact that they both act as permutations.) The new model reduces to the spin-$\frac{1}{2}$ XXX model in the sector with only vacancies and localons, and similarly in the (half-filled) sector with one single electron on each site.

Let us now briefly discuss some physical aspects of the new model. We first remark that we can always (for general lattices in an arbitrary number of dimensions) construct a number of exact eigenstates of the Hamiltonian which show off-diagonal long-range order (ODLRO), which is characteristic for superconductivity [10]. For this we follow the construction which was developed for the Hubbard model by Yang in [2]. The state $\Psi_N = (\eta^\dagger)^N|0\rangle$ is an eigenstate of the Hamiltonian with energy $E = 2\mu N + UL/4 - M$, where M is the total number of nearest-neighbor links $\langle jk\rangle$ in the lattice. Following [2], we compute the following off-diagonal matrix element ($k \neq l$) of the reduced density matrix ρ_2:

$$\langle (k,\downarrow)(k,\uparrow)|\rho_2|(l,\uparrow)(l,\downarrow)\rangle = \frac{\langle 0|\eta^N c^\dagger_{k,\downarrow}c^\dagger_{k,\uparrow}c_{l,\uparrow}c_{l,\downarrow}(\eta^\dagger)^N|0\rangle}{\langle 0|\eta^N(\eta^\dagger)^N|0\rangle} = \frac{N(L-N)}{L(L-1)}. \tag{16}$$

The fact that this off-diagonal matrix element is constant for large distances $|j-k|$ establishes the property of ODLRO for the states Ψ_N.

An important observation is that for the attractive case ($U<0$) with zero magnetic field ($h=0$), the ground state in the sector with an even number $2N$ of electrons is precisely the state $\Psi_N = (\eta^\dagger)^N|0\rangle$. It can be rigorously shown that within each of these sectors (positive density corresponds to negative μ) this ground state is unique. We may thus conclude that in the attractive case our new model exhibits superconductivity.

The local electron pairs that participate in the η pairing have momentum zero. However, the model also admits localons that move with arbitrary momentum. This follows from the fact that $H^0_{j,k}$ acts as a permutation of the electronic states (1) on neighboring sites: Because of this localons cannot decay and move coherently. On a d-dimensional square lattice (with lattice spacing a) the wave function $\sum_{\mathbf{x}}\exp(i\mathbf{x}\cdot\mathbf{k})c^\dagger_{\mathbf{x},\uparrow}c^\dagger_{\mathbf{x},\downarrow}|0\rangle$, which describes a single localon of momentum $\mathbf{k}$ over the bare vacuum, is an exact eigenfunction of the Hamiltonian (6) of energy

$$E = 2d - 2\sum_{m=1}^{d}\cos(k_m a) + UL/4 + 2\mu - M. \tag{17}$$

Multilocalon wave functions, as well as wave functions with single electrons, exist but cannot easily be written down for higher-dimensional lattices. However, in one dimension the model is exactly solvable by Bethe ansatz (BA), and we can obtain explicit expressions for general eigenstates of the Hamiltonian. We think that it is worthwhile to study this exact solution, and that this will lead to a better understanding of the higher-dimensional model as well.

We will here briefly summarize the results of the exact solution in one dimension; the details are deferred to a separate publication [11]. The exact solution starts from the observation that the Hamiltonian is a graded permutation (14) of four states, of which two are fermionic and two are bosonic. The BA analysis for Hamiltonians which are graded permutations was first considered by Sutherland in [8]; see also [12]. The method of solution is the algebraic version of the "nested Bethe ansatz" [13] (for an introduction to the algebraic BA, see [14]). Each step of the nesting involves the introduction of a set of spectral parameters, which are in our case λ_j, $\lambda_k^{(1)}$, and $\lambda_l^{(2)}$, where $j=1,\ldots,(N_e+N_l)$, $k=1,\ldots,N_e$, and $l=1,\ldots,N_\downarrow$.

For each choice of a set of rapidities we can construct an eigenstate of the Hamiltonian H^0 in the sector specified by N_l, N_e, and $N_\downarrow$, with energy E^0 given by

$$E^0 = \sum_{j=1}^{N_e+N_l} 1/(\lambda_j^2 + \tfrac{1}{4}) - L. \tag{18}$$

The boundary conditions for these eigenstates lead to the following set of Bethe equations for the rapidities λ_j, $\lambda_k^{(1)}$, and $\lambda_l^{(2)}$:

$$\left[\frac{\lambda_j-i/2}{\lambda_j+i/2}\right]^L=\prod_{\substack{j'=1\\j'\neq j}}^{N_e+N_l}\frac{\lambda_j-\lambda_{j'}-i}{\lambda_j-\lambda_{j'}+i}\prod_{k'=1}^{N_e}\frac{\lambda_{k'}^{(1)}-\lambda_j-i/2}{\lambda_{k'}^{(1)}-\lambda_j+i/2},$$

$$\prod_{j'=1}^{N_e+N_l}\frac{\lambda_k^{(1)}-\lambda_{j'}+i/2}{\lambda_k^{(1)}-\lambda_{j'}-i/2}=\prod_{l'=1}^{N_l}\frac{\lambda_k^{(1)}-\lambda_{l'}^{(2)}+i/2}{\lambda_k^{(1)}-\lambda_{l'}^{(2)}-i/2}, \quad (19)$$

$$\prod_{\substack{l'=1\\l'\neq l}}^{N_l}\frac{\lambda_{l'}^{(2)}-\lambda_l^{(2)}+i}{\lambda_{l'}^{(2)}-\lambda_l^{(2)}-i}=\prod_{k'=1}^{N_e}\frac{\lambda_{k'}^{(1)}-\lambda_l^{(2)}+i/2}{\lambda_{k'}^{(1)}-\lambda_l^{(2)}-i/2}.$$

These equations, together with the expression (18) for the energy, guarantee that we shall be able to describe explicitly the ground state of our model for arbitrary density of electrons, coupling constant U, and magnetic field h [11].

This work was supported in part by NSF Grant No. PHY-9107261.

[1] O. J. Heilmann and E. H. Lieb, Ann. N.Y. Acad. Sci. **172**, 583 (1971); E. H. Lieb, Phys. Rev. Lett. **62**, 1201 (1989).
[2] C. N. Yang, Phys. Rev. Lett. **63**, 2144 (1989); C. N. Yang and S. Zhang, Mod. Phys. Lett. B **4**, 759 (1990).
[3] E. Lieb and F. Y. Wu, Phys. Rev. Lett. **20**, 1445 (1968).
[4] P.-A. Bares, G. Blatter, and M. Ogata, Phys. Rev. B **44**, 130 (1991).
[5] S. Sarkar, J. Phys. A **23**, L409 (1990); **24**, 1137 (1991); **24**, 5775 (1991).
[6] J. F. Cornwell, *Group Theory in Physics, Vol. III: Supersymmetries and Infinite-Dimensional Algebras* (Academic, New York, 1989). The algebras SU($n|m$) are discussed on p. 270.
[7] C. K. Lai, J. Math. Phys. **15**, 167 (1974).
[8] B. Sutherland, Phys. Rev. B **12**, 3795 (1975).
[9] P. Schlottmann, Phys. Rev. B **36**, 5177 (1987).
[10] C. N. Yang, Rev. Mod. Phys. **34**, 694 (1962).
[11] F. H. L. Essler, V. E. Korepin, and K. Schoutens (to be published).
[12] P. P. Kulish, J. Sov. Math. **35**, 2644 (1986); C. L. Schultz, Physica (Amsterdam) **122A**, 71 (1983).
[13] C. N. Yang, Phys. Rev. Lett. **19**, 1315 (1967).
[14] V. E. Korepin, G. Izergin, and N. M. Bogoliubov, *Quantum Inverse Scattering Method, Correlation Functions and Algebraic Bethe Ansatz* (Cambridge Univ. Press, Cambridge, 1992).

B. MODELS WITH LONG-RANGE INTERACTION

Another rather recent development in the study of one-dimensional systems of strongly correlated electrons is the emergence of models with long-range interactions. Prompted by the work of F. D. M. Haldane[187,188] and B. S. Shastry[189] on spin-chains, various authors investigated Hubbard and t-J models with modified kinetic energy terms. The hopping matrix elements t_{ij} are taken to be proportional to e.g. $1/\sin^2 \frac{\pi}{L} r_{ij})$, where r_{ij} is the distance between sites i and j and L is the length of the lattice. These models might be of relevance in the theory of fractional statistics and the Quantum Hall Effect[190]. The method of solution is different from the nested Bethe *Ansatz* used to solve electronic models with local interaction. For this reason we include below several reprints on the more fundamental issues of how to solve models with long-range interactions.

The history of exact studies of one dimensional models with long-range interactions goes back to the classic works of B. Sutherland[191–199] and F. Calogero[200–205]. In their papers the quantum mechanical N-body problem was studied for various kinds of potentials. It was known in Refs. 203, 206, 207 that for certain potentials one can find an infinite number of conservation laws by means of the Lax method, and the problem becomes integrable. In [repr.III.B.1 – III.B.2] B. Sutherland showed the ground state wave function to be of product form, and gave a general method for determining the spectrum of quantum many-body problems. The method for determining the spectrum is called Asymptotic Bethe *Ansatz*, and has been reviewed in Refs. 208, 197. It was first applied to lattice models with long range interaction by F. D. M. Haldane[187,188,209] and B. S. Shastry[189,210]. Both authors studied antiferromagnetic Heisenberg models with long-range exchange, and succeeded in deriving the excitation spectrum. There are also many recent works on this subject (e.g. Refs. 211–225). In Ref. 226 N. Kawakami and S.-K. Yang evaluated the conformal dimensions of Bose and (spinless) Fermi systems with long-range $1/r^2$ interaction.

Following F. D. M. Haldane's and B. S. Shastry's work various authors investigated electronic models with long-range interactions. Y. Kuramoto and H. Yokoyama investigated a supersymmetric t-J model with a long-range hopping term[227] [repr.III.B.3] They showed that eigenstates of the hamiltonian can be found by Gutzwiller wave functions, and computed spin and charge susceptibilities and the specific heat coefficient by means of the Luttinger-liquid approach. They showed that in the high-density limit the model exhibits a Mott-Hubbard gap and reduces to the antiferromagnetic Heisenberg chain with long-range hopping studied in Refs. 187, 189. N. Kawakami investigated the Kuramoto-Yokoyama model[228] [repr.III.B.4], the SU(N) generalization of the Haldane-Shastry model, and generalisations of the supersymmetric t-J model with one species of boson and m

species of fermions[229,221] by means of the Asymptotic Bethe *Ansatz*. He computed the excitation spectrum and discussed conformal properties of low-lying excitations.

F. Gebhard and A. E. Rückenstein studied the spectrum and the thermodynamics of a Hubbard model with long-range hopping[230]. Their approach is quite different from the Asymptotic Bethe *Ansatz*, but makes use of a conjecture concerning a transformation of the hamiltonian. They found that the model exhibits a metal-insulator (Mott) transition at half-filling if the value of the Hubbard coupling U is equal to the bandwidth W.

Reprinted from:

JOURNAL OF MATHEMATICAL PHYSICS VOLUME 12, NUMBER 2 FEBRUARY 1971

Quantum Many-Body Problem in One Dimension: Ground State*

BILL SUTHERLAND
Physics Department, University of California, Berkeley, California 94720

(Received 12 April 1970)

We investigate the ground state of a system of either fermions or bosons interacting in one dimension by a 2-body potential $V(r) = g/r^2$. In the thermodynamic limit, we determine the ground state energy and pair correlation function.

INTRODUCTION

We wish to investigate the properties of a 1-dimensional N-body system interacting by the 2-body potential $V(r) = g/r^2$. In particular, we shall be concerned in this paper with the ground state in the thermodynamic limit: $N \to \infty$ and volume $L \to \infty$, with the density $d = N/L$ kept finite. The discussion shall be divided into sections as follows: Section I discusses the history and peculiarities of the g/r^2 potential; Sec. II derives the ground state wavefunction for an N-body system contained in a weak harmonic well; Sec. III identifies the square of this wavefunction as identical with a probability distribution function familiar in the theory of random matrices, enabling many properties to be determined immediately by correspondence.

I. HISTORY AND PECULIARITIES OF THE g/r^2 POTENTIAL

Several recent papers have been concerned with systems of particles interacting in one dimension by a two-body potential $V(r) = g/r^2$. One line of investigation proceeds from Dyson's work[1] on phase transitions of classical systems in one dimension, which indicates that systems with attractive potentials falling off slower than $1/r^2$ have phase transitions, while attractive potentials falling off faster than $1/r^2$ do not. Thus one is particularly interested in whether an attractive $1/r^2$ potential has a phase transition. As this paper deals only with the zero temperature problem, it throws no light on the occurrence of phase transitions; finite temperature properties are to be discussed in a later paper. Dyson's work is recalled to emphasize that g/r^2 in one dimension is a particularly interesting choice.

Second, Calogero,[2] treating the corresponding quantum system, solved exactly the 3-body problem and N-body ground state problem in both free space and with mutual harmonic interactions. Calogero's work does not allow one to determine the behavior of the g/r^2 system in the thermodynamic limit. However, our ground state wavefunction is very similar to that of Calogero; the difference in this investigation and Calogero's is best described as one of viewpoint. It is this shift in viewpoint which allows us to proceed further than Calogero, to discuss the particle density, pair correlation function, and ground state energy at finite density.

The potential g/r^2 is very singular at the origin and requires some care to make physical sense. Classically, the attractive potential is too strong at the origin, requiring, for instance, a finite hard core to prevent "fall to the origin." However, for the quantum system, the zero-point motion acts to keep the particle from the origin, so that no cutoff is required, provided that the potential is not too attractive. Thus, for the 2-body Hamiltonian

$$H = -\frac{\partial^2}{\partial x^2} - \frac{\partial^2}{\partial y^2} + \frac{g}{(x-y)^2}, \tag{1}$$

we are led to the restriction $g \geq -\frac{1}{2}$. This point is discussed by Landau and Lifshitz[3] and is reproduced as an appendix at the end of this paper. Within this range of g, there are no bound states. The unnormalized solutions for the 2-body Hamiltonian of Eq. (1) are

$$\begin{aligned} \psi_{\pm} &= e^{iKR} r^{\frac{1}{2}} J_{\pm a}(kr) \\ &\sim e^{iKR} r^{\frac{1}{2}\pm a}, \quad r \to 0, \\ &\sim e^{iKR} \cos(kr - \tfrac{1}{4}\pi \mp \tfrac{1}{2}\pi a), \quad r \to \infty, \\ E &= \tfrac{1}{2}(k^2 + K^2) \end{aligned} \tag{2}$$

with

$$a = \tfrac{1}{2}(1 + 2g)^{\frac{1}{2}} \geq 0,$$

$$x = R - r, \quad k \geq 0,$$

$$y = R + r, \quad r \geq 0.$$

$J_a(x)$ is a Bessel function. From the considerations of Landau and Lifshitz and the Appendix, we select the unique solution $\psi \equiv \psi_+$ corresponding to the upper sign in Eq. (2). The wavefunction for $r \leq 0$ is given by

$$\psi(-|r|) = \pm\psi(|r|), \tag{3}$$

corresponding to bosons or fermions, respectively. Defining a new parameter $\lambda = \frac{1}{2} + a = \frac{1}{2}[1 + (1 + 2g)^{\frac{1}{2}}]$, we see that the "physically reasonable" solutions are $\lambda \geq \frac{1}{2}$. However, as explained in the Appendix, one can also treat free bosons by including the "unreasonable" solutions $\frac{1}{2} > \lambda \geq 0$, corresponding to the lower sign in Eq. (2). With this understanding, in the expressions which follow, we may take λ to range from 0 to $+\infty$.

II. THE GROUND STATE WAVEFUNCTION IN A HARMONIC WELL

We shall now consider the ground state of a system governed by the following Hamiltonian:

$$H = -\sum_{i=1}^{N} \frac{\partial^2}{\partial x_i^2} + \sum_{i<j} \frac{g}{(x_i - x_j)^2} + \omega^2 \sum_{i=1}^{N} x_i^2. \tag{4}$$

The last term is an external potential to contain the system, replacing the usual box of volume L. Since we will be concerned with the limit $N \to \infty$, we will let $\omega \to 0$ in order to produce a finite density of the system in the interior.

We write the wavefunction ψ as $\psi = \phi\varphi$. We verify that the choice

$$\phi = \prod_{i<j} |x_i - x_j|^{\lambda}, \tag{5}$$

$$\varphi = \exp\left(-\tfrac{1}{2}\omega \sum_{i=1}^{N} x_i^2\right), \tag{6}$$

$$\lambda^2 - \lambda = \tfrac{1}{2}g \quad \text{or} \quad \lambda = \tfrac{1}{2} + a, \tag{7}$$

is a solution, with

$$E = \omega N[1 + \lambda(N-1)]. \tag{8}$$

Since, for each ordering of particles, ψ is nodeless, it is the ground state. This is very much like Calogero's ground state wavefunction; however, our Hamiltonian contains an external harmonic well instead of mutual harmonic interactions.

First, let the kinetic term of the Hamiltonian act

on ψ:

$$-\sum_i \frac{\partial^2}{\partial x_i^2}(\phi\varphi) = -2\phi\varphi(\lambda^2 - \lambda)\sum_{i<j}\frac{1}{(x_i - x_j)^2} - 2\lambda\phi\sum_i\left(\sum_{j(\neq i)}\frac{1}{x_i - x_j}\right)\frac{\partial\varphi}{\partial x_i} - \phi\sum_i\frac{\partial^2\varphi}{\partial x_i^2}. \quad (9)$$

Thus, choosing λ according to Eq. (7), we eliminate the mutual interaction term. We may rewrite the second term of Eq. (9) to give

$$-\sum_i\frac{\partial^2\varphi}{\partial x_i^2} - 2\lambda\sum_{i<j}\frac{1}{x_i - x_j}\left(\frac{\partial}{\partial x_i} - \frac{\partial}{\partial x_j}\right)\varphi + \omega^2\left(\sum_i x_i^2\right)\varphi = E\varphi. \quad (10)$$

Now

$$\frac{\partial\varphi}{\partial x_i} = -\omega x_i\varphi, \quad \frac{\partial^2\varphi}{\partial x_i^2} = [\omega^2 x_i^2 - \omega]\varphi.$$

Substituting into Eq. (10), we verify the eigenvalue equation with E as in Eq. (8).

Let us rewrite ψ in terms of the variables

$$y_i = (\omega/\lambda)^{\frac{1}{2}}x_i; \quad (11)$$

then

$$\psi = C^{\frac{1}{2}}\exp\left(-\tfrac{1}{2}\lambda\sum_i y_i^2\right)\prod_{i<j}|y_i - y_j|^\lambda, \quad (12)$$

$$\psi^2 = C\exp\left(-\tfrac{1}{2}\beta\sum_i y_i^2\right)\prod_{i<j}|y_i - y_j|^\beta, \quad (13)$$

with

$$\beta = 2\lambda = 1 + (1 + 2g)^{\frac{1}{2}}. \quad (14)$$

C is a normalization constant, so that

$$C^{-1} = \int_{-\infty}^{+\infty}\cdots\int dy_1\cdots dy_N\psi^2.$$

III. INTERPRETATION OF ψ^2

It is at this point that we recognize the expression in Eq. (13) to be identical to the joint probability density function for the eigenvalues of matrices from a Gaussian ensemble. Choosing β equal to 1, 2, and 4 corresponds to orthogonal, unitary, and symplectic ensembles, respectively. In our case, the most attractive situation, $g = -\frac{1}{2}$, corresponds to $\beta = 1$; free fermions corresponds to $\beta = 2$.

We may immediately go to the literature[4,5] and find the normalization constant C and the 1- and 2-particle correlation functions.

First, it is conjectured[6] that the normalization constant C is given by

$$C^{-1} = (2\pi)^{\frac{1}{2}N}\beta^{-\frac{1}{2}N-\frac{1}{4}\beta N(N-1)} \times [\Gamma(1+\tfrac{1}{2}\beta)]^{-N}\prod_{j=1}^{N}\Gamma(1+\tfrac{1}{2}\beta_j). \quad (15)$$

This conjecture is verified for $\beta = 1, 2, 4$.

The particle density is given by

$$\sigma(y) = N\int_{-\infty}^{+\infty}\cdots\int\psi^2\,dy_2\cdots dy_N, \quad (16)$$

normalized so that

$$\int_{-\infty}^{+\infty}\sigma(y)\,dy = N. \quad (17)$$

One finds[7]

$$\sigma(y) = \begin{cases}\pi^{-1}(2N - y^2)^{\frac{1}{2}}, & y^2 < 2N,\\ 0, & y^2 > 2N.\end{cases} \quad (18)$$

This is true for all β's. The density of x's, normalized in the same way, is

$$d(x) = \begin{cases}\dfrac{1}{\pi}\left(\dfrac{2N\omega}{\lambda} - \dfrac{\omega^2x^2}{\lambda^2}\right)^{\frac{1}{2}}, & x^2 < \dfrac{2N\lambda}{\omega},\\ 0, & x^2 > \dfrac{2N\lambda}{\omega}.\end{cases} \quad (19)$$

We plot $d(x)$ in Fig. 1, comparing it with the corresponding density for the δ-function interaction problem of Lieb and Liniger.[8] Denoting the density at $x = 0$ by d, we find

$$d = \pi^{-1}(2N\omega/\lambda)^{\frac{1}{2}} \quad (20)$$

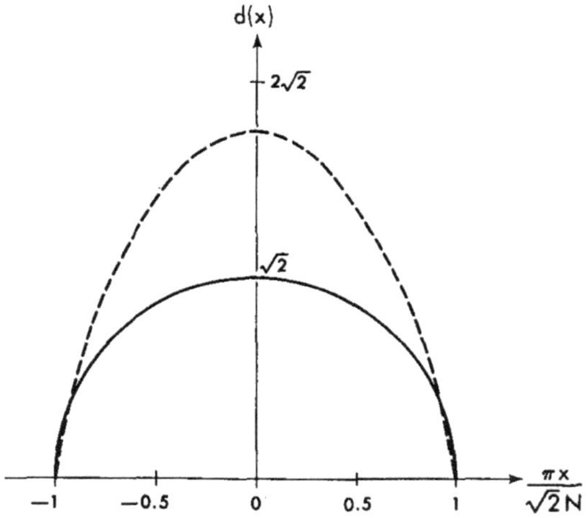

FIG. 1. Particle density $d(x)$, for $\lambda = \frac{1}{2}$ in a well $\omega = \pi^2/2N$, is shown by the solid curve. For comparison, the density of a boson system, interacting by $V(r) = 2\delta(r)$ in one dimension at the same chemical potential in the same well, is shown by the dashed curve.

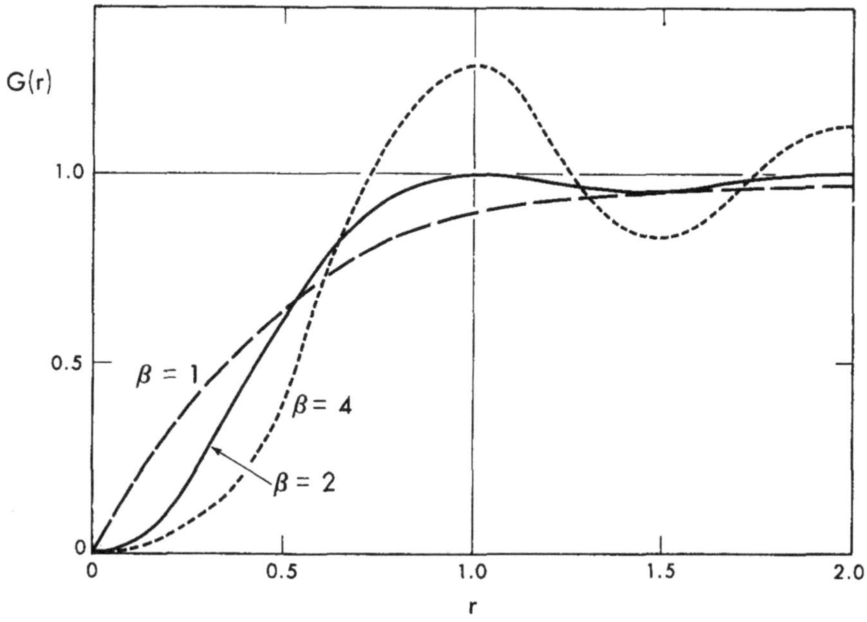

FIG. 2. The pair correlation function is shown for the three values $\beta = 1, 2, 4$; we have taken $d = 1$.

or

$$\omega = \pi^2 d^2 \lambda / 2N. \tag{20'}$$

Thus, we see that, for fixed d, letting $N \to \infty$ means that $\omega \to 0$ as N^{-1}.

We may extract from Eq. (19) the ground state energy for a system at density d in the more conventional box of volume L, instead of a harmonic well. The energy density must be intensive and have dimensions L^{-3}; thus the only choice is

$$E/L = d^3 e(\lambda), \tag{21}$$

where $e(\lambda)$ is a dimensionless function of the coupling constant. The chemical potential is then

$$\mu = 3d^2 e(\lambda). \tag{22}$$

Placing this system in a harmonic well, the chemical potential becomes

$$\mu = 3d^2(x) e(\lambda) + \omega^2 x^2$$

or

$$d(x) = \begin{cases} [(\mu - \omega^2 x^2)/3e(\lambda)]^{\frac{1}{2}}, & x^2 < \mu/\omega^2, \\ 0, & x^2 > \mu/\omega^2. \end{cases} \tag{23}$$

Comparing Eq. (23) with the previous equation (19), we find

$$e(\lambda) = \tfrac{1}{3}\lambda^2\pi^2,$$
$$E/L = \tfrac{1}{3} d^3 \lambda^2 \pi^2. \tag{24}$$

Likewise, we may determine the thermodynamic properties at zero temperature. Considering pressure P as a function of temperature T, fugacity Z, and coupling constant λ, we find that P must be of the form

$$P = T^{\frac{3}{2}} p(Z, \lambda). \tag{25}$$

Thus the energy density is given by

$$\frac{E}{L} = \frac{T^2 \partial}{\partial T}\left(\frac{P}{T}\right)$$
$$= \tfrac{1}{2} T^{\frac{3}{2}} p(Z, \lambda) = \tfrac{1}{2} P. \tag{26}$$

The equation of state at $T = 0$ is therefore found to be

$$P = \tfrac{1}{3} 2 d^3 \lambda^2 \pi^2. \tag{27}$$

We now consider the pair correlation function $G(r)$. Since the density factors out trivially, we take $d = 1$ in the following equations. Then $G(r) \to 1$ as $r \to \infty$, and we write

$$G(r) = 1 - Y(r) \tag{28}$$

and

$$b(k) = \int_{-\infty}^{+\infty} dr\, Y(r) e^{2\pi i k r}. \tag{29}$$

Then we find the following:

(a) For $\beta = 2$,

$$Y(r) = [s(r)]^2 \equiv [(\sin \pi r)/\pi r]^2, \tag{30}$$

$$b(k) = \begin{cases} 1 - |k|, & |k| \le 1, \\ 0, & |k| \ge 1. \end{cases} \tag{31}$$

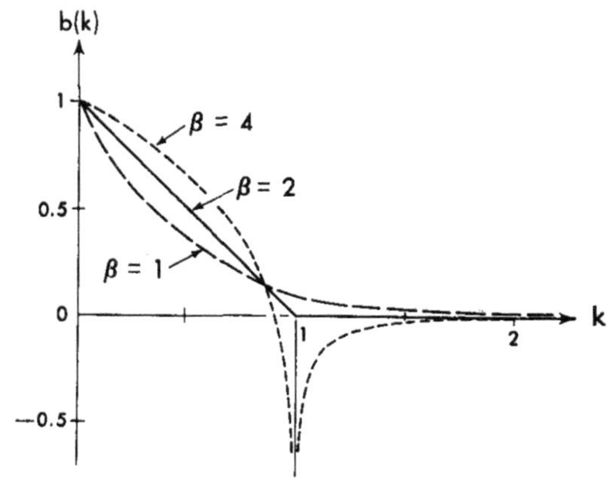

FIG. 3. The negative Fourier transform of the pair correlation function is shown for the same three values of β; again $d = 1$.

(b) For[9] $\beta = 1$,

$$Y(r) = \left(\int_r^\infty s(Z)\,dZ\right)\left(\frac{ds(r)}{dr}\right) + [s(r)]^2, \tag{32}$$

$$b(k) = \begin{cases} 1 - 2|k| + |k|\ln(1 + 2|k|), & |k| \le 1, \\ -1 + |k|\ln\left(\dfrac{2|k| + 1}{2|k| - 1}\right), & |k| \ge 1. \end{cases} \tag{33}$$

(c) For $\beta = 4$, it is conjectured[10] that

$$Y(r) = [s(2r)]^2 - \left(\int_0^{2r} s(Z)\,dZ\right)\left(\frac{1}{2}\frac{ds(2r)}{dr}\right), \tag{34}$$

$$b(k) = \begin{cases} 1 - \frac{1}{2}|k| + \frac{1}{4}|k|\ln[|(|k| - 1)|], & |k| \le 2, \\ 0, & |k| \ge 2. \end{cases} \tag{35}$$

These are the only cases for which the pair correlation function can be evaluated. We notice the system becomes more nearly ordered on a lattice as β increases or, equivalently, as the interaction becomes more repulsive. Figures 2 and 3 show $G(r)$ and $b(k)$ for these three values of β.

APPENDIX

We wish to cut off the g/r^2 potential for $r < b$, replacing it by a less singular potential. Then, letting $b \to 0$, we hope for a unique limiting solution ψ. Two unbiased choices for the potential V when $r < b$ are:

(i) $V = g/b^2$, $r < b$—flattening out the singularity;

(ii) $V = +\infty$, $r < b$—a hard core.

The solution for $r \ge b$ will be

$$\psi = \psi_+ + A\psi_- \tag{A1}$$

$\psi_\pm$ are defined in Eq. (2). We take kb to be small, and assume $V = c^2/b^2$, $r < b$, with c real or imaginary. This includes both cases (i) and (ii).

We match logarithmic derivatives at b, obtaining the equation

$$T = \frac{(\frac{1}{2} + a)b^a k^a + A(\frac{1}{2} - a)b^{-a}k^{-a}}{k^a b^a + A b^{-a}k^{-a}} \tag{A2}$$

with

$$T = \begin{cases} c\tanh(c), & \text{bosons}, \\ c\coth(c), & \text{fermions}. \end{cases} \tag{A3}$$

T is simply a number, equal to $(g^{\frac{1}{2}}\tanh(g^{\frac{1}{2}}), g^{\frac{1}{2}}\coth(g^{\frac{1}{2}}))$ for cutoff (i) and $(+\infty, +\infty)$ for cutoff (ii). Solving for A, we obtain

$$A = -\frac{\frac{1}{2} + a - T}{\frac{1}{2} - a - T}\,b^{2a}k^{2a}. \tag{A4}$$

Thus if $g \ge \frac{1}{2}$, $a > 0$ real, then, as $b \to 0$, $A \to 0$, with the exception of the case

$$T = \tfrac{1}{2} - a. \tag{A5}$$

This is a resonance condition and is not fulfilled for cutoffs (i) and (ii).

For $g < \frac{1}{2}$, a is pure imaginary, and A does not approach a limit. Upon closer examination, one sees that there is no lowest energy eigenstate for the Hamiltonian in this case.

There is some precedent[11] for using the solutions ψ_-, although as we have seen, this is surely artificial. For instance, one wishes to have two independent solutions in the scattering of nonidentical particles by a g/r^2 potential. We take the viewpoint that the mathematical equations are easily extended to the "unphysical" ψ-solutions with no extra effort; thus these cases are included in this paper.

* Work supported in part by National Science Foundation Grant GP 13632.

[1] F. J. Dyson, Commun. Math. Phys. **12**, 91 (1969); **12**, 212 (1969).
[2] F. Calogero, J. Math. Phys. **10**, 2191, 2197 (1969).
[3] L. D. Landau and E. M. Lifshitz, *Quantum Mechanics* (Pergamon, New York, 1958).
[4] C. Porter, *Statistical Theories of Spectra* (Academic, New York, 1965). This reference reprints the original papers of the field.
[5] M. L. Mehta, *Random Matrices* (Academic, New York, 1967). All mathematical techniques used in the present paper are presented in this reference.
[6] F. J. Dyson, J. Math. Phys. **4**, 713 (1963).
[7] E. P. Wigner, Ann. Math. **62**, 548 (1955).
[8] E. Lieb and W. Liniger, Phys. Rev. **130**, 1605 (1963).
[9] M. L. Mehta, Nucl. Phys. **18**, 395 (1960); M. L. Mehta and M. Gaudin, *ibid.*, 420 (1960); M. Gaudin, *ibid.* **25**, 447 (1960); F. J. Dyson, J. Math. Phys. **3**, 166 (1962).
[10] This is identical to the theorem for circular ensembles of F. J. Dyson, J. Math. Phys. **4**, 714 (1963).
[11] F. L. Scarf, Phys. Rev. **112**, 1137 (1958).

Reprinted from:

JOURNAL OF MATHEMATICAL PHYSICS VOLUME 12, NUMBER 2 FEBRUARY 1971

Quantum Many-Body Problem in One Dimension: Thermodynamics*

BILL SUTHERLAND
Physics Department, University of California, Berkeley, California 94720

(Received 21 June 1970)

We continue our investigation of a system of either fermions or bosons interacting in one dimension by a 2-body potential $V(r) = g/r^2$. We first present an approximation for the eigenstates of a general 1-dimensional quantum many-body system. We then apply this approximation to the g/r^2 potential, allowing complete determination of the thermodynamic properties. Finally, comparing the results with those properties known exactly, we conjecture that the approximation is, in fact, exact for the g/r^2 potential.

INTRODUCTION

This paper continues an investigation of the properties of a 1-dimensional quantum N-body system interacting by the 2-body potential $V(r) = g/r^2$. As before, we shall be concerned with the thermodynamic limit: $N \to \infty$ and volume $L \to \infty$, with density $d = N/L$ kept finite. The first paper[1] treated the ground state exactly; this paper will develop an approximation for the excited states, and hence the thermodynamics. Thus, the first section of the present paper introduces this approximation, while the second section applies it to the g/r^2 potential. Finally, the third section compares the results with those properties known exactly: zero temperature thermodynamics, second and third virial coefficients, and the limit of zero interaction. It is found that in all cases our approximation reproduces the exact results. Thus we are finally led to conjecture that the approximation is, in fact, exact for the g/r^2 potential.

I. APPROXIMATE EIGENVALUES OF A 1-DIMENSIONAL N-BODY SYSTEM

We now present an approximation for the energy eigenvalues and thermodynamics of a system of N fermions or bosons interacting in one dimension. Later we shall apply the method specifically to the g/r^2 potential, but we expect it to apply much more generally in one dimension whenever a potential has a phase shift and no bound states. This restriction to potentials with no bound states results because the approximation uses an asymptotic wavefunction which neglects diffraction effects from the simultaneous interaction of three or more particles. The results will then depend on the 2-body Hamiltonian through the exact 2-body phase shift. The scheme is interesting, for it gives the thermodynamics in an approximate but closed form, in contrast to a systematic series expansion. On the other hand, it is difficult to estimate the error. The approximation is modeled on the exact solution of the 1-dimensional δ-function boson problem.[2,3]

Consider the 2-body problem governed by a Hamiltonian:

$$H = -\frac{\partial^2}{\partial x_1^2} - \frac{\partial^2}{\partial x_2^2} + V(|x_1 - x_2|). \qquad (1)$$

For $x_1 \ll x_2$, the asymptotic wavefunction is

$$\psi(x_1, x_2) = e^{i(k_1x_1+k_2x_2)} - e^{i\theta(k_1-k_2)}e^{i(k_2x_1+k_1x_2)}. \qquad (2)$$

We assume that $k_1 > k_2$. Then $S(k) = -\exp[-i\theta(k)]$ is the S matrix and $\theta(k)$ is the 2-body phase shift; $\theta(k)$ is odd in k. For bosons or fermions, respectively, we have

$$\psi\,(x_1 \gg x_2) = \pm\psi\,(x_1 \ll x_2).$$

The energy is given by $E = k_1^2 + k_2^2$. In the center of mass coordinates [Eq. (I.2)], the asymptotic wavefunction becomes

$$\psi = (2/i)e^{iKR}\sin[kr - \tfrac{1}{2}\theta(k)]. \qquad (3)$$

For later reference, we list $\theta(k)$ for the following examples:

$$\delta\text{ function, } V(r) = 2c\delta(r)\text{: } \theta(k) = -2\tan^{-1}(k/c), \qquad (4a)$$

$$\text{hard rod, } v(r) = +\infty,\ r \le b\text{: } \theta(k) = kb, \qquad (4b)$$

$$V(r) = g/r^2\text{: } \theta(k) = (\pi k/2\,|k|)[(1 + 2g)^{\frac{1}{2}} - 1], \qquad (4c)$$

$$\text{free fermions: } \theta(k) = 0, \qquad (4d)$$

$$\text{free bosons: } \theta(k) = -k\pi/|k|. \qquad (4e)$$

Our basic approximation will be to assume that the asymptotic N-body wavefunction is that given by 2-body scattering alone. Therefore, there is no diffraction, and momenta are exchanged in pairs. The asymptotic wavefunction for the region $x_1 \ll x_2 \ll \cdots \ll x_N$ is a sum of $N!$ terms corresponding to the $N!$ permutations P of the N k's, $k_1 > k_2 > \cdots > k_N$:

$$\psi = \sum_P A(P)\exp(i\textstyle\sum k_{Pj}x_j). \qquad (5)$$

The coefficients $A(P)$ are related by 2-body scatterings:

$$A(\cdots k', k\cdots)/A(\cdots k, k'\cdots) = -e^{-i\theta(k-k')}. \qquad (6)$$

For other orderings of the particles, we simply use either Fermi or Bose statistics. This wavefunction is of the form known as Bethe's ansatz.[4]

We are now ready to apply periodic boundary conditions to the wavefunction, determining a unique set of k's for each quantum state. The energy is given by

$$E = \sum k_i^2. \tag{7}$$

For a ring of circumference L, we find the following equation for the k's:

$$e^{-ikL} = (-1)^{N-1} \exp\left(i \sum_{k'} \theta(k' - k)\right). \tag{8}$$

Upon taking the logarithm, we obtain

$$kL = 2\pi I(k) + \sum_{k'} \theta(k - k'). \tag{9}$$

The $I(k)$'s are either integers or half odd integers which come from $\log(\pm 1)/2\pi$ and, in fact, serve as quantum numbers for the problem. They may be taken as free fermion k vectors. Thus, for example, the ground state is given by the $I(k)$'s densely packed about zero.

We now adopt Yang and Yang's derivation for the thermodynamics of the δ-function Bose gas,[3] without rederiving. The pressure is given as a function of temperature T and chemical potential μ by

$$P(\mu, T) = \frac{T}{2\pi} \int_{-\infty}^{\infty} dk \ln(1 + e^{-\epsilon(k)/T}). \tag{10}$$

$\epsilon(k)$ depends upon μ and T and satisfies the integral equation

$$\epsilon(k) = -\mu + k^2 + \frac{T}{2\pi} \int_{-\infty}^{\infty} dk' \theta'(k - k') \times \ln(1 + e^{-\epsilon(k')/T}), \tag{11}$$

with $\theta'(k) = d\theta(k)/dk$, $\theta(k)$ being the 2-body phase shift. As usual, the density d is given by

$$d = \frac{\partial P}{\partial \mu}. \tag{12}$$

We now consider simple examples.

(a) δ-function potential: As our approximation is modeled on this problem, substituting the phase shift (4a) into the above equations gives the exact equations of Yang and Yang.

(b) Hard rod of radius b: The phase shift is given by (4b). Equation (9) for the k's is

$$kL(1 - db) = 2\pi I(k) - bK, \tag{13}$$

$$K = \sum k = \frac{2\pi}{L} \sum I(k). \tag{14}$$

Thus the k's for $K = 0$ are like free fermions in a volume reduced by the hard cores.

We have

$$\epsilon(k) = -\mu + k^2 + bP, \tag{15}$$

$$P = \frac{T}{2\pi} \int_{-\infty}^{\infty} dk \ln\left[1 + \exp\left(\frac{\mu - bP - k^2}{T}\right)\right]. \tag{16}$$

If we write $P_0(\mu, T)$ as the pressure for free fermions, then $P(\mu, T) = P_0(\mu - bP, T)$. We may invert to find $\mu(P) = \mu_0(P) + bP$. Differentiating with respect to P gives the specific volume $v \equiv 1/d$,

$$v(P, T) = v_0(P, T) + b. \tag{17}$$

This clearly shows that the effect of the potential is only to create an excluded volume, and the system otherwise behaves as a system of free fermions. These results are exact.

II. APPLICATION OF THE APPROXIMATION TO THE g/r^2 POTENTIAL

We now apply the approximation of the previous section to the g/r^2 potential. Using the phase shift of Eq. (4c), we find the kernel of the integral equation (11) to be

$$\theta'(k) = 2\pi\gamma\delta(k), \tag{18}$$

with the following definitions:

$$a = \tfrac{1}{2}(1 + 2g)^{\frac{1}{2}}, \quad \lambda = a + \tfrac{1}{2}, \quad \gamma = a - \tfrac{1}{2}. \tag{19}$$

Thus the integral equation simply reduces to the transcendental equation

$$\epsilon = -\mu + k^2 + T\gamma \ln(1 + e^{-\epsilon/T}). \tag{20}$$

Before proceeding with the finite T results, let us first investigate the ground state and low lying excitations.

A. Ground State

In the case of the ground state, we have the k's distributed with a density $\rho(k)$,

$$2\pi\rho(k) = \begin{cases} 1 - 2\pi\gamma\rho(k), & |k| < k_0, \\ 0, & |k| > k_0, \end{cases} \tag{21}$$

where k_0 is a Fermi momentum determined by

$$\int_{-k_0}^{k_0} \rho(k)\, dk = d. \tag{22}$$

The ground state energy density $E_0/L \equiv u_0$ is given by

$$u_0 = \int_{-k_0}^{k_0} \rho(k) k^2\, dk. \tag{23}$$

These equations are easily solved to yield

$$\rho(k) = 1/2\pi\lambda, \tag{21'}$$

$$d = k_0/\pi\lambda, \tag{22'}$$

$$u_0 = k_0^3/3\pi\lambda = \tfrac{1}{3}\pi^2\lambda^2 d^3. \tag{23'}$$

Differentiating u_0 with respect to d gives the chemical potential

$$\mu = (\pi\lambda d)^2. \tag{24}$$

We recall that these $T = 0$ results agree with the exact results of Paper I.

B. Excitations near the Ground State

Excitations near the ground state are finite numbers of hole–particle pairs obtained by taking particles from states k_α below the Fermi surface to states k'_α above the Fermi surface. Then the energy and momentum are given by

$$E - E_0 = \sum [\epsilon(k'_\alpha) - \epsilon(k_\alpha)],$$
$$P = \sum [p(k'_\alpha) - p(k_\alpha)]. \tag{25}$$

We easily find

$$\epsilon(k) = \begin{cases} (k^2 - k_0^2)/\lambda + k_0^2, & |k| < k_0, \\ k^2, & |k| > k_0, \end{cases} \tag{26a}$$

$$p(k) = \begin{cases} k/\lambda, & |k| < k_0, \\ k - k_0 + k_0/\lambda, & |k| > k_0. \end{cases} \tag{26b}$$

Defining $p_0 = k_0/\lambda = \pi d$, we find the dispersion curve $\epsilon(p)$ of the excitations to be

$$\epsilon(p) = \begin{cases} \lambda[p^2 + \gamma p_0^2], & |p| < p_0, \\ [p + \gamma p_0]^2, & |p| > p_0. \end{cases} \tag{27}$$

Note that the derivative is continuous across p_0 and equal to $2\lambda p_0 = 2k_0$.

C. Thermodynamics

Let us now return to Eq. (20). In terms of the following quantities,

$$Z = \text{fugacity} = e^{\mu/T},$$
$$\alpha = \exp[(\mu - k^2)/T] = Ze^{-k^2/T}, \tag{28}$$
$$\zeta = e^w = 1 + e^{-\epsilon/T},$$

Eq. (20) becomes

$$\alpha = \zeta^\lambda - \zeta^\gamma \equiv 2e^{aw} \sinh \tfrac{1}{2}w, \tag{29}$$

with pressure given by

$$P = \frac{T}{2\pi}\int_{-\infty}^{\infty} dk \ln \zeta = \frac{T}{2\pi}\int_{-\infty}^{\infty} dk\, w(k). \tag{30}$$

Writing P as a function of Z, T, and a, we note the symmetry $P(T, -Z, -a) = P(T, Z, a)$. Further, P as a function of Z has the following singularities:

(i) $\lambda \geq 1$, a branch cut along the negative real axis beginning at the branch point

$$Z_0 = -\frac{2}{(4a^2 - 1)^{\frac{1}{2}}}\left(\frac{2a - 1}{2a + 1}\right)^a = \frac{1}{(\lambda\gamma)^{\frac{1}{2}}}\left(\frac{\gamma}{\lambda}\right)^a; \tag{31a}$$

(ii) $1 > \lambda > 0$, two branch cuts extending to infinity from the following two branch points:

$$Z_0 = e^{\pm i\pi\lambda}\frac{2}{(1 - 4a^2)^{\frac{1}{2}}}\left(\frac{1 - 2a}{1 + 2a}\right)^a$$
$$= e^{\pm i\pi\lambda}\frac{1}{(-\lambda\gamma)^{\frac{1}{2}}}\left(-\frac{\gamma}{\lambda}\right)^a. \tag{31b}$$

The location of these branch points in the complex Z plane as a function of λ is shown in Fig. 1. $\lambda = 0$ is the case of free bosons. We therefore conclude from the above that there are no singularities on the positive real Z axis for $\lambda > 0$ and hence no phase transitions in these models.

D. Power Series Expansion for P

We wish to find a power series expansion for P in Z of the form

$$\frac{P}{T} = \frac{1}{2}\left(\frac{T}{\pi}\right)^{\frac{1}{2}}\sum_{n=1}^{\infty} Z^n B_n. \tag{32}$$

We first invert Eq. (29) to give

$$w = \ln \zeta = \sum_{n=1}^{\infty} \alpha^n C_n. \tag{33}$$

Then we perform the k integration on Eq. (30) to give

$$B_n = C_n/n^{\frac{1}{2}}, \tag{34}$$

independent of T. Therefore, the energy per volume u is given by the expression $u = \frac{1}{2}P$, as found in Eq. (I.26).

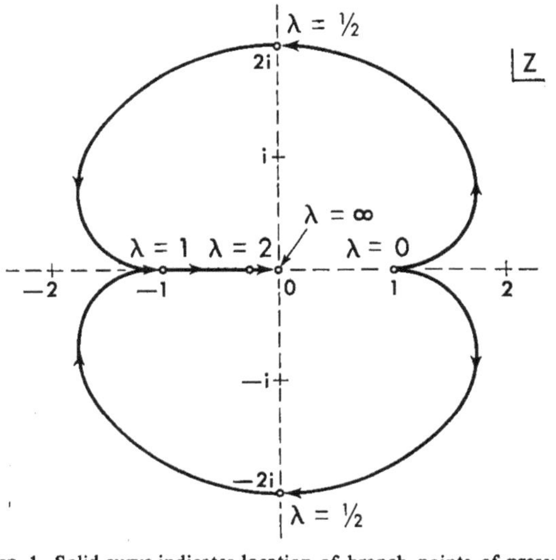

FIG. 1. Solid curve indicates location of branch points of pressure in the complex Z plane as a function of $\lambda \equiv \frac{1}{2}[1 + (1 + 2g)^{\frac{1}{2}}]$.

Upon inverting Eq. (29), we find the coefficients in the expansion of Eq. (33) to be

$$C_n = \frac{(-1)^{n+1}}{n!}(n\gamma + 1)(n\gamma + 2)\cdots(n\gamma + n - 1)$$
$$= \frac{(-1)^{n+1}}{n(\gamma + 1)} \frac{[n(\gamma + 1)]!}{n!\,(n\gamma)!}. \qquad (35)$$

Thus

$$B_n = \frac{(-1)^{n+1}}{n^{\frac{3}{2}}(\gamma + 1)} \frac{[n(\gamma + 1)]!}{n!\,(n\gamma)!}, \qquad (36)$$

and the radius of convergence of the power series of Eq. (32) is

$$R = |\gamma^{\gamma}/\lambda^{\lambda}|, \qquad (37)$$

in agreement with Eq. (31). For $Z \geq R$, P is given by the analytic continuation of the power series of Eq. (32).

E. Classical Limit

The classical limit exists only for the repulsive case, for which Eq. (32) becomes

$$\frac{P}{T} = -\left(\frac{T}{2\pi g}\right)^{\frac{1}{2}} \sum_{n=1}^{\infty} \left(\frac{n^{n-1}}{n!\,n^{\frac{1}{2}}}\right)[-Z(\tfrac{1}{2}g)^{\frac{1}{2}}]^n. \qquad (38)$$

This expression as a function of Z has a branch cut along the negative real axis beginning at

$$Z_0 = -(1/e)(2/g)^{\frac{1}{2}}. \qquad (39)$$

F. Special Cases

We may now consider as examples the special cases investigated in Paper I, with $\beta = 1, 2, 4$. These may all be expressed in closed form.

(a) $\beta = 2, g = 0, \gamma = 0$: $\zeta = 1 + \alpha$,

$$\frac{P}{T} = \frac{1}{2\pi} \int_{-\infty}^{\infty} dk \ln\,[1 + Ze^{-k^2/T}]. \qquad (40)$$

This is the familiar expression for free fermions in one dimension.

(b) $\beta = 1, g = -\frac{1}{2}, \gamma = -\frac{1}{2}$:

$$\ln \zeta = w = 2 \sinh^{-1} \tfrac{1}{2}\alpha,$$

$$\frac{P}{T} = \pi^{-1} \int_{-\infty}^{\infty} dk \sinh^{-1}(\tfrac{1}{2}Ze^{-k^2/T}), \qquad (41)$$

$$d = \pi^{-1} \int_{-\infty}^{\infty} dk(1 + 4Z^{-2}e^{2k^2/T})^{-\frac{1}{2}}. \qquad (41')$$

(c) $\beta = 4, g = 4, \gamma = 1$:

$$\zeta^2 - \zeta = \alpha \quad \text{or} \quad \zeta = \tfrac{1}{2}[1 + (1 + 4\alpha)^{\frac{1}{2}}],$$

$$\frac{P}{T} = (2\pi)^{-1} \int_{-\infty}^{\infty} dk \ln\,[\tfrac{1}{2}(1 + 4Ze^{-k^2/T})^{\frac{1}{2}} + \tfrac{1}{2}], \qquad (42)$$

$$d = \pi^{-1} \int_{-\infty}^{\infty} dk\, Ze^{-k^2/T}[1 + 4Ze^{-k^2/T} + (1 + 4Ze^{-k^2/T})^{\frac{1}{2}}]^{-1}. \qquad (42')$$

III. COMPARISON OF RESULTS WITH KNOWN THERMODYNAMIC PROPERTIES

In this final section, we shall test the approximation of the first section as applied to the g/r^2 potential in the second section. As has been remarked, this potential presents a unique opportunity to test, since the 3-body problem is separable and hence exactly soluble. This enables us to calculate the third virial coefficient, or equivalently the third coefficient B_3 in the fugacity expansion of Eq. (32).

Preliminary to evaluating B_3 exactly, let us first modify Calogero's solution[5] for the 3-body Hamiltonian

$$H = -\sum_{i=1}^{3} \frac{\partial^2}{\partial x_i^2} + \sum_{i<j=1}^{3} \frac{g}{(x_i - x_j)^2} + \omega^2 \sum_{i=1}^{3} x_i^2. \qquad (43)$$

We first transform to the variables

$$\begin{aligned} R &= \tfrac{1}{3}(x_1 + x_2 + x_3), \\ x &= 2^{-\frac{1}{2}}(x_1 - x_2), \\ y &= 6^{-\frac{1}{2}}(x_1 + x_2 - 2x_3), \\ r^2 &= x^2 + y^2, \\ \tan \phi &= x/y. \end{aligned} \qquad (44)$$

The Schrödinger equation becomes

$$H_R\psi + \left(-\frac{\partial^2}{\partial r^2} - \frac{1}{r}\frac{\partial}{\partial r} + \omega^2 r^2 + \frac{M_\phi}{r^2}\right)\psi = E\psi, \qquad (45)$$

$$H_R = -\frac{1}{3}\frac{\partial^2}{\partial R^2} + 3\omega^2 R^2, \qquad (46)$$

$$M_\phi = -\frac{\partial^2}{\partial \phi^2} + \frac{g}{2}\left(\frac{1}{\sin^2} + \frac{1}{\sin^2(\phi + \frac{2}{3}\pi)} + \frac{1}{\sin^2(\phi + \frac{4}{3}\pi)}\right). \qquad (47)$$

Thus the problem is separable.

Only the operator H_R of Eq. (46) differs from Calogero's problem. We seek a solution $\psi = \psi_\phi(\phi)\psi_r(r)\psi_R(R)$. First we solve $M_\phi\psi_\phi = m\psi_\phi$ to find

$$\begin{aligned} \psi_\phi(n) &= (\sin 3\phi)^\lambda C_n^\lambda(\cos \phi), \\ m &= [3(n + \lambda)]^2, \quad n = 0, 1, 2, \cdots. \end{aligned} \qquad (48)$$

C_n^λ is a Gegenbauer polynomial. We then solve for ψ_r finding

$$\psi_r^{(l,n)} = r^{3(l+\lambda)}e^{-\frac{1}{2}\omega r^2}L_n^{3(l+\lambda)}(\omega r^2),$$
$$E_{n,l} = 2\omega[2n + 1 + 3(l + \lambda)], \quad n = 0, 1, 2, \cdots. \qquad (49)$$

L_n^j is a Laguerre polynomial. Finally, the center of mass equation (46) has a solution

$$\begin{aligned} \psi_R^{(m)} &= e^{-\frac{3}{2}\omega R^2}\,\mathrm{Hm}\,[(3\omega)^{\frac{1}{2}}R], \\ Em &= 2\omega(m + \tfrac{1}{2}), \quad m = 0, 1, 2, \cdots. \end{aligned} \qquad (50)$$

Hm is a Hermite polynomial. The total energy E is given by

$$\begin{aligned} E &= E_{n,l} + E_m \\ &= 2\omega[3l + 2n + m + \tfrac{9}{2}] + 6\omega\gamma. \end{aligned} \tag{51}$$

The first term in this expression is simply the free fermion expression, while the last term displaces the entire free fermion spectrum by $6\omega\gamma$. This is a surprisingly simple result. Generally, Calogero made a conjecture, equivalent to the conjecture in our case, that for any N, the spectrum is that of free fermions in a harmonic well ω displaced by $\gamma\omega N(N-1)$. This is verified for $N \leq 3$.

We now wish to compare these exact results with the calculation of the previous section. We do this by directly calculating a fugacity expansion for a g/r^2 system in a macroscopic harmonic well. At the same time, we may easily place the system of Sec. 2 in the same harmonic well, so that we may compare the first three coefficients in the fugacity expansion.

Let us write the grand canonical ensemble as

$$\begin{aligned} Q(Z, \omega, T) &= e^{\Omega} \\ &= \sum_{N=1}^{\infty} Z^N Q_N(\omega, T). \end{aligned} \tag{52}$$

Ω is an extensive variable. We find the coefficients of a fugacity expansion

$$\Omega = \omega^{-1} \sum_{n=1}^{\infty} B_n'(T) Z^n, \tag{53}$$

by the usual derivation of cluster expansions, to be

$$\begin{aligned} B_1' &= \lim_{\omega\to 0} \omega Q_1, \\ B_2' &= \lim_{\omega\to 0} \omega[Q_2 - Q_1^2/2], \\ B_3' &= \lim_{\omega\to 0} \omega[Q_3 - Q_1Q_2 + \tfrac{1}{3}Q_1^3], \quad \text{etc.} \end{aligned} \tag{54}$$

If q_N denotes the partition function for the free fermions in a harmonic well ω, then Calogero's conjecture gives

$$Q_N = e^{-\beta\omega\gamma N(N-1)} q_N. \tag{55}$$

This is correct for $N \leq 3$. Using the expressions for q_N, we find it an easy task to evaluate B_1', B_2', and B_3' from Eq. (54). We leave this task for the reader, quoting the values

$$\begin{aligned} B_1' &= \tfrac{1}{2}T, \quad B_2' = \tfrac{1}{8}T(2\gamma + 1), \\ B_3' &= \tfrac{1}{36}T(3\gamma + 1)(3\gamma + 2). \end{aligned} \tag{54'}$$

On the other hand, the expression of Eq. (32) allows us to calculate Ω as

$$\begin{aligned} \Omega &= \int_{-\infty}^{\infty} \frac{P(x)}{T} dx \\ &= \frac{1}{2}\left(\frac{T}{\pi}\right)^{\frac{1}{2}} \sum_{n=1}^{\infty} B_n Z^n \int_{-\infty}^{\infty} e^{-n\omega^2 x^2/T} dx \\ &= \frac{T}{2\omega} \sum_{n=1}^{\infty} \frac{B_n Z^n}{n^{\frac{1}{2}}}, \end{aligned} \tag{56}$$

where B_n is given by Eq. (36).

Comparing Eq. (54′) and (56), we see that they are the same; our approximation gives the exact second and third virial coefficients.

In all instances where exact results exist, our approximation agrees. We collect these cases:

(i) free fermions or bosons when $g = 0$;
(ii) second and third virial coefficients;
(iii) zero temperature thermodynamics.

The last two points indicate agreement in opposite regimes. We are thus led to the following.

Conjecture: All results of the second section are exact.

It is very likely, although not proven, that Calogero's conjecture implies our conjecture through Eqs. (53) and (56); the converse cannot be true. However, it is believed that the approach of this paper reveals the physical basis of the conjecture to be the absence of diffraction effects in many-body scattering.

We include as an appendix another amusing instance where an approximation applied to the g/r^2 potential yields the exact solution.

APPENDIX

In this appendix, we wish to present a very simple example where again an approximation applied to the g/r^2 potential problem yields an exact result. The example, being a 1-body problem, has little to do with the more complicated many-body problem, yet it gives insight into the peculiarities of the g/r^2 potential and supports the conjecture made in this paper.

We consider a single particle interacting with a 1-dimensional rigid lattice of g/r^2 potentials; the lattice constant is b. However, we first present an approximation, reminiscent of the approximation in the text of the paper, valid for any potential in one dimension which has a phase shift. This approximation is well known in solid state physics.

Let the Schrödinger equation be

$$\left(-\frac{1}{2}\frac{d^2}{dx^2} + U(x)\right)\psi = E\psi, \quad \text{(A1)}$$

with

$$U(x) = \sum_{n=-\infty}^{+\infty} V(x - nb). \quad \text{(A2)}$$

$V(x)$ is the potential between one lattice point and the particle; we assume it to be symmetrical. $U(x)$ is periodic with period b.

We now divide the lattice into equal cells, the nth being $(n - \frac{1}{2})b$ to $(n + \frac{1}{2})b$. Let b be so large that there is no overlap of potentials. We write the solution in the nth cell as

$$\psi(x) = \psi_1^{(n)}\phi_1^{(n)}(kx) + \psi_2^{(n)}\phi_2^{(n)}(kx). \quad \text{(A3)}$$

$\phi_1^{(n)}(kx)$ is a solution in the nth cell corresponding to a plane wave of momentum k, incident from the right on a scatterer at the origin, with amplitude chosen to make the amplitude of the outgoing wave on the left unity. $\phi_2^{(n)}(kx)$ is the time reversal of $\phi_1^{(n)}(kx)$. We may then define a transfer matrix M, acting on the vector

$$\psi^{(n)} \equiv (\psi_1^{(n)}, \psi_2^{(n)}), \quad \text{so that} \quad M\psi^{(n)} = \psi^{(n+1)}. \quad \text{(A4)}$$

In terms of the transmission and reflection amplitudes T and R, we find

$$M = \begin{vmatrix} e^{ibk}/T & R^*/T^* \\ R/T & e^{-ikb}/T^* \end{vmatrix}. \quad \text{(A5)}$$

We shall parametrize the transmission amplitude by

$$T = \exp[i\delta(k)] \cos\theta(k). \quad \text{(A6)}$$

To obtain a band structure, we require the solution to be periodic after N sites. This in turn requires $M^N = I$, and, with $\det M = 1$, we find

$$\operatorname{Tr} M = 2\cos(2\pi n/N), \quad n = 1, 2, \cdots, N,$$

or

$$\cos(bk - \delta)/\cos\theta = \cos(2\pi n/N). \quad \text{(A7)}$$

We solve this transcendental equation for k as a function of n and a band index, and calculate the energy by

$$E = \tfrac{1}{2}k^2. \quad \text{(A8)}$$

Our approximation consists of assuming Eqs. (A7) and (A8) to hold for *all* b. Obviously, for the δ-function potential, our results will be exact since point interactions cannot overlap.

We next apply the previous approximation to the g/r^2 potential, when $0 \geq g \geq -\frac{1}{2}$. First, we see that Eq. (A2) gives

$$U(x) = g \sum_{n=-\infty}^{\infty} (x - nb)^{-2} = g\pi^2 b^{-2} \sin^{-2}(\pi x/b). \quad \text{(A9)}$$

The resulting differential Eq. (A1) is identical with Eq. (47) of the text; this band problem has been solved exactly by Scarf.[6] We shall use Scarf's matching condition through the singular potential, where the most singular solution ψ_- is taken symmetrical and the least singular solution ψ_+ is taken antisymmetrical. As discussed in Paper I, this is an artificial choice, yet necessary for a nontrivial band problem. We easily find the transmission and reflection amplitudes to be

$$T = \sin(\pi a), \quad R = -i\cos(\pi a). \quad \text{(A10)}$$

They are independent of k. Equation (A7) gives

$$\cos(kb) = \sin(\pi a)\cos(2\pi n/N)$$

or

$$k = \frac{\cos^{-1}[\sin(\pi a)\cos(2\pi n/N)]}{b}. \quad \text{(A11)}$$

Thus the allowed bands of k values are of equal width and equally spaced. The energy is given by Eq. (A8) to be

$$E = \frac{\{\cos^{-1}[\sin(\pi a)\cos(2\pi n/N)]\}^2}{2b^2}. \quad \text{(A12)}$$

Upon comparing this result with Scarf's exact solution, we find them to be identical for *all* b, whether the potentials overlap or not. Thus, once again, an approximation applied to the g/r^2 potential, where it would clearly seem to be inappropriate, has in fact yielded exact results.

Addendum: After this paper was submitted, two preprints have appeared which may be used to supply a proof of our final conjecture. In the first preprint, F. Calogero[7] proves his own conjecture, our Eq. (55). One may then easily evaluate the grand canonical ensemble of Eqs. (52) and (53) by saddle point methods, as done by C. Marchioro and E. Presutti[8] in the second preprint. The result is then seen to be identical to our Eq. (56); actually it is easiest to compare the average number of particles $\bar{N} = \partial\Omega/\partial \ln Z$.

* Work supported in part by National Science Foundation Grant GP 13632.

[1] B. Sutherland, J. Math. Phys. **12**, 246 (1971). This paper will be referred to as I, and equation numbers preceded by I.
[2] E. Lieb and W. Liniger, Phys. Rev. **130**, 1605 (1963); E. Lieb, *ibid.*, 1616 (1963).
[3] C. N. Yang and C. P. Yang, J. Math. Phys. **10**, 1115 (1969).
[4] H. A. Bethe, Z. Physik **71**, 205 (1931).
[5] F. Calogero, J. Math. Phys. **10**, 2191, 2197 (1969).
[6] F. L. Scarf, Phys. Rev. **112**, 1137 (1958).
[7] F. Calogero, J. Math. Phys. **12**, March, 1971 (to be published).
[8] C. Marchioro and E. Presutti, "Exact Statistical Model in One Dimension," preprint.

VOLUME 67, NUMBER 10 PHYSICAL REVIEW LETTERS 2 SEPTEMBER 1991

Exactly Soluble Supersymmetric t-J-Type Model with Long-Range Exchange and Transfer

Y. Kuramoto and H. Yokoyama
Department of Applied Physics, Tohoku University, Sendai 980, Japan
(Received 9 May 1991)

The Gutzwiller wave function is shown to be the exact solution of a supersymmetric t-J-type model. The model realizes a Fermi-liquid state in one dimension with a discontinuity in the momentum distribution. Analytic results are obtained for spin and charge susceptibilities, and the specific-heat coefficient with the help of the Luttinger-liquid theory. In the high-density limit the model exhibits a Mott-Hubbard gap and reduces to an antiferromagnetic spin chain with long-range exchange solved by Haldane and Shastry.

PACS numbers: 71.30.+h, 05.30.−d, 74.65.+n, 75.10.Jm

Exactly soluble one-dimensional fermion models such as the Tomonaga-Luttinger model [1], the Hubbard model [2,3], and the supersymmetric t-J model [4-7] show power-law singularities in the momentum distribution. This feature is in marked contrast to the discontinuity at the Fermi surface in Fermi liquids. In this Letter we present an interacting-fermion model that is exactly soluble and shows a discontinuity in the momentum distribution. This model is the first example that realizes a Fermi-liquid state with spin $\frac{1}{2}$ in one dimension. The model includes in the high-density limit the antiferromagnetic Heisenberg chain with long-range exchange which has been solved by Haldane [8] and Shastry [9]. We show that the Gutzwiller wave function is the exact solution of the model. The resultant Fermi-liquid state is identified as a free Luttinger liquid [1]. With this identification we obtain analytic results for most fundamental thermodynamic quantities such as the charge susceptibility, the spin susceptibility, and the low-temperature specific heat. The charge susceptibility indicates the presence of a Mott-Hubbard gap.

The t-J-type Hamiltonian is given by

$$\mathcal{H}=\mathcal{P}\sum_{i\neq j}\left[t_{ij}\sum_{\sigma}c_{i\sigma}^{\dagger}c_{j\sigma}+J_{ij}(\mathbf{S}_i\cdot\mathbf{S}_j-\tfrac{1}{4}n_in_j)\right]\mathcal{P},$$

where $\mathcal{P}$ is the projection operator to exclude the double occupation at each site. Other notations are standard ones. We treat a finite system and impose the periodic boundary condition. Namely, we work with a ring of length N in units of the lattice spacing. We choose N even and require

$$t_{ij}=J_{ij}=tD(x_i-x_j)^{-2},$$

with $D(x_i-x_j)=(N/\pi)\sin[\pi(x_i-x_j)/N]$ and $t>0$. In the macroscopic limit $-t$ reduces to the nearest-neighbor transfer.

The Gutzwiller state $|G\rangle$ is related to the free Fermi sea $|F\rangle$ by $|G\rangle=\mathcal{P}|F\rangle$. Following Ref. [10] we choose the fully polarized state $|N\uparrow\rangle$ of N electrons as the reference state and represent $|G\rangle$ as

$$|G\rangle=\sum_{\{x\},\{y\}}\Psi_G(\{x\},\{y\})\prod_{i\in\{x\}}c_{i\downarrow}^{\dagger}c_{i\uparrow}\prod_{j\in\{y\}}c_{j\uparrow}|N\uparrow\rangle.$$

Here $\{x\}$ denotes the set of coordinates for M down-spin electrons and $\{y\}$ denotes that of Q holes. Thus we have $N=2M+Q$. In order to remove the degeneracy we choose M odd. The amplitude $\Psi_G(\{x\},\{y\})$ is given, apart from a normalization factor, by

$$\Psi_G(\{x\},\{y\})=\exp\left[-i\pi\left(\sum_{\alpha}x_\alpha+\sum_{l}y_l\right)\right]\prod_{\alpha>\beta}D(x_\alpha-x_\beta)^2\prod_{l>m}D(y_l-y_m)\prod_{\alpha,l}D(x_\alpha-y_l). \quad (1)$$

This $\Psi_G(\{x\},\{y\})$ is a generalization of wave functions treated by Sutherland [11], who considered in the continuum space both a boson system and a fermion one, but not their mixtures. Under exchange of coordinates, holes behave as fermions and down-spin electrons behave as hard-core bosons which are called spin bosons hereafter. We note that the hard-core repulsion between holes and spin bosons leads to antisymmetry of $\Psi_G(\{x\},\{y\})$ under exchange of x_α and y_l. In this representation the system of spin bosons and holes has a total momentum of $-\pi(M+Q)$, which in fact is compensated by that of up spins constituting the vacuum.

The singlet nature of $|G\rangle$ appears as [10]

$$\sum_{x_1}\Psi_G(\{x\},\{y\})=0, \quad (2)$$

where x_1 is the first coordinate in $\{x\}$. In applying the transfer operator in $\mathcal{H}$ to $|G\rangle$ we consider the up-spin part $T_\uparrow$ and the down-spin one $T_\downarrow$ separately. $T_\uparrow$ changes $\{y\}$ in $\Psi_G(\{x\},\{y\})$ but leaves $\{x\}$ intact. On the other hand, $T_\downarrow$ causes exchange between pairs of x_α and y_l and is harder to treat. A crucial simplification occurs since $T_\uparrow|G\rangle$ is still a singlet. This follows because Eq. (2) is valid for any $\{y\}$ and thus after application of $T_\uparrow$ the summation over x_1 still gives zero for the wave function. By a rotation in spin space which does not change the singlet state, we can transform $T_\uparrow$ into $T_\downarrow$ and prove the equality $T_\uparrow|G\rangle=T_\downarrow|G\rangle$.

Let us rewrite the Hamiltonian in terms of the spin-boson operator b_i, where $b_i^{\dagger}b_i=n_i/2-S_i^z$, and the hole operator h_i defined by $n_i = 1 - h_i^\dagger h_i$. Using the singlet property described above we obtain

$$\mathcal{H}|G\rangle = \mathcal{P}\sum_{i,j} t_{ij}(2h_i^\dagger h_j + b_i^\dagger b_j + m_i m_j - \tfrac{1}{2} n_i n_j)|G\rangle\,, \quad (3)$$

with $m_i = b_i^\dagger b_i$. We represent the transfer operator for holes in Eq. (3) as T_h and that for spin bosons as T_b. Let $\Psi_{bG}(\{x\},\{y\})$ denote the coordinate representation of $T_b|G\rangle$, and $\Psi_{hG}(\{x\},\{y\})$ that of $T_h|G\rangle$. The ratio Ψ_{bG}/Ψ_G at $(\{x\},\{y\})$ is given by

$$\Psi_{bG}/\Psi_G = t\sum_{n=1}^{N-1} z^{-nN/2} D(n)^{-2} \sum \prod_{\alpha} \prod_{\beta(\neq\alpha)} B_{\alpha\beta}^{(n)} \prod_{l} F_{\alpha l}^{(n)}\,, \quad (4)$$

where $z = \exp(2\pi i/N)$ and

$$B_{\alpha\beta}^{(n)} = 1 - [(1-z^n)Z_\alpha^2 + (1-z^{-n})Z_\beta^2]/(Z_\alpha - Z_\beta)^2\,, \quad (5)$$

$$F_{\alpha l}^{(n)} = \cos(\pi n/N) + \sin(\pi n/N)\cot\Theta_{\alpha l}\,. \quad (6)$$

with $Z_\alpha = \exp(2\pi i x_\alpha/N)$ and $\Theta_{\alpha l} = \pi(x_\alpha - y_l)/N$. We use Greek indices for spin bosons and Latin ones for holes. The first term in $B_{\alpha\beta}^{(n)}$ and that in $F_{\alpha l}^{(n)}$ do not have coordinates of particles. With this in mind we expand Eq. (4) by the use of Eqs. (5) and (6) and classify terms according to the number of particles involved. Then we find that all terms with more than three particles vanish after summation over n. A similar observation has been made in Refs. [8] and [9] in the absence of holes.

For $\Psi_{hG}(\{x\},\{y\})/\Psi_G(\{x\},\{y\})$ we obtain

$$\Psi_{hG}/\Psi_G = 2t\sum_{n=1}^{N-1} z^{-nN/2} D(n)^{-2} \sum \prod_{l} \prod_{m} F_{lm}^{(n)} \prod_{\alpha} F_{l\alpha}^{(n)}\,. \quad (7)$$

Here again terms with more than three particles vanish. The three-body terms in Eqs. (4) and (7) consist of four types depending on the number of spin bosons involved. In the case of three spin bosons the three-body terms combine to a constant owing to the identity

$$\cot\Theta_{\alpha\beta}\cot\Theta_{\alpha\gamma} + \cot\Theta_{\beta\gamma}\cot\Theta_{\beta\alpha} + \cot\Theta_{\gamma\alpha}\cot\Theta_{\gamma\beta} = -1\,.$$

Similar reduction occurs in the case of three holes and that of two spin bosons and one hole.

In the case of two holes and one spin boson the three-body terms do not combine to a constant because of different numerical factors. However, by using the singlet property of Ψ_G we can transform the residual three-body term into

$$\sum_{\alpha}\sum_{l\neq m} \cot\Theta_{\alpha l}\cot\Theta_{\alpha m}\Psi_G = \left[\frac{1}{6}Q(Q-1)(Q-2-3N) + \sum_{l\neq m} \sin^{-2}\Theta_{lm}\right]\Psi_G\,.$$

The details of the algebra will be reported elsewhere.

We have thus seen that the sum of Eqs. (4) and (7) reduces to a constant plus two-body terms. The two-body terms turn out to be just equal to minus the interaction part in Eq. (3). This means that the Gutzwiller wave function $|G\rangle$ is an eigenstate of $\mathcal{H}$. The eigenvalue E is given in terms of $n_e = 2M/N$ by

$$\frac{E}{N\pi^2 t} = -\frac{n_e(n_e^2 - 3n_e + 4)}{12} - \frac{1 - 2n_e/3}{N^2}\,, \quad (8)$$

By the nature of the method of solution it is hard to exclude the possibility of lower-energy states other than $|G\rangle$. Nevertheless there is strong evidence in favor of $|G\rangle$ being the ground state of the system. First, in the dilute limit, E given by Eq. (8) tends to that of the free Fermi sea. Thus if Ψ_G is not the ground state for finite n_e, a phase change should occur as the density is increased. This, however, is unlikely in view of the known properties of related models such as the Hubbard and supersymmetric t-J models. Second, in the high-density limit $n_e = 1$, E agrees with the result of Refs. [8] and [9] with due account of the $-n_i n_j/4$ term in $\mathcal{H}$. In this limit Haldane [8] has confirmed by exact diagonalization up to twelve sites that Ψ_G is indeed the ground state.

The charge susceptibility χ_c in the macroscopic limit is derived from Eq. (8) as

$$\frac{1}{\chi_c} = \frac{\partial^2(E/N)}{\partial n_e^2} = \frac{\pi^2 t(1-n_e)}{2}\,.$$

The divergence of χ_c as n_e approaches 1 is consistent with the formation of the Mott-Hubbard gap, as in the Hubbard model [2] and the t-J model [7]. Let us compare χ_c with the susceptibility $\chi_c^{(0)}$ in the free model with the single-particle energy

$$\epsilon(k) = \pi t\{|k|(1 - |k|/2\pi) - \pi(1 - N^{-2})/3\}$$

for momentum k. The long range of t_{ij} makes its Fourier transform $\epsilon(k)$ dependent on the size of the system. We obtain $1/\chi_c^{(0)} = \pi^2 t(1 - n_e/2)/2$. The ratio

$$\tilde{\chi}_c = \frac{\chi_c}{\chi_c^{(0)}} = \frac{1 - n_e/2}{1 - n_e} \quad (9)$$

is a measure of the many-body effect. Interestingly the right-hand side of Eq. (9) agrees with the inverse of the discontinuity in the momentum distribution obtained by the Gutzwiller *approximation* for models with infinite repulsion.

We now show that the exact solution is consistent with the Luttinger-liquid theory [1]. For this purpose we shift for each spin σ the momentum distribution in the Slater determinant of $|G\rangle$ by $\pi J_\sigma/N$, with J_σ an even integer. Let us first consider the case where only the charge-current excitations are involved: $J_\uparrow = J_\downarrow$. Upon application of transfer operators in Eq. (3) to the shifted wave function, terms with more than three particles vanish as long as $|J_\uparrow| \le M+1$, and the resultant state is shown to be an exact solution. We introduce the charge velocity v_c by $v_c = 2/\pi\chi_c$ which reduces to the Fermi velocity in the noninteracting case. The increment of energy from that of Eq. (8) is calculated to be $\pi v_c J_c^2/2N$, where the charge current J_c is defined [12] by $(J_\uparrow + J_\downarrow)/\sqrt{2}$. This result shows that v_c agrees with the charge-current velocity. The agreement leads to identification of the Fermi-liquid state $|G\rangle$ as a free Luttinger liquid [1]. Then the ratio of Eq. (9) also represents the enhancement factor of the effective mass for the charge current.

We next consider the case $J_\downarrow = 2J_\uparrow$. In this case both charge and spin excitations are involved. The resultant wave function is obtained by replacing in Eq. (1) the momentum $-\pi$ of each particle by $-\pi(1+K/N)$, with $K=J_\uparrow$. Although the state with $K\neq 0$ is not a singlet, close inspection shows that the transfer operator $T_\downarrow$ has the same effect as that of $T_\uparrow$. Therefore the same effective Hamiltonian as in Eq. (3) can be used. With the condition that $|J_\uparrow - J_\downarrow| \leq 2$, terms with more than three particles vanish upon application of transfer operators in Eq. (3) and the shifted state is shown to be an eigenstate. The condition for the exact solution is rather strict in this case. We notice that this is a sufficient condition and suggest that the necessary condition is weaker.

The increment of energy for the case $J_\downarrow = 2J_\uparrow$ is calculated to be $K^2\pi^2 t(1-3n_e/4)/N$. By introducing the spin current [12] $J_s = (J_\uparrow - J_\downarrow)/\sqrt{2}$ and identifying the coefficient of J_s^2, we obtain the spin-current velocity v_s as

$$v_s = \pi t ,$$

which is independent of n_e. Using the property of the free Luttinger liquid we can derive the spin susceptibility χ_s from v_s: $\chi_s = 2/\pi v_s$. At $n_e = 1$ the result is consistent with that of Refs. [8] and [9] derived from the increment of the energy against changing the number M of spin bosons. We note that χ_s is smaller than the noninteracting one $\chi_s^{(0)}$ $(=\chi_c^{(0)})$. Namely, we have the ratio

$$\tilde{\chi}_s = \chi_s/\chi_s^{(0)} = 1 - n_e/2 .$$

The reduction of the homogeneous susceptibility is also present in the supersymmetric t-J model [7] and is due to the antiferromagnetic correlation.

The results for χ_c and χ_s are consistent with correlation-function exponents [13,14] for the Gutzwiller wave function, where no anomalous dimensions appear for either spin or charge. Namely, we have the exponents $K_\rho = K_\sigma = 1$ in the notation of Ref. [12]. We can derive the low-temperature specific heat with the aid of the formula obtained by the conformal field theory [3,7]. The specific-heat coefficient $\tilde{\gamma}$ normalized by the noninteracting one is given by

$$\tilde{\gamma} = \frac{\tilde{\chi}_c + \tilde{\chi}_s}{2} = \frac{(1-n_e/2)^2}{1-n_e} .$$

The many-body effect appears only at $O(n_e^2)$, in contrast to $\tilde{\chi}_c, \tilde{\chi}_s$ and the discontinuity $(1-n_e)^{1/2}$ of the momentum distribution [13] where the effect appears at $O(n_e)$. We note that $\tilde{\gamma}$ diverges as n_e approaches unity. The origin of divergence is the divergent density of states for charge excitations at the edge of the Mott-Hubbard gap [7]. Note that $\tilde{\chi}_c$ vanishes at exactly $n_e = 1$, and we obtain $\tilde{\chi}_s/\tilde{\gamma} = 2$ as in the half-filled case of the Hubbard model.

A peculiar feature of the present model is that the finite-size correction in Eq. (8) contains a nonuniversal contribution in addition to the universal one related to $v_c + v_s$, which is described by the conformal field theory [3,7]. This peculiarity comes from the size dependence in $\epsilon(k)$. Furthermore, the finite-size correction in Eq. (8) does not vanish in the dilute limit. This is not a problem since the exact solution in the form of Eq. (1) is valid only for odd M, which means that n_e has the minimum $2/N$. In fact, with $n_e = 2/N$ the energy is reduced to $2\epsilon(k=0)$, which is indeed exact as can be checked by solving the two-electron problem.

The reduction to the free fermion state in the dilute limit suggests a close relation to the supersymmetric t-J model [7]. In the latter model Yokoyama and Ogata [15] have observed by a numerical study that the Gutzwiller wave function is an excellent approximation not only for the ground-state energy but for structure factors and the momentum distribution at any density. However, a discrepancy appears in the exponent for correlation functions. It has been recognized [8,9] that in contrast to the Heisenberg model the long-range exchange model does not contain marginally irrelevant logarithmic corrections and represents the fixed-point model for the singlet spin liquid. The absence of logarithmic corrections holds for any density in the present supersymmetric model as can be seen in the explicit solutions for the correlation function for the Gutzwiller wave function [13,14]. In this sense the long-range supersymmetric model is regarded as a fixed-point model for Fermi liquids.

With slight modification of parameters in $\mathcal{H}$ the Fermi-liquid fixed point in one dimension should flow toward a Luttinger liquid with no discontinuity in the momentum distribution. The nature of the stable state depends sensitively on the direction of modification. If, for example, the parameters are such that the spin-dimer state is realized [8] in the high-density limit, introduction of holes may lead to a superconducting state. On the other hand, with perturbations such as interchain interactions which increase the dimensionality of the system, the Fermi-liquid fixed point should be greatly stabilized. The present model seems to be a useful reference model to study phase diagrams in the parameter space.

In conclusion, we emphasize that the simplicity of the Gutzwiller wave functions, suitably generalized to describe excited states as well, gives us a unique opportunity to study the behavior of the supersymmetric model without being restricted to the asymptotic regime. It has been pointed out in the high-density limit that there are enormous degeneracies in the excitation spectrum [8]. These degeneracies are called supermultiplets and interpreted in terms of "free spinons" [8]. In a preliminary study we have found in the two-electron system, which represents a dilute limit, that many singlet and triplet excited states are degenerate. Thus the present model poses further intriguing problems, such as whether the supermultiplet structures are present at any density, and what is the physical meaning of the degeneracy.

We thank N. Kawakami for useful discussions on Luttinger liquids, and T. Watanabe for helpful comments on the manuscript.

[1] F. D. M. Haldane, J. Phys. C **14**, 2585 (1980); Phys. Rev. Lett. **45**, 1358 (1980).
[2] N. Kawakami and S.-N. Yang, Phys. Lett. A **148**, 359 (1990).
[3] H. Frahm and V. E. Korepin, Phys. Rev. B **42**, 10553 (1990).
[4] B. Sutherland, Phys. Rev. B **12**, 3795 (1975).
[5] P. Schlottman, Phys. Rev. B **36**, 5177 (1987).
[6] P.-A. Bares and G. Blatter, Phys. Rev. Lett. **64**, 2567 (1990).
[7] N. Kawakami and S.-N. Yang, Phys. Rev. Lett. **65**, 2309 (1990).
[8] F. D. M. Haldane, Phys. Rev. Lett. **60**, 635 (1988); **66**, 1529 (1991).
[9] B. S. Shastry, Phys. Rev. Lett. **60**, 639 (1988).
[10] P. W. Anderson, B. S. Shastry, and D. Hristpulos, Phys. Rev. B **40**, 8939 (1989).
[11] B. Sutherland, Phys. Rev. A **4**, 2019 (1971).
[12] H. J. Schulz, Phys. Rev. Lett. **64**, 2831 (1990).
[13] W. Metzner and D. Vollhardt, Phys. Rev. B **37**, 7382 (1988).
[14] F. Gebhard and D. Vollhardt, Phys. Rev. B **38**, 6911 (1988).
[15] H. Yokoyama and M. Ogata (to be published).

RAPID COMMUNICATIONS

PHYSICAL REVIEW B VOLUME 45, NUMBER 13 1 APRIL 1992-I

Asymptotic Bethe ansatz: Application to the one-dimensional t-J model with long-range exchange and transfer

Norio Kawakami*
Max-Planck-Institut für Festkörperforschung, 7000 Stuttgart 80, Germany
(Received 13 January 1992)

An asymptotic Bethe-ansatz solution is obtained for the one-dimensional t-J model with the $1/r^2$ long-range exchange and transfer, for which the ground-state wave function was obtained exactly by Kuramoto and Yokoyama. The present approach makes it possible to study analytically the energy spectrum as well as bulk quantities. The calculated quantities reproduce the exact results known for low-temperature properties and correlation functions. This solution represents an application of the asymptotic Bethe ansatz to correlated electron systems.

Almost 20 years ago Sutherland proposed the *asymptotic* Bethe-ansatz (ABA) method, which may be exactly applicable to some one-dimensional (1D) quantum systems with long-range interactions.[1,2] The essence of his idea is that even when one cannot seek for a Bethe-ansatz (BA) solution due to the long-range nature of the interactions, there may exist some quantum systems in which the wave function is written down like the BA: $\psi = \Sigma_P A(Q;P)\exp(i\Sigma_j k_{Pj} x_{Qj})$ in the *asymptotic* region $Q = (x_{Q1} \ll x_{Q2} \ll \cdots \ll x_{QN})$, where Q (P) expresses one of $N!$ permutations for the coordinate (momentum) configurations. A typical example is the 1D boson system with inverse-square $(1/r^2)$ long-range interactions for which the ground-state wave function is given by the product of two-body functions (Jastrow-type wave function). Sutherland has shown that the ABA solution constructs the exact spectrum for this system.[1,2] ABA-type equations have also been obtained microscopically by Haldane for the spin systems with inverse-square exchange interactions (Haldane-Shastry model[3,4]) and the full excitation spectrum has been investigated precisely.[5] Furthermore the critical exponents for correlation functions have been studied with the help of conformal field theory.[5,6] In spite of this great success, it remains open whether this method can be generalized to the electron systems, which possess internal symmetry due to the spin degrees of freedom.

Quite recently Kuramoto and Yokoyama have found the exact ground-state wave function for the 1D t-J model with long-range exchange and transfer, which is an extension of the Haldane-Shastry model to electron systems.[7] Though they have successfully investigated ground-state and low-temperature properties, it remains difficult to study systematically the energy spectrum as well as the thermodynamics. It is noted that various correlation functions can be calculated following the work on the Gutzwiller wave function.[8] Motivated by their work, in this paper I generalize Sutherland's ABA method to electron systems and obtain the ABA solution of the above-mentioned t-J model with long-range interactions.

I consider the 1D t-J type model:[7]

$$H = -\sum_{i<j,\sigma} t_{ij} c^\dagger_{i\sigma} c_{j\sigma} + \sum_{i<j} J_{ij}\left(\mathbf{S_i}\cdot\mathbf{S_j} - \frac{1}{4} n_i n_j\right), \quad (1)$$

where both of the transfer and the exchange are assumed to be of inverse-square long-range type:

$$t_{ij} = J_{ij} = (\pi^2 J/N^2)\{\sin[\pi(x_i - x_j)/N]\}^{-2}$$

with $J > 0$, which are chosen to satisfy periodic boundary conditions on N lattice sites. Here the double occupation of every site is strictly forbidden due to the strong electron correlation. Kuramoto and Yokoyama have shown that the ground-state wave function of this Hamiltonian is given exactly by the Gutzwiller projected wave function, i.e., by the product of two-body functions.[7] This characteristic aspect of the wave function is very crucial to apply ABA because it strongly suggests that the many-body S matrix may be decomposed into two-body S matrices as in the case for the Sutherland model[1] and the Haldane-Shastry model.[3,4] Henceforth I will set $J = 1$ for simplicity.

In order to outline the strategy, I begin with the half-filling case.[3,4] Taking the fully polarized up-spin state as the vacuum state,[9] consider the scattering of two spin waves with the momenta p_i and p_j. The resultant S matrix is easily found as $S_{ij} = -\exp[-i\theta(p_i - p_j)]$ with the phase shift function $\theta(p) = \pi\mathrm{sgn}(p)$.[5] In ABA the many-body S matrix is assumed to be decomposed into two-body S matrices in the asymptotic region. Strictly speaking, the asymptotic region cannot be realized for a lattice system with high densities of particles. The idea of ABA still works, however, if the many-body S matrix for the particle moving from one end of the chain to the other can be decomposed into two-body matrices self-consistently. If this is assumed to be the case, one obtains the ABA equation under the periodic boundary conditions in 1D systems with N sites: $p_j N = 2\pi J_j + \Sigma_l \theta(p_j - p_l)$ with J_j being an integer or a half integer, which coincides with that obtained by Haldane microscopically.[5]

Upon doping holes, the charge excitation provides new scattering channels. Therefore one has to deal with the Schrödinger equation with the mixture of down spins and holes in the background of up-spin electrons.[7,9] Consider the two-body scattering in the metallic phase. It is quite instructive to observe the nature of the Gutzwiller function for two particles:[7] $\psi_G \propto f(x_\alpha - x_\beta)$, where $f(x) = (N^2/\pi^2)\sin^2(\pi x/N)$ for two down-spin electrons and $f(x) = (N/\pi)\sin(\pi x/N)$ otherwise. As understood from this wave function in the asymptotic region ($|x_\beta - x_\alpha| \gg 1$), two down spins are scattered similarly as in the half-filling case; hence the resultant S matrix is given by $S_{ij} = -\exp[-i\theta(p_i - p_j)]$. Furthermore the scattering between holes takes place as if they are spinless free fermions. This is also the case for the down spin and the hole. In these cases the S matrix is simply given by $S_{ij} = -1$. Therefore if the wave function for the excited states may be written down in the asymptotic region as $\psi = \sum_P A(Q;P)\exp(i\Sigma_j p_{Pj} x_{Qj})$, Yang's operator $Y_{ij}^{\alpha\beta}$ defined by

$$Y_{ij}^{\alpha\beta} A(x_\alpha, x_\beta; p_i, p_j) = A(x_\alpha, x_\beta; p_j, p_i)$$

(Ref. 10) should satisfy $Y_{ij}^{\alpha\beta} = -\exp[-i\theta(p_i - p_j)]$ for two down-spins and $Y_{ij}^{\alpha\beta} = -1$ otherwise. These results are simplified by introducing the permutation operator $P_{\alpha\beta}$, which interchanges the coordinates x_α and x_β. Notice that when this operator acts on the Gutzwiller wave function it takes the following eigenvalues: $P_{\alpha\beta} = 1$ ($\alpha = \beta =$ down spins), $P_{\alpha\beta} = -1$ ($\alpha = \beta =$ holes), and $P_{\alpha\beta} = -1$ ($\alpha, \beta =$ down spin and hole).[7,9] Consequently, $Y_{ij}^{\alpha\beta}$ is written down simply in terms of the permutation operator $P_{\alpha\beta}$,

$$Y_{ij}^{\alpha\beta} = \lim_{\eta\to 0} \frac{(p_i - p_j)P_{\alpha\beta} + i\eta}{p_i - p_j - i\eta}. \quad (2)$$

Note that $Y_{ij}^{\alpha\beta}$ corresponds to the two-body S matrix. It is easily confirmed that $Y_{ij}^{\alpha\beta}$ satisfies the factorization equation,

$$Y_{jk}^{\alpha\beta} Y_{ik}^{\beta\gamma} Y_{ij}^{\alpha\beta} = Y_{ij}^{\beta\gamma} Y_{ik}^{\alpha\beta} Y_{jk}^{\beta\gamma}, \quad (3)$$

which has the same form as the ordinary Yang-Baxter relation in BA.[10] I refer to this as the *asymptotic* Yang-Baxter relation. Now consider the many-body scattering. It may be quite legitimate to assume that ABA is valid for the present system, based on the observation that the ground-state wave function consists of the product of two-body functions.[7] In this case it is supposed that the many-body S matrix for the particle moving from one end of the chain to the other is decomposed into two-body S matrices. On applying periodic boundary conditions one thus encounters the problem familiar in BA of diagonalizing the product of the $Y_{ij}^{\alpha\beta}$ operators such as $\prod_i P_{ij} Y_{ij}^{ij}$.[10] Fortunately the diagonalization can be performed self-consistently with the help of (3), following the method developed in the nested BA.[10] For the present case it is convenient to make use of the technique applied to the mixture of bosons and fermions.[11] By introducing the pseudomomentum for the charge degree of freedom, q_α, I finally get the generalized ABA equations

$$\exp(ip_j N) = \prod_{\alpha=1}^{N_h} F(q_\alpha - p_j) \prod_{l=1}^{N_h+N_\downarrow} F(p_j - p_l),$$

$$j = 1, 2, \ldots, N_h + N_\downarrow, \quad (4)$$

$$\prod_{j=1}^{N_h+N_\downarrow} F(q_\alpha - p_j) = 1, \quad \alpha = 1, 2, \ldots, N_h, \quad (5)$$

where N_h ($N_\downarrow$) is the number of holes (down-spins) and $F(x) = \lim_{\eta\to 0}(x - i\eta)/(x + i\eta)$. By taking the logarithm (4) and (5) reduce to the transcendental equations,

$$p_j N = 2\pi J_j - \sum_{\alpha=1}^{N_h} \theta(p_j - q_\alpha) + \sum_{l=1}^{N_h+N_\downarrow} \theta(p_j - p_l), \quad (6)$$

$$\sum_{j=1}^{N_h+N_\downarrow} \theta(q_\alpha - p_j) = 2\pi I_\alpha, \quad (7)$$

with $\theta(x) = \pi\,\mathrm{sgn}(x)$, where J_j (I_α) is an integer or half integer characterizing the spin (charge) excitations. The total energy is expressed simply in the noninteracting formula,[3,4,7] because all the interaction effects are incorporated into p_j through phase shifts:

$$E = \pi^2(1 - N^{-2})N_h/6 + \sum_{j=1}^{N_h+N_\downarrow} p_j(p_j - 2\pi)/4.$$

Formulas (6) and (7) are the generalized ABA equations.

I begin with ground-state properties. The density function $\rho_s(p)$ [$\rho_c(q)$] for the spin (charge) pseudomomentum is introduced in the thermodynamic limit. For later convenience the origin of the pseudomomenta is shifted as $(p - \pi) \to p$ and $(q - \pi) \to q$. In this notation the pseudomomenta for the ground state distribute symmetrically around the origin. From (6) and (7) the density functions are obtained as $\rho_s(p) = 1/(2\pi)$ for $|p| < Q$, $\rho_s(p) = 1/(4\pi)$ for $Q < |p| < B$, and $\rho_c(q) = 1/(2\pi)$ for $|q| < Q$. A careful treatment of the limit $\eta \to 0$ by the Wiener-Hopf method gives the values at Fermi points as $\rho_s(\pm B) = 1/(4\sqrt{2}\pi)$ and $\rho_c(\pm Q) = 1/(2\sqrt{2}\pi)$, which are responsible for low-energy elementary excitations. The cutoff parameters ($Q \le B$) are determined by the hole concentration $n_h = N_h/N$ and the magnetization $s_z = (N - N_h - 2N_\downarrow)/(2N)$. One finds $n_h = \int_{-Q}^{Q} \rho_c(q)dq = Q/\pi$ and $s_z = \frac{1}{2}(1 + n_h) - \int_{-B}^{B} \rho_s(p)dp = (\pi - B)/(2\pi)$. Note that $Q = 0$ corresponds to half filling, $B = \pi$ to the singlet state, and $Q = B$ to the fully spin-polarized state. It is easy to calculate bulk quantities with these density functions. All the ground-state quantities agree with those calculated directly by using the Gutzwiller ground-state function.[7] For example, from the second derivative of the ground-state energy with respect to the hole concentration, $(\partial^2/\partial n_h^2)[\int_{-\pi}^{\pi} \frac{1}{4}(p^2 - \pi^2)\rho_s dp + \frac{\pi^2}{6}\int_{-Q}^{Q} \rho_c dq]$, one obtains the compressibility $\chi_c = 4[\pi^2 n_h]^{-1}$, which exhibits a divergence behavior near half filling due to the metal-insulator transition. Futhermore, it is easy to obtain the magnetization in the present formalism: $s_z = \frac{1}{2} - \sqrt{1/4 - H/\pi^2}$ for magnetic fields $H \leq \pi^2(1-n_h^2)/4$.

Now I evaluate the finite-size corrections to the excitation spectrum in order to study the low-energy behavior of correlation functions. Let us introduce basic quantum numbers that specify the low-energy charge and spin excitations: N_c (N_s) labels the change of the hole (down-spin) number, and D_c (D_s) denotes the number of charge (spin) excitations that carry the $4k_F$ ($2k_F$) momentum transfer, where $k_F = \pi(1-n_h)$. Besides these basic numbers, non-negative integers $N_c^{\pm}$ ($N_s^{\pm}$) stand for simple particle-hole-type excitations of the charge (spin) sectors. The elementary excitation energy ΔE is classified in terms of these quantum numbers as

$$\Delta E = \frac{2\pi v_s}{N} x_s + \frac{2\pi v_c}{N} x_c, \tag{8}$$

where $v_s = \pi(1-2s_z)/2$ and $v_c = \pi n_h/2$ are the spin and charge velocites, and x_s (x_c) is the scaling dimension for the spin (charge) sector given by

$$x_s = \frac{1}{2}\left(N_s - \frac{N_c}{2}\right)^2 + \frac{1}{2}D_s^2 + N_s^+ + N_s^-, \tag{9}$$

$$x_c = \frac{1}{8}N_c^2 + 2\left(D_c + \frac{1}{2}D_s\right)^2 + N_c^+ + N_c^-. \tag{10}$$

The quantum numbers satisfy the selection rule $D_s = N_c/2 \bmod 1$ and $D_c = (N_s + N_c)/2 \bmod 1$ owing to the antisymmetric nature of the electron wave function. According to conformal field theory,[12,13] the formula for scaling dimensions verifies that both of the charge and spin sectors are described by $c = 1$ conformal theories, which implies that the low-energy behavior is classified as the Luttinger liquid.[14] Luttinger liquid properties of this model have been already discussed briefly in Ref. 7.

Based on a standard technique in conformal field theory,[13] all the critical exponents for the correlation functions can be read from (9) and (10). Now I evaluate correlation exponents and compare them with exact results for the Gutzwiller wave function.[8] I follow the method adopted in the Hubbard model[15] and subsequently applied to the t-J model.[16] For example, consider the long-distance behavior of the spin-correlation function: $< S^z(r)S^z(0) > \simeq < S^z >^2 + r^{-\alpha_s} \cos(2k_F r)$. By choosing the quantum numbers as $(N_c, N_s, D_c, D_s) = (0,0,0,1)$ one finds the $2k_F$ spin-density wave exponent $\alpha_s = 2x_s + 2x_c = 2$. Similarly other exponents are readily obtained: $2k_F$ charge-density wave (CDW) exponent $\alpha_s' = 2$ by $(N_c, N_s, D_c, D_s) = (0,0,0,1)$, $4k_F$ CDW exponent $\alpha_c = 4$ by $(N_c, N_s, D_c, D_s) = (0,0,1,0)$, and singlet and triplet pairing exponents $\beta_s = \beta_t = 2$ by $(N_c, N_s, D_c, D_s) = (2,1,1/2,-1)$ and (2,2,0,0). Notice that all these values, which are characteristic of the noninteracting free-electron systems,[17] coincide with the exact results for the Gutzwiller wave function.[8] Among others, it is quite instructive to observe the critical exponent for the momentum distribution function. If the momentum distribution is assumed to be written down around the Fermi point as

$$n_k = n_{k_F} - \text{const}|k - k_F|^\theta \text{sgn}(k - k_F), \tag{11}$$

the critical exponent θ is calculated as $\theta = 2x_c + 2x_s - 1 = 0$ by $(N_c, N_s, D_c, D_s) = (1,1,0,1/2)$. Furthermore, it is deduced from (6) and (7) that there is no logarithmic correction to the correlation function. It is thus concluded that $\theta = 0$ implies the existence of a discontinuity at the Fermi point in the momentum distribution, as has been pointed out in Ref. 7 with the Gutzwiller wave function.[8] The present model is assigned as a specific example of the Luttinger liquid, which has a noninteracting fixed point.[7]

Finally let us briefly discuss finite-temperature properties. If one applies a standard technique in BA method straightforwardly,[18] the following expression for the free energy may be deduced at thermal equilibrium:

$$f(T) = -\mu - \frac{H}{2} - \frac{T}{2\pi}\int_{-\pi}^{\pi} \log\{1 + \exp[-\epsilon_s(p)/T]\}dp, \tag{12}$$

where the pseudoenergy functions $\epsilon_s(p)$ and $\epsilon_c(q)$ are given by

$$\epsilon_s(p) = \epsilon_0(p) + H - T\log\{1 + \exp[-\epsilon_c(p)/T]\} + T\log\{1 + \exp[-\epsilon_s(p)/T]\}, \tag{13}$$

$$\epsilon_c(q) = \frac{\pi^2}{6} + \mu - \frac{H}{2} - T\log\{1 + \exp[-\epsilon_s(q)/T]\}, \tag{14}$$

and $\epsilon_0(p) = (p^2 - \pi^2)/4$. By comparing this expression with the exact result for half filling,[5] one notices that there is a discrepancy in two expressions and that (12) does not give a correct answer. Hence, an ordinary procedure for thermodynamics in BA cannot be applied straightforwardly to the present case. This shortage may be due to the missing states in ABA equations, as already pointed out for half-filling case.[5] In order to obtain the full thermodynamics, therefore, the idea of supermultiplet proposed by Haldane[5] should be necessarily introduced. Unfortunately I have not found such an elegant way to construct the supermultiplet for the present model, though it is to be clarified in a future study. Nevertheless it is seen that low-temperature properties of the system can be described correctly by (12)–(14). In fact, the low-temperature expansion of the free energy gives the specific-heat coefficient $\gamma = (\pi/3)(1/v_s + 1/v_c)$, which agrees exactly with the result obtained in Ref. 7 based on the Luttinger-liquid theory introduced by Haldane.[14,19]

In summary, I have derived the generalized ABA solution for the 1D t-J model with the long-range exchange and transfer. This method is a generalization of Sutherland's idea to electron systems, which has enabled systematic studies of energy spectrum as well as bulk quantities. It is suspected that the generalized ABA may provide the exact spectrum of the present system, although the rigorous proof has not been obtained yet. An interesting problem in the future is to clarify the supermultiplet structure of the excitation spectrum[5] in order to obtain the full thermodynamics. Furthermore, ABA conjecture is to be confirmed from the microscopic point of view.

I thank P. Fulde for his hospitality and useful discussions. Fruitful discussions with S.-K. Yang, Y. Kuramoto, and P. Horsch are gratefully acknowleged. This work was partly supported by Yamada Science Foundation.

---

*On leave from Yukawa Institute for Theoretical Physics, Kyoto University, Kyoto 606, Japan.

[1] B. Sutherland, J. Math. Phys. **12**, 246 (1971); **12**, 251 (1971); Phys. Rev. A **4**, 2019 (1971); **5**, 1372 (1971).
[2] B. Sutherland, *Exactly Solvable Problems in Condensed Matter and Relativistic Field Theory, Lecture Notes in Physics*, Vol. 242, edited by B. S. Shastry, S. S. Jha, and V. S. Singh (Springer-Verlag, Berlin, 1985), p. 1.
[3] F. D. M. Haldane, Phys. Rev. Lett. **60**, 635 (1988).
[4] B. S. Shastry, Phys. Rev. Lett. **60**, 639 (1988).
[5] F. D. M. Haldane, Phys. Rev. Lett. **66**, 1529 (1991).
[6] N. Kawakami and S.-K. Yang, Phys. Rev. Lett. **67**, 2493 (1991).
[7] Y. Kuramoto and H. Yokoyama, Phys. Rev. Lett. **67**, 1338 (1991).
[8] W. Metzner and D. Vollhardt, Phys. Rev. Lett. **59**, 121 (1987); F. Gebhard and D. Vollhardt, *ibid.* **59**, 1472 (1987).
[9] P. W. Anderson, B. S. Shastry, and D. Hristopulos, Phys. Rev. B **40**, 8939 (1989).
[10] C. N. Yang, Phys. Rev. Lett. **19**, 1312 (1967).
[11] B. Sutherland, Phys. Rev. B **12**, 3795 (1975).
[12] H. W. Bröte, J. L. Cardy, and M. P. Nightingale, Phys. Rev. Lett. **56**, 742 (1986); I. Affleck, *ibid.* **56**, 746 (1986).
[13] J. L. Cardy, Nucl. Phys. **B270** [FS16], 186 (1986).
[14] F. D. M. Haldane, Phys. Rev. Lett. **45**, 1358 (1980); **47**, 1840 (1981); J. Phys. C **14**, 2585 (1981); Phys. Lett. **81A**, 153 (1981).
[15] H. Frahm and V. E. Korepin, Phys. Rev. B **42**, 10 553 (1990).
[16] N. Kawakami and S.-K. Yang, Phys. Rev. Lett. **65**, 2309 (1990); J. Phys. Condens. Matter **3**, 5983 (1991).
[17] J. Solyom, Adv. Phys. **28** 201 (1979).
[18] C. N. Yang and C. P. Yang, J. Math. Phys. **10**, 1115 (1969).
[19] H. J. Schulz, Phys. Rev. Lett. **64**, 2831 (1990).

REFERENCES

Papers reprinted in this volume are marked by a •.

[1] P. W. Anderson, *Science* **235** (1987) 1196.

[2] F. H. L. Eßler, V. E. Korepin, K. Schoutens, *Phys. Rev. Lett.* **70** (1993) 73.

[3] • C. N. Yang, *Phys. Rev. Lett.* **19** (1967) 1312.

[4] M. Gaudin, *Phys. Lett.* **24**A (1967) 55.

[5] A. M. Tsvelick and P. B. Wiegmann, *Adv. in Phys.* **32** (1983) 453.

[6] N. Andrei, K. Furuya, J. H. Lowenstein, *Rev. Mod. Phys.* **55** (1983) 331.

[7] R. J. Baxter, *Exactly Solved Models in Statistical Mechanics*, New York: Academic Press, (1982).

[8] M. Gaudin, La Fonction d'Onde de Bethe, Masson, 1983.

[9] V. E. Korepin, G. Izergin and N. M. Bogoliubov, *Quantum Inverse Scattering Method, Correlation Functions and Algebraic Bethe Ansatz*, Cambridge University Press, 1993.

[10] Yu. A. Izyumov and Yu. N. Skryabin, *Statistical Mechanics of Magnetically Ordered Systems*, Engl. Transl.; New York: Plenum, (1990).

[11] L. D. Faddeev and L. A. Takhtajan, *Hamiltonian Methods in the Theory of Solitons*, Springer-Verlag, (1987).

[12] M. Jimbo, *Yang-Baxter Equation in Integrable Systems*, Singapore: World Scientific, (1990).

[13] C. Itzykson, H. Saleur, J. B. Zuber, *Conformal Invariance and Applications to Statistical Mechanics*, Singapore: World Scientific, (1986).

[14] E. H. Lieb, D. C. Mattis (eds.), *Mathematical Physics in One Dimension*, New York: Academic Press, (1966).

[15] B. M. McCoy and T. T. Wu, *The Two-Dimensional Ising Model*, Cambridge, Mass.: Harvard University Press, (1973).

[16] B. S. Shastry, S. S. Jha, V. Singh (eds.), *Exactly Solvable Problems in Condensed Matter and Relativistic Field Theory*, Berlin: Springer-Verlag, (1985).

[17] J. Hubbard, *Proc. Roy. Soc.* **276** (1963) 238.

[18] J. Hubbard, *Proc. Roy. Soc.* **277** (1964) 237.

[19] J. Hubbard, *Proc. Roy. Soc.* **284** (1964) 401.

[20] M. C. Gutzwiller, *Phys. Rev. Lett.* **10** (1963) 159.

[21] • E. H. Lieb and F. Y. Wu, *Phys. Rev. Lett.* **20** (1968) 1445.

[22] • A. A. Ovchinnikov, *Sov. Phys. JETP* **30** (1970) 1160.

[23] G. V. Uimin and S. V. Fomichev, *Sov. Phys. JETP* **36** (1973) 1001.

[24] • C. F. Coll, *Phys. Rev.* B**9** (1974) 2150.

[25] • F. Woynarovich, *J. Phys.* C**15** (1982) 85.
[26] • F. Woynarovich, *J. Phys.* C**15** (1982) 97.
[27] • F. Woynarovich, *J. Phys.* C**16** (1983) 5293.
[28] F. Woynarovich, *J. Phys.* C**16** (1983) 6593.
[29] T. C. Choy and W. Young, *J. Phys.* C**15** (1982) 521.
[30] • A. Klümper, A. Schadschneider, J. Zittartz, *Z. Phys.* B**78** (1990) 99.
[31] A. Schadschneider and J. Zittartz, *Z. Phys.* B**82** (1991) 387.
[32] N. Kawakami, A. Okiji, *Phys. Rev.* B**40** (1989) 7066.
[33] F. H. L. Eßler and V. E. Korepin, *Scattering Matrix and Excitation Spectrum of the Hubbard Model, preprint.*
[34] • M. Takahashi, *Prog. Theor. Phys.* **42** (1969) 1098.
[35] M. Takahashi, *Prog. Theor. Phys.* **43** (1970) 860.
[36] M. Takahashi, *Prog. Theor. Phys.* **43** (1970) 1619.
[37] H. Shiba, *Phys. Rev.* B**6** (1972) 930.
[38] F. Woynarovich and T. B. Bahder, *Phys. Rev.* B**33** (1986) 2114.
[39] • F. Woynarovich and K. Penc, *Z. Phys.* B**85** (1991) 269.
[40] F. Woynarovich, *Phys. Rev.* B**43** (1991) 11448.
[41] J. Carmelo, P. Horsch, P. A. Bares, A. A. Ovchinnikov, *Int. J. Mod. Phys.* B**5** (1991) 3.
[42] • M. Takahashi, *Prog. Theor. Phys.* **47** (1972) 69.
[43] • M. Takahashi, *Prog. Theor. Phys.* **52** (1974) 103.
[44] T. Koma, *Prog. Theor. Phys.* **78** (1987) 1213.
[45] T. Koma, *Prog. Theor. Phys.* **81** (1989) 783.
[46] • T. Koma, *Prog. Theor. Phys.* **83** (1990) 655.
[47] N. Kawakami, T. Usuki, A. Okiji, *Phys. Lett.* **137**A (1989) 287.
[48] J. Carmelo, P. Horsch, P. A. Bares, A. A. Ovchinnikov, *Phys. Rev.* B**44** (1991) 9967.
[49] M. Takahashi, *Prog. Theor. Phys.* **46** (1971) 401.
[50] H. Bethe, *Z. Phys.* **71** (1931) 205.
[51] L. D. Faddeev and L. Takhtajan, Zap.Nauch.Semin LOMI Vol 109 (1981) p. 134.
[52] A. A. Vladimirov, *Phys. Lett.* **105**A (1984) 418.
[53] J. Carmelo and A. A. Ovchinnikov, *J. Phys.* C**3** (1991) 757.
[54] J. Carmelo and P. Horsch, *Phys. Rev. Lett.* **68** (1992) 871.
[55] T. Usuki, N. Kawakami, A. Okiji, *Phys. Lett.* **135**A (1989) 476.
[56] F. H. L. Eßler, V. E. Korepin, K. Schoutens, *Nucl. Phys.* B**372** (1992) 559.
[57] • F. H. L. Eßler, V. E. Korepin, K. Schoutens, *Nucl. Phys.* B**384** (1992) 431.
[58] F. H. L. Eßler, V. E. Korepin, K. Schoutens, *Phys. Rev. Lett.* **67** (1991) 3848.

[59] I. Affleck in talk given at the Nato Advanced Study Institute on *Physics, Geometry and Topology*, Banff, August 1989.
[60] O. J. Heilmann and E. H. Lieb, *Ann. New York Acad. Sci.* **172** (1971) 583.
[61] C. N. Yang and S. Zhang, *Phys. Lett.* B4 (1990) 759.
[62] C. N. Yang, *Phys. Rev. Lett.* **63** (1989) 2144.
[63] M. Pernici, *Europhys. Lett.* **12** (1990) 75.
[64] J. Carmelo and D. Baeriswyl, *Phys. Rev.* B**37** (1988) 7541.
[65] A. L. Kholodenko, A. L. Beyerlein, *Phys. Rev.* B**36** (1987) 409.
[66] B. S. Shastry, *Phys. Rev. Lett.* **56** (1986) 1529.
[67] B. S. Shastry, *Phys. Rev. Lett.* **56** (1986) 2453.
[68] • B. S. Shastry, *J. Stat. Phys.* **50** (1988) 57.
[69] R. Z. Bariev, *Phys. Lett.* **138**A (1989) 173.
[70] R. Z. Bariev, *Phys. Lett.* **147**A (1990) 261.
[71] M. Barma and B. S. Shastry, *Phys. Lett.* **61**A (1977) 15.
[72] B. S. Shastry, *private communication.*
[73] R. Z. Bariev, *Theor. Mat. Phys.* **49** (1982) 1021.
[74] T. T. Truong and K. D. Schotte, *Nucl. Phys.* B**230** (1984) 1.
[75] K. D. Schotte, S. Iwabuchi, T. T. Truong, *Z. Phys.* B**60** (1985) 255.
[76] C. Destri, H. J. deVega, *Phys. Rev. Lett.* **69** (1992) 2313.
[77] C. Destri, H. J. deVega, *Nucl. Phys.* B**290** (1987) 363.
[78] C. Destri, H. J. deVega, *J. Phys.* A**22** (1989) 1329.
[79] J. Suzuki, Y. Akutsu, M. Wadati, *J. Phys. Soc. Japan* **59** (1990) 2667.
[80] J. Suzuki, T. Nagao, M. Wadati, *Int. J. Mod. Phys.* B6 (1992) 1119.
[81] A. Klümper, *Thermodynamics of the Anisotropic Spin-1/2 Heisenberg Chain and Related Quantum Chains, preprint.*
[82] M. Takahashi, *Phys. Rev.* B**43** (1991) 5788.
[83] M. Takahashi, *Phys. Rev.* B**44** (1991) 12382.
[84] • H. Tsunetsugu, *J. Phys. Soc. Japan* **60** (1991) 1460.
[85] • A. M. Finkel'shteĭn, *Sov. Phys. JETP Lett.* **25** (1977) 73.
[86] M. Ogata and H. Shiba, *Phys. Rev.* B**41** (1990) 2326.
[87] M. Ogata, T. Sugiyama, H. Shiba, *Phys. Rev.* B**43** (1990) 8401.
[88] M. Ogata and H. Shiba, Proceedings of *ICTP Adriatic Conference and workshop on Strongly Correlated Electron Systems* (Trieste, June 1990).
[89] A. Parola, S. Sorella, *Phys. Rev. Lett.* **64** (1990) 1831.
[90] F. D. M. Haldane, *Phys. Rev. Lett.* **45** (1981) 1358.
[91] • F. D. M. Haldane, *J. Phys.* C**14** (1981) 2585.
[92] • F. D. M. Haldane, *Phys. Lett.* **81**A (1981) 153.
[93] F. D. M. Haldane, *Phys. Rev. Lett.* **47** (1981) 1840.

[94] F. D. M. Haldane, in *Electron Correlation and Magnetism in Narrow-Band Systems*, eds T. Moriya (Springer-Verlag, Berlin, 1981) p. 150.
[95] L. Kadanoff, A. C. Brown, *Ann. Phys. (N.Y.)* **121** (1979) 318.
[96] H. J. Schulz, *Phys. Rev. Lett.* **64** (1990) 2831.
[97] • H. J. Schulz, *Int. J. Mod. Phys.* B**5** (1991) 57.
[98] P. W. Anderson and Y. Ren, *Repulsive Hubbard Model in One Dimension*, preprint.
[99] M. Brecht, J. Voigt, H. Büttner, *Europhys. Lett.* **12** (1990) 289.
[100] J. Carmelo, P. Horsch, A. A. Ovchinnikov, *Phys. Rev.* B**45** (1992) 7899.
[101] H. Frahm and V. E. Korepin, *Phys. Rev.* B**42** (1990) 10533.
[102] • H. Frahm and V. E. Korepin, *Phys. Rev.* B**43** (1991) 5653.
[103] A. A. Belavin, A. A. Polyakov, A. B. Zamolodchikov, *Nucl. Phys.* B**241** (1984) 333.
[104] J. L. Cardy, *Nucl. Phys.* B**270** (1986) 186.
[105] H. W. J. Blöte, J. L. Cardy, M. P. Nightingale, *Phys. Rev. Lett.* **56** (1986) 742.
[106] I. Affleck, *Phys. Rev. Lett.* **56** (1986) 746.
[107] H. J. deVega and F. Woynarovich, *Nucl. Phys.* B**251** (1985) 439.
[108] N. M. Bogoliubov, A. G. Izergin, V. E. Korepin, *Nucl. Phys.* B**275** (1986) 687.
[109] N. M. Bogoliubov, A. G. Izergin, N. Yu. Reshetikhin, *J. Phys.* A**20** (1987) 5361.
[110] • F. Woynarovich and H. P. Eckle, *J. Phys.* A**20** (1987) L97.
[111] F. Woynarovich and H. P. Eckle, *J. Phys.* A**20** (1987) L443.
[112] N. M. Bogoliubov and V. E. Korepin, *Mod. Phys. Lett.* B**1** (1988) 349.
[113] • N. M. Bogoliubov and V. E. Korepin, *Int. J. Mod. Phys.* B**3** (1989) 427.
[114] A. G. Izergin, V. E. Korepin, N. Yu. Reshetikhin, *J. Phys.* A**22** (1989) 2615.
[115] • F. Woynarovich, *J. Phys.* A**22** (1989) 4243.
[116] N. Kawakami and S. K. Yang, *Phys. Rev. Lett.* **65** (1990) 2309.
[117] N. Kawakami and S. K. Yang, *Phys. Lett.* **148**A (1990) 359.
[118] T. Giamarchi and H. J. Schulz, *Phys. Rev.* B**39** (1989) 4620.
[119] K. B. Efetov and A. I. Larkin, *Sov. Phys. JETP* **39** (1974) 1129.
[120] A. Luther and I. Peschel, *Phys. Rev.* B**12** (1975) 3906.
[121] V. N. Popov, *Phys. JETP* **31** (1980) 526.
[122] • B. S. Shastry and B. Sutherland, *Phys. Rev. Lett.* **65** (1990) 243.
[123] • M. Fowler and N. Yu, *Phys. Rev.* B**45** (1992) 11795.
[124] J. Carmelo, P. Horsch, D. Campbell, *Charge and spin dynamical form factors of the Hubbard chain in a magnetic field, preprint.*
[125] M. J. Martins and R. M. Fye, *J. Stat. Phys.* **64** (1991) 271.

[126] F. C. Zhang and T. M. Rice, *Phys. Rev.* B**37** (1988) 3759.

[127] A. B. Harris and R. V. Lange, *Phys. Rev.* **157** (1967) 295; W. F. Brinkmann and T. M. Rice, *Phys. Rev.* B**2** (1970) 1324; C. Castellani, C. di Castro, D. Feinberg, *Phys. Rev. Lett.* **43** (1979) 1957; J. E. Hirsch, *Phys. Rev. Lett.* **54** (1985) 1317, C. Gros, R. Joynt, T. M. Rice, *Phys. Rev.* B**36** (1986) 381.

[128] V. J. Emery, A. S. Kivelson, H. Q. Liu, *Phys. Rev. Lett.* **64** (1990) 475.

[129] M. Ogata, M. U. Luchini, S. Sorella, F. F. Assaad, *Phys. Rev. Lett.* **66** (1991) 2388.

[130] D. Förster, *Z. Phys.* B**82** (1991) 329.

[131] J. F. Cornwell, *Group Theory in Physics, Vol III: Supersymmetries and Infinite-Dimensional Algebras*, Academic Press (1989). The algebras $SU(n|m)$ are discussed on page 270.

[132] M. Scheunert, W. Nahm, V. Rittenberg, *J. Math. Phys.* **18** (1977) 146.

[133] M. Scheunert, W. Nahm, V. Rittenberg, *J. Math. Phys.* **18** (1977) 155.

[134] I. Bars and M. Günaydin, *Comm. Math. Phys.* **91** (1983) 31.

[135] D. Förster, *Phys. Rev. Lett.* **63** (1989) 2140.

[136] P. B. Wiegmann, *Phys. Rev. Lett.* **60** (1988) 821.

[137] C. K. Lai, *J. Math. Phys.* **15** (1974) 1675.

[138] • Bill Sutherland, *Phys. Rev.* B**12** (1975) 3795.

[139] B. Sutherland, *Phys. Rev. Lett.* **20** (1968) 98.

[140] F. H. L. Eßler, V. E. Korepin, *Phys. Rev.* B**46** (1992) 9147.

[141] • P. Schlottmann, *Phys. Rev.* B**36** (1987) 5177.

[142] • P. A. Bares, G. Blatter, *Phys. Rev. Lett.* **64** (1990) 2567.

[143] P. A. Bares, G. Blatter, M. Ogata, *Phys. Rev.* B**44** (1991) 130.

[144] P. A. Bares, PhD thesis, ETH Zürich (1991).

[145] S. Sarkar, *J. Phys.* A**23** (1990) L409, *J. Phys.* A**24** (1991) 1137.

[146] • S. Sarkar, *J. Phys.* A**24** (1991) 5775.

[147] M. Quaisser, A. Schadschneider, J. Zittartz, *J. Phys.* A**25** (1992) L1127.

[148] • N. Kawakami, S. K. Yang, *J. Phys.* C**3** (1991) 5983.

[149] P. Schlottmann, *Phys. Rev.* B**41** (1990) 4164.

[150] P. P. Kulish, *J. Soviet Math.* **35** (1985) 2648.

[151] P. P. Kulish and E. K. Sklyanin, *J. Soviet Math.* **19** (1982) 1596.

[152] P. P. Kulish, N. Yu. Reshetikhin, *J. Phys.* A**16** (1983) L591.

[153] P. P. Kulish, N. Yu. Reshetikhin, *Sov. Phys. JETP* **53** (1981) 108.

[154] L. Takhtajan, *J. Soviet Math.* **23** (1983) 2470.

[155] A. Förster and M. Karowski, *Phys. Rev.* B**46** (1992) 9234.

[156] • A. Förster and M. Karowski, *Nucl. Phys.* B**396** (1993) 611.

[157] A. Klümper, A. Schadschneider, J. Zittartz, *J. Phys.* A**24** (1991) L955.

[158] R. Z. Bariev, *Exact Solution of a one-dimensional anisotropic t-J model, preprint.*

[159] R. Z. Bariev, *Exact Solution of a generalized t-J model in one dimension, preprint.*

[160] A Foerster and M. Karowski, *The Supersymmetric t-J model with Quantum Group Invariance, preprint.*

[161] R. Z. Bariev, *Correlation functions of a one-dimensional anisotropic t-J model, preprint.*

[162] P. A. Bares, J. M. P. Carmelo, J. Ferrer, P. Horsch, *Phys. Rev.* B**46** (1992) 14624.

[163] E. H. Lieb and W. Liniger, *Phys. Rev.* **130** (1963) 1605.

[164] E. H. Lieb and W. Liniger, *Phys. Rev.* **130** (1963) 1611.

[165] J. B. McGuire, *J. Math. Phys.* **6** (1965) 432.

[166] M. Flicker, E. H. Lieb, *Phys. Rev.* **161** (1967) 179.

[167] C. N. Yang, *Phys. Rev.* **168** (1968) 1920.

[168] J. B. McGuire, *J. Math. Phys.* **5** (1964) 622.

[169] F. A. Berezin and V. N. Sushko, *Sov. Phys. JETP* **21** (1965) 865.

[170] E. Brezin and J. Zinn-Justin, *Compt. Rend. Acad. Sci. Paris* B**263** (1966) 670.

[171] • M. Takahashi, *Prog. Theor. Phys.* **46** (1971) 1388.

[172] C. K. Lai, *Phys. Rev. Lett.* **26** (1971) 1472.

[173] J. M. Luttinger, *J. Math. Phys.* **4** (1963) 1154.

[174] R. Heidenreich, R. Seiler, D. A. Uhlenbrock, *J. Stat. Phys.* **22** (1980) 27.

[175] • D. C. Mattis and E. H. Lieb, *J. Math. Phys.* **6** (1965) 304.

[176] • R. Z. Bariev, *J. Phys.* A**24** (1991) L549.

[177] • R. Z. Bariev, A. Klümper, A. Schadschneider, J. Zittartz, *J. Phys.* A**26** (1993) 1249.

[178] • F. H. L. Eßler, V. E. Korepin, K. Schoutens, *Phys. Rev. Lett.* **68** (1992) 2960.

[179] J. E. Hirsch, *Physica* C**158** (1990) 326.

[180] K. A. Penson and M. Kolb, *Phys. Rev.* B**33** (1986) 1663.

[181] M. Kolb and K. A. Penson, *J. Stat. Phys.* **44** (1986) 129.

[182] I. Affleck, J. B. Marston, *J. Phys.* C**21** (1988) 2511.

[183] F. H. L. Eßler, V. E. Korepin, K. Schoutens, *Exact Solution of an Electronic Model of Superconductivity in* $1+1$ *Dimensions, preprint.*

[184] K. Okumura, *Mod. Phys. Lett.* B**7** (1993) 251.

[185] K. Okumura, *J. Phys. Soc. Japan* **62** (1993) 1922.

[186] P. Schlottmann, *Phys. Rev. Lett.* **69** (1992) 2396.

[187] F. D. M. Haldane, *Phys. Rev. Lett.* **60** (1988) 635.

[188] F. D. M. Haldane, *Phys. Rev. Lett.* **66** (1991) 1529.
[189] B. S. Shastry, *Phys. Rev. Lett.* **60** (1988) 639.
[190] N. Kawakami, *Phys. Rev. Lett.* **71** (1993) 275.
[191] B. Sutherland, *Phys. Rev.* **A4** (1971) 2019.
[192] B. Sutherland, *Phys. Rev.* **A5** (1972) 1372.
[193] • B. Sutherland, *J. Math. Phys.* **12** (1971) 251.
[194] • B. Sutherland, *J. Math. Phys.* **12** (1971) 246.
[195] B. Sutherland, *Phys. Rev. Lett.* **34** (1975) 1083.
[196] B. Sutherland, *Phys. Rev. Lett.* **35** (1975) 185.
[197] B. Sutherland, *Rocky Mountain Journal of Mathematics* **8** (1978) 413.
[198] B. Sutherland, *Phys. Rev.* **B38** (1988) 6689.
[199] B. Sutherland, in *Interacting Electrons in Reduced Dimensions*, D. Baeriswyl, D. K. Campbell (eds.) NATO ASI Series B**213**, 1989, Plenum, New York.
[200] F. Calogero, *J. Math. Phys.* **10** (1969) 2191.
[201] F. Calogero, *J. Math. Phys.* **10** (1969) 2197.
[202] F. Calogero, *J. Math. Phys.* **12** (1971) 419.
[203] F. Calogero, *Lett. Nuovo Cimento* **13** (1975) 401.
[204] F. Calogero, *Lett. Nuovo Cimento* **13** (1975) 507.
[205] F. Calogero, O. Ragnisco, C. Marchioro, *Lett. Nuovo Cimento* **13** (1975) 383.
[206] P. D. Lax, *Comm. Pure Appl. Math.* **21** (1968) 467.
[207] J. Moser, *Advances in Math.* **16** (1975) 197.
[208] B. Sutherland, in *Exactly Solvable Problems in Condensed Matter and Relativistic field Theory*, B. S. Shastry, S. S. Jha, V. Singh (eds.) Berlin: Springer Verlag, (1985), p. 1.
[209] F. D. M. Haldane, Z. N. C. Ha, J. C. Talstra, D. Bernard, V. Pasquier, *Phys. Rev. Lett.* **69** (1992) 2021.
[210] B. S. Shastry, *Phys. Rev. Lett.* **69** (1992) 164.
[211] A. P. Polychronakos, *Phys. Rev. Lett.* **69** (1992) 703.
[212] B. Sutherland, B. S. Shastry, *Solution of Some Integrable One-Dimensional Quantum Systems, preprint.*
[213] B. S. Shastry, B. Sutherland, *Super Lax Pairs and Infinite Symmetries in the* $1/r^2$ *System, preprint.*
[214] R. A. Römer and B. Sutherland, *Test of Conformal Invariance in One-Dimensional Quantum Liquid with Long-Range Interaction, preprint.*
[215] H. Ujino, K. Hikami, M. Wadati, *J. Phys. Soc. Japan* **61** (1992) 3425.
[216] K. Hikami and M. Wadati, *Integrable Spin-1/2 Particle Systems with Long-Range Interactions, preprint.*
[217] Z. N. C. Ha and F. D. M. Haldane, *Phys. Rev.* **B46** (1992) 9359.
[218] B. Sutherland, *Phys. Rev.* **B45** (1992) 907.

[219] D. Bernard, M. Gaudin, F. D. M. Haldane, V. Pasquier, *Yang-Baxter Equation in Spin Chains with Long Range Interactions, preprint.*

[220] H. Kiwata, Y. Akutsu, *J. Phys. Soc. Japan* **61** (1991) 1441.

[221] N. Kawakami, *Phys. Rev.* B**46** (1992) 3191.

[222] P. J. Forrester, *J. Phys.* A**25** (1992) 5447.

[223] P. J. Forrester, *Nucl. Phys.* B**338** (1992) 671.

[224] B. D. Simons, P. A. Lee, B. L. Altshuler, *Phys. Rev. Lett.* **70** (1993) 4122.

[225] H. Frahm, *J. Phys.* A**26** (1993) L473.

[226] N. Kawakami and S. K. Yang, *Phys. Rev. Lett.* **67** (1991) 2493.

[227] • Y. Kuramoto and H. Yokoyama, *Phys. Rev. Lett.* **67** (1991) 1338.

[228] • N. Kawakami, *Phys. Rev.* B**45** (1992) 7525.

[229] N. Kawakami, *Phys. Rev.* B**46** (1992) 1005.

[230] F. Gebhard and A. E. Rückenstein, *Phys. Rev. Lett.* **68** (1992) 244.

www.ingramcontent.com/pod-product-compliance
Ingram Content Group UK Ltd.
Pitfield, Milton Keynes, MK11 3LW, UK
UKHW052114170926
8624IPUK00026B/402

9 789810 2153